Emerging Technologies towards Energy Cooperation between Smart Grid and Microgrids

Emerging Technologies towards Energy Cooperation between Smart Grid and Microgrids

Editor

Mohamed A. Mohamed

MDPI

Basel • Beijing • Wuhan • Barcelona • Belgrade • Novi Sad • Cluj • Manchester

Editor
Mohamed A. Mohamed
Minia University
Minia
Egypt

Editorial Office
MDPI
St. Alban-Anlage 66
4052 Basel, Switzerland

This is a reprint of articles from the Special Issue published online in the open access journal *Sustainability* (ISSN 2071-1050) (available at: https://www.mdpi.com/journal/sustainability/special_issues/Energy_Cooperation_Smart_Grid_Microgrids).

For citation purposes, cite each article independently as indicated on the article page online and as indicated below:

Lastname, A.A.; Lastname, B.B. Article Title. *Journal Name* **Year**, *Volume Number*, Page Range.

ISBN 978-3-7258-0087-2 (Hbk)
ISBN 978-3-7258-0088-9 (PDF)
doi.org/10.3390/books978-3-7258-0088-9

Contents

About the Editor

Mohamed A. Mohamed

Mohamed A. Mohamed (SMIEEE) received B.Sc. and M.Sc. degrees in electrical engineering from Minia University, Minia, Egypt, in 2006 and 2010, respectively. He received his Ph.D. degree in electrical engineering from King Saud University, Riyadh, Saudi Arabia, in 2016. He joined the College of Electrical Engineering & Automation, Fuzhou University, China, as a Postdoctoral Research Fellow in 2018. He is currently a Professor in the Department of Electrical Engineering, College of Engineering, Minia University, Minia, Egypt. He has supervised multiple M.Sc. and Ph.D. theses, worked on a number of technical projects, and published various papers and books. He has also joined the editorial board of several scientific journals and the steering committees of several international conferences. His research interests include power system analysis, renewable energy integration, energy management, transportation electrification, electrical vehicles, optimization, cyber-physical systems, smart islands, smart cities, and smart grids. Dr. Mohamed has received numerous local and international awards for his contribution to conducting international scientific research, published and refereed in specialized scientific journals (Q1) and classified according to the Science Citation Index (SCI, EI), which aims to apply modern technologies for energy management, smart grids, and the use of renewable energy to enhance sustainability, reduce carbon dioxide emissions and pollution, achieve optimal use of natural resources, improve energy efficiency and proposing alternative solutions to the energy crisis.

Preface

The growing propagation of microgrids and their remarkable effects on the operation of the smart grid is leading to the development of a sustained environment which is moving away from traditional frameworks. Therefore, tending to microgrid systems to increase their range of benefits can play a significant role in outlining an effective negotiation framework for the microgrids that are connected to the smart grid.

In recent years, numerous research and development projects have been carried out to design energy transactions and economic models and implement local control platforms for manufacturers, consumers, and microgrids. Furthermore, the attention paid to peer-to-peer constructions for energy exchanges and management has grown significantly, with several startups from research and development projects emerging to explore energy trading with a focus on trading surplus energy in a way that allows producers and consumers to exchange surplus energy with their neighbors and on supplying energy in such a way that producers and consumers can directly produce locally renewable products.

This Reprint highlights and discusses the appropriate negotiation structures to maximize the benefits of microgrids that are connected to the smart grid and contributes to the derivation of sustainable future energy systems. In addition to modern techniques for managing uncertainty parameters regarding the microgrid and smart grid, this Reprint also examines multilateral economic distribution frameworks that need to be implemented easily and efficiently without the need for a central agent, with a limited exchange of information comprising the amount and price of the energy exchange. Special attention is given to studies on emerging technologies, such as machine learning and artificial intelligence, etc., for solving emergent challenges surrounding renewable energy integration into the smart grid and uncertainty-aware peer-to-peer energy management and for mitigating smart grids' vulnerability against faults and cyberattacks.

Mohamed A. Mohamed
Editor

sustainability

MDPI

Article

A Novel Control Method for Active Power Sharing in Renewable-Energy-Based Micro Distribution Networks

Wael J. Abdallah [1], Khurram Hashmi [2,3], Muhammad Talib Faiz [4], Aymen Flah [5], Sittiporn Channumsin [6,*], Mohamed A. Mohamed [7,*] and Denis Anatolievich Ustinov [1]

1 Department of Electric Power and Electromechanics, St. Petersburg Mining University, 199106 St. Petersburg, Russia
2 Department of Electrical Engineering, University of Engineering and Technology, Lahore 54890, Pakistan
3 School of Electrical and Electronic Engineering, University College Dublin, D04 V1W8 Dublin, Ireland
4 Department of Electronics and Information Engineering, The Hong Kong Polytechnic University, Kowloon, Hong Kong, China
5 Energy Processes Environment and Electrical Systems Unit, National Engineering School of Gabes, University of Gabes, Gabes 6072, Tunisia
6 Space Technology Research Center, Geo-Informatics and Space Technology Development Agency (GISTDA), Chonburi 20230, Thailand
7 Department of Electrical Engineering, Faculty of Engineering, Minia University, Minia 61519, Egypt
* Correspondence: sittiporn@gistda.or.th (S.C.); dr.mohamed.abdelaziz@mu.edu.eg (M.A.M.)

Abstract: The microgrid is an emerging trend in modern power systems. Microgrids consist of controllable power sources, storage, and loads. An elaborate control infrastructure is established to regulate and synchronize the interaction of these components. The control scheme is divided into a hierarchy of several layers, where each layer is composed of multi-agents performing their dedicated functions and arriving at a consensus of corrective values. Lateral and horizontal interaction of such multi-agents forms a comprehensive hierarchical control structure that regulates the microgrid operation to achieve a compendium of objectives, including power sharing, voltage, and frequency regulation. The success of a multi-agent-based control scheme is dependent on the health of the communication media that is used to relay measurements and control signals. Delays in the transmission of control signals result in an overall deterioration of the control performance and non-convergence. This paper proposes novel multi-agent moving average estimators to mitigate the effect of latent communication links and establishes a hierarchical control scheme incorporating these average estimators to accurately arrive at system values during communication delays. Mathematical models are established for the complete microgrid system to test the stability of the proposed method against conventional consensus-based methods. Case-wise simulation studies and lab-scale experimental verification further establish the efficacy and superiority of the proposed control scheme in comparison with other conventionally used control methods.

Keywords: smart-grid; microgrid control; distribution networks; power electronics

Citation: Abdallah, W.J.; Hashmi, K.; Faiz, M.T.; Flah, A.; Channumsin, S.; Mohamed, M.A.; Ustinov, D.A. A Novel Control Method for Active Power Sharing in Renewable-Energy-Based Micro Distribution Networks. *Sustainability* **2023**, *15*, 1579. https://doi.org/10.3390/su15021579

Academic Editor: Alberto-Jesus Perea-Moreno

Received: 9 December 2022
Revised: 4 January 2023
Accepted: 7 January 2023
Published: 13 January 2023

1. Introduction

Renewable and sustainable energy resource (RES)-based distributed energy units (DGUs) have a variable power yield since RESs display a stochastic nature. DGUs, energy storage units (ESU), hybrid electric vehicles (HEV), data acquisition (DAQ) centers, supervisory control strategies, and local and centralized control structures form a microgrid control framework that manages the micro-network [1,2]. A smart micro-distribution network or microgrid (MG) can be described as an autonomous energy transmission and distribution network capable of self-regulation. The prime objective of the MG is to meet the requirements of connected load demands through effectively controlling its generation, energy storage, and transmission resources. It can, therefore, provide a viable solution for renewable energy integration and utilization [3–5].

Power converters process variable output from energy resources and regulate voltage and currents through local control loops. For power sharing among various nodes, "primary" controls are employed [6]. Several power-sharing techniques are discussed in the literature, such as communication-based optimized control, P-f, Q-V droop, proportional droop, and so on [6–8]. Decentralized droop control schemes are widely utilized in the literature, and in these techniques, proportional reduction has been noted in the frequency and voltage by active and reactive power sharing errors at each distributed node. Several variations have been made in the basic droop techniques so that limitations of conventional droop can be addressed and greater power sharing accuracy can be achieved. However, the limitations of conventional droop control cause a mismatch between the output voltage and frequency and their nominal values. A secondary restorative control layer observes and corrects deviations in frequency and voltage, so that a nominal range of these parameters can be achieved [6,9–16].

Consensus-based control schemes use observers to converge on estimates of values measured at distributed nodes, along with the restoration of voltage and frequency synchronization [1,6,17]. These schemes rely on the transmission of data in the control scheme. These methods can be employed either as a centralized or as a distributed hierarchical control system. Centralized control schemes aim to achieve power sharing in the MG system by utilizing a bidirectional communication structure with a central controller.

The flexibility of the decentralized control techniques for new and innovative control schemes as multi-agent systems is discussed in [6,8,18–22]. Such control methods can combine resources at every node agent to aid in the convergence to the global set point. In such scenarios, the MG control and its operation can be implemented by making an individual power converter an agent within a larger multi-agent-based system. In [8,23,24], MG system objectives, such as restoration of the grid voltage and frequency and power sharing, have been considered as tracker synchronization problems, and all participating system nodes make their efforts for a consensus framework for the corrective values.

Microgrid control and management can be laid out as a multi-level control problem. In this control structure, layer "zero" is responsible for control current and voltage, primary control regulates power-sharing, and secondary control regulates frequency and voltage deviations. Tertiary control performs energy management functions [18,25–27]. Secondary controls work on the principle of consensus among all the participating nodes, making restoration of the voltage and frequency a tracker synchronization problem. A consensus-based control structure may be suitable for realizing power-sharing among participating nodes [28]. Reactive power sharing can also be realized at the primary control layer by adopting a distributed control methodology, which also has the ability to regulate the voltage [29,30]. The MG system's stability is studied in the literature using small-signal modeling with a distributed control scheme to realize the control objectives of the AC grid [17], which highlights substitutes for a secondary centralized control structure and a droop-based primary controller. To regulate different parameters of the AC grid, such as system voltages, active and reactive power regulators have been designed.

The conventional consensus-based controls are heavily reliant upon the efficacy of communication links [7,31]. Fast transmission of measured values, estimated values, and control signals cause early convergence and improve regulation capability, but limitations such as latent communication network and intermittency deteriorate the microgrid's stability and control performance. However, in order to keep the control simple, and for ease of problem formulation, a fault-free communication network has been adopted in the literature without disrupted or latent communication links [18,25,26,29,32]. The information network characteristics demonstrated in these control approaches are time invarying. In [17,28,33–36], researchers have considered a communication network with faults so that controller performance can be analyzed in such scenarios.

In networked energy systems, some communication intermittencies in the control and cyber-network layer are probable. Transmission of values and multi-agent consensus-based controls may be affected due to latencies and delay in information exchanges, and the performance of the MG system will deteriorate. To address these challenges, rational and innovative control schemes are required.

In this paper, an improved control technique that addresses latent communication links in a multilevel control scheme for an islanded AC microgrid (MG) network is presented. To realize active power sharing in the microgrid from all distributed connected nodes, a new power sharing technique has been suggested in this study. A degree of resilience to the communication network latencies has been attained by the estimation of average MG parameters at every node. To verify the viability and effectiveness of the proposed control structure, a comprehensive mathematical analysis and detailed simulation study have been conducted, and results are given to validate the proposed method. In order to realize the restoration of the voltage and frequency, a multi-agent-based consensus control layer has been added to the secondary control of the AC grid.

The main contributions and advantages of this study can be summarized as follows:

1. A novel distributed multi-agent moving average estimation technique is proposed that pre-estimates measured values at every node.
2. A hierarchical distributed control structure has been adopted, which incorporates the multi-agent moving average method proposed in this paper, to achieve active power sharing and regulation of the system-wide voltage and frequency values.
3. The viability and effectiveness of the investigated approach is tested for scenarios where the network layer is suffering from communication link latencies.

This paper is arranged as follows: problem formulation, along with particulars of the AC MG model that have been considered in this work, is discussed in Section 2. The proposed hierarchical control scheme is elaborated in Section 3. Section 4 describes the consensus-based observers. Section 5 presents the voltage and frequency restoration controls schemes. The mathematical model derivation for the mentioned control structure is illustrated in Section 6. Section 7 presents the stability analysis of the propsed control scheme in comparison with the existing consensus-based technique, followed by MG system performance assessment with simulation study in Section 8. Section 9 provides the experimental results carried out for verification of the proposed method. Section 10 compares the proposed method to conventional consensus-based methods. Finally, Section 11 concludes the paper.

2. Materials and Methods: Problem Formulation

This section provides the mathematical details required to develop the control problem. The micro-grid network can be imagined as two overlapped systems, namely the power network and the control and communication network. Each individual DGU can be assumed to be a participating node in the multi-agent-based microgrid. To formulate the problem and introduce new control schemes, the component theories are systematically described with the basic concepts of graph theory and a steady-state mathematical model for the power system with the essential mathematical formulation. A brief explanation of communication link latencies has been provided in the following paragraph as a foreword to the work.

2.1. Microgrid Network Discription

A three-phase, three-wire radial distribution network, shown in Figure 1, is used in this work. As can be seen, the power converters at buses 1 through 4 are interfaced with the coupling inductors and LC filters. The varying RL loads have been linked to bus 2 through bus 4. Bus 1 is not feeding the load demand. Bus 5 offers a coupling point that is common with the power grid. The MG system may be managed independently in the "islanded" mode. MG system-rated parameters are presented in Table 1, and system loads are provided in Table 2.

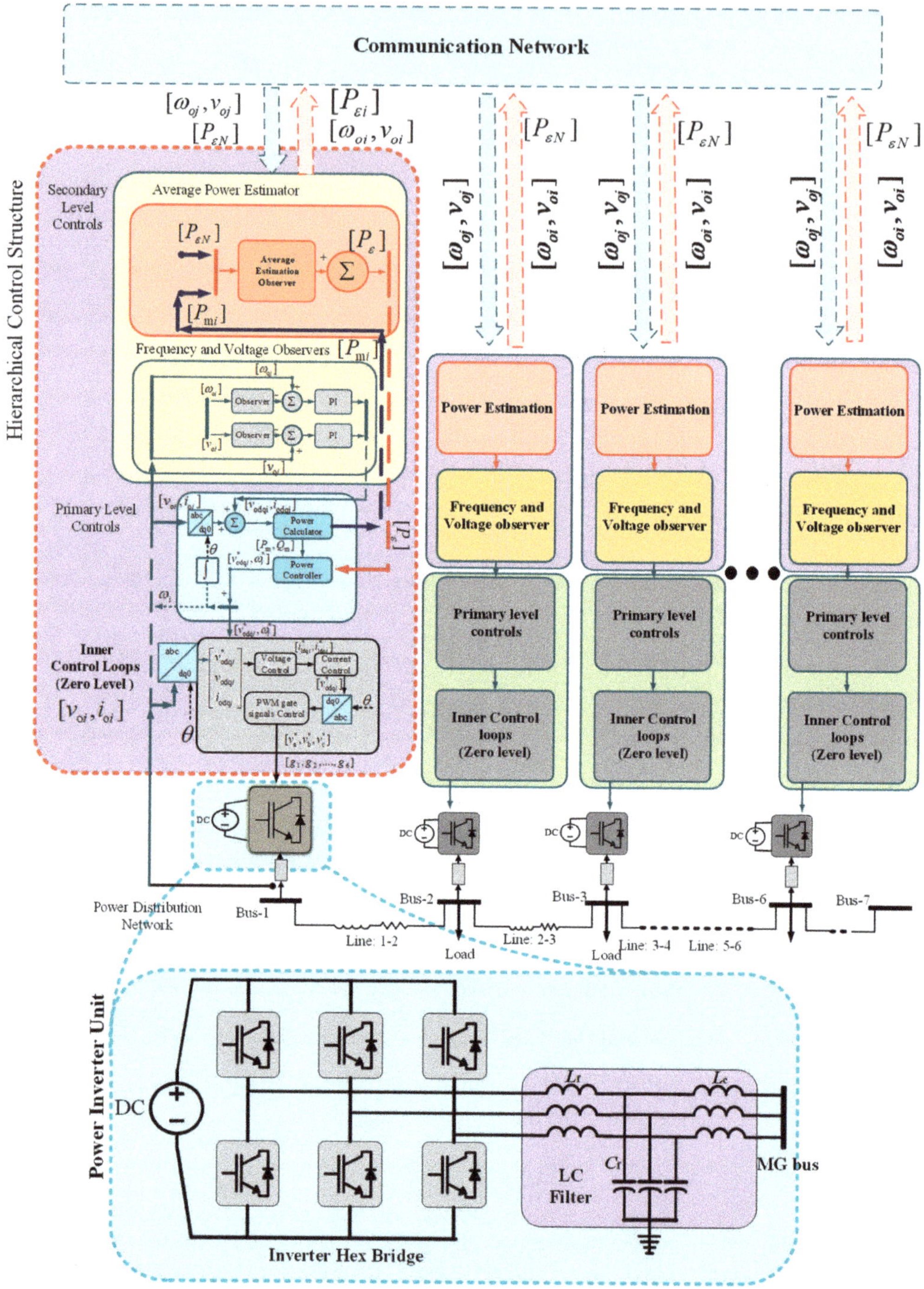

Figure 1. Control and power distribution network of the AC microgrid.

Table 1. Microgrid parameters.

Parameter	Value	Parameter	Value
Lf	1.35 mH	mp	4.4×10^{-6}
Rf	0.1 Ω	nq	1.1×10^{-6}
Cf	25 µF	kpf	0.4
Lc	1.35 mH	kif	0.5
Rc	0.05 Ω	kpV	0.6
Rline	0.2 Ω	kiV	0.3
Lline	0.6 mH	F	0.6
fnom	50 Hz	ωc	31.4
Vnom	380 VL-L		

Table 2. System loads.

Bus No.	Loads Connected (p.u)	
	P	Q
1	0	0
2	0.3	0.3
3	0.25	0.25
4	0.25	0.25
5	0	0

The reactive and active powers being injected from the individual distributed system may be expressed in Equations (1) and (2), respectively. The power network's simple steady-state model is presented in Equation (3). The Y_{busMG} represents the admittance matrix of the bus for the distribution system.

$$P_i = \sum_{k=1}^{N} |Y_{ik} V_i V_k| \cos(\theta_{ik} + \delta_k - \delta_i) \quad (1)$$

$$Q_i = -\sum_{k=1}^{N} |Y_{ik} V_i V_k| \sin(\theta_{ik} + \delta_k - \delta_i) \quad (2)$$

$$[Y_{busMG}] \begin{bmatrix} V_1 \\ V_2 \\ V_3 \\ V_4 \\ V_5 \end{bmatrix} = \begin{bmatrix} I_{s1} \\ I_{s2} \\ I_{s3} \\ I_{s4} \\ I_{s5} \end{bmatrix} \quad (3)$$

where $|Y_{ik}|$ shows admittance connected among i-th and the k-th buses, while $|V_i|$ represents the magnitude of the voltage at the particular i-th inverter, and $|V_k|$ is the magnitude of the voltages at a particular k-th bus; θ_{ik} represents the angle of admittance among the i-th and the k-th buses, δ_k symbolizes the voltage angle at k-th bus, and δ_i represents the voltage angle at the i-th bus.

2.2. Hierarchical Controls and Cyber Network

Figure 1 provides a graphical representation of the layered hierarchical control of the complete microgrid system. The zero-level controls are localized at every node and regulate the voltages and currents for each power converter. Above these is the primary control that handles power-sharing based on a modified multi-agent-based technique with the coefficient of the power controller being adopted as (m_P*, n_Q*). The secondary/tertiary control falls above the primary and secondary layers and contains observers for power, voltage, and frequency values that calculate the above-mentioned constraints based on data from surrounding agent nodes ($\mathrm{U}_{k\in N}\omega_{oi}$, $\mathrm{U}_{k\in N}v_{oi}$, $\mathrm{U}_{k\in N}P_{\varepsilon i}$). Corrective terms are generated by the secondary controllers for power supplied to load, voltage, and frequency values (P_{refi}, ω_{ni}, V_{ni}).

2.2.1. Graph Theoretic Formulation

The measured quantities and the corresponding control signals are transmitted though the communication network layer as illustrated in Figure 2a, which can be exhibited as a digraph as $\Psi = (\Phi_g, E_g, A_g)$, consisting of a non-empty finite set of the N number of agents, located at vertices defined as $\Phi_g = \{\phi_1, \phi_2, \phi_3, \ldots, \phi_N\}$. The arcs that join the distributed vertices are represented by $E_g \subset \Phi_g \times \Phi_g$. The adjacency matrix can be described as $A_g = [a_{ij}] \in R^{N\times N}$. For a microgrid structure, the agent nodes describe system nodes for the di-graph, and the communication networks that connect them can be shown as arcs [37]. Figure 2a explains the communication among the secondary and primary controls of individual nodes. Correction terms for voltage and frequency (V_{nj}, ω_{nj}) have been produced by the observer to achieve the required references of voltage and frequency $\left(v_{ref}, \omega_{ref}\right)$ and neighbor estimates (v_{oj}, ω_{oj}). In addition, $\left(P_{refi}\right)$ calculates the average reference for the injected power based on the average power references $\varepsilon(P_{avg})$, as shown later in the mathematical analysis presented in Section 3.

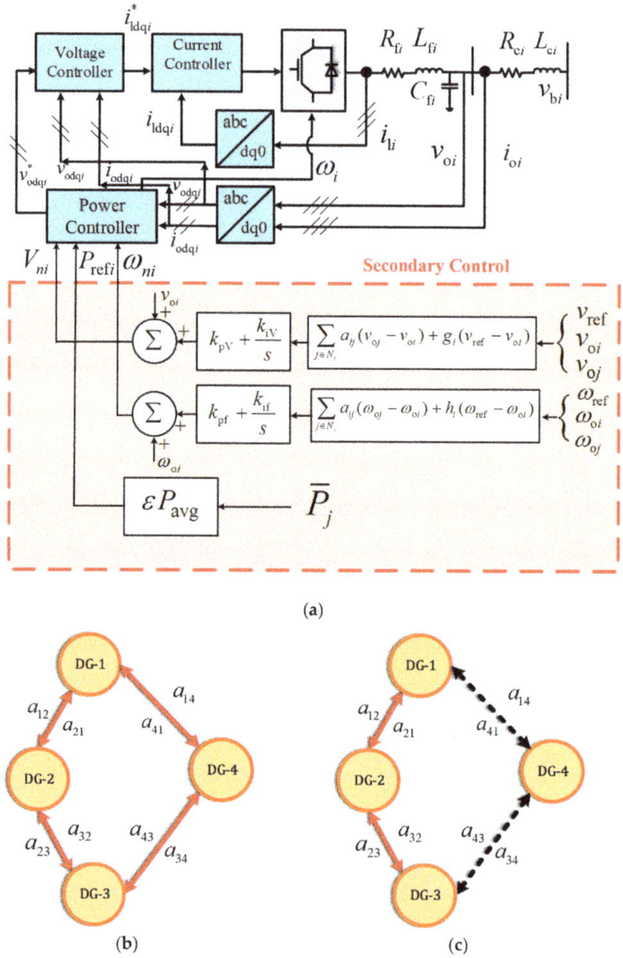

Figure 2. Representation of the distributed control scheme. (**a**) Secondary regulation; (**b**) Fully connected bidirectional communication structure; (**c**) Partially connected communication structure with latencies.

A stable, time-invariant communication network is considered as shown in Figure 2b. To simplify the estimation process, transmission noises can be neglected. A time-invariant di-graph represents the communication network, that is, A_g is considered as a constant for each run of the experiment and simulation. An arc that emanates from the node j and is directed towards the node i can be represented as (ϕ_j, ϕ_i); the node j collects data from the node i. The weight a_{ij} represents the strength of the communication link connecting v_i to v_j, and $a_{ij} > 0$ if $(\phi_j, \phi_i) \in E_g$, otherwise $a_{ij} = 0$. Node i is known as a neighbor of j if the arc $(\phi_i, \phi_j) \in E_g$. A set of the neighboring nodes to the i-th node v_i may be represented as $N_i = \{\phi_j \in V_g : (\phi_i, \phi_j) \in E_g\}$. The Laplacian matrix can be defined as, $\Lambda_g = (l_{ij})_{N \times N}$, where $l_{ij} = -a_{ij}, i \neq j$ and $l_{ij} = \sum_{j-1}^{N} a_{ij}$ for $i = 1, \ldots, N$, such that $\Lambda_{1N} = 0$, with $1_N = (1, \ldots, 1)^T \in R^N$. The in-degree matrix may be represented as $D_g^{\text{in}} = \text{diag}\{d_i^{\text{in}}\}$, where $d_i^{\text{in}} = \sum_{j \in N_i}(a_{ji})$, and the out-degree matrix as $D_g^{\text{out}} = \text{diag}\{d_i^{\text{out}}\}$, where $d_i^{\text{out}} = \sum_{i \in N_i}(a_{ij})$.

A multi-agent observer function has been employed through the control network causing the states of system $\mathbf{x}$ to converge over some delay, which can be expressed as Equations (4) and (5).

$$\dot{\mathbf{x}} = -C(\mathbf{D}_g - \mathbf{A}_g)\mathbf{x} = -C\mathbf{\Lambda}_g\mathbf{x} \tag{4}$$

where

$$\mathbf{\Lambda}_g = \mathbf{D}_g - \mathbf{A}_g \tag{5}$$

In addition, $\mathbf{\Lambda}_g$ is the Laplacian matrix calculated for the communication structure, which is dependent on the adjacency matrix $\mathbf{A}_g$ and the degree matrix $\mathbf{D}_g$. The system states that are represented by $\mathbf{x}$ and $\dot{\mathbf{x}}$ denote a vector containing values obtained through the algorithm of the consensus. The convergence factor has been denoted by C, and its value is dependent on the parameters of the network [38]. Moreover, the $\mathbf{\Lambda}_g$, $\mathbf{D}_g$, and $\mathbf{A}_g$ matrices utilized for the communication model have been illustrated in Appendices B.1–B.3, respectively.

2.2.2. Cyber-Network Link Latencies

In this study, a communication system is considered. Due to the presence of network latencies, three participating nodes have been influenced directly (i.e., DGU 1, 3, and 4), as depicted in Figure 3c. The communication links, represented by dotted lines, experience latencies, while the DGUs are regarded as 1 to 4. DGU 4 is dually influenced by the link deterioration. The communication link latencies for the networked control system can be expressed as Equation (6):

$$\left.\begin{array}{l} x_p(k+1) = A.x_p + B.u_p \\ y_p = C.x_p \end{array}\right\}$$

$$y_p = y_c(k - t_d) \tag{6}$$

where the matrices A, B, and C express the state-space matrices of a networked control system in discrete time, x_p is system state vector adopted in modeling, and u_p and y_p represent the input and outputs of the system, respectively [39]. In this study, communication delay has been approximated as the uniform unit delay t_d [38]. The value of the time delay has been incrementally varied in the discrete equal steps, so that increasing delay for the different scenarios can be emulated.

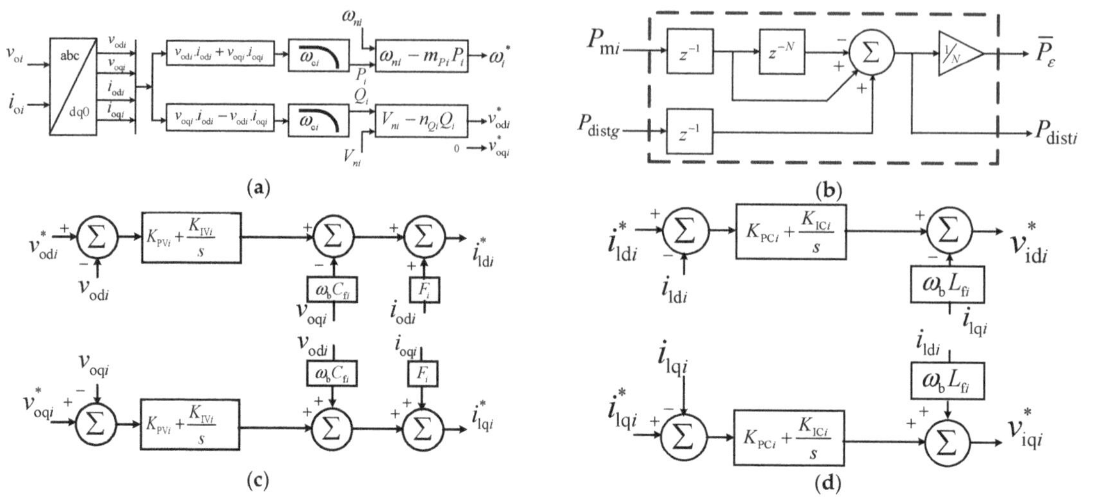

Figure 3. Control levels. (**a**) Power controller (primary control); (**b**) Distributed multi-agent moving average power estimator (secondary control); (**c**) Voltage control loop (zero level control); (**d**) Current control loop (zero level control).

3. Distributed Multi-Agent Moving-Average-Based Control

This section outlines the proposed control method for distributed value estimation and active power sharing. An overview of the method is provided in Section 3.1, followed by a detailed elaboration of the proposed estimators in Section 3.2.

3.1. Overview of the Power Estimation Method

A new control technique is devised for power control and presented here. Considering the concepts of finite and infinite impulse response filters (IIR and FIR) [40], distributed multi-agent moving average estimators have been implemented on each DGU and are described in Figures 2a and 3b. All participating DGUs calculate the average system reference on the calculations received from the neighboring agent. The average power estimates are considered as a correction term for the localized droop control by referring to Equation (7), where the frequency is reduced proportionally to obtain active power sharing accordingly. As shown in Figure 3a, Equations (8) and (9) provide the droop controller. In addition, $\mathbf{x}_{m1}$ through $\mathbf{x}_{mN}$ represent locally measured values, whereas $\mathbf{x}_1$ through $\mathbf{x}_N$ represent average estimates, and the term z^{-1} represents a transmission unit delay of one processing cycle.

$$\omega_i^* = \omega_i - m_{Pi}(P_{mi} - \overline{P}_i) \tag{7}$$

$$x_i(k+1) = \frac{x_{mi}(k-1) - x_{mi}(k-(N+1)) + x_{distg}(k-1)}{N} \tag{8}$$

$$\begin{cases} \Delta x_{mi} = x_{mi}(k-1) - x_{mi}(k-(N+1) \\ x_{disti} = \Delta x_{mi} + x_{distg}(k-1) \\ x_i(k+1) = \frac{\Delta x_{mi} + x_{distg}(k-1)}{N} \end{cases} \tag{9}$$

N represents the number of participating nodes in the network, i denotes the node currently being considered. Exchanged average power has been represented by x_{dist} within nodes: received average power from the adjacent connected nodes is shown by x_{distg}, and injected average power from node i is represented as x_{disti}. By applying the proposed method, all the nodes will converge at a "moving average" of values based on the estimates received from neighboring nodes.

3.2. Distributed Method for the Estimation of Power Injected

If the local measured values of active power at node ***i*** are denoted by x_{mi}, where $x_{mi} = [P_{mi}]$, the estimate of all the active power injected computed at neighboring nodes is given by x_{distg}, where $x_{distg} = [P_{disti}]$.

For a system consisting of *N* DGU nodes, with a multi-agent moving average observer as shown, in matrix representation, the equations can be written as follows:

$$\begin{bmatrix} x_{dist1}(z) \\ x_{dist2}(z) \\ \vdots \\ x_{distN}(z) \end{bmatrix} = \begin{bmatrix} x_{m1}(z) \\ x_{m2}(z) \\ \vdots \\ x_{mN}(z) \end{bmatrix} z^{-1}(1-z^{-N}) + \begin{bmatrix} 0 & 0 & 0 & z^{-1} \\ z^{-1} & 0 & 0 & 0 \\ 0 & \ddots & 0 & 0 \\ 0 & 0 & z^{-1} & 0 \end{bmatrix} \begin{bmatrix} x_{dist1}(z) \\ x_{dist2}(z) \\ \vdots \\ x_{dist(N-1)}(z) \end{bmatrix} \tag{10}$$

where the matrices are

$$[x_m]_{N\times 1} = \begin{bmatrix} x_{m1}(z) \\ x_{m2}(z) \\ \vdots \\ x_{mN}(z) \end{bmatrix}; \ [x_{dist}]_{N\times 1} = \begin{bmatrix} x_{Tdist1}(z) \\ x_{Tdist2}(z) \\ \vdots \\ x_{TdistN}(z) \end{bmatrix};$$
$$[Z]_{N\times N} = \begin{bmatrix} 0 & 0 & 0 & z^{-1} \\ z^{-1} & 0 & 0 & 0 \\ 0 & \ddots & 0 & 0 \\ 0 & 0 & z^{-1} & 0 \end{bmatrix} \tag{11}$$

Therefore, we may write the above system as:

$$[x_{dist}] = [Z][x_{dist}] + [x_m]z^{-1}(1-z^{-N}) \tag{12}$$

We can write:

$$[x_{dist}] - [Z][x_{dist}] = [x_m]z^{-1}(1-z^{-N}) \tag{13}$$

By simplifying, the distributed power estimation may be given as:

$$x_{disti} = \sum_{i=0}^{N} (x_{mi}) \tag{14}$$

The value of power injected at any node in the system may be estimated as:

$$\overline{x}_i = \frac{\sum_{i=0}^{N} (x_{disti})}{N} \tag{15}$$

where $\overline{x}_i = [\overline{P}_i]$ gives an estimate for the active power injected at the ith node. The total number of nodes is given by *N*. From the above analysis, it is shown that the proposed method can estimate the power injected at every node in the system.

4. Distributed Consensus-Based Controls

In [17,41], a consensus technique dependent on the observer has been described for comparison, which calculates the average power injected, $\overline{\mathbf{x}}_i = [\overline{P}_i]$, by observing the previous and current neighborhood measurements.

$$x_i(t) = x_i(t) + \int_0^t \sum_{j\in N_i} a_{ij}(\overline{x}_j(\tau) - \overline{x}_i(\tau)) \cdot d\tau \tag{16}$$

Equation (16) can be further modified by considering communication network delays t_d, in addition to the weight factors in the adjacency matrix a_{ij}.

$$\mathrm{x_i}(t) = \mathrm{x_i}(t) + \int_0^t \sum_{j \in N_i} a_{ij}\overline{\mathrm{x}}_j(\tau - t_d) - \overline{\mathrm{x}}_i(\tau)) \cdot d\tau \tag{17}$$

By considering time derivatives, the system dynamics can be obtained as:

$$\dot{\overline{\mathrm{x}}}_\mathrm{i} = \dot{\mathrm{x}}_\mathrm{i} + \sum_{j \in N_i} a_{ij}(\overline{\mathrm{x}}_j(t - t_d) - \overline{\mathrm{x}}_\mathrm{i}) = \dot{\mathrm{x}}_\mathrm{i} + \sum_{j \in N_i} a_{ij}\overline{\mathrm{x}}_\mathrm{j}(t - t_d) - d_i^{in}\mathrm{x_i} \tag{18}$$

On further simplifying, we obtain:

$$\overline{\mathrm{x}}_\mathrm{i} = \dot{\mathrm{x}}_\mathrm{i} + A_\mathrm{g} \cdot \overline{\mathrm{x}}_\mathrm{j}(t - t_d) - D_\mathrm{g}^{in} \cdot \overline{\mathrm{x}}_\mathrm{i} \tag{19}$$

where Equation (19) represents the observer-based consensus structure dynamics observers; $\mathbf{x_i} = [x_1, x_2, \ldots, x_N]^T$ depicts the measured power vector at all the connected nodes; and $\overline{\mathbf{x}}_\mathbf{i} = [\overline{x}_1, \overline{x}_2, \ldots, \overline{x}_N]^T$ represents the estimation power vector, which is attained through the averaging of the consensus structure. Through Laplace transformation, these equations can be converted into the frequency domain as:

$$s\overline{\mathrm{X}}_\mathrm{i} = s\mathrm{X_i} + sA_\mathrm{g} \cdot e^{-t_d s} \cdot \overline{\mathrm{X}}_\mathrm{j} - D_\mathrm{g}^{in}\overline{\mathrm{X}}_\mathrm{i} \tag{20}$$

With mathematical manipulation, we can obtain:

$$\overline{\mathrm{X}} = s(sI_N + A_g e^{-t_d s} - D_\mathrm{g}^{in})^{-1}\mathrm{X} = H_{\mathrm{obs}}\mathrm{X} \tag{21}$$

where $I_N \in \mathrm{R}^{N \times N}$ is the identity matrix, and H_{obs} represents the transfer function of the observer. If the interval of time is small (i.e., $t_d \to 0$), then $e^{t_d s} \to 1$. Equation (22) can be expressed as:

$$\overline{\mathrm{X}} = s(s\mathbf{I_N} + \mathbf{A_g} - \mathbf{D_g^{in}})^{-1}\mathrm{X} \tag{22}$$

With further simplifications, we obtain:

$$\overline{\mathrm{X}} = s(s\mathbf{I_N} + \mathbf{\Lambda_g})^{-1}\mathrm{X} \tag{23}$$

where the Laplacian matrix can be written as $\mathbf{\Lambda_g} = \mathbf{D_g} - \mathbf{A_g}$. If the $\mathbf{\Lambda_g}$ is a balanced matrix, the components of $\overline{\mathbf{x}}$ meet the requirements of a global reference that presents estimates for the averaged power.

5. Voltage and Frequency Restoration

At the secondary layer, voltage and frequency regulations are achieved using the neighbor multi-agent consensus [35]. The voltage restoration, along with the distributed frequency control approach, is depicted in Figure 2a. The mathematical formulation for the frequency regulation method is given by Equation (24).

$$\begin{cases} \delta\omega_i(k+1) = k_{pf}e_{\omega i}(k) + k_{if}\sum\limits_{i=ko}^{k} e_{\omega i}(k) \\ e_{\omega i}(k+1) = \sum\limits_{j \in N_i} a_{ij}(\omega_{oj}(k) - \omega_{oi}(k)) + h_i\left(\omega_{ref}(k) - \omega_{oi}(k)\right) \end{cases} \tag{24}$$

where the nominal frequency that serves as a reference is given by ω_{ref}. The system frequency ω_{oj} is calculated for the individual nodes in the neighbor of the i-th node. From Figure 2a, k_{pf} and k_{if} are the PI controller gains utilized for secondary frequency restoration. Moreover, k_{pv} and k_{iv} represent the PI controller gains for voltage restoration. The correction term utilized for the frequency set point of the i-th node can be written as $\delta\omega_\mathrm{i}$, and h_i is the pinning gain, which has a non-zero value for the primary node. The voltage regulation method is as follows:

$$\begin{cases} \delta V_i(k+1) = k_{pv}e_{vi}(k) + k_{iv}\sum\limits_{i=ko}^{k} e_{vi}(k) \\ e_{vi}(k+1) = \sum\limits_{j\in N_i} a_{ij}(v_{oj}(k) - v_{oi}(k)) + h_i\left(v_{ref}(k) - v_{oi}(k)\right) \end{cases} \quad (25)$$

where v_{nom} represents the nominal value of voltage references for the MG system per unit (p.u.), and v_{oj} represents the system voltage across for the participating nodes adjacent to node ***i***. The voltage correction items δV_i are added with the i-th inverter's reference. The pinning gain is presented as g_i, which has a non-zero value for the primary node.

6. Microgrid Modelling under Secondary Control Time-Delays

The proposed distributed control strategy has been analyzed using mathematical analysis of the control structure [38,39,42], as depicted in Figure 1. A small-signal model of the MG is constructed with the perturbation of large-signal equations. The MG system model components that have been utilized for the system design and for the stability analysis of the control structure are discussed in this section.

6.1. Primary Power Sharing Control

Figure 3a describes the power sharing scheme, whose small-signal model can be written as:

$$\begin{gathered} \begin{bmatrix} \Delta\dot{\delta} \\ \Delta\dot{P} \\ \Delta\dot{Q} \end{bmatrix} = A_P \cdot \begin{bmatrix} \Delta\delta \\ \Delta P \\ \Delta Q \end{bmatrix} + B_P \begin{bmatrix} \Delta i_{ldq} \\ \Delta v_{odq} \\ \Delta i_{odq} \end{bmatrix} + B_{P\omega com}[\Delta\omega_{com}] \\ \begin{bmatrix} \Delta\omega \\ \Delta v^*_{odq} \end{bmatrix} = \begin{bmatrix} C_{P\omega} \\ C_{Pv} \end{bmatrix} \cdot \begin{bmatrix} \Delta\delta \\ \Delta P \\ \Delta Q \end{bmatrix} \end{gathered} \quad (26)$$

where the matrices A_p, $B_{p\omega com}$, B_p, C_{pv}, and $C_{p\omega}$ have additional variables, particulars of which are offered in the Appendices A and B. Power controllers provide the operating frequency (ω_i) and the voltage references (as components: $v_{odi}{}^*$, $v_{oqi}{}^*$) for the distributed node, which feed the voltage controls, where $v_{oqi}{}^*$ has zero reference [18]. The error generated from the calculated and the estimated average value at any time instant can be expressed as a variation in injected power.

$$[\Delta P_i] = [\overline{P}_i] - [P_i] \quad (27)$$

where $\overline{P}_i$ is the averaged power estimates for the active power at a particular node i, while P_i represents the active power that has been calculated by the power control loop.

The variations in line current, load current, and inverter measurements with the effect of these perturbations are now written as:

$$\begin{aligned} [\Delta\dot{i}_{\text{linedq}}] &= A_{\text{net}}[\Delta i_{\text{linedq}}] + B_{1\text{net}}[\Delta\dot{v}_{\text{dq}i}] + B_{1\text{net}}[\Delta\dot{\omega}_i] \\ [\Delta\dot{i}_{\text{loaddq}}] &= A_{\text{load}}[\Delta i_{\text{linedq}}] + B_{1\text{load}}[\Delta\dot{v}_{\text{dq}i}] + B_{2\text{load}}[\Delta\dot{\omega}_i] \end{aligned} \quad (28)$$

The variations in parameters are updated as:

$$\begin{gathered} [\Delta\dot{x}_{\text{inv}i}] = A_{\text{inv}i}[\Delta x_{\text{inv}i}] + B_{\text{inv}i}[\Delta\dot{v}_{\text{dq}i}] + B_{i\omega\text{com}}[\Delta\dot{\omega}_i + \Delta\omega_{\text{com}}] \\ \begin{bmatrix} \Delta\dot{\omega}_i + \Delta\omega_{\text{com}} \\ \Delta i_{\text{odq}i} \end{bmatrix} = \begin{bmatrix} C_{\text{inv}\omega i} \\ C_{\text{invc}i} \end{bmatrix}[\Delta x_{\text{inv}i}] \end{gathered} \quad (29)$$

The state vector is as follows:

$$[\Delta x_{inv}] = \left[\Delta\delta_i \quad \Delta P_i \quad \Delta Q_i \quad \Delta\phi_{dqi} \quad \Delta\varsigma_{dqi} \quad \Delta i_{ldqi} \quad \Delta v_{odqi} \quad \Delta i_{odqi}\right]^T \tag{30}$$

where the matrices $A_{1\text{net}}$, $B_{1\text{net}}$, A_{load}, $B_{1\text{load}}$, $B_{2\text{load}}$, $A_{\text{inv}i}$, $B_{\text{inv}i}$, $B_{i\omega\text{com}}$, $C_{\text{in}\omega i}$, and $C_{\text{invc}i}$ are provided in the Appendices A and B.

6.2. Composite Complete Microgrid Model

Equation (31) represents the joint model for N number of power-electronics converters that are interfaced to the microgrid.

$$\begin{aligned} [\Delta x_{inv}] &= A_{inv} \cdot [\Delta x_{inv}] + B_{inv} \cdot [\Delta v_{bDQ}] \\ [\Delta i_{oDQ}] &= C_{invc} \cdot [\Delta x_{inv}] \end{aligned} \tag{31}$$

where the state vector is $[x_{inv}] = \left[\Delta x_{inv1} \quad \Delta x_{inv2} \quad \ldots \quad \Delta x_{invN}\right]^T$ and $[\Delta v_{bDQ}] = \left[\Delta v_{bDQ1} \quad \Delta v_{bDQ2} \quad .. \quad \Delta v_{bDQN}\right]^T$.

The microgrid distribution network and system loads can be modelled using KCL and KVL laws. With regards to the line currents and the node voltages, this model is given in Equations (32)–(34).

$$\begin{aligned} [\Delta \dot{i}_{\text{lineDQ}}] &= A_{NET}[\Delta i_{\text{line}DQ}] + B_{1NET}[\Delta \dot{v}_{\varepsilon dqi}] + B_{1NET}[\Delta \dot{\omega}_{\varepsilon i}] \\ [\Delta \dot{i}_{\text{loadDQ}}] &= A_{LOAD}[\Delta i_{\text{load}DQ}] + B_{1LOAD}[\Delta \dot{v}_{\varepsilon dqi}] + B_{2LOAD}[\Delta \dot{\omega}_{\varepsilon i}] \end{aligned} \tag{32}$$

where

$$\begin{cases} \Delta i_{\text{lineDQ}} = [\Delta i_{\text{lineDQ1}}, \Delta i_{\text{lineDQ2}}, \ldots \Delta i_{\text{lineDQ}n}]^T \\ \Delta i_{\text{loadDQ}} = [\Delta i_{\text{loadDQ1}}, \Delta i_{\text{loadDQ2}}, \ldots \Delta i_{\text{loadDQ}p}]^T \\ \Delta v_{\text{bDQ}} = [\Delta v_{\text{bDQ1}}, \Delta v_{\text{bDQ2}}, \ldots \Delta v_{\text{bDQ}m}]^T \\ \Delta\omega = \Delta\omega_{\text{com}} \end{cases} \tag{33}$$

and

$$\left.\begin{aligned} A_{Net} &= \text{Diag}\left[\ A_{Net1}, \quad A_{Net2} \quad .. \quad A_{NetN}\ \right]_{2n\times 2n} \\ B_{2Net} &= \left[\ B_{2Net1}, \quad B_{2Net2} \quad . \quad B_{2NetN}\ \right]^T_{2n\times 1} \\ B_{1Net} &= \left[\ B_{1Net1}, \quad B_{1Net2} \quad . \quad B_{1NetN}\ \right]^T_{2n\times 2m} \end{aligned}\right\} \tag{34}$$

The aforementioned component models are linked to obtain an entire small-signal model, which is elaborated in Equations (35) and (36). The details of the system used can be described as follows: s = four DGUs, m = five nodes, n = five lines, and p = three loads. Simulink and MATLAB are used to solve the system.

$$[\Delta v_{bDQ}] = (M_{Load}[\Delta \dot{i}_{loadDQ}] + M_{net}[\Delta \dot{i}_{lineDQ}] + M_{inv}[\Delta \dot{i}_{oDQ}])R_N \tag{35}$$

$$\begin{bmatrix} \Delta \dot{x}_{\text{inv}} \\ \Delta \dot{i}_{\text{lineDQ}} \\ \Delta \dot{i}_{\text{loadDQ}} \end{bmatrix} = A_{MG} \begin{bmatrix} \Delta x_{\text{inv}} \\ \Delta i_{\text{lineDQ}} \\ \Delta i_{\text{loadDQ}} \end{bmatrix} \tag{36}$$

where Equations (35) and (36) describe the complete system model. A_{MG} represents the system matrix, which is discussed in the Appendices A and B. It should be noted that the time delays are considered to be larger than the sampling time period and are integral multiple of the sampling time.

7. MG System Stability

An analysis of the stability of the MG is undertaken by varying the communication latencies and gains. Moreover, the limits for the tested MG network with the developed control technique are obtained by using this model. The time delays for designated communication link pairs between nodes (a_{14}, a_{41}) and (a_{34}, a_{43}) are changed incrementally to measure the impact on the MG poles and zeros.

The effect of variations in n_{Qi} and m_{Pi}, the reactive and active power control gains, respectively, have been demonstrated in Figure 4a–d. The impacts of m_{Pi} and n_{Qi} variation under the investigated distributed averaging approach can be observed in Figure 4a,b. Figure 4c,d shows the variation effect under a conventional consensus-based approach.

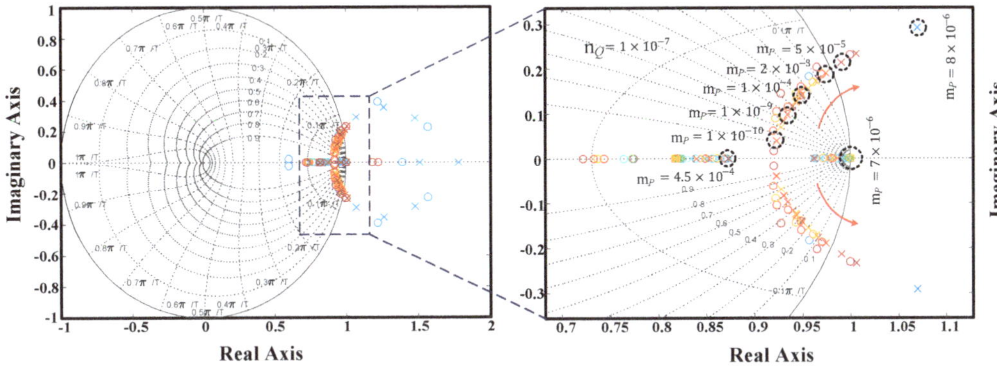

(a)

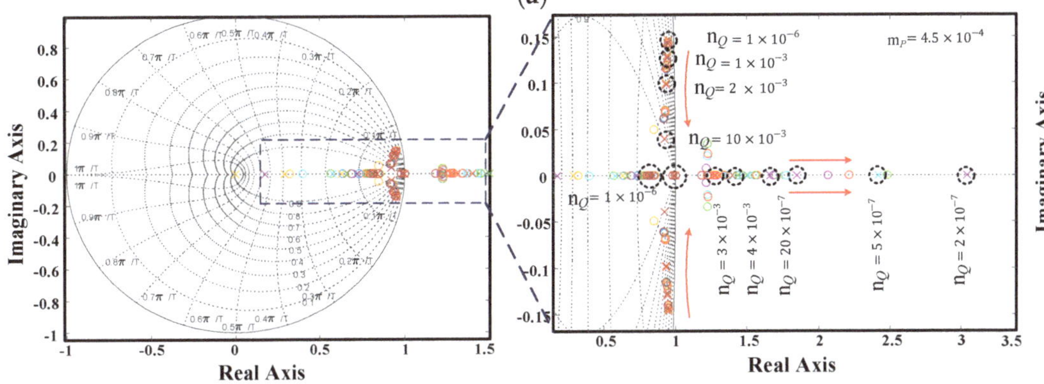

(b)

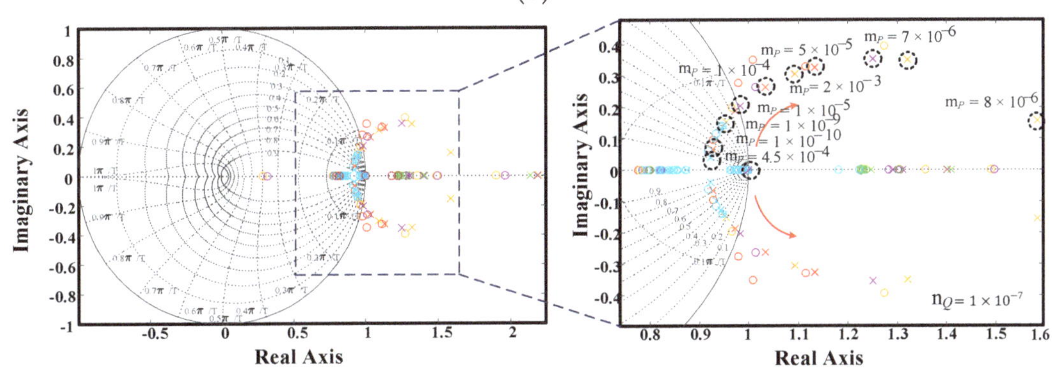

(c)

Figure 4. *Cont.*

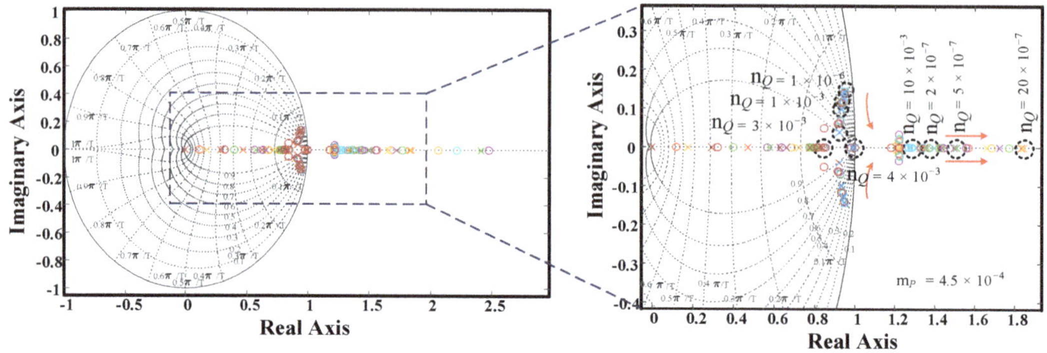

(d)

Figure 4. A comparison of stability under the proposed and other methods. (**a**) Effect of m_P variation with proposed average estimators; (**b**) n_Q variation with proposed average estimators; (**c**) m_P variation with consensus-based observers; (**d**) n_Q variation with consensus-based observers.

The control variables (m_{Pi}, n_{Qi}), where system poles exist in the range of the unit circle border, are the maximum allowable limit for defining the stability of the system. Thus, the control gain and system performance can be predicted by the location of the pole and zeros. Compared to the traditional consensus-based schemes, the investigated technique significantly increases the stability of the system under gain variations of primary-level control. It has been found that the system is more sensitive towards the variations in $\boldsymbol{n_{Qi}}$ than $\boldsymbol{m_{Pi}}$, which causes the reactive power controls to work on a narrower stability margin. The operational limits are tabulated in Table 3.

Table 3. Range of variation for controllers with time delays.

Serial. No.	Control Parameters		
	Power Controller	Minimum	Maximum
1.	Active power: m_p Reactive power: n_q	1×10^{-10} 1×10^{-7}	1×10^{-3} 1×10^{-3}
	Frequency regulation		
2.	k_{pf} k_{if}	0.45 0.14	2.55 0.53
	Voltage regulation		
3.	k_{pV} k_{iV}	0.51 0.15	3.52 0.53
4.	Communication time delay: τ_{delay}	0	2 s

The MG system's behavior with the proposed technique and existing consensus-based power-sharing techniques with the variation in the values of time delays are shown through the pole and zero traces in Figure 5a–d. It can be noted that, compared to a fundamentally consensus-based control, the proposed scheme adds system stability when the MG system is facing communication latencies. Therefore, it can be established that the proposed method using distributed averages provides a greater degree of stability to the MG system compared to the existing control when the system is facing failure of the communication and latencies.

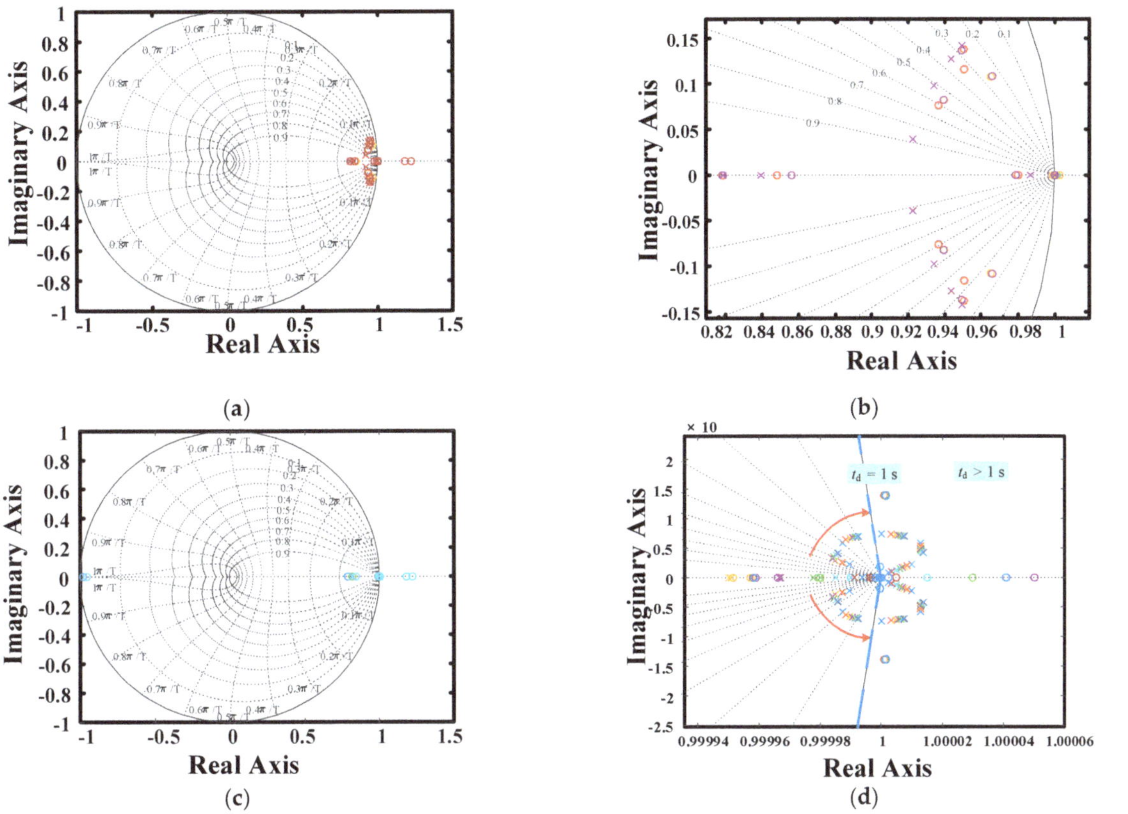

Figure 5. Comparative effect of the time delay on system stability. (**a**) Proposed distributed averaging method; (**b**) Distributed averaging method (magnified); (**c**) Estimation of values based on consensus; (**d**) Estimation of values method based on consensus (magnified).

8. Case Studies

To analyze the MG system with regard to stability, rigorous simulation-based studies have been conducted in MATLAB and Simulink, and the simulation results are shown in this section. Two different time delay cases are evaluated in two links that connect DGU-4 to simulate communication latencies. The control structure strives to achieve MG control of the restoration of system voltage and frequency and also equal power sharing among nodes, as shown in Figures 6 and 7. These results show that the proposed distributed-average-based control approach performs better than the traditional method. As with the consensus-based control, time delays lead to significant deviation in the controlled parameters. However, with the same delays, the proposed method ensures effective power sharing between the system nodes. The results are presented in the following subsections.

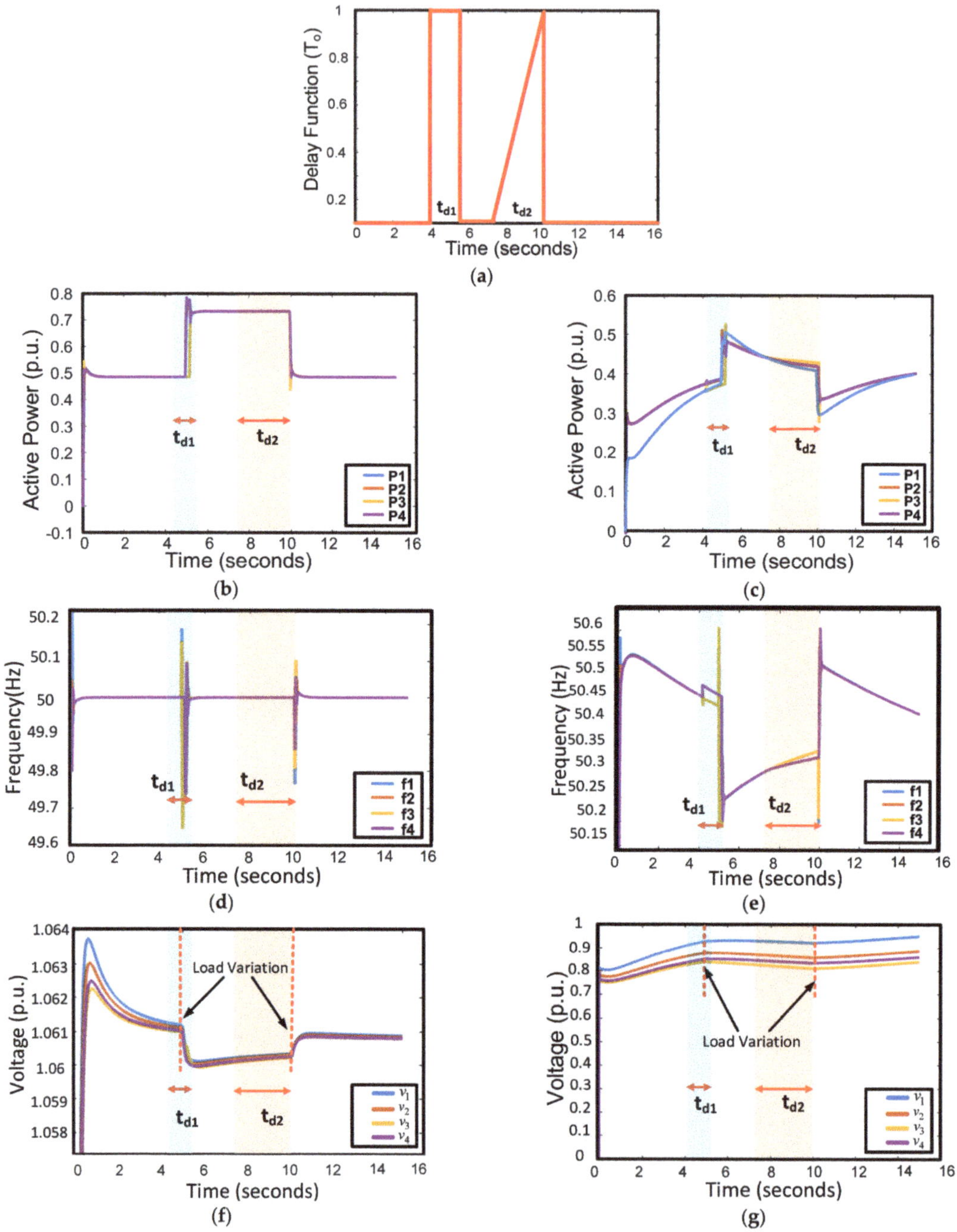

Figure 6. Control schemes' performance under time delays. (**a**) Time delays; (**b**) Distribution of active power using the proposed method; (**c**) Distribution of active-power using consensus method; (**d**) Frequency restoration with proposed method; (**e**) Frequency restoration with consensus-based method; (**f**) Voltage restoration with proposed method; (**g**) Voltage restoration with consensus method.

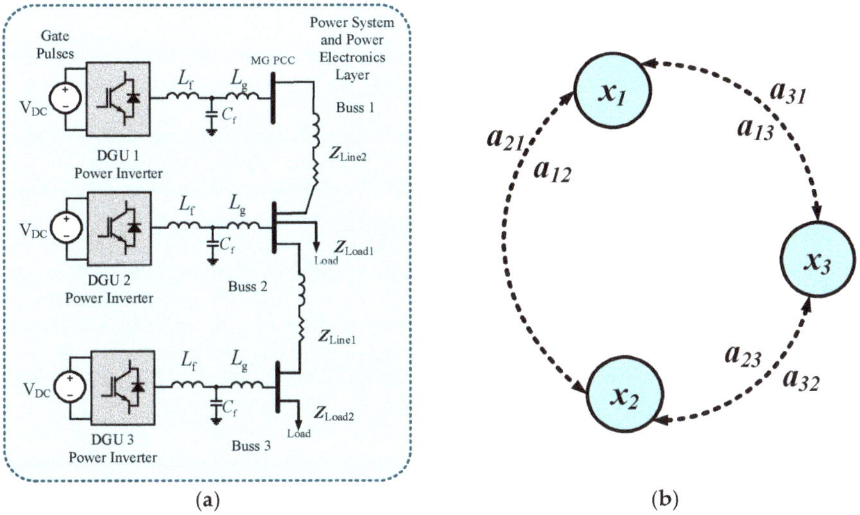

Figure 7. Experimental layout. (**a**) Experimental set-up; (**b**) Communication network structure.

8.1. Active Power Sharing

The active power-sharing approach is assessed in comparison with a consensus observer scheme, as presented in Equation (20). Figure 2b shows the communication nodes, where all the nodes are connected with at least two participating nodes. Failure of the communication link or latencies have been considered in the four links a_{14}, a_{41}, a_{34}, and a_{43}, which leads to a delay in the data transmitted and directed towards and from DGU-4, as depicted in Figure 2c. The results for active power sharing with the investigated scheme with communication link latencies and consensus-based control are presented in Figure 6a,b. The proposed approach shows lesser inaccuracies for a shorter period and converges the system to correct values within finite time (1 s compared to 5 s in the case of the existing consensus-based techniques), whereas a significant deviation in the injected power can be noticed in the case of the consensus approach.

8.2. Frequency Regulation

Figure 6d,e shows the outcomes of the frequency restoration under the proposed method and the consensus-based control, respectively. It has been observed that with the proposed control scheme, the frequency restoration, at its desired value, can be achieved in much less time compared to the existing consensus-based techniques.

8.3. Voltage Regulation

Figure 6f,g demonstrate the voltage restoration results with the proposed scheme and the consensus-based method, respectively. The figures demonstrate that the consensus-based scheme shows greater divergence in node voltages, while the voltage restoration can be achieved without prominent node voltage deviations by the multi-agent moving average estimation method.

In the control scheme suggested here, all participating converters in the MG system maintain a consensus among nodes so that desired corrective values for $P_{ref.i}$, ω_i, and V_i direct the MG system to realize the control objective in a finite time. For the completely consensus-based technique, the MG system takes a larger time to converge the voltage and frequency. It can be noted from the presented simulation results that the suggested technique has shown a better resilience when the system is suffering from communication delays, and the MG control objectives of accurate power sharing among nodes and frequency and voltage restoration can be realized in a finite time.

8.4. Time-Varying Delays in Microgrid Communication Network

The impact of the time-varying delay in the communication network, along with other transients, is discussed in this section. Multiple communication network delays affecting the communication network's DGU-4 (i.e., a_{14}, a_{41}, a_{34}, and a_{43}) are taken into consideration. One of the participating node has been periodically disconnected from its neighboring agent nodes. Load transients are examined at three of the four system buses. The time variations are given. The information signal has been delayed by $T_o(t)$, where T_o represents the time-varying delay, as depicted in Figure 3a. A generalized delay function, to represent the delayed signal, is presented such that y(t) = u(t-To(t)). Here, u(t-To(t)) is the input-delayed function for the system, and the output-delayed function is represented by y(t) [43]. Two kinds of variable delays are considered: t_{d1} represents the step delay, which spans one second and starts at the time t = 4.15 s; at t = 5.15 s, the communication links have been reconnected. In addition, t_{d2} represents the ramp function beginning at the time t = 7.5 s, which then attains the maximum at the time t = 10 s, before abruptly declining to zero. The change in the load of 0.3 p.u. has been added at the time t = 5 s on bus 2 and bus 3. At t = 10 s, these additional loads are removed. The simulation results of the active power sharing obtained with these tests using the suggested approach and those obtained by using conventional consensus-based control are shown in Figure 4b,c. Figure 4d,e compares the frequency restoration results that have been achieved by employing the proposed technique and conventional consensus-based techniques, respectively. Figure 4f,g elaborates the outcomes of the voltage restoration. By comparing the above-mentioned figures, it can be observed that the robustness of the proposed control strategy is superior to the existing consensus-based schemes when the MG system is under communication latencies and link-failure scenarios. In these scenarios, the proposed control scheme can efficiently share the active power between nodes, and the convergence time is much less, in order to achieve the MG nominal values of voltage and frequency.

9. Experimental Results

This section presents experimental results and verifications for the multi-agent moving average estimators control strategy outlined in this paper. We implement a lab-scale micro-grid set-up with DC-AC converters and adjustable R-L loads. Simplified R-L impedances are added to emulate transmission lines, as shown in Figure 7. The set-up is composed of three power inverters, each rated at 60 KVA; DC power supplies; and resistive and inductive loads. For experimental safety purposes, and keeping equipment protection in view, these were operated only at a maximum level of 100 volt-peak and 20 A-peak. To avoid leakage currents, the neutral point is kept floating. The converters are controlled through a customized control board, which embeds a TMS320F28346 Delfino micro-controller unit, EPM570 ALTERA complex programmable logic device, and AD converters. An information network layer is emulated through Ethernet links (IEEE 802.3) and a network switch. With reasonable trade-off and no great loss of generality, the distributed co-operative control schemes are emulated by implementing these in a centralized controller. The communication link delays are emulated in LabVIEW® software, which handles the higher-level controls and information flowing through the network. The power distribution, voltage, and frequency regulation results have been obtained using LabVIEW software, whereas voltage and current traces at inverter terminals have been obtained using Tek MDO3000 and RIGOL DS1052E digital oscilloscopes. The system layout is shown in Figure 7a, and the communication layout emulated in the software is shown in Figure 7b. The parameters used in the experiment are given in Table 4 and the physical layout is shown in Figure 8.

Table 4. Experimental control parameters.

Parameters		**Symbol**	**Values**
System frequency (nominal)		f^*	50 Hz
System voltage (nominal)		V^*	100 V
Switching frequency		f_s	16 kHz
DC link voltage		V_{DC}	150 V
Zero level controllers	Voltage loop controller	K_{pV1}	23
		K_{iV1}	55
	Current loop controller	K_{pC1}	42
		K_{iC1}	110
Primary controllers	Active power controller	m_{p1}	0.0035 rad /Watt
	Reactive power controller	n_{Q1}	10^{-4} rad /VA
Secondary controllers	Voltage restorative controller	K_{pVr}	2.5
		K_{iVr}	0.5
	Frequency restorative controller	K_{pfr}	3.5
		K_{ifr}	0.8
Quasi-average observer		a	0.7
Communication delay		t_d	10 ms (min)–2000 ms (max)

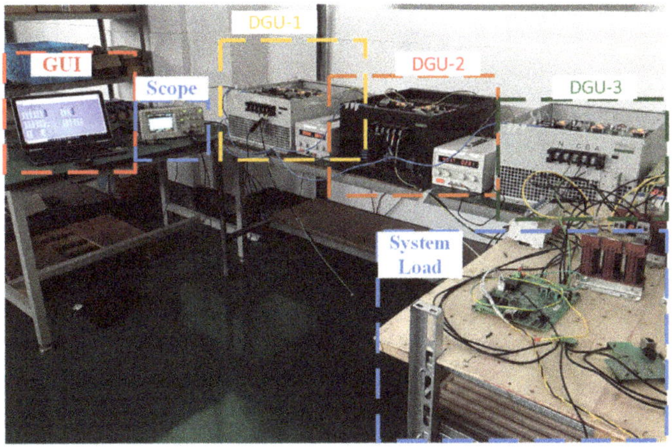

Figure 8. Experimental set-up.

9.1. Active Power Distribution

The performance of the active power distribution controller between DGUs is obtained using ethernet modules through a control and observation platform developed in LabVIEW. The system loads are initially drawing 300 watts (100 watts/DGU). At t = 6 s, the active power demand is increased to around 1360 watts (453 watts/DGU). It is observed that using droop and conventional consensus-based methods, the power sharing is achieved in a greater time span, as shown in Figure 9a. Conversely, using the proposed estimation observers scheme, the power sharing is achieved in a shorter time span and is more accurate, as shown in Figure 9b. Using both control methods, active power sharing is achieved. However, with the droop-based methods, the active power sharing between DGUs is achieved in a greater time span and is more sensitive to communication latencies, whereas using the proposed control methods, the active power sharing is achieved in a shorter time span.

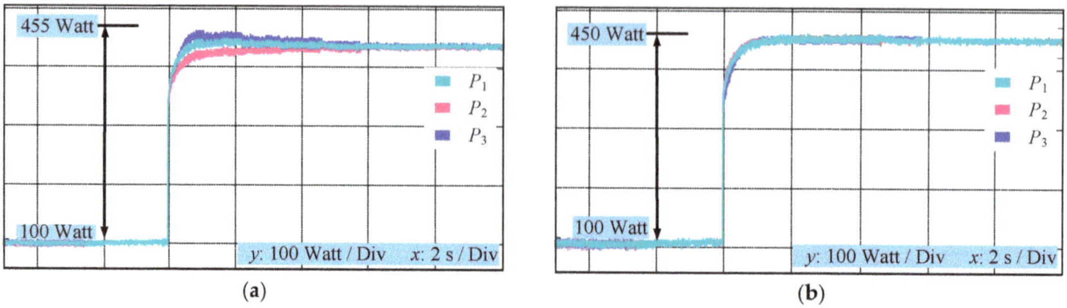

Figure 9. Performance of power distribution controllers. (**a**) Consensu-based controls; (**b**) Proposed observer-based control.

9.2. Frequency Restoration

This section presents the result of frequency restoration for the proposed control method, as compared against a conventional consensus-based control method. A time delay of td = 250 ms is emulated using the network emulated in LabVIEW. Figure 10a presents the results of frequency regulation using the proposed method. It is observed that due to the action of the proposed average estimators in the control scheme, the dynamics of frequency restoration are vastly improved. Following the load change transient that causes this frequency deviation, the frequency is restored back to the nominal value of 50 Hz within 300 ms. Figure 10b gives the results of frequency regulation using a conventional consensus-based control scheme. It can be seen that the system frequency suffers from deviations and inaccuracies with the presence of communication delays. The transient created by the load variation further deteriorates the frequency restoration.

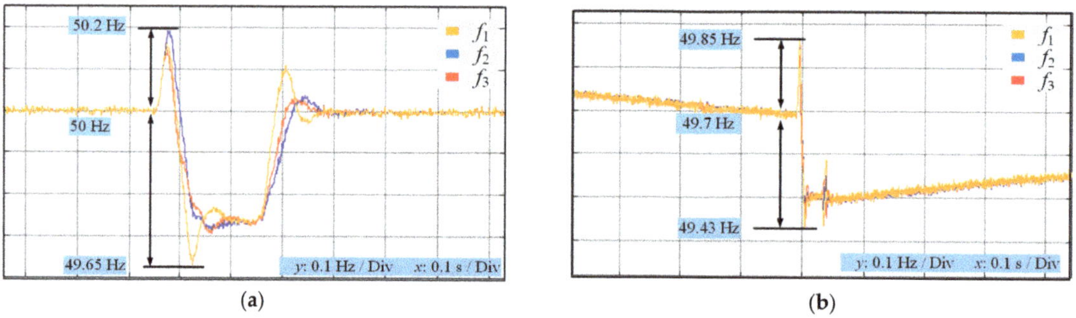

Figure 10. Frequency performance of the controllers. (**a**) Frequency restoration using proposed control method; (**b**) Frequency restoration using consensus controls.

10. Comparisons with Other Control Strategies

To verify the proposed scheme's viability, a comprehensive comparison of the suggested control scheme with the conventional consensus-based approach is presented in this section. Multi-agent-based consensus control structures, as discussed in [19,26,27,30,33], need a mutual agreement between observers of the participating node to satisfy the objective of cooperative control. All of these above-mentioned techniques are highly dependent on the communication network's health, so that desired convergence can be achieved in the MG parameters. Consensus-based control schemes are based on integral functions that employ a collective minimization of errors. Communication latencies cause localized errors that get magnified due to the integral effect of the consensus controllers, and the system values diverge, as may be observed in Figure 6b,d,f and Figure 7c,e,g. Conversely, the proposed scheme in this work takes the value of average estimates as the nominal value of references that are calculated with the aid of distributed averaging observers. The suggested estimation technique is resilient to communication latencies in transmission, as can be seen in Figure 6a,c,e and Figure 7b,c,e. In scenarios such as communication link failure and latency, the performance of the suggested control strategy is compared with existing control schemes, and it is observed that convergence to stable states in these scenarios can be achieved in a short duration of time, which shows that the proposed scheme is more robust. The proposed control scheme has exhibited higher system reliability in several aspects compared to the conventional consensus-based controls, as demonstrated in Table 5.

Table 5. Comparisons of control schema under communication latencies.

Comparison Parameters	Proposed Multi-Agent Moving Average Method	Consensus-Based Methods [18,25,26,29,32]
Active power sharing	Convergence achieved in small time interval	Convergence in larger time interval
Voltage variations	Small variations have been observed that decay in small period of time	Larger variations that decay in longer period of time
Frequency variations	Small variations that decay in small period of time	Larger variations that decay in longer period of time
Convergence: frequency	Achieved in small time	Achieved in medium time
Convergence: voltage	Achieved in small time	Achieved in larger time

11. Conclusions

This work proposes a novel multi-agent moving average estimator to observe distributed system values and thereby reduce the effect of communication latencies in the communication and control network. These estimators are incorporated into a hierarchical control structure that regulates system parameters with active power sharing and voltage and frequency regulation. Mathematical models are derived for the MG system under the proposed method and other conventional consensus-based methods. The viability and superiority of the proposed scheme over other conventional methods is demonstrated using stability analyses derived from these mathematical models. Detailed case-wise simulation studies are carried out in MATLAB, and the results for these control schemes are compared. Furthermore, a lab-scale experimental test bench is implemented to verify the results from the simulation studies. The adopted verification methods collectively establish the efficacy of the proposed control scheme and its superiority over other conventional control schemes. The proposed control method is more resilient to disturbances caused by communication latencies and shows earlier convergence than conventional consensus-based methods.

Author Contributions: Conceptualization, W.J.A., K.H., M.T.F., A.F. and M.A.M.; Methodology, W.J.A., K.H., M.T.F., A.F. and M.A.M.; Validation, W.J.A.; Formal analysis, K.H., M.T.F., A.F. and M.A.M.; Investigation, W.J.A., K.H. and S.C.; Writing—original draft, W.J.A., K.H., M.T.F., A.F. and M.A.M.; Writing—review & editing, W.J.A., K.H., M.T.F., A.F., S.C., M.A.M. and D.A.U.; Supervision, M.A.M.; Funding acquisition, S.C. All authors have read and agreed to the published version of the manuscript.

Funding: This research received no external funding.

Institutional Review Board Statement: Not applicable.

Informed Consent Statement: Not applicable.

Data Availability Statement: Not applicable.

Conflicts of Interest: The authors declare no conflict of interest.

Appendix A. System Matrices

$$A_{NETi}=\begin{bmatrix}\frac{-r_{linei}}{L_{linei}} & \omega_0\\ -\omega_0 & \frac{-r_{linei}}{L_{linei}}\end{bmatrix}$$

$$A_{Loadi}=\begin{bmatrix}\frac{-R_{Loadi}}{L_{Loadi}} & \omega_0\\ -\omega_0 & \frac{-R_{Loadi}}{L_{Loadi}}\end{bmatrix}$$

$$B_{NETi}=\begin{bmatrix}I_{lineQi}\\ -I_{lineQi}\end{bmatrix}$$

$$B_{1NETi}=\begin{bmatrix}\cdots & \frac{1}{L_{linei}} & 0 & \cdots\frac{-1}{L_{linei}} & 0 & \cdots\\ \cdots & 0 & \frac{1}{L_{linei}} & \ldots 0 & \frac{-1}{L_{linei}} & \cdots\end{bmatrix}_{2\times 2m}$$

$$B_{1Loadi}=\begin{bmatrix}\cdots & \frac{1}{L_{Loadi}} & 0 & \cdots\frac{-1}{L_{Loadi}} & 0 & \cdots\\ \cdots & 0 & \frac{1}{L_{Loadi}} & \ldots 0 & \frac{-1}{L_{Loadi}} & \cdots\end{bmatrix}_{2\times 2m}$$

$$B_{2Loadi}=\begin{bmatrix}I_{LoadQi}\\ -I_{LoadDi}\end{bmatrix}$$

$$A_{MG}=\begin{bmatrix}A_{mg1} & B_{inv}R_NM_{Net} & B_{inv}R_NM_{Load}\\ A_{mg2} & A_{Net}+B_{1Net}R_NM_{Net} & B_{1Net}R_NM_{Load}\\ A_{mg3} & B_{1Load}R_NM_{Net} & A_{load}+B_{1Load}R_NM_{Load}\end{bmatrix}$$

$$A_{mg1}=A_{inv}+B_{inv}R_NM_{inv}C_{invc}$$

$$A_{mg2}=B_{1Net}R_NM_{inv}C_{invc}+B_{2Net}C_{inv\omega}$$

$$A_{mg3}=B_{1Load}R_NM_{inv}C_{invc}+B_{2Load}C_{inv\omega}$$

$$rp_{ki}=\frac{\partial\lambda_i}{\partial a_{kk}}$$

$$A_{invi}=\begin{bmatrix}A_{Pi} & 0 & 0 & B_{Pi}\\ B_{V1i}C_{Pvi} & 0 & 0 & B_{V2i}\\ B_{C1i}D_{V1i}C_{Pvi} & B_{C1i}C_{Vi} & 0 & B_{C1i}D_{V2i}+B_{C2i}\\ B_{LCL1i}D_{C1i}D_{V1i}C_{Pvi}+ & B_{LCLli}D_{Cli}C_{Vi} & B_{LCLi}C_{Ci} & A_{LCLi}+\\ B_{LCL2i}[T_{Vi}^{-1}00] & & & \\ B_{LCL3i}C_{Pwi} & & & B_{LCLli}(D_{C1i}D_{V2i}+D_{C2i})\end{bmatrix}_{13\times 13}$$

$$B_{i\omega com}=\begin{bmatrix}B_{P\omega com} & 0 & 0 & 0\end{bmatrix}^T_{13\times 1}$$

$$C_{INV\omega i}=\begin{cases}\begin{bmatrix}C_{p\omega} & 0 & 0 & 0\end{bmatrix}_{1x13}; i=1\\ \begin{bmatrix}0 & 0 & 0 & 0\end{bmatrix}_{1x13}; i\neq 1\end{cases}$$

$$A_P=\begin{bmatrix}0 & -m_p & 0\\ 0 & -\omega_c & 0\\ 0 & 0 & -\omega_c\end{bmatrix}, B_{P\omega com}=\begin{bmatrix}-1\\ 0\\ 0\end{bmatrix} B_P=\begin{bmatrix}0 & 0 & 0 & 0 & 0 & 0\\ 0 & 0 & \omega_c I_{od} & \omega_c I_{oq} & \omega_c V_{od} & \omega_c I_{oq}\\ 0 & 0 & \omega_c I_{oq} & -\omega_c I_{od} & -\omega_c V_{oq} & \omega_c V_{od}\end{bmatrix}$$

$$C_{P\omega}=\begin{bmatrix}0 & -m_P & 0\end{bmatrix}, C_{Pv}=\begin{bmatrix}0 & 0 & -n_Q\\ 0 & 0 & 0\end{bmatrix}$$

where A_{invi}, B_{invi}, $C_{INV\omega i}$, and $B_{i\omega com}$ represent sub-matrices.

Appendix B

Appendix B.1. Adjacency Matrix

$$A_g = \begin{pmatrix} 0 & 1 & 0 & 1 \\ 1 & 0 & 1 & 0 \\ 0 & 1 & 0 & 1 \\ 1 & 0 & 1 & 0 \end{pmatrix}$$

Appendix B.2. Degree Matrix

$$D_g = \begin{pmatrix} 2 & 0 & 0 & 0 \\ 0 & 2 & 0 & 0 \\ 0 & 0 & 2 & 0 \\ 0 & 0 & 0 & 2 \end{pmatrix}$$

Appendix B.3. Laplacian Matrix

$$\Lambda_g = D_g - A_g = \begin{pmatrix} 2 & -1 & 0 & -1 \\ -1 & 2 & -1 & 0 \\ 0 & -1 & 2 & -1 \\ -1 & 0 & -1 & 2 \end{pmatrix}$$

References

1. Guerrero, J.M.; Chandorkar, M.; Lee, T.L.; Loh, P.C. Advanced Control Architectures for Intelligent Microgridspart i: Decentralized and Hierarchical Control. *IEEE Trans. Ind. Electron.* **2013**, *60*, 1254–1262. [CrossRef]
2. Ramos, F.; Pinheiro, A.; Nascimento, R.; de Araujo Silva Junior, W.; Mohamed, M.A.; Annuk, A.; Marinho, M.H. Development of Operation Strategy for Battery Energy Storage System into Hybrid AC Microgrids. *Sustainability* **2022**, *14*, 13765. [CrossRef]
3. Mohamed, M.A. A relaxed consensus plus innovation based effective negotiation approach for energy cooperation between smart grid and microgrid. *Energy* **2022**, *252*, 123996. [CrossRef]
4. Abouzeid, S.I.; Guo, Y.; Zhang, H. Coordinated Control of the Conventional Units, Wind Power, and Battery Energy Storage System for Effective Support in the Frequency Regulation Service. *Int. Trans. Electr. Energy Syst.* **2019**, *29*, e2845. [CrossRef]
5. Bharti, D.; De, M. Framework for Multipoint Optimal Reactive Power Compensation in Radial Distribution System with High Distributed Generation Penetration. *Int. Trans. Electr. Energy Syst.* **2019**, *29*, e12007. [CrossRef]
6. Han, H.; Hou, X.; Yang, J.; Wu, J.; Su, M.; Guerrero, J.M. Review of Power Sharing Control Strategies for Islanding Operation of AC Microgrids. *IEEE Trans. Smart Grid* **2016**, *7*, 200–215. [CrossRef]
7. Han, Y.; Zhang, K.; Li, H.; Coelho, E.A.A.; Guerrero, J.M. MAS-Based Distributed Coordinated Control and Optimization in Microgrid and Microgrid Clusters: A Comprehensive Overview. *IEEE Trans. Power Electron.* **2018**, *33*, 6488–6508. [CrossRef]
8. Han, Y.; Li, H.; Shen, P.; Coelho, E.A.A.; Guerrero, J.M. Review of Active and Reactive Power Sharing Strategies in Hierarchical Controlled Microgrids. *IEEE Trans. Power Electron.* **2017**, *32*, 2427–2451. [CrossRef]
9. Guerrero, J.M.; Matas, J.; de Vicuna, L.G.; Castilla, M.; Miret, J. Decentralized Control for Parallel Operation of Distributed Generation Inverters Using Resistive Output Impedance. *IEEE Trans. Ind. Electron.* **2007**, *54*, 994–1004. [CrossRef]
10. Guerrero, J.M.; De Vicuña, L.G.; Matas, J.; Miret, J.; Castilla, M. Output Impedance Design of Parallel-Connected UPS Inverters. *IEEE Int. Symp. Ind. Electron.* **2004**, *2*, 1123–1128. [CrossRef]
11. Guerrero, J.M.; Matas, J.; De Vicuña, L.G.; Castilla, M.; Miret, J. Wireless-Control Strategy for Parallel Operation of Distributed-Generation Inverters. *IEEE Trans. Ind. Electron.* **2006**, *53*, 1461–1470. [CrossRef]
12. Guerrero, J.M.; Vásquez, J.C.; Matas, J.; Castilla, M.; García de Vicuna, L. Control Strategy for Flexible Microgrid Based on Parallel Line-Interactive UPS Systems. *IEEE Trans. Ind. Electron.* **2009**, *56*, 726–736. [CrossRef]
13. De Brabandere, K.; Bolsens, B.; Van Den Keybus, J.; Woyte, A.; Driesen, J.; Belmans, R. A Voltage and Frequency Droop Control Method for Parallel Inverters. *IEEE Trans. Power Electron.* **2004**, *4*, 2501–2507. [CrossRef]
14. Alizadeh, E.; Birjandi, A.M.; Hamzeh, M. Decentralised Power Sharing Control Strategy in LV Microgrids under Unbalanced Load Conditions. *IET Gener. Transm. Distrib.* **2017**, *11*, 1613–1623. [CrossRef]
15. Xia, Y.; Peng, Y.; Wei, W. Triple Droop Control Method for Ac Microgrids. *IET Power Electron.* **2017**, *10*, 1705–1713. [CrossRef]
16. Hossain, M.A.; Pota, H.R.; Hossain, M.J.; Blaabjerg, F. Evolution of Microgrids with Converter-Interfaced Generations: Challenges and Opportunities. *Int. J. Electron. Power Energy Syst.* **2019**, *109*, 160–186. [CrossRef]
17. Bidram, A.; Nasirian, V.; Davoudi, A.; Lewis, F.L. Droop-Free Distributed Control of AC Microgrids. *IEEE Trans. Power Electron.* **2016**, *31*, 1600–1617. [CrossRef]

18. Bidram, A.; Member, S.; Davoudi, A.; Lewis, F.L.; Guerrero, J.M.; Member, S. Distributed Cooperative Secondary Control of Microgrids Using Feedback Linearization. *IEEE Trans. Power Electron.* **2013**, *28*, 3462–3470. [CrossRef]
19. Lu, L.Y.; Chu, C.C. Consensus-Based Secondary Frequency and Voltage Droop Control of Virtual Synchronous Generators for Isolated AC Micro-Grids. *IEEE J. Emerg. Sel. Top. Circuits Syst.* **2015**, *5*, 443–455. [CrossRef]
20. Wang, X.; Zhang, H.; Li, C. Distributed Finite-Time Cooperative Control of Droop-Controlled Microgrids under Switching Topology. *IET Renew. Power Gener.* **2017**, *11*, 707–714. [CrossRef]
21. Sanjari, M.J.; Gharehpetian, G.B. Unified Framework for Frequency and Voltage Control of Autonomous Microgrids. *IET Gener. Transm. Distrib.* **2013**, *7*, 965–972. [CrossRef]
22. Shahid, M.U.; Khan, M.M.; Xu, J.; Hashmi, K.; Habib, S.; Mumtaz, M.A.; Tang, H. A Hierarchical Control Methodology for Renewable Dc Microgrids Supporting a Variable Communication Network Health. *Electronics* **2018**, *7*, 418. [CrossRef]
23. Lewis, F.L.; Qu, Z.; Davoudi, A.; Bidram, A. Secondary Control of Microgrids Based on Distributed Cooperative Control of Multi-Agent Systems. *IET Gener. Transm. Distrib.* **2013**, *7*, 822–831. [CrossRef]
24. Liu, W.; Gu, W.; Xu, Y.; Wang, Y.; Zhang, K. General Distributed Secondary Control for Multi-Microgrids with Both PQ-Controlled and Droop-Controlled Distributed Generators. *IET Gener. Transm. Distrib.* **2017**, *11*, 707–718. [CrossRef]
25. Zuo, S.; Davoudi, A.; Song, Y.; Lewis, F.L. Distributed Finite-Time Voltage and Frequency Restoration in Islanded AC Microgrids. *IEEE Trans. Ind. Electron.* **2016**, *63*, 5988–5997. [CrossRef]
26. Guo, F.; Wen, C.; Mao, J.; Song, Y.D. Distributed Secondary Voltage and Frequency Restoration Control of Droop-Controlled Inverter-Based Microgrids. *IEEE Trans. Ind. Electron.* **2015**, *62*, 4355–4364. [CrossRef]
27. Agundis-Tinajero, G.; Segundo-Ramírez, J.; Visairo-Cruz, N.; Savaghebi, M.; Guerrero, J.M.; Barocio, E. Power Flow Modeling of Islanded AC Microgrids with Hierarchical Control. *Int. J. Electr. Power Energy Syst.* **2019**, *105*, 28–36. [CrossRef]
28. Lu, X.; Yu, X.; Lai, J.; Guerrero, J.M.; Zhou, H. Distributed Secondary Voltage and Frequency Control for Islanded Microgrids With Uncertain Communication Links. *IEEE Trans. Ind. Inform.* **2017**, *13*, 448–460. [CrossRef]
29. Wang, Y.; Wang, X.; Chen, Z.; Blaabjerg, F. Distributed Optimal Control of Reactive Power and Voltage in Islanded Microgrids. *IEEE Trans. Ind. Appl.* **2017**, *53*, 340–349. [CrossRef]
30. Hashmi, K.; Khan, M.M.; Habib, S.; Tang, H. An Improved Control Scheme for Power Sharing between Distributed Power Converters in Islanded AC Microgrids. In Proceedings of the 2017 International Conference on Frontiers of Information Technology (FIT), Islamabad, Pakistan, 18–20 December 2017; IEEE: Piscataway, NJ, USA; pp. 270–275.
31. Hashmi, K.; Mansoor Khan, M.; Jiang, H.; Umair Shahid, M.; Habib, S.; Talib Faiz, M.; Tang, H. A Virtual Micro-Islanding-Based Control Paradigm for Renewable Microgrids. *Electronics* **2018**, *7*, 105. [CrossRef]
32. Schiffer, J.; Seel, T.; Raisch, J.; Sezi, T. Voltage Stability and Reactive Power Sharing in Inverter-Based Microgrids with Consensus-Based Distributed Voltage Control. *IEEE Trans. Control Syst. Technol.* **2016**, *24*, 96–109. [CrossRef]
33. Guan, Y.; Meng, L.; Li, C.; Vasquez, J.; Guerrero, J. A Dynamic Consensus Algorithm to Adjust Virtual Impedance Loops for Discharge Rate Balancing of AC Microgrid Energy Storage Units. *IEEE Trans. Smart Grid* **2017**, *3053*, 4847–4860. [CrossRef]
34. Lai, J.; Zhou, H.; Lu, X.; Yu, X.; Hu, W. Droop-Based Distributed Cooperative Control for Microgrids with Time-Varying Delays. *IEEE Trans. Smart Grid* **2016**, *7*, 1775–1789. [CrossRef]
35. Hashmi, K.; Khan, M.M.; Shahid, M.U.; Nawaz, A.; Khan, A.; Jun, J.; Tang, H. An Energy Sharing Scheme Based on Distributed Average Value Estimations for Islanded AC Microgrids. *Int. J. Electr. Power Energy Syst.* **2020**, *116*, 105587. [CrossRef]
36. Hashmi, K.; Khan, M.M.; Xu, J.; Shahid, M.U.; Habib, S.; Faiz, M.T.; Tang, H. A Quasi-Average Estimation Aided Hierarchical Control Scheme for Power Electronics-Based Islanded Microgrids. *Electronics* **2019**, *8*, 39. [CrossRef]
37. Bidram, A.; Nasirian, V.; Davoudi, A.; Lewis, F.L. *Cooperative Synchronization in Distributed Microgrid Control*, 1st ed.; Grimble, M.J., Ed.; Springer International Publishing: Cham, Switzerland, 2017; ISBN 978-3-319-50807-8.
38. Coelho, E.A.A.; Wu, D.; Guerrero, J.M.; Vasquez, J.C.; Dragičević, T.; Stefanović, Č.; Popovski, P. Small-Signal Analysis of the Microgrid Secondary Control Considering a Communication Time Delay. *IEEE Trans. Ind. Electron.* **2016**, *63*, 6257–6269. [CrossRef]
39. Mahmoud, M.S.; AL-Sunni, F.M. *Control and Optimization of Distributed Generation Systems*, 1st ed.; Springer International Publishing: Berlin/Heidelberg, Germany, 2015; ISBN 978-3-319-16909-5.
40. Shenoi, B.A. *Introduction to Digital Signal Processing and Filter Design*, 1st ed.; John Wiley & Sons, Inc.: Hoboken, NJ, USA, 2006; ISBN 13 978-0-471-46482-2.
41. Nasirian, V.; Moayedi, S.; Davoudi, A.; Lewis, F.L. Distributed Cooperative Control of Dc Microgrids. *IEEE Trans. Power Electron.* **2015**, *30*, 2288–2303. [CrossRef]
42. Mariani, V.; Vasca, F.; Vsquez, J.C.; Guerrero, J.M. Model Order Reductions for Stability Analysis of Islanded Microgrids With Droop Control. *IEEE Trans. Ind. Electron.* **2015**, *62*, 4344–4354. [CrossRef]
43. *Mathworks Simulink Reference*, R2019a ed.; The MatWorks Inc.: Natick, MA, USA, 2019.

sustainability

MDPI

Article

A Multi-Objective Demand/Generation Scheduling Model-Based Microgrid Energy Management System

Ali M. Jasim [1,*], Basil H. Jasim [1], Habib Kraiem [2,*] and Aymen Flah [3]

1 Electrical Engineering Department, University of Basrah, Basrah 61001, Iraq
2 Department of Electrical Engineering, College of Engineering, Northern Border University, Arar 73222, Saudi Arabia
3 National Engineering School of Gabès, Processes, Energy, Environment and Electrical Systems, University of Gabès, LR18ES34, Gabes 6072, Tunisia
* Correspondence: e.alim.j.92@gmail.com (A.M.J.); habib.kraiem@yahoo.fr (H.K.)

Abstract: In recent years, microgrids (MGs) have been developed to improve the overall management of the power network. This paper examines how a smart MG's generation and demand sides are managed to improve the MG's performance in order to minimize operating costs and emissions. A binary orientation search algorithm (BOSA)-based optimal demand side management (DSM) program using the load-shifting technique has been proposed, resulting in significant electricity cost savings. The proposed optimal DSM-based energy management strategy considers the MG's economic and environmental indices to be the key objective functions. Single-objective particle swarm optimization (SOPSO) and multi-objective particle swarm optimization (MOPSO) were adopted in order to optimize MG performance in the presence of renewable energy resources (RERs) with a randomized natural behavior. A PSO algorithm was adopted due to the nonlinearity and complexity of the proposed problem. In addition, fuzzy-based mechanisms and a nonlinear sorting system were used to discover the optimal compromise given the collection of Pareto-front space solutions. To test the proposed method in a more realistic setting, the stochastic behavior of renewable units was also factored in. The simulation findings indicate that the proposed BOSA algorithm-based DSM had the lowest peak demand (88.4 kWh) compared to unscheduled demand (105 kWh); additionally, the operating costs were reduced by 23%, from 660 USD to 508 USD, and the emissions decreased from 840 kg to 725 kg, saving 13.7%.

Keywords: microgrid; binary orientation search algorithm; demand side management; real-time pricing; energy management; multi-objective management; generation power uncertainty; operating cost

Citation: Jasim, A.M.; Jasim, B.H.; Kraiem, H.; Flah, A. A Multi-Objective Demand/Generation Scheduling Model-Based Microgrid Energy Management System. *Sustainability* **2022**, *14*, 10158. https://doi.org/10.3390/su141610158

Academic Editors: Luis Hernández-Callejo and Mohamed A. Mohamed

Received: 25 July 2022
Accepted: 12 August 2022
Published: 16 August 2022

1. Introduction

1.1. Motivation

Stability and proper management of power system networks are crucial for societies and countries. Optimal power network operation significantly affects economic performance and consumer satisfaction [1]. Smart microgrids (SMGs) facilitate two-way communication between producers and consumers. Consequently, to encourage consumers to control their demand, various costs of electrical energy may have to be applied in what is known as a DSM program, which improves the load profile of consumers [2]. Three DSM categories—environmentally motivated type, market-driven type, and network-driven type—are commonly used, according to the literature. The environmental-driven DSM focuses primarily on environmental and social standards, such as greenhouse gas emission reduction. The network-driven type seeks to maintain system reliability, while the market-driven type seeks to save money for providers and customers [3]. Smart pricing tools for the DSM implementation process include dynamic pricing policies such as time of use (ToU) pricing, off-peak low pricing, critical peak pricing, real-time pricing, and day ahead pricing. In this study, customers' electricity consumption is influenced by a real-time price (RTP)-based DSM program.

The economics of MGs and environmentally friendly electricity generation can benefit greatly from the inclusion of locally distributed units [4]. In MG, distributed generation units such as wind turbines (WTs) and solar photovoltaic (PV) panels, as well as diesel generators (DGs), microturbines (MTs), and fuel cells (FCs), are among the most useful technologies. Despite the fact that distributed generation units, especially renewable ones, improve the MG's performance and environmental parameters, the instability of renewable units reduces the stability of the produced power from local sources. Renewable units, such as solar panels and wind turbines, are highly dependent on weather, making these units unstable. Such events in power production have an impact not only on energy availability, but also on the overall stability of the power grid. Because wind and solar resources are unpredictable, a solar-wind energy system is limited in its ability to operate without the addition of backup power sources such as batteries. Because of this, MGs also include electrical storage systems. MGs' excess electrical energy can be stored in the electrical storage batteries and used later [5]. As a result, the MG's reliance on the upstream network is reduced by local energy sources such as distributed generators and electrical storage systems. The operational schedule of local energy resources, of course, has a significant impact on their performance and the MG's efficiency. The distribution of load demand on a number of distributed generators can be as cost-effective as possible through a process known as economic dispatch, which decides the starting and stopping of each distributed generator. Dynamic economic and emission dispatch (DEED) is a critical optimization problem in the control and operation of power systems. Economic dispatch selects which generators to use to meet electricity demand. It resembles clearing the electricity market. The utility creates an overall marginal cost (supply) curve. The economic operation of a power system is investigated by estimating the penalty factor of the nodes of generation using an approximation of the active power losses. By simultaneously minimizing emission and operation costs, the DEED problem provides online generating schedules over a certain predicted load demand period [6–8].

1.2. Literature Review

Our research focused on the three previously mentioned issues: (1) DEED, (2) DSM, and (3) a multi-objective energy management system. To solve the DEED problem, many optimization techniques have been proposed in the literature. For DEED problems, the authors of [9] proposed the enhanced genetic algorithm (E-GA) and enhanced differential evolutionary (E-DE) algorithms. The DEED problem was solved using a combination of GA and DE in [10], with different generating unit combinations such as hydrothermal, solar–thermal, and wind–thermal. The authors of [11] proposed an enhanced PSO-based DEED problem with wind uncertainties. In [12–14], a multi-agent consensus-based distributed energy management system is proposed. This system takes into account the impact of packet losses in order to eliminate real power mismatch and lower the cost of electricity bills. The issues of the multi-objective optimizer and load shifting-based DSM for cost reduction were not presented. The authors of [15] created a hybrid planning model of DGs and distribution automation (DA) to improve economic, reliability, and operation indices. The objective function minimizes operation, investment, energy loss, and reliability costs. This paper used a stochastic programming approach based on a hybrid simulation called Monte Carlo and simultaneous backward approach to model the uncertainty parameters, such as load, energy price, and network equipment availability. The authors of [16] presented a two-layer energy management model in a smart distribution network that considers flexi-renewable virtual power plants. This model contains market price, load, maximum renewable energy source power, and flexible source demand uncertainties, which are modeled using stochastic programming. The model includes a bi-level optimization model solved by Benders decomposition for a fast solution.

The DSM is the second problem associated with MG energy management. DSM is also regarded as an optimization problem. The DSM program should be used with a number of controllable or shiftable devices, each with its own set consumption pattern. As a result, many meta-heuristic and evolutionary optimization algorithms are preferred for dealing with such complexities [17]. The authors of [18–21] only studied an optimal DSM program based on load shifting to reduce the cost of electricity bills. The issues regarding the multi-objective optimizer and economic dispatch were not addressed in these studies.

Nowadays, the emphasis is primarily on the optimization of DEED and DSM in combination. A GA-based DED and DSM combination for efficient energy management in a microgrid environment was proposed in [22]. The authors of [23] proposed a novel model of the DED problem in regional grids that included DR. The authors of [24,25] proposed the DEED problem and a time-of-use dynamic price-based DSM (not optimal program) combination model with high wind penetration. The primary goal of the literature's DEED and DSM combination models was to demonstrate the effects of DSM on the supply side. Some researchers have studied MGs and DSM in recent years. The authors of [26] proposed a cloud-based multi-agent framework for MG DED and monitoring. This paper adopted an optimal DSM-based time-of-use pricing model. Multi-objective optimization and costs of operation and emission issues for multi-DGs were not applied. In other research, the simultaneous minimization of residential peak load and electricity cost has been presented [27]. On the basis of the time-of-use program, the distribution system's DR program has been evaluated. The issue under consideration has been modeled as multi-objective mixed-integer linear programming. The authors of [28] solved the energy management of a MG connected to the utility power system under probabilistic and deterministic conditions, taking into account the variations of load demand, photovoltaic (PV), and wind turbine (WT) systems. Using an equilibrium optimizer (EO) algorithm, the problem of energy management is solved for a multi-objective function that includes cost minimization, stability improvement, and voltage profile improvement. The authors of [29] solved the issue of energy management within the microgrid by combining customer-oriented with utility-oriented DSM strategies. In light of this, a stochastic energy management framework has been developed in order to implement and evaluate the flexible load-shaping DSM strategy with price-based and incentive-based demand response programs (DRPs) in existing non-dispatchable energy resources. Pedro Faria et al. [30] used the PSO algorithm to minimize the operational cost of distributed energy resources by taking network constraints and demand response into account. In [31], a probabilistic model was adopted to search for multi-objective operation for a smart distribution network with renewable resources (wind and solar). For predicting changes in wind speed and solar radiation, the PDF Rayleigh and beta PDFs PDF were employed. Modeling the creation of solar and wind power at the same time were not taken into account. Despite this, pollutants such as SO_2 and NOx are not taken into account when using the three-constraint method to solve problems. ESS and controllable DGs are not included in the MG configuration proposed in [32], which has four objectives of reducing customer peak load, load curve, costs, and emissions fluctuations while including renewable energy and electric vehicles. An economic model for power dispatch to minimize the operating costs in an AC–DC hybrid MG was presented in [33]. This strategy considered the unpredictability of load demand as well as renewable resources. The method of Hong's two-point estimate was utilized in order to model the uncertainties. The PSO and fuzzy logic systems were used in conjunction with one another to solve the economic dispatch problem. In [34], a genetic algorithm was used for power dispatching in a grid-connected MG in order to minimize the operating costs of PV, WT, FC, and MT systems. An optimal DSM program was not adopted in these studies in order to reduce operation and emission costs as much as possible. Table 1 summarizes the prior research that has been applied to DEED, DSM, and multi-objective function optimization strategies.

Table 1. Summary of previous research studies.

Ref.	Resolved Problem(s)	Limitations
[9–16]	DEED	Demand-side scheduling was not optimized, and optimal multi-objective energy management was not applied.
[18–21]	DSM	DEED and multi-objective energy management were not applied.
[22–34]	DSM and DEED	In [24,25], the optimal DSM is not used. In [26,32,34], multi-objective optimization functions are not employed. The authors of [22,23,27–29,31,33] proposed multi-objective optimization with DR only, and load appliances scheduling based on intelligent DSM with a load-shifting program was not implemented.

1.3. Contributions

According to the previous studies, it is possible to assert that the optimal DSM-based multi-objective supply management optimization of a smart MG that considers randomized natural behavior of RERs is the one that has received the least attention. Because of this, multi-objective management systems were researched and investigated in a MG as part of this research. The following are some novel aspects of this study:

1. This study proposed BOSA-based DSM of a SMG with a MOPSO-based DEED to improve economic and environmental issues. DSM improves the MG load pattern by adopting real-time pricing. MG operators can meet system demand with optimal management of WT, PV panel, DG, MT, and FC energy storage systems, and upstream networks.
2. With the stochastic nature of the renewable resources units, a multi-objective supplier/consumer management system for a SMG based on PDFs has been presented to model the behavior of solar and wind systems, as well as a hybrid wind and solar system, in an effort to achieve the best possible results despite the uncertainty of the situation.
3. The proposed optimal DSM is based on real-time dynamic pricing and the first-ever application of the BOSA optimization algorithm, which employs the load-shifting technique.
4. Using a combination of the optimal DSM program, a multi-objective particle swarm optimizer with the Pareto criterion and fuzzy mechanism based nonlinear sorting was used to find the best MG management program.
5. By utilizing algorithms for optimizing usage of the MG sources and loads, an economic dispatch can be achieved with optimally lower operation costs and pollution outcomes.

1.4. Paper Organization

The remaining sections are organized as follows: Section 2 describes the problem statement. Distribution energy resources and stochastic modeling are demonstrated in Section 3. Section 4 describes the proposed optimal DSM program. The proposed multi-objective optimization model is described in Section 5. The adopted smart MG system is illustrated in Section 6. In Section 7, the BOSA-based MOPSO algorithm is discussed. Results and discussion are covered in Section 8. In Section 9, the main conclusion is presented.

2. Problem Statement

Within the scope of this investigation, a probabilistic model is suggested for demand/generation energy management in SMGs in order to reduce emissions and operational costs. Firstly, a BOSA-based optimal DSM program is proposed to minimize the peak energy consumption using the load-shifting technique by managing the electricity demand of various consumers. The main focus is on reducing peak to average energy consumption by running select appliances when grid stress is low. To that end, all customer loads are first classified as shiftable and non-shiftable. Because shiftable loads play an important role in lowering the peak-to-average ratio, we proposed load-shifting technique in our strategy to optimize the load profile. Due to the inherently unpredictable nature of the wind and solar energies, it is impossible to accurately predict their output and they are always attributed with planning errors due to uncertainty for the following day. Consequently, a PDF is used to model the operation of the solar, wind, and hybrid solar–wind power systems in an effort to achieve the optimal possible results despite the inherent uncertainty in these systems, so as to increase conformity between planning and reality. DEED is one of the most important optimization problems in controlling and operating power systems. DEED recommends a model that uses a MOPSO based on non-dominated sorting and a fuzzy optimization tool to minimize the emissions and operation costs. This model is used to create generating schedules for a predicted load demand period. Additionally, a new price-based DSM program is proposed as a way to solve the dilemma of the high peak unscheduled load profile, to save electricity, and reduce the cost. The section that follows is devoted to modeling and introducing the objective functions.

3. Distribution Energy Resources

In this study, the MG has renewable energy resources such as PV panels and WTs, and nonrenewable ones such as DGs, MTs, and FCs. The adopted grid-connected MG also uses ESS. A DG generates electricity by using a diesel engine and an electric generator. The output power can be adjusted in response to network demand. The MT has the unique ability to simultaneously generate both electricity and heat. A distributed generation unit can generate electricity based on the amount of load it must support. In a FC, the chemical energy of a fuel is converted into electrical energy. Using an ESS, electrical energy can be converted to a form that can be stored and used again at a later time. ESS can help mitigate the effects of renewable energy's inconsistency when it is used in conjunction with a MG.

3.1. Renewable Units Stochastic Modeling

The adoption of renewable energy units such as solar panels and WTs can have a significant impact on environmental issues. One of the drawbacks of WTs and PV panels is their stochastic behavior. For this reason, the stochastic behavior of renewable units is modelled by the PDF. First, the modeling of renewable units is explained, and then the stochastic parameter generation method is described.

3.1.1. PV Panel

The energy generated by a PV panel is proportional to the solar irradiance. The variation of solar radiation can be modeled using the beta PDF. The probability function of solar irradiance is mathematically represented by Equation (1) [35,36]. This equation is valid if $0 \leq x \leq 1,\ \alpha \geq 0,\ \beta \geq 0$.

$$f(x) = C.x^{\alpha-1}(1-x)^{\beta-1} = \frac{1}{\int_0^1 u^{\alpha-1}(1-u)^{\beta-1}du} x^{\alpha-1}(1-x)^{\beta-1} \tag{1a}$$

$$f(x) = \frac{\Gamma(\alpha+\beta)}{\Gamma(\alpha)\Gamma(\beta)} x^{\alpha-1}(1-x)^{\beta-1} \tag{1b}$$

The mean (μ) and variance (σ^2) can be used to express the and β and α parameters as follows,

$$\beta = (1-\mu)\left[\frac{\mu(1+\mu)}{\sigma^2} - 1\right] \tag{2}$$

$$\alpha = \mu\left[\frac{\mu(1+\mu)}{\sigma^2} - 1\right] \tag{3}$$

where the gamma function is denoted by $\Gamma(\ )$; C is constant; α and β are beta PDF parameters. The parameter x represents solar irradiance, while $f(x)$ represents the probability of solar irradiance according to the beta function. The active power produced by the photovoltaic panel can be calculated using Equation (4) [37]. Here, A and λ represent the panel's surface area and efficiency, respectively.

$$P_{pv}(x) = A.\lambda.x \quad (4)$$

As a result, when Equation (1) is used, the PDF $(f_{pv}(P_{pv}))$ for the output power of P_{pv} can be written as follows:

$$f_{pv}(P_{pv}) = \frac{\Gamma(\alpha+\beta)}{\Gamma(\alpha)\Gamma(\beta)}(A.\lambda.x)^{\alpha-1}(1 - A.\lambda.x)^{\beta-1} \quad if\ P_{pv} \in [0, P_{pv}(s)] \quad (5)$$

3.1.2. Wind Turbine

In general, the Rayleigh distribution PDF is given by:

$$f(x) = \frac{xe^{-(x^2/2b^2)}}{b^2} \quad (6)$$

where b is the distribution's scale parameter. The CDF is defined as [38]:

$$F(x) = 1 - e^{-\left(\frac{x^2}{2b^2}\right)} \quad (7)$$

where the wind speed is $(x = V_w)$, $x \in [0, \infty)$, and b is the scale parameter.

Wind turbine output is modeled using the Rayleigh distribution based on the behavior of wind speed [39]. The Rayleigh distribution is a subtype of the Weibull distribution with a shape index of 2.

If V_d is a site's average wind speed, then the scale parameter:

$$V_d = b\sqrt{\frac{\pi}{2}} = 1.253b \quad (8)$$

$$b = \frac{1}{\sqrt{\frac{\pi}{2}}}V_d \quad (9)$$

As a result, by substituting α in PDF and CDF, the WT system's Rayleigh model will be achieved as a function of the average wind speed, as shown in (10) and (11).

$$f_V(V_w) = \frac{2}{\pi}\frac{V_w}{{V_d}^2}e^{-(\pi/4)(V_w/{V_d}^2)^2} \quad (10)$$

$$F_V(V_w) = 1 - e^{-\left(\frac{\pi}{4}\right)\left(\frac{V_w}{V_d}\right)^2} \quad (11)$$

The output power of a specific WT system can be defined as follows [40]:

$$P_w(V_w) = \begin{cases} P_{WR} & V_R \le V_w \le V_{co} \\ P_{WR}\frac{V_w - V_{ci}}{V_R - V_{ci}} & V_R \ge V_w \ge V_{ci} \\ 0 & V_R \le V_w \le V_{ci}\ and\ V_w \ge V_{co} \end{cases} \quad (12)$$

In this equation, V_w, V_{ci}, V_R, and V_{co} represent the wind, cut-in, rated, and cut-out speeds of the WT, while P_{WR} represents its power output. The wind turbine adopted in this study is of the type AIR403 [41], with P_{WR} = 15 kW, V_{ci} = 3.5 m/s, V_{co} = 18 m/s, and V_R = 17.5 m/s.

In this paper, the PDF $(f_P(P_w))$ for the output power of a WT system can be obtained by applying the transformation theorem as follows [42]:

$$f_{p_w}(P_w) = \begin{cases} 1 - [F_V(V_{co}) - F_V(V_{ci})] & P_w = 0 \\ \left(\frac{V_R - V_{ci}}{P_R}\right)\left(\frac{\pi}{2{V_d}^2}\right)\left(V_{ci} + (V_R - V_{ci})\frac{P_w}{P_R}\right)e^{-\left(\frac{V_{ci}+(V_R-V_{ci})\frac{P_w}{P_R}}{\frac{2}{\sqrt{\pi}}V_d}\right)^2} & 0 < P_w < P_R \\ F_V(V_{co}) - F_V(V_R) & P_R = P_w \end{cases} \tag{13}$$

Power generation by hybrid WT–solar PV system (P_h) equals the sum of the power output from the WT system and the PV system.

$$P_h = P_w + P_{pv} \tag{14}$$

Assuming that P_w and P_{pv} are performance-independent according to Equations (4) and (12), a random variable's density function P_h is defined as the convolution of density functions P_w and P_{pv} [43]:

$$f(P_h) = f_{p_w}(P_w) * f_{pv}(P_{pv}) \tag{15}$$

4. The Proposed DSM Program

DSM techniques modify customer demand patterns to change the load curve shape in order to minimize the peak consumption [44]. DSM focuses on energy-saving technologies and financial incentives rather than expanding the transmission and distribution grid or generation capacity. Peak periods of the distribution system's load profile can be rescheduled using an appropriate objective and DSM methodology to eliminate system instabilities caused by high load demand. Six DSM techniques alter the load profile curve. Peak-clipping, valley-filling, load-shifting, load shape flexibility, load growth, and load conservation [45,46]. Figure 1 shows DSM strategies.

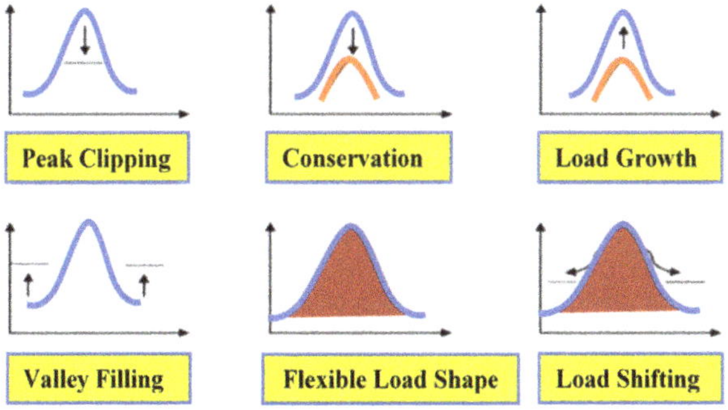

Figure 1. DSM techniques.

Peak clipping involves removing peaks above a certain consumption point to reduce peak demand. Valley filling collects energy storage devices to redistribute loads during off-peak hours [47]. Load shifting shifts on-peak loads to off-peak periods, reducing peak energy demand. Strategic conservation objectives can improve load profiles by reducing customer demand over a daily time. Strategic load growth helps people respond quickly to high demand. Load shape affects SMG reliability [48]. Individuals can participate in the load control strategy, called flexible loads in SMG management. In this paper, a load-shifting technique based optimal DSM using BOSA is proposed to minimize the peak energy consumption. The optimal load reduction profile is applied as an input to the single objective or multi-objective PSO algorithm for DEED purposes. Figure 2 shows the optimal DSM strategy flowchart. In the first step, a survey is conducted to collect load information. After classifying the loads, a load profile comprising both shiftable and non-shiftable loads was generated. With the aid of the load curve, the consumption of peak load and PAR were analyzed. In addition, the durations of on-peak and off-peak hours were determined using the load profile curve. Hourly consumption limits were compared to the pre-determined hourly consumption limit. Depending on the types of appliances that were in use at the time, the excess energy consumption was reduced through load shifting. Load shifting is implemented if shiftable loads were in operation at the time. This entire procedure was observed for 24 h.

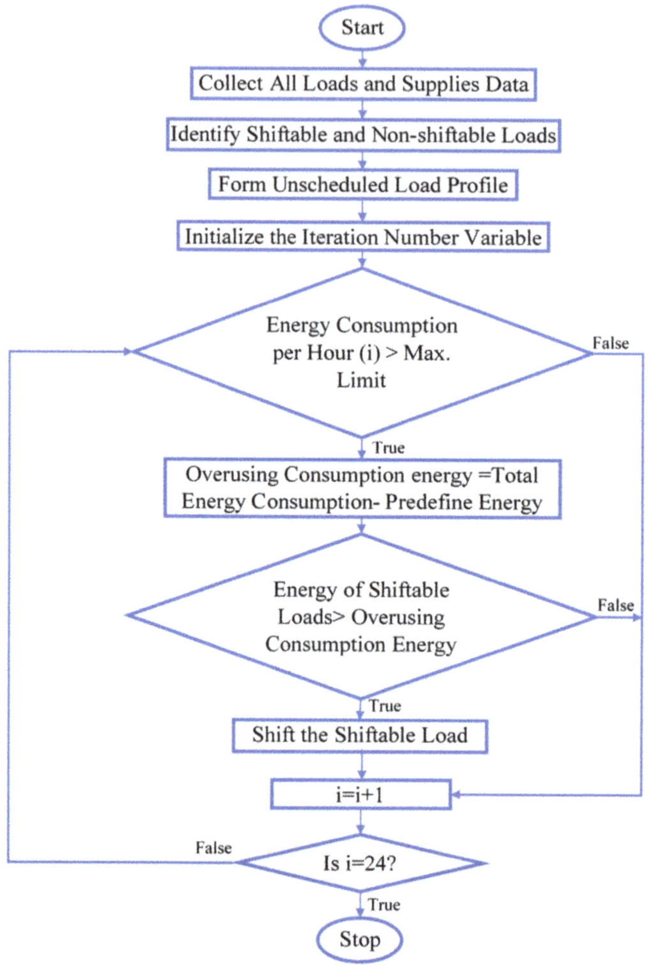

Figure 2. The flowchart of optimal DSM strategy.

4.1. DSM Objective Function

The desired purpose is to reduce the PAR of total energy usage by electrical service providers. The proposed DSM program based on load shifting schedules each of the system's shiftable loads such that the energy consumption curve is as near as feasible to the ideal energy consumption curve. In addition, time slots and shiftable loads are considered movable components. Our objective is to reduce the user's power price while decreasing the PAR to increase the grid's efficiency. The statement of the minimization problem is as follows:

$$Minimize: \sum_{t=1}^{24} \sum_{n=1}^{N} \sum_{m=1}^{M} X(n,m,t)E(n,m)PR(t) \quad (16)$$

$$Subject\ to: \quad P_{Max}(t) \geq \sum_{t=1}^{24} \sum_{n=1}^{N} \sum_{m=1}^{M} X(n,m,t)E(n,m) \tag{17}$$

$$M_n = 24 - l_n \tag{18}$$

where t represents the time slots, n represents the number of appliances, m represents the appliance type, $PR(t)$ represents the price of electricity at time slot t, X represents the ON/OFF state of the device, E represents the energy consumption, and $P_{Max}(t)$ represents the maximum power. Maximum allowable delay of the appliance m is denoted by M_n, and the appliance's duration of operation is l_n. This illustrates that the total energy consumption of N appliances of M kinds during time slot t is equal to or less than the maximum permitted power for decreasing the peak-to-average ratio.

4.2. Constraints

During the process of load scheduling, constraints must be addressed. For instance, the total number of shiftable appliances should surpass the number of loads moved in an hour. Otherwise, we must rein down excessive demand. In addition, there is time-shift restriction for shiftable loads; we may postpone or advance it within the allowed range. Equation (19) states that the number of shiftable appliances at the time step t cannot be fewer than the number of shifted appliances, since DSM can only plan load shiftability. Finally, we impose a time constraint on the shifting of controlled loads

$$S(n,m,t) \leq \sum_{t=1}^{24} H(n,m,t) \qquad \forall -T \leq t \leq T \tag{19}$$

where S represents the shifted appliances, H represents the shiftable appliances, and T represents the maximum time shift.

At all times, the user's power consumption must be less to or equal to the maximum power consumption. This is demonstrated using Equation (20):

$$P^{Demand}(t) \leq P_{max}^{Demand}(t) \qquad \forall\, t \in [0,24] \tag{20}$$

where $P^{Demand}(t)$ is the power demand at hour t-th of the day and $P_{max}^{Demand}(t)$ represents the authorized maximum power demand limit.

To achieve the load duration criteria, all controllable and non-controllable loads are scheduled for the whole 24 h of the day. The load duration is the anticipated operating time for the device type:

$$Subject\ to \sum_{t=1}^{24} \sum_{m=1}^{M} X(m,t) = d_m \tag{21}$$

where d_m represents the operating hours number associated with the load type "m".

Demand for shifted and scheduled loads should be equal to the total daily demand for loads before scheduling:

$$Subject\ to \sum_{t=1}^{24} \sum_{m=1}^{M} B(m,t) = \sum_{t=1}^{24} \sum_{m=1}^{M} A(m,t) \tag{22}$$

where $B(m,t)$ is the total daily demand for the m-th type of load shifting before the t-th hour and $A(m,t)$ is the total daily demand for the m-th type of load shifting after the t-th hour.

The load shifting-based distributed scheduling mechanism addresses the appliances in each time slot cumulatively and generates a full pattern as a result of solving the minimization problem.

4.3. Binary Orientation Search Optimization Algorithm

The BOSA presented in (2019) simulates orientation game rules. Players follow the referee's instructions in this game. The players' initial or starting positions are shown in Equation (23) [49]

$$X_i = \left(x_i^1, \ldots, x_i^d, \ldots, x_i^n\right) \tag{23}$$

where x_i^d represents the position of player i of d-dimension, and n is the number of variables.

In each iteration, the referee is the player with the highest value of the fitness function, as described in Equation (24) [49].

$$\text{Referee} = \begin{cases} Maximization\ problem : location\ of\ \max(f) \\ Minimization\ problem : location\ of\ \min(f) \end{cases} \tag{24}$$

The letter f denotes the value of the fitness function.

The direction in which the referee's hand moves does not always correspond to the direction wherein the referee moves. The referee's hand is the only consideration for players. To simulate the direction, Equations (25) and (26) are used [49]:

$$P_i = 0.8 + 0.2\frac{t}{T} \tag{25}$$

$$Orientation_i^d = \begin{cases} sign\left(\text{Referee}^d - Player_i^d\right) & for\ \ rand < P_i \\ -sign\left(\text{Referee}^d - Player_i^d\right) & else \end{cases} \tag{26}$$

At t iteration, and maximum iteration T.

While each player must move in the referee's direction, some may not be able to. Equations (27) and (28) model this problem.

$$error = 0.2\left(1 - \frac{t}{T}\right) \tag{27}$$

$$x_i^d = \begin{cases} x_i^d + rand * Orientation_i^d * x_{ho}^d & for\ \ rand < error \\ x_{lo}^d + rand * \left(x_{ho}^d - x_{lo}^d\right) & else \end{cases} \tag{28}$$

where x_{ho}^d and x_{lo}^d represent the upper and lower limit.

Particle positions in discrete space are represented by the numbers zero and one for each of the two dimensions. Changing an agent's value from zero to one or from one to zero corresponds to its movement in any dimension. As a result, a probability function is used to calculate the player's displacement in each dimension, and the player's position is then updated as a result. Probability functions in the BOSA are restricted to the intervals [0–1]. In Equation (29) [49], the probability function $S(dX^{j,d}(t))$ is shown.

$$S(dX^{j,d}(t)) = \left|\tanh\left(dX^{j,d}(t)\right)\right| \tag{29}$$

The new position of every player is simulated in accordance with the probability function (30).

$$X^{j,d}(t+1) = \begin{cases} complement\left(X^{j,d}(t)\right) & for\ \ rand < S(dX^{j,d}(t)) \\ X^{j,d}(t) & else \end{cases} \tag{30}$$

A description of the various steps involved in the BOSA is provided as follows

BOSA Algorithm Steps
• Step 1: the system space will be defined, and the initial values will be selected. • Step 2 is the initial positioning of the players. • Step 3 is the evaluation of the players. • In Step 4, the referee will decide which way each player should face. • Step 5: Maintaining an up-to-date position for the referee. • In Step 6, the probability function for displacement is computed, which is the final step. • Step 7 involves bringing the status of the player up to date. • Step 8: repeat Steps 3–7 indefinitely or until the stop condition is satisfied, whichever comes first.

In this study, generation demand-side management is subjected to a multi-objective MG problem. Reducing daily MG operation cost and pollutant production are the main objectives. These indices are optimized with production–consumption power balance and energy storage system constraints. Mathematical explanations of objective functions and constraints follow.

5. Multi-Objective Optimization Model

A stochastic programming model with multiple objectives will be utilized to investigate the effects of existing renewable generation resources, such as wind and solar photovoltaics, on operation costs and pollution emissions. In this study, residential (RL), industrial (IL), and commercial (CL) demand are considered responsive with or without adopting the optimal DSM program. Figure 3 illustrates the proposed optimization model.

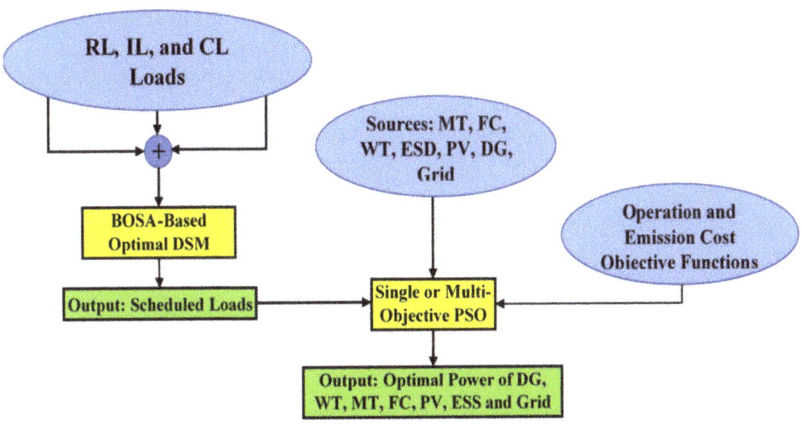

Figure 3. Proposed multi-objective optimization model configuration.

The startup and fixed running costs of distribution generators, the non-spinning and spinning backup costs supplied by distribution generators, and the power purchased/sold from/to the main grid are all considered as operational costs. Uncertain operational costs are determined by realizing and considering the probability of scenario P_{rsc} during the t-th period and sc-th scenario, which are a combination of possible outcomes. Here, the operation function is the cost of running distributed generation units, and the costs related with the expected energy not served (EENS) and the value of lost load (VOLL) for consumers are all included in the operational cost function.

$$Min\ f_A(X) = \sum_{t=1}^{T} F_{cost}(t) = F_S(t) + \sum_{sc=1}^{SC} P_{rsc} \times F_{U,sc}(t) \tag{31}$$

where $F_S(t)$ and $F_{U,\,sc}(t)$ denote certain and uncertain operational cost functions, respectively, P_{rsc} denotes the probability of occurrence of scenario sc. The definitions of the certain and uncertain operational cost functions can be found using Equations (32) and (33), respectively.

$$F_S(t) = \sum_{t=1}^{T}\left\{\sum_{i=1}^{N_{DG}}[P_i(t)OP_i(t)ST_i(t) + SUS_i(t)|ST_i(t) - ST_i(t-1)| + RC_i^{DG}(t)] + P^{ESS}(t)OP_{ESS}(t)ST_{ESS}(t) + SUS_{ESS}(t)|ST_{ESS}(t) - ST_{ESS}(t-1)| + ST_B(t)P_{GB}(t)OP_{GB}(t) - ST_S(t)P_{GS}(t)OP_{GS}(t)\ t\right\} \quad (32)$$

$$F_{U,sc}(t) = \sum_{t=1}^{T}\left\{\sum_{i=1}^{N_{DG}} C_{i,sc}^{DG}(t) + C_{sc}^{ESS}(t) + EENS_{sc}(t) \times VLL(t)\right\} \quad (33)$$

where $P_i(t)$ and $OP_i(t)$ represent the total output power and the price that was being offered for the i-th unit during the t-th period; $ST_i(t)$ is a binary representation of the on mode and off mode of the i-th unit during the t-th period, $P_{ESS}(t)$ and $OP_{ESS}(t)$ denote the total output power and the price that was being offered for the ESS unit during the t-th period; $ST_{ESS}(t)$ is a binary representation of the on mode and off mode of the ESS during the t-th period, $SUS_i(t)$ and $SUS_{ESS}(t)$ are a representation of the costs associated with running and shutting down the i-th and ESS unit during the t-th period; $RC_i^{DG}(t)$ is the reserve costs of the i-th DG during the t-th period. $P_{GB}(t)$ and $P_{GS}(t)$ are the amounts of power exchanged with the utility in period t. $OP_{GB}(t)$ and $OP_{GS}(t)$ represent the offered price for open market power with utility during the t-th period; $C_{i,sc}^{DG}(t)$ and $C_{sc}^{ESS}(t)$ represents the running cost of the i-th DG unit and ESS in the sc-th scenario during the t-th period; and $EENS_s(t)$ and $VLL(t)$ represent the EENS in the sc-th scenario at the t-th period and value of lost load. In Equation (33), $X^T = [X_1, X_2, \ldots, X_T]$ is the state vector of the variables, which includes the active power produced by each DG, the power used in the battery's charge and discharge, and the real power exchanged with the upstream grid.

The quantity of pollution generated by DG units as well as the grid at the energy purchase time are both included in the pollution emissions function. The pollutants include carbon dioxide (CO_2), sulfur dioxide (SO_2), and nitrogen oxides (NO_x), and the mathematical representation model of the pollution emission function can be obtained as follows:

$$Min\ f_B(X) = \sum_{t=1}^{T} F_{Emission}(t) = \sum_{t=1}^{T}[Em_{DG}(t) + Em_{Grid}(t)] \quad (34)$$

The following formula can be used to calculate the average emissions generated by renewable DG units:

$$Em_{DG}(t) = \left\{\sum_{i=1}^{N_{DG}}(E_{CO_2}^{DG}(i) + E_{SO_2}^{DG}(i) + E_{NO_x}^{DG}(i))\right\} \times P_i^{DG}(t) + (E_{CO_2}^{ESS} + E_{SO_2}^{ESS} + E_{NO_x}^{ESS}) \times P^{ESS}(t) \quad (35)$$

where $P_i^{DG}(t)$ is the active power of i-th DG, $E_{CO_2}^{DG}(i)$, $E_{SO_2}^{DG}(i)$, and $E_{NO_x}^{DG}(i)$ denote the amount of CO_2, SO_2, and NO_x pollution resulting from the i-th DG, respectively, and kg/MWh is the unit of measurement. Similarly, the grid pollution can be expressed as follows:

$$Emi_{Grid}(t) = (E_{CO_2}^{Grid}(i) + E_{SO_2}^{Grid}(i) + E_{NO_x}^{Grid}(i)) \times P^{Grid}(t) \quad (36)$$

where $P^{Grid}(t)$ represents the grid power which has maximum and minimum value (± 30 kW), $E_{CO_2}^{DG}(i)$, $E_{SO_2}^{DG}(i)$, and $E_{NO_x}^{DG}(i)$ denote the amount of CO_2, SO_2, and NO_x pollution resulting from the main grid which are assumed as 950 kg/MWh, 0.5 kg/MWh, and 2.1 kg/MWh.

The following constraints are assumed to govern the operation of a typical smart MG.

5.1. Power Balance Constraint

In each interval and scenario, the total amount of electricity made by DGs and utility purchases must match the total amount of demand loads.

$$\sum_{i=1}^{N_{DG}} P_i^{DG}(t) + P^{ESS}(t) + P^{Grid}(t) = \sum_{l=1}^{N_L} P_l^{Optimal.Demand}(t) \quad (37)$$

where $P_l^{Optimal.Demand}(t)$ denotes the total optimal demand power.

5.2. DG Power Constraints

Each unit's maximum and minimum power output is constrained and can be stated as follows:

$$P_{i,min}^{DG}(t) \leq P_i^{DG}(t) \leq P_{i,max}^{DG}(t) \tag{38}$$

$$P_{min}^{ESS}(t) \leq P^{ESS}(t) \leq P_{max}^{ESS}(t) \tag{39}$$

$$P_{min}^{Grid}(t) \leq P^{Grid}(t) \leq P_{max}^{Grid}(t) \tag{40}$$

where $P_{i,min}^{DG}(t)$, $P_{min}^{ESS}(t)$, and $P_{min}^{Grid}(t)$ represent the minimum active power of the i-th DG, the ESS, and the utility, respectively, during period t. The maximum active power units for the period t are $P_{i,max}^{DG}(t)$, $P_{max}^{ESS}(t)$, and $P_{max}^{Grid}(t)$.

5.3. Battery Constraints

For each interval of time, a battery's charging and discharging limitations and equations can be expressed in the following forms [50]:

$$W_{ESS}(t) = W_{ESS}(t-1) + \eta_{ch} P_{ch}(t) I_{ch} - \frac{1}{\eta_{disch}} P_{disch}(t) I_{disch}(t) \tag{41}$$

$$W_{ESS,min} \leq W_{ESS}(t) \leq W_{ESS,max} \tag{42}$$

$$P_{ch}(t) \leq P_{ch,max}(t);\ P_{disch}(t) \leq P_{disch,max} \tag{43}$$

where the quantities $W_{ESS}(t)$ and $W_{ESS}(t-1)$ each represent the amount of energy that is stored in the battery at time t and time $t-1$, respectively; $W_{ESS,min}$ and $W_{ESS,max}$ are the lowest and highest amounts of energy that can be stored in the battery, respectively. $P_{ch,max}/P_{disch,max}$ is the maximum battery charge/discharge power. P_{ch}/P_{disch} refers to the maximum charge/discharge allowed over a given time period. During charging and discharging, η_{ch}/η_{disch} is the battery's efficiency, $I_{ch(t)}$ and $I_{disch}(t)$ are the charge and discharge states of the battery.

6. Proposed Smart MG System

Typically, a MG consists of distribution generators, energy reserves, and loads that can be operated independently or in tandem with the area's primary electrical grid [51–54]. The development of MGs is an aspect of the concept of smart grids; given the benefits of MGs, such as reduced energy costs and enhanced security and system reliability [55–59], it is evident that MGs and smart grids share common goals [60,61]. Moreover, the development of green technologies and the implementation of DSM programs in MGs are contingent on the adoption of smart grid technologies. As seen in Figure 4, a connected grid of residential, commercial, and industrial consumers makes up the SMG under study, in addition to power-generating resources including MTs, DGs, WTs, PV panels, FCs, and an ESS (nickel–metal hydride (NiMH) battery). NiMH batteries have a significantly longer lifespan than lead-acid batteries as well as a significantly higher power and energy density than lead-acid batteries. In addition to not posing any danger, their power output is unaffected by the amount of charge that is currently present in the battery. With the utility, this SMG has the ability to exchange energy. The startup and shutdown costs, DG price offers, the amount of greenhouse gas emissions produced by DGs, as well as the minimum and maximum power generation, are all included in Table 2 (modified from table in [50,62]). Table 3 shows the adopted appliance loads and their attributes.

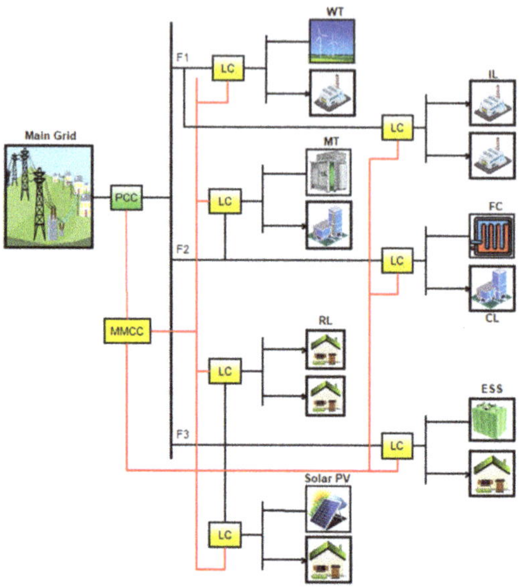

Figure 4. The proposed SMG schematic configuration.

Table 2. Bids and DG source emissions coefficients.

Unit Type	Bid USD/kWh)	SU/SD (USD)	CO_2 (kg/MWh)	SO_2 (kg/MWh)	NO_x (kg/MWh)	P_{min} (kW)	P_{max} (kW)
DG	0.586	0.15	890	0.0045	0.23	30	300
MT	0.457	0.96	750	0.0036	0.1	6	30
FC	0.294	1.65	460	0.003	0.0075	3	30
PV	0.7	0	0	0	0	0	25
WT	0.65	0	0	0	0	0	15
ESS	0.38	0	10	0.0002	0.001	−30	30

Table 3. Numerous appliances and their attributes.

Appliance Number	Appliance Type	Operating Time	IL Rated Power (kW)	CL Rated Power (kW)	RL Rated Power (kW)
1	Nonshiftable	12 AM–12 PM	4	4	2
2	Shiftable	7–9 AM, 11 AM–14 PM & 18 PM–22 PM	0.6	0.6	0.3
3	Shiftable	7 AM–12 PM, 15 PM–20 PM	0.4	0.4	0.2
4	Shiftable	1 AM–10 AM, 15 PM–24 PM	4.8	4.8	2.4
5	Shiftable	7 AM–9 AM, 11 AM–14 PM, 18 PM–22 PM	0.6	0.6	0.3
6	Shiftable	8 AM–18 PM	0.4	0.4	0.2
7	Shiftable	10 AM–15 PM, 18 PM–22 PM	0.4	0.4	0.3
8	Nonshiftable	12 AM–12 PM	2	2	1
9	Nonshiftable	12 AM–12 PM	1.6	1.6	0.8

7. BOSA-Based MOPSO Algorithm

Inequalities and equality constraints must be optimized simultaneously in multi-objective optimization problems because they involve multiple competing objective functions.

$$Min\ F(X) = [f_1(X), f_2(X), \ldots, f_n(X)]^T \tag{44}$$

$$\begin{aligned} Subject\ to \quad & g_i(X) < 0 \quad i = 0, 1, 2, \ldots N_{ueq} \\ & h_i(X) = 0 \quad i = 0, 1, 2, \ldots N_{eq} \end{aligned} \tag{45}$$

where $F(X)$ is a vector that contains the objective functions and X is a vector that contains optimization variables, $f_i(X)$ is the objective function that corresponds to the i-th optimization variable, $g_i(X)$ and $h_i(X)$ are limitations of equality and inequality, and n represents the number of the objective functions that are being considered.

Each pair of solutions in multi-objective optimization can have two distinct relationships with one another: either one solution could dominate the other or no solution can dominate the other.

$$\forall j \in \{1, 2, \ldots, n\}, \quad f_j(X_1) \leq f_j(X_2) \tag{46}$$

$$\exists k \in \{1, 2, \ldots, n\}, \quad f_k(X_1) \leq f_k(X_2) \tag{47}$$

It is possible to use the particle swarm optimization (PSO) algorithm to solve multi-objective problems, and this solution is referred to as multi-objective swarm particle optimization (MOPSO). This is accomplished by applying the concepts of Pareto optimality while utilizing the fundamental principles of PSO. A repository is used to save different solutions when using the MOPSO algorithm. A repository is an external memory in which dominated solutions are stored. This algorithm begins its execution by first working with a collection of random particles. During a series of repeated steps, all of the population particles are compared with one another, and the positions of the particles that dominate the comparison are recorded in the repository. Using the following equation, the new velocity and position of the i-th particle in the dm-th dimension and the $t+1$ repetition are calculated and updated. See [63,64] for more details.

$$v_{idm}^{t+1} = w \times v_{idm}^{t} + c_1 rand_1 \times (p_{bestidm}^{t} - x_{idm}^{t}) + c_2 rand_2 \times (g_{bestidm}^{t} - x_{idm}^{t}) \tag{48}$$

$$x_{idm}^{t+1} = x_{idm}^{t} + v_{idm}^{t+1} \tag{49}$$

The proposed optimal DSM-based MOPSO system applied to the investigated problem can be executed according to the steps outlined below.

1. The required input data are collected at the start of the program, and include: MG structure, utility, and DG operating characteristics PV and WT forecasted output power for every time period under consideration, offering of the real-time price for DGs and utility, the daily demand curve, and pollutant emission coefficients.
2. Set the values for all BOSA parameters.
3. Randomize a population to minimize DSM objective (Equation (16)).
4. For each population within the iteration range, Equations (29) and (30) are used to update positions.
5. Check all the constraints for each population.
6. Initial population of MOPSO, an initial population, is considered based on the problem's limitations and the following relationship:

$$X^0 = [X_1, X_2, \ldots, X_N]^T \tag{50}$$

where X is regarded as the decision variable vector, which consists of the unit's output generation power, the power exchange with the main grid, the amount of load reduction, and the on/off modes in the day ahead vision, which are stated as follows:

$$X = [P_g, U_g]_{1 \times 2nT} \tag{51}$$

$$P_g = \left[P_{dg1}, P_{dg2}, \ldots, P_{dgN_{dg}}, P_{ESS1}, P_{ESS2}, \ldots, P_{ESSN_{ESS}}, P_{grid}, P_{load}\right] \tag{52}$$

$$U_g = \left[U_{dg1}, U_{dg2}, \ldots, U_{dgN_{dg}}, U_{ESS1}, U_{ESS2}, \ldots, U_{ESSN_{ESS}}, U_{grid}, U_{load}\right] \tag{53}$$

$$n = N_{dg} + N_{ESS} + 2 \tag{54}$$

where n represents the number of decision variables, N_{dg} and N_{ESS} represent the total number of generation units and storage units, respectively, and T represents the total number of periods, $\left(P_{dg1}, P_{dg2}, \ldots\right)$, $(P_{ESS1}, P_{ESS2}, \ldots)$, P_{grid}, P_{load} are the vectors of the active power that includes all DGs and storage units; utility grid power and load active power; and U_g is the state vector that indicates whether all units are ON or OFF during period t.

7. For each of the generated populations, the power dispatch algorithm is implemented as shown in Figure 5, and the fitness is calculated using (33) or (36).

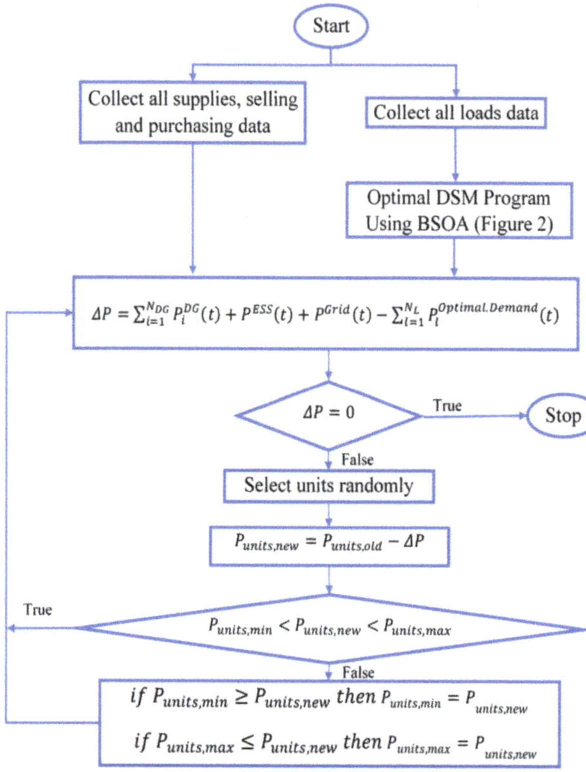

Figure 5. Power dispatch algorithm.

8. Defining non-dominant solutions.
9. Creating a repository for non-dominated solutions.
10. Choosing the best non-dominated solution particle as the leader: the best particle is chosen as the leader by apportioning the search area into equal sections, allocating probability distributions to each part of the identified search space, and finally using the roulette wheel to select the best particle as the leader.
11. Each particle's new velocity and position are calculated using (48) and (49).
12. Modifying the optimal position of every particle:To update each particle's optimal position, the new position of the particle is compared to the position of the particle before.
13. Adding the repository's current non-dominated solutions.
14. The dominated solutions are being removed from the repository.
15. Excessive members will be omitted if the number of individuals in the repository exceeds the pre-specified capacity.
16. The optimization process will end if the maximum number of repetitions is reached; otherwise, return to Step 10.

17. Choosing the best interactive solution: The membership function-based fuzzy logic can be used to select the optimal solution from the optimal Pareto responses. Here, ∂_i^k is the objective function's optimality amount in optimal Pareto response k, which is calculated as follows:

$$\partial_i^k = \begin{cases} 1 & f_i^{min} \geq f_i \\ \frac{f_i^{max} - f_i}{f_i^{max} - f_i^{min}} & f_i^{min} > f_i > f_i^{max} \\ 0 & f_i \geq f_i^{max} \end{cases} \quad (55)$$

where f_i^{max} and f_i^{min} are the objective function's upper and lower limits, respectively. In the method which is being proposed, these values are figured out by using the results of optimizing each objective function. ∂_i^k is between 0 and 1, and a value of 0 indicates that the solution does not meet the designer's objectives, while $\partial_i^k = 1$ indicates that the solution meets the designer's objectives.

8. Results and Discussion

Figure 6a depicts hourly data of the wind speed obtained from the forecast of the weather. The solar PV under consideration is a 25 kW SOLAREX MSX [65]; Figure 6b depicts the average adopted hourly solar irradiance. In a typical system with a 30 kWh battery, the minimum and maximum charges are 10% and 100% of the battery's total capacity, respectively, with a discharge and charge efficiency of 94% [66,67].

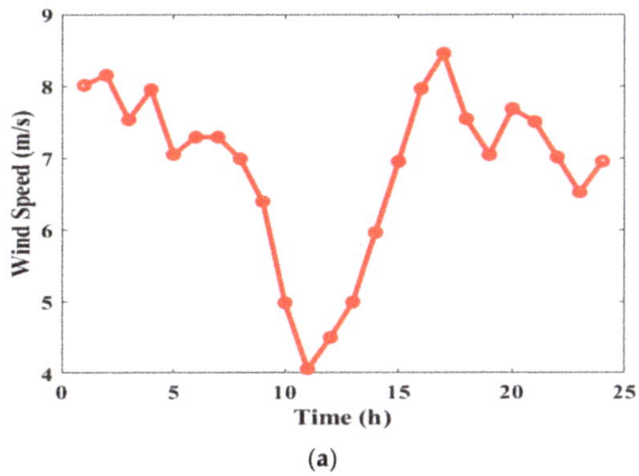

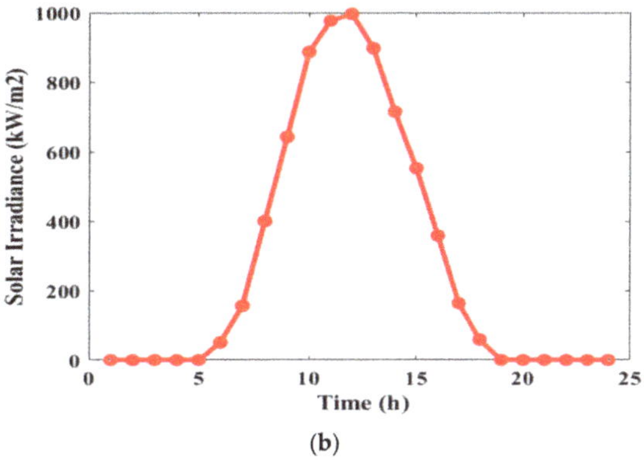

Figure 6. Hourly adopted (**a**) wind speed and (**b**) solar irradiance.

In this paper, the MG connected to the main grid (shown in Figure 4) was used as the test system. Three feeders (F1, F2, and F3), corresponding to medium residential, industrial, and commercial consumers were considered in this system, with the maximum electricity demand assumed to be 40%, 40%, and 20% of the total load energy in the system per time period, respectively.

Figure 7 shows the daily schedule and unscheduled system load demand curve. It illustrates the simulation results of the proposed DSM program for one day (24 h) when the DSM based on the BOSA algorithm is adopted for optimal shifting the controllable (shiftable) loads from on-peak to off-peak hours. Users schedule their peak loads during periods of low electricity costs, resulting in a reduced electric bill. As depicted in Figure 8, proper load scheduling can significantly reduce a user's daily electricity bill. The peak demand value is up to 88.4 kWh using the proposed algorithm-based DSM, which is the lowest in comparison to unscheduled programming (105 kWh). Additionally, the adopted algorithm achieves cost savings (up to 15.7% savings).

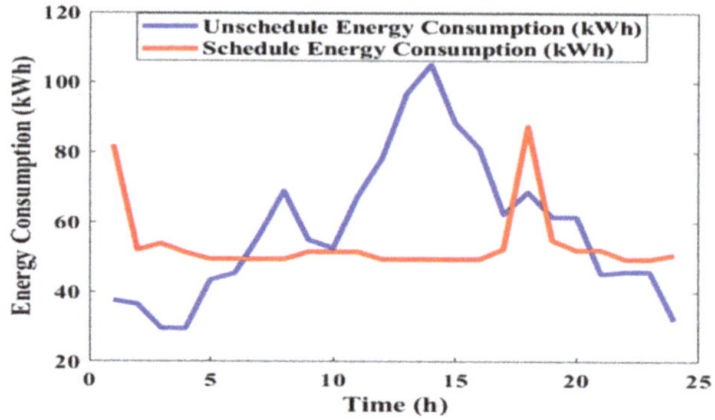

Figure 7. The daily schedule and unscheduled load demand curve.

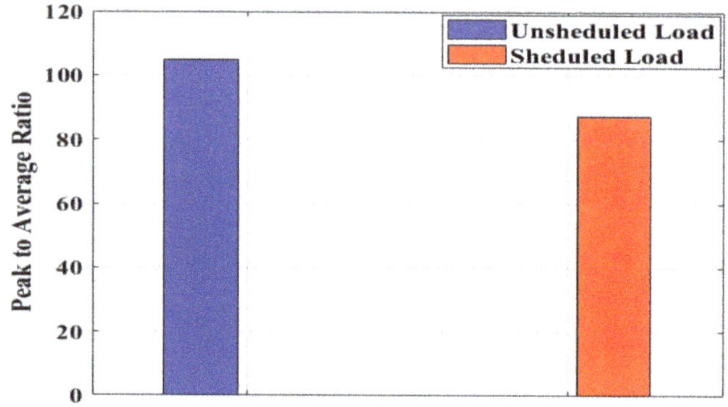

Figure 8. The schedule and non-schedule PAR.

The peak-to-average (PAR) is lowered, which assists in maintaining a healthy equilibrium between the supply and demand of electricity. When it comes to the reduction of PAR, we have placed a higher priority on interruptible appliances. When compared to the non-scheduled load curve, the BOSA technique brings the PAR down to almost 84.3%, as shown in Figure 8. This reduces the electricity cost compared to unscheduled usage. The adopted real-time dynamic price profile is shown in Figure 9.

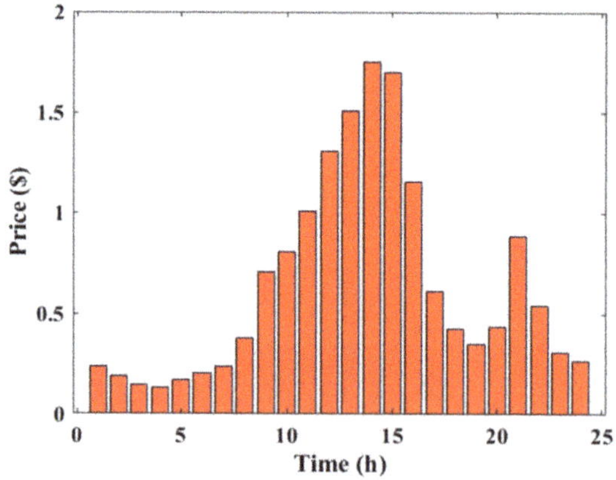

Figure 9. The adopted real-time dynamic price.

Operational costs and pollution emissions were analyzed in order to determine the impact of planning for energy levels, reserves, and optimal DSM on operational costs and environmental impacts, as well as the uncertainty caused by wind and solar resources. The problem was considered under six distinct cases:

- Case #1: single-objective (emission function optimization only) without DSM.
- Case #2: single-objective (emission function optimization only) with DSM.
- Case #3: single-objective (considering operation cost function) without DSM.
- Case #4: single-objective (considering operation cost function) with DSM.
- Case #5: multi-objective (emission and operation cost functions) optimization without DSM.
- Case #6: multi-objective (emission and operation cost functions) optimization with DSM.

8.1. Cases #1 and #2: Emission Function Optimization Only without and with DSM

This section investigated the results of minimizing pollution emissions with and without the DSM. Figures 10 and 11 illustrate the optimal power allocation of the generation units in these two cases, respectively. Figures 10 and 11 reveal that the power production by the WTs and solar PV panels was not considerably different from the predicted amounts presented in Figure 12, which may be related to the fact that these forms of power generation are pollution-free. The data suggested that the utility purchased from the MG for the majority of working periods due to its high level of pollution emissions. In addition, when pollutant emissions were minimized, the DSM had a considerable effect in covering the uncertainties related to wind and solar energy production. As shown in Figure 13a, the power dispatch algorithm has addressed the supply–demand mismatch; however, Case #1 has not resulted in decreased emission because the DSM is not used to minimize power generating requirements. The daily supply–demand power with DSM program implementation based on BOSA is depicted in Figure 13b. It is obvious that the peak-to-average ratio of the load demand has decreased, consequently reducing the necessary generation power, cost, and emissions as indicated in Figure 14. Implementing the DSM program resulted in a 13% reduction in pollutant emissions. The convergence characteristics of the proposed algorithm are shown in Figure 14a,b without and with adoption of DSM, respectively.

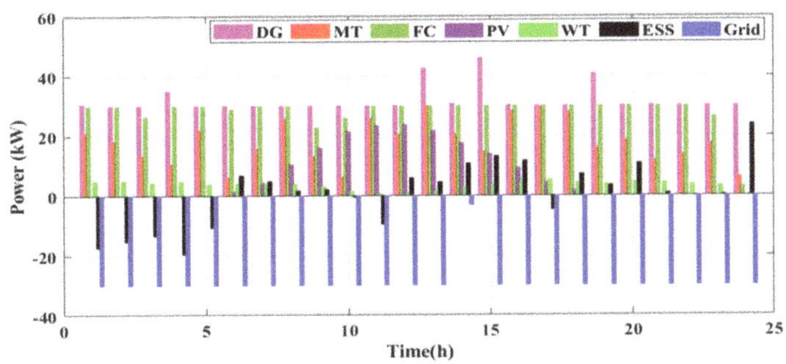

Figure 10. The optimal power allocation of the generation units in Case #1.

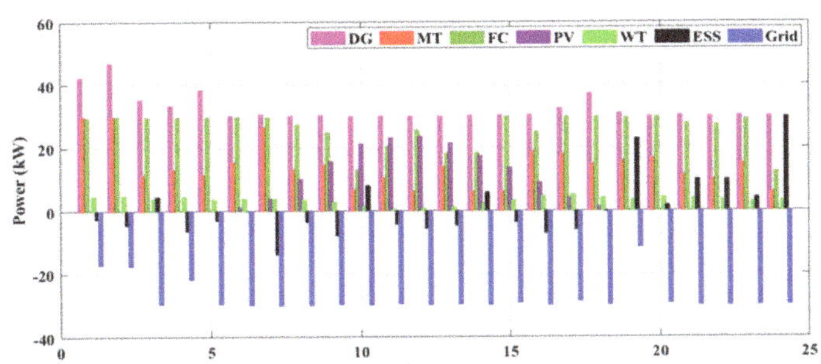

Figure 11. The optimal power allocation of the generation units in Case #2.

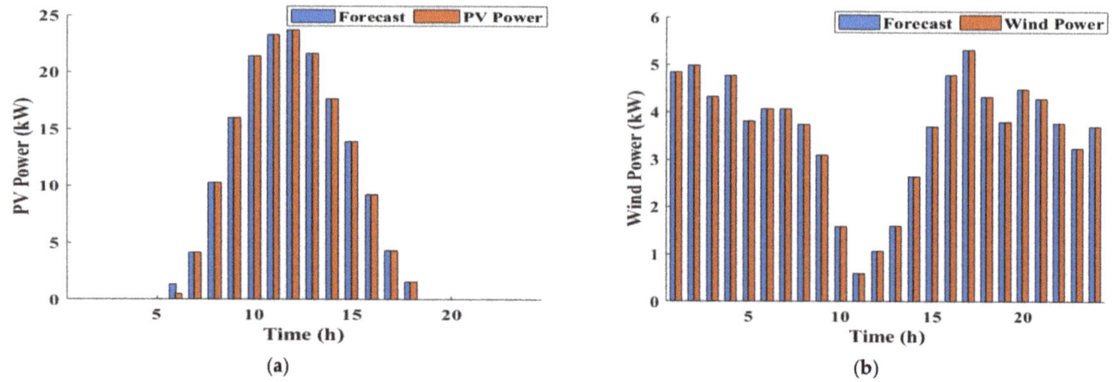

Figure 12. (**a**) PV and (**b**) WT power generation using DSM (Case #2).

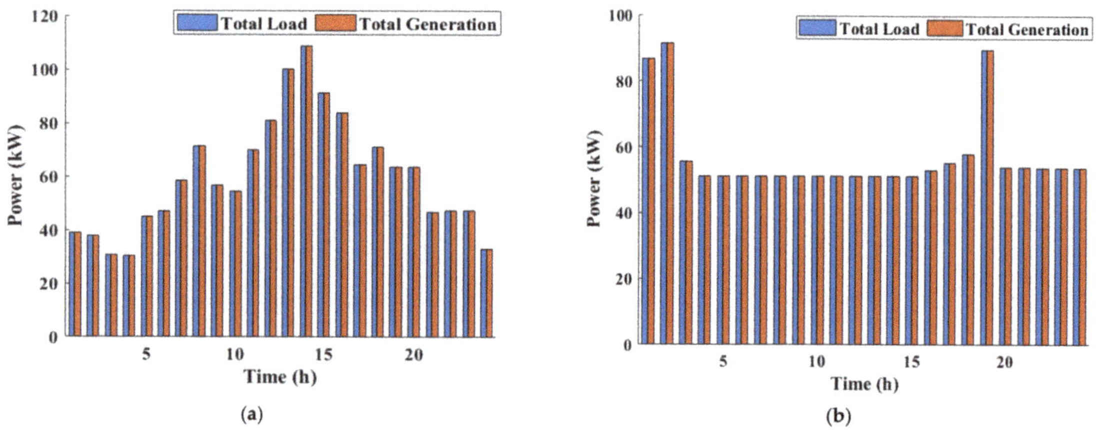

Figure 13. Supply–demand balance of Cases #1 and #2 (**a**) without DSM and (**b**) with DSM.

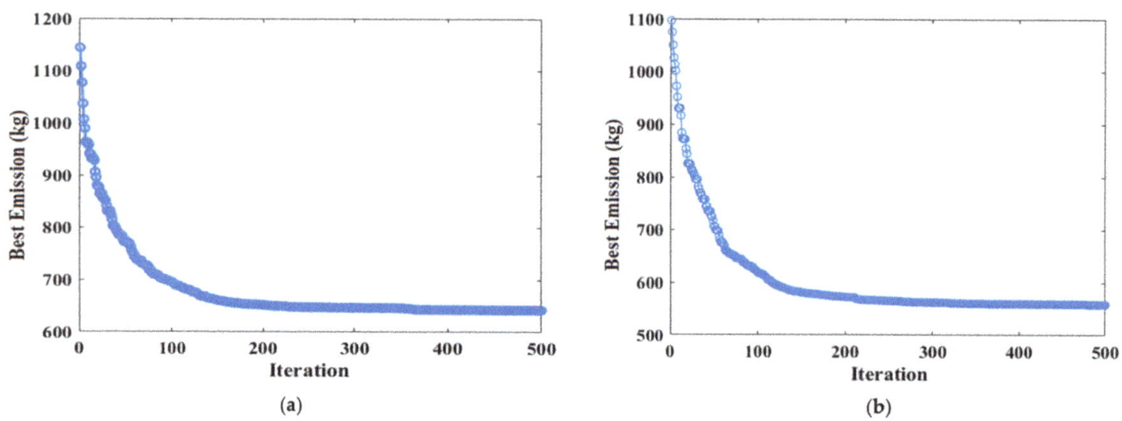

Figure 14. Convergence characteristic of the proposed algorithm (**a**) without DSM and (**b**) with DSM.

8.2. Cases #3 and #4: Operation Cost Function Optimization Only without and with DSM

In these cases, operating cost minimization results were analyzed without and with DSM. Figure 15 illustrates the optimal power generation allocation of the unit without DSM. The simulation's results, which pertain to the case without an optimal DSM program, indicate that charging of the battery started in the earliest hours of the day, when the price has dropped, while the utility purchased electricity from the MG during the peak times of demand, when the price became highest, so that the power consumption could be met by the offering of the price from the distributed generation resources. When there were DSM available in the system (Figure 16), almost the same situation existed. The power generated by the PVs and WTs, especially with adoption of DSM, are depicted in Figure 17a and b respectively. However, because the price of WT and PV resources offered is higher than the price of other generation resources, they cannot gain a great deal of attention when determining the optimal operational cost. As depicted in Figure 17, their power generation is very low or even zero in these cases. As shown in Figure 18a, the supply–demand mismatch has been adjusted by the proposed power dispatch algorithm, but Case#3 has not resulted in lower operating costs due to the fact that the DSM is not adopted to reduce power generation requirements. Figure 18b shows the daily supply–demand power with DSM program adoption based on BOSA. It is evident that the peak-to-average ratio of the load demand is reduced in this case (Case #4), thereby reducing the required generation power and system operation costs as depicted in Figure 19.

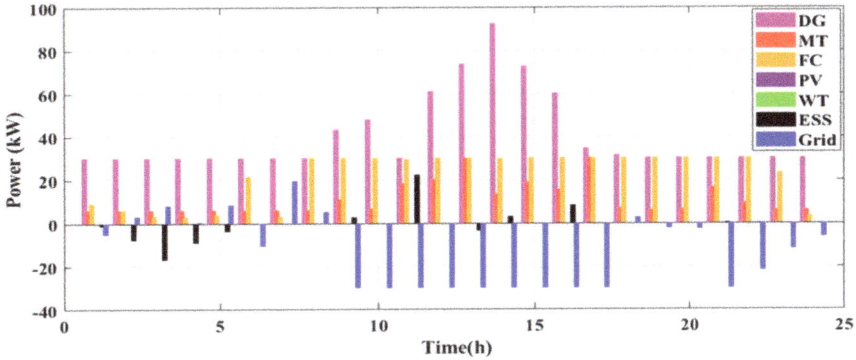

Figure 15. The optimal power distribution of the generating units in Case #3.

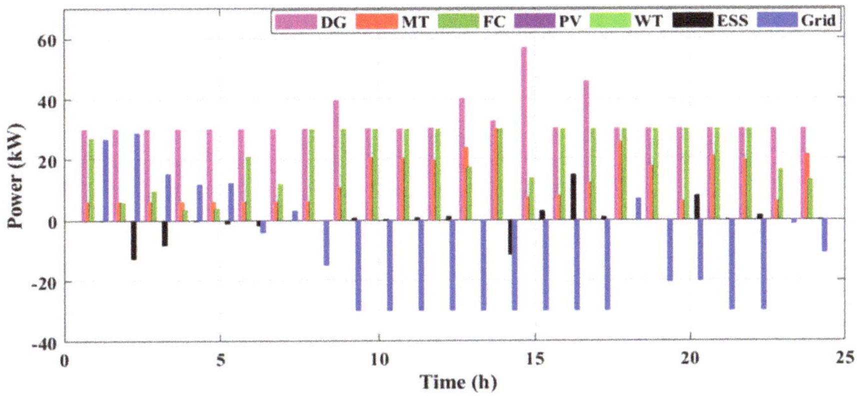

Figure 16. The optimal power distribution of the generating units in Case #4.

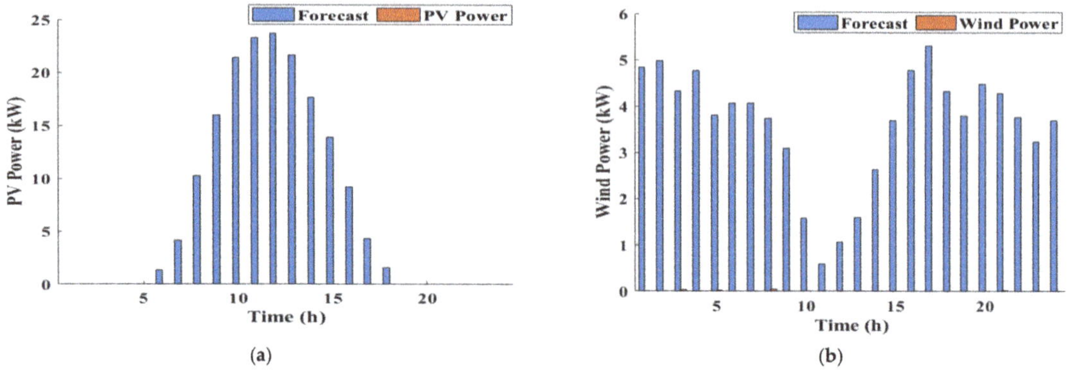

Figure 17. Power generation adopting DSM (**a**) PV and (**b**) WT (Case #4).

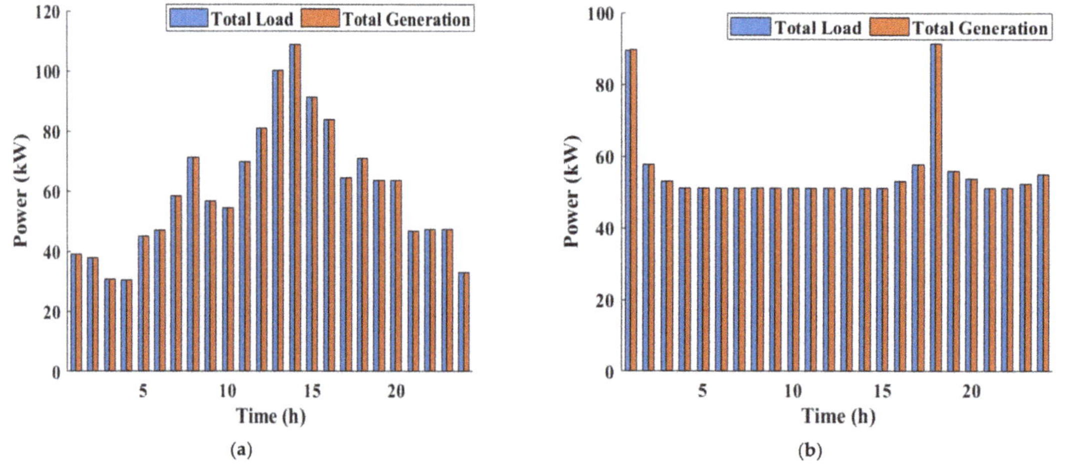

Figure 18. Generation–demand balance of Cases #3 and #4 (**a**) without using DSM and (**b**) with DSM.

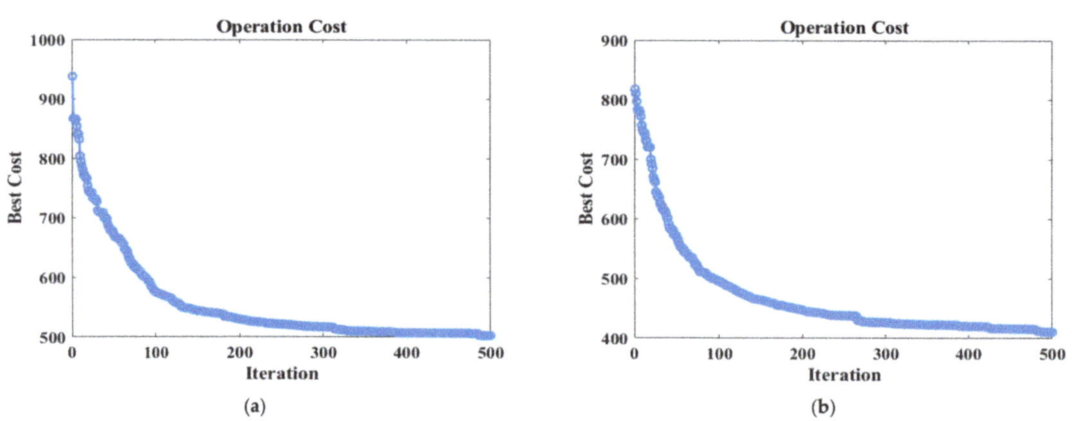

Figure 19. Convergence characteristic of the proposed algorithm (**a**) without DSM and (**b**) with DSM.

8.3. Cases #5 and #6: Emission and Operation Cost Functions Optimization without and with DSM

In these particular cases, the outcomes of minimizing two incompatible functions, namely, pollutant emissions and operating costs, were analyzed, both without and with the optimal DSM program. The optimal power allocation was performed for the simultaneous reduction of operational costs and emissions. Most of the time, the utility purchases power from the SMG as shown in Figure 20 (without DSM) and Figure 21 (with adopting DSM) due to the utility's high pollution levels. Figure 22a,b illustrates the power generation by the WTs and solar PVs takes into consideration the reduction of the operating cost and pollutant emission functions, as well as the simultaneous reduction of the pollutant emission and operating cost functions when DSM was available. As wind and solar power are typically pollution-free, their maximum power generation is achieved when the pollution emission function is considered. These resources, at the same time, do not receive much consideration when it comes to the optimal operational cost because their prices are higher than those of other power generation resources. As a result, in these cases, the energy extracted from WTs and solar PVs results in the lowest operating costs and emissions. The supply–demand mismatch has been fixed by the power dispatch algorithm, but Case #5 has not led to less pollution or lower operating costs overall due to DSM not being adopted to decrease the generation requirements, as shown in Figure 23a. Due to the load demand reduction in Case #6, the supply–demand power mismatch has been resolved by the power dispatch algorithm with lower generation requirements, which has resulted in decreased emissions and lower overall costs of operation (as shown in Figure 23b).

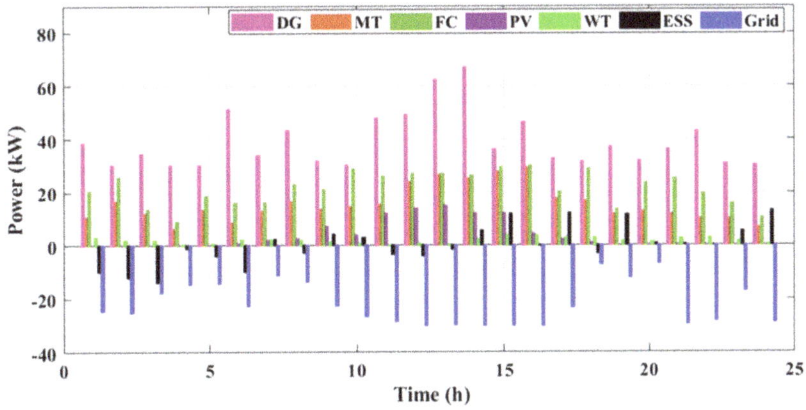

Figure 20. The optimal power distribution of the generating units in Case #5.

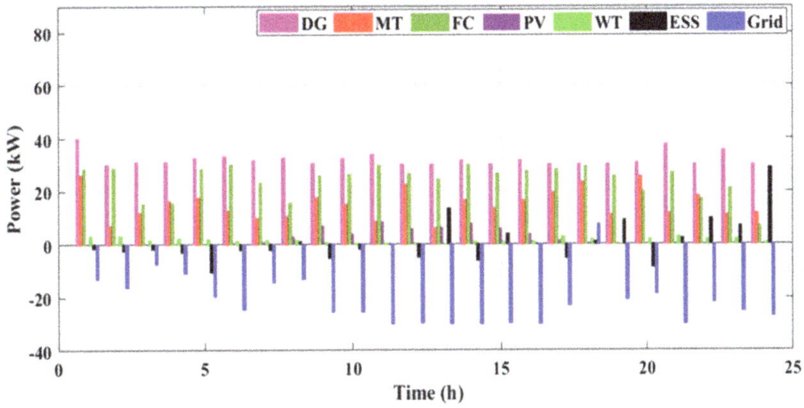

Figure 21. The optimal power distribution of the generating units in Case #6.

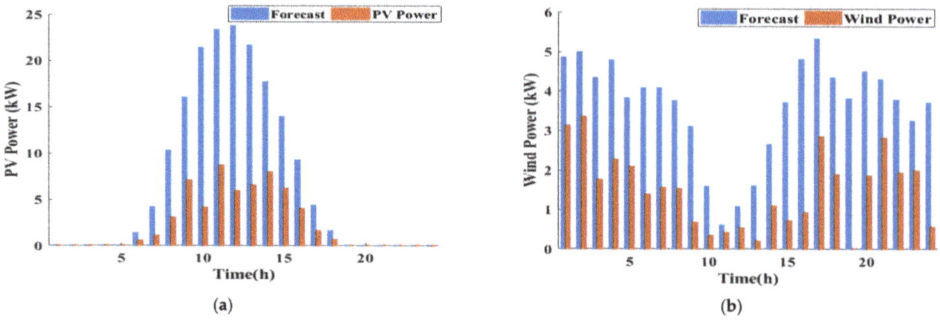

Figure 22. (**a**) Solar PV and (**b**) WT power generation adopting DSM (Case #6).

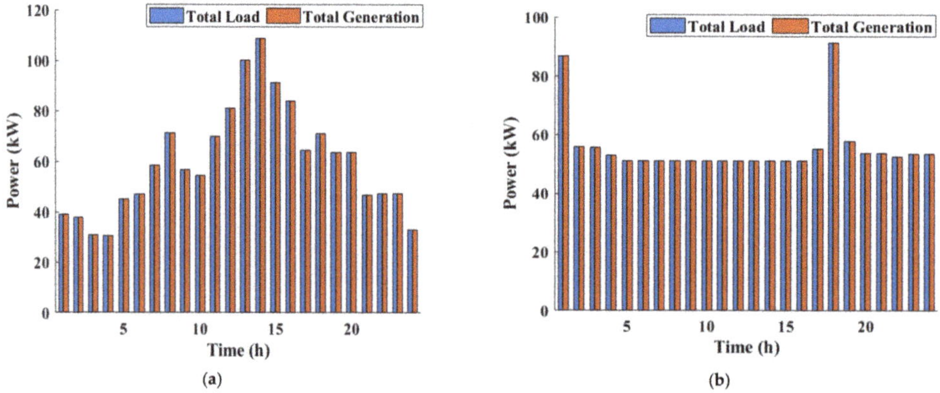

Figure 23. Supply–demand balance of Cases #5 and #6 (**a**) without using DSM and (**b**) with DSM.

According to Figure 24, since operational cost and emissions cost objectives are opposite, traveling from curve initial points to Pareto path endpoints corresponds to a change in operation behavior from low cost and high pollution to high cost and low pollution, with fuzzy mechanisms determining the optimal operation point. Figure 24a shows that without DSM, the optimal operating point cannot be improved to reduce operational costs and pollution emissions. In the case with DSM programs (Figure 24b), they help to improve the optimal operation point, lowering operational costs and pollution emissions by 21% and 7.2%, respectively.

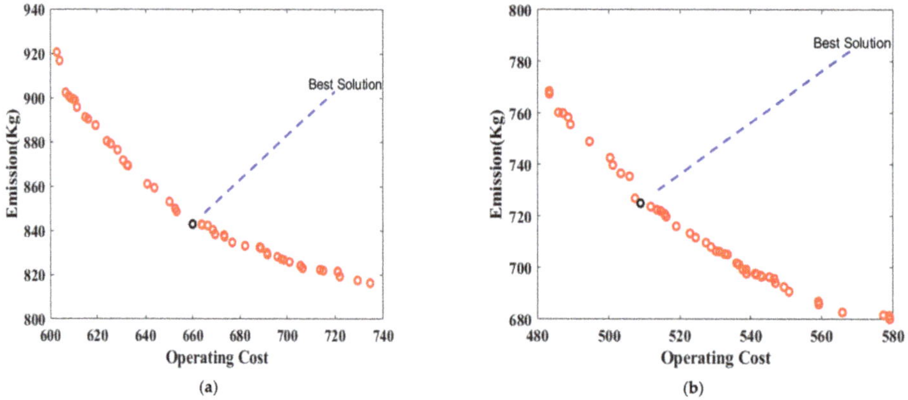

Figure 24. Convergence characteristic of the proposed algorithm (**a**) without DSM and (**b**) with DSM.

8.4. Time Testing Results

These tests are presented to demonstrate the time required to execute the proposed system operation for six adopted scenarios. The elapsed time of the proposed DSM and single-objective or multi-objective optimization algorithms is showed in Table 4. The optimal DSM algorithm is implemented with a maximum iteration number of 300, a population size of 10, maximum limit of 100, and a maximum shift time of 4. The PSO is executed with a maximum iteration number of 500 and a swarm size of 10.

Table 4. The elapsed time of the proposed optimization algorithms.

Operational Condition of MG	Case No.	BOSA (Elapsed Time)	MOPSO (Elapsed Time)	BOSA + MOPSO (Elapsed Time)
Optimal DEED without DSM	#1	-	43.99 s	43.99 s
	#3	-	36.92 s	36.92 s
	#5	-	63.41 s	63.41 s
Optimal DEED with DSM	#2	3.77 s	44.29 s	49.06 s
	#4	3.59 s	37.26 s	40.97 s
	#6	3.7 s	63.72 s	67.85 s

9. Conclusions

This paper proposed a multi-objective DEED combined model with BOSA-based DSM to stay ahead of DSM's utility, generation, and energy consumption reduction benefits. The proposed BOSA-based DSM program adopts a load-shifting program to obtain day-ahead optimal load appliances scheduling. The problems associated with MG operation were addressed by managing the energy according to the analysis of the total operating costs of the MG and the associated pollutant emissions with/without applying optimal DSM based on BOSA. The consumer side has the potential to play an active role in energy production and consumption management. Consumers could participate in RTP-based DSM for consumption management. RTP was implemented to incentivize consumers to better regulate their electricity consumption. In addition, a probabilistic programming technique was employed to model the stochastic behavior of wind and solar cell power generation. The fuzzy-based MOPSO method was used to solve and optimize the DEED proposed model. Simulation results demonstrated that if consumers participate in DSM, it is possible to reduce operational costs and emissions. Among the cases studied, the best results were found when BOSA-based DSM was used to look at operating costs and pollutant emissions at the same time. Additionally, simulation results demonstrated that considering the pollution function as the primary objective made the operational costs increase, so DSM programs were utilized for minimizing the operation cost and vice versa for considering the operation cost as a single objective. This model also demonstrated that if consumers collaborate in DSM, in addition to compensating for production shortages due to the unpredictability of wind and solar power, this results in a decrease in operating costs and system-wide pollution simultaneously. The results show that, when compared to unscheduled peak demand (which was 105 kWh), the proposed BOSA algorithm-based DSM had the lowest peak demand (88.4 kWh). Additionally, operating costs dropped from660 USD to 508USD, a 23% decrease, and emissions dropped from 840 kg to 725 kg, a 13.7% reduction.

Author Contributions: Authors: A.M.J.: original draft, software, methodology, and validation; B.H.J.: supervisor, formal analysis; research resources; investigation; editing; and writing; B.H.J.: validation, H.K. and A.F.; visualization, A.F. and H.K.; project administration, A.F. and H.K.; funding acquisition, H.K. and A.F. All authors have read and agreed to the published version of the manuscript.

Funding: This research was funded by Deputyship for Research & Innovation, Ministry of Education, in Saudi Arabia, project number IF_2020_NBU_451.

Institutional Review Board Statement: Not applicable.

Informed Consent Statement: Not applicable.

Data Availability Statement: Not applicable.

Acknowledgments: The authors extend their appreciation to the Deputyship for Research & Innovation, Ministry of Education, in Saudi Arabia, for funding this research work through project number IF_2020_NBU_451. The authors gratefully thank the Prince Faisal bin Khalid bin Sultan Research Chair in Renewable Energy Studies and Applications (PFCRE) at Northern Border University for their support and assistance.

Conflicts of Interest: The authors declare no conflict of interest.

References

1. Behrangrad, M. A review of demand side management business models in the electricity market. *Renew. Sustain. Energy Rev.* **2015**, *47*, 270–283. [CrossRef]
2. Yan, X.; Ozturk, Y.; Hu, Z.; Song, Y. A review on price-driven residential demand response. *Renew. Sustain. Rev.* **2018**, *96*, 411–419. [CrossRef]
3. Aghaei, J.; Alizadeh, M.; Abdollahi, A.; Barani, M. Allocation of demand response resources: Towards an effective contribution to power system voltage stability. *IET Gener. Transm. Distrib.* **2016**, *10*, 4169–4177. [CrossRef]
4. Aghajani, S.; Kalantar, M. Operational scheduling of electric vehicles parking lot integrated with renewable generation based on bilevel programming approach. *Energy* **2017**, *139*, 422–432. [CrossRef]
5. Akbaria, H.; Browne, M.C.; Ortega, A.; Huang, M.J.; Hewitt, N.J.; Norton, B.; McCormack, S.J. Efficient energy storage technologies for photovoltaic systems. *Sol. Energy* **2019**, *192*, 144–168. [CrossRef]
6. Aydin, D.; Ozyon, S.; Yasar, C.; Liao, T. Artificial bee colony algorithm with dynamic population size to combined economic and emission dispatch problem. *Int. J. Electr. Power Energy Syst.* **2014**, *54*, 144–153. [CrossRef]
7. Rajan, A.; Malakar, T. Optimum economic and emission dispatch using exchange market algorithm. *Int. J. Electr. Power Energy Syst.* **2016**, *82*, 545–560. [CrossRef]
8. Nanjundappan, D.; Nanjundappan, D. Hybrid weighted probabilistic neural network and biogeography based optimization for dynamic economic dispatch of integrated multiple-fuel and wind power plants. *Int. J. Electr. Power Energy Syst.* **2016**, *77*, 385–394. [CrossRef]
9. Zaman, M.; Elsayed, S.M.; Ray, T.; Sarker, R.A. Evolutionary algorithms for dynamic economic dispatch problems. *IEEE Trans. Power Syst.* **2016**, *31*, 1486–1495. [CrossRef]
10. Zaman, F.; Elsayed, S.M.; Ray, T.; Sarker, R.A. Configuring two-algorithm-based evolutionary approach for solving dynamic economic dispatch problems. *Eng. Appl. Artif. Intell.* **2016**, *53*, 105–125. [CrossRef]
11. Aghaei, J.; Niknam, T.; Azizipanah-Abarghooee, R.; Arroyo, J.M. Scenario-based dynamic economic emission dispatch considering load and wind power uncertainties. *Int. J. Electr. Power Energy Syst.* **2013**, *47*, 351–367. [CrossRef]
12. Duan, J.; Chow, M. Robust Consensus-based Distributed Energy Management for Microgrids with Packet Losses Tolerance. *IEEE Trans. Smart Grid* **2019**, *11*, 281–290. [CrossRef]
13. Tajalli, S.; Mardaneh, M.; Taherian-Fard, E.; Izadian, A.; Kavousi-Fard, A.; Dabbaghjamanesh, M.; Niknam, T. DoS-Resilient Distributed Optimal Scheduling in a Fog Supporting IIoT-Based Smart Microgrid. *IEEE Trans. Ind. Appl.* **2020**, *56*, 2968–2977. [CrossRef]
14. Bilal, N.; Jasim, B.H.; Sedhom, B.E.; Guerrero, J.M. Consensus Algorithm-based Coalition Game Theory for Demand Management Scheme in Smart Microgrid. *Sustain. Cities Soc.* **2021**, *74*, 103248. [CrossRef]
15. Pirouzi, S.; Zaghian, M.; Aghaei, J.; Chabok, H.; Abbasi, M.; Norouzi, M.; Shafie-Khah, M.; Catalão, J.P. Hybrid planning of distributed generation and distribution automation to improve reliability and operation indices. *Electr. Power Energy Syst.* **2022**, *135*, 107540. [CrossRef]
16. Asl, S.A.F.; Bagherzadeh, L.; Pirouzi, S.; Norouzi, M.; Lehtonen, M. A new two-layer model for energy management in the smart distribution network containing flexi-renewable virtual power plant. *Electr. Power Syst. Res.* **2021**, *194*, 107085. [CrossRef]
17. Logenthiran, T.; Srinivasan, D.; Shun, T.Z. Demand side management in smart grid using heuristic optimization. *IEEE Trans. Smart Grid* **2012**, *3*, 1244–1252. [CrossRef]
18. Nadeem, J.; Ullah, I.; Akbar, M.; Iqbal, Z.; Khan, F.A.; Alrajeh, N.; Alabed, M.S. An Intelligent Load Management System With Renewable Energy Integration for Smart Homes. *IEEE Access* **2017**, *5*, 13587–13600. [CrossRef]
19. Banala, V.; Sankaramurthy, P.; Chokkalingam, B.; Mihet-Popa, L. Managing the Demand in a Micro Grid Based on Load Shifting with Controllable Devices Using Hybrid WFS2ACSO Technique. *Energies* **2022**, *15*, 790.
20. Sharma, A.; Saxena, A. A demand side management control strategy using Whale optimization algorithm. *SN Appl. Sci.* **2019**, *1*, 703–714. [CrossRef]
21. Kumar, K.; Saravanan, B. Day ahead scheduling of generation and storage in a micro grid considering demand Side management. *J. Energy Storage* **2019**, *21*, 78–86. [CrossRef]
22. Arif, A.; Javed, F.; Arshad, N. Integrating renewables economic dispatch with demand side management in micro-grids: A genetic algorithm-based approach. *Energy Effic.* **2014**, *7*, 271–284. [CrossRef]
23. Chung, H.M.; Su, C.L.; Wen, C.K. Dispatch of generation and demand side response in regional grids. In Proceedings of the 2015 IEEE 15th International Conference on Environment and Electrical Engineering (EEEIC) IEEE, Rome, Italy, 10–13 June 2015; pp. 482–486. [CrossRef]

24. Alham, M.; Elshahed, M.; Ibrahim, D.K.; El Zahab, E.E.D. A dynamic economic emission dispatch considering wind power uncertainty incorporating energy storage system and demand side management. *Renew. Energy* **2016**, *96*, 800–811. [CrossRef]
25. Alham, M.; Elshahed, M.; Ibrahim, D.K.; El Zahab, E.E.D. Optimal operation of power system incorporating wind energy with demand side management. *Ain Shams Eng. J.* **2015**, *8*, 1–7. [CrossRef]
26. Bilal, N.; Jasim, B.; Rahman, Z.-A.; Siano, P. A Novel Robust Smart Energy Management and Demand Reduction for Smart Homes Based on Internet of Energy. *Sensors* **2021**, *21*, 4756. [CrossRef]
27. Shakouri, H.; Kazemi, A. Multi-objective cost-load optimization for demand side management of a residential area in smart grids. *Sustain. Cities Soc.* **2017**, *32*, 171–180. [CrossRef]
28. Deyaa, A.; Ebeed, M.; Ali, A.; Alghamdi, A.; Kamel, S. Multi-Objective Energy Management of a Micro-Grid Considering Stochastic Nature of Load and Renewable Energy Resources. *Electronics* **2021**, *10*, 403. [CrossRef]
29. Seshu, R.; Raghav, L.P.; Raju, D.K.; Singh, A.R. Impact of multiple demand side management programs on the optimal operation of grid-connected microgrids. *Appl. Energy* **2021**, *301*, 117466. [CrossRef]
30. Faria, P.; Soares, J.; Vale, Z.; Morais, H.; Sousa, T. Modified particle swarm optimization applied to integrated demand response and DG resources scheduling. *IEEE Trans. Smart Grid* **2013**, *4*, 606–616. [CrossRef]
31. Zakariazadeh, A.; Jadid, S.; Siano, P. Stochastic multi-objective operational planning of smart distribution systems considering demand response programs. *Electr. Power Syst. Res.* **2014**, *111*, 156–168. [CrossRef]
32. Zheng, Y.; Li, S.; Tan, R. Distributed model predictive control for onconnected microgrid power management. *IEEE Trans. Control. Syst. Technol.* **2018**, *26*, 1028–1039. [CrossRef]
33. Maulik, A.; Das, D. Optimal power dispatch considering load and renewable generation uncertainties in an AC-DC hybrid microgrid. *IET Gener. Transm. Distrib.* **2019**, *13*, 1164–1176. [CrossRef]
34. Yng, L.; Fan, X.; Cai, Z.; Bing, Y. Optimal active power dispatching of microgrid and DistributionNetwork based on model predictive control. *Tsinghua Sci. Technol.* **2018**, *23*, 266–276. [CrossRef]
35. Atwa, Y.; El-Saadany, E.F.; Salama, M.M.A.; Seethapathy, R. Optimal renewable resources mix for distribution system energy loss minimization.,IEEE Transactions on. *Power Syst.* **2010**, *25*, 360–370. [CrossRef]
36. Ettoumi, F.Y.; Mefti, A.; Adane, A.; Bouroubi, M. Statistical analysis of solar measurements in Algeria using beta distributions. *Renew Energy* **2002**, *26*, 47–67. [CrossRef]
37. Deshmukh, M.; Deshmukh, S. Modeling of hybrid renewable energy systems. *Renew. Sustain. Energy Rev.* **2008**, *12*, 235–249. [CrossRef]
38. Athanasios, P.; Pillai, S. *Probability, Random Variables and Stochastic Processes*, 4th ed.; McGraw-Hill: New York, NY, USA, 2002; p. 852.
39. Rahimi, E.; Rabiee, A.; Aghaei, J.; Muttaqi, K.M.; Nezhad, A.E. On the management of wind power intermittency. *Renew. Sustain. Energy Rev.* **2013**, *28*, 643–653. [CrossRef]
40. Gen, B. Reliability and Cost/Worth Evaluation of Generating Systems Utilizing wind and Solar Energy. Ph.D. Thesis, University of Saskatchewan, Saskatoon, SK, Canada, 2005.
41. Owner's Manual of the AIR403 Wind Turbine Made by Southwest Wind Power Inc. Available online: www.nooutage.com/pdf/swwp_air403_landman.pdf (accessed on 16 May 2022).
42. Abouzahr, I.; Ramakumar, R. An approach to assess the performance of utilityinteractive wind electric conversion systems. Energy Conversion. *IEEE Trans.* **1991**, *6*, 627–638.
43. Tina, G.; Gagliano, S.; Raiti, S. Hybrid solar/wind power system probabilistic modelling for long-term performance assessment. *Sol. Energy* **2006**, *80*, 578–588. [CrossRef]
44. Awais, M.; Javaid, N.; Shaheen, N.; Iqbal, Z.; Rehman, G.; Muhammad, K.; Ahmad, I. An Efficient Genetic Algorithm Based Demand Side Management Scheme for Smart Grid. In Proceedings of the 2015 18th International Conference on Network-Based Information Systems, Taipei, Taiwan, 2–4 September 2015.
45. Kothari, D. *Modern Power System Analysis*; Tata McGraw-Hill: New Delhi, India, 2003.
46. Alessandro, A.; De Pascale, G.; Detti, P.; Vicino, A. Load Sceduling for Household Energy Consumption Optimization. *IEEE Trans. Smart Grid* **2013**, *4*, 2364–2373.
47. Shengan, S.; Manisa, P.; Saifur, R. Demand Response as a Load Shaping Tool in an Intelligent Grid With Electric Vehicles. *IEEE Trans. Smart Grid* **2011**, *2*, 624–631.
48. Yi, P.; Dong, X.; Iwayemi, A.; Zhou, C.; Li, S. Real-time Oppertunistic Scheduling for Residential Demand Response. *IEEE Trans. Smart Grid* **2018**, *4*, 227–234. [CrossRef]
49. Mohammad, D.; Montazeri, Z.; Malik, O.P.; Dhiman, G.; Kumar, V. BOSA: Binary Orientation Search Algorithm. *Int. J. Innov. Technol. Explor. Eng. (IJITEE)* **2019**, *9*, 5306–5310.
50. Amjad, A.; Seifi, A.; Niknam, T.; Pahlavani, M.R.A. Multi-objective operation management of a renewable MG (micro-grid) with back-up micro-turbine/fuel cell/battery hybrid power source. *Energy* **2011**, *36*, 6490–6507. [CrossRef]
51. Ali, M.; Basil, H. Grid-Forming and Grid-Following Based Microgrid Inverters Control. *Iraqi J. Electr. Electron. Eng.* **2022**, *18*, 111–131. [CrossRef]
52. Kiani, H.; Hesami, K.; Azarhooshang, A.; Pirouzi, S.; Safaee, S. Adaptive robust operation of the active distribution network including renewable and flexible sources. *Sustain. Energy Grids Netw.* **2021**, *26*, 100476. [CrossRef]
53. Aghaei, J.; Bozorgavari, S.A.; Pirouzi, S.; Farahmand, H.; Korpås, M. Flexibility Planning of Distributed Battery Energy Storage Systems in Smart Distribution Networks. *Iran. J. Sci. Technol. Trans. Electr. Eng.* **2019**, *44*, 1105–1121. [CrossRef]

54. Bilal, N.; Jasim, B.H.; Sedhom, B.E. Distributed secondary consensus fault tolerant control method for voltage and frequency restoration and power sharing control in multi-agent microgrid. *Int. J. Electr. Power Energy Syst.* **2021**, *133*, 107251.
55. Mahmoudi, C.; Flah, A.; Sbita, L. An Overview of Electric Vehicle Concept and Power Management Strategies. In Proceedings of the International Conference on Electrical Sciences and Technologies in Maghreb (CISTEM), Tunis, Tunisia, 3–6 November 2014. [CrossRef]
56. Mahmoudi, C.; Flah, A. A Novel Energy Optimization Approach for Electrical Vehicles in a Smart City. *Energies* **2019**, *12*, 929. [CrossRef]
57. Mohamed, A.; Awwad, E.M.; El-Sherbeeny, A.M.; Nasr, E.A.; Ali, Z.M. Optimal scheduling of reconfigurable grids considering dynamic line rating constraint. *IET Gener. Transm. Distrib.* **2020**, *14*, 1862–1871. [CrossRef]
58. Mohamed, A. A relaxed consensus plus innovation based effective negotiation approach for energy cooperation between smart grid and microgrid. *Energy* **2022**, *252*, 123996. [CrossRef]
59. Mohamed, A.; Almalaq, A.; Abdullah, H.M.; Alnowibet, K.A.; Alrasheedi, A.F.; Zaindin, M.S.A. A distributed stochastic energy management framework based-fuzzy-PDMM for smart grids considering wind park and energy storage systems. *IEEE Access* **2021**, *9*, 46674–46685. [CrossRef]
60. Bilal, N.; Jasim, B.H.; Issa, W.; Anvari-Moghaddam, A.; Blaabjerg, F. A New Robust Control Strategy for Parallel Operated Inverters in Green Energy Applications. *Energies* **2020**, *13*, 3480. [CrossRef]
61. Alhasnawi, B.; Jasim, B.; Siano, P.; Guerrero, J. A Novel Real-Time Electricity Scheduling for Home Energy Management System Using the Internet of Energy. *Energies* **2021**, *14*, 3191. [CrossRef]
62. Sonam, P.; Swarnkar, A.; Niazi, K.R.; Gupta, N. Multiobjective optimal sizing of battery energy storage in grid-connected microgrid. *J. Eng.* **2019**, *18*, 5280–5283. [CrossRef]
63. Coello, C.A.; Lechuga, M.S. MOPSO: A proposal for multiple objective particle swarm optimization. In Proceedings of the 2002 Congress on Evolutionary Computation. CEC'02, Honolulu, HI, USA, 12–17 May 2002.
64. Kraiem, H.; Flah, A.; Mohamed, N.; Alowaidi, M.; Bajaj, M.; Mishra, S.; Sharma, M.K.; Sharma, S.K. Increasing Electric Vehicle Autonomy Using a Photovoltaic System Controlled by Particle Swarm Optimization. *IEEE Access* **2021**, *9*, 72040–72054. [CrossRef]
65. The Solar Power Group Company. Available online: https://thesolarpowergroup.com.au (accessed on 8 April 2022).
66. Chen, C.; Duan, S.; Cai, T.; Liu, B.; Hu, G. Smart energy management system for optimal microgrid economic operation. *IET Renew. Power Gener.* **2011**, *5*, 258–267. [CrossRef]
67. Clement-Nyns, K.; Haesen, E.; Driesen, J. The impact of charging plug-in hybrid electric vehicles on a residential distribution grid. *IEEE Trans. Power Syst.* **2010**, *25*, 371–380. [CrossRef]

sustainability

MDPI

Article

A Multi-Objective Planning Strategy for Electric Vehicle Charging Stations towards Low Carbon-Oriented Modern Power Systems

Hassan Yousif Ahmed [1,*], Ziad M. Ali [1,2], Mohamed M. Refaat [3] and Shady H. E. Abdel Aleem [4]

1 Department of Electrical Engineering, College of Engineering in Wadi Alddawasir, Prince Sattam Bin Abdulaziz University, Wadi Alddawasir 11991, Saudi Arabia
2 Electrical Engineering Department, Aswan Faculty of Engineering, Aswan University, Aswan 81542, Egypt
3 Photovoltaic Cells Department, Electronics Research Institute, Cairo 11843, Egypt
4 Electrical Engineering Department, Valley Higher Institute of Engineering and Technology, Science Valley Academy, Qalyubia 44971, Egypt
* Correspondence: h.ahmed@psau.edu.sa

Abstract: This paper proposes a multi-objective planning framework for electric vehicle (EV) charging stations in emerging power networks that move towards green transportation electrification. Four cases are investigated to study the impacts of EV integration on environmental and economic requirements. In order to facilitate the installation of EV charging stations, the proposed model is formulated to combine the planning models of renewable energy systems, energy storage systems (ESSs), thyristor-controlled series compensators, and transmission lines into the EV-based planning problem. The first objective function aims to maximize EVs' penetration by increasing the networks' capacity to supply charging stations throughout the day, whereas the second objective, on the other hand, emphasizes lowering the carbon dioxide emissions from fossil fuel-based generation units in order to benefit the environment. The third objective is to meet the financial requirements by lowering the initial investment and operating costs of the installed devices. The proposed model is written as a multi-objective optimization problem that is solved using the multi-objective version of the Gazelle optimization algorithm (MGOA). The efficiency of the MGOA was tested by solving a set of four benchmark test functions and the proposed problem. The obtained results demonstrated the MGOA's superiority in solving multi-objective optimization problems when compared to some well-known optimization algorithms in terms of robustness and solution quality. The MGOA's robustness was between 20% and 30% and outperformed other algorithms by 5%. The MGO was successful in outperforming the other algorithms in providing a better solution. The Egyptian West Delta Network simulations revealed a 250 MWh increase in the energy supplied to EVs when energy storage was not used. However, storage systems were necessary for shifting EV charging periods away from high solar radiation scenarios. The use of ESS increased greenhouse gas emissions. When ESS was installed with a capacity of 1116.4 MWh, the carbon emissions increased by approximately 208.29 million metric tons. ESS's role in improving the EV's hosting capacity grows as more renewables are added to the network. ESS's role in improving the EV's hosting capacity rises as more renewables are added to the network.

Keywords: electric vehicles; energy storage systems; transmission lines planning; renewable energy sources; multi-objective Gazelle optimization algorithm; smart grid

Citation: Ahmed, H.Y.; Ali, Z.M.; Refaat, M.M.; Aleem, S.H.E.A. A Multi-Objective Planning Strategy for Electric Vehicle Charging Stations towards Low Carbon-Oriented Modern Power Systems. *Sustainability* **2023**, *15*, 2819. https://doi.org/10.3390/su15032819

Academic Editors: Mohamed A. Mohamed and George Kyriakarakos

Received: 20 December 2022
Revised: 5 January 2023
Accepted: 28 January 2023
Published: 3 February 2023

1. Introduction

Carbon dioxide emissions have risen steadily over the past decades. Energy-related CO_2 emissions were approximately 35 billion metric tons in 2020 and are expected to exceed 40 billion metric tons by 2050 [1]. Globally, governments worldwide have adopted many policies to tackle this issue and generally deal with the problems of climate change. One of these policies is the electrification of the transportation sector. It is planned that by the year 2050, electric vehicles (EVs) will contribute to reducing CO_2 emissions and represent 31% of overall passenger vehicles [1].

The massive use of EVs adversely impacts the power network's inertia, reliability, and frequency regulation capability and is exacerbated if renewable sources are heavily installed. However, these issues can be turned into a solution to many power grid issues if EV penetration is well planned. EVs can provide power grid ancillary services, including peak clipping, load shifting, and frequency management [2]. Many planning strategies for integrating and optimizing EVs in power networks have been proposed in the literature. Static and multi stages-based planning frameworks are commonly presented to model the problem. For example, Abdi-Siab et al. [3] developed a bi-level optimization strategy for the expansion of distribution systems in the presence of plug-in EVs. The upper level aimed to calculate the annual investment, prospective production, and maintenance costs. At the same time, the lower level was concerned with minimizing the cost of the energy purchased from the main grid and optimizing the daily schedule. The results showed that the smart charging strategy led to a lower investment cost and demonstrated the superiority of the proposed solving strategy compared to the strategies that adopted a separate optimization problem. A carbon-oriented planning model that combines the planning of natural gas systems, fast-charging stations for EVs, PV plants, fuel cells, and carbon capture and storage units was proposed by Wu et al. [4]. Moreover, the suggested model was recruited to consider the future scenarios' anticipated adaptation costs and the initial planning costs. The simulations showed that carbon emissions could greatly affect the planning results. When the carbon emission cap was lowered, the planning costs went up, and more infrastructure was installed. Quijano et al. [5] presented a new strategy for improving the hosting capacity of modern distribution networks in the presence of plug-in EVs. The strategy sought to identify the maximum penetration of wind plants, considering the forecast basis, the dynamic operation of EVs, voltage regulation units, and the distributed generators (DGs) penetration. The suggested approach was formulated as a two-stage linear optimization problem. The first stage maximizes the capacity of the installed distributed generation units, while the second minimizes energy losses. The obtained results demonstrated that, by controlling the dispatched power to the EV aggregators, the hosting capacity of the DGs was significantly increased compared to the uncontrolled case.

Commonly, the planning models of the diverse projects that have used transformer taps, energy storage systems (ESSs), transmission lines, and voltage regulators are incorporated into the EV planning problem to facilitate the installation of EV stations. For instance, Mehrjerdi et al. [6] provided a model to determine the nominal level of facilities required, such as fast, intermediate, and slow speed chargers, and to calculate the capacity of EV charging stations. Further, the model was extended to determine the size and operational scheduling of ESSs optimally. The line reinforcement strategy was also employed to strengthen the grid. The results showed an increase in the rated power of fast-speed chargers compared to the intermediate and slow ones. Furthermore, the reduction of line reinforcement resulted in an increase in the battery power installed. Fachrizal et al. [7] introduced an integrated model that facilitated the combination of PV units and EVs and provided an assessment tool to evaluate the hosting capacity of LV distribution networks. To this end, four scenarios were executed, and the system performance was analyzed. The results found that EV smart charging significantly improved the hosting capacity of EVs in residential areas, while PV curtailment significantly increased the hosting capacity of PVs. The authors in [8] also presented a framework to optimally determine the size of the PV units and EVs for charging stations, taking into account load-matching indices. The suggested approach used the self-consumption-sufficiency balance as a new score based on a principle similar to the F1 score in machine learning, which combines a classifier's precision and recall into a single metric by taking their harmonic mean. The high value of the suggested score referred to the fact that the system was close to a self-sufficient state and could not export or curtail a large amount of local generation. The results showed that a self-sufficient state could be achieved with a more extensive combination of PV modules and EVs.

Considering EVs' uncertainty, EVs' behavior is stochastic, and its modeling is critical [9]. A precise formulation for the dynamic operation of EVs is required for their secure integration into the power networks. Stochastic and robust-based methods are commonly used to formulate EV-related uncertainties. A long-term expansion planning model for microgrids was investigated by Mehrjerdi et al. [10]. The objective was to expand the capacity of generation sources, such as the micro-turbine, PV modules, wind station, and ESS, where the EV charging stations appeared as load centers or dispatchable units. The short- and long-term uncertainties were considered. The findings revealed that considering the short-term operation of the dispatchable units reduced the calculated cost to 28% and contributed to minimizing the cost of the long-term plan. Manríquez et al. [11] analyzed the effects of the massive use of EVs on the Chilean national electric network expansion for the year 2030. Data from representative days was used to formulate uncertainties and model the hourly dynamics of the network. Further, the demand for EVs was investigated by considering five scenarios that differed in the number of EVs and the charging strategies employed. The results demonstrated that smart charging resulted in a 2.4% increase in solar power-based generation units and a 2.5% decrease in conventional generating units. Baringo et al. [12] proposed an adaptive, robust optimization framework for expanding the distribution system while accounting for short-term variability and long-term uncertainties. The expansion model was formulated and extended to involve planning renewable energy sources (RESs), ESSs, and EV charging stations. On some representative days inspired by different dynamic operational conditions, short-term uncertainties in load behavior, RESs, EVs, and electricity prices were modeled. Long-term uncertainties were also investigated by considering the uncertainties associated with future peak loads, the electricity future outlook on the transmission network, and the number of EVs. The simulations proved the superiority of the proposed robust approach in terms of computational performance compared to the stochastic approach.

The overall increase in EV usage may deteriorate the power system's reliability and security if it is not appropriately considered and no counteractions are taken [13]. Several studies were carried out to handle this issue. The planning problem of power systems and EVs' charging infrastructure was addressed by Yao et al. [14] to improve power network resilience when subjected to extreme weather conditions. The *N-k* reliability criterion was considered to enhance the security of the systems. The simulation results demonstrated the EVs' ability to improve power network resilience by reducing the amount of load shedding required. Further, a reliability-constrained methodology that considered demand response programs was proposed by Kamruzzaman et al. [13] to maximize EV penetration in modern power systems. The proposed method was carried out by considering the demand response measurements and EVs' load profiles and employing the Monte Carlo simulation to assess some well-known reliability indices. The results revealed that the proposed method could significantly address the negative impacts of EVs and thus maximize EV penetration.

Generally, EV planning is complex, especially if the multi-objective frameworks are formulated to manage several components. Recently, several efficient metaheuristic algorithms have been broadly developed and used to solve EV planning problems. A multi-objective nonlinear planning model was developed by Fan et al. [15]. The aim was to reduce investment costs and maximize charging stations' utilization. The suggested model was solved via a multi-objective Tchebysheff decomposition-based evolutionary algorithm. The results demonstrated the efficiency of the applied methodology in obtaining high-quality solutions. The multi-objective particle swarm optimizer was used by Sadeghi et al. [16] to optimally identify the size of the microgrid's resources, such as PV, wind, and EVs. The problem was formulated to minimize the lifecycle cost of the network, which included the initial, maintenance, and replacement costs, and reduce the probability of power supply loss. An algorithm that relied on the greedy randomized adaptive search and tabu search algorithms was developed by Da Silva et al. [17] to estimate the nominal level of DGs and EVs that can be hosted in distribution networks. The results demonstrated the capability of the proposed algorithm to obtain high-quality solutions and reach high-quality solutions in a few iterations. A bi-level approach based on metaheuristics was developed by Ali et al. [18] to optimally allocate PV and wind power units considering EV parking lots in microgrids. The multi-objective anti-lion optimizer was used to formulate the proposed solving strategy. Ali et al. [19] introduced a two-layer metaheuristic algorithm based on the grey wolf optimizer to improve the hosting capacity of PV-powered plants while dealing with constraints. In addition to maximizing the PV hosting capacity, the proposed model was formulated to harmonize the use of transformers, VAr resources, and EVs. The inner level was concerned with optimizing the charging and discharging power of EVs, transformer taps, and VAr resources, while the outer level focused on maximizing PV penetration.

Aside from metaheuristics, mathematical methods have been steadily employed to solve EV-based optimization problems. What impedes the mathematical methods is that their efficiency dramatically deteriorates when they are applied to solve large-scale nonlinear problems. Many heuristic and decomposition approaches were developed in the literature to simplify the problem. In addition, diverse linearization strategies were employed to linearize nonlinear formulations. For example, Zhang et al. [20] proposed a two-stage optimization model to identify the nominal hosting capacity levels for PV units incorporating soft open points and EVs in distribution networks. The model was converted to mixed-integer second-order cone programming using a convex relaxation strategy. The column-and-constraint generation-based solving algorithm was developed to solve the problem. The suggested solving methodology achieved an acceptable relaxation gap of 0.0573%. The authors in [21] developed a heuristic approach to calculate to what extent domestic EVs can penetrate low-voltage networks, considering different scenarios for heat pump penetration. A new zonal methodology was also employed for EV optimization. The problem was formulated as a linear optimization problem and solved using the CPLEX solver within OATS optimization software.

Based on the above discussion, a hosting capacity model that combines the planning models of many infrastructures as well as provides system operators with a comprehensive view of the impacts of EV integration into emerging networks is rarely presented. Furthermore, testing and developing new solution algorithms is essential to expanding the toolbox of methods. In order to address these issues, this study presents a multi-objective planning framework for increasing the hosting capacity of EVs in power systems while maintaining the environmental requirements for realistic emerging networks. The planning model is extended to combine the planning models of PV plants, ESSs, thyristor-controlled series compensators (TCSCs), and transmission lines to improve the installation of EV charging stations. The main contributions of this paper are summarized as follows:

- This paper investigates a planning model for emerging networks to enable the increase in the use of EVs.
- A multi-objective planning approach for EV charging stations, ESSs, PV plants, TCSCs, and transmission lines is formulated. It makes it easier to determine the maximum number of EVs that can be integrated into the main grid over a day.
- Four scenarios are suggested to identify the environmental impacts of the uncounted penetration of EVs into the grid.
- The multi-objective version of the Gazelle optimization algorithm was developed to solve the proposed problem. It is an extension of the single-objective Gazelle optimization algorithm (GOA), which was first developed by Agushaka et al. [22] in 2022.
- The proposed planning model and adopted solving algorithm were tested using the realistic Egyptian West Delta Network (WDN).

The remainder of the paper is organized as follows. The problem is formulated in Section 2. The multi-objective Gazelle optimization algorithm is presented in Section 3. The simulation results are presented and discussed in Section 4. The conclusions and direction of future works are summarized in Section 5.

2. Mathematical Model

2.1. Objective Functions

The proposed problem has three objectives that are formulated as follows: The first objective (F_1) aims to maximize EVs' penetration, as demonstrated in (1). The target is to calculate to what extent the systems can host EVs, considering the worst-case scenario. This work considers EV charging scenarios (i.e., the grid-to-vehicle scenario) but ignores smart vehicles' ability to support the grid during peak hours or in contingencies (i.e., the vehicle-to-grid scenario).

$$F_1 = max\left(\sum_{h\in H}\sum_{i\in B} P_{EV}^{i,h}\right) \quad (1)$$

The EVs' hoped-for environmental benefits cannot be achieved if fossil fuel-based power plants mainly provide their charging power. Therefore, the second objective (F_2) plans to reduce the carbon dioxide emitted from fossil fuel-based generation units, as shown in (2). The objective is to control the use of EVs such that they do not lead to high carbon emissions from the generation units needed to cover this penetration. The total carbon emitted from the generation units (Car_{CO2}) is estimated by (3), where γ is expressed in ton/MWh.

$$F_2 = min(Car_{CO2}) \quad (2)$$

$$Car_{CO2} = \sum_{h\in H}\sum_{i\in B} \gamma\, P_{Th}^{i,h} \quad (3)$$

The third objective (F_3) focuses on minimizing the investment and operation costs of the devices installed, as described in (4). The first term in (4) covers the cost of the transmission lines and TCSCs required to facilitate the integration of EV charging stations into networks, as calculated in (5). While the second and third terms describe, as explained by (6) and (7), the investment and operating costs of generation stations and ESSs. Accounting for the cost of ESSs is an essential step in developing energy-saving planning strategies, which can help promote the use of ESSs in many applications [23].

$$F_3 = min\left(C^T + C^G + C^{ESS}\right) \quad (4)$$

$$C^T = \sum_{h\in H}\sum_{\substack{i,j\in B\\ i\neq j}} \frac{\lambda(1+\lambda)^y}{(1+\lambda)^y - 1}\left(\alpha_L^{ij}\left(L_{ij}^h - L_{ij}^{h-1}\right) + \alpha_{TCSC}^{i,j}\left(S_{ij}^{TCSC,h} - S_{ij}^{TCSC,h-1}\right)\right) \quad (5)$$

$$\begin{aligned} C^G = \sum_{h\in H}\sum_{i\in B} \tfrac{\lambda(1+\lambda)^y}{(1+\lambda)^y-1}(&\alpha_{g\_cap}^i(P_{Th}^{i,h} - P_{Th}^{i,h-1}) + \alpha_{ren\_cap}^i(P_{PV}^{i,h} - P_{PV}^{i,h-1}) \\ &+365\,\alpha_{g\_op}^i P_{Th}^{i,h} + 365\,\alpha_{ren\_op}^i P_{PV}^{i,h}) \end{aligned} \quad (6)$$

$$C^{ESS} = \sum_{h\in H}\sum_{i\in B} \alpha_{ESS\_cap}^h + \alpha_{ESS\_rep}^h + \alpha_{ESS\_el}^h + \alpha_{ESS\_op}^h \quad (7)$$

where $\frac{\lambda(1+\lambda)^y}{(1+\lambda)^y-1}$ is the capital recovery factor. λ and y denote the interest rate and the planning period, respectively. α_L^{ij} and $\alpha_{TCSC}^{i,j}$ are the investment cost of transmission circuits and TCSCs installed at the route *i*-*j*. $\alpha_{g\_cap}^i$ and $\alpha_{g\_op}^i$ are the capital and operating costs of the thermal generation unit installed at node *i*. $\alpha_{ren\_cap}^i$ and $\alpha_{ren\_op}^i$ are the capital and operating costs of the renewable units built at node *i*. $\alpha_{ESS\_cap}^h$, $\alpha_{ESS\_rep}^h$, $\alpha_{ESS\_el}^h$, and $\alpha_{ESS\_op}^h$ are the capital, replacement, end-life, and operating costs of ESSs. They are calculated as illustrated in [24].

2.2. Operational and Project Constraints

The intensive use of RESs and EVs drives power grids closer to their stability limits, making instability more likely. To safely host EVs and RESs, it is important to use a simple and accurate planning model that considers the power flow reliability and stability constraints [25,26]. The DC power flow is used in this work to formulate the constraints as explained by (8)–(21) [24,27]. Constraint (8) describes the active power flow balance formulation. It maintains the balance between the injected and absorbed nodes' powers. The total power injected into any bus should equal the active power absorbed into the same bus. Equation (9) ensures that the power flow through new and existing transmission lines and TCSCs in each scenario should not exceed the maximum thermal limits. The constraint shown in (10) controls the nodes' voltage angles within the permissible limits.

$$P_{Th}^{i,h} + P_{PV}^{i,h} - P_{d,i}^{h} - P_{ch,i}^{h} + P_{dch,i}^{h} = \sum_{\substack{i,j \in N \\ i \neq j}} P_{ij}^{h} \tag{8}$$

$$-P_{ij}^{max} \leq \frac{L_{ij}{}^{h}\left(\theta_i^s - \theta_j^s\right)}{X_L^{ij}\left(1 - \lambda_{TCSC}^{ij,h}\right)} \leq P_{ij}^{max} \tag{9}$$

$$\theta_i^{min} \leq \theta_i^h \leq \theta_i^{max} \tag{10}$$

Equations (11) and (12) govern the active power generated from PV and fossil fuel-based generation stations. The power dispatched from any generation unit should not violate the designed rating values. Additionally, ESSs are constrained by (13) and (14) to maintain charging and discharging powers less than or equal to their maximum values. The ESSs' storage capacity installed should be lower than nominal levels, as illustrated by (15).

$$P_{PV\_min}^{i} \leq P_{PV}^{i,h} \leq P_{PV\_max}^{i,h} \tag{11}$$

$$P_{Th\_min}^{i} \leq P_{Th}^{i,h} \leq P_{Th\_max}^{i} \tag{12}$$

$$0 \leq P_{dch}^{i,h} \leq P_{dch\_max}^{i,h} \tag{13}$$

$$0 \leq P_{ch}^{i,h} \leq P_{ch\_\max}^{i,h} \tag{14}$$

$$0 \leq E_{ESS,i}^{h} \leq E_{ESS,i}^{max} \tag{15}$$

Each storage system's maximum charging and discharging power is calculated by (16) and (17). The state of charge (SOC) of the ESS is determined by (18).

$$0 \leq P_{dch}^{i,h} \leq min\left(P_{ESS\_max}, \frac{SOC_{ESS}^{i,h} - E_{ESS\_\max}^{i}}{\Delta t}\right) \tag{16}$$

$$0 \leq P_{ch}^{i,h} \leq min\left(P_{ESS\_max}, \frac{E_{ESS\_\max}^{i} - SOC_{ESS}^{i,h}}{\Delta t}\right) \tag{17}$$

$$SOC_{ESS,i}^{h} = SOC_{ESS,i}^{h-1} + \eta_{ESS}^{ch} P_{ch,i}^{h} - \frac{P_{dch,i}^{h}}{\eta_{bat}^{dch}} \tag{18}$$

Equations (19) and (20) guarantee that the number of transmission lines and ESSs does not exceed the maximum number. They imply that projects can be built at any time to be ready for use. Equation (21) governs the size of the TCSCs installed between routes.

$$L_{ij}^{h-1} \leq L_{ij}{}^{h} \leq L_{ij}^{max} \tag{19}$$

$$N_{ESS}^{i,h-1} \leq N_{ESS}^{i,h} \leq N_{ESS\_max}^{i} \tag{20}$$

$$\lambda_{TCSC\_min}^{ij} \leq \lambda_{TCSC}^{ij,h} \leq \lambda_{TCSC\_max}^{ij} \tag{21}$$

This study adopts a scenario-based approach to formulate the model uncertainties. Realistic data for solar radiation and load consumption are used to simulate the stochastic behavior of the PV units and loads. The successive scenarios strategy is used to estimate the capacity of the ESS required accurately and the EV levels that can penetrate across all scenarios, where the problem is solved in each scenario by taking the previous scenario's ESS behavior into account to define the maximum permissible charging and discharging powers that the ESS can support regarding the grid and accommodate its function.

3. Optimization Algorithm

3.1. Gazelle Optimization Algorithm

The GOA is a population-based optimizer. It was first developed by Agushaka et al. [22] in 2022. The GOA simulates the capability of the gazelles to survive in environments where a predator dominates. Gazelles are considered one of the most common prey items for predators. Similar to any population-based algorithm, the mechanism of the GOA's operation is described in two phases—exploitation and exploration. The next phase to be conducted is identified based on a randomly generated number. If it is less than 0.5, the exploitation phase is executed; otherwise, the exploration phase is carried out. The operating mechanism of the GOA is explained in Figure 1.

The exploitation phase mimics gazelles' behavior when they graze peacefully or their behavior when the predator stalks them. In this phase, the Brownian motion, characterized by uniform and controlled steps, effectively covers the domain's neighborhood areas. The exploitation phase can be modeled as follows:

$$X^{t+1} = X^t + S \times R \times R_B \times \left(X^t_{elite} - R_B \times X^t\right) \tag{22}$$

where X^t and X^{t+1} are the gazelles' current and next positions, respectively. X^t_{elite} is a matrix in which the top gazelle vector is replicated N times. S is the gazelles' grazing speed. R_B and R are vectors that, respectively, represent random numbers for Brownian motion and random numbers in [0,1]. The exploration phase simulates the behavior of gazelles when they suddenly spot the predators and the behavior of the predators when they chase the gazelles. The population is divided into two equal groups. The first group represents gazelles, and the second group represents predators. The behavior of gazelles can be derived as follows:

$$X^{t+1} = X^t + S^{top} \times \mu \times CF \times R_B \times \left(X^t_{elite} - R_L \times X^t\right) \tag{23}$$

where S^{top} is the maximum speed that gazelles can reach, and R_L is a vector of random numbers calculated using the Lévy distribution [22]. The behavior of the predators can be formulated by:

$$X^{t+1} = X^t + S^{top} \times \mu \times R \times R_L \times \left(X^t_{elite} - R_L \times X^t\right) \tag{24}$$

where CF regulates the predators' movement, it varies at each iteration and is calculated by:

$$CF = \left(1 - \frac{t}{T}\right)^{2\frac{t}{T}} \tag{25}$$

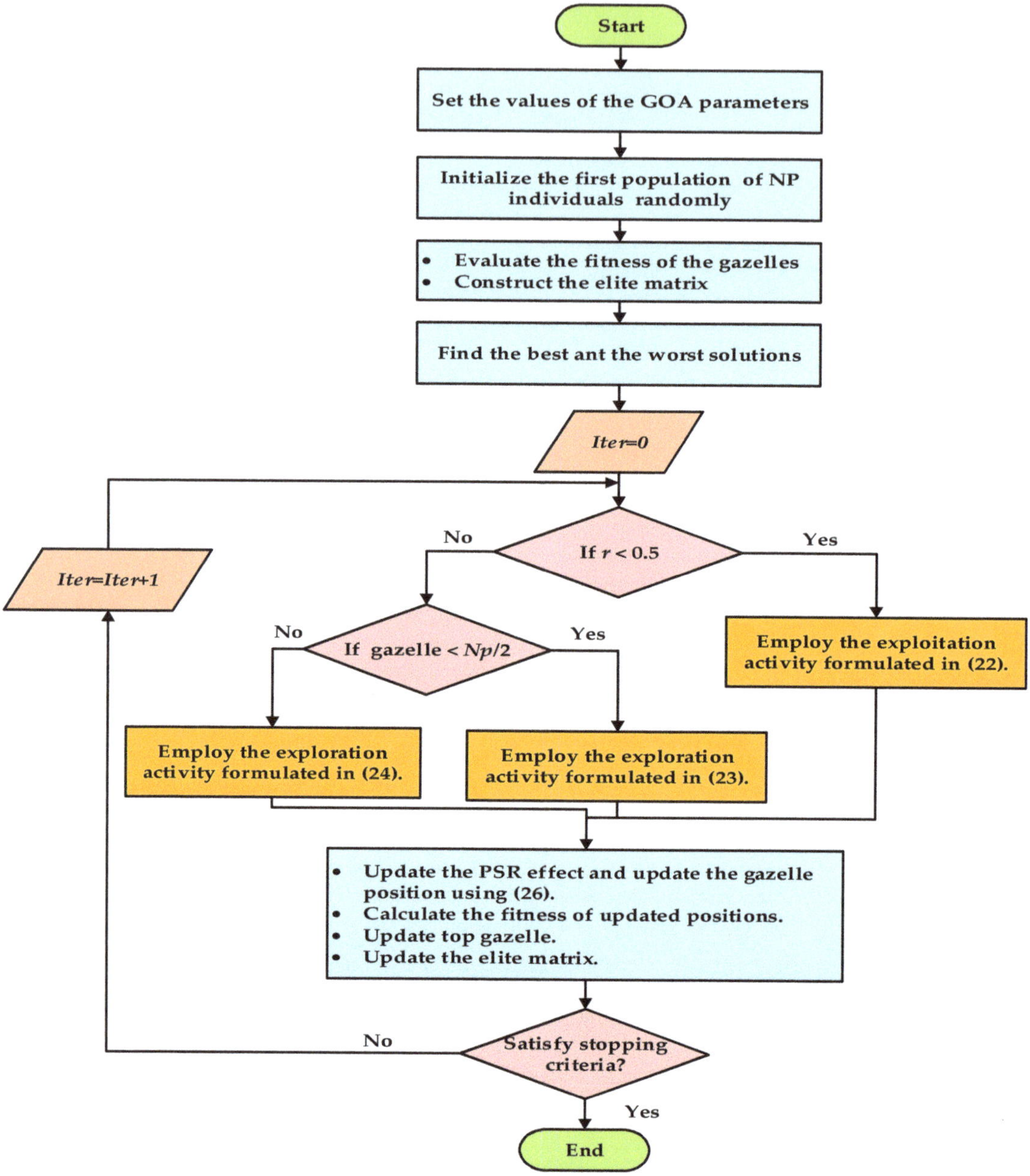

Figure 1. Flowchart of the single-objective version of the GOA [22].

In order to avoid trapping at local optima, the algorithm uses the effect of the predator success rates (*PSR*) to improve the quality of the solutions obtained as follows:

$$X^{t+1} = \begin{cases} X^t + CF \times (LB - R \times (UB - LB)) \times U, \; if \; r \leq PSR \\ X^t + (PSR(1-r) + r) \times \left(X^{r1} - X^{r2}\right), \; if \; r > PSR \end{cases} \tag{26}$$

where r_1 and r_2 are random indices of two individuals in the population. U equals 0 or 1 based on the value of r. More details about the GOA can be found in [22].

3.2. Multi-Objective Gazelle Optimization Algorithm

The current version of the GOA is not able to deal with multi-objective optimization problems. The current version only stores a single solution as the best solution. It is not able to save many solutions as the best solutions for multi-objective problems. The priori and posteriori approaches are commonly used for solving multi-objective problems [28]. Each approach has its advantages and disadvantages. In the priori approach, the multi-objective functions are converted to a single-objective function by aggregating the objectives using a set of weights that is decided based on the significance of each objective. After that, a single-objective solver can be implemented to detect the optimal solution. The posteriori method can maintain the multi-objective problem formulation, and the Pareto optimal solution can be found in a single run [18,28]. Further, its accuracy does not depend on the weights set by experts. The Pareto optimal method preserves a set of dominant solutions that can strike a balance between objectives. This work uses the Pareto optimal approach to formulate the GOA multi-objective version (MGOA). The MGOA can be summarized in Algorithm 1 as follows:

1. The population is first initialized by:

$$X = rand \times (UB - LB) + LB \tag{27}$$

2. After that, in each iteration, the gazelles' position is updated using (22)–(26).
3. A dominating formula is derived for two or more objectives, as follows. The solution X_1 dominates the solution X_2 on if

$$\forall_o \in \{1, 2, \ldots\ldots k\} : f_o(X_1) \leq f_o(X_1) \tag{28}$$

$$\exists_o \in \{1, 2, \ldots\ldots k\} : f_o(X_1) < f_o(X_1) \tag{29}$$

4. In each iteration, the non-dominated solutions are stored in a repository similar to the archives in the multi-objective particle swarm optimizer (MPSO) [29] and the multi-objective salp swarm algorithm (MSSA) [28].
5. The discovered target space is tabulated.
6. Individuals' optimal memory location is updated.
7. The members of the current population that have been dominated are added to the repository, thereby increasing its size.
8. The repository's collected members are re-checked, and the members dominated are eliminated. This level reduces the repository's population.
9. If the number of individuals exceeds the repository's maximum capacity, a portion of the population is eliminated, and the tabulation process is restarted.
10. If the termination criterion is not met, the algorithm returns to the second step and repeats the remaining steps; otherwise, the optimization process terminates.

Algorithm 1: Pseudo-code of MGOA.

Define the problem's dimension, population size (N_p), upper (UB) and lower (LB) bounds of the decision-making variables.
Set the stop criteria (T).
Set MGOA parameters.
Initialize the population (X) by (27).
Compute the cost functions, and check the problem constraints.
Assign a high penalty value to solutions that violate the constraints.
Define the top gazelle.
Construct the elite matrix.
While $(t \leq T)$
  For $i = 1{:}\ N_p$
    If $r < 0.5$
    Update the positions of the gazelles using MGOA's exploitation equation (22).
    Else
  If $i < N_p/2$
    Update the positions of the gazelles using MGOA's exploration equation (23).
  Else
    Update the positions of the gazelles using MGOA's exploration equation (24).
  End If
  End if
    Update The *PSR* effect and update the positions of the gazelles using (26).
  End for
      Calculate the fitness of the objective functions.
      Check the problem constraints and add a high penalty cost to the solution that violates the constraints.
      Define the non-dominant solution.
      Update the MGOA's archive.
      Update the top gazelle and the elite matrix.
 End While

4. Results and Discussion

The proposed planning model and solution algorithm were applied to the Egyptian West Delta network. It is a sub-transmission network in Egypt. It has 52 buses, 8 generation units, and 55 transmission routes, each with two circuits. By 2030, it is planned to install a new generation unit at bus 53, as depicted in Figure 2, to supply loads of 2195.8 MW. The transmission and generation data are presented in [30]. This study arbitrarily suggested that the candidate locations for EV charging stations, PV systems, and ESSs were buses 10, 20, 40, and 50. Figure 3 [31,32] describes the stochastic behaviors of the PV units and load centers.

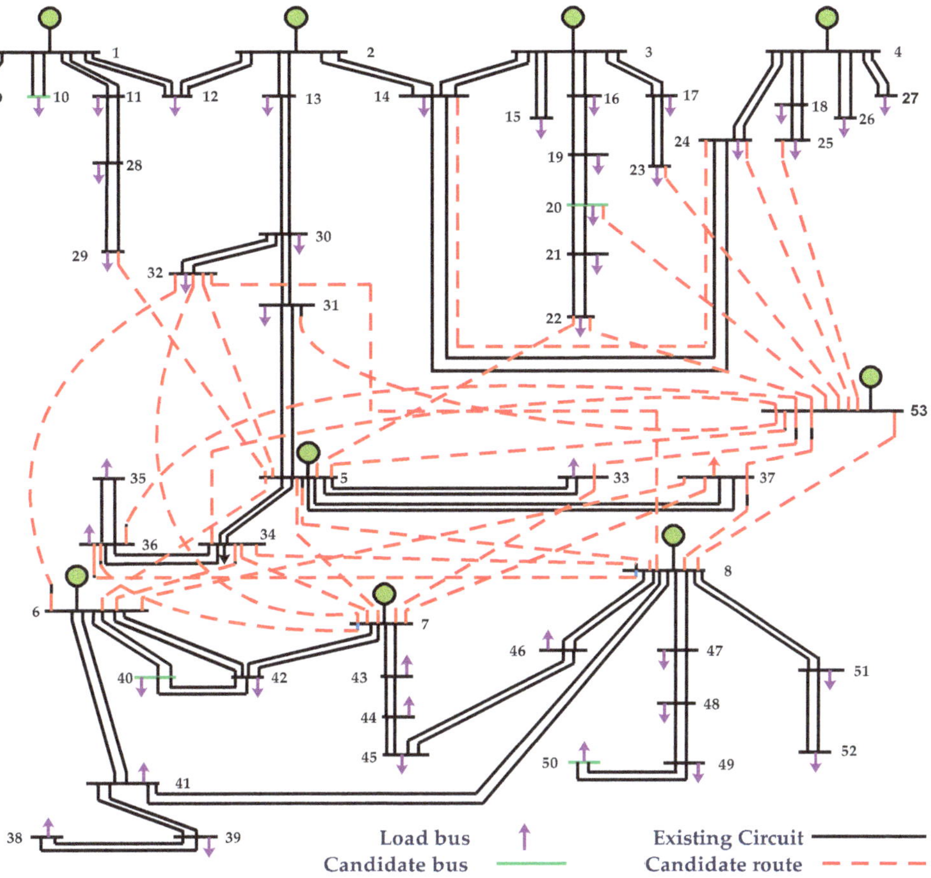

Figure 2. Initial configuration of WDN, adopted from [30].

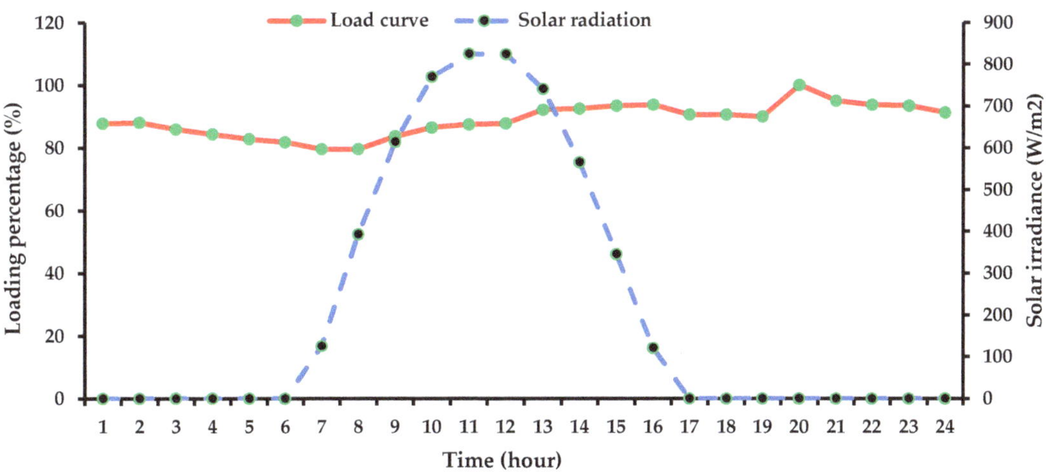

Figure 3. Solar irradiance profile and load behavior during the day.

The sodium-sulfur battery was used in this work. Its technical and economic characteristics were defined in [33]. The investment and operating costs of power plants were given in [34]. The combined cycle-generating unit was selected.

Four planning cases were executed to study the environmental and economic impacts of the integration of EV charging stations into emerging networks. The four cases are:

- Case #1: The WDN was planned to consider the environmental requirements as a priority, increasing EV penetration, and the financial requirements as second and third priorities, respectively. The ESSs were forced to charge in scenarios of high solar irradiance.
- Case #2: As in Case #1, the environmental requirements were the priority, but ESSs were not used.
- Case #3: The WDN was planned to consider the increasing EV penetration as a priority, then the environmental and economic requirements as second and third priorities, respectively. The ESSs were also forced to charge in scenarios of high solar irradiance.
- Case #4: The WDN was planned with EV penetration as a priority, like Case #3. However, the ESSs were not planned in this case.

4.1. *Testing the Performance of MGOA*

The efficacy of the MGOA was experimentally tested by solving a set of four benchmark-challenging test functions, named ZDT, developed by Zitzler et al. [35], in addition to the multi-objective planning problem investigated in this work. The mathematical models of the ZDT functions are given in [36].

The performance of the MGOA was compared to the multi-objective salp swarm algorithm (MSSA) [28] and the multi-objective particle swarm optimizer (MPSO) [29]. The MSSA and MOPSO are well-known algorithms in the literature for solving multi-objective optimization problems. It is worth stating that the population size is set to 200 for the ZDT functions and 40 for the planning problem. The maximum number of iterations applied in this work is 1000 for solving the ZDT functions and 500 for solving the proposed problem. The results are obtained through 30 independent runs for the ZDT functions and 20 for the proposed planning problem.

4.1.1. ZDT Test Problems

Table 1 shows that the MGOA performed better than the MSSA and MPSO on most ZDT functions. All algorithms were compared using the inverted generational distance metric (IGD) as a performance indicator. The MGOA was superior in terms of the average (*Ave*), standard deviation (*Std*), median, and best and worst solutions.

Table 1. Results of MGOA, MSSA, and MPSO for solving the ZDT test functions (using IGD).

Function	Algorithm	Ave.	Std.	Median	Best	Worst
ZDT1	MGOA	0.00237	0.00060811	0.001714	0.002078	0.0031
	MSSA [28]	0.00286	0.000841427	0.0025	0.0023	0.0043
	MPSO [28]	0.00422	0.003103	0.0037	0.0015	0.0101
ZDT2	MGOA	0.00117576	0.000057544	0.00107459	0.0010384	0.0014706
	MSSA [28]	0.0037	0.00130958	0.0044	0.0015	0.0047
	MPSO [28]	0.00156	0.000174	0.0017	0.0013	0.0017
ZDT3	MGOA	0.02403	0.0007621	0.0250	0.0234	0.0298
	MSSA [28]	0.02986	0.000898888	0.0296	0.0291	0.0314
	MPSO [28]	0.03782	0.006297	0.0362	0.0308	0.0497
ZDT1 (Linear front)	MGOA	0.0024589	0.0004557	0.004532	0.001905	0.0027835
	MSSA [28]	0.0033	0.000731437	0.0034	0.0025	0.0041
	MPSO [28]	0.00922	0.005531	0.0098	0.0012	0.0165

By inspecting the set of Pareto optimal points in Figure 4, it may be determined that the solutions obtained by the MGOA were uniformly distributed and nearly similar to those obtained by the MSSA and MPSO. The results also demonstrated the MGOA's ability to approximate the true front of these functions efficiently. In all functions, the MGOA's coverage was very competitive, and its convergence was very high. Although these functions are diverse in complexity and difficulty, the MGOA succeeded in driving solutions to explore different regions of the true Pareto optimal front and avoid trapping in one region.

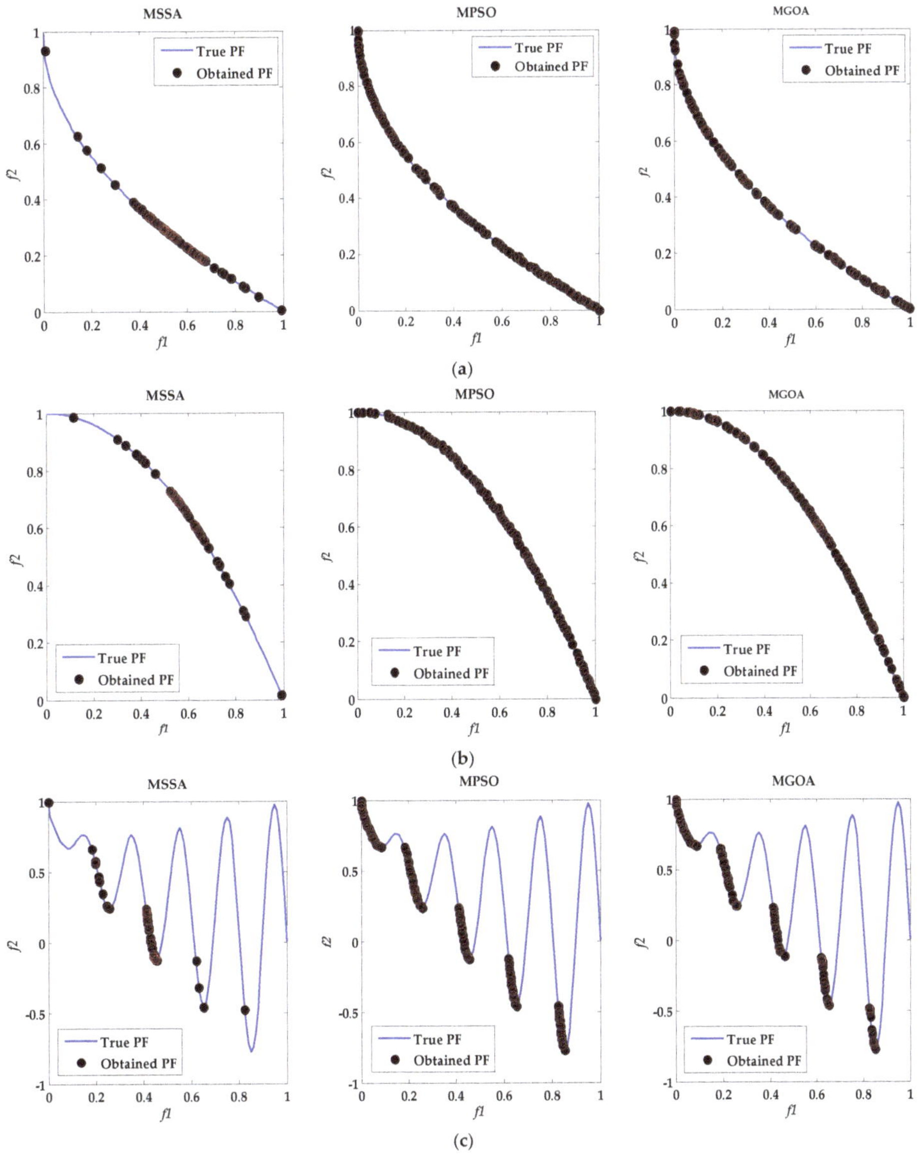

Figure 4. *Cont.*

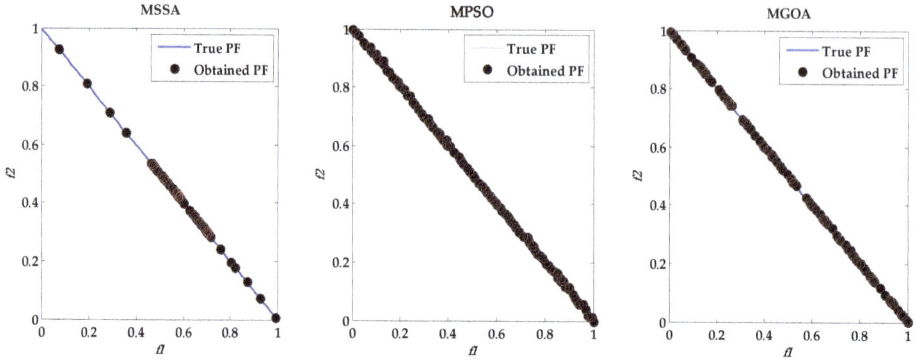

(d)

Figure 4. Best Pareto front determined by MGOA, MSSA, and MPSO on: (**a**) ZDT1, (**b**) ZDT2, (**c**) ZDT3, and (**d**) ZDT1 with a linear front.

4.1.2. Proposed Problem

Table 2 summarizes the results obtained by the MGOA, MSSA, and MPSO when employed to solve the problem in Scenario number 7. Figure 5 depicts the dominant solutions (F1, F2, and F3) for Cases #1, #2, #3, and #4. By inspecting the results in Table 2, it can be observed that the MGOA had merits over the MSSA and MPSO in terms of robustness and obtaining high-quality solutions. The robustness of the MGOA ranged between 20 and 30%, while the MSSA and MPSO varied by about 20–25% and 15–25%, respectively.

Table 2. Results of MGOA, MSSA, and MPSO in solving the problem (Scenario number 7).

Case Number	Solver	Best Solution			Robustness	Time (s)
		- F1	F2	F3		
Case #1	GOA	0	1600.945	689.645	5	1010.47
	SSA	0	1600.945	696.366	4	641.22
	PSO	0	1600.945	693.550	5	805.03
Case #2	GOA	0	1459.345	682.454	6	1081.47
	SSA	0	1459.345	688.538	6	721.31
	PSO	0	1459.345	683.835	5	826.12
Case #3	GOA	−387.0318	1943.468	731.529	4	1017.55
	SSA	−386.862	1943.468	733.230	4	644.66
	PSO	−387.0318	1943.468	733.194	3	821.87
Case #4	GOA	−547.031	1943.468	736.929	6	1027.31
	SSA	−546.975	1943.468	742.478	5	650.26
	PSO	−547.031	1943.468	740.695	5	845.23

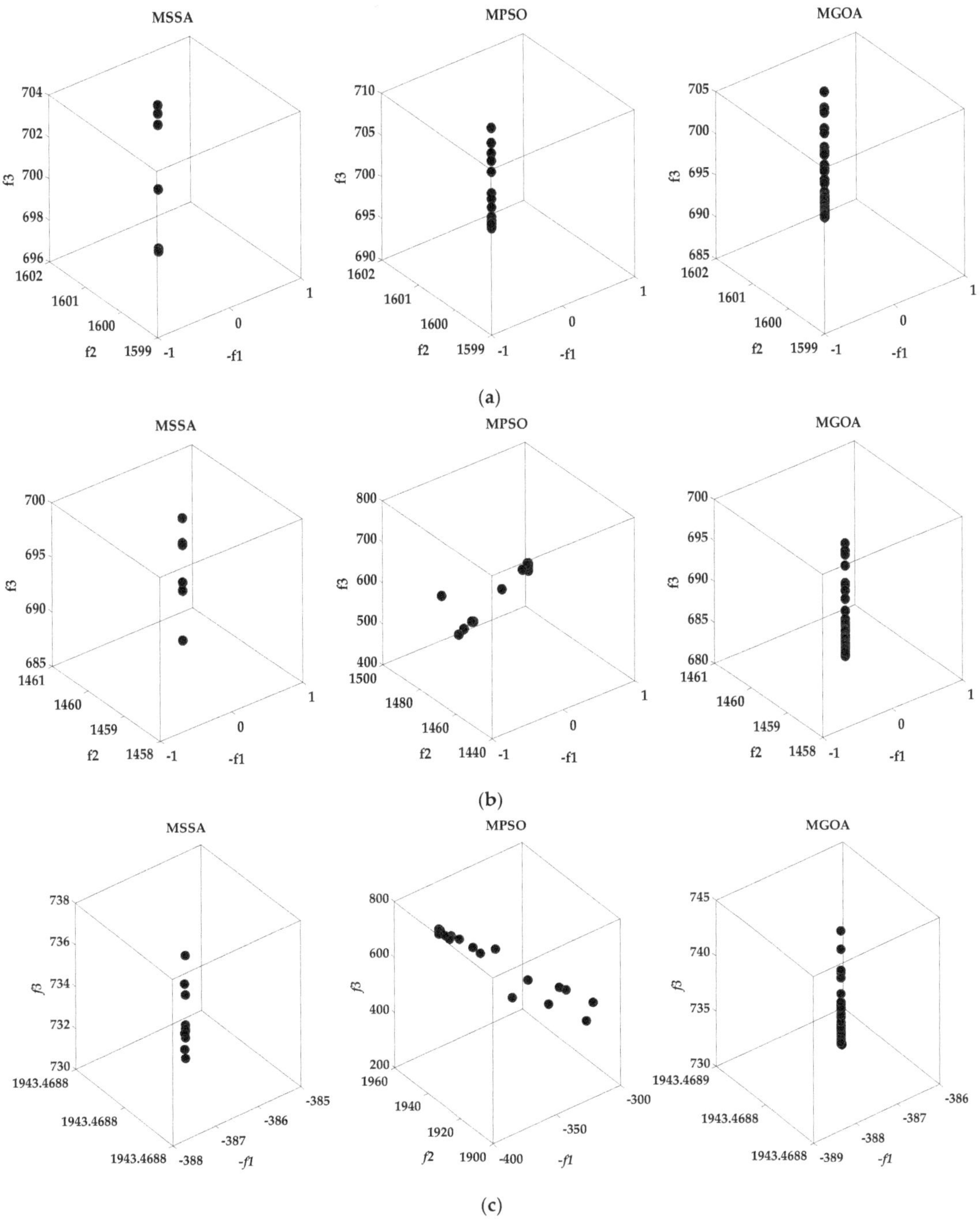

Figure 5. *Cont.*

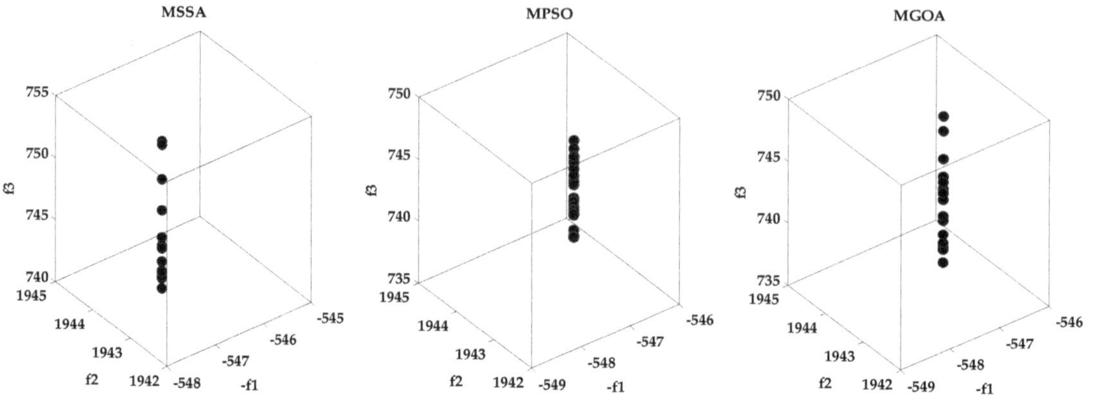

(d)

Figure 5. Best Pareto front determined by MGOA, MSSA, and MPSO on: (**a**) Case #1, (**b**) Case #2, (**c**) Case #3, and (**d**) Case #4.

Moreso, Table 2 indicates that the MSSA consumed less time than the MGOA and MPSO. For example, the MSSA saved about 33.3% and 12.6% of the time required for solving Case # 2 by the MGOA and MPSO, respectively.

4.2. WDN Planning

This sub-section analyzes the impacts of different planning strategies for EV charging stations on environmental requirements and EV penetration size. As mentioned before, the network was planned in the first strategy, considering that achieving the environmental requirements was the priority through minimizing the amount of carbon emitted from fossil fuel-based units. In the second strategy, the network was planned while neglecting ESSs, with plans to discuss the technical effects of ESSs on the emission of CO_2. The WDN was planned in the third strategy, increasing EV penetration as the highest priority while still addressing environmental concerns. Finally, in the fourth strategy, the ESS planning model was ignored in order to estimate how much the use of ESSs affects the size of the EV charging stations installed. Table 3 presents the energy charged by EVs, the amount of carbon emitted, the size of the ESSs installed, and the total planning cost in each case.

Table 3. Results obtained in Cases #1,2,3, and 4.

Item	Case #1	Case #2	Case #3	Case #4
Energy charged by EVs (GWh)	0	0.005964	9.89	10.14
Carbon emission (million metric tons)	37,882.81	37,674.52	46,643.2	46,643.25
Size of ESS (MWh)	1116.4	0	1116.4	0
Total cost (million USD)	1249.7	830.71	1278.9	848.73

Figures 6 and 7 describe the amount of carbon emitted in each case. It may be observed that the general use of EVs led to high carbon emissions from the generation stations to cover this penetration, as shown in Cases #3 and #4. The network could technically accommodate the estimated size of vehicles between 9.89 and 10.14 GWh, as described in Figures 8 and 9. This size caused an increase of approximately 23% in CO_2 emitted compared to Case #1, as described in Figure 7.

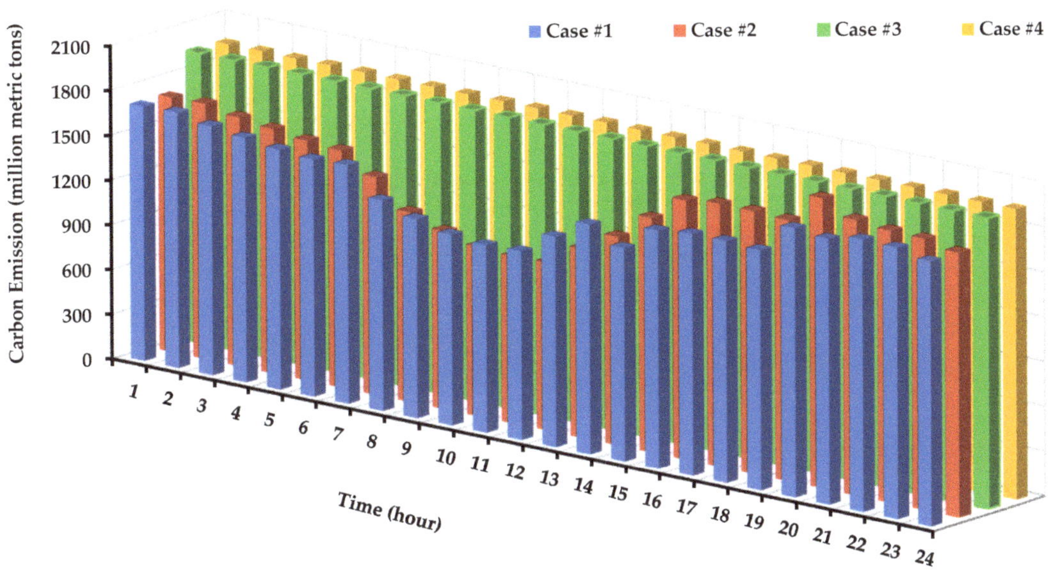

Figure 6. Carbon emissions in each case during the day.

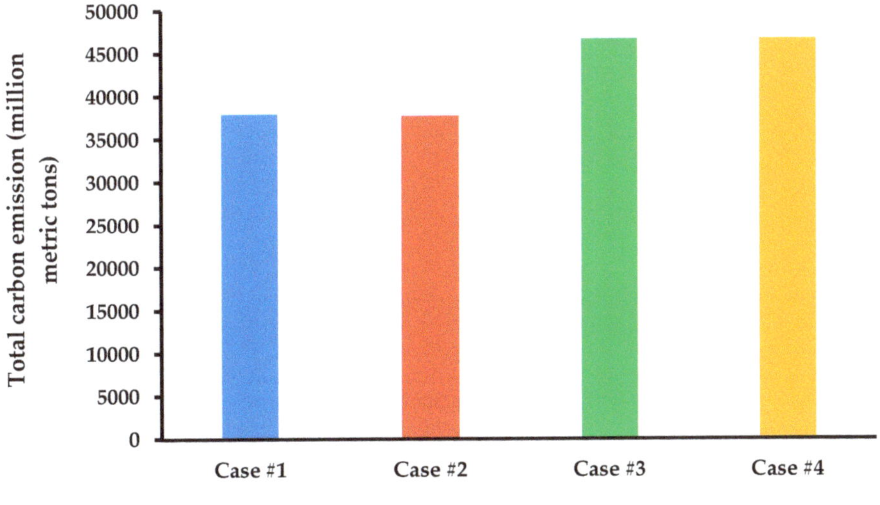

Figure 7. Comparison of total carbon emitted emissions in each case.

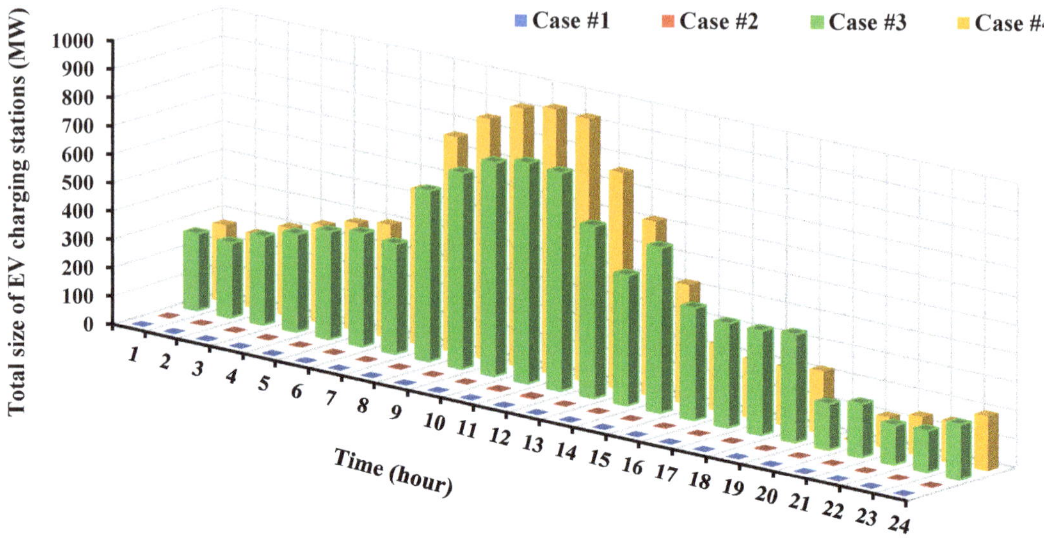

Figure 8. EV hosting capacity in each case during the day.

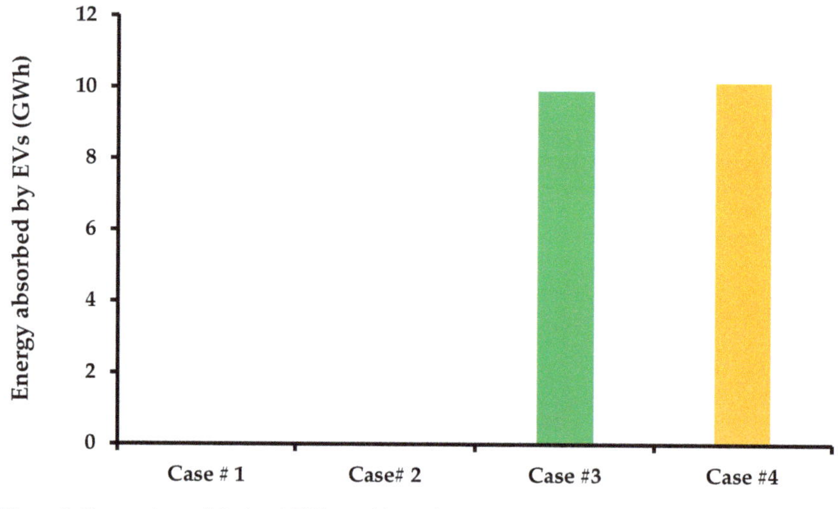

Figure 9. Comparison of the total EV hosted in each case.

In Case #1, the total amount of carbon was about 37,882.81 million metric tons and decreased to 37,674.52 million metric tons when ESSs were not planned and did not share in the power consumed by PV units. This decrease totaled about 208.29 million metric tons. This implies that, in Case #1, the ESS partially relied on fossil fuel power plants to obtain the required power for charging, which harmed the environmental conditions for which they were primarily designed. The energy dispatched from the PV units was insufficient to cover the energy required by the ESSs and loads. Load centers consume all the energy dispatched by the PV units. It can be stated that the ESS's role grows as more RES is added to the network. It is worth remarking that EVs were not allowed for integration into the network when environmental concerns were considered the priority, as illustrated in Figures 8 and 9. Adding electric vehicles would have caused an extra burden on the network. The fossil fuel power plants would have to increase their generation to compensate for this. The PV unit's output was insufficient to cover the energy required by the loads and EVs.

The results also showed that in Case #1 and Case #2, the scenarios in which carbon emissions declined were correlated with periods when solar radiation was as high as that observed in the scenario numbers 7, 8, 9, 10, 11, 12, 13, 14, 15, 16, and 17 (see Figure 6). The mix of several renewable energy sources, which have different production profiles during the day, may positively influence low carbon emissions over the day.

When EV use was considered the priority (see Cases #3 and #4), EV penetration during the day was about 9.89 and 10.14 GWh, respectively, as depicted in Figure 9. The results also showed that the penetrating ability of EVs can be increased during periods of high solar radiation, as noticed in hours 7, 8, 9, 10, 11, 12, 13, 14, 15, 16, and 17 (see Figure 8). In Case #3, the ESSs participated in shifting the EV charging periods away from high solar radiation scenarios by storing energy during these hours and discharging it during other hours as needed. The findings also revealed an increase in carbon emissions due to an increase in the EV hosting capacity. An increase of about 8760.4 million metric tons was calculated in Case #3 compared to Case #1. As a result, any increase in EV hosting capacity should be accompanied by an increase in RES to avoid reliance on fossil fuels to compensate for this increase.

Figure 10 depicts the planning cost for each case. The planning costs in each case were about USD 1249.7, 830.71, 1278.9, and 848.73 million, respectively. It can be noticed that the ESS planning cost constituted a large portion of the planning cost, as illustrated in Cases #2 and 4. A reduction of about 33.5% in the planning cost was observed due to the non-use of ESSs.

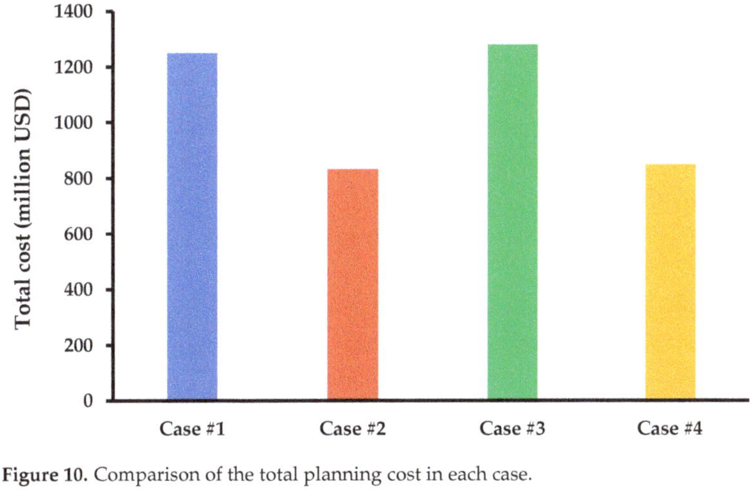

Figure 10. Comparison of the total planning cost in each case.

5. Conclusions

The integration of EVs into power networks commonly affects the network's reliability and operation. Furthermore, if widely adopted, the uncounted use of EVs may negatively impact the environmental requirements and result in high carbon emissions from power plants to cover this penetration. In this work, a planning strategy that safely increases the use of EVs while considering the environmental and technical requirements was developed. The proposed strategy was formulated as an optimization problem with the three objectives of improving the EV host's capacity, reducing carbon dioxide emissions, and decreasing the investment and operating costs of the devices installed. The planning models of RESs, ESSs, TCSCs, and transmission lines were combined into the EV charging station planning model to optimally facilitate EVs' integration from an environmental and technical point of view. The problem was a challenging multi-objective optimization problem. In order to solve it, the multi-objective version of the GOA was derived. Four case studies were carried out on the Egyptian West Delta Network to demonstrate the efficiency of the suggested model. In the first case, achieving the environmental requirements was the first priority, regardless of the EV penetration levels. In the second case, the network was planned while neglecting the ESSs, with plans to discuss the technical effects of the ESSs on the emission of CO_2. The Egyptian West Delta Network was planned in the third case, while increasing EV penetration was the highest priority. Finally, in the fourth case, the ESS planning model was ignored in order to estimate how much the use of ESSs affects the size of the EV charging stations installed. The main findings of this work can be summarized as follows:

- The MGOA was superior in comparison with the MSSA and MPSO in most of the ZDT functions investigated.
- The MGOA outperformed the MSSA and MPSO in terms of robustness and obtaining high-quality solutions when applied to solve the proposed problem. The robustness of the MGOA was about 20–30%, while the MSSA and MPSO were about 20–25% and 15–25%, respectively.
- The MSSA was faster than the MGOA and MPSO in solving the proposed problem. Approximately 12.6–36.7% of the time executed was saved by the MSSA.
- The results demonstrated that the scenarios in which carbon emissions declined were correlated with periods when solar radiation was high.
- According to the case studies, the simulations indicated a 23.2–23.8% increase in carbon emissions due to the widespread use of EVs.
- The total amount of carbon emitted decreased by about 208.29 million metric tons when the ESSs were not planned and did not share in the power consumed by the PV units. The ESS's role rises as more RESs are added to the network.

One point that was out of the scope of this study was the investigation of the impact of reliability constraints, such as the *N*-1 security and short-circuit current requirements, on the penetration of EV charging stations. Moreover, further work is needed to consider the ESSs' degradation in grid-connected operation and their effects on the hosting capacity levels of EVs and RES. The EV's behavior is stochastic; therefore, more analysis is required to investigate their consequences on networks' steady-state and transient stability.

Author Contributions: H.Y.A., M.M.R. and S.H.E.A.A. designed the problem under study; Z.M.A. and M.M.R. performed the simulations and obtained the results; H.Y.A. and Z.M.A. analyzed the obtained results; M.M.R. and S.H.E.A.A. wrote the paper, which was further reviewed by H.Y.A. and Z.M.A. All authors have read and agreed to the published version of the manuscript.

Funding: The authors extend their appreciation to the Deputyship for Research and Innovation, Ministry of Education in Saudi Arabia, for funding this research work through project number (IF2/PSAU/2022/01/21870).

Institutional Review Board Statement: Not applicable.

Informed Consent Statement: Not applicable.

Data Availability Statement: The data presented in this study are available on request from the corresponding author.

Conflicts of Interest: The authors declare no conflict of interest.

Nomenclature

DGs	Distributed generators
ESS	Energy storage system
EV	Electric vehicle
GOA	Gazelle optimization algorithm
MGOA	Multi-objective Gazelle optimization algorithm
MPSO	Multi-objective particle swarm optimizer
MSSA	Multi-objective salp swarm algorithm
NaS	Sodium-sulfur battery
PV	Photovoltaic
SOC	State of charge
RESs	Renewable and sustainable energy sources
TCSC	Thyristor-controlled series compensator
WDN	Egyptian West Delta Network
Data and Indices	
$h \in H$	Index and set for scenarios
$i \in B$	Index and set for buses
C^T, C^G, C^{ESS}	Total cost of transmission projects, generation units, and energy storage systems
$L_{ij}^h - L_{ij}^{h-1}$	Total number of circuits exists between buses i and j at scenario h
$N_{ESS}^{i,h}, N_{ESS\_max}^i$	Number of batteries installed at scenario h and maximum number of batteries can be installed at bus i
$P_{EV}^{i,h}$	Estimated EV power at scenario h and bus i
$P_{Th}^{i,h}, P_{PV}^{i,h}$	Output active power in MW of the thermal unit and PV unit, respectively at scenario h and bus i
$P_{d,i}^h$	Active power consumed by the load at bus i (MW)
$P_{ch,i}^h, P_{dch,i}^h$	Charging and discharging active power of batteries at bus i (MW)
$P^{bat,r}$	Rated power of the selected ESS
P_{ij}^h	Active power flow in a route between bus i and j (MVA)
P_{ij}^{max}	Maximum rated of power flow in a route between bus i and j (MW)
$SOC_{ESS,i}^h$	SOC of ESS at bus i and scenario h
θ_i^h	Voltage angle at bus i (p.u)
y	Planning period
$\eta_{bat}^{ch}, \eta_{bat}^{dch}$	Charging and discharging efficiencies of ESS
$\alpha_L^{ij}, \alpha_{TCSC}^{i,j}$	Cost of circuits and TCSC device installed between bus i and bus j
$\alpha_{ren\_cap}^i, \alpha_{ren\_op}^i$	Capital and operating costs of RES installed at bus i
$\alpha_{g\_cap}^i, \alpha_{g\_op}^i$	Capital and operating costs of a fossil fuel-based generation unit installed at bus i
$\alpha_{ESS_{cap}}^h, \alpha_{ESS_{rep}}^h, \alpha_{ESS_{el}}^h, \alpha_{ESS\_op}^h$	Capital cost, replacement cost, end-life cost, and the operating cost of ESSs at bus i
β_{ij}	Susceptance of the route between bus i and j
λ	Interest rate

References

1. International Energy Outlook 2021—U.S. Energy Information Administration (EIA). Available online: https://www.eia.gov/outlooks/ieo/ (accessed on 31 January 2022).
2. Scarabaggio, P.; Carli, R.; Cavone, G.; Dotoli, M. Smart Control Strategies for Primary Frequency Regulation through Electric Vehicles: A Battery Degradation Perspective. *Energies* **2020**, *13*, 4586. [CrossRef]
3. Abdi-Siab, M.; Lesani, H. Distribution Expansion Planning in the Presence of Plug-in Electric Vehicle: A Bilevel Optimization Approach. *Int. J. Electr. Power Energy Syst.* **2020**, *121*, 106076. [CrossRef]
4. Wu, T.; Wei, X.; Zhang, X.; Wang, G.; Qiu, J.; Xia, S. Carbon-Oriented Expansion Planning of Integrated Electricity-Natural Gas Systems with EV Fast-Charging Stations. *IEEE Trans. Transp. Electrif.* **2022**, *8*, 2797–2809. [CrossRef]
5. Quijano, D.A.; Melgar-Dominguez, O.D.; Sabillon, C.; Venkatesh, B.; Padilha-Feltrin, A. Increasing Distributed Generation Hosting Capacity in Distribution Systems via Optimal Coordination of Electric Vehicle Aggregators. *IET Gener. Transm. Distrib.* **2021**, *15*, 359–370. [CrossRef]

6. Mehrjerdi, H.; Hemmati, R. Stochastic Model for Electric Vehicle Charging Station Integrated with Wind Energy. *Sustain. Energy Technol. Assess.* **2020**, *37*, 100577. [CrossRef]
7. Fachrizal, R.; Ramadhani, U.H.; Munkhammar, J.; Widén, J. Combined PV–EV Hosting Capacity Assessment for a Residential LV Distribution Grid with Smart EV Charging and PV Curtailment. *Sustain. Energy Grids Netw.* **2021**, *26*, 100445. [CrossRef]
8. Fachrizal, R.; Shepero, M.; Åberg, M.; Munkhammar, J. Optimal PV-EV Sizing at Solar Powered Workplace Charging Stations with Smart Charging Schemes Considering Self-Consumption and Self-Sufficiency Balance. *Appl. Energy* **2022**, *307*, 118139. [CrossRef]
9. Ebeed, M.; Aleem, S.H.E.A. Overview of Uncertainties in Modern Power Systems: Uncertainty Models and Methods. In *Uncertainties in Modern Power Systems*; Zobaa, A.F., Abdel, A., Shady, H.E., Eds.; Academic Press: Cambridge, MA, USA, 2021; pp. 1–34. ISBN 978-0-12-820491-7.
10. Mehrjerdi, H. Dynamic and Multi-Stage Capacity Expansion Planning in Microgrid Integrated with Electric Vehicle Charging Station. *J. Energy Storage* **2020**, *29*, 101351. [CrossRef]
11. Manríquez, F.; Sauma, E.; Aguado, J.; de la Torre, S.; Contreras, J. The Impact of Electric Vehicle Charging Schemes in Power System Expansion Planning. *Appl. Energy* **2020**, *262*, 114527. [CrossRef]
12. Baringo, L.; Boffino, L.; Oggioni, G. Robust Expansion Planning of a Distribution System with Electric Vehicles, Storage and Renewable Units. *Appl. Energy* **2020**, *265*, 114679. [CrossRef]
13. Kamruzzaman, M.D.; Benidris, M. A Reliability-Constrained Demand Response-Based Method to Increase the Hosting Capacity of Power Systems to Electric Vehicles. *Int. J. Electr. Power Energy Syst.* **2020**, *121*, 106046. [CrossRef]
14. Yao, F.; Wang, J.; Wen, F.; Tseng, C.L.; Zhao, X.; Wang, Q. An Integrated Planning Strategy for a Power Network and the Charging Infrastructure of Electric Vehicles for Power System Resilience Enhancement. *Energies* **2019**, *12*, 3918. [CrossRef]
15. Fan, V.H.; Dong, Z.; Meng, K. Integrated Distribution Expansion Planning Considering Stochastic Renewable Energy Resources and Electric Vehicles. *Appl. Energy* **2020**, *278*, 115720. [CrossRef]
16. Sadeghi, D.; Hesami Naghshbandy, A.; Bahramara, S. Optimal Sizing of Hybrid Renewable Energy Systems in Presence of Electric Vehicles Using Multi-Objective Particle Swarm Optimization. *Energy* **2020**, *209*, 118471. [CrossRef]
17. Da Silva, E.C.; Melgar-Dominguez, O.D.; Romero, R. Simultaneous Distributed Generation and Electric Vehicles Hosting Capacity Assessment in Electric Distribution Systems. *IEEE Access* **2021**, *9*, 110927–110939. [CrossRef]
18. Ali, A.; Mahmoud, K.; Lehtonen, M. Optimization of Photovoltaic and Wind Generation Systems for Autonomous Microgrids With PEV-Parking Lots. *IEEE Syst. J.* **2022**, *16*, 3260–3271. [CrossRef]
19. Ali, A.; Mahmoud, K.; Lehtonen, M. Maximizing Hosting Capacity of Uncertain Photovoltaics by Coordinated Management of OLTC, VAr Sources and Stochastic EVs. *Int. J. Electr. Power Energy Syst.* **2021**, *127*, 106627. [CrossRef]
20. Zhang, S.; Fang, Y.; Zhang, H.; Cheng, H.; Wang, X. Maximum Hosting Capacity of Photovoltaic Generation in SOP-Based Power Distribution Network Integrated With Electric Vehicles. *IEEE Trans. Ind. Inform.* **2022**, *18*, 8213–8224. [CrossRef]
21. Edmunds, C.; Galloway, S.; Dixon, J.; Bukhsh, W.; Elders, I. Hosting Capacity Assessment of Heat Pumps and Optimised Electric Vehicle Charging on Low Voltage Networks. *Appl. Energy* **2021**, *298*, 117093. [CrossRef]
22. Agushaka, J.O.; Ezugwu, A.E.; Abualigah, L. *Gazelle Optimization Algorithm: A Novel Nature-Inspired Metaheuristic Optimizer*; Springer: London, UK, 2022; Volume 6, ISBN 0123456789.
23. Yan, G.; Liu, D.; Li, J.; Mu, G. A Cost Accounting Method of the Li-Ion Battery Energy Storage System for Frequency Regulation Considering the Effect of Life Degradation. *Prot. Control Mod. Power Syst.* **2018**, *3*, 4. [CrossRef]
24. Rawa, M.; AlKubaisy, Z.M.; Alghamdi, S.; Refaat, M.M.; Ali, Z.M.; Aleem, S.H.E.A. A Techno-Economic Planning Model for Integrated Generation and Transmission Expansion in Modern Power Systems with Renewables and Energy Storage Using Hybrid Runge Kutta-Gradient-Based Optimization Algorithm. *Energy Rep.* **2022**, *8*, 6457–6479. [CrossRef]
25. Yao, M.; Molzahn, D.K.; Mathieu, J.L. An Optimal Power-Flow Approach to Improve Power System Voltage Stability Using Demand Response. *IEEE Trans. Control Netw. Syst.* **2019**, *6*, 1015–1025. [CrossRef]
26. Scarabaggio, P.; Carli, R.; Dotoli, M. Noncooperative Equilibrium Seeking in Distributed Energy Systems Under AC Power Flow Nonlinear Constraints. *IEEE Trans. Control Netw. Syst.* **2022**, *9*, 1731–1742. [CrossRef]
27. Almalaq, A.; Alqunun, K.; Refaat, M.M.; Farah, A.; Benabdallah, F.; Ali, Z.M.; Aleem, S.H.E.A. Towards Increasing Hosting Capacity of Modern Power Systems through Generation and Transmission Expansion Planning. *Sustainability* **2022**, *14*, 2998. [CrossRef]
28. Mirjalili, S.; Gandomi, A.H.; Mirjalili, S.Z.; Saremi, S.; Faris, H.; Mirjalili, S.M. Salp Swarm Algorithm: A Bio-Inspired Optimizer for Engineering Design Problems. *Adv. Eng. Softw.* **2017**, *114*, 163–191. [CrossRef]
29. Coello Coello, C.A.; Pulido, G.T.; Lechuga, M.S. Handling Multiple Objectives with Particle Swarm Optimization. *IEEE Trans. Evol. Comput.* **2004**, *8*, 256–279. [CrossRef]
30. Fathy, A.A.; Elbages, M.S.; El-Sehiemy, R.A.; Bendary, F.M. Static Transmission Expansion Planning for Realistic Networks in Egypt. *Electr. Power Syst. Res.* **2017**, *151*, 404–418. [CrossRef]
31. Egyptian Electricity Holding Company Annual Reports. Available online: http://www.moee.gov.eg/english_new/report.aspx (accessed on 31 January 2022).
32. Renewables.Ninja. Available online: https://www.renewables.ninja/ (accessed on 31 January 2022).
33. Mostafa, M.H.; Abdel Aleem, S.H.E.; Ali, S.G.; Ali, Z.M.; Abdelaziz, A.Y. Techno-Economic Assessment of Energy Storage Systems Using Annualized Life Cycle Cost of Storage (LCCOS) and Levelized Cost of Energy (LCOE) Metrics. *J. Energy Storage* **2020**, *29*, 101345. [CrossRef]

34. Saboori, H.; Hemmati, R. Considering Carbon Capture and Storage in Electricity Generation Expansion Planning. *IEEE Trans. Sustain. Energy* **2016**, *7*, 1371–1378. [CrossRef]
35. Zitzler, E.; Deb, K.; Thiele, L. Comparison of Multiobjective Evolutionary Algorithms: Empirical Results. *Evol. Comput.* **2000**, *8*, 173–195. [CrossRef]
36. Mirjalili, S. Dragonfly Algorithm: A New Meta-Heuristic Optimization Technique for Solving Single-Objective, Discrete, and Multi-Objective Problems. *Neural Comput. Appl.* **2016**, *27*, 1053–1073. [CrossRef]

sustainability

MDPI

Article

An Energy Storage Assessment: Using Frequency Modulation Approach to Capture Optimal Coordination

Wan Chen [1,*], Baolian Liu [1], Muhammad Shahzad Nazir [1,*], Ahmed N. Abdalla [2,3], Mohamed A. Mohamed [4], Zujun Ding [1], Muhammad Shoaib Bhutta [5] and Mehr Gul [6]

1 Faculty of Automation, Huaiyin Institute of Technology, Huai'an 223003, China; xinyu970521@163.com (B.L.); dingzujun@hyit.edu.cn (Z.D.)
2 Faculty of Electronic Information Engineering, Huaiyin Institute of Technology, Huai'an 223003, China; ahmed@hyit.edu.cn
3 Jiangsu Foreign Expert Workshop, Jiangsu Provincial Department of Sciences and Technology, Huai'an 223003, China
4 Electrical Engineering Department, Faculty of Engineering, Minia University, Minia 61519, Egypt; dr.mohamed.abdelaziz@mu.edu.eg
5 School of Automotive and Transportation Engineering, Guilin University of Aerospace Technology, Guilin 541004, China; shoaibbhutta@hotmail.com
6 Department of Electrical Engineering, Balochistan University of Information Technology, Engineering and Management Sciences (BUITEMS), Quetta 1800, Pakistan; mehr.gul@buitms.edu.pk
* Correspondence: calvinchen@hyit.edu.cn (W.C.); msn_bhutta88@yahoo.com (M.S.N.)

Abstract: To reduce the allocation of energy storage capacity in wind farms and improve economic benefits, this study is focused on the virtual synchronous generator (synchronverter) technology. A system accompanied by wind power, energy storage, a synchronous generator and load is presented in detail. A brief description of the virtual synchronous generator control strategy is given. The capacity allocation is based on different optimization goals and the optimal energy storage capacity configuration of the coordinated frequency modulation (FM) control strategy. The detail of the dual-loop control strategy is carried out by establishing the grid-connected transfer function model of the synchronverter energy storage and a theoretical model of life cycle cost is established. The optimal control strategy of coordinated FM for wind storage is implemented using MATLAB software. The simulation showed that the proposed strategy provided the energy storage capacity at high wind speed, which is configured to be 5.9% of the installed capacity of the wind turbine, marking a reduction of 26% compared with the 8% capacity required for independent support. In addition, the proposed method has improved the energy storage capacity configuration of the coordinated FM control strategy.

Keywords: energy storage system (ESS); synchronverter; wind energy; frequency modulation (FM); capacity configuration

Citation: Chen, W.; Liu, B.; Nazir, M.S.; Abdalla, A.N.; Mohamed, M.A.; Ding, Z.; Bhutta, M.S.; Gul, M. An Energy Storage Assessment: Using Frequency Modulation Approach to Capture Optimal Coordination. *Sustainability* **2022**, *14*, 8510. https://doi.org/10.3390/su14148510

Academic Editor: Andrea Nicolini

Received: 4 June 2022
Accepted: 7 July 2022
Published: 12 July 2022

1. Introduction

To protect the ecological environment and ensure sustainable economic development, the use of wind, solar and other new energy sources to generate electricity has become the focus of authorities [1,2]. Furthermore, with the development of technology and the economy, people's demand for electricity is rapidly increasing, which leads to the scarcity of traditional energy. New energy sources do not need to consume increasingly exhausted fossil energy, nor will they cause environmental pollution, ensuring sustainable production [3]. While the synchronization characteristic of a synchronous motor can ensure its automatic synchronization with a large power grid, the inverter can only achieve synchronization with the power grid through phase-locked control [4]. In a large power network, the output impedance of the synchronous motor is high due to its winding, which has a strong ability to suppress current disturbance [5]. The inverter has weak resistance to current and is prone to overcurrent. The synchronous motor has a strong ability to cope with external disturbances due to its structural characteristics, while the inverter topology is mainly composed of power electronic devices, which have limited bearing capacity and poor resistance in case of system failure [6]. Therefore, the characteristics of the synchronous motor can ensure the stable operation of the power system. The characteristics of large synchronous motor impedance and large inertia can thus guarantee the stable operation of the power system. A new control method of inverter synchronverter control is proposed to improve the stability of power grid operation [7,8]. The excellent technical characteristics, such as fast output and stable operation, of the ESS are utilized to compensate for the uncontrollable shortcomings of wind turbines and to support the frequency of the system in a stable state in time [9]. The intermittent ability to generate electricity from either a single wind turbine or wind farms, as well as the inherent challenges of supplying and sustaining connectivity to the electrical grid, point to the necessity for energy storage [10]. Energy storage offers a way to collect and balance wind energy as it is produced and store it for use at a later time when demand might outstrip supply. The most practicable or advantageous energy storage technologies for wind generation are said to fall into four categories: battery storage, flywheel storage, compressed air storage, and pumped hydro storage. However, considering the high cost of ESS, it is necessary to study the optimal control strategy of coordinated FM for wind energy storage [10].

In this study, an ESS model is established to conduct parameter tuning and deeply analyze the influence of parameter changes on the stability of each link of the ESS, thereby laying a foundation for the subsequent ESS to be connected to the wind energy system and providing a stable output. Under the condition that the original wind storage coordination FM strategy cannot meet the demand, a wind storage coordination frequency optimization strategy is proposed. Therefore, the optimal allocation of energy storage capacity is studied. Furthermore, due to the high cost of energy storage, we also considered the characteristics and economics of FM technology and proposes an optimal allocation method for energy storage capacity. According to the requirements of FM, the design method of the energy storage rated power and the rated capacity is determined.

2. Background Study

The basic idea of a synchronverter has been provided [11,12], which makes the grid-connected inverter resemble the operational characteristics of SG in various ways. Researchers [13] proposed a synchronverter scheme that reflects the proper operating characteristics of the synchronous machine, whereas the technique reflects the swing equation of conventional SG. Furthermore, operational metrics such as Q and P_f can be compared between the standard SG and synchronverter [14]. However, a current-controlled synchronverter is the same as a current source, providing voltage and frequency support for the system. In [15], a voltage-controlled synchronverter technique is proposed to alleviate the shortcomings of the current-controlled synchronverter. The goal of voltage-controlled synchronverter techniques is to simulate the rotor inertia and system frequency modulation characteristics of SG in frequency control to improve the system's frequency stability [16]. The reactive voltage relationship is primarily considered in voltage control to control the stable voltage output [17]. The synchronverter performs power management and frequency modulation functions thanks to the power controller and voltage frequency controller [18]. The synchronverter is a system that simulates the inertia of a traditional power system by combining control algorithms, renewable energy sources, energy storage devices, and power electronics [19]. The synchronverter is a system that connects various storage units, generation units, and the utility grid. In today's grid systems, variable wind turbines are employed, and these turbines are connected with back-to-back inverters, allowing for total decoupling of inertia from the utility grid [18]. The AC to DC converters and an additional inverter at the front end connect the energy storage and wind systems. This system is unresponsive to changes in inertia [12]. According to the literature, the main model concepts for many topologies are identical; however, the implementation of each topology model differs. Only a few topologies use mathematical equations to fully simulate synchronous generator behavior, and only a few topologies use swing equations to copy the synchronous generator's inconsistent performance. The simple structure of the synchronverter-based wind energy storage system is presented in Figure 1.

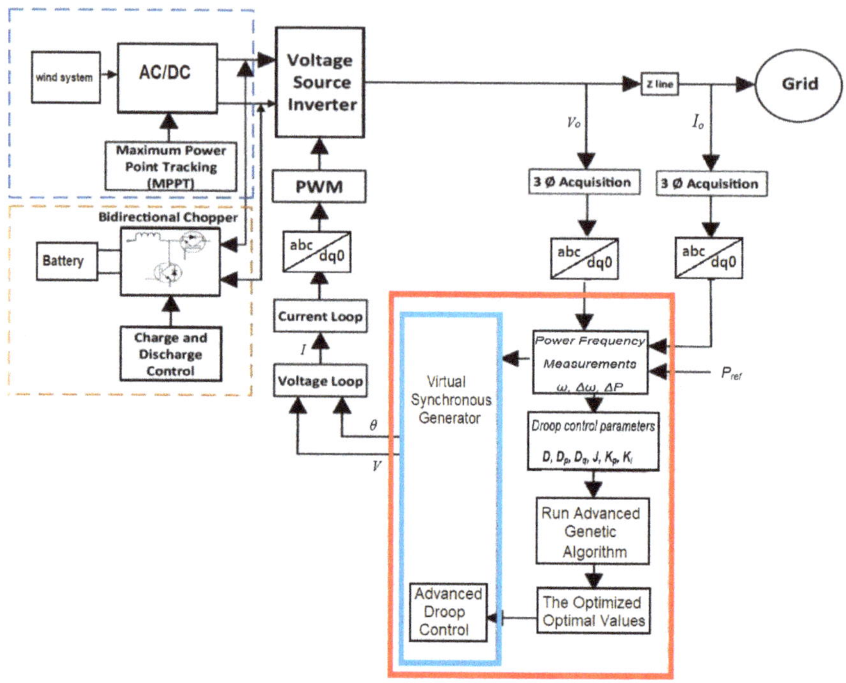

Figure 1. The simple structure of the synchronverter-based wind energy storage system.

2.1. Virtual Synchronous Generator

Synchronverter technology allows embedding the mechanical and electrical transient equation of a traditional synchronous generator into the control strategy of the external power electronic converter of wind power or photovoltaic unit, to simulate the external electromagnetic and mechanical motion characteristics of the synchronous motor [17]. To avoid too complex modelling and give consideration to practicability, the second-order synchronous motor model as the simulation modelling target is used. The motion equation of the second-order rotor of a synchronous motor controlled by VSG can be expressed as

$$\begin{aligned} T_m - T_e - D(\omega - \omega_{ref}) &= J\tfrac{d\omega}{dt} \\ \tfrac{d\theta_1}{dt} &= \omega \end{aligned} \quad (1)$$

In Equation (1), T_m and T_e represent the mechanical and electromagnetic torque of the prime mover, respectively; θ_1 represents the angle of work; ω and ω_{ref} represent the rotor angular velocity and rated angular velocity, respectively; the J parameter is the rotational inertia coefficient of the rotor; and D is the damping coefficient of the rotor.

The stator electrical equation can be expressed as

$$u_{abc} = e_{abc} - C_s i_{abc} - L_s \frac{di_{abc}}{dt} \quad (2)$$

In Equation (2), u_{abc} is the stator side-induced electromotive force; e_{abc} is the three-phase output voltage of the stator side; and L_s and C_s are armature inductors and capacitors, respectively.

Using the above two models, the modelling process is relatively simple, but also can ensure the inertia characteristics of the synchronous motor simulated by the inverter.

Low- and high-frequency oscillation phenomena exist in the synchronverter grid-connected system due to external interference and factors of the synchronverter itself [19]. The introduction of virtual synchronization makes the system have low-frequency oscillation characteristics similar to that of a synchronous motor [20]. Furthermore, the combined action of some functions of the synchronverter and external interference will stimulate the original mode change of the inverter and cause the high-frequency oscillation of the grid-connected system [14]. At present, most scholars seldom present the grid-connection stability and mostly use the small-signal model. Researchers [15] have discussed the small-signal model adopted to study the stability of a single machine incorporated into the power grid. Some [16] studied the situation of two machines connected to a microgrid under small-signal interference and also studied the influence of the change in the control parameters and circuit parameters on the stability of the system. Others [18] have derived the synchronous frequency resonance phenomenon excited by the synchronverter access power grid in detail and proposed corresponding measures to suppress the oscillation through analyzing its influencing factors. In [19], the authors adopted the classical method of stability analysis in a power system, taking the power angle and frequency of synchronverter as the state variables, establishing the state–space equation, and used the characteristic root locus method to study the influence of parameter changes on the stability of the system. The authors in [20] studied the stability of a synchronverter connected to a large power grid, establishing the transfer function between output power and input power, and analyzed the dynamic characteristics of the system.

2.2. Frequency Modulation Coordination

The rotor inertia control has a short maintenance time, and the response speed is slow when the rotor pitch control is adopted, so the power of the frequency recovery cannot be provided in time [21]. Therefore, it is difficult to keep the power system frequency in a stable state solely by relying on the wind turbine's FM. Researchers [22] constructed an FM model covering energy storage and wind farms. It is clear that the addition of energy storage is beneficial to ease the frequency fluctuation of the power grid, but the waste of energy storage is caused to some extent due to the failure to consider the FM method used on wind farms. Therefore, considering only the FM means of energy storage and ignoring the FM means used in wind farms is not conducive to its application in engineering. The authors in [23] used ESS and wind control systems to assist power grid FM successively. Without considering the coordination function of the two, the economic efficiency of frequency regulation is reduced by making full use of their complementary characteristics. Therefore, isolated consideration of wind and energy storage participating in the FM system causes economic loss. This study adopts the coordinated FM control strategy of wind storage and builds the simulation model of the wind storage system and the conventional power system. It analyses the frequency change of the system and improves the technical economy and engineering practicality of the frequency regulation. Meanwhile, coordinated FM of wind storage restrains the secondary frequency drop phenomenon in the process of wind power speed recovery, which is of great significance to transform wind power from FM with the ability to provide an auxiliary service to the participating system and improve the safety and stability of the power grid.

2.3. Optimal Energy Storage Capacity

Energy storage is connected to the power system, which can optimize the FM effect of the wind farm. However, considering the high cost of energy storage, how to optimize the configuration of energy storage capacity and improving the efficiency of wind farm FM has become an urgent problem to be solved [24]. The energy storage capacity allocation methods used to calm the stroke power of wind farms include the economic index optimization method, considering the economy; the frequency-domain analysis method, using spectrum analysis; and the probability statistical method, based on distribution value allocation [25]. The frequency-domain analysis method is described in [26]. Fourier transform is adopted to analyze the output data, to determine the target value to be stabilized and the frequency band where the energy storage is located, and to determine the maximum and minimum value of the accumulated energy storage in the selected period. When the SOC of the ESS is within the allowable range and can maintain the normal operation of the system, the capacity that the ESS needs to be configured to is determined through simulation analysis [27]. When adopting this method, it is necessary to collect the historical data of wind farms, which requires a high degree of sample selection and has a strong dependence on the data. Meanwhile, the energy storage capacity configured by this method is relatively large. Authors [28] have discussed that a probabilistic statistical method be adopted, and that capacity allocation also be carried out based on historical data. Firstly, the component values of the two types of hybrid batteries are separated by the wavelet decomposition method. Then, according to the component value statistics, one determines its distribution law [29]. Different confidence levels are used to determine the capacity of the energy storage configuration. Researchers [28] studied the optimization method of the economic indicators adopted. The optimal economic effect is taken as the objective function, and the mathematical algorithm of chance-constrained programming is adopted to seek the optimal value of the energy storage capacity and configure the capacity needed for economic optimization [29]. A comprehensive optimal allocation method for energy storage capacity is proposed, which is constrained by the operating control energy of the ESS and targeted at the optimization of the FM effect and economic synthesis. By allocating a proper energy storage, the wind farm can adjust the frequency variation of the power grid in time just like traditional power supply, thereby improving the engineering applicability and economy of the combined FM of wind storage.

3. Proposed Methods

3.1. Virtual Synchronous Generator Model

A voltage and current-controlled energy storage synchronverter adopts a double-loop control strategy. The outer loop is an active and reactive power link, through which amplitude and phase are generated. The inner loop is the current loop, which generates the voltage reference value of the pulse modulation signal by inductance current control. In the current-control-type energy storage virtual synchronous machine, the outer loop uses power control to generate the reference current, and the inner loop uses current control to generate the voltage reference signal. According to the control strategy of the virtual synchronous machine for energy storage, the transmission model of the ESS is established. The parameter setting and stability analysis are carried out for the energy storage system, to provide a stable output value for subsequent participation in wind power FM. When the voltage-controlled energy storage of the synchronverter is connected to the grid, the system has two coordinate systems: the synchronverter itself and the rotation coordinate system of the grid. The proposed dual-loop control strategy is depicted in Figure 2, while Figure 3 presents the simulation model of the proposed system, where batteries are used to maximize the full life cycle value of the energy storage.

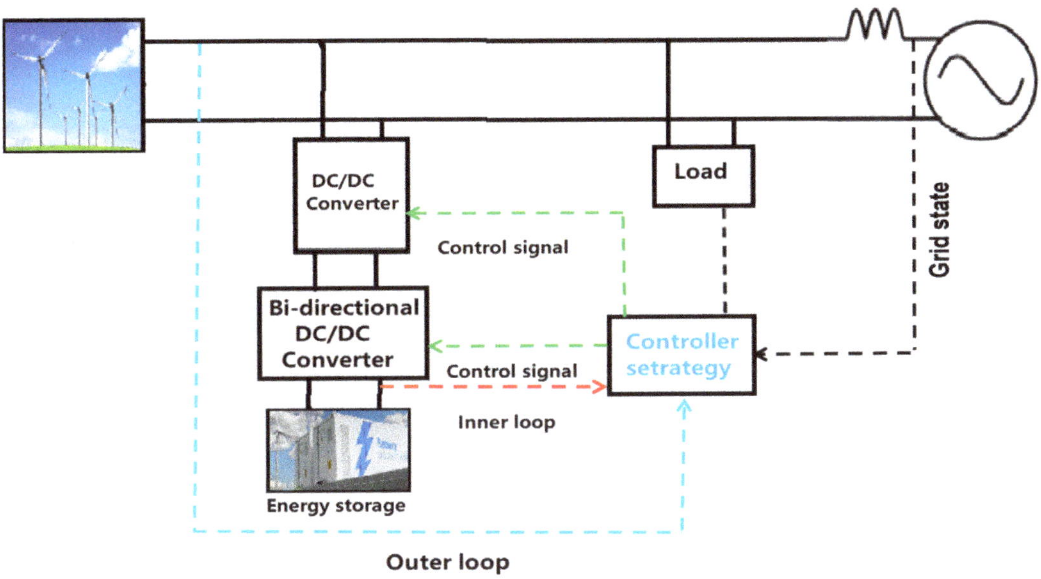

Figure 2. Proposed dual-loop control strategy.

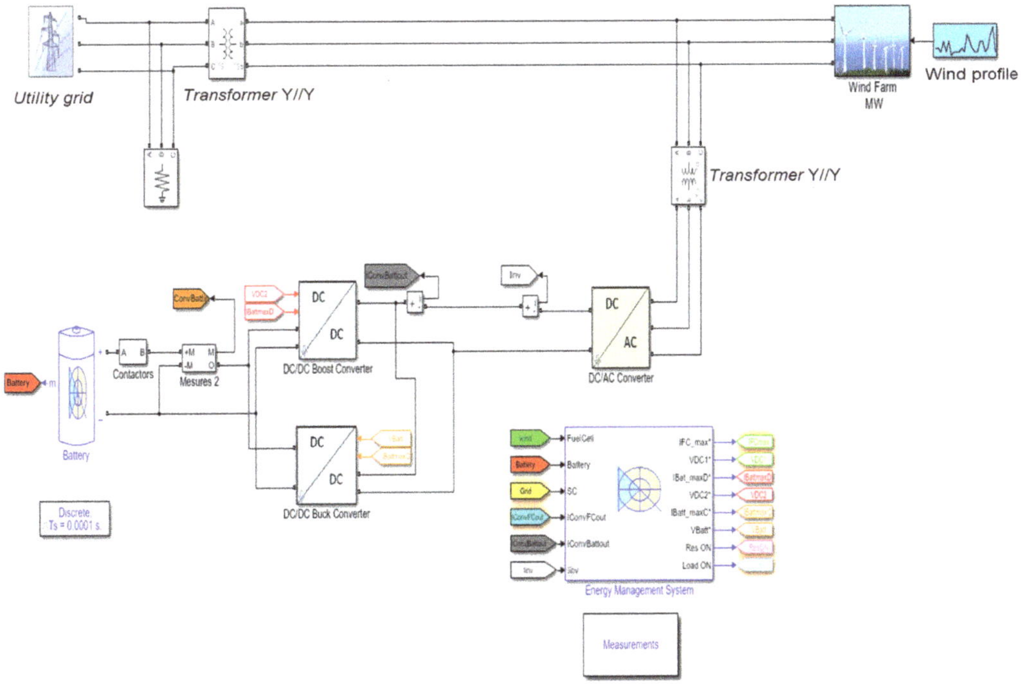

Figure 3. Simulation model of the proposed system.

3.2. Optimized Control Strategy

The optimized control strategy of coordinated FM of wind energy storage is presented in Figure 4. For the output FM power (P_f) of the whole wind resource, the field is calculated and then allocated to the energy storage and wind. According to the principle of energy storage priority in distribution, if the energy storage capacity is greater than the FM power output required by the whole wind field, the FM task will be undertaken by the energy storage. If the energy storage capacity is less than the FM power output required by the whole wind field, the energy storage will be full, and the remaining FM task is undertaken by the wind. The calculation of the P_f regards the entire wind farm output needs, and then allocate it to energy storage and wind turbines. The allocation is based on the principle of energy storage priority, as in the energy storage capacity has to be greater than the FM power that the entire wind farm needs to output. The FM task is undertaken by energy storage; if the energy storage capacity is less than the FM power, the entire wind farm needs the output and the full energy storage, while the remaining frequency adjustment tasks are undertaken by the wind. The energy storage capacity optimization and configuration are depicted in Figure 5.

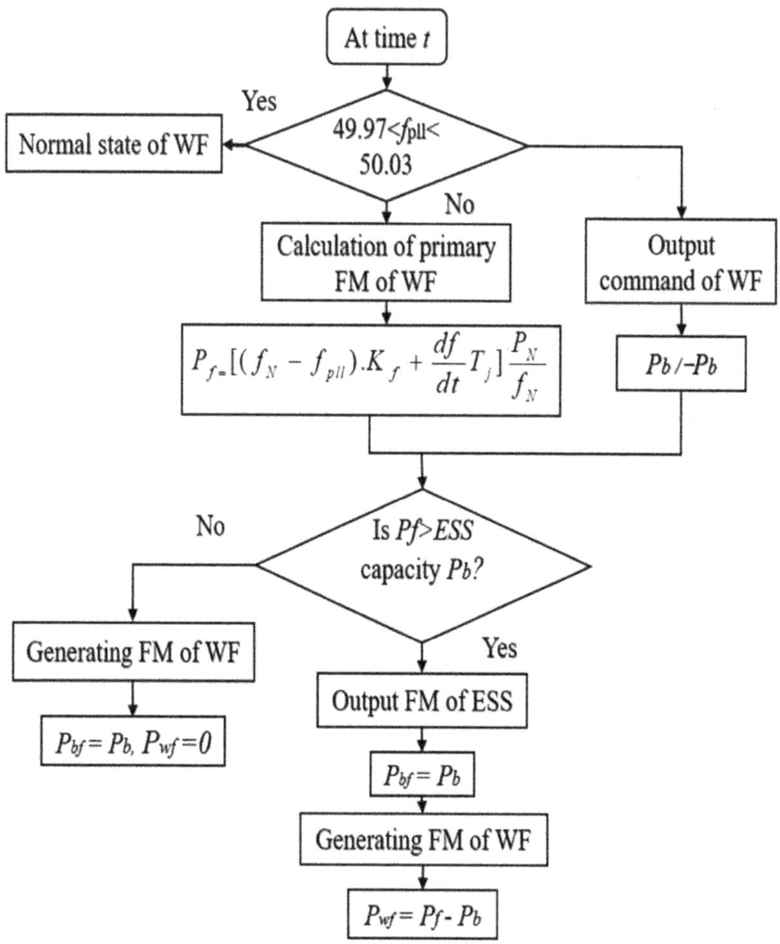

Figure 4. Optimization control strategy.

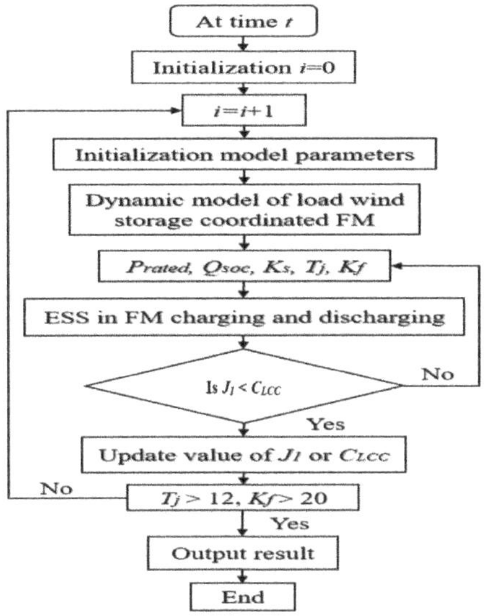

Figure 5. Energy storage capacity optimization and configuration.

In these figures, P_{bf} is the FM power of the energy storage system; P_f is the FM power of the wind field; P_{wf} is the FM power of the wind turbine; P_{bref} is the given power value; T is the FM duration of the wind; P_b is the energy storage capacity; N_f is the rated frequency of the system; $\frac{df_{pll}}{dt}$ is the collected system frequency; df_{pll}/dt is the rate of change in system frequency; and T_j and K_f are the inertial time constant and active power FM coefficient of the synchronverter, respectively.

3.3. Optimal Economic Capacity

The life cycle cost of the ESS can reflect the average cost of the energy storage power station during the life cycle. The investment cost mainly includes the investment cost in the early stage and the replacement cost of the later device. The operation and maintenance costs mainly include fixed operation and maintenance costs determined by PCS and variable operation and maintenance costs determined by ESS charging and discharging.

The ESS device cost is mainly composed of energy storage devices, power conversion systems, and some auxiliary equipment:

$$C_{sys} = C_{bat} + C_{pcs} + C_{bop} \tag{3}$$

where C_{bat} is the energy storage device cost; C_{pcs} is the power conversion system cost; and C_{bop} is the auxiliary equipment cost.

The cost of energy storage device C_{bat} expressed as

$$C_{bat} = \frac{C_E E_{rated}}{\eta} \tag{4}$$

where E_{rated} is the rated power of the ESS (kW.h); η is the conversion efficiency of the ESS (%); C_E is the unit power price (\$/(kW.h)); P_{rated} is the rated power of the ESS (KW); and t is the discharge time (h) of the energy storage system.

The power conversion system cost is expressed as

$$C_{pcs} = C_P P_{rated} \tag{5}$$

where C_p is the unit power price of P_{CS} ($/kW).

$$C_{bop} = C_B E_{rated} \tag{6}$$

where C_B is the unit electricity price of auxiliary equipment ($/(kW.h)).

$$C_{LCC} = \overline{C}_{rep} + \overline{C}_{sys} + C_{POM} + C_{VOM} \tag{7}$$

$$\overline{C}_{sys} = (\frac{C_E E_{rated}}{\eta} + C_P P_{rated} + C_B E_{rated})\frac{i(1+i)^N}{(1+i)^N - 1} \tag{8}$$

where i is the discount rate (%); and N is the project period (years).

When the project cycle is greater than the life cycle of the energy storage system, the ESS needs to replace equipment. P_{CS} and auxiliary equipment generally have a service life of ten years.

$$\overline{C}_{rep} = \frac{C_E E_{rated}}{\eta} \sum_{\beta=1}^{k} \frac{(1-\alpha)^{\beta n}}{(1+i)^{\beta n}} \frac{i(1+i)^N}{(1+i)^N - 1} \tag{9}$$

where C_{rep} is the average annual reduction ratio of the cost of energy storage devices; k is the number of battery replacements, $k = N/n - 1$, and n is the battery life (years); β is the βth replacement of the battery in the energy storage system.

$$C_{VOM} = C_e = \frac{C_e t D}{\eta} P_{rated} \tag{10}$$

$$C_{POM} = C_f P_{rated} \tag{11}$$

where C_f is the operation and maintenance cost per unit of power (USD/(kW·year)). Variable operation and maintenance costs mainly consider electricity cost, C_e (USD/year), whereas C_{e_P} is the average annual electricity cost of the ESS per unit of power (USD/(kW·year)).

4. Results and Discussion

4.1. Results

Figure 6 presents the wind storage coordination and FM control strategy based on the frequency outer loop of the energy storage compensation. Figure 6 depicts wind power does not participate in FM, but solely energy storage is used for frequency recovery. Where the grid has 20% wind power installed capacity, load disturbance is added, and the load takes up 5% of the system capacity. In Figure 6, the energy storage supports the FM control strategy of the wind farm, and the frequency of the system after stabilization is 49.878 Hz. The energy storage capacity that needs to be configured to restore the frequency to the stable value accounts for 8% of the wind field capacity. While this method is adopted, the lowest frequency of the system increases by 41% and the steady-state frequency increases by 20% compared with the wind farm without FM capacity.

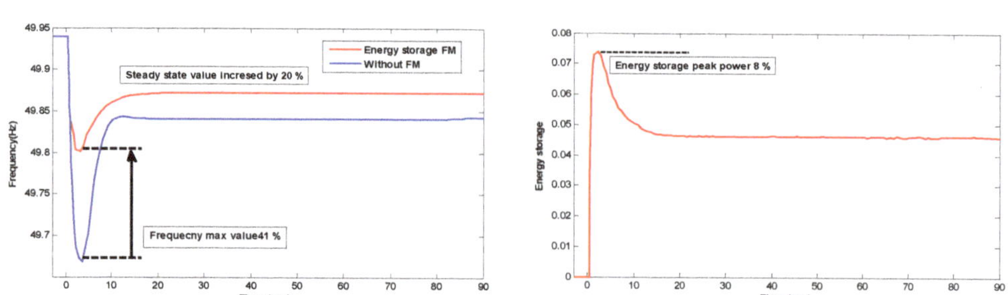

Figure 6. FM control strategy of the wind farm supported by energy storage.

An independent FM mode of energy storage is adopted and the original control strategy of wind energy storage to coordinate the FM by adding an appropriate amount of energy storage is presented in Figure 6. At the grid with 20% wind power installed, load disturbance of 5% system capacity is applied. Figure 7 depicts a wind speed of 8.6 m/s, the frequency characteristics of the system when solely the energy storage FM was adopted, and the original strategy of wind energy storage coordination FM. The FM effect is similar to that when the energy storage participates in the FM solely, and compared with wind energy storage, the frequency characteristics are greatly improved.

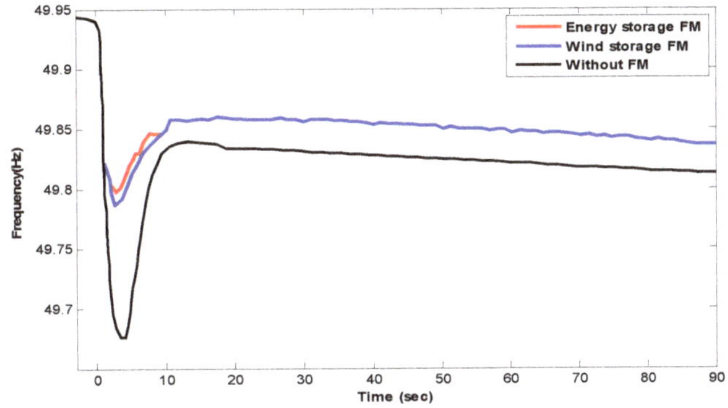

Figure 7. The FM effect of the original strategy of wind energy storage coordination control.

Figure 8 compares the output of the two control strategies of a single FM for energy storage and coordinated FM for wind storage. When the wind speed is 8.6 m/s, compared with the FM mode of wind energy storage, wind farms are supported when the original control strategy of wind storage coordinating FM is adopted. The energy storage output is found relatively smooth and the peak value of output is calculated smaller. Similarly, the total FM output of the wind and the energy storage participating system is the same under the two strategies. Under the condition that the FM effect is consistent, the energy storage alone support strategy is adopted and the energy storage capacity to be configured is 8% of the wind field capacity; thus, the energy storage capacity to be configured when the original control strategy of wind storage coordination and FM is adopted under 5.7% of the wind field capacity, which is 28% less than the energy storage alone support strategy.

When the wind speed of the wind is 11.2 m/s, the frequency characteristics of the system are consistent with those in Figure 9 when solely the energy storage FM is adopted, and with the original control strategy of the wind storage coordinated FM. Similarly, in the grid with 20% wind power installed capacity, load disturbance is added, and the load takes up 5% of the system capacity. In Figure 9, when the wind speed is 11.2 m/s, the energy storage capacity that needs to be configured is 13.9%, exceeding the capacity configuration (8%) when the energy storage supports wind farm FM. This is because, when the wind speeds are higher, the power drop amplitude after the wind exits the FM is observed to be more serious. Under a constant speed range, when the wind speed is high, the power drop range is observed to be deeper after the wind exits the FM, and the deeper the wind's power has dropped, the more high-power energy storage needed to compensate. Therefore, under high wind speed, the short-time supporting power of the wind is larger. The original control strategy that energy storage solely compensates for power drop still needs to be optimized.

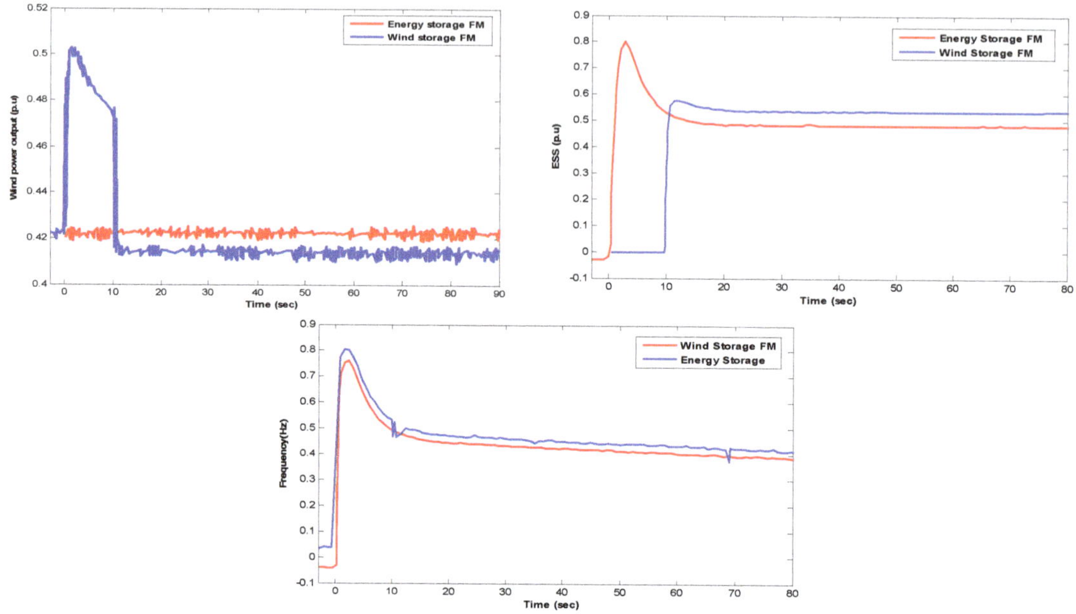

Figure 8. Comparison of the output of the original control strategy of the FM coordination energy storage and wind energy storage.

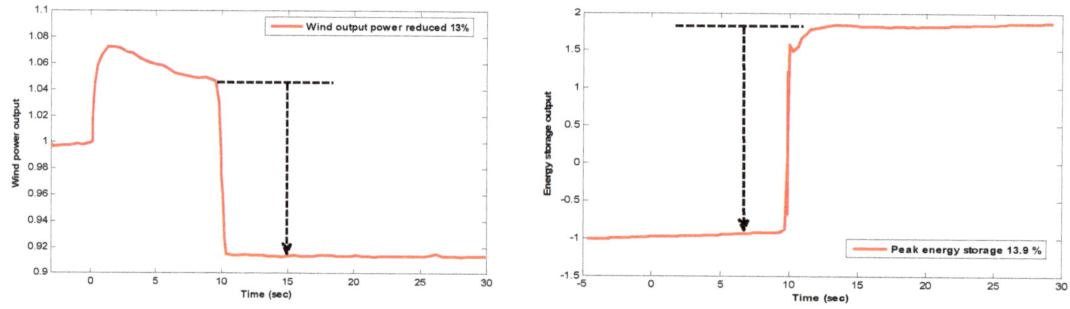

Figure 9. FM effect of the original control strategy for wind storage coordination.

Figure 10 depicts the coordinated control strategy of wind power inertia release and steady-state support of energy storage requiring 5.9% energy storage. The wind speed is 11.2 m/s, at which it requires 8% of the energy storage phase for independent support compared with a reduction of 26%.

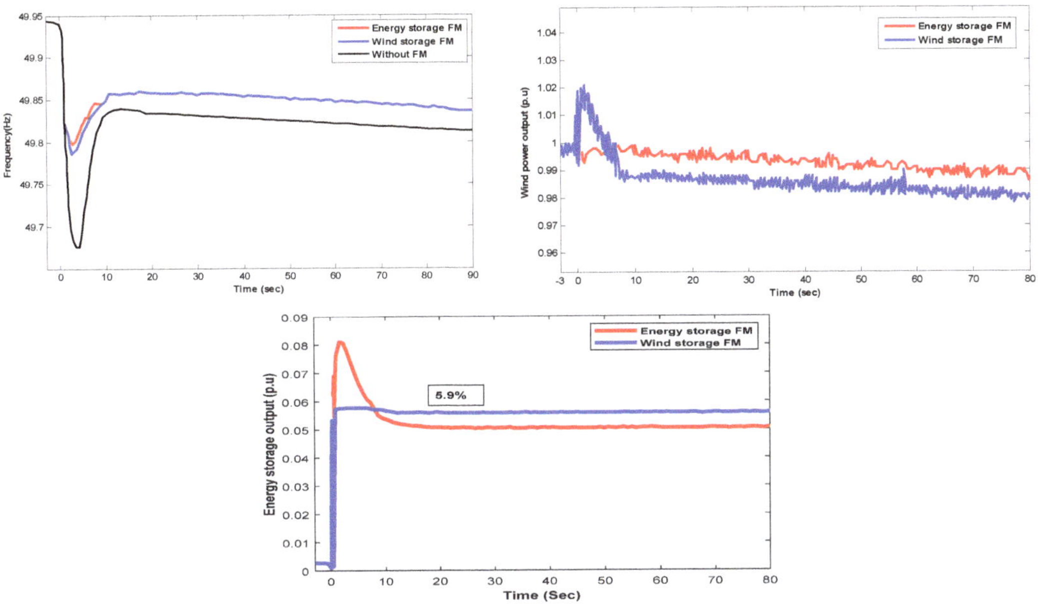

Figure 10. Energy storage compensation steady-state strategy under a 11.2 m/s wind speed.

4.2. Discussion

The specific parameters of a lithium battery are given in Table 1. This strategy is adopted, and the energy storage capacity at high wind speeds is configured to be 5.9% of the installed capacity of the wind turbine. It is a reduction of 26% compared with the 8% capacity required for independent support of energy storage and improves the economics of energy storage participating in primary frequency regulation. Considering the battery's development and commercial conditions, this study selects lithium batteries for capacity configuration. The lithium batteries have a strong ability to withstand high power and extreme temperatures and need not consume water resources in the production process. On the other hand, the service life, cost, energy storage, and mass production conditions of lithium batteries are relative to other batteries. There are fewer restrictions, making them widely used in commerce. Considering these two aspects comprehensively, the lithium battery is used for analysis.

Table 1. Energy storage cost analysis.

Type of Battery	Unit Capacity Price C_E (USD/kW·h)	Unit Power Price C_p (USD/kW)	O&M Cost C_f (USD)/(kW·h)	Charging Electricity Price C_c (USD/kW.h)	Conversion Efficiency (%)	Life Time Period (Year)
Lithium battery	21,600	7270	1040	3.5	0.85	10

Table 2 represents the wind storage coordination strategy's energy storage capacity requirements under different wind speeds. Energy storage gets similar, supporting 8% under 6.2 m/s and 8.6% at 11.2 m/s wind speeds, while optimized and original strategies get different values under various wind speeds. The synchronverter of wind power has an energy reserve to participate in the primary FM of the grid; but, when the FM exits and the speed is restored, it causes the second drop in the grid frequency and deteriorates the system frequency dynamics. To improve the secondary frequency drop problem based on the comprehensive speed recovery strategy, the parameters of the FM support strategy are optimized, but the secondary frequency drop was not eliminated. To eliminate the secondary frequency drop, an ESS is added to the power generation side to cooperate with wind power generating units and traditional generating units to participate in grid frequency adjustment to maintain the grid frequency in a stable state. When a similar FM effect is achieved, the coordinated and optimized control strategy of wind power inertia release and energy storage steady-state support is the optimal strategy for wind storage coordinated FM.

Table 2. Energy storage capacity requirements under different wind speeds.

Wind Speed	6.2 m/s	8.6 m/s	11.2 m/s
With energy storage supports	8%	8%	8%
Original strategy: secondary fall compensation	6.7%	5.6%	13.9%
Optimization strategy: energy storage and compensation	5.5%	5.2%	5.9%

The design method of the same rated power and capacity the operating performance of the ESS and meeting the FM are the constraints. The ESS has the lowest average annual cost per unit of power to obtain greater benefits for capacity allocation. The obtained control parameters, FM effect evaluation index, economic evaluation index, rated power P_{rated} and rated capacity E_{rated} are shown in Table 3. According to the maximum frequency deviation of the grid and considering the operating performance of the energy storage system. The rated power P_{rated} of the ESS is determined by optimizing the control parameters of the energy storage system and the rated capacity E_{rated} of the ESS is determined according to the state of charge (SOC) of the ESS. Meanwhile, the average annual cost per unit power of the ESS C_{LCC} is calculated according to the life cycle cost model. The capacity configuration is based on the minimum primary FM effect J as the optimization objective.

Table 3. Capacity allocation is based on the robust economic model, comprehensive optimal capacity allocation and control variable.

Model	Parameter	Value
Robust economic model	Control variable (T_j)	4
	Control variable (K_f)	5
	Economic evaluation index (J)	0.131
	Economic evaluation index (C_{LCC}) (10^3 USD)	66.30
	Power (%)	1.4
Optimal capacity allocation	Control variable (T_j)	8
	Control variable (K_f)	13
	Economic evaluation index (J)	0.098
	Economic evaluation index (C_{LCC}) (10^3)	150
	Power (%)	3.7
Control variable	Control variable (T_j)	12
	Control variable (K_f)	20
	Economic evaluation index (J)	0.096
	Economic evaluation index (C_{LCC}) (10^3)	23,200
	Power (%)	5.9

5. Conclusions

This study discussed the characteristics of FM and configured the wind energy storage capacity by optimizing the controlling parameters. The main conclusions are as follows:

- One can improve the economics of energy storage by determining the design method of its rated power and the capacity according to the FM requirements.
- The energy storage capacity under high wind speeds is configured to be 5.9% of the installed capacity, which is a reduction of 26% compared with the 8% capacity required for independent support of energy storage.
- The comprehensive optimal energy storage capacity configuration of the coordinated FM control strategy is improved.

In a follow-up study, the coordinated control of the three generators and the coordinated FM of the wind storage and synchronous generators still need to be studied. In addition, comprehensive consideration is needed to ascertain the various factors that affect system resonance.

Author Contributions: Conceptualization, writing—original draft preparation, formal analysis, W.C., M.S.N. and A.N.A.; methodology, investigation, M.S.N., A.N.A., M.S.B., M.A.M. and W.C.; review and editing, B.L., Z.D., M.A.M. and M.G. All authors have read and agreed to the published version of the manuscript.

Funding: This research did not receive any direct funding.

Institutional Review Board Statement: Not applicable.

Informed Consent Statement: Not applicable.

Data Availability Statement: Not applicable.

Acknowledgments: The authors are highly grateful to their affiliated universities and institutes for providing research facilities.

Conflicts of Interest: The authors declare no conflict of interest.

Abbreviations

FM	Frequency modulation
ESS	Energy storage system
SOC	State of charge
P_f	FM power
N_f	Rated frequency of the system
P_{wf}	Power of the wind turbine
K_f	Inertial time constant
P_b	Energy storage capacity
C_{bat}	Energy storage device cost
C_{pcs}	Power conversion system cost
C_{bop}	Auxiliary equipment cost
E_{rated}	Rated power of the ESS
C_B	Unit electricity price
C_f	Operation and maintenance cost
J	Optimization objective

References

1. Ramírez, M.; Castellanos, R.; Calderón, G.; Malik, O. Placement and sizing of battery energy storage for primary frequency control in an isolated section of the Mexican power system. *Electr. Power Syst. Res.* **2018**, *160*, 142–150.
2. Bhutta, M.S.; Sarfraz, M.; Ivascu, L.; Li, H.; Rasool, G.; ul Abidin Jaffri, Z.; Farooq, U.; Ali Shaikh, J.; Nazir, M.S. Voltage Stability Index Using New Single-Port Equivalent Based on Component Peculiarity and Sensitivity Persistence. *Processes* **2021**, *9*, 1849.
3. Rahman, F.S.; Kerdphol, T.; Watanabe, M.; Mitani, Y. Optimization of virtual inertia considering system frequency protection scheme. *Electr. Power Syst. Res.* **2019**, *170*, 294–302.
4. El-Bidairi, K.S.; Nguyen, H.D.; Mahmoud, T.S.; Jayasinghe, S.; Guerrero, J.M. Optimal sizing of Battery Energy Storage Systems for dynamic frequency control in an islanded microgrid: A case study of Flinders Island, Australia. *Energy* **2020**, *195*, 117059.
5. Belila, A.; Amirat, Y.; Benbouzid, M.; Berkouk, E.M.; Yao, G. Virtual synchronous generators for voltage synchronisation of a hybrid PV-diesel power system. *Int. J. Electr. Power Energy Syst.* **2020**, *117*, 1056776.

6. Zhu, T.; Nazir, M.S.; Ali Mokhtarzadeh, A.; Abdalla, A.N.; Nazir, H.M.; Chen, W. Improve performance of induction motor drive using weighting factor approach-based gravitational search algorithm. *Int. J. Electron.* **2022**, *109*, 900–913.
7. Nazir, M.S.; Abdalla, A.N.; MMetwally, A.S.; Imran, M.; Bocchetta, P.; Javed, M.S. Cryogenic-Energy-Storage-Based Optimized Green Growth of an Integrated and Sustainable Energy System. *Sustainability* **2022**, *14*, 5301.
8. Abdalla, A.N.; Nazir, M.S.; Tao, H.; Cao, S.; Ji, R.; Jiang, M.; Yao, L. Integration of energy storage system and renewable energy sources based on artificial intelligence: An overview. *J. Energy Storage* **2021**, *40*, 102811.
9. Das, C.K.; Mahmoud, T.S.; Bass, O.; Muyeen, S.; Kothapalli, G.; Baniasadi, A.; Mousavi, N. Optimal sizing of a utility-scale energy storage system in transmission networks to improve frequency response. *J. Energy Storage* **2020**, *29*, 101315.
10. Nazir, M.S.; Abdalla, A.N.; Sohail, H.M.; Tang, Y.; Rashed, G.I.; Chen, W. Optimal planning and investment of Multi-renewable power generation and energy storage system capacity. *J. Electr. Syst.* **2021**, *17*, 171–181.
11. Soni, N.; Doolla, S.; Chandorkar, M.C. Improvement of transient response in microgrids using virtual inertia. *IEEE Trans. Power Deliv.* **2013**, *28*, 1830–1838.
12. D'Arco, S.; Suul, J.A.; Fosso, O.B. A virtual synchronous machine implementation for distributed control of power converters in smartgrids. *Electr. Power Syst. Res.* **2015**, *122*, 180–197.
13. Sakimoto, K.; Miura, Y.; Ise, T. Stabilization of a power system including inverter type distributed generators by the virtual synchronous generator. *IEEJ Trans. Power Energy* **2012**, *132*, 341–349.
14. Alipoor, J.; Miura, Y.; Ise, T. Power system stabilization using virtual synchronous generator with alternating moment of inertia. *IEEE J. Emerg. Sel. Top. Power Electron.* **2015**, *3*, 451–458.
15. Jongudomkarn, J.; Liu, J.; Ise, T. Comparison of current-limiting strategies of virtual synchronous generator control during fault ride-through. *IFAC-Pap. Line* **2018**, *51*, 256–261.
16. Hirase, Y.; Abe, K.; Sugimoto, K.; Shindo, Y. A grid-connected inverter with virtual synchronous generator model of algebraic type. *Electr. Eng. Jpn.* **2013**, *184*, 10–21.
17. Guerrero, J.M.; Vasquez, J.C.; Matas, J.; de Vicuna, L.G.; Castilla, M. Hierarchical control of droop-controlled ac and dc microgrids-a general approach toward standardization. *IEEE Trans. Ind. Electron.* **2011**, *58*, 158–172.
18. Liu, J.; Miura, Y.; Bevrani, H.; Ise, T. Enhanced virtual synchronous generator control for parallel inverters in microgrids. *IEEE Trans. Smart Grid.* **2015**, *8*, 2268–2277.
19. Tamrakar, U.; Galipeau, D.; Tonkoski, R.; Tamrakar, I. Improving transient stability of photovoltaic-hydro microgrids using virtual synchronous machines. In Proceedings of the 2015 IEEE Eindhoven Power Tech, Eindhoven, The Netherlands, 29 June–2 July 2015; pp. 1–6.
20. Onaolapo, A.K.; Carpanen, R.P.; Dorrell, D.G.; Ojo, E.E. Reliability Evaluation and Financial Viability of an Electricity Power Micro-Grid System with the Incorporation of Renewable Energy Sources and Energy Storage: Case Study of KwaZulu-Natal, South Africa. *IEEE Access* **2021**, *9*, 159908–159924.
21. Kerdphol, T.; Rahman, F.S.; Watanabe, M.; Mitani, Y. Optimization of Virtual Inertia Control Considering System Frequency Protection Scheme. In *Virtual Inertia Synthesis and Control*; Springer: Berlin/Heidelberg, Germany, 2021; pp. 227–247.
22. Chen, J.; Jin, T.; Mohamed, M.A.; Annuk, A.; Dampage, U. Investigating the Impact of Wind Power Integration on Damping Characteristics of Low Frequency Oscillations in Power Systems. *Sustainability* **2022**, *14*, 3841.
23. Nazir, M.S.; Abdalla, A.N.; Zhao, H.; Chu, Z.; Nazir, H.M.J.; Bhutta, M.S.; Javed, M.S.; Sanjeevikumar, P. Optimized economic operation of energy storage integration using improved gravitational search algorithm and dual stage optimization. *J. Energy Storage* **2022**, *50*, 104591.
24. Al-Ghussain, L.; Ahmad, A.D.; Abubaker, A.M.; Mohamed, M.A. An integrated photovoltaic/wind/biomass and hybrid energy storage systems towards 100% renewable energy microgrids in university campuses. *Sustain. Energy Technol. Assess.* **2021**, *46*, 101273.
25. Meng, L.; Zafar, J.; Khadem, S.K.; Collinson, A.; Murchie, K.C.; Coffele, F.; Burt, G.M. Fast frequency response from energy storage systems—A review of grid standards, projects and technical issues. *IEEE Trans. Smart Grid.* **2019**, *11*, 1566–1581.
26. Judge, P.D.; Green, T.C. Modular multilevel converter with partially rated integrated energy storage suitable for frequency support and ancillary service provision. *IEEE Trans. Power Deliv.* **2018**, *34*, 208–219.
27. Arifin, Z.; Firmanto, A. Battery Energy Storage System as Frequency Control at Substation based on Defense Scheme Mechanism. In Proceedings of the 2021 International Seminar on Intelligent Technology and Its Applications (ISITIA), Surabaya, Indonesia, 21–22 July 2021.
28. Alsiraji, H.A.; El-Shatshat, R. Comprehensive assessment of virtual synchronous machine based voltage source converter controllers. *IET Gener. Transm. Distrib.* **2017**, *11*, 1762–1769.
29. Vargas, R.Z.; Lopes, J.C.; Colque, J.C.; Azcue, J.L.; Sousa, T. Energy Storage System Integration with Wind Generation for Primary Frequency Support in the Distribution Grid. *Simpósio Bras. Sist. Elétricos-SBSE* **2020**. [CrossRef]

sustainability

MDPI

Article

Energy Conservation Measures and Value Engineering for Small Microgrid: New Hospital as a Case Study

Saleh Abdulaziz Almarzooq [1], Abdullah M. Al-Shaalan [1], Hassan M. H. Farh [2,*] and Tarek Kandil [3]

1 Electrical Engineering Department, College of Engineering, King Saud University, Riyadh 11421, Saudi Arabia; 439105730@student.ksu.edu.sa (S.A.A.); shaalan@ksu.edu.sa (A.M.A.-S.)
2 Department of Building and Real Estate, Faculty of Construction and Environment, Hong Kong Polytechnic University, Hung Hom, Kowloon, Hong Kong
3 Department of Electrical and Computer Engineering, College of Engineering and Computing, Georgia Southern University, Statesboro, GA 30460, USA; thassankandil@georgiasouthern.edu
* Correspondence: hfarh.hussein@polyu.edu.hk

Abstract: Energy conservation measures can not only improve energy efficiency; it can also enhance microgrid resilience. This paper aims at investigating energy conservation in a small microgrid, using a new hospital in Riyadh city as a case study, to satisfy the Saudi Building Code (SBC part 601) requirement of energy conservation as the first case. The second case study aims to apply and simulate additional advanced energy conservation requirements. The new hospital has considered energy conservation measures uch as upgraded Heating, Ventilation, and Air Conditioning (HVAC), lighting type effect, thermal insulation, and window material. These energy conservation considerations made a difference in the annual energy saving and efficiency of its microgrid. This study used Autodesk Revit software to obtain building information modeling (BIM) and eQUEST to perform energy simulations. The two software programs are integrated together to perform comprehensive energy simulations with detailed building information from the model by Autodesk Green Building Studio (GBS). The energy conservation measures mainly focused on energy management and saving in the building's electrical installations. All utilized equipment in the hospital should follow the Saudi standards issued by the national authorities. The simulation results revealed a noticeable annual energy saving of up to 19.82% for the second case, using a great thermal resistance building envelope, energy-saving lighting system, and highly rated Energy Efficiency Ratio (EER) HVAC system compared to the first case. More than 100,000 SR in yearly energy saving was achieved by implementing the second case study. Applying the Value Methodology (VM) to the proposed hospital in this study saved more than 700,000 SR in the initial cost of the hospital.

Keywords: energy conservation; pareto analysis; energy consumption; eQUEST; Revit; value methodology

Citation: Abdulaziz Almarzooq, S.; Al-Shaalan, A.M.; Farh, H.M.H.; Kandil, T. Energy Conservation Measures and Value Engineering for Small Microgrid: New Hospital as a Case Study. *Sustainability* **2022**, *14*, 2390. https://doi.org/10.3390/su14042390

Academic Editor: Mohamed A. Mohamed

Received: 20 January 2022
Accepted: 17 February 2022
Published: 19 February 2022

1. Introduction

Energy conservation measures can improve not only the energy efficiency of a microgrid but also the resilience and reliability of a microgrid. A microgrid can be comprised of a decentralized photovoltaics, wind turbines, diesel generators, batteries, fuel cells (FCs), and loads that can operate autonomously (island mode) [1–4] or interconnected with the utility grid (connected mode) [5,6]. Energy conservation refers to minimizing energy consumption using more efficient equipment and appliances as well as sensible utilization. Refraining from using products or services becomes a way to save energy, but it is also possible to do this by utilizing various energy conservation measures that reduce energy consumption compared to their conventional counterparts. Energy conservation is an important component of long-term viability and development. In addition, it is a critical issue, as the use of nonrenewable resources has an impact on the environment. For example, the use of fossil fuel contributes to air and water pollution because carbon dioxide is produced when oil, coal, and gas are burned in power plants, heating systems, and automobile engines. As is well-known, carbon dioxide acts as a straightforward layer in the atmosphere and contributes to global warming, a phenomenon which is also known as greenhouse gases. In our atmosphere, global warming has certain negative consequences. Some of these are lethal, such as the spread of various diseases, temperature increase, extended hurricane risk, economic burdens, polar ice melting, and an increased rate and intensity of storms [7,8].

Numerous studies have been conducted for energy conservation and management, strategies which have impacted this study positively. For example, in [9], different energy-saving strategies for hospital and healthcare facilities were reviewed to improve energy efficiency where the hospital's consumption differs based on the nature of use, the year of constitution, the building envelope status, the level of insulation, the climatic nature, the maintenance level, and the age of the mechanical and electrical equipment, and energy management level. Abdullah Al-Shaalan proposed some strategies to mitigate customers' power outages and showed that consumers bear an enormous role in embracing and observing certain energy consumption behaviors when using their equipment [10,11]. The authors in [12,13] revealed that energy consumption can be reduced to reasonable levels by consumer incentive strategies. In [14], the authors showed that continued education and awareness may motivate people to conserve energy through a variety of motivation factors, such as appeals to environmental and parental obligation, lifestyle rewards, incentives, and disincentives. In [15], this study adopted a city-scale consumption for several types of energy conservation measures, such as occupants' behavior and energy-efficient equipment, to evaluate the policies of energy conservation in the domestic sector. A well-designed natural or partial natural ventilation system may reduce the energy requirement for a cooling system in a building [16]. The authors showed the benefit of using a two-stage algorithm to achieve reasonable energy usage in a school campus [17]. In [18,19], the authors showed that using an automated, energy-saving LED luminaires system in a building will effectively reduce the energy consumption of that building. The advantages of using an automated LED lighting system in buildings in terms of their high performance and decreased energy consumption have been proven. Due to the several advantages of LED lighting, their usage has increased dramatically in street lighting, medical lighting, and residential lighting; this is because of its characteristics, such as low maintenance requirements, a long life cycle, and environmental friendliness.

In [20–22], the Saudi Energy Efficiency Center (SEEC) embarked on a campaign that led to the establishment of an energy label, to be compulsorily stuck to major household appliances, such as air-conditioners, washing machines, refrigerators, and freezers. The outcome of this campaign was a reduction in the huge size of electricity demand, as well as the relieving of a sizable part of the tariff from the consumer's burden. According to the American Society of Heating, Refrigerating and Air-Conditioning Engineers (ASHRAE), the first step towards net-zero energy in small hospitals and healthcare facilities is attained by achieving 30% energy savings [23]. There are several recommendations regarding HVAC, lighting and building insulation that may assist effectively in energy savings. A proper HVAC system is designed by choosing a system type which is best for functionality, with a very highly efficient HVAC system, requiring a 10 Energy Efficiency Ratio (EER) as a minimum rating. In addition, proper temperature control based on functionality, time, and Seal Class A for supply and outdoor ducts may improve energy savings positively [23]. Regarding the whole building's lighting power density (10.7 W/m^2), it is recommended that lighting be switched from manual to auto-off in all hospital zones except patient care areas to achieve 30% energy savings. Side lighting capacity can be maximized with carefully designed skylights and fenestration to obtain illumination without unnecessary solar heat gain. For interior lighting, the use of state-of-the-art products, efficient equipment and lighting design techniques are recommended. Using high-performance lighting may reduce the cost and size of the HVAC system, which will reduce the initial cost of electrical installation in the building [24,25]. The authors of [26,27] demonstrate the need for a multi-criteria implementation when selecting energy-efficient materials, which assists in decreasing energy consumption instantly, as it is required for the following years. Real estate developers, consulting companies, and any related professionals interacting together may affect energy conservation positively [28].

Value Methodology (VM) is defined as a systematic process to improve the value of a project, product, or process through the analysis of functions and resources [29]. VM includes two processes, which are value analysis and value engineering. Value analysis applies the VM process to existing applications [30]. Improvement in initial and life cycle costs in construction projects can be achieved by the implementation of VM [31]. Value engineering involves applying the VM process to new products, projects, or processes. The value of a function is defined as the relationship between cost and performance [32].

Based on the literature review mentioned above, this article considers the first study that investigates energy conservation measures in a small microgrid in a healthcare facility, which is an actual hospital design in Riyadh city, Saudi Arabia. This study focused on energy conservation measures including upgraded HVAC, lighting type effect, thermal insulation, and window material, which have not been utilised before, especially in the exceptional climate of Riyadh. This study employed Building Information Modeling (BIM) for the first time with an energy modeling approach to simulate the energy load of a hospital in Riyadh, Saudi Arabia. The weather in Riyadh is exceptional because it is very dry throughout the year, with the temperature normally ranging from 5 °C to 45 °C. Therefore, it is essential to utilize actual weather data and integrate it with BIM to obtain a robust energy conservation result. The applied energy simulation in this study included electrical, mechanical, and medical equipment by integrating the BIM with building energy modeling methodologies for energy reduction in a hospital. This study covered two different cases where both used the same simulation model, which is a 150-bed hospital design model with several enhancements in energy conservation, such as the building envelope and the HVAC system. The first case is to apply the minimum Saudi Building Code (SBC) energy conservation measures, which have been improved gradually in recent years. The second case is to improve the level of energy conservation, which includes more efficient lighting, HVAC, window material, and building insulation systems.

The integration between Revit and eQUEST has been achieved through Green Building Studio (GBS). The integration between the two different software programs aims to produce complete building data across the energy simulation, which assures robust and reliable results. The actual design of the hospital model was used to create an energy simulation. The energy model considered all energy effective factors, such as material, location, and usage. An efficient HVAC system, energy-efficient lighting fixtures, and highly efficient building envelope material are used in the study to reduce energy consumption, taking into consideration the requirements of certain codes, such as the Saudi Central Board for Accreditation of Healthcare Institutions (CBAHI) and SBC, which assure the comfort level in the proposed hospital in Riyadh, Saudi Arabia. In comparison to the first case, the simulation results indicated a significant yearly energy saving of up to 19.82 percent in the second case, which used a high thermal resistance building envelope, energy-efficient lighting, and a high EER HVAC system. Applying value engineering to the hospital has saved more than 700,000 SR.

This study includes five sections, which are as follows: Section 1 contains the introduction. Section 2 describes the proposed hospital under study. Section 3 explains the methodology of the study. Discussion and results are explained in Section 4. The conclusion from the simulation results is discussed in the last section.

2. Description of the Proposed Hospital under Study

The proposed hospital for the case study is a 150-bed hospital based in Riyadh, Saudi Arabia. The total area of the hospital is 28,909 m^2. The building is a three-story building, comprising ground, first and second levels. Within are out-patient clinics, a laboratory, a pharmacy, intensive care units, and in-patient rooms, as well as administration support areas. Revit takes architecture, structure, and the mechanical and electrical designers' implementation as an input. The energy model is the output of Revit and the input for GBS. The outputs of GBS are gbXM and weather data files, which go to eQUEST as an input. Detailed energy results are the output of eQUEST, produced by simulating the gbXML file. Figure 1 shows a schematic diagram of the proposed microgrid, including the actual hospital in Riyadh city. All building material, such as walls, windows, equipment, and sun path throughout the year, are taken by the energy model, as shown in Figure 2. After the simulation process, the output will be detailed energy consumption results. Figure 3 shows the Building Information Modelling (BIM) of the hospital as it contains the integrated architectural, structural, electrical, mechanical, and medical models. Figure 4 shows the material of the wall, which can be adjusted easily to allow several simulation results.

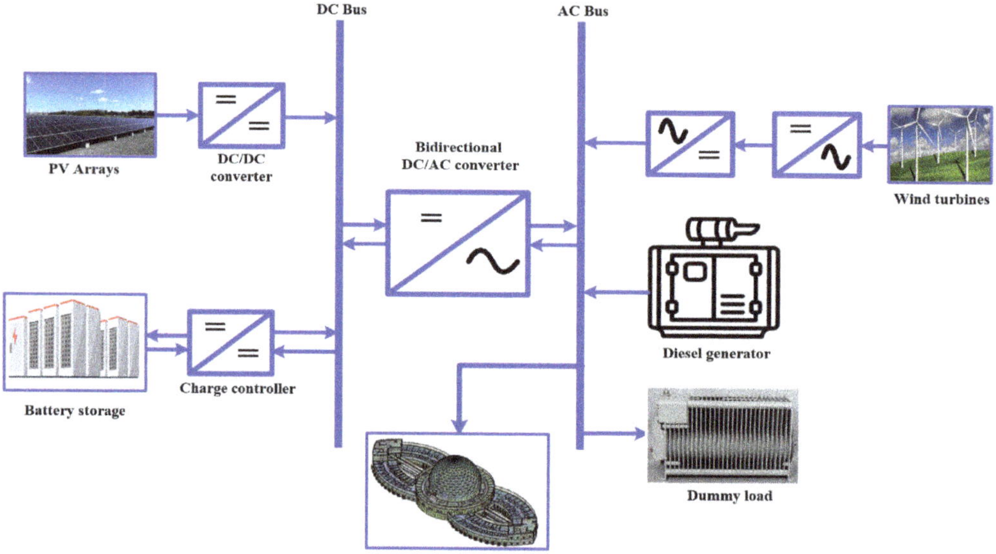

Figure 1. Schematic diagram of the proposed microgrid including actual hospital in Riyadh city.

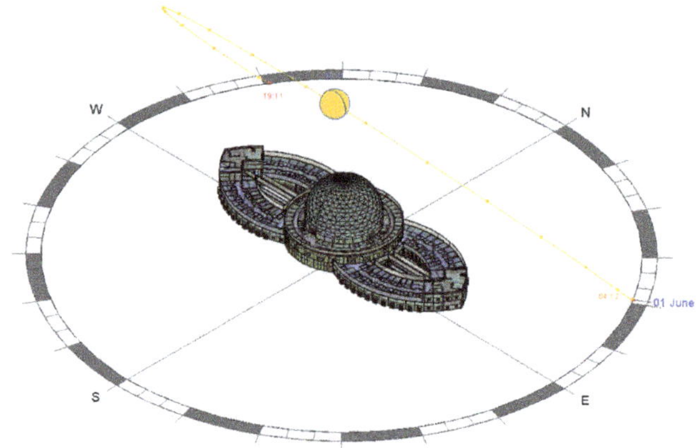

Figure 2. Sun path in the energy model.

Figure 3. Ground floor of the Building Information Modelling (BIM).

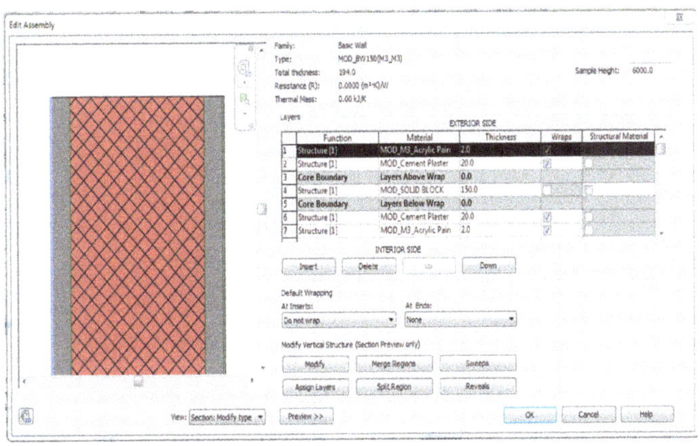

Figure 4. The material and different parameters of wall.

3. Research Methodology

Using Computer-aided design to simulate two different energy consumption cases for the same BIM hospital design is the main purpose of this study. As mentioned in the introduction, the first case is to apply the minimum Saudi Building Code (SBC) energy conservation measures. The second case is to improve the level of energy conservation, which includes measures such as more efficient lighting, HVAC, window material, and building insulation systems. However, a comparison of lighting, window material, and insulation systems has been presented to allow a better understanding of the choices. The Revit and eQUEST software packages were applied in the building of information modeling and energy conservation simulation, which is useful in all design stages. The two computer programs required to perform the energy analysis are Autodesk Revit and eQUEST. Autodesk Revit is a building information modeling software for architects, structural engineers, mechanics, electrical engineers, and plumbers. The software allows users to design a structure and its components in 3D, which can reduce conflict and time. Moreover, the designed electrical system can be analyzed through Revit in various ways, such as safety and energy analysis. Green Building Studio is one of Autodesk's cloud-based software packages, which has a built-in application in Revit that allows the running of building performance simulations to perform energy analysis. eQUEST includes many functions related to the design of HVAC systems that can be applied in residential, healthcare facilities, as well as several other building types. The software calculates the load and design of the systems. It also performs a simulation of building energy, energy audit, evaluation, analysis, and costing. Therefore, these processors can perform a descriptive and quantitative analysis of all components of glass windows, roofs, walls, etc. It can also respond to all HVAC effects from cooling, heating, and ventilation. eQUEST is very popular in providing accurate and sound results with reasonable efforts and time. On the other hand, Revit may communicate with eQUEST by gbXML data models, as shown in Figure 5. gbXML is used to transform geometry data, HVAC system, lighting system, and internal load. A gbXML data file can be created after creating the energy model in Revit.

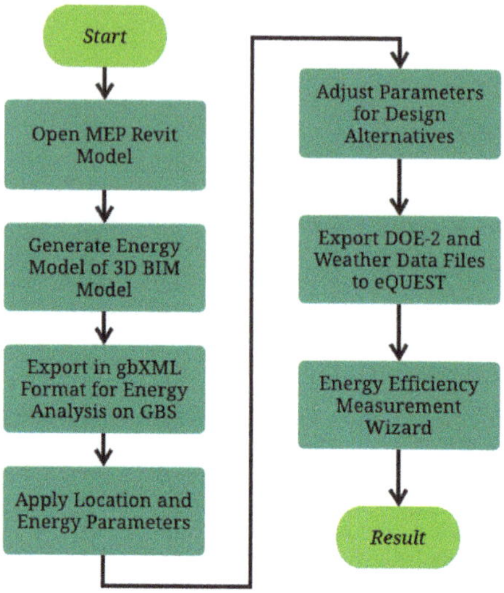

Figure 5. Flow chart of BIM and eQUEST integration procedure.

Calculating the heat transfer (Q) can be used to compare the effectiveness of using different building envelopes, such as window material and building insulation system. It can be calculated by the following equation [33]:

$$Q = U \times A \times \Delta T \tag{1}$$

where U is the overall heat transfer coefficient, A is the area, and ΔT is the temperature difference.

The model requires an equation for proper simulation that shows the interaction between the factors and their levels. Therefore, the total energy of the building was used, and it can be expressed as follows [34].

$$E_{total} = \beta_0 + \left(\sum_{n=1}^{n=m} (\beta_n \times E_n) \right) + \epsilon \tag{2}$$

where β_0 and β_n are areas coefficients, E_n is the energy usage for each area, and ϵ is error tolerance.

For the purpose of setting measures and adopting methodologies towards energy conservation in the Riyadh hospital building, the following procedures have been implemented as follows:

- Compiling all data and information needed for the simulation programming.
- Monitoring conformity with the Saudi Building Code (Part 601: Energy Efficiency) and SASO requirements, as shown in Figure 6 [8].
- Estimating thermal behavior of the hospital building in relation to its outdoor environment.
- Visualizing the impact of daylight and artificial light inside the building.
- Identifying the factors affecting the energy consumption in the hospital building.
- Making detailed analysis of different factors such as lighting, HVAC, thermal insulation material, shading, and envelope considerations using eQUEST.
- Applying VM to reduce cost and improve quality.

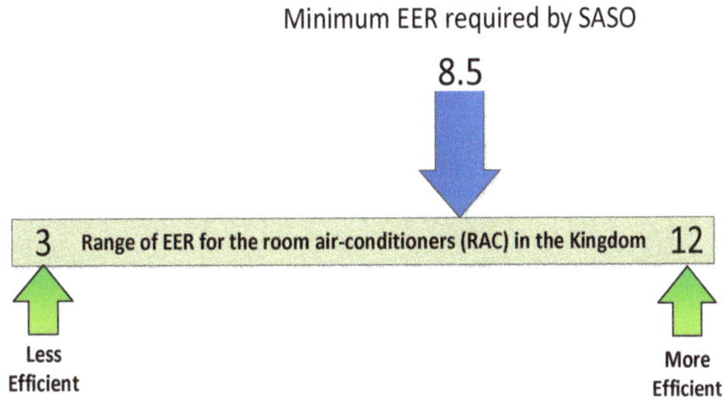

Figure 6. Minimum EER requirement By Saudi Standards [8].

After reviewing these studies and previous works that showed different methods of reducing the consumption of electrical energy in the buildings, which are based on statistical data and computer programs, we find that the main factor in electricity consumption is the HVAC system, followed by lighting in hospitals. Therefore, choosing and installing a high-efficiency HVAC system, characterized by EER, corresponding to those values specified in the Saudi standards issued in this regard, using high-efficiency lighting fixtures, will help in reducing energy usage. In addition, the use of insulating glass in the windows, in order to reduce the amount of light from the window and thus reduce the heat that can leak into the place, is important. In Saudi Arabia, the HVAC system consumes a huge amount of power, since the weather is very hot in summer. The HVAC system is a very complex system because the system must satisfy and work with several functionalities, such as different areas requiring different humidity levels. Therefore, the HVAC system plays a very important role in the hospital. Figure 7 shows that an HVAC system can consume 65% of electric energy in Saudi buildings. This means that more than 50% of total energy consumption in hospitals are from HVAC systems [35].

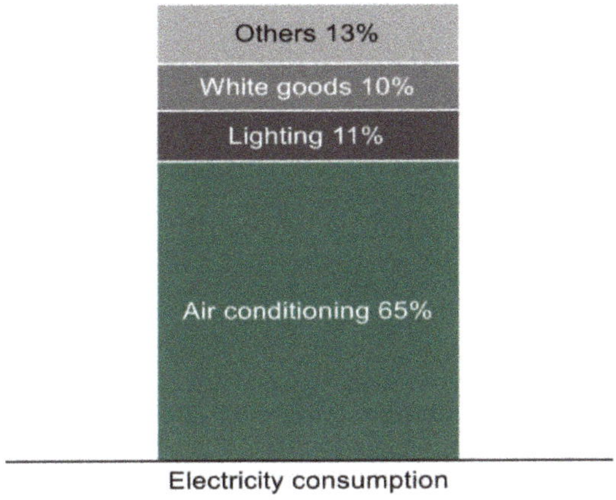

Figure 7. Electricity consumption of Saudi buildings.

It is important to use thermal insulation for the outer shell of the hospital to reduce the heat exchange. Selection and design of glass windows in terms of transparency, shading, visibility, and the amount of natural light entering through them is important. Choosing the orientation of the hospital will play some part in allowing direct sunlight to enter the hospital. The ideal orientation is to face north, to increase comfort at a lower cost. As shown in Figure 8, maximum electricity consumption in Saudi Arabia is increasing in the summer, and hospitals are not different from any building in the sense that most electricity goes to cool the building.

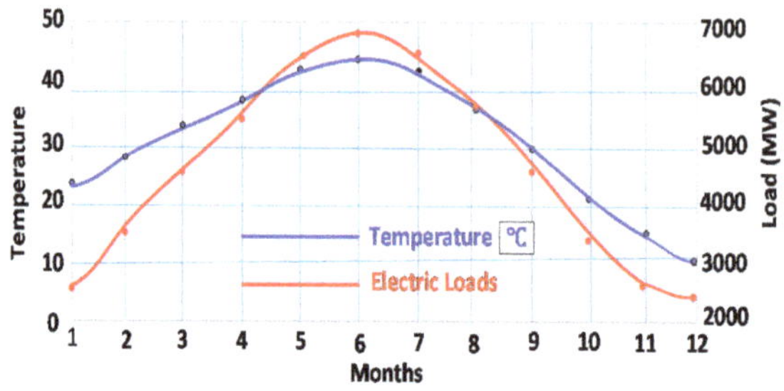

Figure 8. Power consumption change with temperature change in Saudi Arabia.

There are many factors influencing electrical consumption, but most of them depend on the place, the occupants, the functionality of a building, and the electrical equipment in operation. The study investigates major factors that may affect energy consumption in a hospital and contribute to energy conservation. The weather in Riyadh is very hot and dry in the summer and is cold and dry in winter. Over the year, the temperature normally differs from 9 °C to 43 °C and is rarely below 5 °C or above 45 °C. Figure 9 shows the daily average high and low temperature in Riyadh city throughout the year [36]. Riyadh's weather plays a huge role in power consumption through air conditioning.

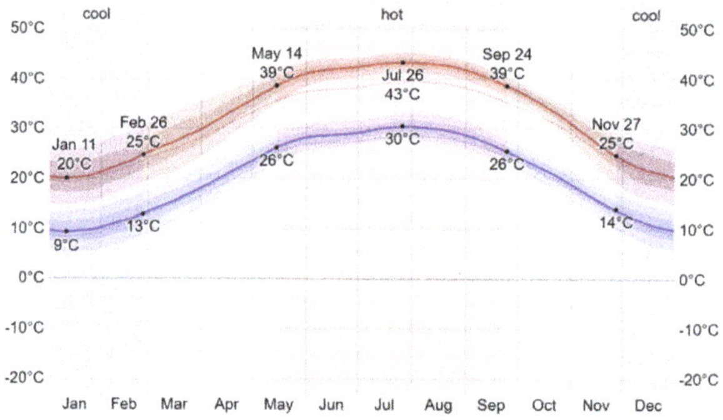

Figure 9. The daily average high (red line) and low (blue line) temperature in Riyadh city.

Value Methodology (VM) is a process to identify unnecessary costs and potential performance improvements, as well as to improve value. Additionally, the process offers alternatives while assuring quality and reliability. However, it is not a process for cost cutting or quality reduction. Value is defined as a fair return or the equivalent in goods, services, or money for something exchanged. There are six phases in a VM job plan, which are: the information phase, the function analysis phase, the creative phase, the evaluation phase, the development phase, and the presentation phase, as shown in Figure 10. The information phase involves obtaining a thorough understanding of project, system, or study item. The function analysis phase involves identifying the greatest opportunities for value improvement. The creative phase involves generating a large quantity of ideas or alternatives to accomplish the functions identified in the previous phase. The evaluation phase involves judging the ideas generated during the creative phase. The development phase involves developing selected ideas into a proposal for change in order to improve value.

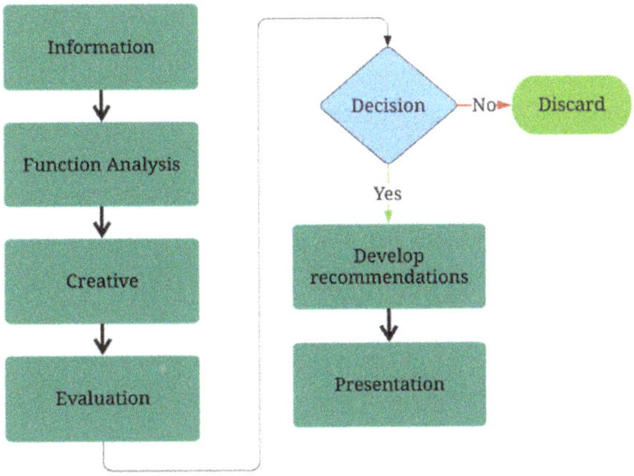

Figure 10. Flow chart of VM job plan.

4. Simulation Results and Discussions

The Revit and eQUEST software packages were applied for building information modeling and energy conservation. The designed electrical system can be analyzed through Autodesk Revit in various ways, such as safety and energy analysis. Revit may communicate with eQUEST through gbXML data models, which can be created after creating the energy model in Revit, as discussed previously in Section 3. Geometry data, HVAC systems, lighting systems, and internal loads are all transformed using gbXML. The comparison of lighting fixtures, window material, building insulation material, and HVAC system have been conducted complying with the Saudi Building Code (SBC part 601) for energy efficiency. There are two cases that have been simulated in this study, using Revit and eQUEST. The first case involved applying the minimum new SBC requirements, such as using two concrete block walls with a 10 cm polyester insulation as a building envelope, LED lighting fixtures, and a 9.2 EER HVAC system. In the second case, a higher thermal resistance insulation system and reflected double glass for windows were used, which can be adjusted in Revit then used to generate the energy model. However, comparing different types of equipment or materials is presented to investigate their effect on the load.

4.1. Lighting Fixture Type Effect on the Load

Lighting can account for up to 25% of energy consumption in a hospital, depending on the types of activities carried out in them. One of the most common energy saving practices for lighting systems in any building is an automatic lighting control system. The lighting control system can be controlled automatically with different controlling options, such as presence, motion, or daylight sensors.

It is unrealistic to compare incandescent lighting fixtures with CFL and LED because outdated lighting technology consumes a lot of power and has a low lifespan, but it is useful for the purpose of comparison to show that LED makes a difference in terms of energy saving and lifespan compared to the CFL. As shown in Table 1 and Figure 11, using LED instead of CFL lighting fixtures results in an energy saving of up to 28%. Therefore, highly efficient lighting fixtures may assist in noticeable energy savings. LED lighting may affect energy conservation positively with low radiated heat and high energy efficiency, since LED lights are more reliable, better in directional lighting, and high in brightness and intensity. Therefore, design engineers can optimize lighting design to be more efficient and directive.

Table 1. Comparison between incandescent, CFL and LED lighting.

	Incandescent	CFL	LED
Lifespan in hours	1000	9000	10,000
Watts's equivalent	60	14	10
12 h daily operation cost (0.18 SR/kWh)	0.129	0.030	0.021
Annual operation cost in SR	1135	11.03	7.88
20 years operation cost in SR	946.08	220.75	157.68

Figure 11. Operation cost of a single fixture over 20 years comparison of lighting type.

4.2. Impact of Window Material on HVAC Load

To justify the idea that the window material may affect the air conditioning load, three different materials are considered in this study, which are single glass, vacuum insulated double glass, and double shade reflector. Table 2 shows the effect of window material on the annual HVAC load (kW) for a 16 m^2 room area with a 3 m^2 window area. Several considerations have been taken to obtain accurate results, which are the exposure outside the wall in a southerly direction and the occupancy of two persons. The load is the actual loads according to Riyadh weather conditions within a 24 h operation. The simulation results shown in Table 1 and Figure 12 reveal that using vacuum insulated double glass instead of single glass will save 14% on annual air conditioning power consumption, while the use of reflective double glass instead of single glass will save 35% on annual air conditioning power consumption.

Table 2. Effect of window material on HVAC load.

Window Material	Heat Transfer Coefficient (W/m^2 Kelvin)	HVAC Load (kW)	Annual HVAC Load (kW)
Reflective double glass	2.13	3.2	11,439
Vacuum insulated double glass	3.8	5.6	15,168
Single glass	5.6	5.7	17,666

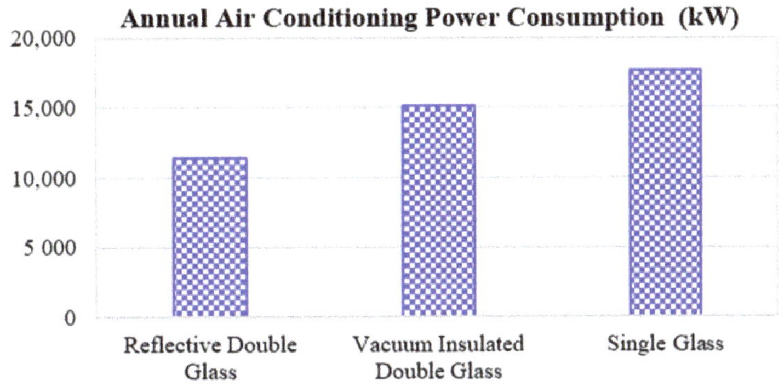

Figure 12. Effect of window material on air conditioning load.

4.3. Impact of Building Thermal Insulation on HVAC Loads

The selection of thermal insulation in buildings participates mainly in reducing the heat exchange between the inside and the outside of the building. Wall and ceiling insulation assist in reducing HVAC loads, which will result in smaller HVAC equipment in terms of size and a lower energy usage. From Table 3 and Figure 13, which are based on a 300 m^2, two-story house building, it is demonstrated that insulation type plays a significant role in energy conservation for the HVAC load, as the cooling capacity can be reduced by up to 73% when comparing the lowest with the highest thermal resistance insulation system. As per the minimum values accepted by the SBC Part 601 regarding energy efficiency, the minimum thermal resistance value (R-value) accepted is 2.9, which means that two concrete block walls with a 10 cm polyester insulation is the minimum accepted insulation type.

Table 3. Impact of the building insulation system on the Energy conservation.

The Building Insulation System Used for the Exterior Wall	Thermal Resistance (R-Value)	Required Cooling Capacity	Conservation Percentage
Hollow concrete block	0.25	290,000	0%
Red clay block	0.42	200,000	30%
Volcanic block	0.43	190,000	33%
Concrete block with insulation (polystyrene filling)	0.55	170,000	40%
Red clay block with insulation (polystyrene filling)	0.62	160,000	45%
Volcanic block with insulation (polystyrene filling)	0.64	155,000	47%
Exterior insulation finishing system (EIFS) 5 cm insulation thickness	2.0	95,000	68%
Two concrete block walls with a 10 cm polyester insulation	2.9	88,000	70%
Two Red clay block walls with a 10 cm polyester insulation	3.1	85,000	71%
ACC block (density 400) 35 cm thickness	3.1	85,000	71%
Two Volcanic block walls with a 10 cm polyester insulation	3.3	82,000	72%
Exterior insulation finishing system (EIFS) 10 cm insulation thickness	3.4	80,000	73%

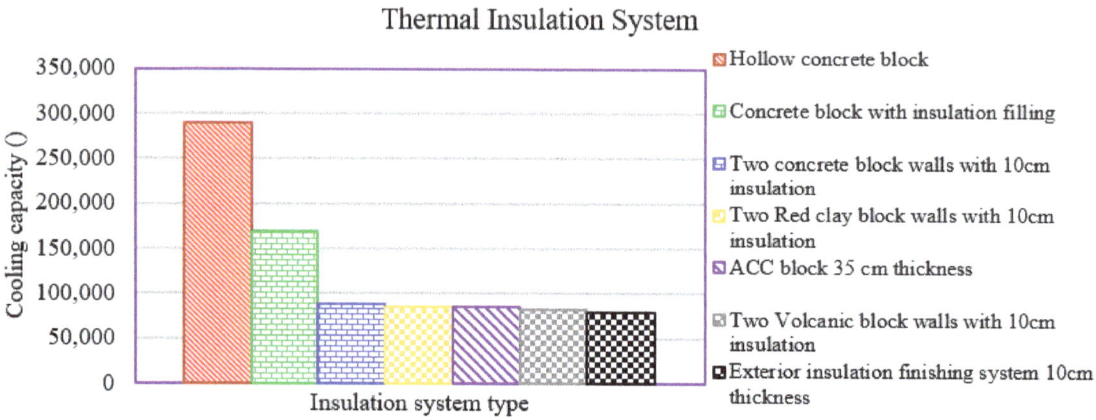

Figure 13. Thermal insulation system of 300 m^2 two story building.

The first case study aims to apply the minimum new SBC requirements, as specified earlier. In order to get this result, the minimum thermal insulation requirement has been met using LED lighting fixtures, an HVAC system rated as 9.2 EER, reflective double glass windows, and two concrete block walls with a 10 cm polyester insulation. Figures 14 and 15 revealed that HVAC (space cool, space heat, and ventilation fans) is the most consuming load, consuming 46%, followed by miscellaneous equipment (28%) and lighting (16%). Therefore, the annual energy cost is 533,703 SR according to the simulation model.

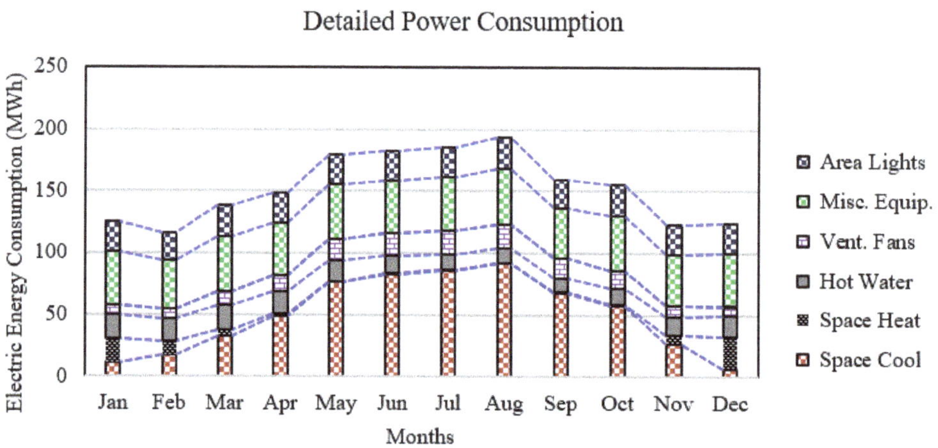

Figure 14. The electric energy consumption over one year for the first case study.

On the other hand, the energy conservation measures specified in the second case have been applied to obtain better hospital energy consumption rates. An HVAC system with better energy consumption, which is rated as 11.8 EER, has been used. Additionally, occupancy and daylight sensors and controls for lighting system in the hospital have been applied, as well as an improved insulation system. Figures 16 and 17 reveal that HVAC (space cool, space heat, and ventilation fans) is the most consuming load, consuming 38%, followed by miscellaneous equipment (34%) and lighting (16%). The annual energy cost is 429,677 SR in the second case.

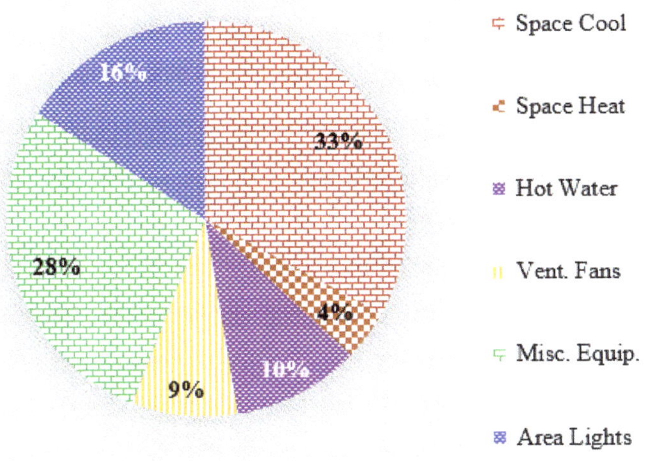

Figure 15. Simulation results of the first case study.

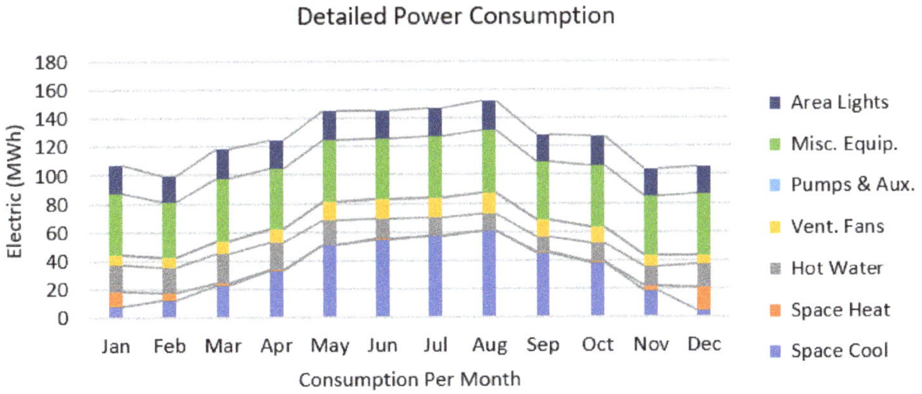

Figure 16. The electric energy consumption over one year for the second case study.

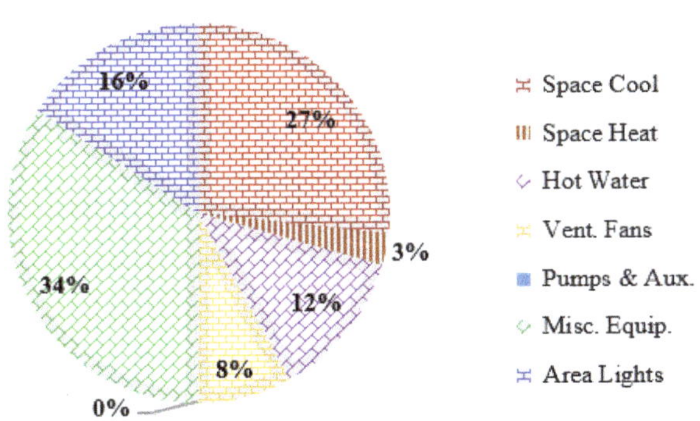

Figure 17. Simulation results of the second case study.

From the results of the first and second case studies, it can be concluded that the second case study result achieves a 19.82% annual energy saving compared to the first case study. The first case study achieved an acceptable energy saving, meeting the minimum code requirements, whereas the best energy saving was achieved in the second case study. This energy saving mostly occurred in the summer months, as shown in Figure 18, which shows the benefit of improving the building envelope and the HVAC systems. The results showed 104,026 SR in annual savings in the second case study. The HVAC load played a huge role in savings through the use of enhanced building envelope systems and higher efficiency systems. The annual HVAC load of the first case is 837.3 MWh, while the second case is 564.9 MWh. Therefore, the annual energy savings regarding HVAC is 32.5% in the second case. Using daylight and occupancy sensors for the same lighting system results in an 18.2 % energy saving in the lighting system.

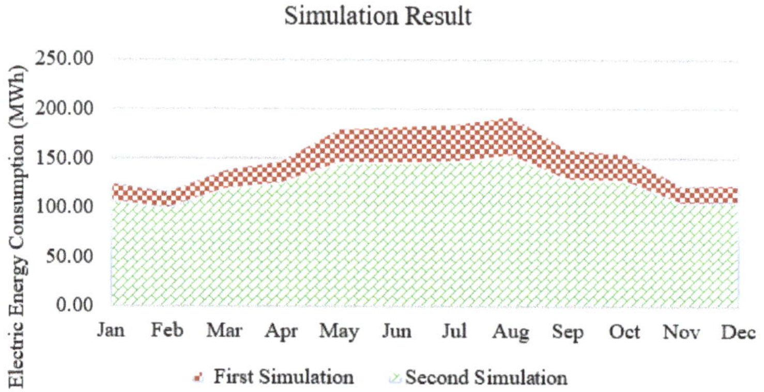

Figure 18. Simulation Results of first and second cases in MWh.

4.4. Value Engineering Analysis

Pareto analysis is a formal technique that can be used when there are numerous possible alternative options competing. It is an innovative way to look at the causes of problems since it stimulates thinking and organizes perspectives. This technique aids in identifying the top 20% of causes that must be addressed to overcome the remaining 80% of problems. From the Pareto analysis in Figure 19, it can be concluded that 80% of the initial cost is from interior lighting, electrical services, distribution panels, feeders, and busways.

A huge number of functions have been generated and analyzed, and only three ideas are further discussed, which are as follows: reducing the number of battery-packed fixtures, using aluminum busways conductors instead of copper conductors, and eliminating the busway system by using cables. The design used approximately 40% of lighting fixtures with a battery packed, reducing the number of battery-packed fixtures to 25%. The emergency lighting system has been designed excessively, which results in unnecessary initial costs and life cycle costs. The proposed idea is to optimize the quantity of the emergency lights, maintaining compliance with the applicable standard. The initial cost saving is shown in Table 4.

The busway system uses copper conductors. Using aluminum conductor busways instead of copper conductor busways maintains the current carrying capacity without having an effect on voltage drop along the entire network. An aluminum conductor busway is larger than a copper conductor busway in size; however, it is better in price while still maintaining the same current carrying capacity. Coordination with architectural drawings is required to verify adequate space above the ceiling and vertical shafts. The advantages of implementing the idea are an initial cost reduction and a lighter weight, while the disadvantage is an increase in size and dimension. The initial cost saving result is shown in Table 5.

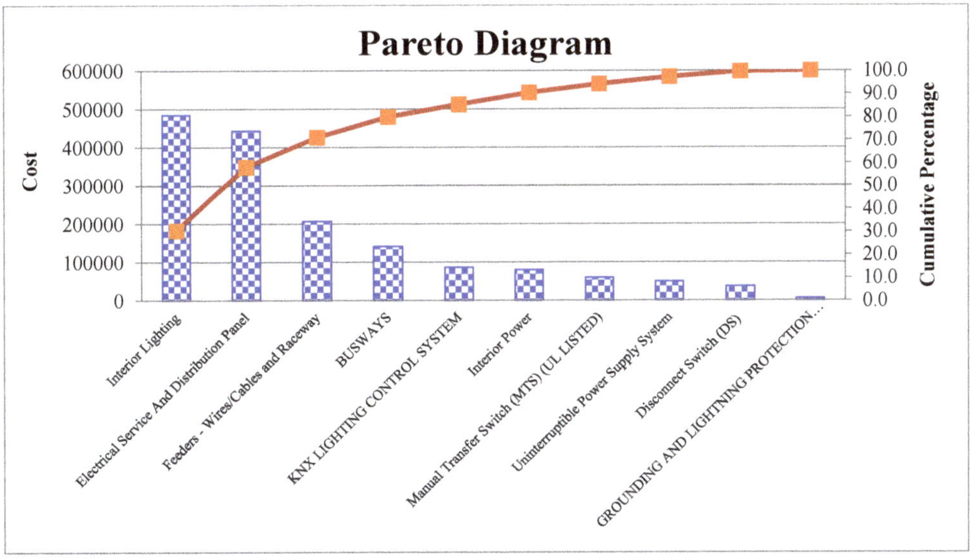

Figure 19. Pareto analysis.

Table 4. Initial cost saving of battery packed fixtures reduction.

VM Stage	Initial Cost
Original Design	1,212,700 SR
Proposed Design	812,586 SR
Potential Savings	309,114 SR

Table 5. Initial cost saving of using aluminum busway instead of copper.

VM Stage	Initial Cost
Original Design	1,413,000 SR
Proposed Design	989,100 SR
Potential Savings	423,900 SR

The high order distribution system uses a busways system. Using a cable distribution system instead of a busway system maintains the same current carrying capacity without having an effect on voltage drop along the entire network. Hence, the busway is better in terms of space, while the cable distribution system has better price for limited length while maintaining the same current carrying capacity and voltage drop. The cable system may require more space in the ceiling void and vertical shaft, and therefore adequacy of the available space should be verified. The initial cost saving result is shown in Table 6. From value engineering analysis, it is better to use an aluminum busway than to eliminate the busway, since it will be more affective in initial cost reduction without affecting the quality of the system. Additionally, reducing the number of battery-packed lighting fixtures will reduce the cost and will keep the required functionality unaffected.

Table 6. Initial cost saving of eliminating busway system.

VM Stage	Initial Cost
Original Design	1,413,000 SR
Proposed Design	1,021,000 SR
Potential Savings	392,000 SR

5. Conclusions

Overall, this work has presented a comparison of two cases in regard to energy conservation and has shown that the second case results in better energy consumption. A significant amount of energy is normally consumed by a hospital's complex electrical loads. This energy can be positively reduced by increasing overall building efficiency through certain energy conservation measures that will improve the efficiency and resilience of the microgrid. Two cases have been thoroughly studied and analyzed. The first case study investigated minimum energy conservation measures to meet Saudi Building Code (SBC part 601) requirements. In the second case study, additional advanced energy conservation measures such as upgraded HVAC, lighting type, building thermal insulation, and window material were applied and simulated. The proposed work applied two different energy conservation cases at an actual hospital in Riyadh city, which has an exceptional climate. The simulation findings revealed that the second case results saved around 20 percent of annual energy consumption by using a great thermal resistance building envelope, an energy-saving lighting system, and a highly rated EER HVAC system compared to the first case study. The first case study achieved promising energy savings since it follows the SBC (601) requirements, while the second case study attained significant energy savings. The benefits of improving the building envelope and HVAC systems are very noticeable during the summer months. Following the SBC (601) requirements guaranteed noticeable energy conservation. The second case study will save a million SR in energy costs every 10 years. However, since technology and systems are improving gradually, it is recommended to push forward with improved energy conservation measures and requirements to capitalize on energy saving in buildings that run for 24 h, such as hospitals, as witnessed in the obtained results. On the other hand, Value Engineering methodology increases the overall hospital facilities and equipment design and reduces the initial, operating, and running costs. Additionally, this study concluded that 733,014 SR in initial cost saving from the 1,354,000 SR of the busway and emergency lighting system can be saved, which is a considerable energy and cost saving. The study focused on energy conservation and VM implementation. As energy conservation will play a huge role in the country's total power consumption, as well as in emissions reduction, healthcare facilities should be designed to promote energy-efficient buildings.

Author Contributions: Conceptualization, S.A.A., H.M.H.F. and T.K.; methodology, S.A.A. and H.M.H.F.; software, S.A.A.; validation, S.A.A.; formal analysis, S.A.A., A.M.A.-S. and H.M.H.F.; investigation, S.A.A. and H.M.H.F.; resources, A.M.A.-S. and T.K.; data curation, S.A.A. and H.M.H.F.; writing—original draft preparation, S.A.A. and H.M.H.F.; writing—review and editing, A.M.A.-S. and H.M.H.F.; visualization, S.A.A. and H.M.H.F.; supervision, A.M.A.-S. and H.M.H.F.; project administration, A.M.A.-S. and H.M.H.F.; funding acquisition, A.M.A.-S. All authors have read and agreed to the published version of the manuscript.

Funding: This research received no external funding.

Institutional Review Board Statement: Not applicable.

Informed Consent Statement: Not applicable.

Acknowledgments: This work was supported by the Researchers Supporting Project number (RSP-2021/337), King Saud University, Riyadh, Saudi Arabia.

Conflicts of Interest: The authors declare no conflict of interest.

References

1. Omotoso, H.O.; Al-Shaalan, A.M.; Farh, H.M.; Al-Shamma'a, A.A. Techno-Economic Evaluation of Hybrid Energy Systems Using Artificial Ecosystem-Based Optimization with Demand Side Management. *Electronics* **2022**, *11*, 204. [CrossRef]
2. Alturki, F.A.; Al-Shamma'a, A.A.; Farh, H.M.; AlSharabi, K. Optimal sizing of autonomous hybrid energy system using supply-demand-based optimization algorithm. *Int. J. Energy Res.* **2021**, *45*, 605–625. [CrossRef]
3. Alnowibet, K.; Annuk, A.; Dampage, U.; Mohamed, M.A. Effective Energy Management via False Data Detection Scheme for the Interconnected Smart Energy Hub–Microgrid System under Stochastic Framework. *Sustainability* **2021**, *13*, 11836. [CrossRef]

4. Gong, X.; Dong, F.; Mohamed, M.A.; Abdalla, O.M.; Ali, Z.M. A secured energy management architecture for smart hybrid microgrids considering PEM-fuel cell and electric vehicles. *IEEE Access* **2020**, *8*, 47807–47823. [CrossRef]
5. Farh, H.M.H.; Eltamaly, A.M.; Al-Shaalan, A.M.; Al-Shamma'a, A.A. A novel sizing inherits allocation strategy of renewable distributed generations using crow search combined with particle swarm optimization algorithm. *IET Renew. Power Gener.* **2021**, *15*, 1436–1450. [CrossRef]
6. Farh, H.M.; Al-Shaalan, A.M.; Eltamaly, A.M.; Al-Shamma'a, A.A. A novel severity performance index for optimal allocation and sizing of photovoltaic distributed generations. *Energy Rep.* **2020**, *6*, 2180–2190. [CrossRef]
7. Iwaro, J.; Mwasha, A. A review of building energy regulation and policy for energy conservation in developing countries. *Energy Policy* **2010**, *38*, 7744–7755. [CrossRef]
8. Wulfinghoff, D.R. The modern history of energy conservation: An overview for information professionals. *Electron. Green J.* **2000**, *1*. [CrossRef]
9. Kolokotsa, D.; Tsoutsos, T.; Papantoniou, S. Energy conservation techniques for hospital buildings. *Adv. Build. Energy Res.* **2012**, *6*, 159–172. [CrossRef]
10. Al-Shaalan, A.M. Proposed Measures to Mitigate Energy Interruptions in Residential Sector. *J. Power Energy Eng.* **2017**, *5*, 50. [CrossRef]
11. Al-Shaalan, A.M. Investigating Practical Measures to Reduce Power Outages and Energy Curtailments. *J. Power Energy Eng.* **2017**, *5*, 21. [CrossRef]
12. Gabbar, H.A. *Energy Conservation in Residential, Commercial, and Industrial Facilities*; Gabbar, H.A., Ed.; John Wiley & Sons: Hoboken, NJ, USA, 2018.
13. Marshall, M.J. Social-Psychological Correlates of Household Energy Conservation Activities. Ph.D. Dissertation, The Claremont Graduate University, Claremont, CA, USA, 1987.
14. McMakin, A.H.; Malone, E.L.; Lundgren, R.E. Motivating residents to conserve energy without financial incentives. *Environ. Behav.* **2002**, *34*, 848–863. [CrossRef]
15. Shimoda, Y.; Asahi, T.; Taniguchi, A.; Mizuno, M. Evaluation of city-scale impact of residential energy conservation measures using the detailed end-use simulation model. *Energy* **2007**, *32*, 1617–1633. [CrossRef]
16. De Dear, R.; Brager, G.S. The adaptive model of thermal comfort and energy conservation in the built environment. *Int. J. Biometeorol.* **2001**, *45*, 100–108. [CrossRef] [PubMed]
17. Ho, Y.F.; Chang, C.C.; Wei, C.C.; Wang, H.L. Multi-objective programming model for energy conservation and renewable energy structure of a low carbon campus. *Energy Build.* **2014**, *80*, 461–468. [CrossRef]
18. Gobinath, J.; Kumar, M.V.; Lenin, M. Recent approach for energy conservation and user fulfillment using automated led lighting system in buildings through wireless network. In Proceedings of the 2013 International Conference on Green Computing, Communication and Conservation of Energy (ICGCE), Chennai, India, 12–14 December 2013; pp. 28–33.
19. Wang, Y.; Alonso, J.M.; Ruan, X. High-performance LED drivers. *IEEE Trans. Ind. Electron.* **2017**, *64*, 5751–5753. [CrossRef]
20. Saudi Energy Efficiency Center (SEEC). Mandatory Application of Energy Efficiency Ratio (EER) on major electrical equipment. In *Annual Report*; SEEC: Riyadh, Saudi Arabia, 2020.
21. Belaid, F.; Al, M. The Role of Residential Energy Efficiency in Shaping the Energy Transition in Saudi Arabia: Key challenges and initiatives. In *IAEE Energy Foru*; International Association for Energy Economics: Cleveland, OH, USA, 2021; pp. 19–23.
22. AlGhamdi, A. *Saudi Arabia Energy Report*; (No. ks—2020-dp25); KAPSARC: Riyadh, Saudi Arabia, 2020.
23. Athalye, R. Through the past decade: How advanced energy design guides have influenced the design industry. *ASHRAE Trans.* **2015**, *121*, 1C.
24. Alzubaidi, S.; Soori, P.K. Energy efficient lighting system design for hospitals diagnostic and treatment room—A case study. *J. Light Vis. Environ.* **2012**, *36*, 23–31. [CrossRef]
25. Fujii, E.T.; Mak, J. A model of household electricity conservation behavior. *Land Econ.* **1984**, *60*, 340–351. [CrossRef]
26. Omer, M.A.; Noguchi, T. A conceptual framework for understanding the contribution of building materials in the achievement of Sustainable Development Goals (SDGs). *Sustain. Cities Soc.* **2020**, *52*, 101869. [CrossRef]
27. Passer, A.; Cresnik, G.; Schulter, D.; Maydl, P. Life Cycle Assessment of buildings comparing structural steelwork with other construction techniques. In *International Conference on Life Cycle Management*; ETH: Zürich, Switzerland, 2007; Volume 3.
28. Zhu, J.; Cao, Y.; Zhai, J.; Zhao, X.; Zhao, Y.; Kang, S. Analysis on synergies and trade-offs in green building development: From the perspective of SDG 11. *Chin. J. Popul. Resour. Environ.* **2019**, *17*, 341–351. [CrossRef]
29. Younker, D. *Value Engineering: Analysis and Methodology*; CRC Press: Boca Raton, FL, USA, 2003; Volume 30.
30. Al Amri, T.; Marey-Pérez, M. Value engineering as a tool for sustainability in the construction industry of Oman. *Value Eng.* **2020**, *29*, 7433–7444.
31. Othman, I.; Kineber, A.F.; Oke, A.E.; Zayed, T.; Buniya, M.K. Barriers of value management implementation for building projects in Egyptian construction industry. *Ain Shams Eng. J.* **2021**, *12*, 21–30. [CrossRef]
32. Dahooie, J.H.; Dehshiri, S.J.H.; Banaitis, A.; Binkytė-Vėlienė, A. Identifying and prioritizing cost reduction solutions in the supply chain by integrating value engineering and gray multi-criteria decision-making. *Technol. Econ. Dev. Econ.* **2020**, *26*, 1311–1338. [CrossRef]
33. Chhabra, R.P. *CRC Handbook of Thermal Engineering*; Chhabra, R.P., Ed.; CRC Press: Boca Raton, FL, USA, 2017.
34. Gross, M.; Mautz, R. *Renewable Energies*; Routledge: England, UK, 2014.

35. Saudi Energy Efficiency Center (SEEC). *Introducing the New Requirements of Air Conditioners Standard SASO-2663 the Seasonal Energy Efficiency Ratio (SEER)*; SEEC: Riyadh, Saudi Arabia, 2020.
36. Diebel, J.N.O.K.J. "Weather Spark," Cedar Lake Ventures, Inc. 31 December 2016. Available online: https://weatherspark.com/ (accessed on 26 January 2021).

sustainability

MDPI

Article

Optimal Location and Sizing of Distributed Generators in Power System Network with Power Quality Enhancement Using Fuzzy Logic Controlled D-STATCOM

Prashant [1], Anwar Shahzad Siddiqui [1], Md Sarwar [1,*], Ahmed Althobaiti [2] and Sherif S. M. Ghoneim [2,*]

1 Department of Electrical Engineering, Faculty of Engineering & Technology, Jamia Millia Islamia, New Delhi 110025, India; prashant.pacificcold@gmail.com (P.); assiddiqui@jmi.ac.in (A.S.S.)
2 Electrical Engineering Department, College of Engineering, Taif University, P.O. Box 11099, Taif 21944, Saudi Arabia; ahmed.althobaiti@tu.edu.sa
* Correspondence: msarwar@jmi.ac.in (M.S.); s.ghoneim@tu.edu.sa (S.S.M.G.)

Abstract: This article presents the selection of location and sizing of multiple distributed generators (DGs) for boosting performance of the radial distribution system in the case of constant power load flow and constant impedance load flow. The consideration of placing and sizing of DGs is to meet the load demand. This article tries to overcome the limitations of existing techniques for determining the appropriate location and size of DGs. The selection of DG location is decided in terms of real power losses, accuracy, and sensitivity. The size of DG is measured in terms of real and reactive power. Both positioning and sizing of DG are analyzed with the genetic algorithm and the heuristic probability distribution method. The results are compared with other existing methods such as ant-lion optimization algorithm, coyote optimizer, modified sine-cosine algorithm, and particle swarm optimization. Further, the power quality improvement of the network is assessed by positioning D-STATCOM, and its location is decided on the basis of the nearby bus having poor voltage profile and high total harmonic distortion (THD). The switching and controlling of D-STATCOM are assessed with fuzzy logic controller (FLC) for improving the performance parameters such as voltage profile and THD at that particular bus. The proposed analytical approach for the system is tested on the IEEE 33 bus system. It is observed that the performance of the system with the genetic algorithm gives a better solution in comparison to heuristic PDF and other existing methods for determining the optimal location and size of DG. The introduction of D-STATCOM into the system with FLC shows better performance in terms of improved voltage profile and THD in comparison to existing techniques.

Keywords: distributed generator; D-STATCOM; fuzzy logic controller; genetic algorithm; power quality

Citation: Prashant; Siddiqui, A.S.; Sarwar, M.; Althobaiti, A.; Ghoneim, S.S.M. Optimal Location and Sizing of Distributed Generators in Power System Network with Power Quality Enhancement Using Fuzzy Logic Controlled D-STATCOM. *Sustainability* **2022**, *14*, 3305. https://doi.org/10.3390/su14063305

Academic Editor: Mohamed A. Mohamed

Received: 4 February 2022
Accepted: 4 March 2022
Published: 11 March 2022

1. Introduction

The current activities of the advanced power system have become very complicated, which needs to necessarily satisfy the increasing energy needs in an efficient manner [1,2]. The civil, fiscal, and other substantial considerations warrant the site of generation centers being placed at places distant from load centers. The reorganization and deregulation of power companies have led to making system governance unpredictable. The factors which are to be considered while carrying out extension of the transmission system are the following: cutback stability margins, chances of tripping outages, and rising power cuts. The distributed generator (DG) installation would be of the utmost benefit where the installation of new transmission lines and setting up of new power-generating units are not feasible [3–5]. It is also reasonable to believe that the selection of the right DG technology, including the optimal location and scale of the DG, would help in decreasing the losses in such a system. The purpose of optimally placing DG in a power system network is for achieving correct operation of the network with system error minimization and voltage profile enhancement inclusive of improved stability, reliability, and load ability [6,7]. In contrast to the classical approach, smart techniques for optimal DG sizing and positioning are simple and have good convergence characteristics for application to composite power system networks [8–10]. DG is also capable of significantly mitigating harmonics and improving the voltage profile along with acquiesced transmission and distribution investment [11–13]. DG is flexible enough to locate anywhere in the system as per requirement of minimum transmission losses [14–16]. There have been many techniques proposed to perform such activities as load durability test, exact loss formula, Newton–Raphson, Gauss–Seidal method etc., [17–19]. At present, in the era of soft computing technique, every system is trying to obtain fast convergence region so as to reduce the minimum losses with lesser component involved [20,21].

The biggest challenge is to find the optimal size of DG to fulfil the load requirements. It is observed that there are four types of DG given as:

(a) DG1: It supports only real power; no component of reactive power at unity power factor. For example: solar PV array.
(b) DG2: It supports both real power and reactive power at 0.8–0.85 power factor leading. For example: wind, tidal geothermal.
(c) DG3: It supports only reactive power; no component of real power at 0 power factor. For example: synchronous condenser, capacitor bank.
(d) DG4: It absorbs the real power, giving the reactive power to the system at 0.8–0.85 power factor lagging. For example: doubly-fed induction generator [22–24].

Recent work has found that there are many advantages to the installation of DGs within the power system; namely raising the efficiency of the output and lowering energy deficit on the electrical grid [25–28]. In this article, only DG2 is used as per system requirements. The size and location of DG were proposed in many articles by using different techniques such as crow bar search method, neural network, evolutionary algorithm, intelligent water droplet, swarm optimization, etc. [29–33]. The advanced algorithms [34,35] or other existing techniques for identifying the size and location of DG have the biggest challenge of improper total harmonic distortion (THD), sensitivity, and accuracy. It is observed in literature that the THD value obtained was in the range of 15–18%, accuracy and sensitivity are also in the range of 10–15% which are quite high and inadequate, due to which voltage profile is distorted. Such issues with existing techniques create the motivation to overcome these problems. These issues are satisfactorily resolved with the proposed scheme. The genetic algorithm (GA) is proposed to determine the sizing and location of DG with improved THD, accuracy, sensitivity, and voltage profile. GA tends to be more highly permissible than heuristic optimization strategies in order to find the optimal results with less effort than absolute search. In this paper, heuristic optimization strategy which is a combination of two methods, namely, Poisson distribution method and normal distribution method, is also utilized to find the performance of the radial distribution system and show the effectiveness of GA in its comparison. Such a heuristic method requires more effort for providing detailed solutions in terms of accuracy, sensitivity, realism, and power loss. In order to have better results with minimal efforts, GA is applied. It is also observed that the proposed technique gives better results in comparison to existing techniques presented in literature [4,9,12,15,17,20,28,30,32,35–45]. However, to analyze the efficacy of the proposed work in this paper, the results achieved are compared with the existing techniques such as ant-lion optimization algorithm [42], coyote optimizer [43], modified sine-cosine algorithm [44], and particle swarm optimization [45]. The GA seems to have the main advantage of utilizing entity definition, i.e., strings, instead of handling such entities themselves. Multiple problems in electricity grid optimization are manifestations in numerical optimization but linear and other nonlinear methods in programming consider them difficult to address. In this article, the heuristic probability distribution method and GA are applied for finding the best location and sizing of DG in the IEEE 33 bus system for the constant power load model (CPLF) and the constant impedance load model (CILF). The performance parameters show the superiority of GA over the heuristic probability distribution method and other existing methods such as ant-lion optimization algorithm [42], coyote optimizer [43], modified sine-cosine algorithm [44], and particle swarm optimization [45] in the IEEE 33 standard bus system. By considering lower real power losses, improved accuracy, and sensitivity, selection criteria of three DGs location are decided, and after that, size of DG is decided in terms of real and reactive power. Afterwards, the impact of the distribution static compensator (D-STATCOM) is assessed for improving the power quality performance at the line near to the bus terminal having higher distortion. In order to improve the power quality of performance parameters, the switching of D-STATCOM is performed by using the fuzzy logic controller (FLC). With the involvement of D-STATCOM, the distortion level has been improved considerably with FLC in comparison to existing techniques. It is preferred to utilize optimal contribution of DG for both the CILF and the CPLF system so as to fulfill the load demand using GA.

2. Optimal Location and Sizing of DG

This paper aims to find the optimal location and size of DG. For this, two approaches are adopted. Firstly, the location and size are found using the heuristic probability distribution method (PDF), and then the GA method is used. Both these methods utilize CPLF and CILF to optimize the DG location and size.

DG is placed at a particular bus to fulfil the load demand. Since load demand is fixed for all buses, the selection of bus for DG placement is assessed in terms of minimum real power losses, improved accuracy, and improved sensitivity factor. The selection of DG location is analyzed by using heuristic PDF and GA.

2.1. Heuristic Probability Distribution Method (PDF)

The location and position of DG are decided by the heuristic probability distribution method which is presented in Equation (1). The heuristic probability distribution method is a combination of the Poisson distribution method and normal distribution and it is denoted by operator '*J*' and is expressed by the following equation:

$$J = \min\left(\sum_{i=1}^{33} X_i |S_i - S_{ref}|^2 + pdf(\lambda)\left(\sum_{i=1}^{33} S_{loss}\right)\right) \tag{1}$$

J is an operator of heuristic probability distribution method for minimization of power losses shown in Equation (2) and λ is the difference between measured and reference power.

$$S_{loss} = \sum_{i=1}^{33} S_i - S_{load} \tag{2}$$

S_{loss} is the complex power loss between two buses and S_{load} is the power consumed by the load.

Putting the value of Equation (2) into Equation (1), Equation (3) is obtained as:

$$J = \min\left(\sum_{i=1}^{33} X_i |S_i - S_{ref}|^2 + pdf(\lambda)\left(\sum_{i=1}^{33} S_i - S_{load}\right)\right) \tag{3}$$

where the measured (S_i) and reference value (S_{ref}) of complex power are given by Equations (4) and (5), respectively:

$$S_i = P_i + jQ_i \tag{4}$$

$$S_{ref} = P_{ref} + jQ_{ref} \tag{5}$$

X_i is the design factor given by Equation (6) to improve the accuracy such that minimum power loss can be attained.

$$X_i = \frac{S_{rated}}{\left|C_1 S_i + C_2 S_{ref}\right|^2} \quad (6)$$

The probability distribution factor in Equation (3) is given by Equation (7):

$$pdf(\lambda) = \frac{e^{-\frac{\lambda^2}{2\sigma^2}}}{\sigma\sqrt{2\pi}} \quad (7)$$

where $\lambda = S_i - S_{ref}$ = difference factor between measured and reference value of power.

As per standard expression of normal distribution and comparing with Equation (7), Equation (8) is obtained:

$$pdf(S_i - S_{ref}) = \frac{e^{-\frac{|S_i - S_{ref}|^2}{2\sigma^2}}}{\sigma\sqrt{2\pi}} \quad (8)$$

where S_{ref} is the mean, σ is the standard deviation, and its corresponding variance is given as σ^2.

Putting the value of Equations (6)–(8) into Equation (3), Equation (9) is obtained:

$$J = \min\left(\sum_{i=1}^{33} \frac{S_{rated}\left|S_i - S_{ref}\right|^2}{\left|C_1 S_i + C_2 S_{ref}\right|^2} + \frac{e^{-\frac{|S_i - S_{ref}|^2}{2\sigma^2}}}{\sigma\sqrt{2\pi}}\left(\sum_{i=1}^{33} S_i - S_{load}\right)\right) \quad (9)$$

where,

C_1 is the accuracy measured in terms of real power measurement. Its range lies between 0.02 and 0.04. The mathematical representation of C_1 is given as ($\Delta P_i/P_i$).

C_2 is the accuracy measured in terms of reactive power measurement. Its range lies between 0.03 and 0.05. The mathematical representation of C_2 is given as ($\Delta Q_i/Q_i$).

Power flow at bus '*i*' is given by Equations (10) and (11) as:

$$\Delta P = P_i - P_{ref} \quad (10)$$

$$\Delta Q = Q_i - Q_{ref} \quad (11)$$

where, P_i and Q_i are real and reactive power flow between two buses and their reference values are given as P_{ref} and Q_{ref} as shown in Equations (12) and (13):

$$P_i = P_{ij} = \sum_{i=1}^{33}\sum_{j=1}^{33} V_i V_j (G_{ij}\cos(\theta_i - \theta_j) + B_{ij}\sin(\theta_i - \theta_j)) \quad (12)$$

$$Q_i = Q_{ij} = \sum_{i=1}^{33}\sum_{j=1}^{33} V_i V_j (G_{ij}\cos(\theta_i - \theta_j) - B_{ij}\sin(\theta_i - \theta_j)) \quad (13)$$

where conductance (G_{ij}) and susceptance (B_{ij}) are given by Equations (14) and (15), respectively:

$$G_{ij} = \frac{R_{ij}}{\sqrt{R_{ij}^2 + X_{ij}^2}} \quad (14)$$

$$B_{ij} = \frac{X_{ij}}{\sqrt{R_{ij}^2 + X_{ij}^2}} \quad (15)$$

The best performance index can be measured by differentiating Equation (9) to obtain the minimal optimal solution as shown in Equation (16):

$$\frac{\partial J}{\partial S_i} = 0; \frac{\partial^2 J}{\partial S_i{}^2} > 0 \quad (16)$$

J is the function of variable S_i which employs $J(S_i)$. For checking the convergence nature, Taylor series expansion for second order is used for 'k' iteration as shown in Equation (17):

$$J(S_i^{k+1}) = J(S_i^k) + h\frac{\partial(J(S_i^k))}{\partial S_i^k} + h^2\frac{\partial^2(J(S_i^k))}{\partial(S_i^k)^2} \quad (17)$$

Using the condition of Equation (16) in Equation (17), Equation (18) is obtained:

$$J(S_i^{k+1}) - J(S_i^k) = h^2\frac{\partial^2(J(S_i^k))}{\partial(S_i^k)^2} \quad (18)$$

Rearranging Equation (18), Equation (19) is obtained:

$$\frac{J(S_i^{k+1}) - J(S_i^k)}{h^2} = \frac{\partial^2(J(S_i^k))}{\partial(S_i^k)^2} > 0 \quad (19)$$

where $h = S_i^{k+1} - S_i^k$. In order to obtain Equation (19), there is a constraint limit to attain the convergence level, which is shown in Equation (20):

$$Magnitude\left|J(S_i^{k+1}) - J(S_i^k)\right| < \varepsilon \quad (20)$$

The complete iterative process under the heuristic PDF method is shown using Figure 1.

The load representations for CPLF and CILF are given by Equations (21)–(23):

$$S_{load} = P_L + jQ_L \quad (21)$$

where

$$P_L = P_i\left(\frac{V_o}{V_i}\right)^x \quad (22)$$

$$Q_L = Q_i\left(\frac{V_o}{V_i}\right)^y \quad (23)$$

V_o is rated voltage of a particular bus while x and y are constant load parameter and its values are shown in Table 1.

Table 1. Constant parameter value for CPLF and CILF.

Load Type	x	y
CPLF	0	0
CILF	1.8	1.8

The constant values of x and y are taken for power load and impedance load; that is why they are called constant power load and constant impedance load.

The sensitivity and accuracy can be represented as $\frac{\partial(J)}{\partial(S_i)}$ in pu and $S_i - S_{ref}$ in pu, respectively.

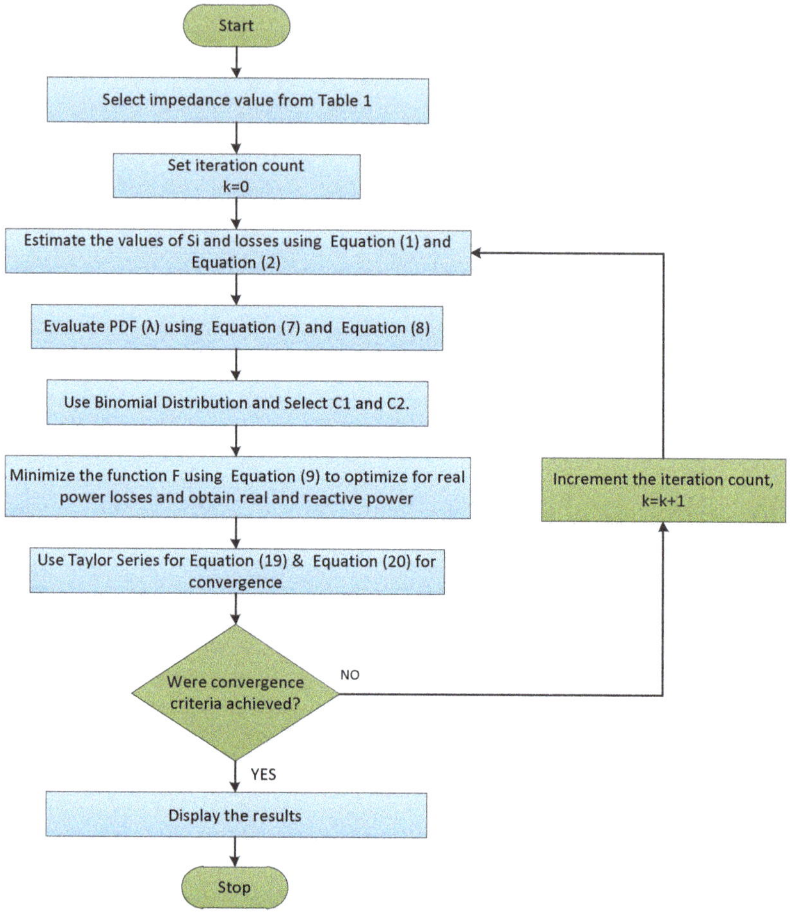

Figure 1. Flowchart of heuristic PDF method.

2.2. Performance Evaluation for Standard IEEE Tested 33 Bus Systems Using Heuristic PDF

The effectiveness of the proposed scheme using the conventional method is being tested initially on the standard IEEE 33 bus test system for CPLF and CILF [36,37]. The typical diagram of the IEEE 33 bus test system is shown in Figure 2 [37].

The IEEE 33 bus test system is modelled in MATLAB with location of fault from 0–90 km. This system consists of 33 buses and 32 lines and has 12.66 kV, 3.715 MW, and 2.3 MVar load size. Thirty percent of the entire load is the size of the source unit used. The source unit voltage is 12.66 kV and the system's lower and upper voltage is set between 0.95 pu and 1.05 pu with standard internal parameters as given in [37]. Bus 1 is considered as the slack bus or main substation. The internal parameters of the IEEE 33 bus test system are shown in [36].

The proposed design illustrated from Equations (1)–(23) is applied on the standard IEEE 33 bus test system for estimation of location of DG using the heuristic PDF method. By using these equations, the parameters such as real power loss, accuracy, and sensitivity are estimated as shown in Table 2. The design part has already been discussed in the previous section.

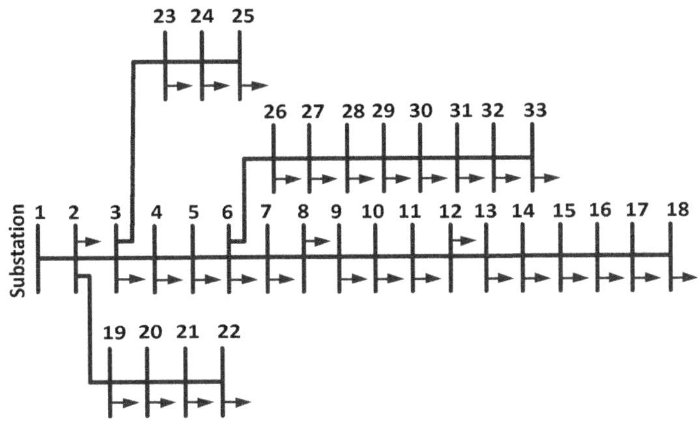

Figure 2. IEEE 33 bus test system.

Table 2. Performance parameter of DG at different locations for the IEEE 33 bus system with the heuristic PDF method for CPLF and CILF.

Bus No.	CPLF			CILF		
	Real Power Losses (pu)	Accuracy (pu)	Sensitivity (pu)	Real Power Losses (pu)	Accuracy (pu)	Sensitivity (pu)
1	0.31	0.031	0.056	0.32	0.071	0.066
2	0.29	0.029	0.047	0.31	0.087	0.065
3	0.27	0.021	0.057	0.29	0.078	0.061
4	0.21	0.019	0.056	0.30	0.079	0.055
5	0.29	0.025	0.049	0.31	0.080	0.061
6	0.28	0.031	0.054	0.27	0.079	0.059
7	0.27	0.049	0.055	0.26	0.080	0.057
8	0.31	0.050	0.061	0.25	0.090	0.056
9	0.32	0.071	0.054	0.30	0.069	0.055
10	0.31	0.072	0.049	0.31	0.091	0.053
11	0.29	0.098	0.095	0.41	0.098	0.052
12	0.31	0.098	0.075	0.29	0.084	0.049
13	0.27	0.084	0.085	0.28	0.072	0.045
14	0.29	0.087	0.084	0.31	0.073	0.044
15	0.31	0.088	0.045	0.33	0.074	0.042
16	0.27	0.086	0.039	0.32	0.078	0.043
17	0.25	0.075	0.039	0.27	0.069	0.041
18	0.27	0.069	0.042	0.29	0.081	0.048
19	0.31	0.058	0.025	0.39	0.079	0.047
20	0.30	0.047	0.024	0.40	0.081	0.046
21	0.27	0.059	0.034	0.28	0.083	0.057
22	0.29	0.098	0.036	0.31	0.087	0.051
23	0.31	0.094	0.054	0.32	0.081	0.050
24	0.27	0.095	0.045	0.33	0.079	0.065
25	0.29	0.089	0.078	0.27	0.074	0.079
26	0.27	0.075	0.084	0.31	0.072	0.065
27	0.31	0.061	0.084	0.30	0.077	0.054
28	0.33	0.072	0.094	0.32	0.076	0.051
29	0.34	0.065	0.018	0.33	0.075	0.053
30	0.35	0.050	0.045	0.31	0.091	0.064
31	0.42	0.064	0.061	0.41	0.094	0.091
32	0.41	0.042	0.082	0.43	0.080	0.085
33	0.39	0.084	0.083	0.39	0.078	0.084

2.3. Genetic Algorithm

GA is used to find the best location of DG in the IEEE 33 bus system. It involves reproduction, crossover, and mutation. Its procedure can be illustrated by the flowchart as shown in Figure 3. The process begins with selection of a binary string as shown by Equation (24), the parameters for which are assumed as follows:

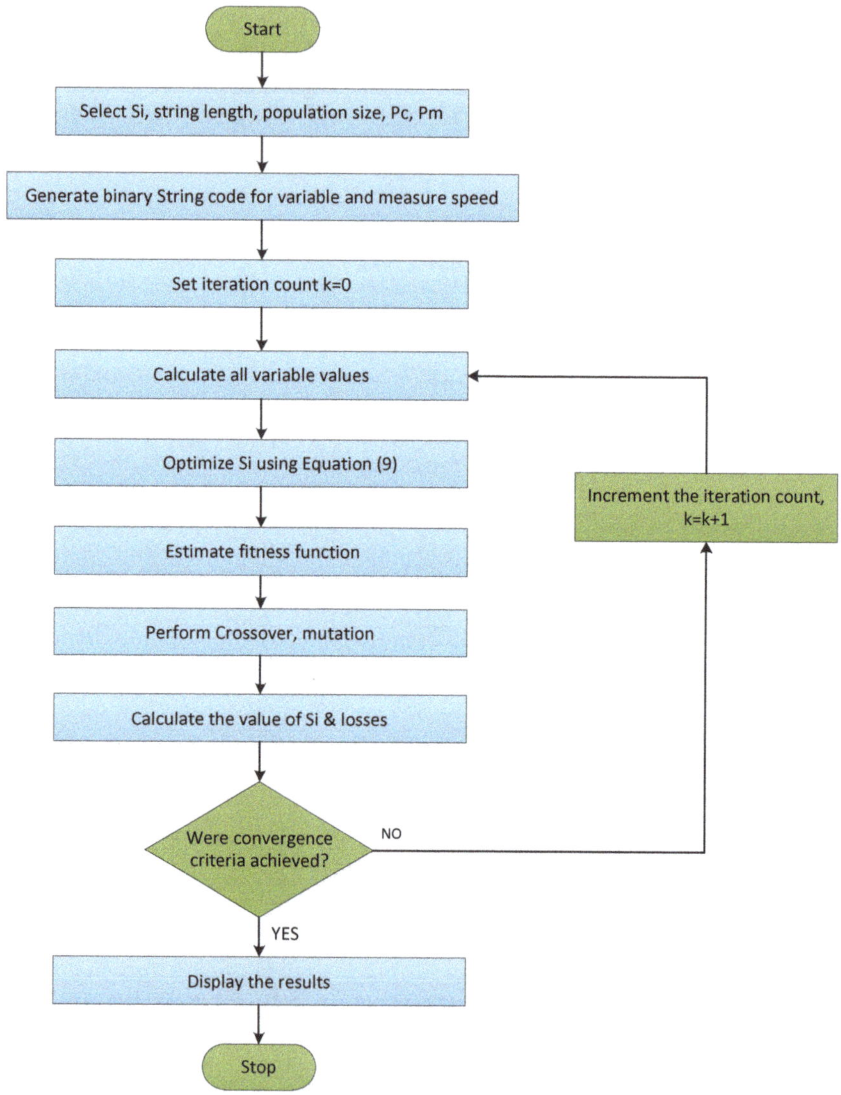

Figure 3. Flowchart representing the GA.

Population size: 6, length of the complete string: 8, crossover probability, $P_c = 0.9$, mutation probability, $P_m = 0.02$.

$$S_i^{\,r} = S_i^{\,\min} + \frac{S_i^{\,\max} - S_i^{\,\min}}{2^l - 1} y^r \tag{24}$$

where r is the iteration count. The application of iteration in Equation (24) results in Equations (25) and (26):

$$f^r = \frac{1}{(1 + \frac{\alpha \varepsilon^r}{S_i^r})} \tag{25}$$

α is step size which is equal to 0.5.

$$\varepsilon^r = S_i^r - S_i^{ref} \tag{26}$$

Figures 4 and 5 show real power loss at bus 17 for CPLF and CILF using GA. The performance parameter for finding the DG location can be analyzed from Tables 2 and 3 at different DG locations in the IEEE 33 bus system using heuristic PDF and GA methods for both CPLF and CILF. Figures 6 and 7 provide the performance parameter comparison of GA with the heuristic method at different buses for CPLF and CILF. Moreover, in order to have comprehensive analysis and comparison for the DG location with GA and heuristic pdf, the results are compared with those obtained with the ant-lion optimization algorithm [42], coyote optimizer [43], modified sine-cosine algorithm [44], and particle swarm optimization [45]. The performance parameter evaluation for the estimation of DG location by using the ant-lion optimization algorithm [42], coyote optimizer [43], modified sine-cosine algorithm [44], and particle swarm optimization [45] is shown in Tables 4–7 for CPLF and CILF type of load.

From Tables 2–7, it is observed that the selection of DGs location is better found under GA for CPLF and CILF in comparison to other methods. By comparing the real power loss, sensitivity, and accuracy, it can be concluded that bus 17 is the best location for the placement of DG for both CPLF and CILF. In the sequence manner, bus 3 and bus 4 are the second and third position of placing the next two DGs. The comparative results for positioning of three DGs with different methods are shown in Table 8.

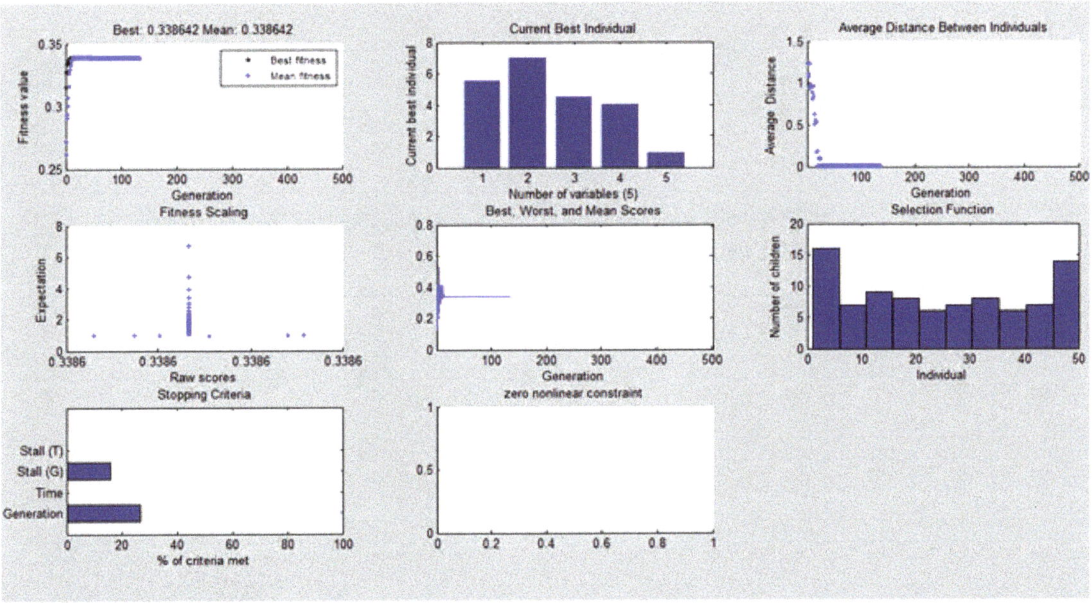

Figure 4. Real power loss at bus 17 after 150 iterations using GA for CPLF.

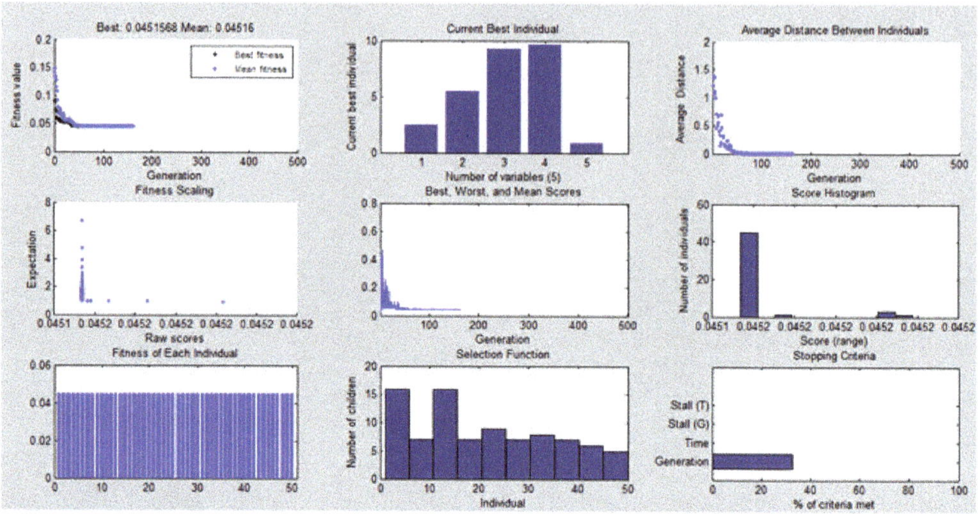

Figure 5. Real power loss at bus 17 after 155 iterations using GA for CILF.

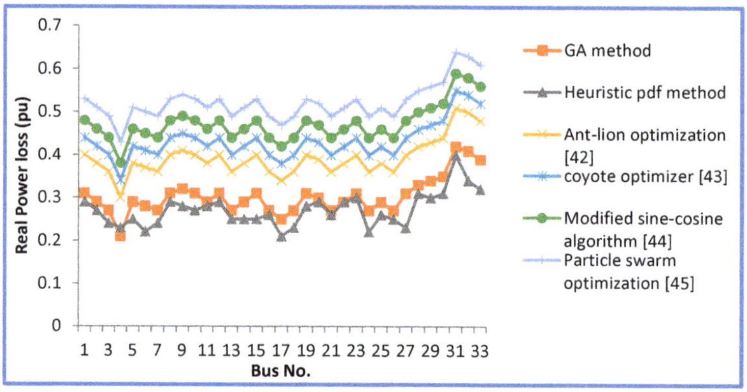

Figure 6. Real power loss comparison for CPLF.

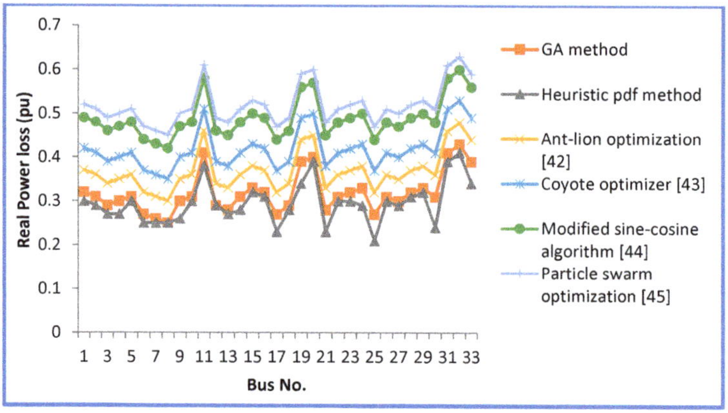

Figure 7. Real power loss comparison for CILF.

Table 3. Performance parameter of DG at different locations for the modified IEEE 33 bus system with GA for CPLF and CILF.

Bus No.	CPLF			CILF		
	Real Power Losses (pu)	Accuracy (pu)	Sensitivity (pu)	Real Power Losses (pu)	Accuracy (pu)	Sensitivity (pu)
1	0.29	0.029	0.052	0.30	0.069	0.061
2	0.27	0.027	0.044	0.29	0.082	0.059
3	0.24	0.019	0.050	0.27	0.076	0.060
4	0.23	0.017	0.051	0.27	0.078	0.052
5	0.25	0.016	0.042	0.30	0.077	0.058
6	0.22	0.024	0.051	0.25	0.078	0.055
7	0.24	0.040	0.050	0.25	0.072	0.049
8	0.29	0.045	0.052	0.25	0.084	0.051
9	0.28	0.055	0.050	0.26	0.061	0.049
10	0.27	0.056	0.048	0.30	0.089	0.050
11	0.28	0.081	0.047	0.38	0.094	0.050
12	0.29	0.089	0.069	0.29	0.082	0.045
13	0.25	0.079	0.079	0.27	0.070	0.044
14	0.25	0.079	0.078	0.28	0.071	0.043
15	0.25	0.079	0.031	0.32	0.072	0.041
16	0.26	0.076	0.037	0.31	0.075	0.041
17	0.21	0.070	0.034	0.23	0.061	0.039
18	0.23	0.065	0.032	0.28	0.070	0.041
19	0.28	0.051	0.021	0.34	0.071	0.045
20	0.29	0.045	0.022	0.39	0.072	0.041
21	0.26	0.052	0.031	0.23	0.077	0.051
22	0.29	0.091	0.035	0.30	0.081	0.047
23	0.30	0.090	0.051	0.30	0.079	0.047
24	0.22	0.091	0.044	0.29	0.071	0.047
25	0.26	0.081	0.071	0.21	0.069	0.074
26	0.25	0.070	0.081	0.30	0.069	0.064
27	0.23	0.089	0.080	0.29	0.071	0.051
28	0.31	0.090	0.079	0.31	0.072	0.049
29	0.30	0.091	0.015	0.32	0.071	0.047
30	0.31	0.081	0.035	0.24	0.085	0.059
31	0.40	0.089	0.051	0.39	0.084	0.082
32	0.34	0.081	0.073	0.41	0.077	0.071
33	0.32	0.082	0.075	0.34	0.071	0.077

Table 4. Performance parameter of DG at different locations for the IEEE 33 bus system with the ant-lion optimization algorithm [42] for CPLF and CILF.

Bus No.	CPLF			CILF		
	Real Power Losses (pu)	Accuracy (pu)	Sensitivity (pu)	Real Power Losses (pu)	Accuracy (pu)	Sensitivity (pu)
1	0.4	0.042	0.064	0.37	0.081	0.075
2	0.38	0.04	0.055	0.36	0.097	0.074
3	0.36	0.032	0.065	0.34	0.088	0.07
4	0.3	0.03	0.064	0.35	0.089	0.064
5	0.38	0.036	0.057	0.36	0.09	0.07
6	0.37	0.042	0.062	0.32	0.089	0.068
7	0.36	0.06	0.063	0.31	0.09	0.066
8	0.4	0.061	0.069	0.3	0.1	0.065
9	0.41	0.082	0.062	0.35	0.079	0.064
10	0.4	0.083	0.057	0.36	0.101	0.062
11	0.38	0.109	0.103	0.46	0.108	0.061
12	0.4	0.109	0.083	0.34	0.094	0.058
13	0.36	0.095	0.093	0.33	0.082	0.054

Table 4. *Cont.*

Bus No.	CPLF			CILF		
	Real Power Losses (pu)	Accuracy (pu)	Sensitivity (pu)	Real Power Losses (pu)	Accuracy (pu)	Sensitivity (pu)
14	0.38	0.098	0.092	0.36	0.083	0.053
15	0.4	0.099	0.053	0.38	0.084	0.051
16	0.36	0.097	0.047	0.37	0.088	0.052
17	0.34	0.086	0.047	0.32	0.079	0.05
18	0.36	0.08	0.05	0.34	0.091	0.057
19	0.4	0.069	0.033	0.44	0.089	0.056
20	0.39	0.058	0.032	0.45	0.091	0.055
21	0.36	0.07	0.042	0.33	0.093	0.066
22	0.38	0.109	0.044	0.36	0.097	0.06
23	0.4	0.105	0.062	0.37	0.091	0.059
24	0.36	0.106	0.053	0.38	0.089	0.074
25	0.38	0.1	0.086	0.32	0.084	0.088
26	0.36	0.086	0.092	0.36	0.082	0.074
27	0.4	0.072	0.092	0.35	0.087	0.063
28	0.42	0.083	0.102	0.37	0.086	0.06
29	0.43	0.076	0.026	0.38	0.085	0.062
30	0.44	0.061	0.053	0.36	0.101	0.073
31	0.51	0.075	0.069	0.46	0.104	0.1
32	0.5	0.053	0.09	0.48	0.09	0.094
33	0.48	0.095	0.091	0.44	0.088	0.093

Table 5. Performance parameter of DG at different locations for the IEEE 33 bus system with the coyote optimizer [43] for CPLF and CILF.

Bus No.	CPLF			CILF		
	Real Power Losses (pu)	Accuracy (pu)	Sensitivity (pu)	Real Power Losses (pu)	Accuracy (pu)	Sensitivity (pu)
1	0.44	0.051	0.068	0.42	0.088	0.082
2	0.42	0.049	0.059	0.41	0.104	0.081
3	0.4	0.041	0.069	0.39	0.095	0.077
4	0.34	0.039	0.068	0.4	0.096	0.071
5	0.42	0.045	0.061	0.41	0.097	0.077
6	0.41	0.051	0.066	0.37	0.096	0.075
7	0.4	0.069	0.067	0.36	0.097	0.073
8	0.44	0.07	0.073	0.35	0.107	0.072
9	0.45	0.091	0.066	0.4	0.086	0.071
10	0.44	0.092	0.061	0.41	0.108	0.069
11	0.42	0.118	0.107	0.51	0.115	0.068
12	0.44	0.118	0.087	0.39	0.101	0.065
13	0.4	0.104	0.097	0.38	0.089	0.061
14	0.42	0.107	0.096	0.41	0.09	0.06
15	0.44	0.108	0.057	0.43	0.091	0.058
16	0.4	0.106	0.051	0.42	0.095	0.059
17	0.38	0.095	0.051	0.37	0.086	0.057
18	0.4	0.089	0.054	0.39	0.098	0.064
19	0.44	0.078	0.037	0.49	0.096	0.063
20	0.43	0.067	0.036	0.5	0.098	0.062
21	0.4	0.079	0.046	0.38	0.1	0.073
22	0.42	0.118	0.048	0.41	0.104	0.067
23	0.44	0.114	0.066	0.42	0.098	0.066
24	0.4	0.115	0.057	0.43	0.096	0.081
25	0.42	0.109	0.09	0.37	0.091	0.095
26	0.4	0.095	0.096	0.41	0.089	0.081
27	0.44	0.081	0.096	0.4	0.094	0.07

Table 5. *Cont.*

Bus No.	CPLF			CILF		
	Real Power Losses (pu)	Accuracy (pu)	Sensitivity (pu)	Real Power Losses (pu)	Accuracy (pu)	Sensitivity (pu)
28	0.46	0.092	0.106	0.42	0.093	0.067
29	0.47	0.085	0.03	0.43	0.092	0.069
30	0.48	0.07	0.057	0.41	0.108	0.08
31	0.55	0.084	0.073	0.51	0.111	0.107
32	0.54	0.062	0.094	0.53	0.097	0.101
33	0.52	0.104	0.095	0.49	0.095	0.1

Table 6. Performance parameter of DG at different locations for the IEEE 33 bus system with the modified sine-cosine algorithm [44] for CPLF and CILF.

Bus No.	CPLF			CILF		
	Real Power Losses (pu)	Accuracy (pu)	Sensitivity (pu)	Real Power Losses (pu)	Accuracy (pu)	Sensitivity (pu)
1	0.48	0.059	0.075	0.49	0.096	0.089
2	0.46	0.057	0.066	0.48	0.112	0.088
3	0.44	0.049	0.076	0.46	0.103	0.084
4	0.38	0.047	0.075	0.47	0.104	0.078
5	0.46	0.053	0.068	0.48	0.105	0.084
6	0.45	0.059	0.073	0.44	0.104	0.082
7	0.44	0.077	0.074	0.43	0.105	0.08
8	0.48	0.078	0.08	0.42	0.115	0.079
9	0.49	0.099	0.073	0.47	0.094	0.078
10	0.48	0.1	0.068	0.48	0.116	0.076
11	0.46	0.126	0.114	0.58	0.123	0.075
12	0.48	0.126	0.094	0.46	0.109	0.072
13	0.44	0.112	0.104	0.45	0.097	0.068
14	0.46	0.115	0.103	0.48	0.098	0.067
15	0.48	0.116	0.064	0.5	0.099	0.065
16	0.44	0.114	0.058	0.49	0.103	0.066
17	0.42	0.103	0.058	0.44	0.094	0.064
18	0.44	0.097	0.061	0.46	0.106	0.071
19	0.48	0.086	0.044	0.56	0.104	0.07
20	0.47	0.075	0.043	0.57	0.106	0.069
21	0.44	0.087	0.053	0.45	0.108	0.08
22	0.46	0.126	0.055	0.48	0.112	0.074
23	0.48	0.122	0.073	0.49	0.106	0.073
24	0.44	0.123	0.064	0.5	0.104	0.088
25	0.46	0.117	0.097	0.44	0.099	0.102
26	0.44	0.103	0.103	0.48	0.097	0.088
27	0.48	0.089	0.103	0.47	0.102	0.077
28	0.50	0.100	0.113	0.49	0.101	0.074
29	0.51	0.093	0.037	0.50	0.1	0.076
30	0.52	0.078	0.064	0.48	0.116	0.087
31	0.59	0.092	0.08	0.58	0.119	0.114
32	0.58	0.07	0.101	0.6	0.105	0.108
33	0.56	0.112	0.102	0.56	0.103	0.107

Table 7. Performance parameter of DG at different locations for the IEEE 33 bus system with the particle swarm optimization [45] for CPLF and CILF.

Bus No.	CPLF			CILF		
	Real Power Losses (pu)	Accuracy (pu)	Sensitivity (pu)	Real Power Losses (pu)	Accuracy (pu)	Sensitivity (pu)
1	0.53	0.062	0.077	0.52	0.099	0.094
2	0.51	0.06	0.068	0.51	0.115	0.093
3	0.49	0.052	0.078	0.49	0.106	0.089
4	0.43	0.05	0.077	0.50	0.107	0.083
5	0.51	0.056	0.07	0.51	0.108	0.089
6	0.50	0.062	0.075	0.47	0.107	0.087
7	0.49	0.08	0.076	0.46	0.108	0.085
8	0.53	0.081	0.082	0.45	0.118	0.084
9	0.54	0.102	0.075	0.50	0.097	0.083
10	0.53	0.103	0.07	0.51	0.119	0.081
11	0.51	0.129	0.116	0.61	0.126	0.08
12	0.53	0.129	0.096	0.49	0.112	0.077
13	0.49	0.115	0.106	0.48	0.1	0.073
14	0.51	0.118	0.105	0.51	0.101	0.072
15	0.53	0.119	0.066	0.53	0.102	0.07
16	0.49	0.117	0.06	0.52	0.106	0.071
17	0.47	0.106	0.06	0.47	0.097	0.069
18	0.49	0.1	0.063	0.49	0.109	0.076
19	0.53	0.089	0.046	0.59	0.107	0.075
20	0.52	0.078	0.045	0.6	0.109	0.074
21	0.49	0.09	0.055	0.48	0.111	0.085
22	0.51	0.129	0.057	0.51	0.115	0.079
23	0.53	0.125	0.075	0.52	0.109	0.078
24	0.49	0.126	0.066	0.53	0.107	0.093
25	0.51	0.12	0.099	0.47	0.102	0.107
26	0.49	0.106	0.105	0.51	0.1	0.093
27	0.53	0.092	0.105	0.5	0.105	0.082
28	0.55	0.103	0.115	0.52	0.104	0.079
29	0.56	0.096	0.039	0.53	0.103	0.081
30	0.57	0.081	0.066	0.51	0.119	0.092
31	0.64	0.095	0.082	0.61	0.122	0.119
32	0.63	0.073	0.103	0.63	0.108	0.113
33	0.61	0.115	0.104	0.59	0.106	0.112

Table 8. Performance comparison among different methods.

Technique	Bus No.	CPLF			CILF		
		Real Power Loss (pu)	Accuracy (pu)	Sensitivity (pu)	Real Power Loss (pu)	Accuracy (pu)	Sensitivity (pu)
Heuristic method	17	0.25	0.075	0.039	0.27	0.069	0.041
	3	0.27	0.021	0.057	0.29	0.078	0.061
	4	0.21	0.019	0.056	0.30	0.079	0.055
Genetic Algorithm	17	0.21	0.070	0.034	0.23	0.061	0.039
	3	0.24	0.019	0.050	0.27	0.076	0.060
	4	0.23	0.017	0.051	0.27	0.078	0.052
Ant-lion optimization algorithm [42]	17	0.34	0.086	0.047	0.32	0.079	0.05
	3	0.36	0.032	0.065	0.34	0.088	0.07
	4	0.3	0.03	0.064	0.35	0.089	0.064
Coyote optimizer [43]	17	0.38	0.095	0.051	0.37	0.086	0.057
	3	0.4	0.041	0.069	0.39	0.095	0.077
	4	0.34	0.039	0.068	0.4	0.096	0.071

Table 8. *Cont.*

Technique	Bus No.	CPLF			CILF		
		Real Power Loss (pu)	Accuracy (pu)	Sensitivity (pu)	Real Power Loss (pu)	Accuracy (pu)	Sensitivity (pu)
Modified sine-cosine algorithm [44]	17	0.42	0.103	0.058	0.44	0.094	0.064
	3	0.44	0.049	0.076	0.46	0.103	0.084
	4	0.38	0.047	0.075	0.47	0.104	0.078
Particle Swarm optimization [45]	17	0.47	0.106	0.06	0.47	0.097	0.069
	3	0.49	0.052	0.078	0.49	0.106	0.089
	4	0.43	0.05	0.077	0.50	0.107	0.083

2.4. Optimal Sizing of DG

After deciding the optimal location of three DGs, it is also required to check the DG size to confirm its placement at bus 17. DG size is estimated from Tables 9–11 by using heuristic pdf, GA, ant-lion optimization algorithm [42], coyote optimizer [43], modified sine-cosine algorithm [44], and particle swarm optimization [45] for both CPLF and CILF type of load. It is observed that the size of DG when placed at bus 17 is found to be the minimum. The relative comparison of DG size for bus 17, 4, and 3 is shown in Table 12. The minimal size of DG is obtained under GA as compared to the heuristic method. Figures 8–10 illustrate the real power and Figure 11 illustrates the reactive power at bus 17 for both CPLF and CILF. From the results, it is confirmed that bus 17 is the best choice for the location and optimum sizing of DG is also obtained here. Subsequently, bus 3 and bus 4 have the appropriate size of DGs, which assures the confirmation of the DG location. The location and size of three DGs at the allotted location in the IEEE 33 bus system are shown in Figure 12.

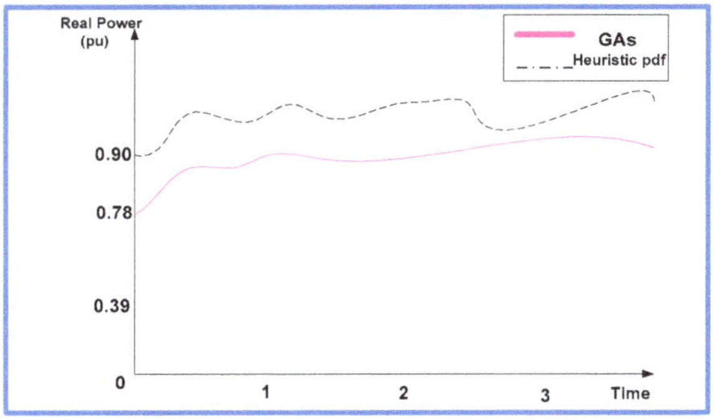

Figure 8. Real power at bus 17 after150 iterations for CPLF.

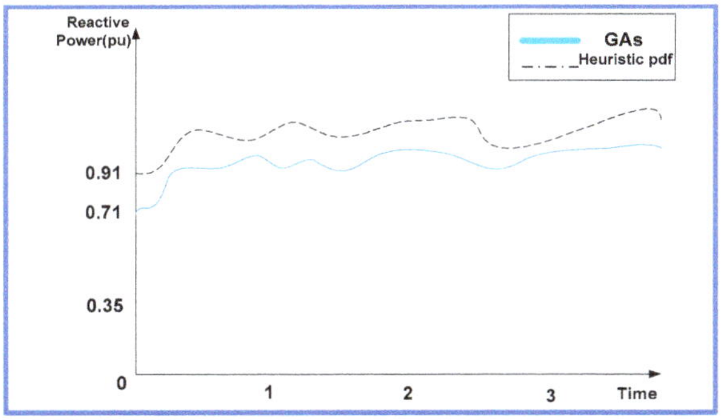

Figure 9. Reactive power at bus 17 after 150 iterations using GA for CPLF.

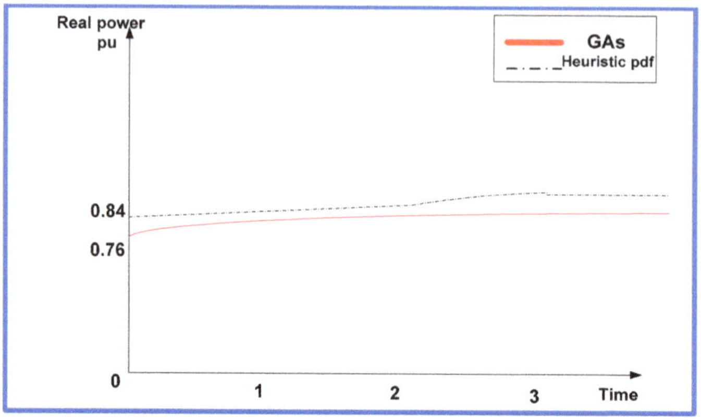

Figure 10. Real power at bus 17 after 155 iterations using GA for CILF.

Table 9. DG size comparison for CPLF and CILF using GA and heuristic PDF.

Bus No.	CPLF				CILF			
	Heuristic PDF		GA		Heuristic PDF		GA	
	P (pu)	Q (pu)	P (pu)	Q (pu)	P (pu)	Q (pu)	P (pu)	Q (pu)
1	1.01	0.98	1.02	0.99	1.09	1.02	0.99	1.02
2	1.02	0.94	1.10	0.96	1.03	0.99	1.01	0.99
3	1.05	0.85	0.85	0.78	1.10	1.01	0.90	0.89
4	1.23	0.99	0.91	0.81	1.09	0.95	0.81	0.91
5	1.12	1.01	1.31	0.75	1.11	0.99	1.29	0.76
6	1.03	0.94	1.05	0.73	1.05	1.02	1.01	0.85
7	1.04	0.96	1.06	0.74	1.07	0.95	1.10	0.72
8	0.91	0.95	0.95	0.91	1.02	0.97	0.99	0.84
9	0.89	0.81	0.97	0.82	0.95	0.95	0.94	0.92
10	1.02	0.95	1.03	0.76	0.98	0.92	1.01	0.96
11	0.99	1.00	0.99	1.02	1.02	1.05	0.98	1.01
12	0.98	1.01	0.99	1.03	0.99	1.03	0.92	1.01
13	1.89	1.03	1.95	1.04	1.71	1.02	1.84	1.00
14	0.85	1.05	0.85	1.05	0.97	1.05	1.12	1.01
15	0.94	1.09	0.94	1.10	1.00	0.99	0.99	1.10
16	0.94	1.11	0.98	1.01	0.99	1.02	0.94	1.08
17	0.89	0.90	0.78	0.71	0.84	0.98	0.76	0.70
18	0.95	1.13	0.96	1.04	1.05	1.10	0.99	1.01
19	0.94	1.84	0.99	1.88	0.92	1.21	0.94	1.08
20	1.00	1.24	1.07	1.44	1.10	1.12	1.02	1.40
21	1.00	1.11	1.08	1.21	1.09	1.01	1.02	1.01
22	0.95	1.24	1.98	1.34	1.01	1.10	1.91	1.21
23	0.80	1.00	1.89	1.44	1.02	1.08	1.21	1.01
24	0.81	0.99	1.85	1.00	0.95	1.01	1.01	1.09
25	0.79	1.20	1.84	1.22	0.89	1.12	1.00	1.12
26	0.84	1.21	1.85	1.23	0.94	1.12	1.02	1.11
27	0.99	0.99	2.00	1.02	0.96	0.96	1.01	1.10
28	0.98	0.84	1.75	0.86	1.03	1.01	1.05	0.95
29	0.99	0.95	1.82	0.98	1.21	0.99	1.51	0.97
30	1.33	0.99	1.72	0.99	1.23	0.92	1.10	0.94
31	0.99	1.00	1.81	1.01	1.42	1.01	1.02	1.08
32	1.20	1.01	1.51	1.05	1.47	1.11	1.01	1.06
33	1.05	1.12	1.31	1.15	1.25	1.10	1.01	1.01

Table 10. DG size comparison for CPLF and CILF using the ant-lion optimization algorithm [42] and coyote optimizer [43].

Bus No.	CPLF				CILF			
	Ant-Lion Optimization Algorithm [42]		Coyote Optimizer [43]		Ant-Lion Optimization Algorithm [42]		Coyote Optimizer [43]	
	P (pu)	Q (pu)	P (pu)	Q (pu)	P (pu)	Q (pu)	P (pu)	Q (pu)
1	1.1	1.04	1.08	1.07	1.18	1.11	1.07	1.09
2	1.11	1	1.16	1.04	1.12	1.08	1.09	1.06
3	1.14	0.91	0.91	0.86	1.19	1.1	0.98	0.96
4	1.32	1.05	0.97	0.89	1.18	1.04	0.89	0.98
5	1.21	1.07	1.37	0.83	1.2	1.08	1.37	0.83
6	1.12	1	1.11	0.81	1.14	1.11	1.09	0.92
7	1.13	1.02	1.12	0.82	1.16	1.04	1.18	0.79
8	1	1.01	1.01	0.99	1.11	1.06	1.07	0.91
9	0.98	0.87	1.03	0.9	1.04	1.04	1.02	0.99
10	1.11	1.01	1.09	0.84	1.07	1.01	1.09	1.03
11	1.08	1.06	1.05	1.1	1.11	1.14	1.06	1.08
12	1.07	1.07	1.05	1.11	1.08	1.12	1	1.08
13	1.98	1.09	2.01	1.12	1.8	1.11	1.92	1.07
14	0.94	1.11	0.91	1.13	1.06	1.14	1.2	1.08
15	1.03	1.15	1	1.18	1.09	1.08	1.07	1.17
16	1.03	1.17	1.04	1.09	1.08	1.11	1.02	1.15
17	0.98	0.96	0.84	0.79	0.93	1.07	0.84	0.77
18	1.04	1.19	1.02	1.12	1.14	1.19	1.07	1.08
19	1.03	1.9	1.05	1.96	1.01	1.3	1.02	1.15
20	1.09	1.3	1.13	1.52	1.19	1.21	1.1	1.47
21	1.09	1.17	1.14	1.29	1.18	1.1	1.1	1.08
22	1.04	1.3	2.04	1.42	1.1	1.19	1.99	1.28
23	0.89	1.06	1.95	1.52	1.11	1.17	1.29	1.08
24	0.9	1.05	1.91	1.08	1.04	1.1	1.09	1.16
25	0.88	1.26	1.9	1.3	0.98	1.21	1.08	1.19
26	0.93	1.27	1.91	1.31	1.03	1.21	1.1	1.18
27	1.08	1.05	2.06	1.1	1.05	1.05	1.09	1.17
28	1.07	0.9	1.81	0.94	1.12	1.1	1.13	1.02
29	1.08	1.01	1.88	1.06	1.3	1.08	1.59	1.04
30	1.42	1.05	1.78	1.07	1.32	1.01	1.18	1.01
31	1.08	1.06	1.87	1.09	1.51	1.1	1.1	1.15
32	1.29	1.07	1.57	1.13	1.56	1.2	1.09	1.13
33	1.14	1.18	1.37	1.23	1.34	1.19	1.09	1.08

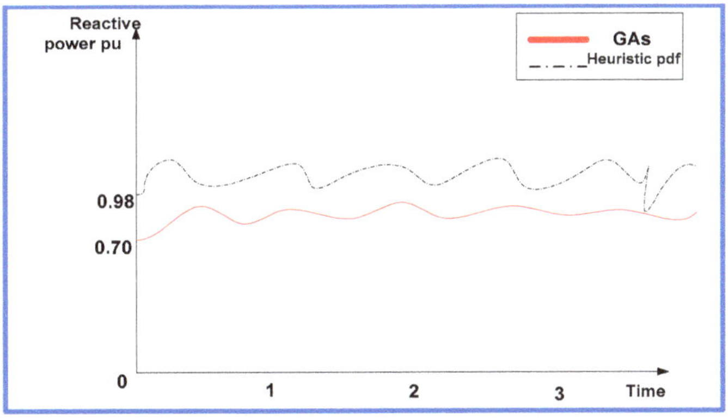

Figure 11. Reactive power at bus 17 after 155 iterations using GA for CILF.

Table 11. DG size comparison for CPLF and CILF using the modified sine-cosine algorithm [44] and particle swarm optimization [45].

Bus No.	CPLF				CILF			
	Modified Sine-Cosine Algorithm [44]		Particle Swarm Optimization [45]		Modified Sine-Cosine Algorithm [44]		Particle Swarm Optimization [45]	
	P (pu)	Q (pu)	P (pu)	Q (pu)	P (pu)	Q (pu)	P (pu)	Q (pu)
1	1.19	1.1	1.14	1.15	1.27	1.2	1.15	1.16
2	1.2	1.06	1.22	1.12	1.21	1.17	1.17	1.13
3	1.23	0.97	0.97	0.94	1.28	1.19	1.06	1.03
4	1.41	1.11	1.03	0.97	1.27	1.13	0.97	1.05
5	1.3	1.13	1.43	0.91	1.29	1.17	1.45	0.9
6	1.21	1.06	1.17	0.89	1.23	1.2	1.17	0.99
7	1.22	1.08	1.18	0.9	1.25	1.13	1.26	0.86
8	1.09	1.07	1.07	1.07	1.2	1.15	1.15	0.98
9	1.07	0.93	1.09	0.98	1.13	1.13	1.1	1.06
10	1.2	1.07	1.15	0.92	1.16	1.1	1.17	1.1
11	1.17	1.12	1.11	1.18	1.2	1.23	1.14	1.15
12	1.16	1.13	1.11	1.19	1.17	1.21	1.08	1.15
13	2.07	1.15	2.07	1.2	1.89	1.2	2	1.14
14	1.03	1.17	0.97	1.21	1.15	1.23	1.28	1.15
15	1.12	1.21	1.06	1.26	1.18	1.17	1.15	1.24
16	1.12	1.23	1.1	1.17	1.17	1.2	1.1	1.22
17	1.07	1.02	0.9	0.87	1.02	1.16	0.92	0.84
18	1.13	1.25	1.08	1.2	1.23	1.28	1.15	1.15
19	1.12	1.96	1.11	2.04	1.1	1.39	1.1	1.22
20	1.18	1.36	1.19	1.6	1.28	1.3	1.18	1.54
21	1.18	1.23	1.2	1.37	1.27	1.19	1.18	1.15
22	1.13	1.36	2.1	1.5	1.19	1.28	2.07	1.35
23	0.98	1.12	2.01	1.6	1.2	1.26	1.37	1.15
24	0.99	1.11	1.97	1.16	1.13	1.19	1.17	1.23
25	0.97	1.32	1.96	1.38	1.07	1.3	1.16	1.26
26	1.02	1.33	1.97	1.39	1.12	1.3	1.18	1.25
27	1.17	1.11	2.12	1.18	1.14	1.14	1.17	1.24
28	1.16	0.96	1.87	1.02	1.21	1.19	1.21	1.09
29	1.17	1.07	1.94	1.14	1.39	1.17	1.67	1.11
30	1.51	1.11	1.84	1.15	1.41	1.1	1.26	1.08
31	1.17	1.12	1.93	1.17	1.6	1.19	1.18	1.22
32	1.38	1.13	1.63	1.21	1.65	1.29	1.17	1.2
33	1.23	1.24	1.43	1.31	1.43	1.28	1.17	1.15

Table 12. DG size comparison at the three best bus locations.

Load Type	Bus No.	Heuristic Method		Genetic Algorithm		Ant-Lion Optimization Algorithm [42]		Coyote Optimizer [43]		Modified Sine-Cosine Algorithm [44]		Particle Swarm Optimization [45]	
		P (pu)	Q (pu)	P (pu)	Q (pu)	P (pu)	Q (pu)	P (pu)	Q (pu)	P (pu)	Q (pu)	P (pu)	Q (pu)
CPLF	17	0.89	0.9	0.78	0.71	0.98	0.96	0.84	0.79	0.98	0.96	0.84	0.79
CILF		0.84	0.98	0.76	0.7	0.93	1.07	0.84	0.77	0.93	1.07	0.84	0.77
CPLF	3	1.05	0.85	0.85	0.78	1.14	0.91	0.91	0.86	1.14	0.91	0.91	0.86
CILF		1.1	1.01	0.9	0.89	1.19	1.1	0.98	0.96	1.19	1.1	0.98	0.96
CPLF	4	1.23	0.99	0.91	0.81	1.32	1.05	0.97	0.89	1.32	1.05	0.97	0.89
CILF		1.09	0.95	0.81	0.91	1.18	1.04	0.89	0.98	1.18	1.04	0.89	0.98

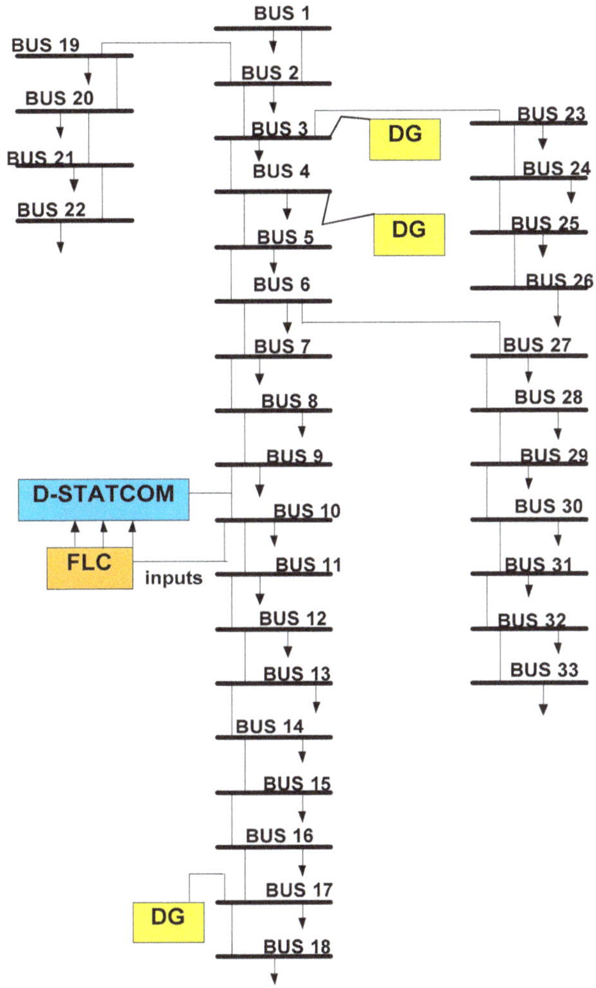

Figure 12. Layout of the IEEE 33 bus system with allocated positioning and sizing of DG and D-STATCOM.

3. Positioning and Impact of D-STATCOM

The positioning and impact of D-STATCOM in the IEEE-33 bus system have to be assessed for improving the power quality performance parameters such as THD and voltage profile. The rating of D-STATCOM is 1 pu reactive power. In order to improve power quality performance, control of D-STATCOM is decided by using the fuzzy logic controller. In this article, only one D-STATCOM is used, so it has to be connected at that bus where large distortions are present. In order to define the inputs and output of FLC, Equation (9) is further used. Now, we differentiate Equation (9) to obtain the optimal solution.

$$\text{Let's define } E = S_i - S_{ref} \tag{27}$$

$$\frac{d(J)}{dE} = 0 \tag{28}$$

Equation (27) enables us to differentiate J with respect to E as given in Equation (28). After differentiating, Equation (29) is obtained:

$$e^{-\frac{|E|^2}{2\sigma^2}} = \frac{S_{rated}\sigma\sqrt{2\pi}}{\left|C_1S_i + C_2S_{ref}\right|2(S_i - S_{load})} \tag{29}$$

By arranging the terms in Equation (29) and by taking log on both sides, the value of E is obtained as given by Equation (30).

$$E = \sqrt{2\sigma^2 \log(\frac{\left|C_1S_i + C_2S_{ref}\right|2(S_i - S_{load})}{\left|S_{rated}\sigma\sqrt{2\pi}\right|})} \tag{30}$$

Equation (30) gives the optimal solution. Now, the inputs to FLC are error (E) and change in error (ΔE) which are shown in Equation (31) as:

$$\Delta E = E(t) - E(t-1) \tag{31}$$

The detailed structure of the D-STATCOM-based FLC controller is shown in Figures 13 and 14. In Figure 13, distorted real and reactive power is measured and passes through the PID controller which generates the measured complex power. The measured complex power is compared with its reference value which produces error. The error and derivative of error act as inputs to FLC. The actual design of FLC for D-STATCOM switching is shown in Figure 14. The output of FLC is reference voltage which is compared with reference value and generates pulses for switching the D-STATCOM. In the same pattern, ΔE can be expressed in standard form.

In order to design rules of the fuzzy set, a 7 × 7 matrix is taken so that the model will have more precise and better results. This means both inputs have seven membership functions. Using Cramer's product rule, FLC rule is designed as shown in Figure 15 and its surface view is shown in Figure 16. Tables 13 and 14 show the notification of membership function for first input, as initial error E is around 0.12 and mapping of input with output, respectively.

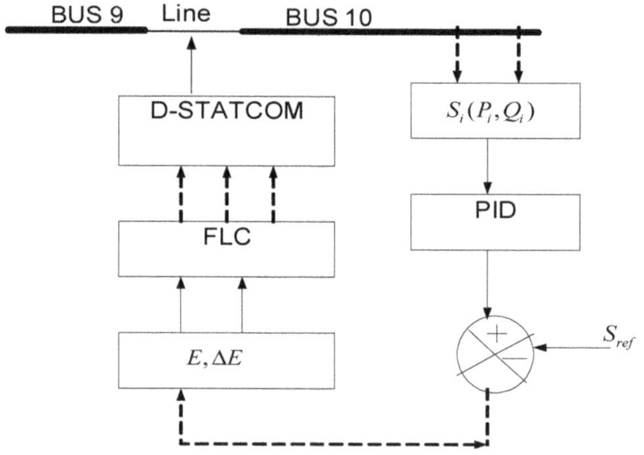

Figure 13. Structure of D-STATCOM-based FLC controller.

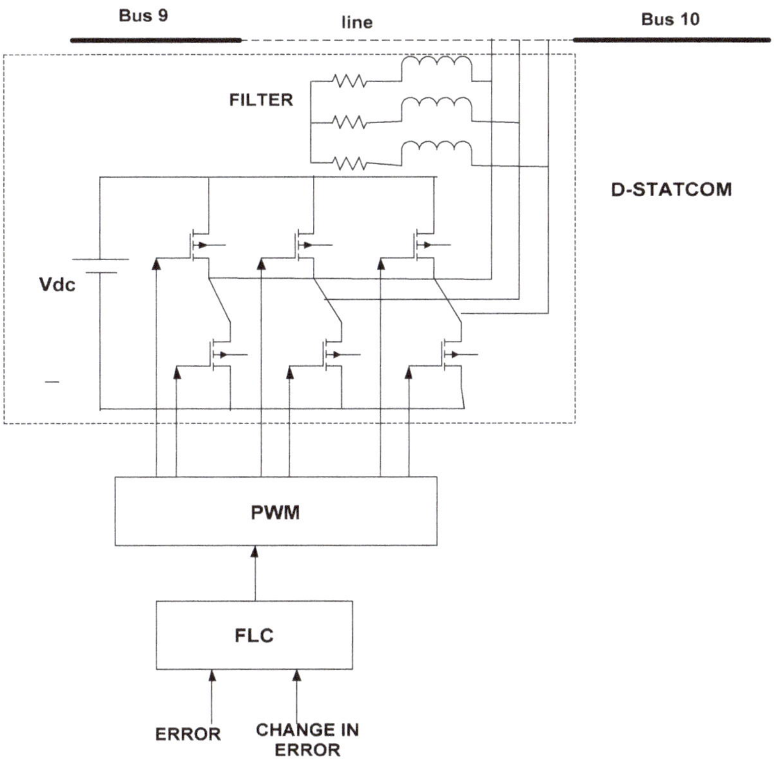

Figure 14. Switching of D-STATCOM.

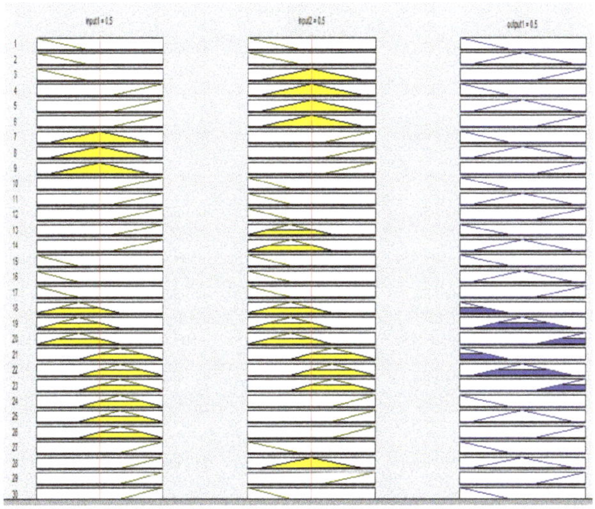

Figure 15. Fuzzy rules of FLC.

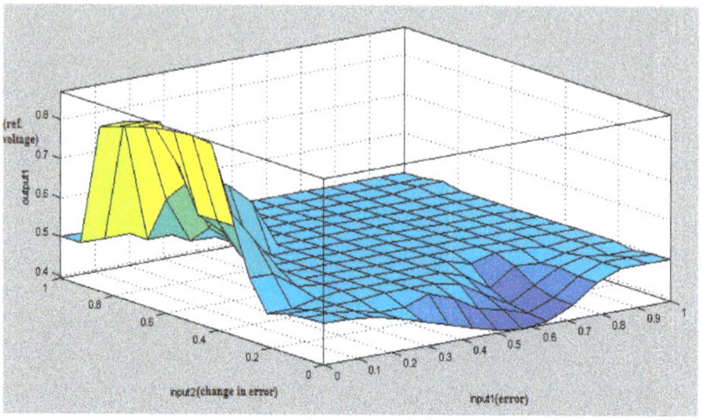

Figure 16. Surface view of FLC between inputs and output.

Table 13. Mathematical notation of input and output.

E/Δ*E*		**NB**	**NM**	**NS**	**ZS**	**PS**	**PM**	**PB**
		0.15	**0.40**	**0.65**	**0.9**	**0.65**	**0.40**	**0.15**
NB	0.25	0.25	0.40	0.65	0.9	0.65	0.40	0.25
NM	0.50	0.50	0.50	0.65	0.9	0.65	0.50	0.50
NS	0.75	0.75	0.75	0.75	0.9	0.75	0.75	0.75
ZS	1	1	1	1	1	1	1	1
PS	0.75	0.75	0.75	0.75	0.9	0.75	0.75	0.75
PM	0.50	0.50	0.50	0.65	0.9	0.65	0.50	0.50
PB	0.25	0.25	0.40	0.65	0.9	0.65	0.40	0.25

Table 14. Mapping of output with input.

E/Δ*E*	**NB**	**NM**	**NS**	**ZS**	**PS**	**PM**	**PB**
NB	NM	NS	ZS	PS	PM	PB	PM
NB	NB	NM	NS	ZS	PS	PM	PB
NM	NM	NM	NS	ZS	PS	NM	NM
NS	NS	NS	NS	ZS	NS	NS	NS
ZS	ZS	ZS	ZS	ZS	ZS	ZS	ZS
PS	PS	PS	PS	ZS	PS	PS	PS
PM	PM	PM	NS	ZS	PS	PM	PM

From Tables 13 and 14, it can be demonstrated that, under certain parameters, a precise control of objectives can be achieved using fuzzy observations, mapping, and control, if the observations become sufficiently accurate as the goal is approached.

The seven membership functions corresponding to the first input are shown in Equation (32) and these seven membership functions are taken from x_1 to x_7. The seven membership functions are chosen in order to have better and more precise results. The nature of the membership function is triangular and input is split into seven membership functions. Each membership function from x_1 to x_7 is assigned with values in the numerator as shown in Equation (32).

$$E = \frac{0.03}{x_1} + \frac{0.06}{x_2} + \frac{0.09}{x_3} + \frac{0.12}{x_4} + \frac{0.09}{x_5} + \frac{0.06}{x_6} + \frac{0.03}{x_7} \tag{32}$$

Dividing Equation (32) by 0.12, Equation (33) is obtained as:

$$E = \frac{0.25}{x_1} + \frac{0.50}{x_2} + \frac{0.75}{x_3} + \frac{1}{x_4} + \frac{0.75}{x_5} + \frac{0.50}{x_6} + \frac{0.25}{x_7} \tag{33}$$

Notification of membership function for the second input is given in Equation (34):

$$\Delta E = \frac{0.15}{x_1} + \frac{0.40}{x_2} + \frac{0.65}{x_3} + \frac{0.9}{x_4} + \frac{0.65}{x_5} + \frac{0.40}{x_6} + \frac{0.15}{x_7} \tag{34}$$

From the Cartesian product rule:

$$A = EB = \Delta E,\ A\ U\ B = \max\ (A\ or\ B)$$

Further membership function is taken as:

$$x_1 = NB, x_2 = NM, x_3 = NS, x_4 = ZS, x_5 = PS, x_6 = PM, x_7 = PB$$

Table 15 gives the comparative analysis of voltage profile at different buses with different existing techniques and D-STATCOM-based FLC for both CPLF and CILF load; whereas computational expression of THD is given by Equation (35).

Table 16 gives the comparative analysis of THD (%) of real power different buses with different existing techniques and D-STATCOM-based FLC for both CPLF and CILF load.

$$THD = \sqrt{\frac{1}{g^2} - 1} \tag{35}$$

where 'g' is distortion factor which is defined as ratio of rms fundamental harmonic value to rms value of voltage.

Table 15. Voltage (pu) comparisons at different buses with D-STATCOM-based FLC and existing methods.

Bus No.	Load Type	D-STATCOM-Based FLC	Harmony Search Algorithm [4]	Ant Colony Algorithm [9]	Big-Bang–Crunch Method [30]
1	CPLF	0.86	0.81	0.78	0.74
	CILF	0.87	0.82	0.79	0.75
2	CPLF	0.86	0.81	0.78	0.74
	CILF	0.85	0.8	0.79	0.75
3	CPLF	0.87	0.82	0.81	0.77
	CILF	0.86	0.81	0.8	0.76
4	CPLF	0.85	0.8	0.78	0.74
	CILF	0.87	0.82	0.8	0.76
5	CPLF	0.86	0.81	0.77	0.73
	CILF	0.85	0.8	0.79	0.75
6	CPLF	0.85	0.8	0.79	0.75
	CILF	0.9	0.85	0.83	0.79
7	CPLF	0.9	0.85	0.83	0.79
	CILF	0.91	0.86	0.84	0.8
8	CPLF	0.91	0.86	0.83	0.79
	CILF	0.91	0.86	0.82	0.78
9	CPLF	0.91	0.86	0.84	0.8
	CILF	0.91	0.86	0.84	0.8
10	CPLF	1	0.98	0.97	0.93
	CILF	1.01	0.99	0.97	0.93
11	CPLF	0.9	0.85	0.82	0.78
	CILF	0.92	0.87	0.85	0.81
12	CPLF	0.76	0.71	0.87	0.83
	CILF	0.79	0.74	0.72	0.68

Table 15. *Cont.*

Bus No.	Load Type	D-STATCOM-Based FLC	Harmony Search Algorithm [4]	Ant Colony Algorithm [9]	Big-Bang–Crunch Method [30]
13	CPLF	0.8	0.75	0.73	0.69
	CILF	0.81	0.76	0.74	0.7
14	CPLF	0.79	0.74	0.72	0.68
	CILF	0.8	0.75	0.74	0.7
15	CPLF	0.8	0.75	0.73	0.69
	CILF	0.74	0.69	0.65	0.61
16	CPLF	0.89	0.84	0.81	0.77
	CILF	0.9	0.85	0.83	0.79
17	CPLF	0.99	0.97	0.94	0.9
	CILF	0.98	0.95	0.9	0.86
18	CPLF	0.95	0.9	0.89	0.85
	CILF	0.93	0.88	0.86	0.81
19	CPLF	0.97	0.92	0.91	0.87
	CILF	0.95	0.9	0.87	0.83
20	CPLF	0.96	0.91	0.89	0.85
	CILF	0.9	0.85	0.83	0.79
21	CPLF	0.91	0.86	0.84	0.8
	CILF	0.93	0.88	0.81	0.77
22	CPLF	0.94	0.89	0.84	0.8
	CILF	0.76	0.71	0.7	0.66
23	CPLF	0.75	0.7	0.69	0.65
	CILF	0.74	0.69	0.68	0.64
24	CPLF	0.73	0.68	0.67	0.63
	CILF	0.72	0.67	0.66	0.62
25	CPLF	0.71	0.66	0.64	0.6
	CILF	0.7	0.65	0.64	0.6
26	CPLF	0.71	0.66	0.64	0.6
	CILF	0.69	0.64	0.63	0.59
27	CPLF	0.7	0.65	0.62	0.58
	CILF	0.71	0.66	0.64	0.6
28	CPLF	0.72	0.67	0.65	0.61
	CILF	0.72	0.67	0.64	0.6
29	CPLF	0.72	0.67	0.65	0.61
	CILF	0.75	0.7	0.69	0.65
30	CPLF	0.76	0.71	0.7	0.66
	CILF	0.76	0.71	0.7	0.66
31	CPLF	0.81	0.76	0.74	0.7
	CILF	0.77	0.72	0.71	0.67
32	CPLF	0.78	0.73	0.72	0.68
	CILF	0.79	0.74	0.73	0.69
33	CPLF	0.8	0.75	0.74	0.7
	CILF	0.78	0.73	0.72	0.68

Table 16. THD (%) of real power comparison at different buses with D-STATCOM-based FLC and existing methods.

Bus No.	Load Type	D-STATCOM-Based FLC	Harmony Search Algorithm [4]	Ant Colony Algorithm [9]	Big-Bang–Crunch Method [30]
1	CPLF	6.81	7.81	8.21	8.82
	CILF	6.82	7.82	7.91	8.52
2	CPLF	7.81	8.81	9.12	9.73
	CILF	6.8	7.8	7.91	8.52
3	CPLF	7.71	8.71	9.11	9.72
	CILF	6.81	7.81	8.81	9.42
4	CPLF	6.8	7.8	9.11	9.72
	CILF	6.82	7.82	8.11	8.72
5	CPLF	6.81	7.81	8.1	8.71
	CILF	8.9	9.9	9.21	9.82
6	CPLF	6.8	7.8	9.17	9.78
	CILF	8.5	8.5	9.12	9.73
7	CPLF	8.4	8.4	9.14	9.75
	CILF	6.86	7.86	8.98	9.59
8	CPLF	6.86	7.86	9.12	9.73
	CILF	6.86	7.86	8.65	9.26
9	CPLF	7.5	8.5	8.88	9.49
	CILF	7.4	8.4	8.78	9.39
10	CPLF	5.2	9.12	9.89	10.5
	CILF	6.3	9.65	9.65	10.26
11	CPLF	7.6	8.6	8.89	9.5
	CILF	7.5	8.5	8.92	9.53
12	CPLF	6.71	7.71	7.87	8.48
	CILF	7.1	8.1	8.55	9.16
13	CPLF	7.6	8.6	8.77	9.38
	CILF	7.7	8.7	8.88	9.49
14	CPLF	7.1	8.1	8.9	9.51
	CILF	7.5	8.5	8.87	9.48
15	CPLF	6.75	7.75	7.87	8.48
	CILF	6.69	7.69	7.99	8.6
16	CPLF	6.84	7.84	7.98	8.59
	CILF	6.85	7.85	8.02	8.63
17	CPLF	6.2	7.25	9.02	9.63
	CILF	7.3	7.95	8.22	8.83
18	CPLF	6.9	7.9	7.99	8.6
	CILF	6.91	7.91	7.95	8.56
19	CPLF	6.92	7.92	8.25	8.86
	CILF	6.9	7.9	8.65	9.26
20	CPLF	6.91	7.91	8.14	8.75
	CILF	6.85	7.85	9.02	9.63
21	CPLF	6.86	7.86	9.22	9.83
	CILF	6.88	7.88	8.99	9.6
22	CPLF	6.89	7.89	8.55	9.16
	CILF	6.71	7.71	8.42	9.03
23	CPLF	6.7	7.7	8.11	8.72
	CILF	6.69	7.69	7.87	8.48

Table 16. *Cont.*

Bus No.	Load Type	D-STATCOM-Based FLC	Harmony Search Algorithm [4]	Ant Colony Algorithm [9]	Big-Bang–Crunch Method [30]
24	CPLF	6.68	7.68	7.98	8.59
	CILF	6.67	7.67	7.84	8.45
25	CPLF	6.66	7.66	7.92	8.53
	CILF	6.65	7.65	7.83	8.44
26	CPLF	6.66	7.66	7.8	8.41
	CILF	6.64	7.64	7.75	8.36
27	CPLF	6.65	7.65	7.79	8.4
	CILF	6.66	7.66	7.85	8.46
28	CPLF	6.67	7.67	7.87	8.48
	CILF	6.67	7.67	7.9	8.51
29	CPLF	6.67	7.67	7.75	8.36
	CILF	6.7	7.7	7.98	8.59
30	CPLF	6.69	7.69	7.87	8.48
	CILF	6.61	7.61	7.9	8.51
31	CPLF	6.66	7.66	7.96	8.57
	CILF	6.72	7.72	7.92	8.53
32	CPLF	6.73	7.73	7.93	8.54
	CILF	6.65	7.65	7.95	8.56
33	CPLF	6.5	7.5	7.84	8.45
	CILF	6.73	7.73	7.83	8.44

Now, a suitable location for D-STATCOM at a particular bus has to be decided. From Figures 9 and 10, it can be inferred that voltage profile and THD (%) of the real power at bus 10 are found to be worst with existing methods such as harmony search algorithm [4], ant colony algorithm [9], and big-bang–crunch method [30] for both CPLF and CILF. Therefore, D-STATCOM is placed at bus 10. Figures 17 and 18 show the voltage (pu) comparisons at bus 10 for CPLF and CILF. It is also observed that switching of D-STATCOM with FLC improves its voltage profile and THD (%). This means that switching of D-STATCOM with FLC helps to improve the performance parameters in comparison to existing methods. The comparison of voltage profile and THD (%) of real power at bus 10 with existing techniques and D-STATCOM-based FLC is shown in Tables 17 and 18 and its graphical comparison is shown in Figures 19 and 20.

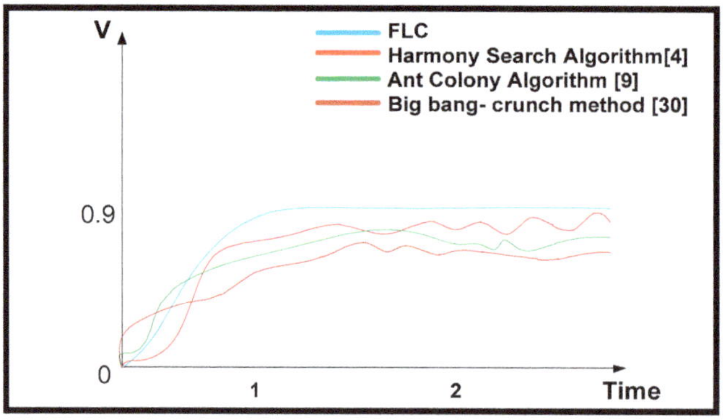

Figure 17. Voltage comparison at bus 10 for CPLF.

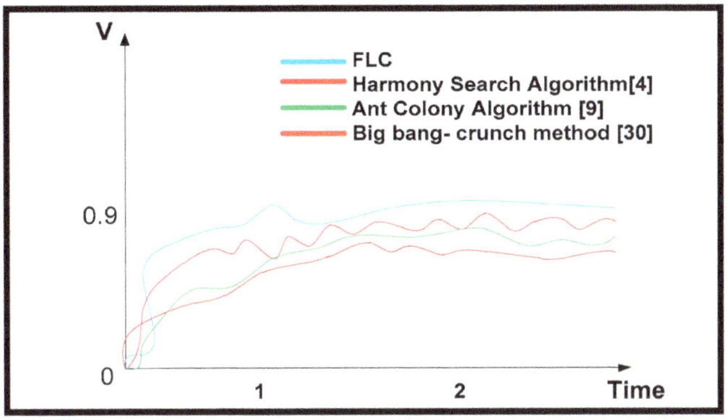

Figure 18. Voltage comparison at bus 10 for CILF.

Table 17. Comparative analysis for voltage (pu) at bus 10 for both types of load.

Load Type	FLC	D-STATCOM-Based FLC	Harmony Search Algorithm [4]	Ant Colony Algorithm [9]
CPLF	1	0.98	0.97	0.93
CILF	1.01	0.99	0.97	0.93

Table 18. Comparative analysis for THD (%) of real power at bus 10 for both types of load.

Load Type	FLC	D-STATCOM-Based FLC	Harmony Search Algorithm [4]	Ant Colony Algorithm [9]
CPLF	5.2	9.12	9.89	10.5
CILF	6.3	9.65	9.65	10.26

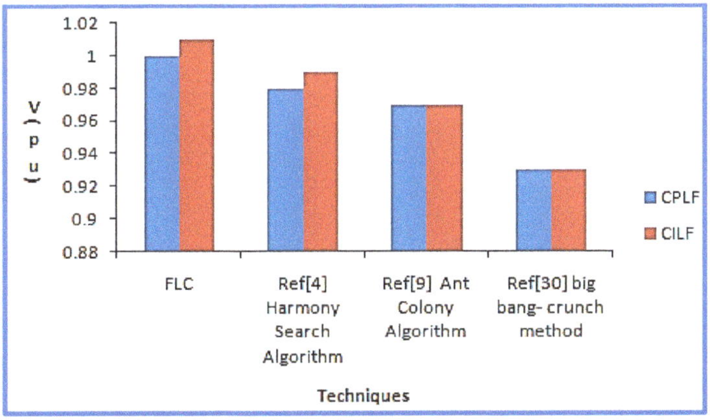

Figure 19. Graphical voltage comparison at bus 10 with different methods for CPLF and CILF.

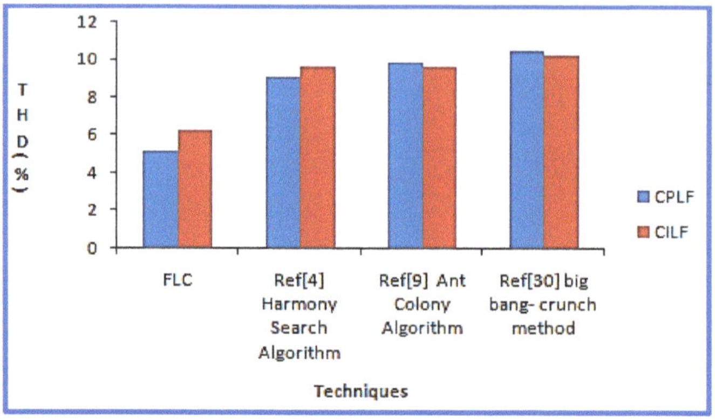

Figure 20. Graphical THD (%) of real power comparison at bus 10 with different methods for CPLF and CILF.

4. Result Summary

This article shows location and sizing of three DGs for CPLF and CILF types of load using GA and heuristic PDF method. The location of the DGs was obtained based on three different parameters such as line power losses, accuracy, and sensitivity. The application of GA method shows that all three parameters of DG placement are improved in comparison with the heuristic PDF method as well as existing methods such as ant-lion optimization algorithm, coyote optimizer, modified sine-cosine algorithm, and particle swarm optimization. It is also observed that after obtaining optimal location for DG placement at bus 17, optimal sizing for DG has been determined among all buses and the most optimal solution turns out to be at bus 17 in terms of minimum real and reactive power. It is found that determination of size of DG is quite satisfactorily observed under GA in comparison to the heuristic PDF method, ant-lion optimization algorithm, coyote optimizer, modified sine-cosine algorithm, and particle swarm optimization. Subsequently, bus 3 and bus 4 are the second and third best locations for placement of DGs with optimal size. Now, positioning of D-STATCOM on the IEEE 33 bus system is assessed in such a way that the bus with the worst performance in terms of THD and voltage profile has to be detected. On the basis of the analysis performed, bus 10 is found to be most suitable for locating the D-STATCOM. Afterwards, in order to improve the THD and voltage profile, switching of D-STATCOM is performed through FLC, consequently showing superiority over other existing methods. The current research will provide considerable expertise and also acts as a guide for researchers, including utility engineers, regarding problems to be addressed in order to optimize the size and position of DG units within electrical power systems. The metaheuristic computation approaches recently unveiled could be implemented for optimal design and fitting of DG in network delivery in future.

5. Conclusions

This article presents the optimal location and sizing of three DGs in the IEEE 33 bus test system by using the heuristic PDF method and GA for CPLF and CILF. Associated bus locations are examined for analysis of the impact of optimal placement and size of DG. The optimal locations of DGs are selected in terms of performance parameters such as line power losses, sensitivity, and accuracy while sizing of DG is obtained in terms of real and reactive power. It is evident that the parameters such as voltage profile, line power loss, accuracy, sensitivity, THD, etc. are improved with the GA method as compared to the heuristic PDF method and other existing methods such as the ant-lion optimization algorithm, coyote optimizer, modified sine-cosine algorithm, and particle swarm optimization. It is also confirmed that determining DG size is resolved quite satisfactorily with the GA method in terms of real power and reactive power, rather than heuristic PDF and other existing methods. Further positioning of D-STATCOM is being decided on the basis of bus having the worst voltage profile and THD (%) of real power. The D-STATCOM is controlled with FLC which gives improved voltage and less THD of real power in comparison to existing techniques.

Author Contributions: Conceptualization, P. and M.S.; formal analysis, P., A.A. and S.S.M.G.; funding acquisition, S.S.M.G.; investigation, A.A.; methodology, P., M.S. and A.A.; project administration, P., A.A. and S.S.M.G.; resources, S.S.M.G.; software, P. and M.S.; supervision, P., A.S.S. and M.S.; validation, P. and M.S.; visualization, P., A.S.S., M.S. and A.A.; writing—original draft, P., A.S.S. and M.S.; writing—review and editing, P., A.A. and S.S.M.G. All authors have read and agreed to the published version of the manuscript.

Funding: This research was funded by TAIF UNIVERSITY RESEARCHERS SUPPORTING PROJECT, grant number TURSP-2020/34", Taif University, Taif, Saudi Arabia, and "The APC was funded by SHERIF GHONEIM".

Data Availability Statement: All data generated or analysed during this study are included in this research article and any relevant information related to the current study are available from the corresponding author on reasonable request.

Acknowledgments: The authors would like to acknowledge the financial support received from Taif University Researchers Supporting Project Number (TURSP-2020/34), Taif University, Taif, Saudi Arabia.

Conflicts of Interest: The authors declare no conflict of interest.

Abbreviations

DG	Distributed generator	δ_i	Load angle
CPLF	Constant power load flow	θ_i	Impedance angle at '*i*' bus
CILF	Constant impedance load flow	θ_j	Impedance angle at '*j*' bus
FLC	Fuzzy logic controller	ε	Tolerance limit
GA	Genetic Algorithm	y^r	Initial code of string
D-STATCOM	Distribution static compensator	λ	Difference between measured and reference power
THD	Total harmonic distortion	σ	Standard deviation
PDF	Probability distribution method	E	Error
PWM	Pulse width modulation	NB	Negative big
S_{ij}	Complex power between 2 buses *i* & *j*	NM	Negative medium
P_{ij}	Real power between 2 buses *i* & *j*	NS	Negative small
Q_{ij}	Reactive power between 2 buses *i* & *j*	PB	Positive big
S_{loss}	Complex power *loss*	PM	Positive medium
G_{ij}	Conductance between *i* & *j* bus	PS	Positive small
B_{ij}	Susceptance between '*i*' & '*j*' bus	ZS	Zero

References

1. Bernardon, D.P.; Mello, A.P.C.; Pfitscher, L.L.; Canha, L.N.; Abaide, A.R.; Ferreira, A.A. Real-time reconfiguration of distribution network with distributed generation. *Electr. Power Syst. Res.* **2014**, *107*, 59–67. [CrossRef]
2. Chicco, G.; Mazza, A. An overview of the probability-based methods for optimal electrical distribution system reconfiguration. In Proceedings of the Fourth International Symposium on Electrical and Electronics Engineering (ISEEE), Galați, Romania, 11–13 October 2013; pp. 1–10.
3. Chan, C.-M.; Liou, H.-R.; Lu, C.-N. Operation of distribution feeders with electric vehicle charging loads. In Proceedings of the 2012 IEEE 15th International Conference on Harmonics and Quality of Power (ICHQP), Hong Kong, China, 17–20 June 2012; pp. 695–700.
4. Ganguly, S.; Samajpati, D. Distributed generation allocation on radial distribution networks under uncertainties of load and generation using genetic algorithm. *IEEE Trans. Sustain. Energy* **2015**, *6*, 688–697. [CrossRef]
5. Martins, V.F.; Borges, C.L.T. Active distribution network integrated planning incorporating distributed generation and load response uncertainties. *IEEE Trans. Power Syst.* **2011**, *26*, 2164–2172. [CrossRef]

6. Salman, N. Practical mitigation of voltage sag in distribution networks by combining network reconfiguration and DSTATCOM. In Proceedings of the IEEE International Conference on Power and Energy (PECON2010), Kualalumpur, Malaysia, 29 November–1 December 2010.
7. Li, G.; Shi, D.; Duan, X.; Li, H.; Yao, M. Multiobjective optimal network reconfiguration considering the charging load of PHEV. In Proceedings of the Power and Energy Society General Meeting, San Diego, CA, USA, 22–26 July 2012; IEEE: New York, NY, USA, 2012; pp. 1–8.
8. Pandi, V.R.; Zeineldin, H.H.; Xiao, W. Determining optimal location and size of distributed generation resources considering harmonic and protection coordination limits. *IEEE Trans. Power Syst.* **2013**, *28*, 1245–1254. [CrossRef]
9. Swarnkar, A.; Gupta, N.; Niazi, K.R. Optimal placement of fixed and switched shunt capacitors for large-scale distribution systems using genetic algorithms. In Proceedings of the Innovative Smart Grid Technologies Conf. Europe (ISGT Europe), Gothenburg, Sweden, 11–13 October 2010; IEEE: New York, NY, USA, 2010; pp. 1–8.
10. Farahani, V.; Vahidi, B.; Abyaneh, H.A. Reconfiguration and capacitor placement simultaneously for energy loss reduction based on an improved reconfiguration method. *IEEE Trans. Power Syst.* **2012**, *27*, 587–595. [CrossRef]
11. Tuladhar, S.R.; Singh, J.G.; Ongsakul, W. Multi-objective approach for distribution network reconfiguration with optimal DG power factor using NSPSO. *IET Gener. Transm. Distrib.* **2016**, *10*, 2842–2851. [CrossRef]
12. Jasthi, K.; Das, D. Simultaneous distribution system reconfiguration and DG sizing algorithm without load flow solution. *IET Gener. Transm. Distrib.* **2017**, *12*, 1303–1313. [CrossRef]
13. Kalambe, S.; Agnihotri, G. Loss minimization techniques used in distribution network: Bibliographical survey. *Renew. Sust. Energy Rev.* **2014**, *29*, 184–200. [CrossRef]
14. Muhammad, M.A.; Mokhlis, H.; Naidu, K.; Franco, J.F.; Illias, H.A.; Wang, L. Integrated data base approach in multi-objective network reconfiguration for distribution system using discrete optimisation techniques. *IET Gener. Transm. Distrib.* **2018**, *12*, 976–986. [CrossRef]
15. Lalitha, M.P.; Reddy, V.V.; Usha, V. Optimal Dg placement for minimum real power loss in radial distribution system using PSO. *J. Theor. Appl. Inf. Technol.* **2010**, *13*, 107–116.
16. Gözel, T.; Hocaoglu, M.H. An analytical method for the sizing and siting of distributed generators in radial systems. *Electr. Power Syst. Res.* **2009**, *79*, 912–918. [CrossRef]
17. Zhang, S.; Cheng, H.; Li, K.; Bazargan, M.; Yao, L. Optimal siting and sizing of intermittent distributed generators in distribution system. *IEEJ Trans. Electr. Electron. Eng.* **2015**, *10*, 628–635. [CrossRef]
18. Kumar, K.S.; Jayabarathi, T. Power system reconfiguration and loss minimization for a distribution systems using bacterial foraging optimization algorithm. *Int. J. Electr. Power Energy Syst.* **2012**, *36*, 13–17. [CrossRef]
19. Kaur, M.; Ghosh, S. Network reconfiguration of unbalanced distribution networks using fuzzy-firefly algorithm. *Appl. Soft Comput.* **2016**, *49*, 868–886. [CrossRef]
20. Hung, D.Q.; Mithulananthan, N. Multiple distributed generator placement in primary distribution networks for loss reduction. *IEEE Trans. Ind. Electron.* **2013**, *60*, 1700–1708. [CrossRef]
21. Angelim, J.H.; Affonso, C.M. Impact of distributed generation technologyand location on power system voltage stability. *IEEE Latin Am. Trans.* **2016**, *14*, 1758–1765. [CrossRef]
22. Wu, Y.-K.; Lee, C.-Y.; Liu, L.-C.; Tsai, S.H. Study of reconfiguration for the distribution system with distributed generators. *IEEE Trans. Power Deliv.* **2010**, *25*, 1678–1685. [CrossRef]
23. Liu, K.-Y.; Sheng, W.; Liu, Y.; Meng, X. A network reconfiguration method considering data uncertainties in smart distribution networks. *Energies* **2017**, *10*, 618. [CrossRef]
24. Rao, R.; Ravindra, K.; Satish, K.; Narasimham, S.V.L. Power loss minimization in distribution system using network reconfiguration in the presence of distributed generation. *IEEE Trans. Power Syst.* **2013**, *28*, 317–325. [CrossRef]
25. Prabha, D.R.; Jayabarathi, T.; Umamageswari, R.; Saranya, S. Optimal location and sizing of distributed generation unit using intelligent water drop algorithm. *Sustain. Energy Technol. Assess.* **2015**, *11*, 106–113. [CrossRef]
26. Rezk, H.; Abdelkareem, M.A.; Ghenai, C. Performance evaluation and optimal design of stand-alone solar PV-battery system for irrigation in isolated regions: A case study in Al Minya (Egypt). *Sustain. Energy Technol. Assess.* **2019**, *36*, 100556. [CrossRef]
27. El-Zonkoly, A.M. Optimal placement of multi-distributed generation units including different load models using particle swarm optimization. *Swarm Evol. Comput.* **2011**, *1*, 50–59. [CrossRef]
28. Naik, S.G.; Khatod, D.K.; Sharma, M.P. Optimal allocation of combined DG and capacitor for real power loss minimization in distribution networks. *Int. J. Electr. Power Energy Syst.* **2013**, *53*, 967–973. [CrossRef]
29. Sandeep, K.; Ganesh, K.; Jaydev, S. A MINLP technique for optimal placement of multiple DG units in distribution systems. *Electr. Power Energy Syst.* **2014**, *63*, 609–617.
30. Othman, M.M.; Walid, E.; Yasser, G.H.; Almoataz, Y.A. Optimal placement and sizing of distributed generators in unbalanced distribution systems using supervised big bang-big crunch method. *IEEE Trans. Power Syst.* **2015**, *30*, 911–919. [CrossRef]
31. Naresh, A.; Mahat, P.; Mithulananthan, N. An analytical approach for DG allocation in primary distribution network. *Int. J. Electr. Power Energy Syst.* **2006**, *28*, 669–678.
32. Mithulananthan, N.; Oo, T.; Phu, L.V. Distributed generator placement in power distribution system using genetic algorithm to reduce losses. *TIJSAT* **2004**, *9*, 55–62.

33. Hamed, S.H. Intelligent water drops algorithm: A new optimization method for solving the multiple knapsack problem. *Int. J. Intell. Comput. Cybern.* **2008**, *1*, 193–212.
34. Muqbel, A.; Elsayed, A.H.; Abido, M.A.; Mantawy, A.A.; Al-Awami, A.T.; El-Hawary, M. Optimal Sizing and Location of Solar Capacity in an Electrical Network Using Lightning Search Algorithm. *Electr. Power Compon. Syst.* **2020**, *47*, 1247–1260. [CrossRef]
35. Siahbalaee, J.; Rezanejad, N.; Gharehpetian, G.B. Reconfiguration and DG Sizing and Placement Using Improved Shuffled Frog Leaping Algorithm. *Electr. Power Compon. Syst.* **2020**, *47*, 1475–1488. [CrossRef]
36. Baran, M.E.; Wu, F.F. Network reconfiguration in distribution systems for loss reduction and load balancing. *IEEE Trans. Power Deliv.* **1989**, *4*, 1401–1407. [CrossRef]
37. Bagherinezhad, A.; Palomino, A.D.; Li, B.; Parvania, M. Spatio-Temporal Electric Bus Charging Optimization with Transit Network Constraints. *IEEE Trans. Ind. Appl.* **2020**, *56*, 5741–5749. [CrossRef]
38. Elmetwaly, A.H.; Eldesouky, A.A.; Sallam, A.A. An Adaptive D-FACTS for Power Quality Enhancement in an Isolated Microgrid. *IEEE Access* **2020**, *8*, 57923–57942. [CrossRef]
39. Castiblanco-Pérez, C.M.; Toro-Rodríguez, D.E.; Montoya, O.D.; Giral-Ramírez, D.D. Optimal Placement and Sizing of D-STATCOM in Radial and Meshed Distribution Networks Using a Discrete-Continuous Version of the Genetic Algorithm. *Electronics* **2021**, *10*, 1452. [CrossRef]
40. Yuvaraj, T.; Devabalaji, K.; Ravi, K. Optimal Placement and Sizing of DSTATCOM Using Harmony Search Algorithm. *Energy Procedia* **2015**, *79*, 759–765. [CrossRef]
41. Rukmani, D.K.; Thangaraj, Y.; Subramaniam, U.; Ramachandran, S.; Madurai Elavarasan, R.; Das, N.; Baringo, L.; Imran Abdul Rasheed, M. A New Approach to Optimal Location and Sizing of DSTATCOM in Radial Distribution Networks Using Bio-Inspired Cuckoo Search Algorithm. *Energies* **2020**, *13*, 4615. [CrossRef]
42. Moayedi, H.; Mosavi, A. Synthesizing Multi-Layer Perceptron Network with Ant lion Biogeography-Based Dragonfly Algorithm Evolutionary Strategy Invasive Weed and League Champion Optimization Hybrid Algorithms in Predicting Heating Load in Residential Buildings. *Sustainability* **2021**, *13*, 3198. [CrossRef]
43. Kamel, S.; Amin, A.; Selim, A. Application of coyote optimizer for Optimal DG Placement in Radial Distribution Systems. In Proceedings of the 2019 International Conference on Computer, Control, Electrical, and Electronics Engineering ICCCEEE, Khartoum, Sudan, 21–23 September 2019.
44. Qu, C.; Zeng, Z.; Dai, J.; Yi, Z.; He, W. A Modified Sine-Cosine Algorithm Based on Neighborhood Search and Greedy Levy Mutation. *Comput. Intell. Neurosci.* **2018**, *2018*, 4231647. [CrossRef]
45. Al-Masri, H.M.K.; Al-Sharqi, A.A.; Magableh, S.K.; Al-Shetwi, A.Q.; Abdolrasol, M.G.M.; Ustun, T.S. Optimal Allocation of a Hybrid Photovoltaic Biogas Energy System Using Multi-Objective Feasibility Enhanced Particle Swarm Algorithm. *Sustainability* **2022**, *14*, 685. [CrossRef]

sustainability

MDPI

Article

Robust Power System State Estimation Method Based on Generalized M-Estimation of Optimized Parameters Based on Sampling

Yu Shi [1], Yueting Hou [2], Yue Yu [2], Zhaoyang Jin [2,*] and Mohamed A. Mohamed [3]

1 Department of Science, Shandong Jiaotong University, Jinan 250353, China
2 Department of Electrical Engineering, Shandong University, Jinan 250100, China
3 Department of Electrical Engineering, Faculty of Engineering, Minia University, Minia 61519, Egypt
* Correspondence: zhaoyang.jin@sdu.edu.cn; Tel.: +86-178-6299-7751

Abstract: Robustness is an important performance index of power system state estimation, which is defined as the estimator's capability to resist the interference. However, improving the robustness of state estimation often reduces the estimation accuracy. To solve this problem, this paper proposes a power system state estimation method for generalized M-estimation of optimized parameters based on sampling. Compared with the traditional robust state estimator, the generalized M-estimator based on projection statistics improves the robustness of state estimation, and the proposed optimized parameter determination method improves the overall accuracy of state estimation by appropriately adjusting its robustness. Considering different degrees of non-Gaussian distributed measurement noises and bad data, the estimation accuracy the proposed method is demonstrated to be up to 23% higher than the traditional generalized M-estimator through MATLAB simulations in IEEE 14, 118 bus test systems, and Polish 2736 bus system.

Keywords: Gaussian distribution; M-estimator; power system state estimation; precision; robustness; weighted least square method

Citation: Shi, Y.; Hou, Y.; Yu, Y.; Jin, Z.; Mohamed, M.A. Robust Power System State Estimation Method Based on Generalized M-Estimation of Optimized Parameters Based on Sampling. *Sustainability* **2023**, *15*, 2550. https://doi.org/10.3390/su15032550

Academic Editor: Noradin Ghadimi

Received: 25 November 2022
Revised: 15 January 2023
Accepted: 17 January 2023
Published: 31 January 2023

1. Introduction

Modern power systems need to grasp the real-time operation of the power system comprehensively and quickly. It can more accurately analyze and predict the system's operation trends to ensure the power system's economy and safety. Power system state estimation is an effective means to detect the real-time operation data of the power system which was first proposed by the Schweppe in 1970 [1]. Abundant research has been conducted on power system state estimation to optimize the real-time redundant measurement data of the power system, to realize the real-time and reliable monitoring of the power system, and to ensure the safe and reliable operation of the system [2–8].

The weighted least squares (WLS) method has been widely adopted since 1970, with the advantages of good convergence performance and accurate estimation results. Its disadvantage is that it is not sufficiently robust to measured noise that follows the Gaussian distribution without bad data interference. In the real power network, not all measurement noise can be represented strictly by a Gaussian distribution, and occasionally significant errors occur due to various known telemetry noises or faults. There may also be other types of outliers that strongly affect the estimated state but may or may not carry bad data. All of these factors can potentially affect the estimator's estimation accuracy. A series of studies have been conducted to improve the robustness of state estimation, which is summarized in Table 1:

Table 1. Reference summary.

Authors	Origin	Purpose	Advantages and Disadvantages
L. Mili, M. G. Cheniae, N. S. Vichare and P. J. Rousseeuw	USA	To describe a fast and robust method for identifying the leverage points [9].	The method is very fast and compatible with real-time applications, but it does not apply to all forms of lever points.
J. Zhao and L. Mili	USA	To develop a robust dynamic state estimator of a cyber-physical system [10].	The H-infinity filter is able to handle large system uncertainties as well as suppress outliers, but the estimation efficiency of this method is low.
M. B. Djukanovic, M. H. Khammash and V. Vittal	USA	To present a framework for robust stability assessment in multimachine power systems [11].	The proposed method significantly reduces computational complexity and at the same time preserves the accuracy in predicting stability robustness.
Z. Lyu, H. Wei, X. Bai, D. Xie, L. Zhang and P. Li	CHN	To propose an norm estimator [12].	The proposed estimator has high computational efficiency and robustness.
E. Kyriakides, S. Suryanarayanan and G. T. Heydt	USA	To demonstrate the Huber function technique in a power engineering application [13].	This technique reduces large residuals but not accuracy.
M. Göl and A. Abur	TR	To develop a PMU placement strategy [14].	This method can improve the stability and accuracy of estimation.
M. Netto, J. Zhao and L. Mili	USA	To develop a robust extended Kalman filter [15].	The robust extended Kalman filter exhibits good tracking capabilities under Gaussian process and observation noise while suppressing observation outliers, even in position of leverage. However, it presents poor performance under non-Gaussian noise.
I. Akingeneye, J. Wu and J. Yang	USA	To develop PMU placement algorithms to improve the power grid state estimation [16].	The performance of the low complexity algorithms approach that of the exhaustive search algorithm, but with a much lower complexity.
G. Wang, G. B. Giannakis and J. Chen	USA	To put forward a novel LAV estimator leveraging recent algorithmic advances in composite optimization [17].	The algorithm efficiently deals with the non-convexity and non-smoothness of LAV-based PSSE, but it relies on solving a sequence of convex quadratic subproblems.
M. Huang, Z. Wei, G. Sun and H. Zang	CHN	To propose a hybrid SE for distribution systems [18].	The estimator method provides more reliable estimation results with a limited number of SCADA measurements, while biased estimated results can exist if some buses are far away from the measuring points.
C. H. Ho, H. C. Wu, S. C. Chan and Y. Hou	CHN	To present a robust statistical approach [19].	The proposed approach outperforms conventional approaches using the ADMM with L1 outlier detection in state estimation accuracy and convergence speed.
J. Zhao, M. Netto and L. Mili	USA	To develops a robust iterated extended Kalman filter based on the generalized maximum likelihood approach [20].	GM-IEKF can achieve both robustness and statistical efficiency, but its vulnerability to system parameter and topology errors.

The problem of existing robust state estimators is that robustness of the state estimator is usually achieved at the expense of estimation accuracy. The reason is that the existing robust state estimators use fixed parameter settings, which cannot adjust the state estimator's robustness and estimation accuracy in different scenarios. In [21], it is indicated that the measurement noise in PMU is likely to be non-Gaussian, which leads to the increased probability of outliers (measurements that significantly departs from their true values). Moreover, the types, parameters, and proportions of the non-Gaussian measurement distribution are different in different systems. Therefore, existing robust state estimators will be less robust when bad data and outliers are more frequent and will have insufficient accuracy when bad data and outliers are less frequent. To overcome this problem, this paper first evaluates the impact of bad data and non-Gaussian measurement noises on existing generalized

M-estimation parameters based on extensive Monte Carlo simulations. Finally, this paper proposes a generalized M-estimator of optimized parameters based on sampling which can adaptively select the appropriate parameters according to the probability distribution of the system measurement noise, accurately identify the outliers, and significantly reduce the effects of non-Gaussian measurement noise and bad data.

The contributions of this paper are summarized as follows:

- Demonstrate that higher robustness does not necessarily improve the estimation accuracy of the state estimator, and the best accuracy can be achieved if the robustness is tuned at an appropriate level;
- Propose a new robust power system state estimation method that can adaptively tune its robustness according to different levels of non-Gaussian distributed measurement noise and bad data.

The paper is structured as follows. Section 2 introduces the existing robust state estimation methods. Section 3 describes the drawbacks of existing robust state estimation methods. Section 4 presents the proposed generalized M state estimator of optimal parameters based on sampling. Section 5 presents and discusses the simulation results. Finally, Section 6 concludes this paper and identifies opportunities for future developments.

2. Existing Robust State Estimators for Power System Estimation

This section presents two widely used robust state estimators for power system state estimation, M-estimator (including Huber estimator and LAV estimator) and generalized M-estimator, which is improved based on the M-estimator.

2.1. M-Estimation of the Static State Estimation Method

The concept of M-estimation was first used by Huber for robust estimation of distribution centers and subsequently generalized to regression [22]. In general, an M-estimator is a maximum-likelihood estimator. It minimizes an objective function that is expressed as a function $\rho(\mathbf{r})$ for measuring the residue, according to the constraints given by the measurement equation:

Objective function:

$$\sum_{i=1}^{m} \rho(r_i) \tag{1}$$

Constraint condition:

$$\mathbf{z} = \mathrm{h}(\mathbf{x}) + \mathbf{r} \tag{2}$$

where $\rho(r_i)$ is a selected function that can measure the residual r_i, z is the measurement vector, $\mathbf{x}$ is a state vector and, $\mathrm{h}(\mathbf{x})$ is a measurement function.

The Huber estimator objective function is expressed as:

$$J(\mathbf{x}) = \sum_{i=1}^{m} \omega_i^2 \rho(r_i) \tag{3}$$

where the Huber function, $\rho(r_i)$, can be defined as:

$$\rho(r_i) = \begin{cases} r_i{}^2/2 & \text{for } |r_i| \le \beta \\ \beta|r_i| - \beta^2/2 & \text{for } |r_i| > \beta \end{cases}. \tag{4}$$

The first partial derivative $\psi(r_i)$ of r_i is expressed as:

$$\psi(r_i) = \begin{cases} r_i, & |r_i| < \beta \\ \beta * sign(r_i), & else \end{cases}. \tag{5}$$

The least absolute value (LAV) function can be defined as:

$$\rho(r_i) = |r_i| \tag{6}$$

where r_i is the normalized residual, the parameter β is a fixed value, generally set to 1.5. In fact, if β is set to infinite, the Huber estimator is equal to WLS estimator parameters, which has the highest accuracy. If β is set to 0, the Huber estimator turns into a LAV estimator, which has the highest robustness. Therefore, the Huber estimator can be regarded as an estimator which strikes a balance between the advantages of the WLS estimator and the LAV estimator.

The loss functions of the Huber and the LAV estimator are shown in Figure 1:

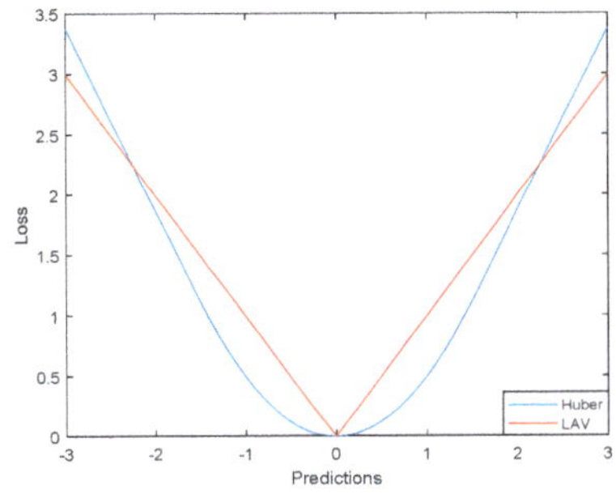

Figure 1. Loss functions of the Huber and LAV estimators.

2.2. Generalized M-Estimator

The external interference of the system causes outliers. There are multiple types of external interference. For example, there can be errors in sampling, such as recorded bias, calculation errors, etc. It can also be caused by various accidental and abnormal factors. The presence of outliers affects the accuracy of state estimation. Although the Huber estimator and the LAV estimator have better robustness compared to the WLS estimator, it is still obviously affected by outliers. Therefore, early robust estimators suppress the effect of outliers by reducing their weights according to their normalized residuals.

The normalized residual is based on Mahalanobis distance, which is the weighted average distance between each point. The disadvantage is that the average point is significantly affected by outliers, leading to an inaccurate judgment of outliers. In contrast, the median is hardly affected by the outliers. However, the variance of each point to the median cannot be directly defined. It is more reasonable to use the projection statistics method. The idea of the projection statistics method is to study those coming from the coordinated median **M**, and through each data point direction v. Formally, we have v = $\mathbf{l}_i - \mathbf{M}$, where:

$$\mathbf{M} = \left[\underset{j}{median}\mathbf{l}_{j1}, \cdots, \underset{j}{median}\mathbf{l}_{m}\right]^T \tag{7}$$

where j is obtained from 1 to m. The resulting distance calculated with (7) will be referred to as the projection statistics, or the PS_i for short. Then, the i-th calculated projection statistic, PS_i, is compared to a given threshold. The labeled outliers are then degraded using the following weight function: $\omega_i^2 = \min(1, d^2/PS_i^2)$, where d = 1.5 is set to produce good statistical efficiency.

The generalized M-estimator combines outlier detection and the Huber loss function, whose objective function is expressed as:

$$J(\mathbf{x}) = \sum_{i=1}^{m} \omega_i^2 \rho(r_s). \tag{8}$$

To solve the objective function, we calculated the first derivative about **x**:

$$\frac{\partial J(x)}{\partial x} = \sum_{i=1}^{m} -\frac{\omega_i a_i}{s} \psi(r_{si}) = 0 \tag{9}$$

where $s = 1.4826\ median_i\ |r_i|$.

It is clear that this is a set of equations which can be solved by iterated re-weighted least square (IRLS) algorithm [23].

3. Disadvantages of Existing Robust State Estimators for Power System State Estimation

Existing robust estimators mainly consider the suppression performance on bad data, while the assessment of their outlier suppression performance caused by non-Gaussian measurement noise is insufficient. This might result in a decreased estimation accuracy of the state estimator in the presence of non-Gaussian measurement noise. The two most common types of non-Gaussian measurement noise are bimodal Gaussian distribution and Laplace distribution, which are introduced in detail in Section 3.1. Section 3.2 discusses the impact of non-Gaussian measurement noise on the performance of existing robust estimators. Section 3.3 presents the purpose of this study.

3.1. Non-Gaussian Distributed Measurement Noises

3.1.1. Bimodal Gaussian Distribution

In practice, the errors of voltage and current measurements might follow bimodal Gaussian mixture (BGM) distribution [24]. The probability density function can be obtained by the superposition of two Gaussian probability density functions:

$$PDF_{BGM} = \prod_{i=1}^{2} \omega_i f_{N(\hat{\mu}_i,\hat{\sigma}_i^2)}(y) \tag{10}$$

where $f_N(y)$ represents the Gaussian probability density function, $N(\mu,\sigma^2)$ represents the normal distribution, and ω represents the weights corresponding to the combination of Gaussian components. The symbol "^" represents the estimated quantity, and the subscript i represents the i-th Gaussian component combination. The weight of each Gaussian component combination is the product of all the Gaussian component weights involved in that combination, satisfying:

$$\omega_1 + \omega_2 = 1. \tag{11}$$

For the bimodal mixed Gaussian distribution in the method, the mean and variance are as follows:

$$\mu_{12} = \omega_1\mu_1 + \omega_2\mu_2 \tag{12}$$

$$\sigma_{12}^2 = \omega_1\sigma_1^2 + \omega_2\sigma_2^2 + \omega_1\omega_2(\mu_1 - \mu_2)^2 \tag{13}$$

where ω represents the weight, μ represents the mean value, σ represents the variance, subscripts 1 and 2 indicate the components 1 and 2 of the BGM, respectively, and subscript 12 indicates the resulting BGM distribution. In the proposed method, μ_{12} is set to 0, which results in

$$\frac{\mu_1}{\mu_2} = -\frac{\omega_2}{\omega_1} = k. \tag{14}$$

Combine (12) and assume $\sigma_1 = \sigma_2 = \sigma$, we have,

$$\mu_1 = -k\omega_2\sigma \tag{15}$$

$$\mu_2 = k\omega_1\sigma. \tag{16}$$

Therefore, the distribution of the BGM measurement noise can be changed by adjusting the parameter, k.

3.1.2. Laplace Distribution

The Laplace distribution is also called bi-exponential distribution because it can be seen as the combination of two exponential distributions at different positions. The probability density function of the Laplace distribution is:

$$f(x|\mu,b) = \frac{1}{2b}\exp(-\frac{|x-\mu|}{b}) = \frac{1}{2b}\begin{cases}\exp(-\frac{x-\mu}{b}), if\ x \geq \mu \\ \exp(-\frac{\mu-x}{b}), if\ x < \mu\end{cases} \tag{17}$$

where μ is the position parameter, $b > 0$ is the scaling parameter. If $\mu = 0$, then the positive half happens to be an exponential distribution of scale 1/2. The difference between Laplace distribution and Gaussian distribution is that the Gaussian distribution represents the square of the difference relative to the mean, while the Laplace distribution is represented by the absolute value relative to the difference [25]. Thus, the tail Laplace distribution is much flatter than that of the Gaussian distribution.

3.2. Effect of Non-Gaussian Measurement Noises on the Performance of Existing Robust State Estimators

State estimation aims to determine the most likely states in the system based on the measurements. One way to achieve this is through maximum likelihood estimation (MLE). Assuming that all measurement noises have known probability distributions, all measurements' joint probability density function can be written with these unknown parameters, called the likelihood function, which will peak when the unknown parameter is chosen closest to its actual value. When the measurement noise follows the Gaussian distribution, the deduced optimal solution method is the WLS [26].

As shown in Figure 2, WLS is no longer the optimal solution when the measurement noise does not follow the Gaussian distribution. This is because, compared to the Gaussian distribution, the tail convergence rate of the bimodal Gaussian distribution and the Laplace distribution are faster than that of the Gaussian distribution tail convergence rate (so they are also known as a heavy tail distributions). Therefore, they have higher outlier probability than the Gaussian distribution, i.e., their absolute value in the white part is greater than that of the Gaussian distribution. If WLS is still used in the existence of non-Gaussian distributed noises, outliers will be assigned with large weights, resulting in reduced estimation accuracy.

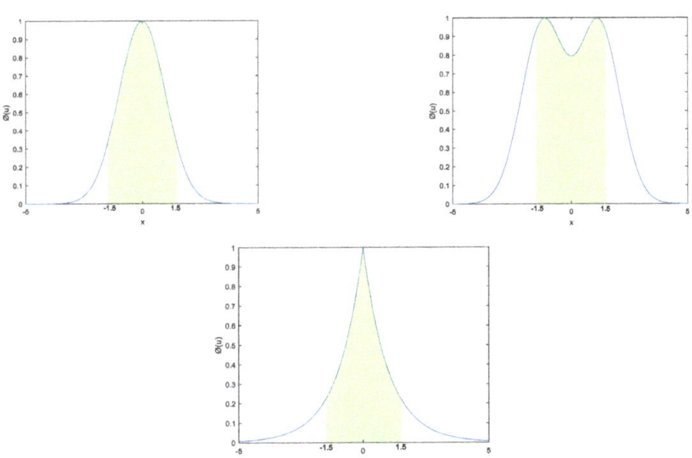

Figure 2. Probability density functions of Gaussian, bimodal Gaussian, and Laplacian distributions.

In discussing the effect of non-Gaussian measurement noise on the performance of existing robust state estimators, this subsection only discusses M-estimators since the generalized M-estimator is still developed based on the M-estimators. As shown in (5), the value of the target function of LAV is proportional to the absolute value of the deviation value, so when the deviation value is large when the outlier appears, the weight is much less than the WLS, so it has high robustness for the outliers. Although there may be significant proportion of non-Gaussian measurement noises, the Gaussian measurement noise still constitutes the main part for real systems. Therefore, the LAV estimator might not achieve the highest estimation accuracy.

The Huber estimator combines the advantages of both WLS and LAV according to (4). The size of the parameter β determines Huber's emphasis on WLS and LAV: when $\beta \rightarrow \infty$, Huber estimator is equivalent to WLS; when $\beta = 0$, it is equal to LAV. The parameter β is generally set to be the fixed value of 1.5. Therefore, the existing Huber estimators and generalized M-estimators based on fixed β values may have insufficient robustness or accuracy in the existence of the non-Gaussian distribution measurement noise with different proportions and parameters. Therefore, it is urgent to propose a generalized M-estimator (or Huber estimator) that optimizes the value according to the measurement noises.

3.3. Research Purpose of Generalized M State Estimation of Optimization Parameters Proposed

As shown in the above sub-sections, non-Gaussian distributed measurement noises are more likely to cause outliers than the Gaussian distributed measurement noises. Therefore, different levels of robustness of the estimator are required to ensure the best estimation accuracy under different types and proportions of non-Gaussian distributed measurement noises. Traditional generalized M-estimation based on Huber estimation with fixed parameter β can ensure high robustness against bad data and high estimation accuracy under Gaussian distributed noise. Its estimation accuracy is decreased if the measurement noises are non-Gaussian distributed. So, the purpose of this study is to propose a method that can adaptively change the value of β so that the generalized M-estimator can have high estimation accuracy under different non-Gaussian distributed measurement noises while ensuring good robustness against bad data.

4. Generalized M State Estimator of Optimized Parameters Based on Sampling

To overcome the drawback of the traditional generalized M-estimator, a novel generalized M state estimator of optimized parameters based on sampling is proposed in this section, where the random sampling method is described in Section 4.1, and the algorithm process of sampling-based optimization parameter is introduced in Section 4.2.

4.1. Optimized Parameter Selection Method Based on the Random Sampling Method

In Section 3.2, it is explained that non-Gaussian noise has a higher outlier probability than Gaussian noise. Hence, the value of β needs to be reduced to increase the estimator's robustness. Because the optimal value is affected by many uncertain factors, such as the configuration of measurement noise, the probability distribution of measurement noise, and the probability of bad data occurrence, it is almost impossible to obtain the exact optimal value of β. Therefore, this paper proposes the sagging optimization selection method to find a value of β close to its optimal value according to the probability ratio of the outliers:

$$\beta = 1.5 - a\frac{\sum_{i=1}^{N} P_{Truthi}(x \geq 1.5)}{\sum_{i=1}^{N} P_{Gaussi}(x \geq 1.5)}. \tag{18}$$

Define $n_1 = \sum_{i=1}^{N} P_{Truthi}(x \geq 1.5)$, which is the number of times when the actual standardization error ((measurement value-mean)/standard deviation) of all the measurements in the system is greater than 1.5, is the quantity to be determined. Define

$n_2 = \sum_{i=1}^{N} P_{Gaussi}(x \geq 1.5)$, which is the number of times when all the measurements obey the Gaussian distribution, is the known quantity. The variable a is the droop coefficient and the quantity to be determined. This paper takes the following steps to determine n_1 and a:

Step 1: for a system with m measurements, randomly select N measurements;

Step 2: For the N measurements, record L groups of data from their corresponding devices (L is large enough), and record the number of normalized errors greater than 1.5, n_1.

Step 3: Calculate the droop coefficient a by (19):

$$a = (1.5 - \beta_{\min})\frac{n_2}{n_{1,\max}} \tag{19}$$

where $\beta_{\min}$ is the set minimum value. In order to ensure the convergence ability of the estimator, this paper takes $\beta_{\min}$ = 0.1, which is a conservative value. n_1,max is the number of all measurements whose normalized errors are greater than 1.5 that are calculated from the probability density function at the highest non-Gaussian degree.

If most of the non-Gaussian distribution measurements in the system obey a bimodal Gaussian distribution, then the formula for the optimization values can be approximately reduced to:

$$\beta = 1.5 - Akh \tag{20}$$

where A is the droop coefficient to be sought, and k is the average bimodal Gaussian error coefficient, which is defined in Equation (14), as the non-Gaussian measurement ratio. This paper takes the following methods to determine A, k, and h, as shown in Figure 3.

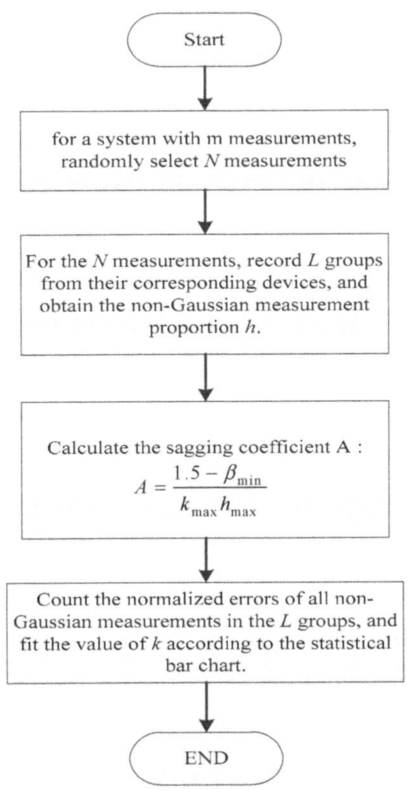

Figure 3. The process defined by A, k, and h.

4.2. The Proposed Generalized M State Estimation Algorithm of Optimized Parameters Based on Sampling

Based on the optimization parameter selection method proposed in the previous section, this paper proposes an improved generalized M state estimation algorithm. The steps are shown in Figure 4.

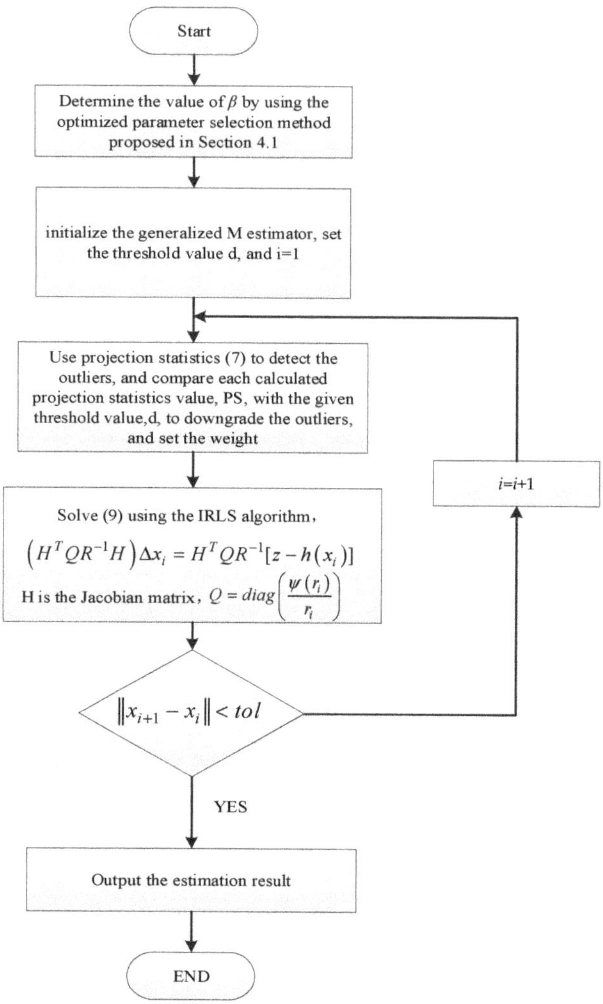

Figure 4. The flowchart of generalized M state estimation algorithm for optimization parameters based on sampling.

5. Simulated Examples

To verify the effectiveness of the proposed method, the IEEE 14 and 118 bus test systems are selected to simulate the performance generalized M-estimator with different parameters. All simulations were conducted in MATLAB, using an Intel Core i7-9750CPU (@2.6 Hz) 16 GB memory computer. Section 5.1 describes the impact of bad data and non-Gaussian measurement noise on generalized M-estimators with different values of β by conducting exhausted Monte Carlo simulations in the IEEE 14 bus test system.

Section 5.2 demonstrates the effectiveness of the proposed generalized M-state estimator of the optimized parameter in a large system, the IEEE 118 bus test system.

5.1. Effect of Bad Data and Measurement Noise on the Performance of the Generalized M-Estimator

To ensure the generality of the results, all of the results shown in this section are obtained by averaging the results over the 500 Monte Carlo simulations. To ensure the validity of the generalized M-state estimation method, a set of redundant conventional measurements and PMU measurements were selected in IEEE 14 bus test system as shown in Figure 5.

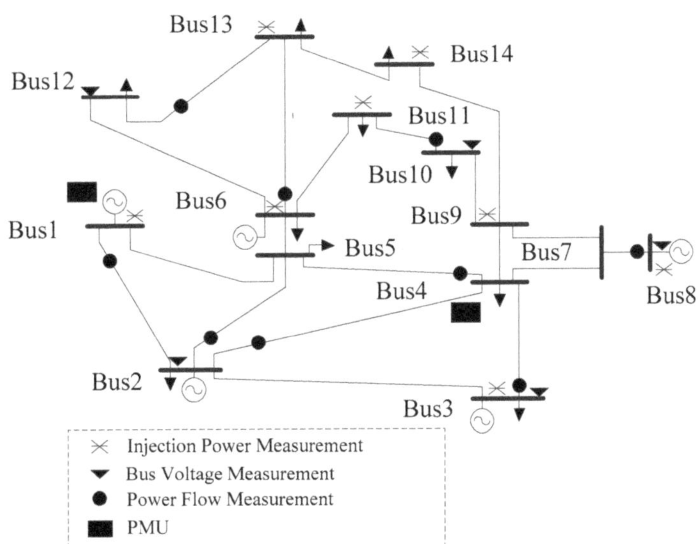

Figure 5. Diagram of IEEE 14 test system and measurement configuration.

For Gaussian errors, the standard deviation is set to a percentage of the measurement according to the type of measurement as follows:

SCADA: P, Q measured value (0.02), voltage (0.002)

PMU: Voltage amplitude (0.002), Voltage phase angle (0.01)

The performance of the generalized M-estimator can be adjusted by varying values of β. In this simulation test, the least value of β is 10^{-2}. With this configuration, the generalized M-estimator can be considered to be an LAV estimator. The largest value of β is set to be 10^4. The corresponding estimator can be considered to be a WLS estimator. In the range of 10^{-2} to 10^4, β increases at the fixed ratio of $10^{1/30}$, all estimators with different values of β are tested in the IEEE 14 bus test system. The root mean square error (RMSE) is used as the performance index of the estimators.

Case 1: Existence of bad data with fixed percentage errors. We compared the robustness of the state estimators by introducing fixed percentage errors to the voltage magnitude measurement at bus 1 and obtaining the RMSE of the estimators when the noise probability density function follows the standard Gaussian distribution. The voltage magnitude measurement at bus 1 is changed to 0, 0.2, 0.4, 0.6, 0.8 of its true value (100%, 80%, 60%, 40%, 20% errors, respectively), and tested separately. The simulation results are shown in Figure 6.

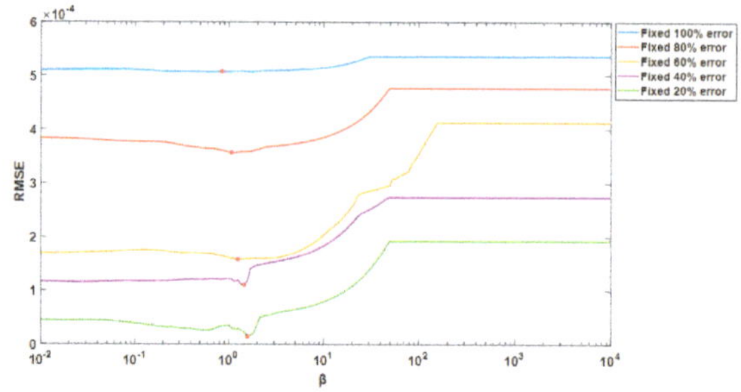

Figure 6. The trend of RMSE of each estimator with different β values at fixed error parameter percentages from 0.2 to 1. The red dot is the nadir.

The following conclusions can be obtained:

(1) After introducing fixed errors of different magnitudes, the trend of RMSE of each curve is roughly the same: the general trend is that RMSE increases as a larger β is selected. It will not change after a certain value;

(2) There is an optimal value of β for each fixed percentage error, and the value decreases with the percentage of the error parameter, as summarized in Table 2.

Table 2. Optimal β values of each estimator with different fixed error parameters.

Percentage of Fixed-Error Parameters	100%	80%	60%	40%	20%
The Optimal β Value	0.3415	0.5412	0.6813	0.8577	2.3263

Case 2: Existence of bad data with random percentage of errors. The robustness of generalized M state estimators with different values of β is compared by five individual simulations with three random bad data points in the measurements and observing their performances. The test results are shown in Figure 7.

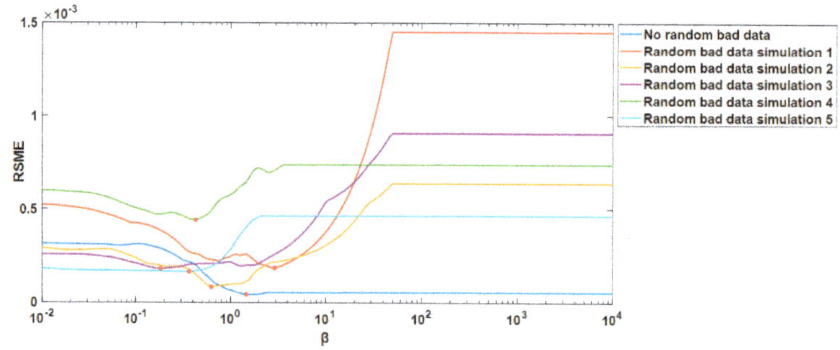

Figure 7. The trend of RMSE of each estimator with different β values with random error parameter percentages. The red dot is the nadir.

The following conclusions can be obtained:

(1) When there are no bad data in the measurements, the trend of RMSE with the change of β is decreasing in general; the value of β which achieves the lowest RMSE is 1.5, which is the exact value of β for normal generalized M-estimator;

(2) After the introduction of three random errors, the trend of the RMSE of each simulation with the change of β is roughly the same: The general trend is increasing. It will not change after a certain value;

(3) There is an optimal β whose corresponding estimator achieves the lowest RMSE for each simulation, which varies between 10^{-1} and 10^{1} with different bad data.

Case 3: Existence of measurements with bimodal Gaussian distribution noise. When the bimodal interval size k (as introduced in Section 3.1) is between 0~5, the trend of the RMSE with estimators with different values of β is shown in Figure 8.

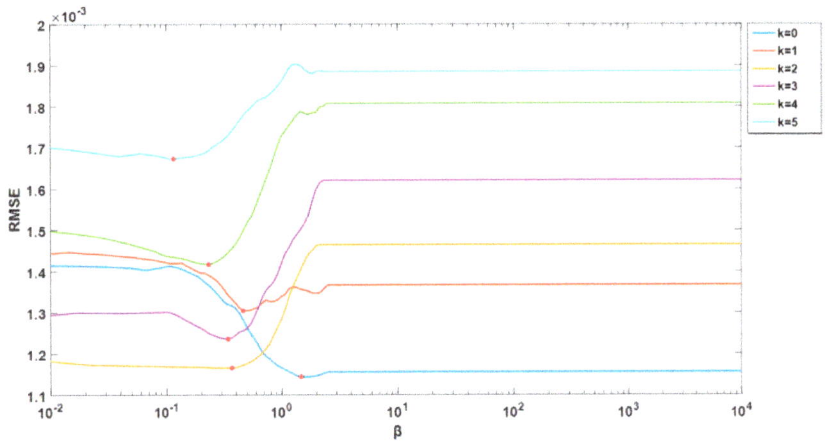

Figure 8. The trend of RMSE of each estimator with different β values in the presence of bimodal Gaussian distribution noise with bimodal intervals. The red dot is the nadir.

The following conclusions can be obtained:

(1) When $k \leq 1$, RMSE decreases with the increase of β, and when $k > 1$, RMSE decreases with the increase of β;

(2) The optimal β whose corresponding estimator achieves the lowest RMSE decreases with the increase of k;

(3) In the presence of non-Gaussian measurement noise, the value range of the optimal β is between 10^{-1} and 10^{1}, which can roughly cover the value range of the optimal β in the presence of bad data;

(4) Since the occurrence of bad data is low and the value range of the optimal β in the bad data case can be roughly covered by that in the non-Gaussian measurement noise case, the optimal β can be determined by only considering the influence of non-Gaussian measurement noise.

The optimal βs obtained by simulation in the presence of measurement noises in bimodal Gaussian distribution with different values of k are summarized in Table 3 and plotted in Figure 9. The optimized βs calculated according to the proposed optimized parameter selection method (20) are also plotted in Figure 9. It can be seen that with the increase of k, the calculated optimized β decreases, and are equal to the optimal β at $k = 0$ and $k \approx 5$. Although the optimized β is larger than the optimal β for $0 < k < 5$, the decreasing trend guarantees significantly better estimation accuracy compared to the estimator with a fixed c value while k increases.

Table 3. Optimal β values in the presence of different bi-modal Gaussian distributed measurement noises.

The Bimodal Interval Size (k)	0	1	2	3	4	5
The Optimal β	1.5000	0.4642	0.2712	0.1849	0.1324	0.1000

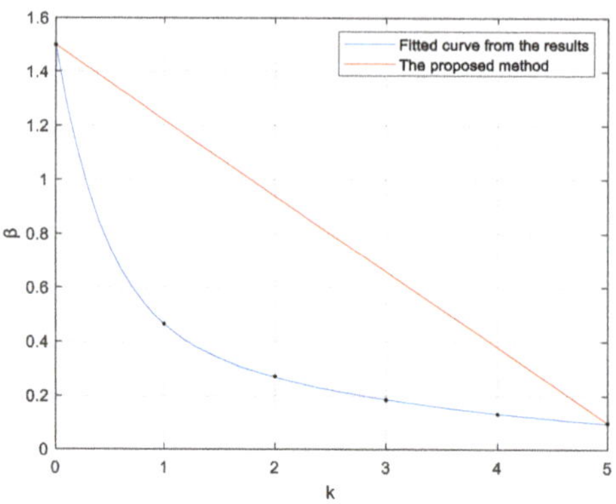

Figure 9. A plot of the optimal value and optimized value of β against the value of k.

5.2. Simulation Examples of the IEEE118 Bus Test System

To verify the proposed generalized M-estimation method of optimized parameters based on sampling, simulations are conducted in the IEEE 118 bus test system with the network diagram and measurement configuration shown in Figure 10.

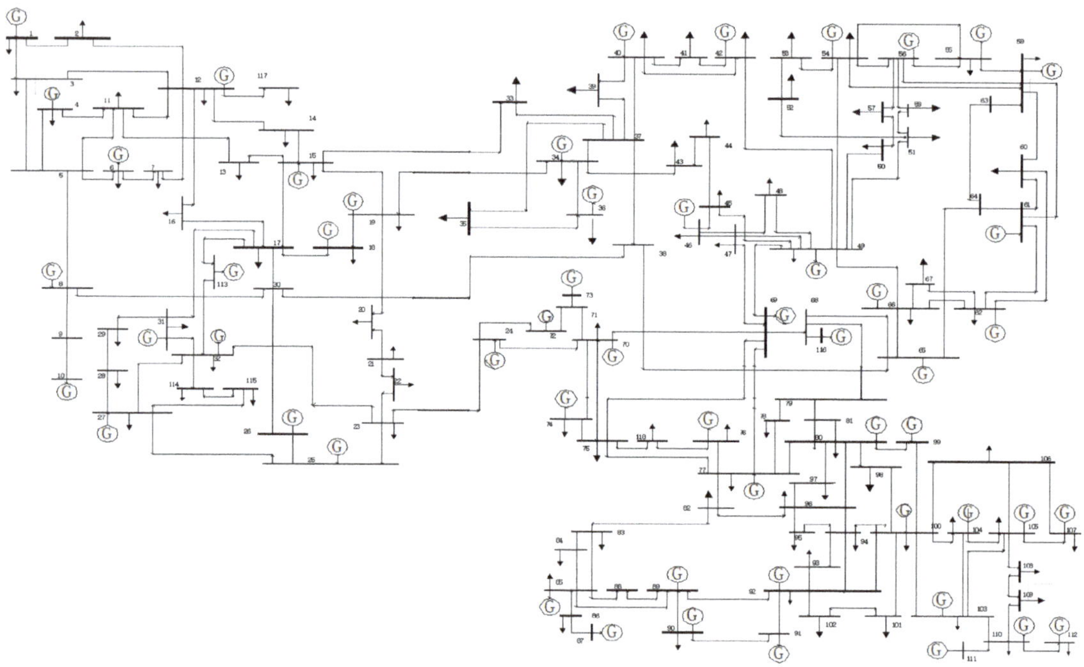

Figure 10. Diagram for IEEE 118 test system.

Among the total 186 measurements in the system, 28 of them have bimodal Gaussian distributed noises with the bimodal interval size (k) randomly set between 0~5. According to the proposed optimized parameter selection method, the optimized β, β opd, is determined as follows:

Step 1: Randomly select 80 measurements;

Step 2: For each measurement selected, test the measurement noises and record 10,000 groups of data; 12 non-Gaussian measurement noises are counted according to the statistics, and the proportion of non-Gaussian measurement, h, is 15%;

Step 3: Calculate the sagging coefficient A:

$$A = \frac{1.5 - \beta_{\min}}{k_{\max} h_{\max}} = \frac{1.5 - 0.1}{5 \times 0.5} = 0.56$$

where $k_{\max}$ is the maximum bimodal Gaussian error coefficient possible, and $h_{\max}$ is the maximum proportion of the bimodal Gaussian distribution measurement noise possible.

Step 4: Find the average k according to the statistic bar chart of all bimodal Gaussian distributed measurements noises. To test the proposed method in different scenarios, 5 different simulations with average $k \approx 1, 2, 3, 4, 5$ are conducted with the statistical bar charts shown in Figure 11.

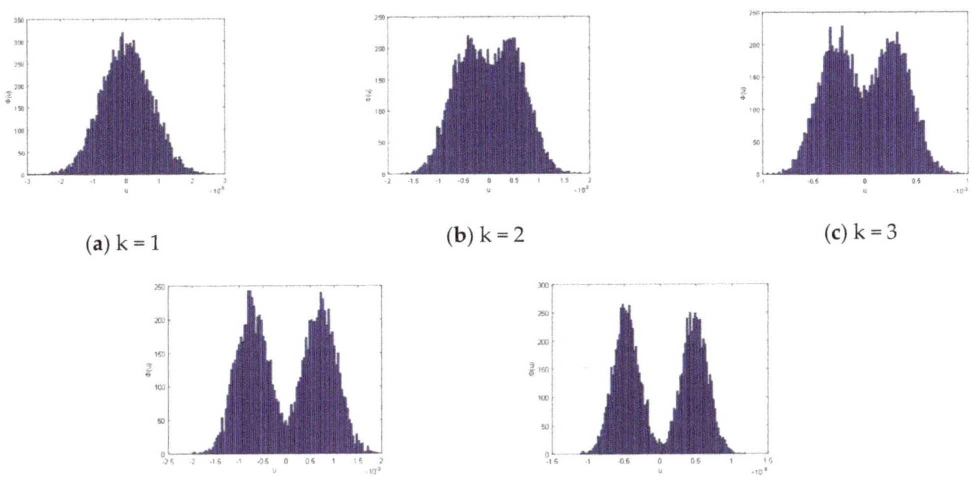

(**a**) k = 1 (**b**) k = 2 (**c**) k = 3

(**d**) k = 4 (**e**) k = 5

Figure 11. Average distribution graph of non-Gaussian measurement.

To verify the effectiveness of the proposed generalized M-estimator of optimized parameters, the RMSEs of the estimators obtained from Monte Carlo simulations for different ks and against β are shown in Figure 12. The optimized β, β_{opd}, can be calculated according to (20) with the obtained A and h. The optimal β, β_{opl}, and the RMSEs at β_{opl}, β_{opd}, $\beta = 1.5$ for different ks can be obtained from Figure 12. The results are summarized in Table 4.

Table 4. Average standard residuals of different estimators in IEEE118 systems at different non-Gaussian noise ratios.

k	5	4	3	2	1
β_{opd}	0.1005	0.3804	0.6603	0.9402	1.2201
β_{opl}	0.1001	0.1467	0.2326	0.3687	0.9261
RMSE at β_{opd}	1.55×10^{-3}	1.49×10^{-3}	1.45×10^{-3}	1.17×10^{-3}	1.15×10^{-3}
RMSE at β_{opl}	1.52×10^{-3}	1.43×10^{-3}	1.26×10^{-3}	1.06×10^{-3}	1.11×10^{-3}
RMSE at β = 1.5	1.82×10^{-3}	1.73×10^{-3}	1.63×10^{-3}	1.32×10^{-3}	1.18×10^{-3}
((RMSE at β = 1.5) − (RMSE at β_{opd}))/RMSE at β = 1.5 × 100%	14.84%	13.87%	11.04%	11.36%	2.54%

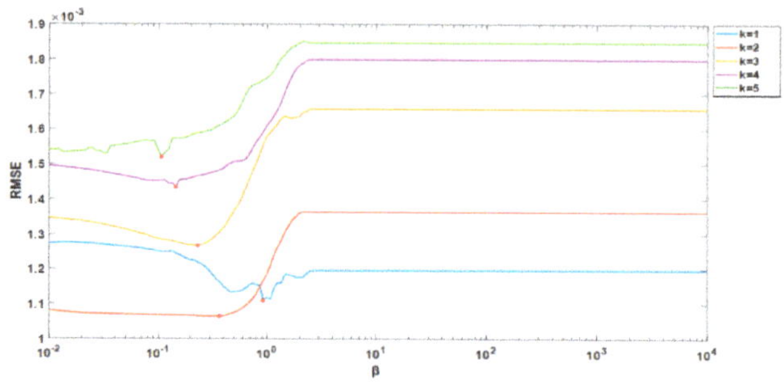

Figure 12. The trend of RMSE of each estimator with the value of β in systems with different bimodal interval size systems. The red dot is the nadir.

The analysis performed in Figure 12 and Table 4 leads to the following conclusions:

(1) β_{opd} is larger than β_{opl}, but the RMSE is not much different from the actual RMSE;

(2) The RMSE at β_{opd} is significantly lower than the RMSE at $\beta = 1.5$.

The conclusions above demonstrate the effectiveness of the proposed method. The underlying reason for these results is that the proposed method reduces the value of β to increase the estimator's robustness when the level of non-Gaussian measurement noise increases. The linear drooping characteristic of β gives the proposed estimator significantly better estimator accuracy than the traditional generalized M-estimator, but slightly lower estimator accuracy than the estimator with β_{opl}, which is acceptable considering the difficulty to obtain β_{opl}.

5.3. Simulation Examples of the Polish 2736 Bus System

To verify that which is proposed, simulations are conducted in the Polish 2736 bus test system [27].

Among the total 3269 measurements in the system, 280 of them have bimodal Gaussian distributed noises with the bimodal interval size (k) randomly set between 0~5. According to the proposed optimized parameter selection method, the optimized β, β opd, is determined as follows:

Step 1: Randomly select 800 measurements;

Step 2: For each measurement selected, test the measurement noises and record 10,000 groups of data; 120 non-Gaussian measurement noises are counted according to the statistics, and the proportion of non-Gaussian measurement, h is 15%;

Step 3: Calculate the sagging coefficient A;

Step 4: Find the average k according to the statistic bar chart of all bimodal Gaussian distributed measurements noises. To verify the effectiveness of the proposed generalized M-estimator of optimized parameters, the RMSEs of the estimators obtained from Monte Carlo simulations for different ks and against β are shown in Table 5.

Table 5. Average standard residuals of different estimators in Polish 2736 systems at different non-Gaussian noise ratios.

k	5	4	3	2	1
β_{opd}	0.6751	0.8330	0.9876	1.167	1.340
β_{opl}	0.1079	0.1467	0.2326	0.3687	0.9261
RMSE at β_{opd}	1.69×10^{-3}	1.54×10^{-3}	1.39×10^{-3}	1.22×10^{-3}	1.10×10^{-3}
RMSE at β_{opl}	1.61×10^{-3}	1.48×10^{-3}	1.34×10^{-3}	1.16×10^{-3}	1.05×10^{-3}
RMSE at $\beta = 1.5$	2.03×10^{-3}	1.96×10^{-3}	1.78×10^{-3}	1.57×10^{-3}	1.43×10^{-3}
((RMSE at $\beta = 1.5$) − (RMSE at β_{opd}))/RMSE at $\beta = 1.5 \times 100\%$	16.74%	21.43%	21.91%	22.29%	23.08%

Analysis performed in Table 5 leads to the following conclusions:

(1) The β value calculated by sagging coefficient is slightly different from the actual β value, but its RMSE is not much different from the actual RMSE;

(2) The RMSE corresponding to the calculated β value is significantly higher than that when $\beta = 1.5$, and the RMSE is significantly higher when the k value is larger.

These conclusions are almost the same with those in Section 5.2, demonstrating the adaptability of the proposed even in very large networks.

6. Conclusions and Future Works

This paper presents a generalized M-estimation method for power system state estimation with sampling-based optimization parameters. The sampling statistics method can set the parameter according to the different non-Gaussian measurement noise parameters and the proportion to adjust its robustness, which increases its overall estimation accuracy. The generalized M-estimation defines standardized residuals based on the median, thus being more robust than ordinary M-estimation methods. Exhaustive simulation results in the IEEE14 bus test system demonstrate the impacts of non-Gaussian measurement noise and bad data on estimation accuracy, which are summarized as follows:

(1) Small β can increase the robustness of the estimator but decrease the estimation accuracy in the normal operation case where no bad data occur, and vice versa. There is an optimal β that can achieve the highest overall estimation accuracy;

(2) The optimal β decreases with the increase of the magnitude of bad data and the non-Gaussian degree of measurement noise;

(3) The selection of the optimal β value mainly depends on the non-Gaussian degree and proportion of the measured noise.

Simulations in the IEEE 118 bus test system and Polish 2736 bus system verify that the proposed method can improve the estimation accuracy by up to 23% compared to the traditional generalized M state estimator in the presence of different degrees of non-Gaussian measurement noise.

Although this paper has reached the expected goal, further research can be carried out by considering more newly developed estimators, adopting more complex grid models, and considering more complex situations.

Author Contributions: Y.S.: Conceptualization, Methodology, Data curation, Investigation, Software, Writing original draft. Y.H.: Methodology, Validation, Visualization, Writing. Y.Y.: Validation, Visualization, Editing. Z.J.: Supervision, Funding acquisition, Project administration, Writing. M.A.M.: Review & editing. All authors have read and agreed to the published version of the manuscript.

Funding: This research was funded by National Natural Science Foundation of China, grant number 51907106.

Institutional Review Board Statement: Not applicable.

Informed Consent Statement: Informed consent was obtained from all subjects involved in the study.

Data Availability Statement: Research data is available at https://drive.google.com/file/d/17QEiRvHyZ38LqEr7hUffad1m9SvLZDn-/view?usp=sharing, https://drive.google.com/file/d/1TDkySmOztwf4i3HJyeMVU16FdUWvm8DZ/view?usp=sharing (accessed on 4 June 2022).

Conflicts of Interest: The authors declare no conflict of interest.

Nomenclature

WLS	Weighted least squares
LAV	Least absolute value
PMU	Phasor measurement unit
IRLS	Iterated re-weighted least square
BGM	Bimodal Gaussian mixture
MLE	Maximum likelihood estimation
SCADA	Supervisory control and data acquisition
RMSE	Root mean square error

References

1. Schweppe, F.C.; Wildes, J. Power System Static-State Estimation, Part I–Part III. *IEEE Trans. Power Appar. Syst.* **1970**, *PAS-89*, 120–125. [CrossRef]
2. Yu, K.; Watson, N.; Arrillaga, J. Error Analysis in Static Harmonic State Estimation: A Statistical Approach. *IEEE Trans. Power Deliv.* **2005**, *20*, 1045–1050. [CrossRef]
3. Jiang, W.; Vittal, V.; Heydt, G.T. A Distributed State Estimator Utilizing Synchronized Phasor Measurements. *IEEE Trans. Power Syst.* **2007**, *22*, 563–571. [CrossRef]
4. Wang, B.; He, G.; Liu, K. A New Scheme for Guaranteed State Estimation of Power System. *IEEE Trans. Power Syst.* **2013**, *28*, 4875–4876. [CrossRef]
5. Guo, Y.; Wu, W.; Zhang, B.; Sun, H. A Fast Solution for the Lagrange Multiplier-Based Electric Power Network Parameter Error Identification Model. *Energies* **2014**, *7*, 1288–1299. [CrossRef]
6. Zhu, K.; Nordstrom, L.; Ekstam, L. Application and analysis of optimum PMU placement methods with application to state estimation accuracy. In Proceedings of the 2009 IEEE Power & Energy Society General Meeting, Calgary, AB, Canada, 26–30 July 2009; pp. 1–7.
7. Wall, P.; Terzija, V. Simultaneous Estimation of the Time of Disturbance and Inertia in Power Systems. *IEEE Trans. Power Deliv.* **2014**, *29*, 2018–2031. [CrossRef]
8. Rostami, M.; Lotfifard, S. Distributed Dynamic State Estimation of Power Systems. *IEEE Trans. Ind. Inform.* **2018**, *14*, 3395–3404. [CrossRef]
9. Mili, L.; Cheniae, M.; Vichare, N.; Rousseeuw, P. Robustification of the least absolute value estimator by means of projection statistics [power system state estimation]. *IEEE Trans. Power Syst.* **1996**, *11*, 216–225. [CrossRef]
10. Zhao, J.; Mili, L. A Theoretical Framework of Robust *H*-Infinity Unscented Kalman Filter and Its Application to Power System Dynamic State Estimation. *IEEE Trans. Signal Process.* **2019**, *67*, 2734–2746. [CrossRef]
11. Djukanovic, M.; Khammash, M.; Vittal, V. Sensitivity based structured singular value approach to stability robustness of power systems. *IEEE Trans. Power Syst.* **2000**, *15*, 825–830. [CrossRef]
12. Lyu, Z.; Wei, H.; Bai, X.; Xie, D.; Zhang, L.; Li, P. Lp Quasi Norm State Estimator for Power Systems. *J. Mod. Power Syst. Clean Energy* **2022**, *10*, 871–882. [CrossRef]
13. Kyriakides, E.; Suryanarayanan, S.; Heydt, G. State Estimation in Power Engineering Using the Huber Robust Regression Technique. *IEEE Trans. Power Syst.* **2005**, *20*, 1183–1184. [CrossRef]
14. Göl, M.; Abur, A. PMU placement for robust state estimation. In Proceedings of the 2013 North American Power Symposium (NAPS), Manhattan, KS, USA, 22–24 September 2013; pp. 1–5.
15. Netto, M.; Zhao, J.; Mili, L. A robust extended Kalman filter for power system dynamic state estimation using PMU measurements. In Proceedings of the 2016 IEEE Power and Energy Society General Meeting (PESGM), Boston, MA, USA, 17–21 July 2016; pp. 1–5.
16. Akingeneye, I.; Wu, J.; Yang, J. Optimum PMU placement for power system state estimation. In Proceedings of the 2017 IEEE Power & Energy Society General Meeting, Chicago, IL, USA, 16-20 July 2017; pp. 1–5.
17. Wang, G.; Giannakis, G.B.; Chen, J. Fast LAV Estimation via Composite Optimization. In Proceedings of the 2019 IEEE Power & Energy Society General Meeting (PESGM), Atlanta, GA, USA, 4–8 August 2019; pp. 1–5.
18. Huang, M.; Wei, Z.; Sun, G.; Zang, H. Hybrid State Estimation for Distribution Systems With AMI and SCADA Measurements. *IEEE Access* **2019**, *7*, 120350–120359. [CrossRef]
19. Ho, C.H.; Wu, H.C.; Chan, S.C.; Hou, Y. A Robust Statistical Approach to Distributed Power System State Estimation With Bad Data. *IEEE Trans. Smart Grid* **2020**, *11*, 517–527. [CrossRef]
20. Zhao, J.; Netto, M.; Mili, L. A Robust Iterated Extended Kalman Filter for Power System Dynamic State Estimation. *IEEE Trans. Power Syst.* **2017**, *32*, 3205–3216. [CrossRef]
21. Wang, S.; Zhao, J.; Huang, Z.; Diao, R. Assessing Gaussian Assumption of PMU Measurement Error Using Field Data. *IEEE Trans. Power Deliv.* **2018**, *33*, 3233–3236. [CrossRef]
22. Zhao, J.; Zhang, G.; Dong, Z.Y.; La Scala, M. Robust Forecasting Aided Power System State Estimation Considering State Correlations. *IEEE Trans. Smart Grid* **2018**, *9*, 2658–2666. [CrossRef]
23. Kotiuga, W.W.; Vidyasagar, M. Bad Data Rejection Properties of Weighted Least Absolute Value Techniques Applied to Static State Estimation. *IEEE Power Eng. Rev.* **1982**, *PER-2*, 32. [CrossRef]

24. Jin, Z.; Zhao, J.; Chakrabarti, S.; Ding, L.; Terzija, V. A hybrid robust forecasting-aided state estimator considering bimodal Gaussian mixture measurement errors. *Int. J. Electr. Power Energy Syst.* **2020**, *120*, 105962.
25. Chen, J.; Jin, T.; Mohamed, M.A.; Annuk, A.; Dampage, U. Investigating the Impact of Wind Power Integration on Damping Characteristics of Low Frequency Oscillations in Power Systems. *Sustainability* **2022**, *14*, 3841. [CrossRef]
26. Radhoush, S.; Bahramipanah, M.; Nehrir, H.; Shahooei, Z. A Review on State Estimation Techniques in Active Distribution Networks: Existing Practices and Their Challenges. *Sustainability* **2022**, *14*, 2520. [CrossRef]
27. Available online: http://www.pserc.cornell.edu/matpower/ (accessed on 4 June 2022).

sustainability

MDPI

Article

Technical and Economic Evaluation for Off-Grid Hybrid Renewable Energy System Using Novel Bonobo Optimizer

Hassan M. H. Farh [1,*], Abdullrahman A. Al-Shamma'a [2,*], Abdullah M. Al-Shaalan [2], Abdulaziz Alkuhayli [2], Abdullah M. Noman [2] and Tarek Kandil [3]

1 Faculty of Construction and Environment, Department of Building and Real Estate, Hong Kong Polytechnic University, Hung Hom, Kowloon, Hong Kong

2 Department of Electrical Engineering, College of Engineering, King Saud University, Riyadh 11421, Saudi Arabia; shaalan@ksu.edu.sa (A.M.A.-S.); aalkuhayli@ksu.edu.sa (A.A.); anoman@ksu.edu.sa (A.M.N.)

3 Department of Electrical and Computer Engineering, College of Engineering and Computing, Georgia Southern University, Statesboro, GA 30460, USA; thassankandil@georgiasouthern.edu

* Correspondence: hfarh.hussein@polyu.edu.hk (H.M.H.F.); ashammaa@ksu.edu.sa (A.A.A.-S.)

Abstract: In this study, a novel bonobo optimizer (BO) technique is applied to find the optimal design for an off-grid hybrid renewable energy system (HRES) that contains a diesel generator, photovoltaics (PV), a wind turbine (WT), and batteries as a storage system. The proposed HRES aims to electrify a remote region in northern Saudi Arabia based on annualized system cost (ASC) minimization and power system reliability enhancement. To differentiate and evaluate the performance, the BO was compared to four recent metaheuristic algorithms, called big-bang–big-crunch (BBBC), crow search (CS), the genetic algorithm (GA), and the butterfly optimization algorithm (BOA), to find the optimal design for the proposed off-grid HRES in terms of optimal and worst solutions captured, mean, convergence rate, and standard deviation. The obtained results reveal the efficacy of BO compared to the other four metaheuristic algorithms where it achieved the optimal solution of the proposed off-grid HRES with the lowest ASC (USD 149,977.2), quick convergence time, and fewer oscillations, followed by BOA (USD 150,236.4). Both the BBBC and GA algorithms failed to capture the global solution and had high convergence time. In addition, they had high standard deviation, which revealed that their solutions were more dispersed with obvious oscillations. These simulation results proved the supremacy of BO in comparison to the other four metaheuristic algorithms.

Keywords: hybrid renewable energy system; bonobo optimizer; annualized system cost; optimal solution; convergence rate; renewable energy fraction; artificial intelligent algorithms

Citation: Farh, H.M.H.; Al-Shamma'a, A.A.; Al-Shaalan, A.M.; Alkuhayli, A.; Noman, A.M.; Kandil, T. Technical and Economic Evaluation for Off-Grid Hybrid Renewable Energy System Using Novel Bonobo Optimizer. *Sustainability* **2022**, *14*, 1533. https://doi.org/10.3390/su14031533

Academic Editor: Mohamed A. Mohamed

Received: 28 December 2021
Accepted: 20 January 2022
Published: 28 January 2022

1. Introduction

Renewable energy resources, unlike conventional generation resources, are inexhaustible and lasting sources of energy. The world requires renewable and reliable energy resources since they are much cleaner and produce energy without the harmful effects of pollution [1,2]. There are different renewable energy generation sources, like wind turbines (WT), solar photovoltaics (PV), biomass energy, geothermal energy, etc. Besides their advantages, such as being environmental friendly, sustainability, etc., wind energy and solar PV energy have been used frequently due to the reduction in their manufacturing cost and growing industrial and residential applications [3]. These sources can be used individually or connected as a hybrid to feed power to the grid. In addition, renewable energy sources can be used to electrify remote areas, which considered off-grid loads and are unable to be supplied from an attainable grid. Due to the non-reliability and excessive sizing of using a single source (e.g., PV) to feed power to off-grid areas, hybrid renewable energy resources are proposed to meet these challenges [4]. Hybrid renewable energy systems (HRES) can comprise PVs, WT, diesel generators, batteries, fuel cells (FCs), etc. However, due to the nonlinear and random operation of renewable energy resources, challenges arise when they are used to electrify off-grid loads [5,6]. The challenges include reduced reliability, more control complexity, design and sizing considerations, unstable and less energy, etc. These challenges represent complex and nonlinear optimization problems. Optimization means reaching the optimal solution and the best design of the HRES with minimum cost [7,8].

There are numerous optimization algorithms that have been used to search for the optimal sizing of hybrid renewable energy systems, including traditional algorithms, artificial intelligence (soft computing) algorithms [9–17], hybrid algorithms [4,6,18], and software tools [19–23]. Traditional algorithms can be classified as analytical algorithms [24–27], graphical algorithms [28,29], probabilistic algorithms [24,30,31], and numerical techniques [32]. Traditional techniques are simple. However, they have certain requirements to define the optimization problem. For example, graphical techniques depend on the solar irradiance and wind speed to determine the sizing of the HRES, which causes over sizing or under sizing [33]. Analytical algorithms cannot deal with many resources and may consume more computational time compared to AI algorithms. On the other hand, other techniques such as soft computing techniques, hybrid, and software tools have no specific requirements like traditional techniques, which leads to solving the optimization problems effectively. Consequently, this literature focuses on software tools, hybrid, and soft-computing techniques.

Spreading the use of HRES to electrify rural on-grid/off-grid areas motivated researchers to explore more technical and economic feasibility issues. A techno-economic analysis was implemented using the HOMER PRO software tool in [20–23]. The authors in [20] studied, analyzed, and designed the techno-economic feasibility of a solar PV–diesel–battery energy system to electrify a village in Pakistan, taking the availability time of the grid as a constraint. The optimal sizing and the techno-economic feasibility were carried out using HOMER PRO software. The findings of this study revealed that the levelized cost of electricity (LCOE) in on-grid HRES is more economical than off-grid HRES. In [21], a study was developed using the HOMER PRO software tool to redesign and refinance a remote HRES (solar PV, diesel, and battery) on a small island in Thailand. The HRES was optimized to attain the lowest cost of electricity. In [22], a study was carried out to study the electrification of a rural community in Benin. The study found that the hybrid solar PV–diesel–battery system attained the lowest cost optimal system.

As mentioned above, optimal sizing of an HRES is unavoidable since over-sizing results in increasing the initial cost and under sizing may result in reducing the shared power from the HRES and consequently reducing system reliability. Recently, great attention has been centered on studying the optimal sizing of an HRES. Some of the published studies focused on a single objective mathematical model and others focused on multi-objective models [13,34–36]. The authors in [34] developed a model based on fuzzy logic to minimize the annualized cost of the HRES that encompasses solar PV–WT–battery. The authors in [13] utilized a genetic algorithm (GA) to find the optimal sizing of an HRES in power distribution networks based on minimizing the power losses and the expected energy not supplied. The authors in [35] used GA for optimal sizing for a hybrid PV–WT–battery system integrated into the energy management strategy. The energy management was attained based on a proposed economic model predictive control approach [12]. The authors in [36] combined a quasi-steady operational method and GA for optimal sizing of a solar PV–pump storage hydroelectric energy system based on investment cost and loss of power supply probability (LPSP) as an objective function. In [37], a MATLAB model was developed for the optimal design of a wind–hydro system based on reduction the cost of energy (COE) and CO_2 emissions. The authors in [38] applied a crow search algorithm for the optimal sizing of an autonomous microgrid including PV–WT to supply a research center based on the annual system cost. In [39], particle swarm optimization (PSO) technique-based Monte Carlo simulation was employed to reach the optimal size of a solar PV–WT–battery based on minimizing the total annual cost (TAC). In [40], GA was used to find a multi-objectives sizing solution for a solar PV–WT–solar collector–battery system based on minimizing the net present cost (NPC). The authors in [41] utilized the Pareto evolutionary algorithm to minimize the LCOE and the CO_2 life cycle emissions (LCE) for a solar PV–WT–DG–battery standalone system. In [42], a water cycle algorithm in addition to moth-flame optimization were used for the techno-economic optimal design of a solar PV–biogas generator–0umped hydro energy storage–battery energy system. The objective function was a minimization of TNPV. The water cycle and moth-flame optimizer techniques were compared and assessed with the GA. The authors in [43] introduced the differential evolution algorithm (DEA) incorporated with the fuzzy technique to optimally design a solar PV–WT–DG–hydrogen–battery energy system based on minimum cost, emissions, and unmet load. An improved fruit fly optimizer technique was proposed by [44] to optimally design a hybrid solar PV–WT–diesel–battery system based on minimizing the TAC and the pollutant emission. The authors in [45] utilized a line-up competition algorithm (LUCA) to determine the optimal design of a solar PV–WT–DG–battery energy system based on minimizing the TAC and CO_2 emissions. The authors in [17] proposed a cuckoo search (CS) technique for the optimal design of three HRESs: solar PV–battery, WT–battery, and solar PV–WT–battery systems, minimizing total system cost. This algorithm was compared to the PSO and GA algorithms, and the results show that CS gave better solutions and faster convergence. The author in [46] utilized the grey wolf optimization algorithm for the optimal design of an HRES encompassing a PV–WT–biomass system based on minimizing the TNPC and LPSP. The findings obtained by GWO were compared to the findings obtained by GA and simulated annealing (SA) algorithms, and the superiority of the GWO was confirmed. The authors in [47] introduced a hybrid Big Bang–Big Crunch (BBBC) technique for the optimal sizing of a solar PV–WT–battery standalone HRES. In [48], the authors utilized the ant colony optimization (ACO) based integer continuous domain programming to optimally size a PV–WT system based on minimizing the TAC.

Based on the preceding literature review, metaheuristic algorithms proved their efficacy and robustness to deal with the optimal design optimization problem of an HRES in comparison to the analytical and graphical ones. Therefore, this study proposes a novel bonobo optimizer (BO) algorithm, which was applied to search for the optimal design of the proposed off-grid HRES to electrify an urban region located in northern Saudi Arabia. The main novelty and contributions of this study are manifested as shown below:

- A novel BO algorithm was applied for the first time in this research to find the optimal design of the proposed off-grid HRES based on minimizing the annualized system cost and enhancing the power system reliability level.
- Technical and economic evaluation were attained for the proposed off-grid HRES that includes PV, WT, diesel, and batteries to electrify an urban region in northern Saudi Arabia, namely, Al Sulaymaniyah.
- To validate the performance soundness and credibility, the BO algorithm was compared to other four metaheuristic algorithms, namely, BBBC [47], GA [35], crow search [38], and the butterfly optimization algorithm (BOA) [49] in terms of optimal and worst solutions captured, mean, standard deviation (STDEV), convergence rate, and oscillations around steady state.
- The simulation results revealed the supremacy performance of the BO algorithm compared to the other four metaheuristic algorithms. It attained the optimal design of the HRES with the minimum ASC (USD 149,977.2), quick convergence time, and fewer oscillations, followed by BOA (USD 150,236.4). The BBBC and GA algorithms failed to capture the global solution and had high STDEV, high oscillations, and high convergence time.

This research paper is organized as follows: Section 2 describes the proposed off-grid HRES components and their mathematical modeling. In Section 3, the problem formulation is covered, including the objectives, constraints, and the proposed BO algorithm for HRES optimal sizing. Section 4 covers the simulation results, analysis, and discussions. Finally, Section 5 epitomizes the conclusions.

2. Description of the Proposed Off-Grid Hybrid Renewable Energy System

The proposed HRES under study is shown in Figure 1. This HRES encompasses two renewable sources—solar PV and WT—in addition to the diesel generator and battery banks for energy storage. Both solar PV arrays and batteries are interconnected directly to the DC busbar, whereas the WT and the diesel are interconnected to the AC busbar, as demonstrated in Figure 1 below. The bidirectional DC/AC converter has two jobs. The first job is converting the AC to DC, where it works as a bridge rectifier in this process. The second job is converting the DC to AC, and it works as an inverter in this job. The batteries are used to cover and supply the load once the renewable resources are not capable of supplying the required load demand. The diesel generators are the second standby generation source, which are used in case the renewables are not available and the energy stored in the batteries is consumed. The mathematical modeling of the HRES components under study are demonstrated in the following sections.

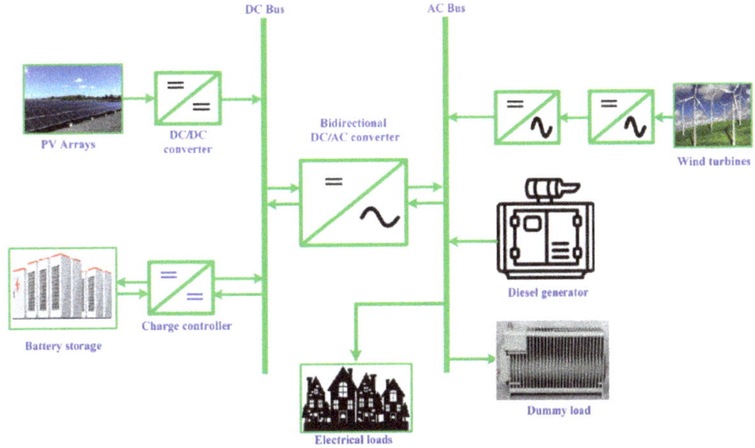

Figure 1. The hybrid renewable energy system (HRES) under study.

2.1. Modeling of PV Generation Source

The output power captured from the PV modules (P_{PV}) as a function of the PV rated power is [50]:

$$P_{PV} = P_r \, f_{PV} \left(\frac{\overline{G_T}}{\overline{G_{T,STC}}} \right) [1 + \alpha_P (T_c - T_{c,STC})] \quad (1)$$

where P_r is the PV rated power; $\overline{G_{T,STC}}$ and $\overline{G_T}$ are the solar PV radiation for standard test conditions (STCs) and normal conditions, respectively; $T_{c,STC}$ and T_c are the temperatures under STCs and normal conditions, respectively; and α_P and f_{PV} are the power temperature coefficient and derating coefficient, respectively. The PV steady-state temperature is formulated as shown below [50].

$$T_c = \frac{T_a + (NOCT - T_{a,NOCT})(1 - 1.11\eta_{MPP}(1 - \alpha_P T_{c,STC}))\left(\frac{\overline{G_T}}{G_{T,NOCT}}\right)}{1 + 1.11(\alpha_P \, \eta_{MPP,STC})(NOCT - T_{a,NOCT})\left(\frac{\overline{G_T}}{G_{T,NOCT}}\right)} \quad (2)$$

where

T_a and $NOCT$ are the cell temperature at ambient and nominal operating, respectively.

$T_{a,NOCT}$ and $G_{T,NOCT}$ are the ambient temperature and solar PV radiation under nominal operating, respectively;

$\eta_{MPP,STC}$ and η_{MPP} are the PV MPP efficiency under STCs and normal conditions, respectively.

Each PV energy system is equipped with a maximum power point tracker (MPPT) device, which is used to extract the maximum power captured from the PV system by controlling the duty cycle of the DC/DC converter. The metaheuristic MPPT techniques proved their efficacy and accuracy in comparison to the conventional techniques to deal with both the normal and partial shading cases. The conventional techniques gave acceptable performance under uniform conditions but may have trapped to the local peak under partial shading.

2.2. Modeling of the WT Genertation System

As demonstrated in Figure 2, the wind turbine starts to generate the output power once the wind speed exceeds the cut-in speed (u_c). At the rated speed (u_r), the wind power reaches its rated value till the cut-off or furling speed (u_f), where the wind turbine stops running if the wind speed exceeds its cut-off speed value. The output power captured by the WT depends on P_r, u_c, u_r, and u_f, which can be formulated as shown below [51].

$$P_{WT}(u) = \begin{cases} 0 & u < u_c \text{ or } u > u_f \\ P_r \times \frac{u^2 - u_c^2}{u_r^2 - u_c^2} & u_c \le u \le u_f \\ P_r & u_r \le u \le u_f \end{cases} \tag{3}$$

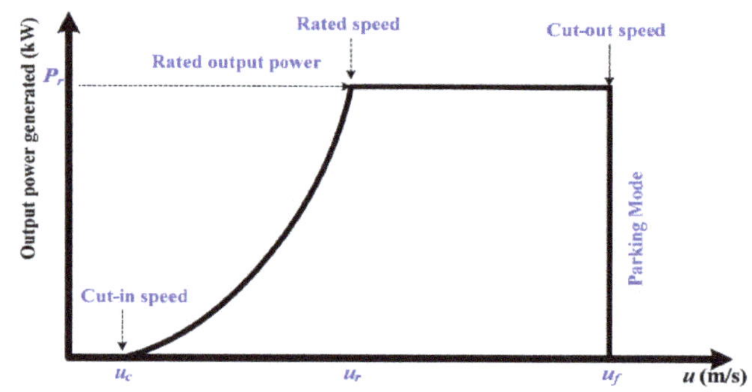

Figure 2. The WT power generated versus the wind speed characteristics.

2.3. Modeling of the Battery Bank as a Storage System

As a result of the intermittent nature of the renewable generation sources, whether WT or PV, the battery bank is considered a compulsory component in the HRES. They act as a storage element for the extra energy and supply it in case of deficiency by the WT and PV. The batteries will be in charging mode if the generated output power from the renewables (WT and PV) is higher than the demand, whereas they will be in discharging mode if the generated output power from the renewable sources is lower than the demand. The state of charge (SOC) or the charging power of the batteries at time t can be calculated with the following formula [52].

$$SOC(t) = SOC(t-1)(1-\sigma) + \left(E_{GA}(t) - \frac{E_L(t)}{\eta_{inv}}\right)\eta_{Bat}. \tag{4}$$

where

σ is the hourly self-discharging rate;
$E_L(t)$ is the total demand;
$E_{GA}(t)$ is the total output power;
η_{inv} is the inverter efficiency;
η_{Bat} is the battery bank efficiency.

During the discharging mode, the SOC must be more than or equal to the minimum allowable limit (SOC_{min}), whereas it must not exceed the maximum allowable limit (SOC_{max}) during the charging mode. These modes or cases are expressed as follows:

$$SOC(t) = \begin{cases} SOC_{min} & SOC(t) < SOC_{min} \\ SOC(t) & SOC_{min} < SOC(t) < SOC_{max} \\ SOC_{max} & SOC(t) > SOC_{max} \end{cases} \quad (5)$$

2.4. Modeling of the Diesel Generator

The diesel generator is considered the secondary standby generation source, which is utilized to cover the load if the renewables (WT and PV) and the batteries are not capable of covering the load demand. The annual fuel cost depends on the diesel rated power and output power generated, which can be formulated as follows [50]:

$$C_{Diesel} = C_F \sum_{t=1}^{8760} A \times P_{Diesel}(t) + B \times P_R \quad (6)$$

where

C_F is the fuel cost per liter;

A and B are the fuel constants;

P_R and $P_{Diesel}(t)$ are the diesel rated power and output power generated at time t, respectively.

3. Problem Formulation: Objectives and Constraints

The proposed BO technique is applied for the optimal design/sizing of the proposed HRES, including solar *PV*, *WT*, diesel, and batteries. The objective of the HRES sizing optimization problem is to minimize the total annualized system cost (*ASC*) at an acceptable limit of reliability, taken as loss of power supply probability (*LPSP*). The objective function, in addition to the constraints of this optimization problem, can be summarized as follows:

$$\begin{array}{c} Min\ ASC, \\ subject\ to: \begin{cases} LPSP \leqslant LPSP_{desired} \\ REF \leqslant REF_{desired} \\ 0 \leq P_{PV} \leq P_{PV,max} \\ 0 \leq P_{WT} \leq P_{WT,max} \\ P_{Bat.,min} \leq P_{Bat.} \leq P_{Bat.,max} \\ P_{Diesel,min} \leq P_{Diesel} \leq P_{Diesel,max} \end{cases} \end{array} \quad (7)$$

where $LPSP_{desired}$ is the predetermined limit of *LPSP*, which gives an indication about the reliability, and $REF_{desired}$ is the predetermined limit of the renewable energy fraction (*REF*).

Based on the minimum *ASC*, the proposed BO algorithm will determine the optimal sizing of the solar *PV*, *WT*, diesel, and battery bank as follows:

$$x = [P_{PV},\ P_{WT}, P_{Diesel}, P_{Bat.}] \quad (8)$$

where

P_{PV}, and P_{WT} are the optimal sizing in kW of *PV* and *WT*, respectively;

P_{Diesel} and $P_{Bat.}$ are the diesel rated power in kW and the capacity of battery in kWh, respectively;

$P_{PV,max}$, $P_{WT,max}$, $P_{Bat.\ ,max}$, and $P_{Diesel,max}$ are the maximum power limit of the *PV*, *WT*, batteries, and diesel, respectively.

The ASC equals the annualized capital costs ($C_{Cap.}$), plus the O&M cost ($C_{O\&M}$) and annualized replacement costs ($C_{Rep.}$), as follows [50]:

$$\text{ASC} = C_{Cap.} + C_{O\&M} + C_{Rep.} \quad (9)$$

The total capital cost is expressed as follows [50]:

$$C_{Cap.} = (C_{Ren.} + C_{Bat.} + C_{Diesel})\frac{i(1+i)^{Y_{proj}}}{(1+i)^{Y_{proj}} - 1} \quad (10)$$

where $C_{Ren.}$, $C_{Bat.}$, and C_{Diesel} are the capital costs of the renewable resources, battery bank, and diesel, respectively.

The annualized replacement cost can be formulated as shown below [50].

$$C_{Rep.} = C_{Rep.}\frac{i}{(1+i)^{Y_{rep}} - 1} \quad (11)$$

where $C_{Rep.}$ is the capital replacement cost and Y_{rep} is the lifetime of each component.

The LPSP can be defined as the power supply loss potential, which means that the HRES is not capable of supplying the load demand. On the other hand, it can be estimated by the percentage of the energy deficit of the total energy production of the HRES. Therefore, the LPSP gives us indication about the performance of the proposed HRES in terms of reliability and can be formulated as follows [50]:

$$LPSP = \frac{\sum_{t=0}^{T} \text{Power Failure Time } (PFT)}{T} \quad (12)$$

The *PFT* is the time interval during which the load cannot be supplied. An LPSP that equals 0% means that the load will be supplied all the time (*T*) and 100% LPSP means that the load cannot supplied all the time (*T*).

The *REF* represents the energy portion submitted to the load demand that was generated by renewable generation sources and can be expressed as follows:

$$REF = \left(1 - \frac{E_{L,Diesel}}{E_{L,served}}\right) \times 100 \quad (13)$$

where $E_{L,Diesel}$ represents the total load supplied by diesel. The REF changes from 0% to 100%. A REF of 0% means that the diesel generator only supplies the load and no renewables, and 100% means that the load is supplied only by renewables. Therefore, the REF changes between 0% and 100%, which means that the power supplied is the result of sharing all renewable and non-renewable generation sources.

4. Application of Bonobo Optimizer Technique for HRES Optimal Sizing

The bonobo optimizer (BO) is a new meta-heuristic technique that is influenced by bonobos' reproductive strategy and social behavior. Das et al. developed a population-based method [53]. For efficient optimization, the algorithm uses bonobos' fission–fusion search approach. Bonobos divide into smaller groups known as fissions for the purpose of locating food and then reuniting (fusion) at night to sleep, as demonstrated in Figure 3. Females are depicted by light forms, whereas males are depicted by dark forms, as shown in Figure 4. This one-of-a-kind method was incorporated into the algorithm to improve the efficiency of the search mechanism. BO is like other heuristics, and each solution in the population is termed a bonobo and the alpha bonobo (αbonobo) is the bonobo with the highest rank in the population's dominance hierarchy. More so, bonobos move forward through the positive phase (pp) and negative phase (np) of their phase probability, which indicates either population diversity or selection pressure (np). Positive phase count (ppc) and negative phase count (npc) are the counts of the consecutive number of iterations of pp and np, respectively. To reproduce young bonobos, the bonobo uses four main mating strategies: promiscuous and restrictive, consortship, and extra-group mating [53].

Figure 3. Bonobo social groups: both fission and fusion. Females are light forms; males are dark.

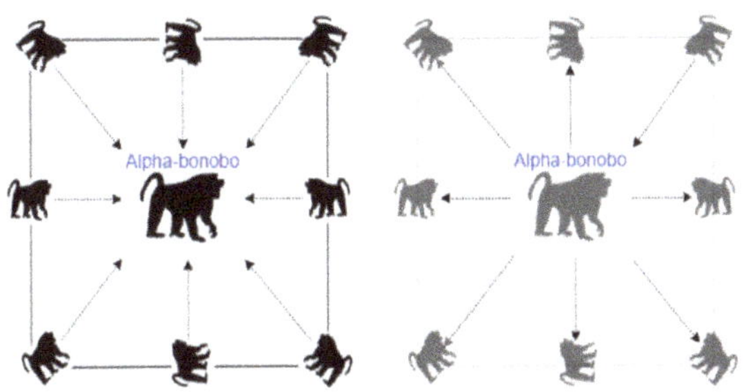

Figure 4. Movement directions of bonobos in pp (dark forms) and np (light forms) with higher probabilities.

The mating tactics alter depending on the phase condition (positive or negative). First, the positive phase (pp) depicts the state of the bonobo community, in which there is enough food, protection from other communities, breeding success, and genetic variation among bonobos. The odds for the first two types of mating, i.e., promiscuous and restricted mating, are higher during this phase. An oestrus female is available for both alpha bonobos (the highest-ranking male among all bonobos in a society) and other lower-ranking males in the promiscuous kind of mating. In the event of restricted mating, however, only the higher-ranking males are allowed to join. Consortship mating and extra-group mating are more likely in the case of negative phase (np), which signals a negative state in the society. A pair is separated from their natal community and spend their time together in a type of mating known as consortship. They rejoin with their community after a few days or weeks. In the case of extra-group mating, a female bonobo is found engaging in mating with males from other communities. Furthermore, compared to the other, the likelihood of extra-group mating is quite low. In the proposed BO, these physical processes are artificially recreated with the help of mathematics for optimization. The flowchart of the proposed bonobo optimizer algorithm is shown in Figure 5.

4.1. Promiscuous and Restrictive Mating Approach

The bonobos' mating approach is determined by the phase probability parameter (pp). The value of pp is set to 0.5 at the start, and it is changed after each iteration. If a random number r with a value between 0 and 1 is discovered to be less than or equal to pp, a new bonobo is born, as indicated in the following formula [53].

$$n\_b_j = b_j^i + r_1 s^{\alpha}\left(\alpha_j^b - b_j^i\right) + (1 - r_1)s^s flag\left(b_j^i - b_j^p\right) \tag{14}$$

where b is bonobo; n_b_j and α_j^b are the new offspring and α bonobo jth variables, respectively; j is an integer that ranges from 1 to d (variables number); and the variables b_j^i and b_j^p represent the ith and pth bonobo variable values, respectively. A value in the range of

0 to 1 is represented by r_1. The sharing coefficients for the αbonobo and pth bonobos are s^α and s^s, respectively. The $flag$ argument has a value between −1 and 1. When the optimal solution of the ith bonobo produces a better result than the pth bonobos, this is known as promiscuous mating. In this case, the $flag$ is given one point. Restrictive mating is a different term for the same thing. The $flag$ and αbonobo are given −1 in this regard.

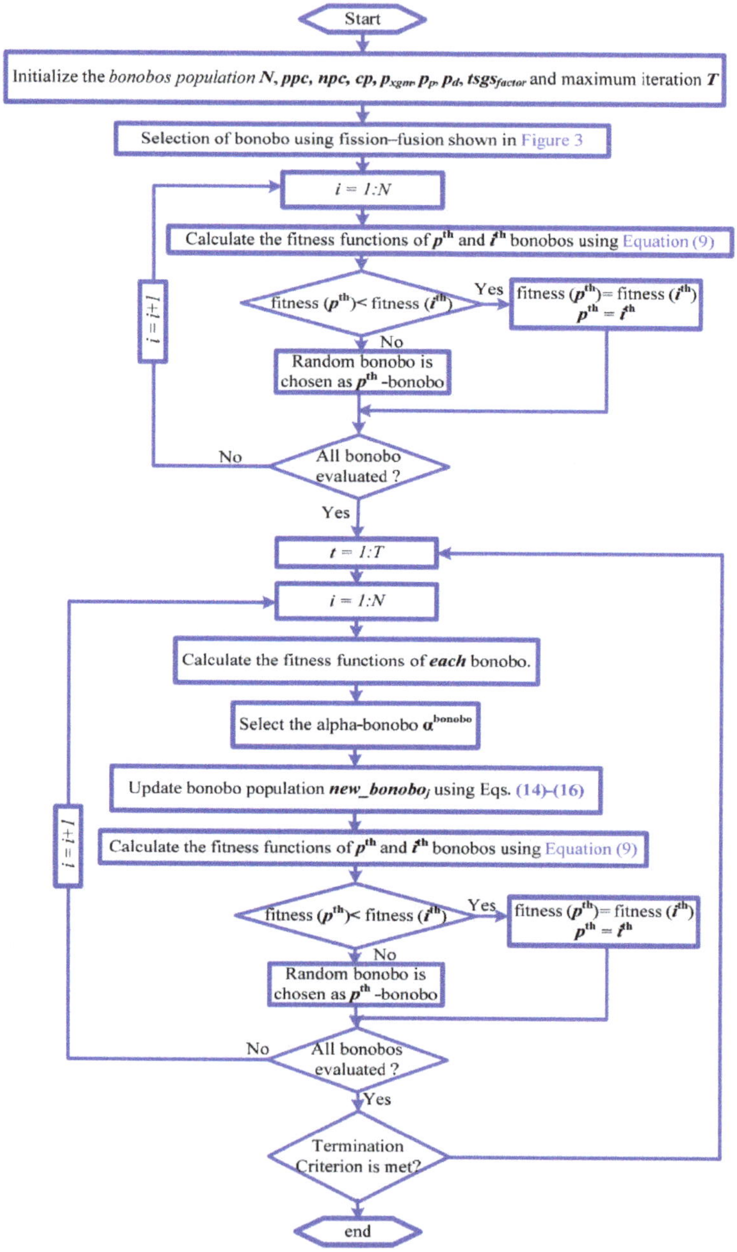

Figure 5. The bonobo optimizer algorithm flowchart.

4.2. Consortship and Extra-Group Mating Strategies

When phase p_p is less than the random integer r, this type of mating occurs. If r_2 is equal to or less than the probability of extra-group mating (p_{xgm}), the solution is updated by extra-group mating [53].

$$n\_b_j = \begin{cases} b_j^i + e^{(r_3^2+r_3-2r_3^{-1})}\left(Var\_max_j - b_j^i\right) & \alpha_j^{bonobo} \geq b_j^i \\ b_j^i - e^{(-r_4^2+2r_4-2r_4^{-1})}\left(b_j^i - Var\_min_j\right) & r_3 \leq p_d \\ b_j^i - e^{(r_3^2+r_3-2r_3^{-1})}\left(b_j^i - Var\_min_j\right) & \alpha_j^{bonobo} \leq b_j^i \\ b_j^i + e^{(-r_4^2+2r_4-2r_4^{-1})}\left(Var\_max_j - b_j^i\right) & r_3 \geq p_d \end{cases} \quad (15)$$

The p_d is set to 0.5 and is gradually updated based on the nature of evolution. The p_d optimizes the search process to find the best result. The lower and higher boundaries of the jth variable are represented by Var_min_j and Var_max_j respectively.

When the value of r_2 is greater than the value of p_{xgm}, a new offspring is produced utilizing the consortship mating strategy, which is as follows [53]:

$$n\_b_j = \begin{cases} n\_b_j + e^{r_5} flag(1+r_1)\left(b_j^i - b_j^p\right) & r_6 \leq p_d \\ b_j^p & otherwise \end{cases} \quad (16)$$

where r_1, r_2, r_3, r_4, and r_5 are random numbers ranging from 0 to 1.

5. Simulation Results and Discussion

The bonobo optimizer (BO) was proposed and applied to the optimal design/sizing of the proposed hybrid energy system including solar PV, WT, diesel, and batteries to electrify an urban area called Al Sulaymaniyah village in Arar in the northern area of Saudi Arabia. To validate the performance of the BO, it was compared to four other artificial intelligence algorithms—BBBC, crow search, GA, and BOA—to search for the optimal solution of the proposed HRES with a quick convergence rate. MATLAB R2019b/Windows 10/64-bit was used for the implementation of this optimization problem with 500 iterations and 50 runs for all five artificial intelligence algorithms. The average hourly solar irradiance and wind speed values were considered in this study. The parameters used in this study for all artificial intelligence algorithms are summarized in Table 1.

Table 1. The parameters used in this study for all four artificial intelligence algorithms.

Algorithm	Main Parameters	Value
BO	Rate of change in phase probability	0.0035
	Sharing coefficient for selected bonobo	1.3
	Sharing coefficient for alpha bonobo	1.25
	Extra-group mating probability	0.001
BBBC	Energy of rabbit	2
	Population size	100
Crow search	Awareness probability	0.1
	Flight length	2
GA	Population	100
	Selection	Roulette wheel
	Mutation rate	0.2
	Crossover rate	0.8
BOA	Sensory modality	0.01
	Power exponent	0.1

The simulation results of the proposed BO compared to the four artificial intelligence algorithms are shown in Table 2 for achieving the optimal design/sizing of the proposed

HRES with an LPSP of 0%. The proposed HRES contains PV, WT, a diesel generator, and batteries. As shown in Table 2, the BO had the best performance, followed by BOA, compared to the other three artificial intelligence algorithms (BBBC, crow, and GA), where it achieved the optimal solution/sizing of the proposed HRES with the lowest ASC. This is also demonstrated in Figure 6, which shows the optimal ASC using BO compared to the other four metaheuristic techniques for LPSP = 0%. In addition, the worst solution of the BOA was better than the optimal solution of the BBBC, crow, and GA algorithms. The optimal design/sizing of the PV, WT, diesel generator, and batteries for all five metaheuristic algorithms is shown in Figure 7. Moreover, Figure 8a,b shows the five performance indicators (optimal solution, worst solution, mean, median, and STDEV) that were used to evaluate the performance of all five metaheuristic algorithms. As revealed in Figure 8, BO had the best performance compared to the other four metaheuristic algorithms in terms of optimal solution, worst solution, mean, median, and STDEV. The BO and crow search algorithms achieved the smallest STDEV for the proposed HRES. A low STDEV means that the optimal solutions are concentrated around the mean. Therefore, they have fewer oscillations around steady state and follow the optimal solution quickly, taking less time to converge. On the other hand, BBBC and GA had the lowest performance based on these five performance indicators. They had high standard deviation. This means that the optimal solutions were highly dispersed. Therefore, they had obvious oscillations around the optimal solution and took more time to converge.

Table 2. Simulation results of the proposed BO compared to the other four artificial intelligence techniques for achieving the optimal design/sizing of the proposed HRES with 0% LPSP.

Algorithm	Performance Indicators	P_{PV} (kW)	P_{WT} (kW)	$P_{Bat.}$ (kW)	P_{Diesl} (kW)	ASC (USD/year)	REF (%)
BO	Optimal	341.3	403.9	499.8	215.9	149,977.2	82.7
	Worst	342	353.7	500	220.12	150,320.8	81.04
	Mean	340.62	400.36	499.83	215.84	150,033.1	82.61
	STDEV	5.97	12.75	0.2996	2.894	95.397	0.4021
BBBC	Optimal	397.85	348.4	923.9	229.1	150,883.5	64.55
	Worst	293.95	434.95	1436.3	248.06	158,884.5	70.52
	Mean	372.75	364.27	844.21	239.71	154,138.5	86.67
	STDEV	72.14	69.41	293.67	7.45	2007.99	10.379
Crow Search	Optimal	341.3	403.98	499.22	215.93	151,618.8	82.74
	Worst	358.7	405.75	498.86	221.74	151,832.6	83.046
	Mean	341.84	405.04	499.11	216	151,640.4	82.774
	STDEV	4.568	2.931	0.3925	0.8840	50.772	0.0984
GA	Optimal	370.21	311.51	1027.5	210.8	150,718	85.22
	Worst	496.29	260.45	1355.8	204.24	154,770	88.27
	Mean	375.56	346.95	940.64	219.09	151,944.6	85.55
	STDEV	45.92	52.921	161.35	21.43	927.45	1.6267
BOA	Optimal	361.28	373.70	950.01	227.34	150,236.4	63.32
	Worst	352.07	331.31	657.54	228.07	151,968.8	61.28
	Mean	343.87	353.63	850.27	214.06	151,130.5	66.93
	STDEV	26.359	27.982	161.531	11.461	422.106	6.2367

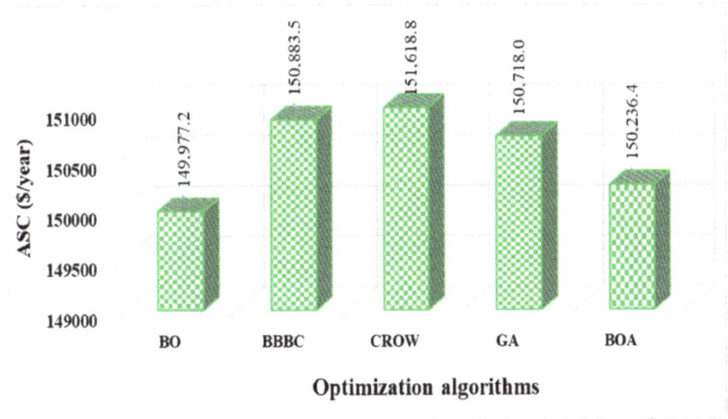

Figure 6. The optimal ASC using BO compared to the other four metaheuristic techniques for LPSP = 0%.

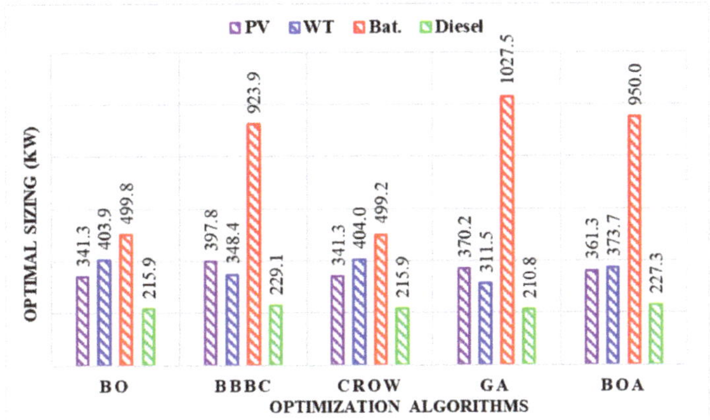

Figure 7. The optimal sizing of the solar PV, WT, diesel generator, and batteries for all five metaheuristic algorithms.

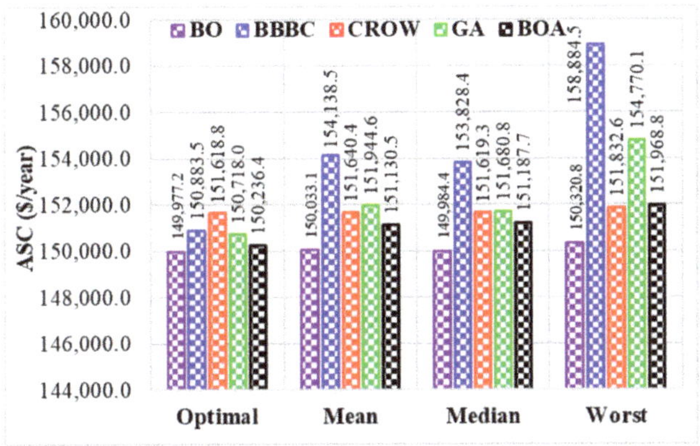

(a)

Figure 8. *Cont.*

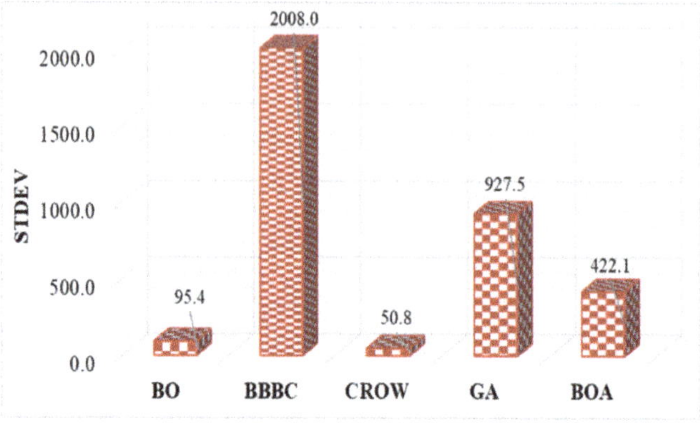

(b)

Figure 8. The performance indicators with all eight metaheuristic algorithms: (**a**) optimal, worst, mean, median, and (**b**) STDEV.

Figure 9 presents the ASC versus the run number for the BO technique in comparison to the other four metaheuristic techniques. This graph proved that BO followed by BOA and crow search followed the global solution and achieved the optimal sizing of the HRES with less ASC and fewer oscillations. This is due to the fact that their standard deviation was low. This means that the optimal solutions were concentrated around the mean. Therefore, they had fewer oscillations around steady state, as shown in Figure 9, and followed the optimal solution quickly, taking less time to converge, as shown in Figure 10. On the other hand, BBBC and GA were trapped to the local solution with obvious oscillations around the steady state, as shown in Figure 9. On the other hand, Figure 10 presents the convergence rate of all five metaheuristic algorithms: BO, BBBC, crow search, GA, and BOA. This figure shows that the BO followed the global solution with a faster convergence rate than the other four metaheuristic algorithms (BBBC, crow search, GA, and BOA). This figure also emphasizes that both BBBC and GA may have been trapped to the local solution and had the lowest convergence rate compared to the other metaheuristic techniques.

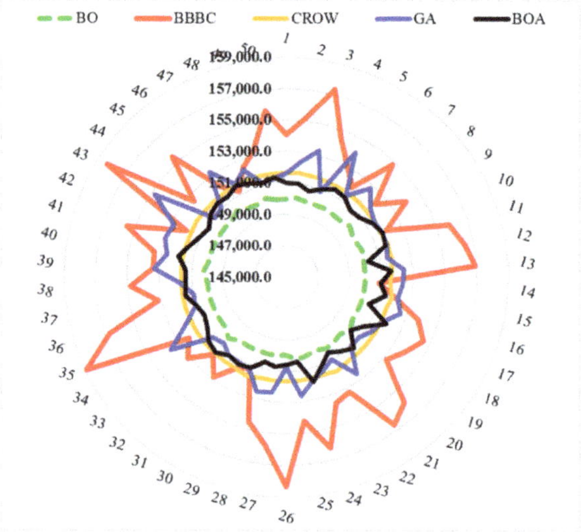

Figure 9. The ASC versus run number using BO compared to the other four metaheuristic techniques.

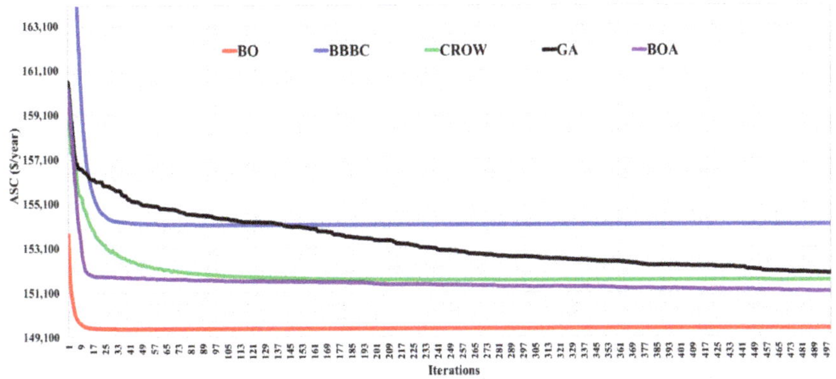

Figure 10. The convergence time of BO compared to the other four metaheuristic techniques.

Under different values of diesel prices (USD/liter), the simulation findings of the proposed BO are introduced in Table 3 for achieving the optimal design/sizing of the proposed HRES that includes solar PV, WT, a diesel generator, and batteries. As the diesel price increased, the optimal design/sizing of the diesel generator decreased, and at the same time, the renewable sizes (WT and PV) and REF% increased, as shown in Table 3. This is also evident in Figure 11a,b. On the other hand, the CO_2 emissions (kg/yr) reduced dramatically from 454,729.3 to 94,232.7 kg/yr with the increase in diesel prices from 0.1 to 1.5 USD/liter, as shown in Figure 11c. In addition, both ASC and fuel cost doubled as the diesel price increased from 0.1 to 1.5 USD/liter.

Table 3. Simulation results of the proposed BO algorithm under different diesel prices for achieving the optimal sizing of the proposed HRES.

Diesel Price (USD/Liter)	PV	WT	Bat.	Diesel	REF	Surplus	Diesel Hours	Fuel cost (USD/yr)	CO_2 (kg/yr)	ASC (USD/yr)
0.1	315.7	7.1	109.5	289.3	43.1	6.6	6021.0	30,660.6	454,729.3	98,185.6
0.2	275.9	255.7	287.1	226.6	71.6	11.0	3520.0	29,593.7	229,691.9	116,353.0
0.3	288.8	312.6	493.1	215.8	77.8	68.9	2669.0	33,386.3	178,479.0	129,834.4
0.4	341.0	354.9	499.8	222.8	81.0	54.9	2242.0	38,296.3	152,589.4	140,686.4
0.5	341.3	403.9	499.8	215.9	82.7	3.5	2020.0	42,722.8	138,529.7	149,977.2
0.6	327.8	422.3	499.8	202.2	83.1	20.3	1992.0	48,897.7	135,260.3	158,364.8
0.7	347.3	455.7	499.5	187.7	84.5	56.3	1844.0	50,872.3	123,745.5	166,060.2
0.8	372.6	482.5	482.0	176.1	85.5	15.7	1724.0	52,893.3	115,260.1	173,061.2
0.9	372.6	482.5	482.0	176.1	85.5	17.8	1724.0	59,505.0	115,260.1	179,672.8
1.0	402.8	496.8	489.2	134.7	87.1	16.2	1659.0	54,689.9	101,971.3	185,830.4
1.1	402.5	496.9	489.0	134.7	87.1	55.9	1660.0	60,189.7	102,002.2	191,319.3
1.2	403.1	497.7	487.3	134.5	87.1	21.7	1657.0	65,512.1	101,843.4	196,821.0
1.3	404.9	498.9	499.8	126.6	87.5	64.2	1634.0	67,803.4	98,737.1	202,402.6
1.4	451.4	499.7	500.0	121.4	88.1	2.5	1583.0	69,060.1	94,232.8	207,520.9
1.5	451.4	499.7	500.0	121.4	88.1	88.3	1583.0	73,992.8	94,232.7	212,453.8

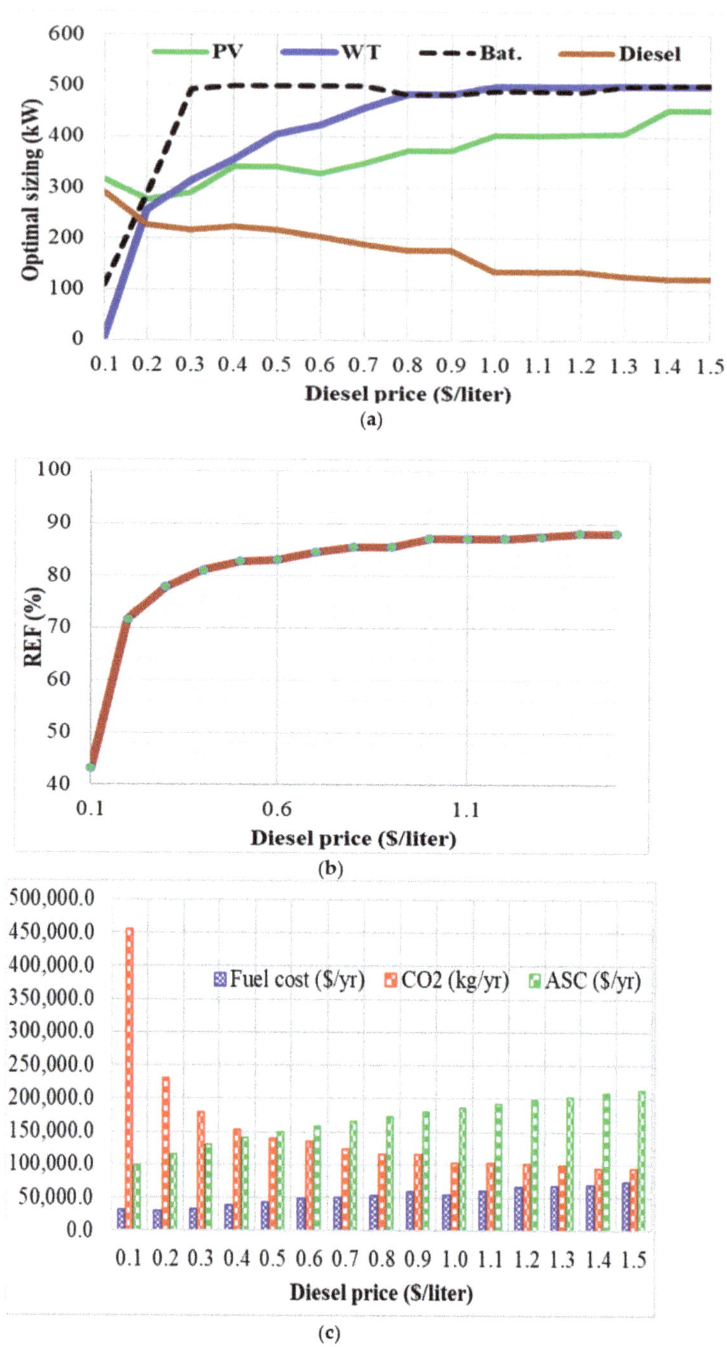

Figure 11. Simulation results using the proposed BO algorithm under different diesel prices of (**a**) optimal design/sizing of the solar PV, WT, diesel generator, and batteries; (**b**) REF %; (**c**) ASC, fuel cost, and CO_2 emissions.

6. Conclusions

The bonobo optimizer (BO) was proposed and applied for the optimal design/sizing of a hybrid renewable energy system (HRES) including PV, WT, diesel, and batteries to electrify an urban area called Al Sulaymaniyah village, in Arar in the northern part of Saudi Arabia. For the validation purposes, the BO was compared to four metaheuristic algorithms—BBBC, crow search, GA, and BOA—to find the optimal solution for the HRES with a quick convergence rate. These performance indicators (optimal solution, worst solution, mean, median, STDEV, and convergence rate) were used to discern the most appropriate performance among these five metaheuristic algorithms. The simulation findings revealed that the BO outperformed the other four metaheuristic algorithms—BBBC, crow search, GA, and BOA—where it achieved the optimal HRES solution/sizing with minimum ASC (USD 149,977.2), quick convergence time, and fewer oscillations around steady state. Both the BBBC and GA algorithms trapped into the local solution and failed to capture the global solution. In addition, they had high standard deviation, which means that the optimal solutions were highly dispersed. Hence, they had obvious oscillations around the optimal solution and took longer to converge. On the other hand, both BO and crow search had low standard deviation, which means that the optimal solutions were concentrated around the mean. These results prove the efficacy and robustness of the proposed BO algorithm compared to the other four metaheuristic optimization algorithms.

Author Contributions: Conceptualization, H.M.H.F. and A.A.A.-S.; methodology, H.M.H.F. and A.A.A.-S.; software, H.M.H.F. and A.A.A.-S.; validation, H.M.H.F. and A.A.A.-S.; formal analysis, H.M.H.F., A.A.A.-S., A.M.A.-S., A.A., A.M.N. and T.K.; investigation, H.M.H.F. and A.A.A.-S.; resources, A.M.A.-S., A.M.N., and A.A.; data curation, H.M.H.F. and A.A.A.-S.; writing—original draft preparation, H.M.H.F.; writing—review and editing, A.M.A.-S., A.A., A.M.N. and T.K.; visualization, H.M.H.F., A.A.A.-S. and A.M.N.; supervision, A.M.A.-S.; project administration, A.M.A.-S.; funding acquisition, A.M.A.-S. All authors have read and agreed to the published version of the manuscript.

Funding: This research received no external funding.

Acknowledgments: The authors would like to acknowledge the Researchers Supporting Project number (RSP-2021/337), King Saud University, Riyadh, Saudi Arabia.

Conflicts of Interest: The authors declare no conflict of interest.

References

1. Eltamaly, A.M.; Farh, H.M.H.; Al Saud, M.S. Impact of PSO reinitialization on the accuracy of dynamic global maximum power detection of variant partially shaded PV systems. *Sustainability* **2019**, *11*, 2091. [CrossRef]
2. Alturki, F.A.; Al-Shamma'a, A.A.; Farh, H.M.H. Simulations and dSPACE Real-Time Implementation of Photovoltaic Global Maximum Power Extraction under Partial Shading. *Sustainability* **2020**, *12*, 3652. [CrossRef]
3. Gao, K.; Wang, T.; Han, C.; Xie, J.; Ma, Y.; Peng, R. A Review of Optimization of Microgrid Operation. *Energies* **2021**, *14*, 2842. [CrossRef]
4. Lian, J.; Zhang, Y.; Ma, C.; Yang, Y.; Chaima, E. A review on recent sizing methodologies of hybrid renewable energy systems. *Energy Convers. Manag.* **2019**, *199*, 112027. [CrossRef]
5. Acuna, L.G.; Padilla, R.V.; Mercado, A.S. Measuring reliability of hybrid photovoltaic-wind energy systems: A new indicator. *Renew. Energy* **2017**, *106*, 68–77. [CrossRef]
6. Gong, X.; Dong, F.; Mohamed, M.A.; Abdalla, O.M.; Ali, Z.M. A secured energy management architecture for smart hybrid microgrids considering PEM-fuel cell and electric vehicles. *IEEE Access* **2020**, *8*, 47807–47823. [CrossRef]
7. Alnowibet, K.; Annuk, A.; Dampage, U.; Mohamed, M.A. Effective Energy Management via False Data Detection Scheme for the Interconnected Smart Energy Hub–Microgrid System under Stochastic Framework. *Sustainability* **2021**, *13*, 11836. [CrossRef]
8. Belmili, H.; Haddadi, M.; Bacha, S.; Almi, M.F.; Bendib, B. Sizing stand-alone photovoltaic–wind hybrid system: Techno-economic analysis and optimization. *Renew. Sustain. Energy Rev.* **2014**, *30*, 821–832. [CrossRef]
9. Zhao, X.; Wang, C.; Su, J.; Wang, J. Research and application based on the swarm intelligence algorithm and artificial intelligence for wind farm decision system. *Renew. Energy* **2019**, *134*, 681–697. [CrossRef]
10. Chen, H.-C. Optimum capacity determination of stand-alone hybrid generation system considering cost and reliability. *Appl. Energy* **2013**, *103*, 155–164. [CrossRef]
11. Shayeghi, H.; Hashemi, Y. Application of fuzzy decision-making based on INSGA-II to designing PV–wind hybrid system. *Eng. Appl. Artif. Intell.* **2015**, *45*, 1–17. [CrossRef]

12. Zhang, Y.; Ma, C.; Lian, J.; Pang, X.; Qiao, Y.; Chaima, E. Optimal photovoltaic capacity of large-scale hydro-photovoltaic complementary systems considering electricity delivery demand and reservoir characteristics. *Energy Convers. Manag.* **2019**, *195*, 597–608. [CrossRef]
13. Delgado-Antillón, C.; Domínguez-Navarro, J. Probabilistic siting and sizing of energy storage systems in distribution power systems based on the islanding feature. *Electr. Power Syst. Res.* **2018**, *155*, 225–235. [CrossRef]
14. Menshsari, A.; Ghiamy, M.; Mousavi, M.M.; Bagal, H. Optimal design of hybrid water-wind-solar system based on hydrogen storage and evaluation of reliability index of system using ant colony algorithm. *Int. Res. J. Appl. Basic Sci.* **2013**, *4*, 3582–3600.
15. Mohamed, A.F.; Elarini, M.M.; Othman, A.M. A new technique based on Artificial Bee Colony Algorithm for optimal sizing of stand-alone photovoltaic system. *J. Adv. Res.* **2014**, *5*, 397–408. [CrossRef]
16. Maleki, A.; Askarzadeh, A. Optimal sizing of a PV/wind/diesel system with battery storage for electrification to an off-grid remote region: A case study of Rafsanjan, Iran. *Sustain. Energy Technol. Assess.* **2014**, *7*, 147–153. [CrossRef]
17. Sanajaoba, S.; Fernandez, E. Maiden application of Cuckoo Search algorithm for optimal sizing of a remote hybrid renewable energy System. *Renew. Energy* **2016**, *96*, 1–10. [CrossRef]
18. Berrueta, A.; Heck, M.; Jantsch, M.; Ursúa, A.; Sanchis, P. Combined dynamic programming and region-elimination technique algorithm for optimal sizing and management of lithium-ion batteries for photovoltaic plants. *Appl. Energy* **2018**, *228*, 1–11. [CrossRef]
19. Zhou, W.; Lou, C.; Li, Z.; Lu, L.; Yang, H. Current status of research on optimum sizing of stand-alone hybrid solar–wind power generation systems. *Appl. Energy* **2010**, *87*, 380–389. [CrossRef]
20. Farh, H.M.; Al-Shaalan, A.M.; Eltamaly, A.M.; Al-Shamma'A, A.A. A Novel Crow Search Algorithm Auto-Drive PSO for Optimal Allocation and Sizing of Renewable Distributed Generation. *IEEE Access* **2020**, *8*, 27807–27820. [CrossRef]
21. Veilleux, G.; Potisat, T.; Pezim, D.; Ribback, C.; Ling, J.; Krysztofiński, A.; Ahmed, A.; Papenheim, J.; Pineda, A.M.; Sembian, S. Techno-economic analysis of microgrid projects for rural electrification: A systematic approach to the redesign of Koh Jik off-grid case study. *Energy Sustain. Dev.* **2020**, *54*, 1–13. [CrossRef]
22. Odou, O.D.T.; Bhandari, R.; Adamou, R. Hybrid off-grid renewable power system for sustainable rural electrification in Benin. *Renew. Energy* **2020**, *145*, 1266–1279. [CrossRef]
23. Amirkhalili, S.; Zahedi, A. Techno-economic analysis of a stand-alone hybrid wind/fuel cell microgrid system: A case study in Kouhin region in Qazvin. *Fuel Cells* **2018**, *18*, 551–560. [CrossRef]
24. Luna-Rubio, R.; Trejo-Perea, M.; Vargas-Vázquez, D.; Ríos-Moreno, G. Optimal sizing of renewable hybrids energy systems: A review of methodologies. *Sol. Energy* **2012**, *86*, 1077–1088. [CrossRef]
25. Jakhrani, A.Q.; Othman, A.-K.; Rigit, A.R.H.; Samo, S.R.; Kamboh, S.A. A novel analytical model for optimal sizing of standalone photovoltaic systems. *Energy* **2012**, *46*, 675–682. [CrossRef]
26. Hung, D.Q.; Mithulananthan, N.; Bansal, R. Analytical strategies for renewable distributed generation integration considering energy loss minimization. *Appl. Energy* **2013**, *105*, 75–85. [CrossRef]
27. Sanjel, N.; Baral, B.; Acharya, M.; Gautam, S. Analytical modelling for optimized selection between renewable energy systems and the conventional grid expansion. *J. Phys. Conf. Ser.* **2019**, *1266*, 012014. [CrossRef]
28. Sinha, S.; Chandel, S. Review of recent trends in optimization techniques for solar photovoltaic–wind based hybrid energy systems. *Renew. Sustain. Energy Rev.* **2015**, *50*, 755–769. [CrossRef]
29. Markvart, T. Sizing of hybrid photovoltaic-wind energy systems. *Sol. Energy* **1996**, *57*, 277–281. [CrossRef]
30. Clúa, J.G.G.; Mantz, R.J.; De Battista, H. Optimal sizing of a grid-assisted wind-hydrogen system. *Energy Convers. Manag.* **2018**, *166*, 402–408. [CrossRef]
31. Upadhyay, S.; Sharma, M. A review on configurations, control and sizing methodologies of hybrid energy systems. *Renew. Sustain. Energy Rev.* **2014**, *38*, 47–63. [CrossRef]
32. Cabral, C.V.T.; Oliveira Filho, D.; Diniz, A.S.A.C.; Martins, J.H.; Toledo, O.M.; Lauro de Vilhena, B. A stochastic method for stand-alone photovoltaic system sizing. *Sol. Energy* **2010**, *84*, 1628–1636. [CrossRef]
33. Khatib, T.; Ibrahim, I.A.; Mohamed, A. A review on sizing methodologies of photovoltaic array and storage battery in a standalone photovoltaic system. *Energy Convers. Manag.* **2016**, *120*, 430–448. [CrossRef]
34. Giallanza, A.; Porretto, M.; Puma, G.L.; Marannano, G. A sizing approach for stand-alone hybrid photovoltaic-wind-battery systems: A Sicilian case study. *J. Clean. Prod.* **2018**, *199*, 817–830. [CrossRef]
35. Rullo, P.; Braccia, L.; Luppi, P.; Zumoffen, D.; Feroldi, D. Integration of sizing and energy management based on economic predictive control for standalone hybrid renewable energy systems. *Renew. Energy* **2019**, *140*, 436–451. [CrossRef]
36. Mahmoudimehr, J.; Shabani, M. Optimal design of hybrid photovoltaic-hydroelectric standalone energy system for north and south of Iran. *Renew. Energy* **2018**, *115*, 238–251. [CrossRef]
37. Portero, U.; Velázquez, S.; Carta, J.A. Sizing of a wind-hydro system using a reversible hydraulic facility with seawater. A case study in the Canary Islands. *Energy Convers. Manag.* **2015**, *106*, 1251–1263. [CrossRef]
38. Elbaz, A.; Güneşer, M.T. Using crow algorithm for optimizing size of wind power plant/hybrid PV in Libya. In Proceedings of the 2019 3rd International Symposium on Multidisciplinary Studies and Innovative Technologies (ISMSIT), Ankara, Turkey, 11–13 October 2019; pp. 1–4.
39. Maleki, A.; Khajeh, M.G.; Ameri, M. Optimal sizing of a grid independent hybrid renewable energy system incorporating resource uncertainty, and load uncertainty. *Int. J. Electr. Power Energy Syst.* **2016**, *83*, 514–524. [CrossRef]

40. Mayer, M.J.; Szilágyi, A.; Gróf, G. Environmental and economic multi-objective optimization of a household level hybrid renewable energy system by genetic algorithm. *Appl. Energy* **2020**, *269*, 115058. [CrossRef]
41. Dufo-López, R.; Bernal-Agustín, J.L.; Yusta-Loyo, J.M.; Domínguez-Navarro, J.A.; Ramírez-Rosado, I.J.; Lujano, J.; Aso, I. Multi-objective optimization minimizing cost and life cycle emissions of stand-alone PV–wind–diesel systems with batteries storage. *Appl. Energy* **2011**, *88*, 4033–4041. [CrossRef]
42. Das, M.; Singh, M.A.K.; Biswas, A. Techno-economic optimization of an off-grid hybrid renewable energy system using metaheuristic optimization approaches–case of a radio transmitter station in India. *Energy Convers. Manag.* **2019**, *185*, 339–352. [CrossRef]
43. Abedi, S.; Alimardani, A.; Gharehpetian, G.; Riahy, G.; Hosseinian, S. A comprehensive method for optimal power management and design of hybrid RES-based autonomous energy systems. *Renew. Sustain. Energy Rev.* **2012**, *16*, 1577–1587. [CrossRef]
44. Zhao, J.; Yuan, X. Multi-objective optimization of stand-alone hybrid PV-wind-diesel-battery system using improved fruit fly optimization algorithm. *Soft Comput.* **2016**, *20*, 2841–2853. [CrossRef]
45. Shi, B.; Wu, W.; Yan, L. Size optimization of stand-alone PV/wind/diesel hybrid power generation systems. *J. Taiwan Inst. Chem. Eng.* **2017**, *73*, 93–101. [CrossRef]
46. Tabak, A.; Kayabasi, E.; Guneser, M.T.; Ozkaymak, M. Grey wolf optimization for optimum sizing and controlling of a PV/WT/BM hybrid energy system considering TNPC, LPSP, and LCOE concepts. *Energy Sources Part A Recovery Util. Environ. Eff.* **2019**, 1–21. Available online: https://www.tandfonline.com/doi/abs/10.1080/15567036.2019.1668880?journalCode=ueso20 (accessed on 27 December 2021).
47. Ahmadi, S.; Abdi, S. Application of the Hybrid Big Bang–Big Crunch algorithm for optimal sizing of a stand-alone hybrid PV/wind/battery system. *Sol. Energy* **2016**, *134*, 366–374. [CrossRef]
48. Fetanat, A.; Khorasaninejad, E. Size optimization for hybrid photovoltaic–wind energy system using ant colony optimization for continuous domains based integer programming. *Appl. Soft Comput.* **2015**, *31*, 196–209. [CrossRef]
49. Injeti, S.K. Butterfly optimizer-assisted optimal integration of REDG units in hybrid AC/DC distribution micro-grids based on minimum operational area. *J. Electr. Syst. Inf. Technol.* **2021**, *8*, 13. [CrossRef]
50. Alturki, F.A.; Al-Shamma'a, A.A.; Farh, H.M.; AlSharabi, K. Optimal sizing of autonomous hybrid energy system using supply-demand-based optimization algorithm. *Int. J. Energy Res.* **2021**, *45*, 605–625. [CrossRef]
51. Alturki, F.A.; Farh, H.M.H.; Al-Shamma'a, A.A.; AlSharabi, K. Techno-Economic Optimization of Small-Scale Hybrid Energy Systems Using Manta Ray Foraging Optimizer. *Electronics* **2020**, *9*, 2045. [CrossRef]
52. Al-Shamma'a, A.A.; Alturki, F.A.; Farh, H.M. Techno-economic assessment for energy transition from diesel-based to hybrid energy system-based off-grids in Saudi Arabia. *Energy Transit.* **2020**, *4*, 31–43. [CrossRef]
53. Das, A.K.; Pratihar, D.K. A new bonobo optimizer (BO) for real-parameter optimization. In Proceedings of the 2019 IEEE Region 10 Symposium (TENSYMP), Kolkata, India, 7–9 June 2019.

sustainability

MDPI

Article

Economic Viability of NaS Batteries for Optimal Microgrid Operation and Hosting Capacity Enhancement under Uncertain Conditions

Mohammed M. Alhaider [1,*], Ziad M. Ali [1,2,*], Mostafa H. Mostafa [3] and Shady H. E. Abdel Aleem [4]

1 Department of Electrical Engineering, College of Engineering in Wadi Alddawasir, Prince Sattam Bin Abdulaziz University, Wadi Alddawasir 11991, Saudi Arabia
2 Electrical Engineering Department, Aswan Faculty of Engineering, Aswan University, Aswan 81542, Egypt
3 Electrical Power and Machines Department, International Academy for Engineering and Media Science, Cairo 12411, Egypt; m.h.mostafa2050@gmail.com
4 Department of Electrical Engineering, Institute of Aviation Engineering and Technology, Giza 12658, Egypt; shady.abdelaleem@iaet.edu.eg
* Correspondence: m.alhaider@psau.edu.sa (M.M.A.); dr.ziad.elhalwany@aswu.edu.eg (Z.M.A.)

Abstract: Recent developments have increased the availability and prevalence of renewable energy sources (RESs) in grid-connected microgrids (MGs). As a result, the operation of an MG with numerous RESs has received considerable attention during the past few years. However, the variability and unpredictability of RESs have a substantial adverse effect on the accuracy of MG energy management. In order to obtain accurate outcomes, the analysis of the MG operation must consider the uncertainty parameters of RESs, market pricing, and electrical loads. As a result, our study has focused on load demand variations, intermittent RESs, and market price volatility. In this regard, energy storage is the most crucial facility to strengthen the MG's reliability, especially in light of the rising generation of RESs. This work provides a two-stage optimization method for creating grid-connected MG operations. The optimal size and location of the energy storage are first provided to support the hosting capacity (HC) and the self-consumption rate (SCR) of the RESs. Second, an optimal constrained operating strategy for the grid-connected MG is proposed to minimize the MG operating cost while taking into account the optimal size and location of the energy storage that was formerly determined. The charge–discharge balance is the primary criterion in determining the most effective operating plan, which also considers the RES and MG limitations on operation. The well-known Harris hawks optimizer (HHO) is used to solve the optimization problem. The results showed that the proper positioning of the battery energy storage enhances the MG's performance, supports the RESs' SCR (reached 100% throughout the day), and increases the HC of RESs (rising from 8.863 MW to 10.213 MW). Additionally, when a battery energy storage system is connected to the MG, the operating costs are significantly reduced, with a savings percentage rate of 23.8%.

Keywords: economic analysis; hosting capacity; market price; microgrid; bi-level optimization; renewable energy sources; sodium–sulfur batteries; uncertainty

Citation: Alhaider, M.M.; Ali, Z.M.; Mostafa, M.H.; Aleem, S.H.E.A. Economic Viability of NaS Batteries for Optimal Microgrid Operation and Hosting Capacity Enhancement under Uncertain Conditions. *Sustainability* **2023**, *15*, 15133. https://doi.org/10.3390/su152015133

Academic Editor: George Kyriakarakos

Received: 14 September 2023
Revised: 16 October 2023
Accepted: 20 October 2023
Published: 22 October 2023

1. Introduction

Due to the rapid growth of the international economy, electricity demand is increasing dramatically. As a result, pollution of the environment and fuel shortage are ongoing issues [1]. The global electricity market report demonstrates the strong connection between the growth of the global economy and the increase in electricity demand. In 2021, the worldwide gross domestic product increased by 5.9%, while in 2022, it increased by 4.9%. As a result, the global electricity demand increased by 6% in 2021 and 2.4% in 2022 [2]. To meet electrical requirements and reduce pollution, renewable energy sources (RESs) are used [3,4]. Nearly 320 GW of RES capacity was available in 2022: an increase of over 8%.

On the other hand, excessive usage of RESs may result in issues with the power system [5], including overloaded electrical system components, increased power losses, increased transmission line loading, issues with over-voltage, and a higher risk of short circuits [6,7].

Diverse techniques are used in research studies to address these issues and boost the capacity of RESs:

- The conventional grid support technique employs a variety of economically effective autonomous inverter control techniques [8].
- To reserve bus voltage in this manner, a tap changer with a low-voltage transformer is utilized [9].
- Active power curtailment of RESs approach that uses active power curtailment schemes to avoid voltage increases in feeders with high RES penetration [10].
- Demand-side management method: with this approach, users' energy consumption is decreased to diminish the issue of overvoltage [11].
- Reactive power control method: in this method, the relationship between the reactive power and active power of RESs is controlled to support the voltage buses of the electrical system [12].
- Energy storage systems (ESSs) address the issues brought on by high RES generation; the ESS is incorporated into the electrical power system in this approach [13].

In order to support renewable energies, the RES's hosting capacity (HC) without electrical operational issues must be improved [14]. The microgrid (MG) aims to combine RESs for self-consumption, using the energy as it is being produced to make RESs a commoditized alternative for electricity production and more cost-effective [15]. Therefore, RES generation is prioritized for self-consumption. In order to reduce overall operating costs, excess RES output should be added to the distribution or transmission grid when RES power exceeds the electrical load, i.e., to be stored [16–19].

Bearing these considerations in mind, enhancing the HC of RESs becomes crucial for the MGs and electrical power systems in general. Additionally, many factors, including topographical boundaries and intermittency, adversely impact the output power of RESs; managing these factors presents a significant problem [19,20]. A promising piece of equipment is an ESS, which has the potential to increase system reliability, increase its capacity for renewable energy sources, and make it more resilient to interruptions [21].

In order to improve the performance of electrical power systems, such as lowered transmission losses, increased energy efficiency, supported power quality, and reduced environmental pollution [22], MG systems connect ESS units, renewable and non-renewable energy resources, and diverse controllers. In addition, the transformer connecting the main grid and the MG can occasionally experience an overload because of the increased electrical load [23]. The accepted approach thus calls for strengthening the transformer to boost its capacity (reinforcement), but drawbacks include a lack of ability to reinforce the transformer and a low rate of return on investment [24]. As a result, to address the issue of transformer overloading, demand during peak hours must be reduced [25]. As a result, the method used to implement the best ESS allocation and reasonably schedule the output power of distributed generators (DGs), ESSs, and main grids in accordance with conditions for renewable and non-renewable energy resources and electrical demand not only affects the consistency of the energy supply of the electrical power system but also controls the cost and reliability of its operation to a significant extent.

The battery management system (BMS) is essential for controlling the state of power batteries, ensuring their safety, and improving their service efficiency [26]. BMS should keep the battery in appropriate working conditions and protect it from overcharge, over-discharge, and abnormal thermal conditions. A battery is a multistate, complicated nonlinear system. As a result, creating an effective and precise BMS serves as both the foundation for battery control and the key to successful battery management [27]. Battery data gathering, modeling and state estimation, charge and discharge control, problem detection and warning, temperature management, balancing control, and communication are only a few of a BMS's fundamental duties, as depicted in Figure 1 [28]. SOX algorithms improve

BMS and increase battery health and performance by measuring and predicting the state of energy, state of charge, state of power, and state of health of the battery with a high degree of accuracy, better fault tolerance, and robustness [29]. BMS controls the temperature of batteries to keep them working safely and effectively. Low temperatures can cause a reduction in battery capacity and poorer charging/discharging efficiency, while high temperatures can hasten battery aging and pose safety issues [30]. After each charging cycle, the voltages of all the cells in the battery pack are equalized through a process called battery balancing. Either the highest-charged cell is discharged, or the charge is transferred from one cell to another cell. Thus, the equalization management systems, one of the major components of the BMS, are essential to reducing such inter-cell inconsistency by redistributing the energy among the cells [31]. Due to the growing use of batteries in highly complex and powerful applications, fault detection has emerged as a critical function of the BMS. This is performed to guarantee the system's safe and dependable functioning [32].

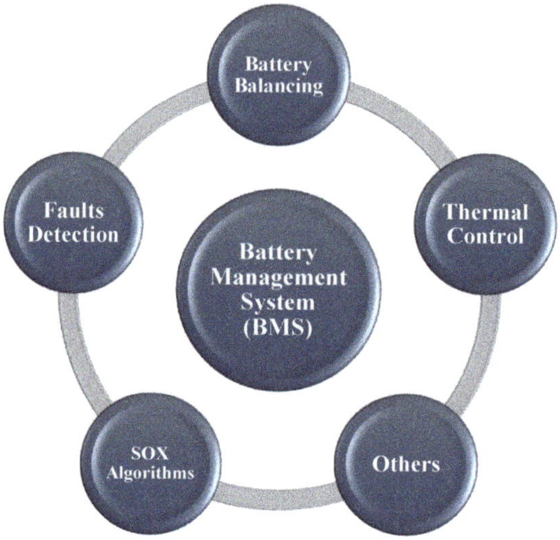

Figure 1. Main functions of a battery management system.

BMS is not the same for all types of batteries or schemes. It is designed specifically for the type of battery it is meant to manage. Different batteries have different chemistries, voltages, and charge/discharge characteristics, so the BMS needs to be tailored to these specifics [33]. For example, lithium-ion batteries require a different BMS compared to lead–acid batteries. The BMS for a lithium-ion battery needs to monitor factors such as cell voltage, temperature, and state of charge to ensure safe and efficient operation. Lead–acid batteries have different parameters that need to be monitored [34]. Additionally, different schemes (such as series or parallel configurations) may require specialized BMS setups to ensure that each cell or battery module is properly balanced and protected. It is crucial to match the BMS with the specific battery chemistry and configuration to optimize performance and safety [35]. The BMS for sodium–sulfur (NaS) batteries is unique to this specific type of battery chemistry. NaS batteries operate on a high-temperature principle, utilizing molten sodium and sulfur as active materials [36]. The BMS for NaS batteries needs to be designed to handle the unique characteristics of this chemistry. It monitors parameters such as temperature, cell voltage, state of charge, and cell balancing. Additionally, it manages the high operating temperatures that NaS batteries require for proper function. Since NaS batteries are commonly used in large-scale energy storage applications, the BMS also plays a critical role in ensuring the safety, performance, and longevity of the battery system. It helps prevent issues such as overcharging, over-discharging, and thermal runaway, which are crucial for maintaining the integrity of the battery and preventing safety hazards. The BMS for NaS batteries is tailored to the specific characteristics and operating conditions of NaS chemistry, ensuring the safe and efficient operation of these high-temperature batteries [37].

In order to improve the operation of the MG, various research papers have been published in recent years to examine the best ESS allocation and schedule the output power of DGs, ESSs, and the main grid. Numerous articles focused solely on the optimal ESS size [38–40]. For instance, active dispatch mode, a novel method suggested by Li et al. [38], allows for optimizing ESS capacity while enhancing electrical power system reliability. Additionally, the developed method considered the uncertainty characteristics and operated the ESS to shift peak demand. To reduce the cost of maintaining the electrical power system, Hou et al. [39] developed an optimal capacity model for wind turbines, photovoltaics, and ESSs. The model considered various ESS kinds, including compressed air, gravity storage, and battery storage. A mixed-integer linear program was suggested by Panuschka and Hofmann [40] to optimize the ESS in large industrial loads and increase flexibility. Additionally, many studies [41–43] focus on investigating how the ESS's location affects the functionality of the MG in addition to its size. In order to reduce the cost of operating the MG, Chen and Duan [41] presented an optimization methodology for determining the best location and size for ESS and DG. Mostafa et al. [42] presented an optimization technique to obtain the best ESS allocation for improving the voltage stability and performance of the MG by lowering power losses. A mathematical model was developed by Qiu et al. [43] to analyze the best ESS and micro-turbine scheduling and energy scheduling for the MG to manage its operation. However, most of the earlier studies had not looked at the effects of ESSs on the SCR, HC of the RESs, and overloading rates of transformers.

This work addresses this research gap by introducing a two-stage optimization approach to develop the operation of grid-connected MG while taking market price volatility, intermittent RES, and fluctuations in electricity demand into account. Firstly, the optimal size and location of the ESS are determined to support the HC and the SCR rate of the RESs. Secondly, the optimal working strategy is executed for the grid-connected MG to minimize the MG working cost, considering the EES's optimal size and location, which was obtained first.

The explained problem is investigated using the Harris hawks optimizer (HHO). During its wildlife rabbit hunting operations, the HHO imitates Harris hawks. With this intelligent strategy, Harris hawks can imitate several hunting attitudes based on different situations and rabbit evasion techniques [44]. HHO is superior to several swarm intelligence optimization algorithms, including Gaussian process optimization, firefly algorithm, biogeography-based optimization, particle swarm optimization, and grey wolf optimization algorithm, according to results verified over a variety of engineering optimization problems and benchmark functions [44]. The outcomes also show that HHO achieves a respectable balance between exploitation and exploration, enhancing the HHO's capacity to produce superior outcomes [44]. Furthermore, [45] demonstrated that the HHO is a potentially significant optimizer that supports the investigation of complex non-linear problems.

The following is a summary of this paper's principal advances:

- This work examines the impact of the ESS on RES hosting capacity and transformer loads connected to the MG with the main grid.
- An optimal ESS allocation is proposed to support the self-consumption rate of the RESs and reduce the overall operating costs of the MG while taking into account the actual operation of the MG with a high penetration level of RESs and taking market price volatility, intermittent RES, and changes in electrical load demand into account.
- The presented operation mode operates ESS units actively to optimally utilize the benefits from the ESS and minimize the operation cost of the various DGs included in the MG, taking into account the various cost factors, efficiency, and lifecycle of ESS. This is performed through operation constraints of the grid-connected MG, real-time modification, and energy management strategies.

The rest of the article is structured as follows: The MG arrangement, HC for RESs, and modeling of uncertainty are clarified in Section 2. The problem's mathematical formulation, HHO, and its application to address the problem are presented in Section 3. In Section 4, we give the simulation findings and discuss them. Section 5 concludes with a brief review of the work completed, the findings of the study, and future research.

2. Microgrid Configuration

The IEEE 33-bus system discussed in [13] has been employed in this study. The MG under-investigation is depicted in Figure 2. The entire data from [13] for each photovoltaic (PV) and wind turbine (WT) unit is provided in Table 1.

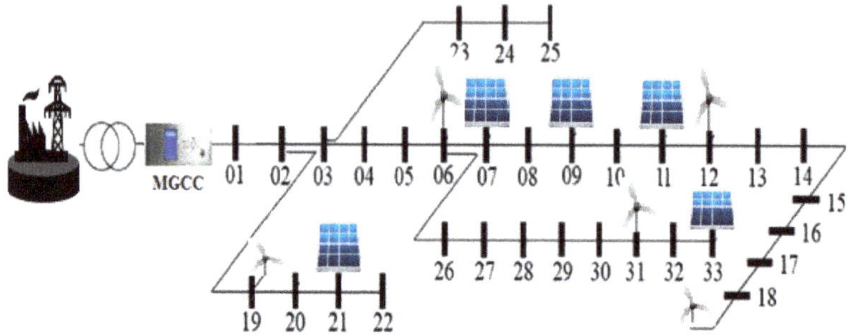

Figure 2. The MG investigated in this study.

Table 1. The entire data for WTs and PVs.

RES Type	Properties	Values				
PV	Location	7	9	11	21	33
	size (MW)	0.24	0.36	0.36	0.36	0.6
WT	Location	6	12	18	19	31
	size (MW)	1.2	0.6	0.6	0.96	1.2

2.1. Wind Turbine Units

The WT units related to wind speed are usually separated into four parts, as expressed in Equation (1) [19].

$$Power_h^{WT} = \begin{cases} 0 & vel_{WT,h} < vel_{WT}^{cut-in} \\ Power_R^{WT}\left(\frac{(vel_{WT,h})^3-(vel_{WT}^{cut-in})^3}{(vel_{RWT})^3-(vel_{WT}^{cut-in})^3}\right) & vel_{WT}^{cut-in} \le vel_{WT,h} < vel_{RWT} \\ Power_R^{WT} & vel_{RWT} \le vel_{WT,h} < vel_{WT}^{cut-out} \\ 0 & vel_{WT,h} \ge vel_{WT}^{cut-out} \end{cases} \quad (1)$$

where $vel_{WT,h}$, vel_{RWT}, $vel_{WT}^{cut-out}$, and vel_{WT}^{cut-in} express the current hourly wind speed, rated WT speed, the cut-out WT speed, and cut-in WT speed, respectively. $Power_h^{WT}$ and $Power_R^{WT}$ denote the output and rated powers of the WT, respectively.

2.2. Photovoltaic (PV) Stations

The output power of a PV station relates to the solar irradiance and the ambient temperature, as represented by Equation (2) [19].

$$Power_h^{PV} = NU^{PV} Power_R^{PV}\left(\frac{GI}{GI_0}\right)\left(1 - TC_{coff}(TC_{ambient} - 25)\right)\eta_v\eta_R \quad (2)$$

where $Power_R^{PV}$ and $Power_h^{PV}$ express the rated power and the output of the PV stations. η_v and η_R express the efficiency of the inverter and the PV relative efficiency. NU^{PV} denotes the number of PV stations. TC_{coff} and $TC_{ambient}$ express the temperature coefficient and the ambient temperature. GI and GI_0 are the global and standard solar irradiance under standard test conditions.

2.3. Operation of the Main Grid

The operation cost of the main grid $\left(Cost^{grid}\right)$ relates to the output of the main grid $\left(Power_h^{grid}\right)$ and market energy price $\left(MP_h^{grid}\right)$ in (USD/kW) considering several market price scenarios (NS) and their probabilities (x_h^s) as represented by Equation (3).

$$Cost^{grid} = \sum_{s=1}^{NS} \left(x_h^s.MP_h^{grid}\cdot Power_h^{grid}\right) \quad (3)$$

2.4. Battery Storage System

Battery energy storage systems (BSSs) come in a variety of forms, including lead–acid, lithium-ion, sodium–sulfur, nickel–cadmium, etc. [21]. The HC value of RES, grid stability, power calculations, peak load reduction, and energy management of the MG are all obviously impacted differently by each type's technological characteristics [19]. The technology of NaS is one of the most promising ones, which uses liquid sodium as the negative electrode and liquid sulfur as the positive electrode and is composed of inexpensive materials. High energy capacity, high efficiency, long cycling life, high operating temperature, and reasonable prices are some of the benefits of NaS batteries [5,15]. Additionally, Mostafa et al. [21] developed a methodology for calculating storage costs that considers both the technical and economic aspects of each storage type's many storage types. The conclusion of the study confirmed that because of its high efficiency, long lifespan, and affordable replacement prices, NaS storage offers the best cost among alternative storage options.

The capital cost (BSS_c) of the BSS is related to its power ($Power^{BS}$) and energy ($Energy^{BS}$) capacities as given in Equation (4).

$$BSS_c = \left(Cost^P\cdot Power^{BS}\right) + \left(Cost^E\cdot Energy^{BS}\right) \quad (4)$$

where $Cost^P$ (USD/kW) and $Cost^E$ (USD/kWh) are the costs related to the power and energy sizes of the storage.

2.5. Renewable Energy Hosting Capacity

The electrical power system used to be one-way, with energy flowing from the main grid to demands. The electrical network should allow for an exchange of power as RESs are increasingly used [46]. The high penetration level of RES, however, could have an adverse impact on how the electrical power grid functions and possibly result in several operational issues, including overloading of electrical system components, increased power losses, increased transmission line loading, overvoltage issues, and an increased risk of short circuits [47]. Researchers must determine how many RESs the power grid can accommodate without surpassing its operational limit. Electrical power system RES capacity may expand despite power grid limits [48]. Supporting the electrical network's HC increases RES penetration without electrical problems [48]. Figure 3 shows RES hosting capacity. Hosting capacity of RESs ($HCRES$) represents the ratio between the injected output power of RESs ($Power_{RESs}$) and the apparent power of demand (S_{load}), as given in Equation (5) [47]. The self-consumption rate of RESs (Ψ) expresses the ratio between the actual energy of RES ($Energy_{RES}$) and the overall produced RES energy ($Energy_{RES}^{rated}$) as given in Equation (6) [48].

$$HCRES\% = \frac{Power_{RESs}}{S_{load}}\cdot 100 \quad (5)$$

$$\Psi = \frac{Energy_{RES}}{Energy_{RES}^{rated}} \tag{6}$$

The power systems need to improve Ψ and *HCRES*. The ESSs are essential electrical network elements that can effectively perform this [49].

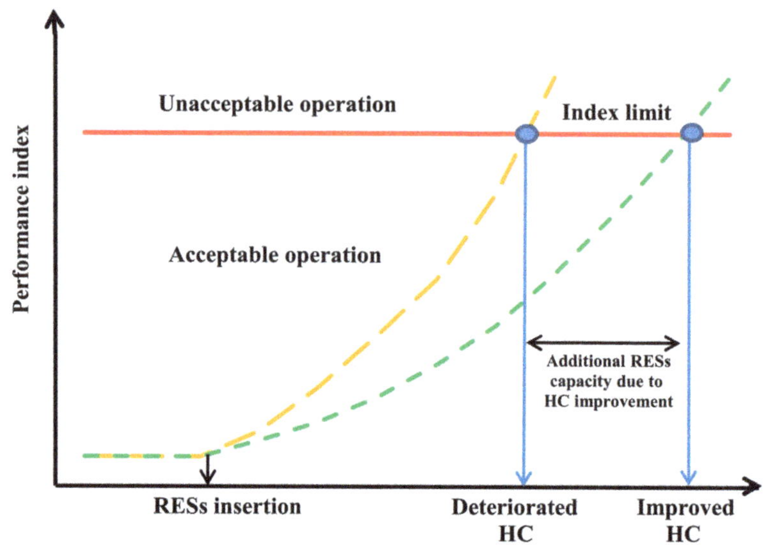

Figure 3. The HC and the importance of its support (as noted by the green dashed line).

2.6. Uncertainty Modeling

In stochastic optimization, creating suitable scenarios is crucial to enabling decisions based on precise assessments of uncertainties [19]. As a result, to accurately reflect the usual measurement, the uncertainties' estimations must use a realistic method. Each uncertainty modeling technique would require a distinct design for the system. Consequently, it is critical to use the appropriate approach when modeling uncertainty. In contrast to the deterministic technique, which relies on precise knowledge of well-known characteristics, uncertainty modeling simulates the volatility in market price, RES output power, and electrical load. In order to imitate the probability properties of the parameters, random distributions are employed as inputs to the random optimization problem [50]. To illustrate the numerous system parameter uncertainties, this study develops a number of scenarios, each with a known probability. Uncertain scenarios for electricity demand, RES capacity for generation, and market pricing are generated using the fuzzy clustering method (FCM). Then, grouping these instances into a more reasonable set is desirable.

FCM is used in this work to split a specific number of data (M) into a specific number of clusters (O), with $O = 10$, as specified in [51]. As the number of scenarios rises, the problem becomes more complex and challenging, requiring a larger processing package.

A matrix Z with a collection of column vectors z_j, where $j = 1, 2, \ldots, M$, collects the data required for clustering. To group Z, FCM requires the elements O and the fuzziness component (d), where $d > k$ and $d > 1$. The procedure is presupposed to end at a predetermined tolerance (*eps*). There are five phases in the FCM clustering algorithm:

- Phase 1: A membership matrix ($k = [k_{ij}]_{O \times M}$) is initialized randomly, where the sum of each column j in k must equal 1. O random centroids are chosen from the data. These centroids are gathered in a vector $= [O_i]_{1 \times O}$.

- Phase 2: compute the new centroids utilizing Equation (7):

$$O_i = \frac{\sum_{i=1}^{M} k_{ij}^{d} \times z_j}{\sum_{i=1}^{M} k_{ij}^{d}} \quad (7)$$

- Phase 3: compute the elements of the membership matrix $\left(k = [k_{ij}]_{O\times M}\right)$ for each element in Z, where:

$$k_{ij} = \frac{1}{\sum_{p=1}^{O}\left(\frac{\|z_j - O_i\|}{\|z_j - O_p\|}\right)^{\frac{2}{d-1}}} \quad (8)$$

- Phase 4: compute $f_{FCM}^{(n)} = \sum_{j=1}^{M}\sum_{i=1}^{O} k_{ij}^{d}\|z_j - O_i\|$, where $f_{FCM}^{(n)}$ represents the objective function value at the nth iteration.
- Phase 5: if $\left\|f_{FCM}^{(n)} - f_{FCM}^{(n-1)}\right\| < eps,\ \forall n$, stop the algorithm; otherwise, repeat the procedure starting from Phase 2.

3. Problem Formulation

Problem formulation can be divided into objective function system constraints and optimization algorithms, as follows.

3.1. Objective Function

Two levels are used to introduce the objective function. First, the BSS's optimal size and location, as specified by OF_1 in Equation (9), are implemented to increase the SCR of all RESs in the MG, where $\Psi(x)$ reflects the self-consumption of all RESs in the MG. Second, using the optimal BSS size and position that was initially determined, an optimal operating strategy is implemented for the grid-connected MG, as shown in Equation (10), to reduce the MG operating cost where OF_2 expresses the total cost of BSS per day ($TCSS$) (USD/day), generation costs of WT and PV (USD/kWh), and the operation cost of the utility (USD/kWh), considering the several PV output power scenarios (MS) and their probabilities $\left(b_h^{pv,s}\right)$, several WT output power scenarios (WS) and their probabilities $\left(b_h^{wt,s}\right)$, and numerous market price scenarios (NS) and their probabilities (x_h^s).

$$OF_1 = \max\Psi(x) = \sum_{h=1}^{H}\left(\frac{Energy_{RES,h}}{Energy_{RES,h}^{rated}}\right) \quad (9)$$

$$OF_2 = \sum_{h=1}^{H}\sum_{s=1}^{NS}\left(x_h^s \cdot MP_h^{grid} \cdot Power_h^{grid}\right) + \sum_{s=1}^{MS}\left(b_h^{pv,s} \cdot B_h^{PV} \cdot Power_h^{PV}\right) + \sum_{s=1}^{WS}\left(b_h^{wt,s} \cdot B_h^{WT} \cdot Power_h^{WT}\right) + TCSS \quad (10)$$

where $Energy_{RES,h}$ and $Energy_{RES,h}^{rated}$ are the total RESs energy. $Power_h^{grid}$, $Power_h^{WT}$, and $Power_h^{PV}$ are the main grid, WT, and PV output powers at each hour h, respectively. MP_h^{grid}, B_h^{WT}, and B_h^{PV} are the kWh price of the main grid, WT, and PV at h, respectively.

Life cycle assessment includes all costs of batteries, such as capital cost, operation and maintenance cost, and replacement cost of batteries [52]. In this research, it was assumed that the purchase price of BSS covers all of its components, including the capital and replacement costs during the course of the project. The capital cost (CBS_c) of the BSS is related to its power ($Power^{BS}$) and energy ($Energy^{BS}$) capacities as given in Equation (11).

$$CBS_c = \left(Cost^P \cdot Power^{BS}\right) + \left(Cost^E \cdot Energy^{BS}\right) \quad (11)$$

where $Cost^P$ (USD/kW) and $Cost^E$ (USD/kWh) are the coefficient cost of BSS function of rated power of BSS and its energy. To obtain the replacement number of BSS, first, the number of cycles achieved over the BSS $(Battery_{cycles})$ is obtained by Equations (12) and (13) to determine the BSS lifetime. Second, Equation (14) used to obtain the lifetime of BSS (L_{BS}) relying on the life cycle of a battery ($Battery_{Lifecycles}$) and $Battery_{cycles}$.

$$n_{Battery}(h,j) = \left(k_{a(h)} - k_{a(h-1)}\right)y_{a(h)}, \quad \forall h \in H, \forall j \in D \tag{12}$$

$$Battery_{cycles} = \sum_{j=1}^{D}\sum_{h=1}^{H} n_{Battery}(h,j) \tag{13}$$

$$L_{BS} = \frac{Battery_{Lifecycles}}{Battery_{cycles}} \tag{14}$$

where $n_{Battery}(h,j)$ expresses the BSS cycles, and D represents the total number of operating days per year. BSS has two statuses: charge and discharge, $k_{a(h)}$ represents the status of BSS at each hour during the operating days per year. Therefore, the BSS replacement number ($RN_{Battery}$) through the project lifetime (Q) is represented by Equation (15).

$$RN_{Battery} = \frac{Q}{L_{BSS}} \tag{15}$$

Accordingly, $TCSS$ (USD/day) can be obtained by using Equation (16) as a function of the interest rate i.

$$TCSS = \frac{1}{D \cdot Q}\left(\frac{i(1+i)^{Q}}{(1+i)^{Q}-1} \cdot CBS_c \cdot RN_{Battery}\right) \tag{16}$$

3.2. Constraints

For the solution to be applicable, the investigation must incorporate many sets of constraints, as follows:

3.2.1. RES Constraints

The generated power by WT must be restricted by its minimum power value $Power_{h,min}^{WT}$ and its maximum power value $Power_{h,max}^{WT}$ as represented in Equation (17). Similarly, the generated power by PV must be restricted by its minimum power value $Power_{h,min}^{PV}$ and its maximum power value $Power_{h,max}^{PV}$ as represented in Equation (18).

$$Power_{h,min}^{WT} \leq Power_{h}^{WT} \leq Power_{h,max}^{WT}, \forall h \tag{17}$$

$$Power_{h,min}^{PV} \leq Power_{h}^{PV} \leq Power_{h,max}^{PV}, \forall h \tag{18}$$

3.2.2. Power Balance

The total produced output power from the different DGs must be equal to the total load scenarios ($P_{load,s,h}$) and their probabilities ($\psi^{s,h}$) at all times during the day, as represented in Equation (19).

$$\begin{aligned}\sum_{s=1}^{NS} x_h^s \cdot Power_h^{grid} &+ \sum_{s=1}^{MS}\left(b_h^{pv,s} \cdot Power_h^{PV}\right) + \sum_{s=1}^{WS}\left(b_h^{wt,s} \cdot Power_h^{WT}\right) \\ &+ Power_{DIS,h}^{BS} \\ &= \sum_{s=1}^{LS}\left(\psi^{s,h} \cdot P_{load,s,h}\right) + Power_{CH,h}^{BS} + \sum_{b=1}^{NR} Power_{b,h}^{losses} \quad \forall h \in H\end{aligned} \tag{19}$$

where $Power_{DIS,h}^{BS}$ and $Power_{CH,h}^{BS}$ are the BSS discharge and charge, respectively. $Power_{b,h}^{losses}$ and NR are the active MG loss of the bth line and the number of lines.

3.2.3. Voltage Limits

The root mean square (rms) value of the bus voltage $\left(Voltage^{bus}\right)$ must not decrease below the minimum voltage $Voltage^{bus}_{min}$ which is set to 0.95 p.u. value, and do not increase over the maximum voltage value $Voltage^{bus}_{max}$ which is set to 1.05 p.u. in this study, as represented in Equation (20).

$$Voltage^{bus}_{min} \le Voltage^{bus} \le Voltage^{bus}_{max} \tag{20}$$

3.2.4. Carrying Current Capacity Limit

The current flowing in each branch (TI^{line}_{RMS}) must not exceed the maximum carrying capacity of the branch ($TI^{line-max}_{RMS}$), as given by Equation (21).

$$TI^{line}_{RMS} \le TI^{line-max}_{RMS} \tag{21}$$

3.2.5. Energy Storing Limits

BSSs have many limits that must be considered in this study, such as the charging power ($Power^{BS}_{CH,h}$) and the discharging power ($Power^{BS}_{DIS,h}$), as represented by Equations (22) and (23).

$$Power^{BS}_{CH,h} \le Power^{BS-max}_{CH,h}, \ \forall h \le H \tag{22}$$

$$Power^{BS}_{DIS,h} \le Power^{BS-max}_{DIS,h}, \ \forall h \le H \tag{23}$$

The state of charge of BSS ($BSOC^h$) must be restricted by its minimum ($BSOC^h_{min}$) and maximum ($BSOC^h_{max}$) thresholds as specified in Equation (24) with respect to the efficiency of charge (η^{Bat}) of the BSS. The current $BSOC^h$ is a function of the previous $BSOC^{h-1}$ and the charge and discharge capacities at h as specified in Equation (25). The initial $BSOC$ $\left(BSOC^{in}\right)$ is considered at h = 1, as specified in Equation (25).

$$BSOC^h_{min} \le BSOC^h \le BSOC^h_{max}, \ \forall h \le H \tag{24}$$

$$BSOC^h = \begin{cases} BSOC^{in} + \Delta h \ \eta^{Bat} \ Power^{BS}_{CH,h} - \Delta h \ Power^{BS}_{DIS,h}, & h = 1 \\ BSOC^{h-1} + \Delta h \ \eta^{Bat} \ Power^{BS}_{CH,h} - \Delta h \ Power^{BS}_{DIS,h}, & \forall h \ge 2, \ h \in H \end{cases} \tag{25}$$

At the end of the day, the $BSOC^h$ should be the same $BSOC^{in}$ to maintain $BSOC^{in}$ is always constant, as represented by Equation (26).

$$BSOC^h = BSOC^{in}, \ h = H \tag{26}$$

Equation (27) demonstrates that when the efficiency η^{Bat} is taken into account, the discharge power is always equal to the charge power.

$$\sum_{t=1}^{T} Power^{BS}_{DIS,h} = \sum_{t=1}^{T} Power^{BS}_{CH,h} \cdot \eta^{Bat} \tag{27}$$

3.3. *Harris Hawks Optimizer*

Heidari et al. presented the Harris hawks optimizer (HHO) in 2019 [53,54], a recent population-based optimization method. A flock of hawks will startle its prey, usually a rabbit, by attacking it from several angles. A leader hawk encircles the victim in this synchronized attack. The hawks' abilities to alter their hunting strategies in response to the hunting environment and the rabbits' struggle to avoid capture. The three stages of Harris hawk hunting are exploration, the transition from exploration to exploitation, and globalization of search (exploitation). The hawks scour the immediate region throughout their excursion, using their outstanding vision to find rabbits. The first tactic relies on all the hawks cooperating to shock the rabbit, whereas the second focuses on having the hawks' leader attack the rabbit following the rabbit's abilities. Hawks may choose where to sit depending on the locations of nearby hawks, as expressed in (28), provided that $\alpha < 0.5$ is fulfilled and each choice is given an equal probability.

$$H\ (t+1) = \begin{cases} H_R(t) - \alpha\ |H_R(t) - 2\tau\ H(t)| & Q \geq 0.5 \\ (H_{Best}(t) - \left(\frac{1}{M}\sum_{i=1}^{M} H_i(t))\right) - \varphi\ (LB + \varnothing\ (UB - LB) & Q < 0.5 \end{cases} \tag{28}$$

where $H(t+1)$ denotes the hawks' location vector at iteration $t+1$, $H(t)$ denotes the Hawks' location vector during iteration t. The place of the prey is represented by $H_{Best}(t)$; the total number of hawks is represented by M; and the random values α, τ, φ, $\varnothing$, and Q are generated from the range [0, 1].

HHO can transition from exploration to exploitation by using rabbit escape energy (E):

$$E = 2\ E_o\left(1 - \frac{t}{T}\right) \tag{29}$$

E_o is the rabbit's initial random energy, which is calculated for each iteration from the range $[-1, 1]$, and T stands for the maximum number of iterations. According to the rabbit escape scenario, Harris hawks can hunt using either a hard besiege or a delicate attachment method. The rabbit attempts to escape the gentle besiege with $p \geq 0.5$ and $|E| \geq 0.5$ but finally fails. These attacks involve the hawks softly around the rabbit to make it more tired before coming in from nowhere. Equations (30) and (31) are used to express this behavior.

$$H(t+1) = \Delta H(t) - E|(J \cdot H_{Best}(t)) - H(t)| \tag{30}$$

$$\Delta H(t) = H_{Best}(t) - H(t) \tag{31}$$

where J is the rabbit's random escape strength and $\Delta H(t)$ is the difference between the $H_{Best}(t)$ and $H(t)$ in iteration t.

The rabbit is worn out and has minimal escape energy throughout the difficult besiege, where $r \geq 0.5$ and $|E| < 0.5$. As a result, the hawks seldom ever surround the rabbit to launch an unexpected assault. This behavior is represented by Equation (32).

$$H(t+1) = H_{Best}(t) - E|\Delta H(t)| \tag{32}$$

Further, more advanced soft and hard siege techniques are possible, as mentioned in [53]. For further information regarding the HHO, the reader could refer to [53,54].

4. Numerical Results and Their Discussion

To more effectively solve the uncertainty parameters and comprehend the implications of parameter uncertainty on the result, the stochastic technique makes use of many scenarios and the corresponding probability. Using historical data, 1000 scenarios have been developed to simulate the uncertainty of each PV and WT, load demand, and market price change. Then, in order to shorten calculation time, the number of PV, WT, demand, and market price scenarios is reduced to the ten most-probable scenarios using a scenario reduction technique based on the FCM clustering algorithm.

Following the scenario reduction technique, the clustering powers of the PV installed in bus 7 and their probabilities are shown in Figure 4. The clustering power of the PV systems installed in buses 9, 11, and 21 is shown in Figure 5, along with their probabilities. In addition, Figure 6 displays the clustered power of PV installed in bus 33 and their probabilities.

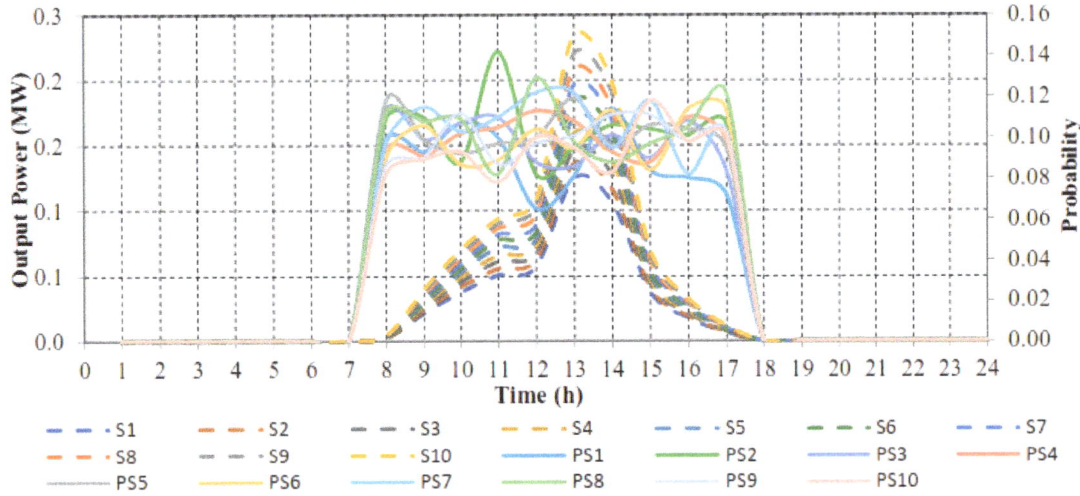

Figure 4. Hourly power scenarios and their probabilities of PV installed in bus 7.

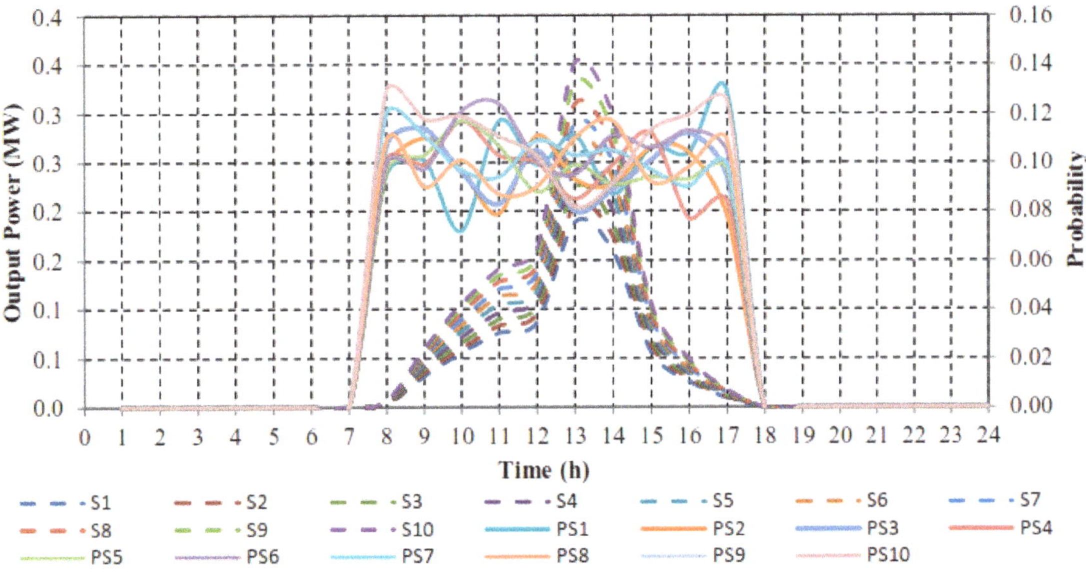

Figure 5. Hourly power scenarios and their probabilities of PVs installed in buses 9, 11, and 21.

The clustering powers of the WTs placed in buses 6 and 31, as well as their probabilities, are presented in Figure 7. The WT systems installed in buses 12 and 18 have the same output power because they have the same historical data and size. The clustering powers of the WT placed in buses 12 and 18 and their probabilities are shown in Figure 8. The clustering powers of WT installed in bus 19 are shown in Figure 9 and their probabilities. The demand for electricity is never uniform; it changes every hour. The maximum load in the MG under study is 3.715 MW; the hourly load is a percentage of the maximum load. The key factor contributing to the complexity of MG management is the inherent variability of the load that customers need. As a result, this analysis takes the electrical load's uncertainty into account. Figure 10 depicts the power load clustering scenarios and their probabilities for each hour of the day. Figure 11 illustrates the market price clustering scenarios and their probabilities for each hour of the day.

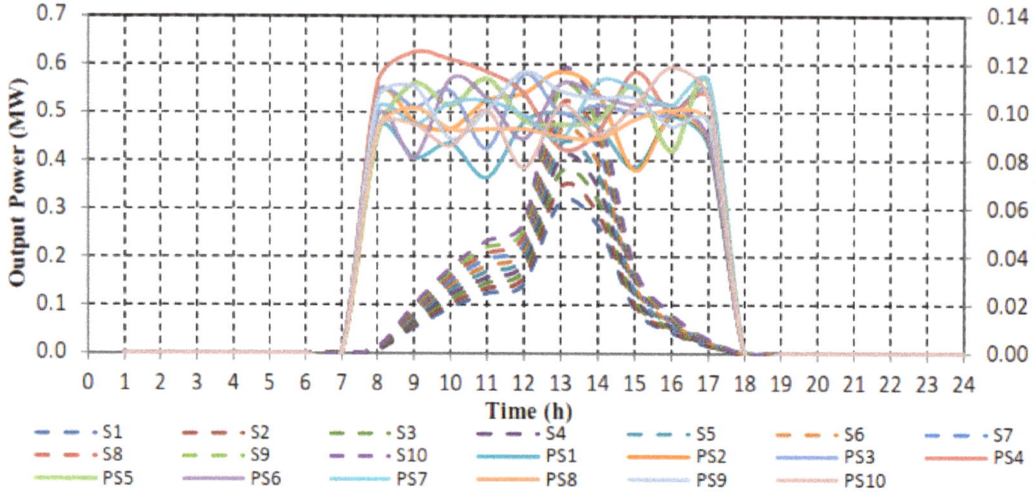

Figure 6. Hourly power scenarios and their probabilities of PV installed in bus 33.

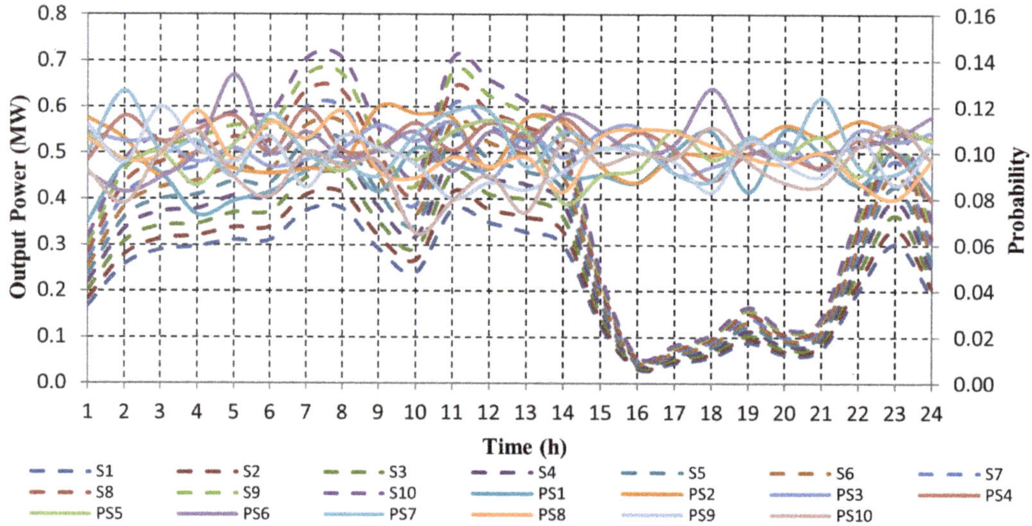

Figure 7. Hourly power scenarios and their probabilities of WTs installed in buses 6 and 31.

The system voltage fluctuations and total power loss will be adversely affected by the placement of the BSS in the MG. Therefore, developing an optimal approach for choosing the appropriate location and size for the BSS is essential. The SCR and HC of RESs must be improved, and the operation cost of the MG must be minimized while taking into account operational microgrid constraints such as PV and WT uncertainty, electrical load variation, market price fluctuations, RES power limits, power balance limits, voltage limits, line capacity constraints, and energy storage limits. In order to maximize SCR to 100% and reduce MG operation cost while considering operational MG limits, BSS is added to each bus, boosting its capacity. This plan was carried out using the HHO, including the following steps:

- Read all RES, BSS, and MG data in Step 1.
- Execute the MG's load flow in Step 2 and store the results.
- In Step 3, start the HHO program.
- Step 4 attaches the BSS to each bus using various power and energy values.
- Step 5: Run the load flow and obtain the value of the objective function for each size.
- Step 6: Repetition of Steps 4 and 5 will help to determine the best BSS size and location for the most remarkable (the best global) objective function.

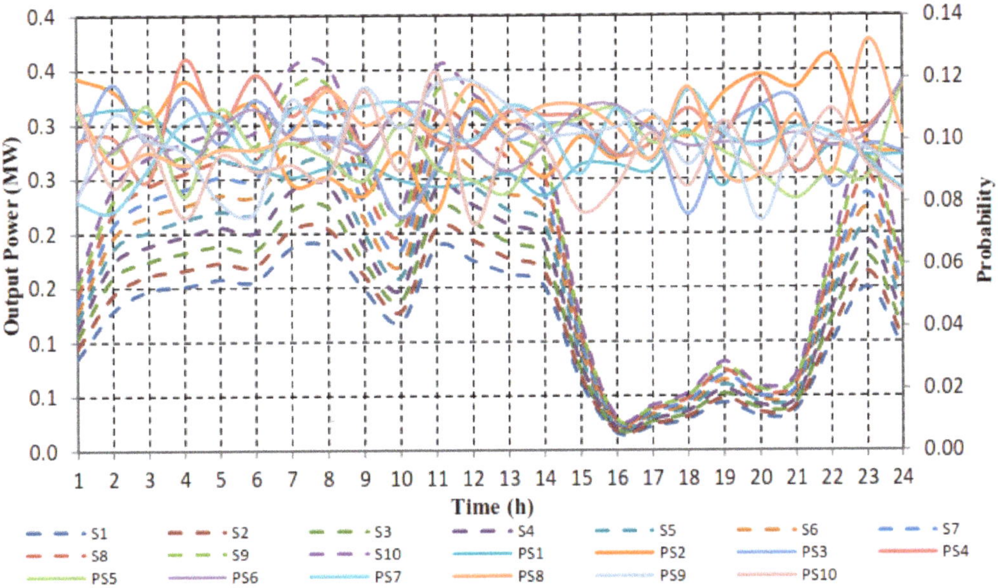

Figure 8. Hourly power scenarios and their probabilities of WTs installed in buses 12 and 18.

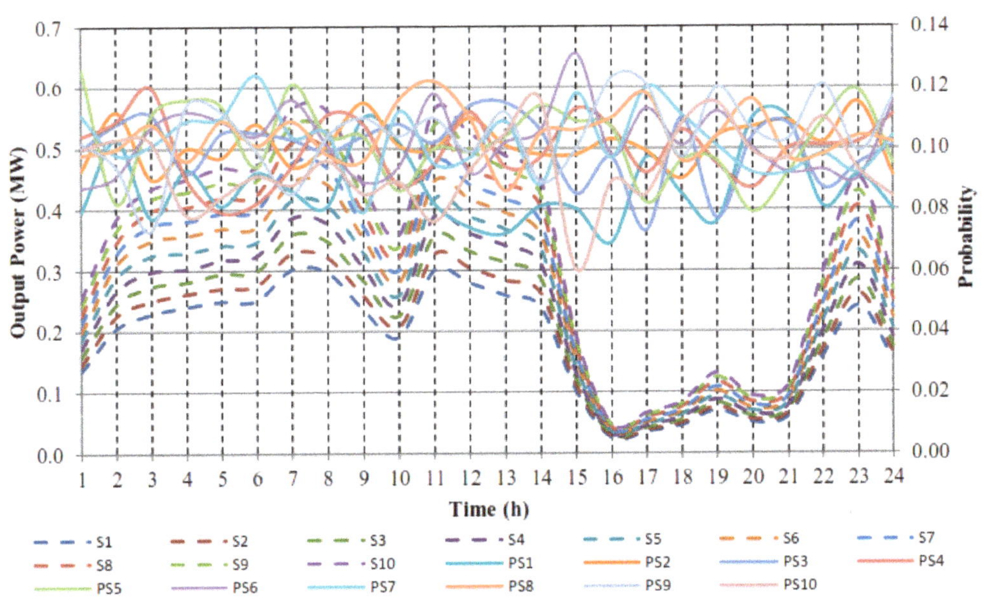

Figure 9. Hourly power scenarios and their probabilities of WTs installed in bus 19.

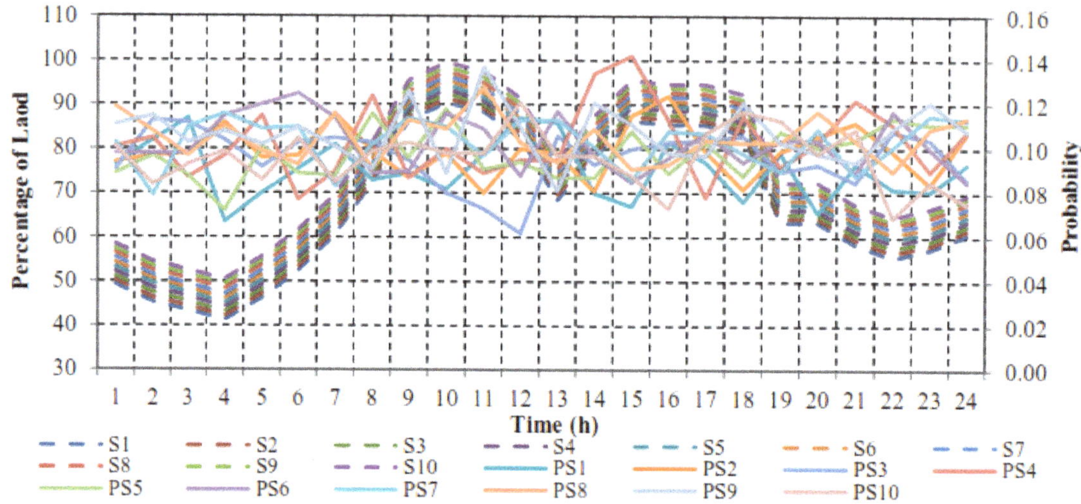

Figure 10. Hourly generated electrical load scenarios and their probabilities.

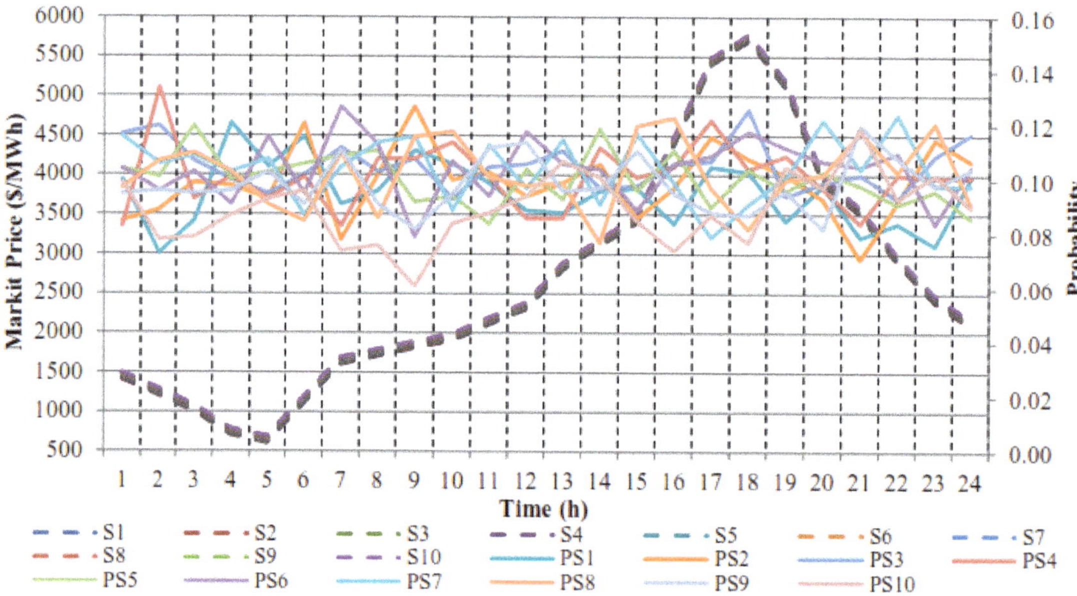

Figure 11. Hourly generated market price scenarios and their probabilities.

The PV's and WT's specified bids are set to 2.8 (USD/kWh) and 1.72 (USD/kWh), respectively [2]. Table 2 provides information on the cost, efficiency, and durability of the NaS battery used in this study [2,13].

Table 2. Efficiency, cost factors, and lifecycle of NaS batteries [2,13].

Battery	Capital Power Cost (USD/kW)	Capital Energy Cost (USD/kWh)	Efficiency (%)	Lifecycle	Lifetime (Years)
NaS	350	300	95	4500	15

The appropriate position and size of the NaS batteries are given in Table 3 in order to maximize SCR and lower the operating cost of the MG. The MG's cost-effectiveness is established by lowering running expenses and maximizing the system's self-consumption of PV and WT after determining the NaS's optimal position.

Table 3. Optimal allocation of the NaS battery.

Battery	Location	Power (MW)	Energy (MWh)
NaS	6	2.06	12.37

Figure 12 shows the SCR of WTs and PVs on an hourly basis without the NaS battery. It is clear that neither of them always consumes themselves entirely. The self-consumption of WT varies between 81.8% and 91.4% from hours 3 to 5, and it equals 90.1% in hour 7. Additionally, the self-consumption of PV varies between 31.8% and 66.2% from hour 13 to hour 14. Undoubtedly, the MG operator wants to increase the WT and PV's self-consumption to 100% at all times of the day in order to make them more commoditized and accessible as options for electricity generation. After adding the NaS battery, the WT and PV self-consumption equals 100% during the entire day. The NaS positively affects RES self-consumption in this regard.

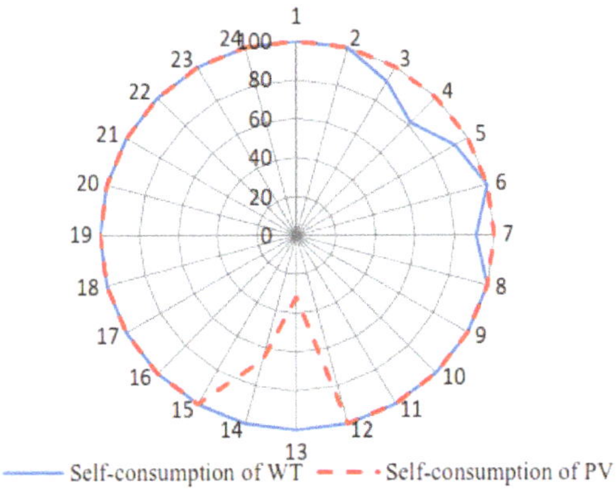

Figure 12. The self-consumption of PV and WT with no NaS battery connected.

Figure 13 shows the hourly HC of RESs in the investigated MG without and with one NaS battery. As shown in this figure, the HC of RESs increases from 8.863 MW in the absence of a NaS battery to 10.213 MW, improving the HC of RESs.

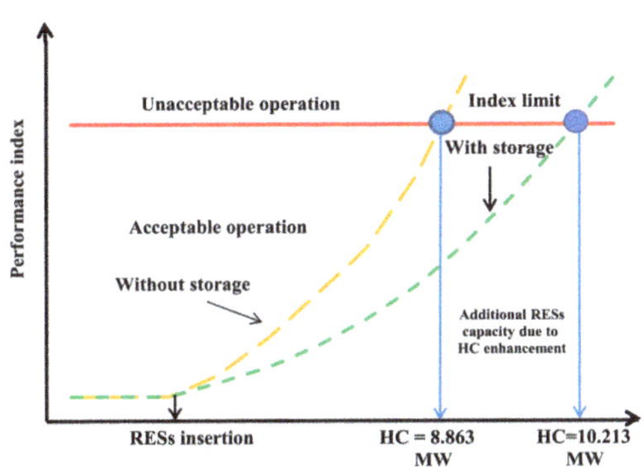

Figure 13. HCRES in the studied MG.

The operational costs of the MG with and without the NaS battery are shown in Table 4, which provides the MG economic analysis after determining the appropriate NaS battery allocation from an economic perspective.

Table 4. Impact of the NaS battery on the MG from an economic perspective.

Case	With No NaS	With NaS
Operation cost (USD/day)	183,645.3	138,550.7
BS_c (USD)	-----	4,432,225.5
RN_{BSS}	-----	3
Cost NaS/day (USD/day)	-----	1457.2
Total operating cost/day (USD/day)	183,645.3	140,007.9
Saving (%)	-----	23.8

The total cost per day for the project includes the capital and replacement costs for the NaS, as shown in Table 4. The project life span in this analysis is 25 years, and the interest rate is 0.08. In order to know the NaS batteries' replacement number over the project's duration, the batteries' expected lifetimes are calculated. The saving percentage is determined with respect to the base scenario while keeping in mind the life cycles of the NaS provided in Table 2 and the overall number of cycles completed through the NaS $\left(Battery_{cycles}\right)$ per year. The results showed that adding NaS to the MG considerably reduces operating expenses.

The optimal output powers for the main grid, PV, WT, and NaS at each hour of the day are presented in Figure 14. Figure 15 shows the NaS battery's SOC for each hour of the day. In order to comply with the MG constraints, Figures 14 and 15 indicate that the battery storage is charged when the energy price is low and the overall load is not high, such as the first periods from hour 1 to hour 7. When the energy market price is high, such as between hours 16 and 21, the battery storage begins to discharge in order to lower the MG's operating costs. Figure 14 clearly shows that the BSS is charged in the early periods because of the low market price and light total load. Figure 16 shows that the SOC of the battery changed from 0% at hour 1 to 100% at hour 14.

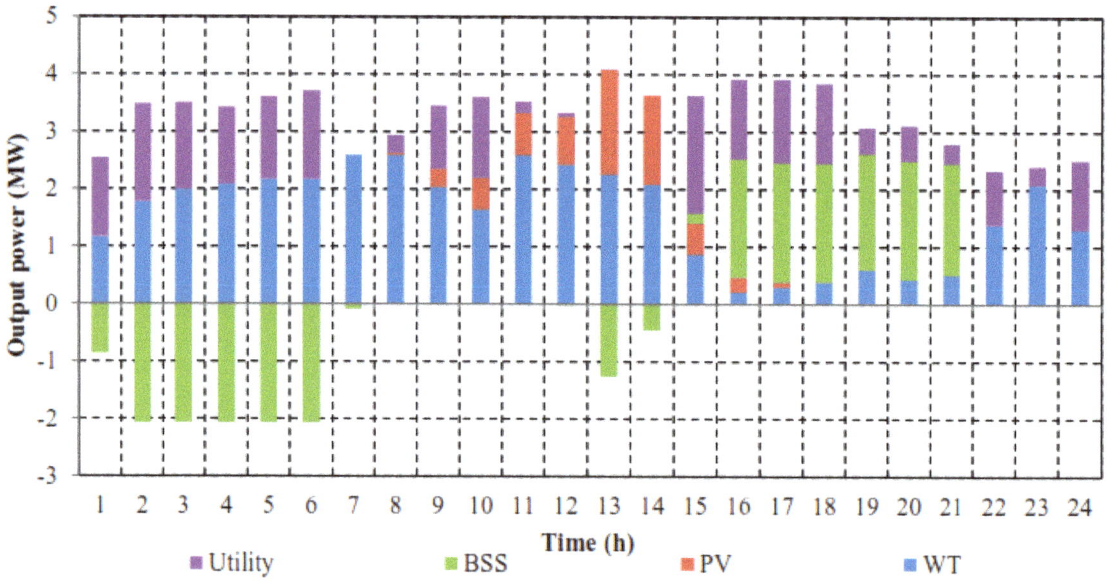

Figure 14. Optimal hourly power of the main grid, WT, PV, and the BSS.

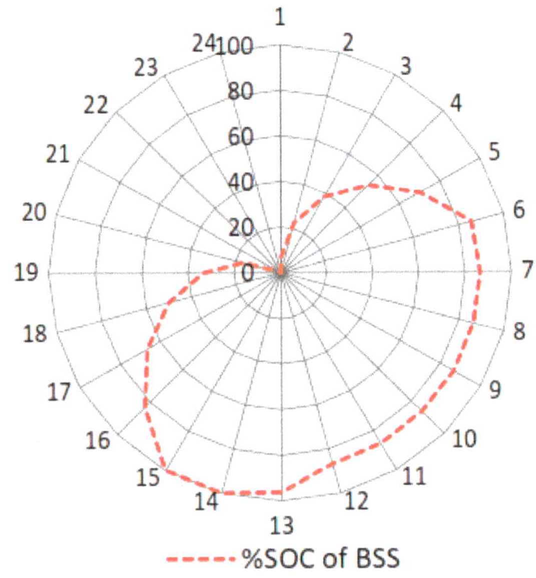

Figure 15. SOC of the BSS.

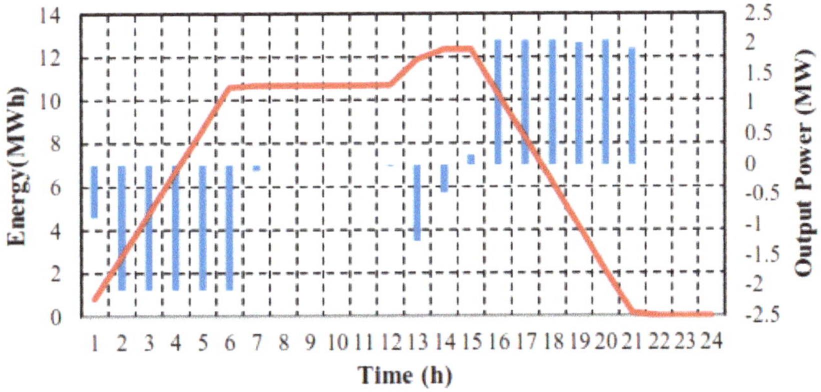

Figure 16. Charge and discharge power and SOC of the BSS.

The transformer between the primary grid and the considered MG provides power for the MG's electrical load. When the output power of the RESs diminishes, more electricity is drawn from the main grid and delivered to the MG via the transformer. The transformer in the MG under study has a rated capacity S_{Tr}^{rated} and rated power $(Power_{Tr}^{rated})$ of 3500 kVA and 2976 kW, respectively. Transformer overloads may occur when the overall output power of the RESs is low and the overall demand is high. Figure 17 shows the transformer's load rate both with and without storage. It also shows the transformer's maximum rated power.

Power reversal occurs when there is a more significant difference between the total output power of all RES and the entire load, such as between hours 12 and 14, where MG operates without storage. Transformer overloads can also occur when there is a high overall demand and a low overall output power from RESs, as shown in Figure 17 for the case where MG runs without storage throughout the hours from hour 16 to hour 18. The transformer load rate does not increase above its rated power after adding the BSS. It is necessary to reduce the transformer load rate because the ESS relocated the load from on-peak to off-peak hours of the day. As a result, it will delay the reinforcement of the transformer size.

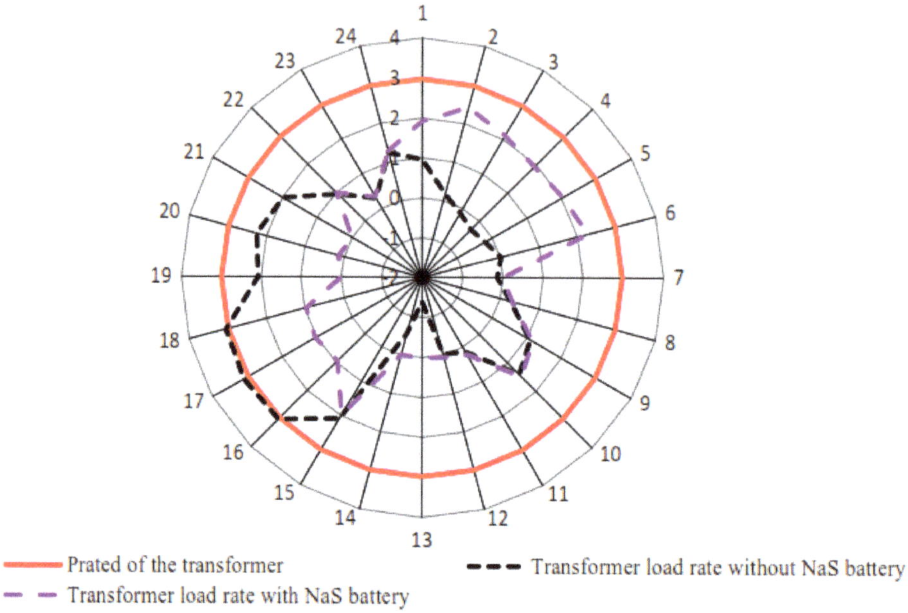

Figure 17. The transformer's rated power and load rate with and without storage.

Additionally, Figure 18 illustrates how the MG's power losses in the early phases increase compared to the basic situation. Another interesting finding from Figure 16 is that the SoC of the BSS is continuous from hour 8 to hour 12. The power losses decreased from hour 15 to hour 19 when the BSS discharged. The MG's overall power losses for the day nevertheless decreased from 1493.2 kW to 1471.1 kW after the addition of the BSS. The BSS affects the MG's overall power losses in this way.

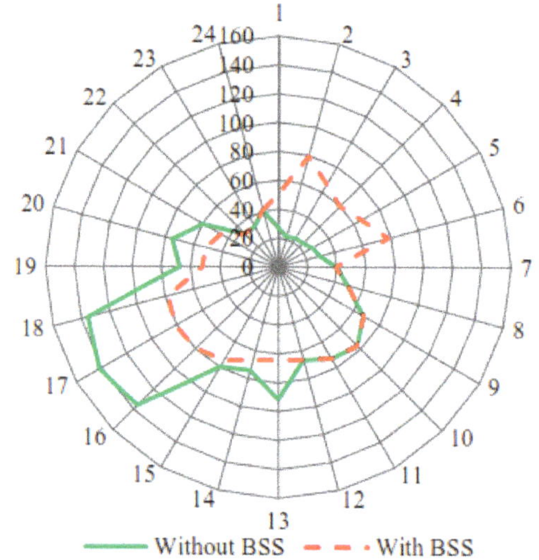

Figure 18. Hourly power losses (kW).

Figure 19 depicts the voltage profile of the MG at four distinct times following the daily load profile: at $h = 4$ (low loading), $h = 10$ (very high loading), $h = 14$ (high loading), and $h = 21$ (medium loading). It is crucial to note from Figure 19a that the voltage profile of the MG at the fourth hour in the absence of a BSS is close to 1 per unit at all buses because of the light load on each bus. Due to low market pricing and loads on each bus during the integration of a BSS, the BSS is charged during the fourth hour, resulting in a lower voltage profile for the MG than in the base scenario while taking into account the voltage limitations on each bus. Figure 19b indicates that the voltage profile of the MG at the 10th hour is the same in all cases due to the integrated BSS's constant SOC. As seen in Figure 19c, when the BSS is charged at this hour, the voltage profile of the MG at $h = 14$ decreases. Figure 19d makes it clear that the MG's voltage profile at $h = 21$ was superior to the base case since the BSS is discharged at this time. In this way, the BSS improved the voltage profile of the MG at these specific instances.

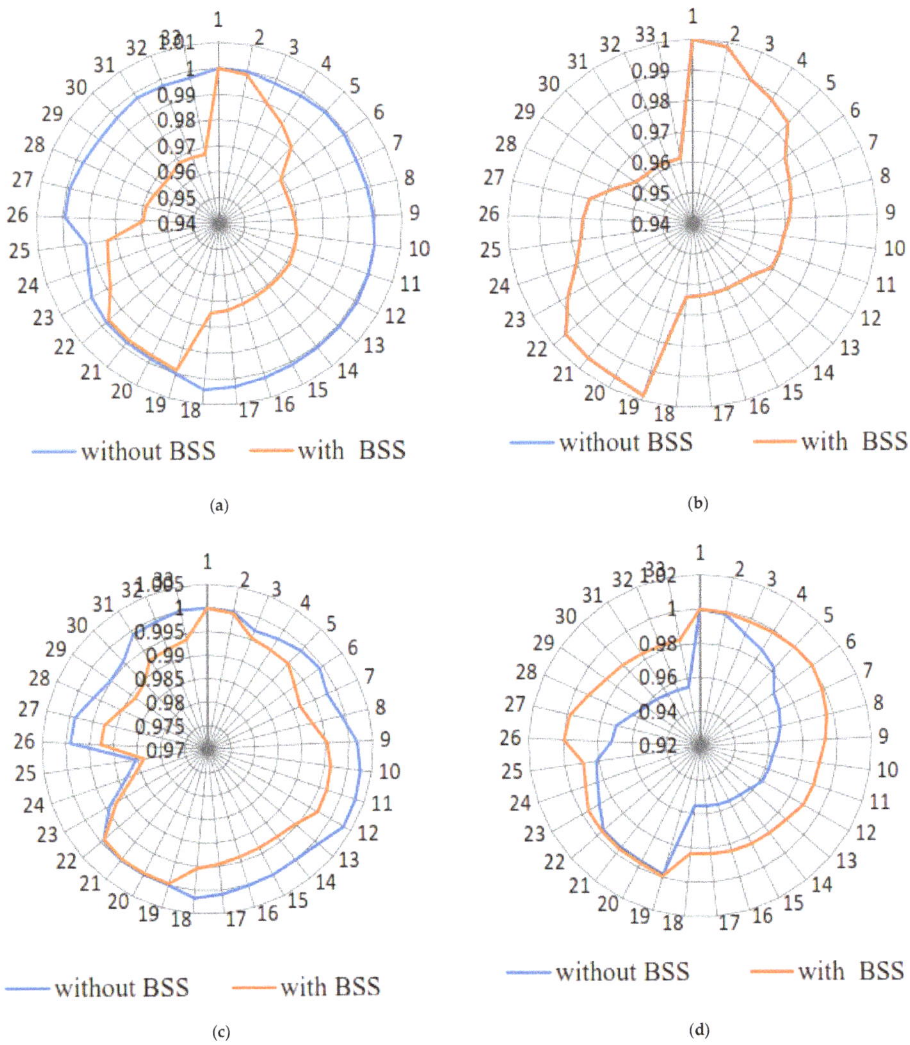

Figure 19. Bus voltage for each bus at four selective periods at: (**a**) $h = 4$, (**b**) $h = 10$, (**c**) $h = 14$, and (**d**) $h = 21$.

To sum up, many research studies [54–62] have examined the economic MG operational costs using different characteristics. Some studies ignored uncertainty parameters and focused on economics. Other researchers considered some uncertainty parameters but ignored others. Table 5 presents a comparison between some studies and the proposed study from where the renewable energy resources included in the MG, the uncertainty parameters, the percentage of saving, and the studied topics discussed in the paper. The methodology used in this study is generic and effective, as seen in Table 5.

Table 5. Overview of recent research works addressing the planning and operation of MGs.

Ref.	Renewables Used		Uncertainty				Topics Studied						
	PV	WT	PV	WT	Demand	Market Price	Total Operation Cost		SCR of RESs	HC	Transformer Load Rate	Voltage Profile	Power Losses
							Studied	Saving%					
[55]	√	√	√	√	X	X	√	7.00	X	X	X	X	X
[56]	√	√	X	X	X	X	√	16.80	√	X	√	√	√
[57]	√	X	√	X	√	X	√	20.00	√	X	X	√	X
[58]	√	√	√	√	X	X	√	6.99	X	X	X	X	X
[59]	√	√	X	X	√	X	√	20.50	X	X	X	X	X
[60]	√	X	X	X	X	X	√	25.10	X	X	X	X	X
[61]	√	√	√	√	√	X	√	10.00	X	X	X	X	X
[62]	X	√	X	√	√	X	√	4.50	X	X	X	√	X
[63]	√	√	X	X	√	X	√	3.20	X	X	X	X	√
Proposed	√	√	√	√	√	√	√	23.80	√	√	√	√	√

5. Conclusions

An optimization model is suggested in this work to identify the optimal location and size of a BSS in the grid-connected MG while taking into account the uncertainty of RESs, variations in electrical loads, and fluctuations of market prices. This is carried out in order to maximize RES self-consumption rate, RES hosting capacity, and MG operating cost minimization. To address the economic sustainability of the BSS and to understand the hosting capacity of RESs in the MGs as well as the load rate of the transformer that connects the main grid and the MG, decision-makers can efficiently use the provided optimization framework. The main conclusions of the paper are summarized as follows:

- The study demonstrated that the BSS has a beneficial effect on the RES' SCR, showing that after adopting the BSS, the RES' SCR achieved 100% at all times of the day.
- Furthermore, it has been shown that adding NaS batteries improves the HC of RESs greatly, as shown by the increase in HCRES from 8.863 MW in the absence of a NaS battery to 10.213 MW.
- The results showed that the optimal placing of the BSS in the MG considerably reduces its overall operating costs in terms of the cost of operation. The MG's operating costs were 183,645.3 (USD/day) before the BSS was installed; the MG's operating costs were 140,007.9 (USD/day) after the BSS was installed. As a result, the savings percentage rate was 23.8%.
- The optimal BSS placement helps reduce the overall active power losses.
- Transformer overloads are likely to occur within a few hours due to the high load and low output power of RESs during these hours. Through comparison, we have found that the BSS could shift the load from the day's on-peak hours to the off-peak hours, which is essential in lowering the transformer load rate. It is, therefore, used to delay reinforcing the transformer.

Finally, future research should focus on the exploration of the hybridization of different BSSs in MGs to be more techno-economically effective while also enhancing the performance of the MG and lowering operational costs. In addition, in future works, all costs will be considered.

Author Contributions: M.M.A., M.H.M. and S.H.E.A.A. designed the problem under study; Z.M.A. and M.M.A. performed the simulations and obtained the results; M.M.A. and M.H.M. analyzed the obtained results; M.M.A., M.H.M. and S.H.E.A.A. wrote the paper, which was further reviewed by Z.M.A. All authors have read and agreed to the published version of the manuscript.

Funding: This project was supported by the Deanship of Scientific Research at Prince Sattam Bin Abdulaziz University under the research project (PSAU2022/01/19413).

Data Availability Statement: The data presented in this study are available on request from the corresponding author.

Acknowledgments: The authors acknowledge the Deanship of Scientific Research at Prince Sattam Bin Abdulaziz University for supporting this research project (PSAU2022/01/19413).

Conflicts of Interest: The authors declare no conflict of interest.

References

1. Debouza, M.; Al-Durra, A.; EL-Fouly, T.H.M.; Zeineldin, H.H. Survey on microgrids with flexible boundaries: Strategies, applications, and future trends. *Energies* **2022**, *205*, 107765. [CrossRef]
2. International Energy Agency (IEA). *Electricity Market Report 2022*; International Energy Agency: Paris, France, 2022.
3. Caparrós Mancera, J.J.; Saenz, J.L.; López, E.; Andújar, J.M.; Segura Manzano, F.; Vivas, F.J.; Isorna, F. Experimental analysis of the effects of supercapacitor banks in a renewable DC microgrid. *Appl. Energy* **2022**, *308*, 118355. [CrossRef]
4. Shang, Z.; Hossain, M.; Wycisk, R.; Pintauro, P. Poly(phenylene sulfonic acid)-expanded polytetrafluoroethylene composite membrane for low relative humidity operation in hydrogen fuel cells. *J. Power Sources* **2022**, *535*, 231375. [CrossRef]
5. Abo-Khalil, A.G.; Alobaid, M. A Guide to the Integration and Utilization of Energy Storage Systems with a Focus on Demand Resource Management and Power Quality Enhancement. *Sustainability* **2023**, *15*, 14680. [CrossRef]
6. Villanueva-Rosario, J.A.; Santos-García, F.; Aybar-Mejía, M.E.; Mendoza-Araya, P.; Molina-García, A. Coordinated ancillary services, market participation and communication of multi-microgrids: A review. *Appl. Energy* **2022**, *308*, 118332. [CrossRef]
7. Xuanyue, S.; Wang, X.; Wu, X.; Wang, Y.; Song, Z.; Wang, B.; Ma, Z. Peer-to-peer multi-energy distributed trading for interconnected microgrids: A general Nash bargaining approach. *Int. J. Electr. Power Energy Syst.* **2022**, *138*, 107892. [CrossRef]
8. Stetz, T.; Marten, F.; Braun, M. Improved Low Voltage Grid-Integration of Photovoltaic Systems in Germany. *IEEE Trans. Sustain. Energy* **2013**, *4*, 534–542. [CrossRef]
9. Kabiri, R.; Holmes, D.G.; McGrath, B.P.; Meegahapola, L.G. LV Grid Voltage Regulation Using Transformer Electronic Tap Changing, With PV Inverter Reactive Power Injection. *IEEE J. Emerg. Sel. Top. Power Electron.* **2015**, *3*, 1182–1192. [CrossRef]
10. Tonkoski, R.; Lopes, L.A.; El-Fouly, T.H. Coordinated active power curtailment of grid connected PV inverters for overvoltage prevention. *IEEE Trans. Sustain. Energy* **2011**, *2*, 139–147. [CrossRef]
11. Yao, E.; Samadi, P.; Wong, V.W.S.; Schober, R. Residential Demand Side Management Under High Penetration of Rooftop Photovoltaic Units. *IEEE Trans. Smart Grid* **2015**, *7*, 1597–1608. [CrossRef]
12. Samadi, A.; Shayesteh, E.; Eriksson, R.; Rawn, B.; Söder, L. Multi-objective coordinated droop-based voltage regulation in distribution grids with PV systems. *Renew. Energy* **2014**, *71*, 315–323. [CrossRef]
13. Rawa, M.; Abusorrah, A.; Al-Turki, Y.; Mekhilef, S.; Mostafa, M.H.; Ali, Z.M.; Aleem, S.H.E. Optimal Allocation and Economic Analysis of Battery Energy Storage Systems: Self-Consumption Rate and Hosting Capacity Enhancement for Microgrids with High Renewable Penetration. *Sustainability* **2020**, *12*, 10144. [CrossRef]
14. Nudell, T.R.; Brignone, M.; Robba, M.; Bonfiglio, A.; Ferro, G.; Delfino, F.; Annaswamy, A.M. Distributed control for polygeneration microgrids: A Dynamic Market Mechanism approach. *Control Eng. Pract.* **2022**, *121*, 105052. [CrossRef]
15. Guibentif, T.M.M.; Vuille, F. Prospects and barriers for microgrids in Switzerland. *Energy Strategy Rev.* **2022**, *39*, 100776. [CrossRef]
16. Alghamdi, B.; Cañizares, C. Frequency and voltage coordinated control of a grid of AC/DC microgrids. *Appl. Energy* **2022**, *310*, 118427. [CrossRef]
17. Khodadoost Arani, A.A.; Gharehpetian, G.B.; Abedi, M. Review on Energy Storage Systems Control Methods in Microgrids. *Int. J. Electr. Power Energy Syst.* **2019**, *107*, 745–757. [CrossRef]
18. Hajiaghasi, S.; Salemnia, A.; Hamzeh, M. Hybrid energy storage system for microgrids applications: A review. *J. Energy Storage.* **2019**, *21*, 543–570. [CrossRef]
19. Mostafa, M.H.; Aleem, S.H.E.A.; Ali, S.G.; Abdelaziz, A.Y.; Ribeiro, P.F.; Ali, Z.M. Robust Energy Management and Economic Analysis of Microgrids Considering Different Battery Characteristics. *IEEE Access* **2020**, *8*, 54751–54775. [CrossRef]
20. Mansouri, S.A.; Ahmarinejad, A.; Nematbakhsh, E.; Javadi, M.S.; Nezhad, A.E.; Catalão, J.P.S. A sustainable framework for multi-microgrids energy management in automated distribution network by considering smart homes and high penetration of renewable energy resources. *Energy* **2022**, *245*, 123228. [CrossRef]
21. Mostafa, M.H.; Aleem, S.H.E.A.; Ali, S.G.; Ali, Z.M.; Abdelaziz, A.Y. Techno-economic assessment of energy storage systems using annualized life cycle cost of storage (LCCOS) and levelized cost of energy (LCOE) metrics. *J. Energy Storage.* **2020**, *29*, 101345. [CrossRef]
22. Bai, H.; Tang, X.; Yuan, Z.; Li, Q.; Pan, S.; Deng, P.; Zhou, C.; Luo, N. A novel three-phase unbalanced power flow solution for islanded microgrids with distributed generations under droop controls. *Energy Rep.* **2022**, *8*, 1438–1447. [CrossRef]
23. Tomin, N.; Shakirov, V.; Kozlov, A.; Sidorov, D.; Kurbatsky, V.; Rehtanz, C.; Lora, E.E.S. Design and optimal energy management of community microgrids with flexible renewable energy sources. *Renew. Energy.* **2022**, *183*, 903–921. [CrossRef]

24. Zhang, L.; Liang, D.; Wen, Q.; Liu, H.; Liu, Y.; Li, D.; Gao, Y.; Xu, H.; Li, W.; Sun, H. Transient optimization control strategy of hybrid capacity regulating transformer. *Electr. Power Syst. Res.* **2022**, *206*, 107803. [CrossRef]
25. Hossain, E.; Perez, R.; Nasiri, A.; Padmanaban, S. A Comprehensive Review on Constant Power Loads Compensation Techniques. *IEEE Access* **2018**, *6*, 33285–33305. [CrossRef]
26. Li, Q. Design and practical application analysis of thermal management system for power battery in new energy vehicles. *Results Phys.* **2023**, *54*, 107063. [CrossRef]
27. Habib, A.; Hasan, M.; Issa, G.; Singh, D.; Islam, S.; Ghazal, T. Lithium-Ion Battery Management System for Electric Vehicles: Constraints, Challenges, and Recommendations. *Batteries* **2023**, *9*, 3. [CrossRef]
28. Li, W.; Rentemeister, M.; Badeda, J.; Schulte, D.; Sauer, D.U. Digital twin for battery systems: Cloud battery management system with online state-of-charge and state-of-health estimation. *J. Energy Storage* **2020**, *30*, 101557. [CrossRef]
29. Yang, S.; Zhang, Z.; Cao, R.; Wang, M.; Cheng, H.; Zhang, L.; Jiang, Y.; Chen, B.; Ling, H.; Lian, Y.; et al. Implementation for a cloud battery management system based on the CHAIN framework. *Energy AI* **2021**, *5*, 100088. [CrossRef]
30. Balasingam, B.; Ahmed, M.; Pattipati, K. Battery management systems-challenges and some solutions. *Energies* **2021**, *13*, 11. [CrossRef]
31. Okay, K.; Eray, S.; Eray, A. Development of prototype battery management system for PV system. *Renew. Energy* **2022**, *181*, 1294–1304. [CrossRef]
32. Lipu, M.H.; Hannan, M.A.; Karim, T.; Hussain, A.; Saad, M.; Miah, M.; Mahlia, I. Intelligent algorithms and control strategies for battery management system in electric vehicles: Progress, challenges and future outlook. *J. Clean. Prod.* **2021**, *292*, 126044. [CrossRef]
33. Raoofi, T.; Yildiz, M. Comprehensive review of battery state estimation strategies using machine learning for battery Management Systems of Aircraft Propulsion Batteries. *J. Energy Storage* **2023**, *59*, 106486. [CrossRef]
34. Hamed, M.; El-Tayeb, A.; Moukhtar, I.; el Dein, A.; Abdelhameed, E. A review on recent key technologies of lithium-ion battery thermal management: External cooling systems. *Results Eng.* **2022**, *16*, 100703. [CrossRef]
35. Ebbs-Picken, T.; da Silva, C.; Amon, C. Design optimization methodologies applied to battery thermal management systems: A review. *J. Energy Storage* **2023**, *67*, 107460. [CrossRef]
36. Sezer, H.; Aygun, M.; Mason, J.; Baran, E.; Celik, I. A Computational Model for Sodium Sulfur Battery Analysis. *ECS Trans.* **2015**, *69*, 91. [CrossRef]
37. Vudata, S.; Bhattacharyya, D. Thermal management of a high temperature sodium sulphur battery stack. *Int. J. Heat Mass Transf.* **2021**, *181*, 122025. [CrossRef]
38. Li, H.; Zhang, C.; Sun, B. Optimal design for component capacity of integrated energy system based on the active dispatch mode of multiple energy storages. *Energy* **2021**, *227*, 120522. [CrossRef]
39. Hou, H.; Xu, T.; Wu, X.; Wang, H.; Tang, A.; Chen, Y. Optimal capacity configuration of the wind-photovoltaic-storage hybrid power system based on gravity energy storage system. *Appl. Energy.* **2020**, *271*, 115052. [CrossRef]
40. Panuschka, S.; Hofmann, R. Impact of thermal storage capacity, electricity and emission certificate costs on the optimal operation of an industrial energy system. *Energy Convers. Manag.* **2019**, *185*, 622–635. [CrossRef]
41. Chen, C.; Duan, S. Optimal allocation of distributed generation and energy storage system in microgrids. *IET Renew. Power Gener.* **2014**, *8*, 581–589. [CrossRef]
42. Mostafa, M.H.; Aleem, A. Optimal Allocation of Energy Storage System for Improving Performance of Microgrid Using Symbiotic Organisms Search. In Proceedings of the 2018 Twentieth International Middle East Power Systems Conference (MEPCON), Cairo, Egypt, 18–20 December 2018; pp. 474–479.
43. Qiu, J.; Zhao, J.; Zheng, Y.; Dong, Z.; Dong, Z.Y. Optimal allocation of BESS and MT in a microgrid. *IET Gener. Transm. Distrib.* **2018**, *12*, 1988–1997. [CrossRef]
44. Tripathy, B.K.; Maddikunta, P.K.R.; Pham, Q.-V.; Gadekallu, T.R.; Dev, K.; Pandya, S.; ElHalawany, B.M. Harris Hawk Optimization: A Survey on Variants and Applications. *Comput. Intell. Neurosci.* **2022**, *2022*, 2218594. [CrossRef] [PubMed]
45. Alabool, H.M.; Alarabiat, D.; Abualigah, L.; Heidari, A.A. Harris hawks optimization: A comprehensive review of recent variants and applications. *Neural Comput. Appl.* **2021**, *33*, 8939–8980. [CrossRef]
46. Kim, S.; Kim, J.; Jin, Y.; Yoon, Y. Optimal Bidding Strategy for Renewable Microgrid with Active Network Management. *Energies* **2016**, *9*, 48. [CrossRef]
47. Arif, S.; Rabbi, A.E.; Ahmed, S.U.; Hossain Lipu, M.S.; Jamal, T.; Aziz, T.; Sarker, M.R.; Riaz, A.; Alharbi, T.; Hussain, M.M. Enhancement of Solar PV Hosting Capacity in a Remote Industrial Microgrid: A Methodical Techno-Economic Approach. *Sustainability* **2022**, *14*, 8921. [CrossRef]
48. Koirala, A.; Van Acker, T.; D'hulst, R.; Van Hertem, D. Hosting capacity of photovoltaic systems in low voltage distribution systems: A benchmark of deterministic and stochastic approaches. *Renew. Sustain. Energy Rev.* **2022**, *155*, 111899. [CrossRef]
49. Yuan, J.; Weng, Y.; Tan, C.-W. Determining maximum hosting capacity for PV systems in distribution grids. *Int. J. Electr. Power Energy Syst.* **2022**, *135*, 107342. [CrossRef]
50. Hazır, Ö.; Ulusoy, G. A classification and review of approaches and methods for modeling uncertainty in projects. *Int. J. Prod. Econ.* **2020**, *223*, 107522. [CrossRef]
51. Zhao, Z.; Zhao, J.; Song, K.; Hussain, A.; Du, Q.; Dong, Y.; Liu, J.; Yang, X. Joint DBN and Fuzzy C-Means unsupervised deep clustering for lung cancer patient stratification. *Eng. Appl. Artif. Intell.* **2020**, *91*, 103571. [CrossRef]

52. Heidary, H.; El-Kharouf, A.; Wilckens, R.; Bozorgmehri, S.; Salimi, M.; Golmohammad, M. Life cycle assessment of solid oxide fuel cell vehicles in a natural gas producing country; comparison with proton electrolyte fuel cell, battery and gasoline vehicles. *Sustain. Energy Technol. Assess.* **2023**, *59*, 103396. [CrossRef]
53. Heidari, A.A.; Mirjalili, S.; Faris, H.; Aljarah, I.; Mafarja, M.; Chen, H. Harris hawks optimization: Algorithm and applications. *Futur. Gener. Comput. Syst.* **2019**, *97*, 849–872. [CrossRef]
54. Almotairi, S.; Badr, E.; Abdul Salam, M.; Dawood, A. Three Chaotic Strategies for Enhancing the Self-Adaptive Harris Hawk Optimization Algorithm for Global Optimization. *Mathematics* **2023**, *11*, 4181. [CrossRef]
55. Shahbazbegian, V.; Dehghani, F.; Shafiyi, M.A.; Shafie-khah, M.; Laaksonen, H.; Ameli, H. Techno-economic assessment of energy storage systems in multi-energy microgrids utilizing decomposition methodology. *Energy* **2023**, *283*, 128430. [CrossRef]
56. Alqarni, M. Sodium sulfur batteries allocation in high renewable penetration microgrids using coronavirus herd immunity optimization. *Ain Shams Eng. J.* **2022**, *13*, 101590. [CrossRef]
57. Tercan, S.M.; Demirci, A.; Gokalp, E.; Cali, U. Maximizing self-consumption rates and power quality towards two-stage evaluation for solar energy and shared energy storage empowered microgrids. *J. Energy Storage* **2022**, *51*, 104561. [CrossRef]
58. Wei, W.; Ye, L.; Fang, Y.; Wang, Y.; Chen, X.; Li, Z. Optimal Allocation of Energy Storage Capacity in Microgrids Considering the Uncertainty of Renewable Energy Generation. *Sustainability* **2023**, *15*, 9544. [CrossRef]
59. Gong, Q.; Fang, J.; Qiao, H.; Liu, D.; Tan, S.; Zhang, H.; He, H. Optimal Allocation of Energy Storage System Considering Price-Based Demand Response and Dynamic Characteristics of VRB in Wind-PV-ES Hybrid Microgrid. *Processes* **2019**, *7*, 483. [CrossRef]
60. Zhao, H.; Xu, J.; Xu, K.; Sun, J.; Wang, Y. Optimal Allocation Method of Source and Storage Capacity of PV-Hydrogen Zero Carbon Emission Microgrid Considering the Usage Cost of Energy Storage Equipment. *Energies* **2022**, *15*, 4916. [CrossRef]
61. Zhong, X.; Sun, X.; Wu, Y. A Capacity Optimization Method for a Hybrid Energy Storage Microgrid System Based on an Augmented ε- Constraint Method. *Energies* **2022**, *15*, 7593. [CrossRef]
62. Kreishan, M.Z.; Zobaa, A.F. Scenario-Based Uncertainty Modeling for Power Management in Islanded Microgrid Using the Mixed-Integer Distributed Ant Colony Optimization. *Energies* **2023**, *16*, 4257. [CrossRef]
63. Battula, A.R.; Vuddanti, S.; Salkuti, S.R. A Day Ahead Demand Schedule Strategy for Optimal Operation of Microgrid with Uncertainty. *Smart Cities* **2023**, *6*, 491–509. [CrossRef]

sustainability

MDPI

Article

A Distribution Network Planning Method Considering the Distributed Energy Resource Flexibility of Virtual Power Plants

Zhichun Yang [1,*], Gang Han [1], Fan Yang [1], Yu Shen [1], Yu Liu [1], Huaidong Min [1], Zhiqiang Zhou [2], Bin Zhou [2], Wei Hu [1] and Yang Lei [1]

1 Electric Power Research Institute of State Grid Hubei Co., Ltd., Wuhan 430037, China; m15872382912@163.com (G.H.); yangf_82@163.com (F.Y.); YShen185467@163.com (Y.S.); liuyu1795@gmail.com (Y.L.); minhuaidong@foxmail.com (H.M.); huweitest@163.com (W.H.); YangL86947852@163.com (Y.L.)

2 State Grid Hubei Electric Power Co., Ltd., Wuhan 430037, China; jianle3600@126.com (Z.Z.); BZhou145871@163.com (B.Z.)

* Correspondence: yangzhichun36000@163.com

Abstract: To solve the overload problem caused by the high proportion of renewable energy into the power system, it is particularly important to find a suitable distribution network planning scheme. Existing studies have effectively reduced the planning cost by incorporating virtual power plants into the distribution planning process, but there is no quantitative analysis of the flexible resources inside the virtual power plant. At the same time, the traditional planning process does not pay much attention to the acquisition of photovoltaic and load data. Therefore, in this paper, we propose a distribution network planning method considering the flexibility of distributed energy resources in virtual power plants. Firstly, taking the distribution network planning including the virtual power plant as the research object, the flexibility of the distributed energy resource of the virtual power plant was quantified. Then, in order to achieve the goal of minimizing the operating cost of system planning, a distribution network planning model considering the flexibility of distributed energy resources in the virtual power plant is established. In this model, the impact of virtual power plants flexibility on the distribution network planning process is mainly considered. Secondly, this paper uses the improved k-means clustering algorithm to obtain the typical data of PV and load. The algorithm effectively overcomes the impact of PV and load output fluctuations on the planning process. Finally, the simulation results show that the proposed planning model can effectively reduce the operation cost of system planning by using distributed energy storage system and distributed energy resource flexibility. At the same time, the PV absorption rate of the PV power station inside the distribution network is improved.

Keywords: virtual power plant; distributed energy resource; distribution network planning; distributed energy storage system; flexibility quantization; improved k-means clustering algorithm

Citation: Yang, Z.; Han, G.; Yang, F.; Shen, Y.; Liu, Y.; Min, H.; Zhou, Z.; Zhou, B.; Hu, W.; Lei, Y. A Distribution Network Planning Method Considering the Distributed Energy Resource Flexibility of Virtual Power Plants. *Sustainability* **2023**, *15*, 14399. https://doi.org/10.3390/su151914399

Academic Editor: Mohamed A. Mohamed

Received: 22 August 2023
Revised: 15 September 2023
Accepted: 28 September 2023
Published: 30 September 2023

1. Introduction

In the context of the global promotion of carbon emission reduction, the proportion of renewable energy such as photovoltaic (PV) in the power system continues to increase. However, the ensuing problem is that the uncertainty of renewable energy output such as photovoltaic will endanger the safety and stability of distribution network operation [1]. With the rapid development of power systems, the level of demand side load is also increasing. Demand-side resources under the new power system will become grid-side friendly and interactive resources, including electric vehicles (EVs), HVAC systems, etc. Due to the wide distribution of these resources, virtual power plant technology should be used to aggregate them to adjust the power load resources on the demand side, thereby improving the flexibility of the power grid [2].

In the existing research on distribution network planning, it is mainly to achieve the optimization objectives of reducing planning costs and peak load reduction. Among the existing methods to reduce the cost of the planning process and achieve peak filling, the most effective way is to increase the interaction between the distribution network and the demand side resources of the virtual power plant. Therefore, the distribution network planning process is closely related to the output characteristics of the source side and the load forecasting characteristics of the load side [3]. Therefore, the fluctuation of PV output and the participation of flexible resources on the demand side will affect the final planning results. Therefore, in order to reduce the planning cost, the existing research often includes the photovoltaic and distributed energy storage resources in the virtual power plant in the distribution network planning process.

At present, there are some literatures on flexible resources in virtual power plants. In order to characterize the flexibility of resources such as air conditioners and electric vehicles in virtual power plants, the authors in [4] established corresponding virtual battery models to achieve this goal. In [5], the membership matrix of flexible resources in each virtual power plant is obtained by cloud model method, and the subjective and objective weights are combined with the membership matrix to evaluate the level of flexible resources. The authors in [6] achieve joint optimization of their participation in power and reserve markets by evaluating the power availability zone of flexible resources within a virtual power plant. In [7], the authors proposed a method to characterize the flexibility of virtual power plants using flexible radar charts. In order to accurately evaluate the power generation characteristics of distributed generation resources in virtual power plants, the authors in [8] puts forward the corresponding evaluation criteria. In order to reduce system congestion, the authors in [9] considers the flexibility of distributed energy resource in virtual power plants and establishes a flexible trading market centered on the distribution network.

In view of the participation of virtual power plants and other flexible resources in the distribution network planning process, relevant studies have been carried out in the literature. In order to realize distribution network planning including distributed energy storage and distributed generation, the authors in [10,11] proposes a multi-objective coordinated planning model that comprehensively considers multiple factors. The authors proposed a multi-stage joint distribution network planning strategy in [12] to address the placement of EV charging stations and energy storage in the system. The authors in [13,14] in order to solve the problem of load fluctuation in the planning process, a distribution network planning method considering new load and distributed generation access is proposed. In [15,16], the authors all took island microgrid planning as the research object and effectively reduced the planning cost by introducing hydrogen storage systems. Based on the above studies, the authors in [17–20] focused on the safety and stability of the operation during the planning process of the distribution network including wind power and electric vehicles. In [21,22], the author integrated PV, energy storage and wind power and established a distributed energy management system. Then, the energy management system is integrated into the distribution network planning process to effectively improve the economy of the planning process. In order to achieve the goal of improving the economy of the planning process, the authors incorporated distributed energy into the planning process in the form of demand response in [23]. By introducing distributed energy storage and photovoltaic power generation in the distribution network planning process, the planned low-carbon economy goal is achieved. In [24,25], the authors incorporated distributed generation and energy storage in virtual power plants into the distribution planning process, which effectively reduced its planning cost. In [26], by integrating the flexibility of distributed energy into the planning process of urban distribution network, the author effectively reduces the cost of line expansion in the planning process by using load flexibility and conventional reinforcement methods.

The above studies included the virtual power plant with distributed energy resources into the distribution network planning, but did not quantitatively analyze the internal flexible resources of the virtual power plant. At the same time, it does not focus on the impact of PV and load output fluctuations on the distribution network planning results. Therefore, this paper proposes a distribution network planning method that considers the distributed energy resource flexibility of virtual power plants. Compared with [10,11,15,16,23], the proposed model considers the fluctuation of PV and load output into the planning process. Furthermore, compared with [17–22], this paper adopts an improved k-means clustering algorithm to deal with the fluctuation of PV and load output. Compared with [12–14,24–26], the proposed model quantifies the flexibility of the demand side resources of the virtual power plant in the planning process of the distribution network. The main contributions of this paper are as follows.

(1) In the distribution network planning process, the improved k-means clustering algorithm is used for the fluctuation of PV and load output.
(2) The distributed energy resource flexibility model of virtual power plant is established to quantify the flexibility of demand side resources in virtual power plant during the planning process.
(3) The proposed programming model is converted into a typical mixed integer linear programming model by a second-order cone optimization technique. The improved IEEE RTS-24 node system is used to verify the superiority of the proposed programming model.

The remainder of this article consists of the following. Section 2 introduces the distribution network planning framework proposed in this paper. Section 3 discusses in detail the distributed energy resource flexibility model for virtual power plants with EVs. Section 4 introduces the programming model proposed in this paper, and gives the method of solving the model. In Section 5, the details of the example simulation are introduced and discussed. Finally, the conclusion is given in Section 6.

2. Distribution Network Planning Framework

The load in the power system always increases with the increase of the distribution network planning year, so it is necessary to invest in new lines or expand the original lines to maintain the stable operation of the system. Due to the increasing proportion of renewable energy access such as photovoltaic, distribution network planning also needs to consider many factors.

In order to solve the problems mentioned above, this paper puts forward a set of suitable distribution network planning schemes. Firstly, the paper brings virtual power plant flexibility resources into the distribution network planning process. Then, considering the investment cost of line expansion, the flexibility compensation cost of virtual power station, the investment cost of the energy storage system and the abandonment cost of the photovoltaic power station, the corresponding distribution network planning model is established. Finally, by solving the model, a distribution network planning scheme is obtained.

Virtual power plants as a new technology, through information collection, control and communication technology, it can aggregate cross-region, multi-type controllable distributed resources, and can effectively play the flexibility and complementarity of distributed resources. At the same time, as a typical class of adjustable load, electric vehicles achieve a flexible balance in all aspects of operation, which brings corresponding economic benefits. By introducing electric vehicles into the system as an effective means of transferring peak load to the system, planning costs can be further reduced. Therefore, this paper mainly focuses on the virtual power plant with distributed energy resource such as EV. Although the existing studies on the flexibility resources of virtual power plant mainly focus on the operation aspect, some studies also take it into consideration in the planning process. There are two main reasons for this. On the one hand, electric vehicles can play a role in cutting peaks and filling valleys. On the other hand, in order to obtain a better planning scheme, the concept of operation is usually introduced in the planning process, the planning and operation are jointly optimized. At same time, in order to obtain better planning schemes, operational concepts are usually introduced in the planning process to obtain the best investment and operation schemes.

In this paper, the flexibility modeling of EV in a virtual power plant is used to describe its up-and-down standby capacity. Then, under the relevant constraints, a certain load adjustment margin is set quantitatively according to the flexibility of electric vehicles. The total power supply capacity of the superior grid, PV and energy storage systems is insufficient during peak power load hours. At this time, price compensation incentives can be used to guide flexible power loads to reduce electricity consumption. When the power load is in the off-peak period, the system guides the flexible load to increase the power consumption due to the surplus of the total power supply.

In order to overcome the impact of PV output fluctuation on the planning process, an improved k-means clustering algorithm is used to obtain the PV output data of 96 h in a typical year. At the same time, considering the characteristics of energy storage, each PV plant is equipped with a corresponding energy storage system. Similar to the acquisition of typical PV data, a modified k-means clustering algorithm was used to obtain typical load data for 96 h from historical data. The process of using the improved k-means clustering algorithm is shown in Figure 1.

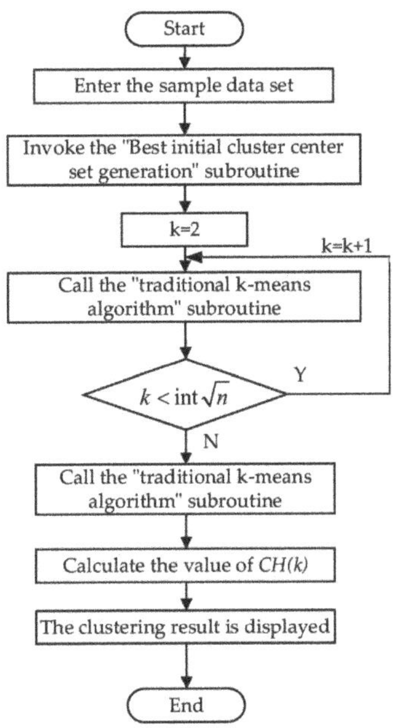

Figure 1. The process of using the improved k-means clustering algorithm.

In the flow chart, $CH(k)$ is the evaluation index of clustering effect of k-means algorithm. The value of $CH(k)$ reflects the effect of clustering, and the larger the value of $CH(k)$ indicates the better effect of clustering. The steps of the "Best initial cluster center set generation" subroutine are as follows.

Step1: Calculate the density parameter of each data and find the data S_n corresponding to the maximum value.

Step2: If S_n is unique, add S_n to the initial cluster center candidate set. If it corresponds to multiple sample data, the data with the smallest cohesion degree is selected according to the cohesion parameter and added to the initial cluster center candidate set.

Step3: Delete the data from the sample set with S_n as the center and d_{mean} (average distance of the sample) as the radius.

Step4: Repeat the process from Step1 to Step3 until an initial cluster center is found.

Finally, Figure 2 shows the proposed distribution network planning framework.

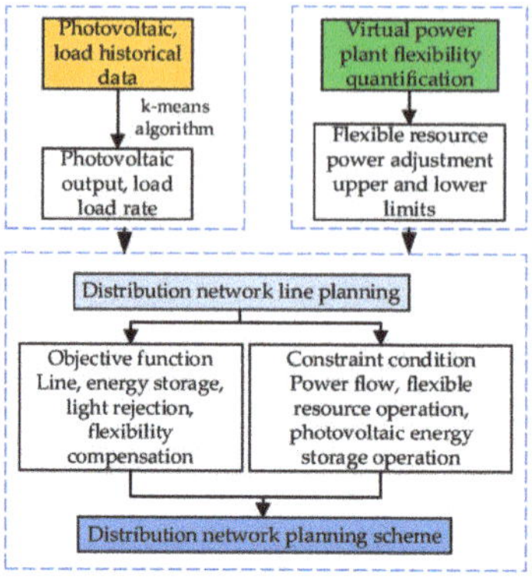

Figure 2. Planning framework.

3. Distributed Energy Resource Flexibility Model of Virtual Power Plant

In order to incorporate virtual power plants and EVs into the distribution network planning process as typical distributed energy sources, a quantitative model of cluster EVs flexibility based on EVs flexibility should be established first. Then, on the basis of this model, the charge and discharge optimization is carried out to improve the economy of Time-of-use electricity price (TOU) cluster electric vehicles [27]. The optimal charging and discharging power is obtained, and the standby capacity of EV cluster involved in centralized regulation is calculated as reference power. Secondly, the calculated results are used as the operating limits of the adjustable power of the flexible load of the virtual power plant in the planning process. Finally, by reducing the power consumption of flexible load during peak load hours, the flexible load during low power consumption periods can be effectively increased, the economy of planning can be improved, and the photovoltaic absorption rate can be improved.

When characterizing the reserve capacity of an electric vehicle, its charge-discharge limits must be observed. On the basis of analyzing the power feasible region of EV, the time axis is discretized, and the scheduling cycle of cluster EV is divided into T period, and the length of each period is Δt. Assume that the charge and discharge power in Δt remains unchanged, where the charge/discharge power of EV is as follows.

$$P_{i,t} = P_{i,t,c}u_{i,t,c}\eta_{i,c} - \frac{P_{i,t,d}u_{i,t,d}}{\eta_{i,d}} \tag{1}$$

$$0 \leq P_{i,t,c} \leq P_{c,\max} \tag{2}$$

$$0 \leq P_{i,t,d} \leq P_{d,\max} \tag{3}$$

$$u_{i,t,c} + u_{i,t,d} \leq 1 \tag{4}$$

where, $P_{i,t,c}$, $P_{i,t,d}$ are respectively the power input and output of electric vehicles. $u_{i,t,c}$, $u_{i,t,d}$ represent the power input and output states of electric vehicles respectively, which are binary variables. $\eta_{i,c}$, $\eta_{i,d}$ are the power input and output efficiency of electric vehicles respectively. $P_{c,\max}$, $P_{d,\max}$, respectively represent the maximum power input and output power of electric vehicles. For this reason, the standby capacity of the EV will change with the change of the real-time power of the EV. Equation (1) represents the relationship between the real-time power of EV and its charging and discharging power. Equations (2) and (3) indicate that the EV charging and discharging power cannot exceed the limit. Equation (4) indicates that EV charging and discharging cannot be carried out simultaneously.

When calculating the charging and discharging power of an EV, it is necessary to consider the limitations of its battery capacity, so the EV should be run as follows.

$$E \geq E_{\min} = \begin{cases} E_{base}, \ t_{base} \leq t \leq t_d \\ E_{\exp} - P_{c,\max}(t_{\exp} - t), \ t > t_d \end{cases} \quad (5)$$

where, $E_{\min}$ is the minimum capacity of EV. E_{base} represents the minimum power level set by the user. $E_{\exp}$ represents the off-grid power of EV. t_{base} indicates the time for EV charging to the minimum charge. t_d is the off-network time of EV.

Under TOU, charge and discharge optimization is carried out with the goal of improving the operation economy of cluster electric vehicles, and the optimization objective function is as follows:

$$\max F_{evag} = \sum_{i \in N} \sum_{t \in T} f_{ev,i,t} \quad (6)$$

$$f_{ev,i,t} = f_{d,i,t} - f_{c,i,t} \quad (7)$$

$$f_{ev,i,t} \geq -P^0_{i,t,c} \pi_{c,t} \quad (8)$$

$$f_{c,i,t} = P_{i,t,c} \pi_{c,t} \quad (9)$$

$$f_{d,i,t} = P_{i,t.d} \pi_{d,t} \quad (10)$$

where, F_{evag} is the aggregate revenue of N EVs centrally controlled by the aggregator. $f_{ev,i,t}$ is the income of a single EV after centralized regulation $f_{d,i,t}$, $f_{c,i,t}$ respectively represent the discharge income and charging cost after EV participates in scheduling. $\pi_{d,t}$, $\pi_{c,t}$ are the discharge and charging prices of EV respectively. $P^0_{i,t,c}$ represents the real-time interactive power of electric vehicles that do not participate in centralized scheduling. Formula (8) ensures that after a single EV participates in centralized regulation, it has benefits compared with the previous individual charging. Equations (9) and (10) represent the charging cost and discharging benefit of a single EV, respectively.

The final optimized power of EV can be calculated by Formulas (1)–(11), and then the standby capacity of each EV can be obtained by considering the influence of power boundary and power boundary.

$$E_{i,t} = E_{i,0} + \gamma_{i,t} \sum_{t=1}^{T} P_{i,t} \quad (11)$$

$$P_{cu,i,t} = \gamma_{i,t} \max\left\{ \min(-P_{c,\max} + P_{i,t}, \frac{E_{i,t} - E_{\min,t+1}}{\Delta t} + P_{i,t}), 0 \right\} \quad (12)$$

$$P_{cd,i,t} = \gamma_{i,t} \max\left\{ \min(P_{d,\max} - P_{i,t}, \frac{E_{\max} - E_{i,t}}{\Delta t} - P_{i,t}), 0 \right\} \quad (13)$$

where, $E_{i,t}$ represent the real-time power of EV battery. $E_{i,0}$ is the initial charge of the EV battery. $\gamma_{i,t}$ indicates the online status of EV. $P_{cu,i,t}$, $P_{cd,i,t}$ are the upper and lower reserve capacities of EV respectively. $P_{d,\max} - P_{i,t}$, $-P_{c,\max} + P_{i,t}$ are the maximum increased charging and discharging power of EV respectively, that is, the power boundary. $(E_{i,t} - E_{\min,t+1})/\Delta t + P_{i,t}$ is the discharge space of EV, that is, the energy boundary. Equation (11) represents the relationship between EV real-time charge and power. For this purpose, the standby capacity of the electric vehicle can be obtained as shown in Figure 3.

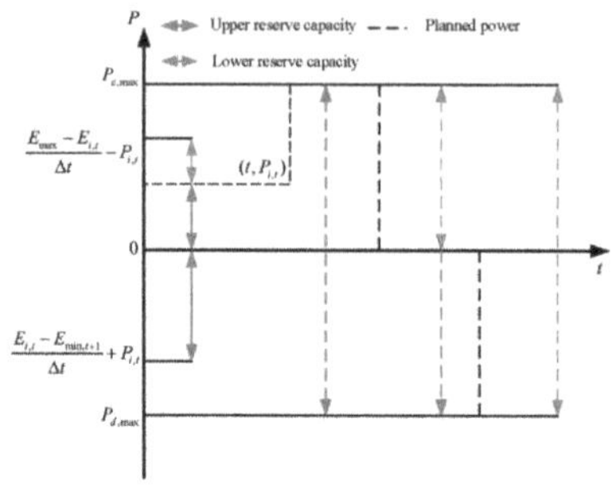

Figure 3. Electric vehicle backup capability.

In the subsequent planning model, the spare capacity of electric vehicles can be used as the adjustable power limit of the virtual power plant flexibility resources planned by the distribution network, as shown in the following equation:

$$P_t^{down,\max} = \sum_{i=1}^{N} P_{cu,i,t} \tag{14}$$

$$P_t^{up,\max} = \sum_{i=1}^{N} P_{cd,i,t} \tag{15}$$

where, $P_t^{down,\max}$, $P_t^{up,\max}$ represent the down and up boundaries of the flexibility resources of the virtual power plant respectively.

4. Distribution Network Planning Model

This section introduces the construction of distribution network planning model, taking into account the factors of peak loading and valley filling and new energy consumption [28]. In this model, all economic costs are converted to average annual investment at the time of planning. Firstly, by solving the model proposed in Section 3, the adjustable power upper and lower limits of 96 h in typical planning year of distribution network are obtained. Then, the flexibility of virtual power plant electric vehicles and distributed energy storage are utilized to reduce the investment cost of line expansion and the cost of light abandonment. Finally, the power balance constraint, power flow constraint, virtual power plant flexible operation constraint, photovoltaic operation constraint, main network exchange power constraint, energy storage operation constraint, voltage and current safety constraint and other related constraints are comprehensively considered for distribution network planning [29].

4.1. Objective Function

The objective function of distribution network planning model consists of four parts. It includes the flexibility compensation cost of virtual power station, the investment cost of energy storage system, the investment cost of distribution network line expansion and the abandonment cost of photovoltaic power station.

$$\min C = c_{line} + c_{dsr} + c_{ess} + c_{cur} \tag{16}$$

where, c_{line} represents the investment cost of distribution network line expansion. c_{dsr} is the flexibility compensation cost of virtual power station. c_{ess} represents the investment cost of energy storage system. c_{cur} is the abandonment cost of photovoltaic power station. Its specific expression forms are as follows.

4.1.1. The Investment Cost of Distribution Network Line Expansion

$$c_{line} = \frac{r(1+r)^n}{(1+r)^n - 1} \sum_{ij \in \theta_l} c_{ij} m_{ij} \tag{17}$$

where, r is the discount rate of line investment. n indicates the planned usage year. In this document, the value is 10. θ_l indicates the branch set to be built. c_{ij} represents the cost of building a single line. m_{ij} indicates the number of new lines.

4.1.2. The Flexibility Compensation Cost of Virtual Power Station

$$c_{dsr} = \sum_{v \in \delta_T} d_v \sum_{t=1}^{T} (c_t^{up} \sum_{j \in \delta_{up}} P_{t,v,j}^{up} + c_t^{down} \sum_{i \in \delta_{down}} P_{t,v,i}^{down}) \tag{18}$$

where, δ_T represents the set of typical days. δ_{up}, δ_{down} represent the set of nodes in the system where the load can be raised and lowered respectively. d_v indicates the number of days in a typical day. c_t^{up}, c_t^{down} respectively represent the unit compensation electricity price that can be raised and lowered in the system. $P_{t,v,j}^{up}$, $P_{t,v,i}^{down}$ indicate that the load capacity in the system can be increased and decreased respectively.

4.1.3. The Investment Cost of Energy Storage System

$$c_{ess} = (1 + r_{op} + r_{ma} + r_{sc}) r_{de} \sum_{i \in \delta_{ess}} (c_p P_{ess,i}^r + c_e E_{ess,i}^r) \tag{19}$$

where, c_p, c_e represent the unit power and capacity cost of the energy storage system respectively. r_{op}, r_{ma}, r_{sc}, r_{de} represent the conversion coefficients of the operation, maintenance, disposal and depreciation costs of the energy storage system respectively. δ_{ess} indicates the node set for configuring the energy storage system. $P_{ess,i}^r$, $E_{ess,i}^r$ are the rated power and capacity of the energy storage system respectively.

4.1.4. The Abandonment Cost of Photovoltaic Power Station

$$c_{cur} = \sum_{i \in \delta_{PV}} \sum_{v \in \delta_T} d_v \sum_{t=1}^{T} c_{PV} (P_{t,v,i}^{PV,\max} - P_{t,v,i}^{PV}) \tag{20}$$

where, δ_{PV} is the set of nodes configured with photovoltaic power stations. c_{PV} represents the unit cost of light abandonment of photovoltaic power stations. $P_{t,v,i}^{PV,\max}$, $P_{t,v,i}^{PV}$ represent the maximum power limit and actual output power of the photovoltaic power station respectively.

4.2. Constraint

4.2.1. Power Balance Constraint

$$P_{t,v}^{buy} + P_{t,v}^{PVess} - A^{new} P_{ij,t,v}^{new} - A^{old} P_{ij,t,v}^{old} = P_{t,v}^{up} + P_{t,v}^{load} - P_{t,v}^{down} \tag{21}$$

$$Q_{t,v}^{buy} + Q_{t,v}^{PVess} - A^{new} Q_{ij,t,v}^{new} - A^{old} Q_{ij,t,v}^{old} = Q_{t,v}^{load} \tag{22}$$

$$P_{t,v}^{PVess} = P_{t,v}^{PV} - P_{t,v}^{ess} \tag{23}$$

$$Q_{t,v}^{PVess} = Q_{t,v}^{PV} - Q_{t,v}^{ess} \tag{24}$$

where, $P_{t,v}^{buy}$, $Q_{t,v}^{buy}$ are the power interaction values of the distribution network and the upper power market, respectively. $P_{t,v}^{PVess}$, $Q_{t,v}^{PVess}$ are active power and reactive power of optical storage combination respectively. $P_{ij,t,v}^{new}$, $Q_{ij,t,v}^{new}$ represent the real-time power of a new line on a typical day. $P_{ij,t,v}^{old}$, $Q_{ij,t,v}^{old}$ respectively represent the real-time power of the existing line on a typical day. A^{new}, A^{old} represent the node association matrix of the new and existing branches respectively. $P_{t,v}^{load}$, $Q_{t,v}^{load}$ are active and reactive power loads on typical days respectively. $P_{t,v}^{ess}$, $Q_{t,v}^{ess}$ respectively represent the real-time power output of the energy storage system. Equations (21) and (22) are the active and reactive power balance of the distribution network, respectively.

4.2.2. Power Flow Constraint of Distribution Network

$$\sum_{i \in r(j)} (P_{ij,t,v} - \frac{P_{ij,t,v}^2 + Q_{ij,t,v}^2}{U_{i,t,v}^2} r_{ij}) = P_{j,t,v} + \sum_{c \in s(j)} P_{jc,t,v} \tag{25}$$

$$\sum_{i \in r(j)} (Q_{ij,t,v} - \frac{P_{ij,t,v}^2 + Q_{ij,t,v}^2}{U_{i,t,v}^2} x_{ij}) = Q_{j,t,v} + \sum_{c \in s(j)} Q_{jc,t,v} \tag{26}$$

$$U_{i,t,v}^2 - U_{j,t,v}^2 = 2(r_{ij}P_{ij,t,v} + x_{ij}Q_{ij,t,v}) - I_{ij,t,v}^2(r_{ij}^2 + x_{ij}^2) \tag{27}$$

where, r_{ij}, x_{ij} are the resistance and reactance of branch ij. $P_{ij,t,v}$, $Q_{ij,t,v}$ represent the real-time power of branch ij respectively. $P_{j,t,v}$, $Q_{j,t,v}$ represent the real-time injected power of node j respectively. $U_{i,t,v}$, $I_{ij.t,v}$ represent the voltage of node i and the current on branch ij, respectively. $r(j)$ represents the first node set with j as the tail node. $s(j)$ represents the set of tail nodes starting with j.

The above power flow constraints are strongly non-convex constraints, and their direct computation is complicated. Therefore, the second order conical convex optimization technique is used to transform the above constraints into linear constraints. The square of the line node voltage and branch current is defined as $U'_{i,t,v}$, $I'_{ij,t,v}$. Therefore, when the square of the branch current is a strictly increasing function of the objective function, and the node load is not set to a limit, there are as follows.

$$I'_{ij,t,v} \geq \frac{P_{ij,t,v}^2 + Q_{ij,t,v}^2}{U_{i,t,v}^2} \tag{28}$$

Therefore, the power flow constraint Formulas (25)–(27) of the distribution network can be converted into the following form after transformation.

$$\sum_{i \in r(j)} (P_{ij,t,v} - I'_{ij,t,v} r_{ij}) = P_{j,t,v} + \sum_{c \in s(j)} P_{jc,t,v} \tag{29}$$

$$\sum_{i \in r(j)} (Q_{ij,t,v} - I'_{ij,t,v} x_{ij}) = Q_{j,t,v} + \sum_{c \in s(j)} Q_{jc,t,v} \tag{30}$$

$$U'_{i,t,v} - U'_{j,t,v} = 2(r_{ij}P_{ij,t,v} + x_{ij}Q_{ij,t,v}) - I'_{ij,t,v}(r_{ij}^2 + x_{ij}^2) \tag{31}$$

$$\left\| \begin{matrix} 2P_{ij,t,v} \\ 2Q_{ij,t,v} \\ I'_{ij,t,v} - U'_{i,t,v} \end{matrix} \right\|_2 \leq I'_{ij,t,v} + U'_{i,t,v} \tag{32}$$

4.2.3. Constraints on the Flexible Operation of Virtual POWER Plants

$$u_{t,v,i}^{down} + u_{t,v,i}^{up} \leq 1 \tag{33}$$

$$0 \leq P_{t,v,i}^{down} \leq P_{t,v,i}^{down,\max} u_{t,v,i}^{down} \tag{34}$$

$$0 \leq P_{t,v,i}^{up} \leq P_{t,v,i}^{up,\max} u_{t,v,i}^{up} \tag{35}$$

$$\sum_{t}^{t+T^{down,\max}} u_{t,v,i}^{down} \leq T^{down,\max} \tag{36}$$

$$\sum_{t}^{t+T^{up,\max}} u_{t,v,i}^{up} \leq T^{up,\max} \tag{37}$$

In the formula, $u_{t,v,i}^{down}$, $u_{t,v,i}^{up}$ represent the state of virtual power plant flexible load reduction and load increase respectively. $P_{t,v,i}^{down,\max}$, $P_{t,v,i}^{up,\max}$ represent the flexible load reduction and upward load boundary of a typical day virtual power plant respectively. $T^{down,\max}$, $T^{up,\max}$ represent the maximum continuous downward adjustment and upward adjustment time of virtual power plant flexibility load respectively.

4.2.4. Constraints on Photovoltaic Operation

$$0 \leq P_{t,v,i}^{PV} \leq P_{t,v,i}^{PV,\max} \tag{38}$$

$$0 \leq Q_{t,v,i}^{PV} \leq Q_{t,v,i}^{PV,\max} \tag{39}$$

where, $P_{t,v,i}^{PV,\max}$, $Q_{t,v,i}^{PV,\max}$ are the maximum power output of the photovoltaic power station.

4.2.5. Constraint on Switching Power of the Main Network

$$0 \leq P_{t,v}^{buy} \leq P^{buy,\max} \tag{40}$$

$$0 \leq Q_{t,v}^{buy} \leq Q^{buy,\max} \tag{41}$$

where, $P^{buy,\max}$, $Q^{buy,\max}$ are the maximum interaction power with the upper power market.

4.2.6. Energy Storage Operation Constraints

$$0 \leq P_{t,v}^{ess,d} \leq P_{t,v}^{ess,d,\max} \tag{42}$$

$$0 \leq P_{t,v}^{ess,c} \leq P_{t,v}^{ess,c,\max} \tag{43}$$

$$P_{t,v}^{ess} = P_{t,v}^{ess,c} \xi_c - \frac{P_{t,v}^{ess,d}}{\xi_d} \tag{44}$$

$$E^{ess,\min} \leq E_{t,v}^{ess} \leq E^{ess,\max} \tag{45}$$

$$E_{t,v}^{ess} + \sum_{\tau=1}^{t} \left[(\lambda_{\mathrm{ds}})^{t-\tau} P_{t,v}^{ess} \right] = E^{ess,0} \tag{46}$$

where, λ_{ds} is the power conversion coefficient of the energy storage. $P_{t,v}^{ess,d}$, $P_{t,v}^{ess,c}$ respectively represent the real-time power output and input of the energy storage. $P_{t,v}^{ess,d,\max}$, $P_{t,v}^{ess,c,\max}$ respectively represent the maximum power output and input of the energy storage. ξ_d, ξ_c respectively represent the maximum power output and input efficiency of the energy storage.

$E_{t,v}^{ess}$, $E^{ess,0}$ represent the storage real-time energy and initial energy respectively. $E^{ess,\min}$, $E^{ess,\max}$ respectively represent the energy storage capacity limits.

4.2.7. Voltage and Current Safety Constraints

$$0 \le U'_{i,t,v} \le U'^{,\max}_{i,t,v} \tag{47}$$

$$0 \le I'_{i,t,v} \le I'^{,\max}_{i,t,v} \tag{48}$$

where, $U'^{,\max}_{i,t,v}$, $I'^{,\max}_{i,t,v}$ are the maximum values of the square of node voltage and the square of branch current, respectively.

In the aspect of model solving, the proposed programming model is transformed into a directly solvable mixed integer linear programming model by using second-order cone optimization technique. This paper intends to use YALMIP on MATLAB platform for mathematical modeling of the model, and call CPLEX solver for solving.

5. Simulation Results and Analysis

5.1. Model Building

In this paper, the superiority of the proposed programming model is verified by simulation on IEEE RTS-24 node system. The relevant data of this node are referred to [30]. Meanwhile, the 96-h photovoltaic output and conventional load data are shown in Figures 4 and 5.

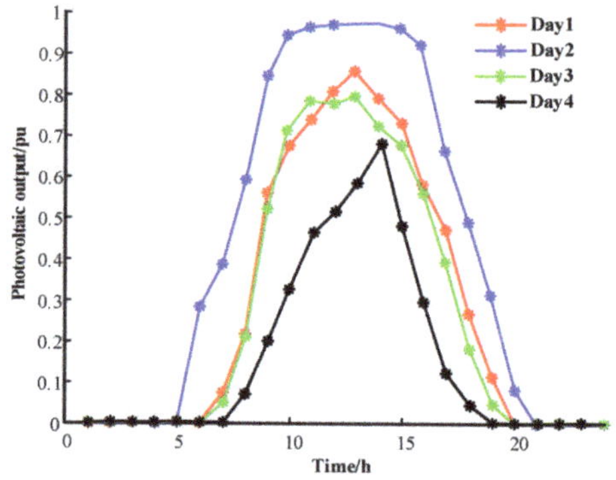

Figure 4. PV output in 96 h.

By solving the model presented in Section 3, the time series load power data and adjustable power range of the EV flexible load in the virtual power plant can be obtained. In this example, the charge and discharge power limit is 3 kW, the battery capacity ranges from 0.1 to 1 kWh during operation, and the battery capacity is set to 30 kWh. The calculated sequential load power data of 10,000 electric vehicles for 96 h and the adjustable power range up and down are shown in Figure 6. Among them, the quantified adjustable power range of electric vehicles will be used as the upward and downward load boundary in planning model.

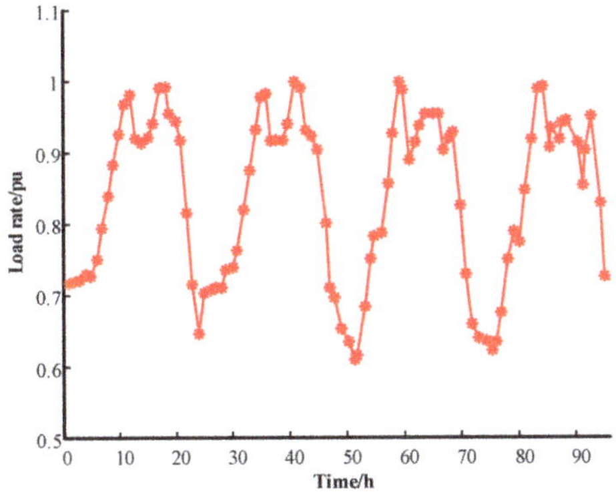

Figure 5. 96 h load rate.

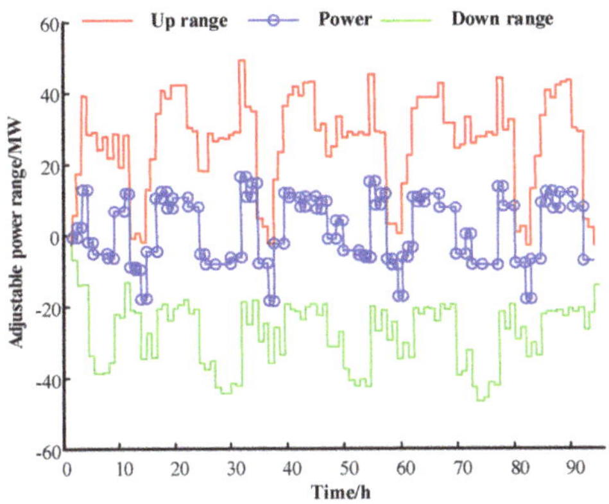

Figure 6. The sequential load power of 10,000 electric vehicles for 96 h and its adjustable power range up and down.

In the example setting, the 60 MW PV power station is connected on nodes 3, 8, 9, 10, 14 and 19 respectively, and the 130 MW PV power station is connected on nodes 7, 21, 22 and 23 respectively. In addition, each photovoltaic power station is equipped with an energy storage system, and its related parameters are shown in Table 1. At the same time, 10,000 electric vehicles are set up at nodes 3, 6, 9, 13, 15 and 18 as demand-side flexibility resources. The time-sharing compensation price of flexible load in virtual power plant is shown in Table 2.

Table 1. Related parameters of energy storage system.

Parameters	Value	Parameters	Value
r_{op}	0.02	r_{sc}	0.01
r_{de}	0.05	c_p (yuan/kW)	300
r_{ma}	0.02	c_e (yuan/kW)	3500

Table 2. Time-of-use compensation price.

Time	09:00–14:00 18:00–22:00	00:00–07:00 22:00–24:00	07:00–09:00 14:00–18:00
c_{down}	0.30	0.10	0.25
c_{up}	0.15	0.35	0.25

Finally, a 5-year distribution network planning work is carried out based on the above example setting, and the planning results are reflected in the economic cost of line expansion and planning.

5.2. Analysis of Result of Example Planning

In order to reflect the improvement of virtual battery flexibility resources on distribution network planning results and new energy consumption, this paper compares the planning results by setting two different cases.

Case 1. The flexibility of the virtual power plant flexibility resources is quantified before being incorporated into the distribution network planning process.

Case 2. The VPP flexibility resources are directly incorporated into the distribution network planning process.

5.2.1. Result Analysis of Line Expansion

In the actual planning process, the initial route scheme of this year is the optimal solution of the previous year's planning scheme. The optimization results of five-year distribution network line planning under the two cases are shown in Tables 3 and 4 respectively, where "6–10(1)" indicates that a line needs to be expanded between nodes 6 and 10 in the planning year.

Table 3. Results of planned line expansion for distribution network in Case 1.

Year	Line Expansion
1	6–10(1), 7–8(2), 13–14(1)
2	1–5(1), 7–8(1)
3	20–22(1)
4	3–22(1), 6–8(1), 10–12(1)
5	2–5(1), 9–11(1), 14–22(1)

Table 4. Results of planned line expansion for distribution network in Case 2.

Year	Line Expansion
1	6–10(1), 7–8(2), 10–11(1), 13–14(1)
2	1–5(1), 7–8(1), 10–12(1)
3	3–22(1), 20–22(1), 10–12(1)
4	3–22(1), 6–8(1), 10–12(1), 10–11(1), 13–14(1)
5	2–5(1), 7–8(1), 15–16(1), 15–23(1), 16–18(1)

As can be seen from Tables 3 and 4, compared with case 2, by adding flexibility resources to distribution network planning in case 1. It can be seen that the required line expansion in each planning year is significantly reduced, effectively reducing the overall planning cost.

5.2.2. Economic Analysis of Planning Results

In order to verify the advantages of the proposed planning model in improving the planning economy, the total planning costs of the two schemes were compared, as shown in Figure 7. The data column on the left of each planning year is the planning result of Case 1, and the data column on the right is the planning result of Case 2. In order to compare the difference between the individual costs of the two cases in detail, the comparison of line investment costs of the two cases is shown in Figure 8. The comparison of the abandoned light cost of photovoltaic power stations in the two cases is shown in Figure 9.

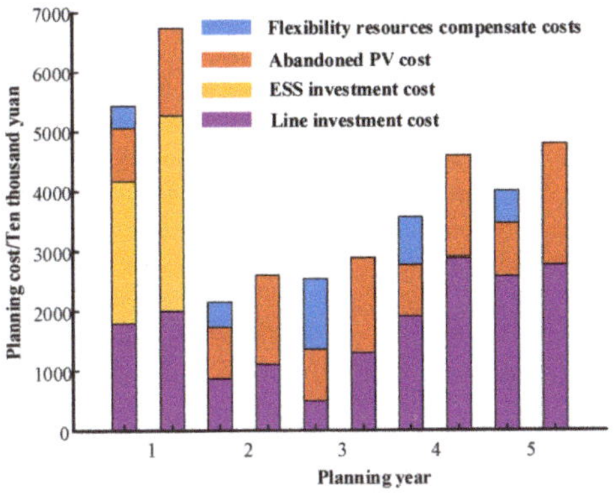

Figure 7. Distribution network planning cost comparison.

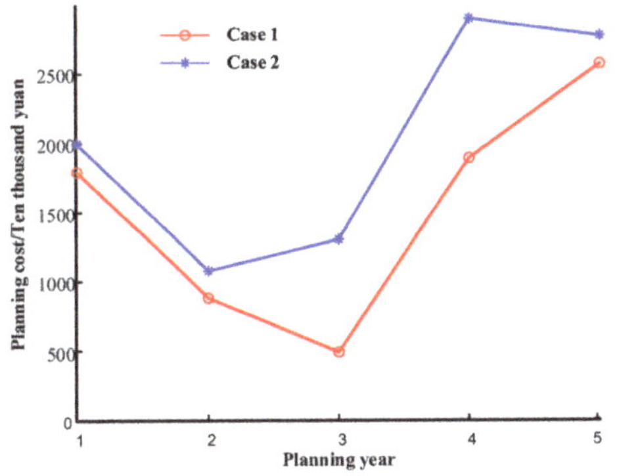

Figure 8. Line investment cost comparison.

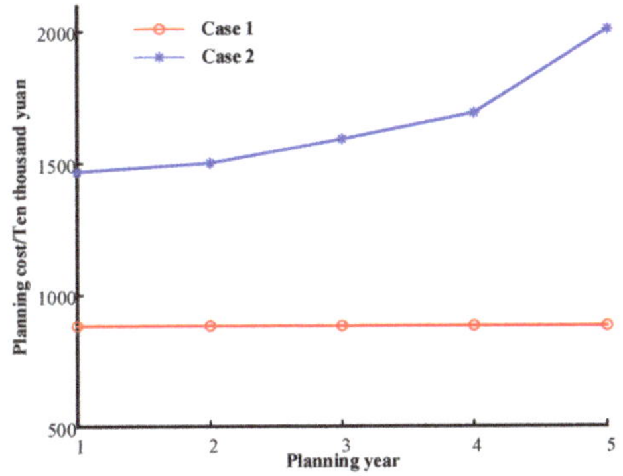

Figure 9. Comparison of abandoned light cost of photovoltaic power station.

As can be seen from Figure 7, by adding flexibility resources to distribution network planning in case 1, the annual total planning cost is lower than that in case 2. Therefore, the planning strategy in Case 1 is chosen every year as the initial planning scenario for the next year. In Figure 7, the first year is the initial planning of the IEEE RTS-24 node system, which requires the initial investment in the energy storage equipment and the initial construction of the line. Therefore, the total cost of the initial year of planning is higher. In the second to fifth years, the planning cost also increases with the increase of the load level.

It can be seen from Figures 8 and 9 that the addition of flexible resources can reduce the investment cost of line expansion in the distribution network. This is because the downlink load in the flexible resource reduces the peak load of the system during operation, which in turn reduces the peak power transmitted by the line. At the same time, because electric vehicles can be used as a flexible energy storage in the virtual power station, excess photovoltaic power generation can be stored, thereby increasing the photovoltaic consumption rate. Therefore, the proposed model can effectively improve the economy of the planning process.

5.2.3. Analysis of Photovoltaic Absorption Results

In order to further verify the effectiveness of the proposed model in improving the PV uptake rate of the system, the comparison of the photovoltaic consumption results of the two cases on the third node of the typical day is shown in Figure 10.

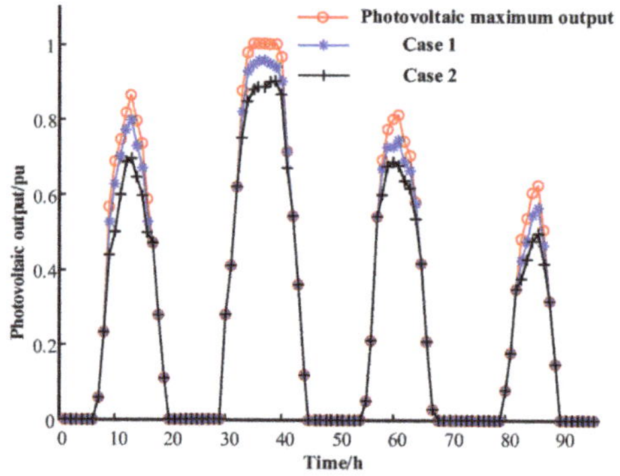

Figure 10. Comparison of PV consumption results on typical day of node 3.

It can be seen from the figure that when flexible resources are involved in planning, the actual PV output of the third node in the first year is more consistent with the maximum output curve than the actual PV output and the maximum output curve. This is because the addition of flexible resources such as distributed energy storage enables the photovoltaic power output of photovoltaic power station nodes to achieve on-site absorption. It reduces discarded light, improves the absorption rate of photovoltaics, and reduces power flow in the grid. While improving the economy of distribution network planning, the "peak cutting and valley filling" of power system is realized to a certain extent.

6. Conclusions

Aiming at the problem of quantifying the flexibility resources in virtual power plant in the existing literature, a distribution network planning method considering the flexibility of distributed energy resource in virtual power plant is proposed. The following conclusions can be drawn after setting corresponding examples for verification.

1. The planning method proposed in this paper analyzes the flexibility resources quantitatively and brings them into the planning process of distribution network. It effectively reduces the investment cost of line development and improves the economy of the planning process.
2. The proposed planning method utilizes the characteristic of flexible resources to transfer the peak load effectively and maintain the stable operation of the system. At the same time, distributed energy resource storage is used to improve the system's photovoltaic consumption rate.

Author Contributions: Conceptualization, Z.Y., G.H. and F.Y.; methodology, Z.Y. and Y.S.; software, Z.Y., Y.L. (Yu Liu) and H.M.; validation, Z.Y., Z.Z. and B.Z.; data curation, W.H. and G.H.; writing—original draft preparation, Z.Y. and G.H.; writing—review and editing, Z.Y., G.H. and Y.L. (Yang Lei). All authors have read and agreed to the published version of the manuscript.

Funding: This research received no external funding.

Institutional Review Board Statement: Not applicable.

Informed Consent Statement: Not applicable.

Data Availability Statement: Due to privacy restrictions, all data sources cannot be shared.

Conflicts of Interest: The authors declare no conflict of interest.

References

1. Li, H.; Ren, Z.; Fan, M.; Li, W.; Xu, Y.; Jiang, Y.; Xia, W. A review of scenario analysis methods in planning and operation of modern power systems: Methodologies, applications, and challenges. *Electr. Power Syst. Res.* **2022**, *205*, 107722. [CrossRef]
2. Dong, L.; Fan, S.; Wang, Z.; Xiao, J.; Zhou, H.; Li, Z.; He, G. An adaptive decentralized economic dispatch method for virtual power plant. *Appl. Energy* **2021**, *300*, 117347. [CrossRef]
3. Fan, H.V.; Dong, Z.; Meng, K. Integrated distribution expansion planning considering stochastic renewable energy resources and electric vehicles. *Appl. Energy* **2020**, *278*, 115720. [CrossRef]
4. Jiechen, W.; Xin, A.I.; Junjie, H. Methods for characterizing flexibilities from demand-side resources and their applications in the day-ahead optimal scheduling. *Trans. China Electrotech. Soc.* **2020**, *35*, 1973–1984.
5. Renhe, Z. *Flexible Resources Value Evaluation Model and Application of High Proportion New Energy Power System*; North China Electric Power University: Beijing, China, 2021.
6. Zhouyang, W.; Xin, A.I.; Junjie, H. Reserve optimization and real- time scheduling off requency regulation ancillary service with participation of flexible resource on demand side. *Autom. Electr. Power Syst.* **2021**, *45*, 148–157.
7. Yasuda, Y.; Gomez-Lazaro, E.; Menemenlis, N. Flexibility chart. evaluation on diversity of flexibility in various areas. In Proceedings of the 12th Wind Integration Workshop, London, UK, 22–24 October 2013. 6p.
8. Ceseña, E.A.M.; Capuder, T.; Mancarella, P. Flexible distributed multi-energy generation system expansion planning under uncertainty. *IEEE Trans Smart Grid.* **2015**, *7*, 348–357. [CrossRef]
9. Zhang, C.; Ding, Y.; Nordentoft, N.C.; Pinson, P.; Østergaard, J. FLECH. a Danish market solution for DSO congestion management through DER flexibility services. *J. Mod. Power Syst. Clean Energy* **2014**, *2*, 126–133. [CrossRef]
10. Shaoyun, G.; Youwei, Z.; Hong, L. Bi-layer expansion programming method for active distribution network considering dynamic grid reconfiguratio. *Power Syst. Technol.* **2018**, *42*, 1526–1536.
11. Wu, Z.; Liu, Y.F.; Gu, W.; Liu, P.X.; Li, J.J.; Li, Z. A modified decomposition method for multistage planning of energy storage, distributed generation and distribution network. *Proc. CSEE* **2019**, *39*, 4705–4715,4973.
12. Jia, L.; Hu, Z.C.; Song, Y.H.; Ding, H.J. Joint planning of distribution networks with distributed energy storage systems and electric vehicle charging stations. *Proc. CSEE* **2017**, *37*, 73–84.
13. Zhong, Q.; Sun, W.; Yu, N.; Liu, C.; Wang, F.; Zhang, X. Load and power forecasting in active distribution network planning. *Proc. CSEE* **2014**, *34*, 3050–3056.
14. Tomasson, E.; Soder, L. Improved importance sampling for reliability evaluation of composite power systems. *IEEE Trans. Power Syst.* **2017**, *32*, 2426–2434. [CrossRef]
15. Li, H.; Ren, Z.; Trivedi, A.; Verma, P.P.; Srinivasan, D.; Li, W. A noncooperative game-based approach for microgrid planning considering existing interconnected and clustered microgrids on an island. *IEEE Trans. Sustain. Energy* **2022**, *13*, 2064–2078. [CrossRef]
16. Li, H.; Ren, Z.; Trivedi, A.; Srinivasan, D.; Liu, P. Optimal planning of dual-zero microgrid on an island towards net-zero carbon emission. *IEEE Trans. Smart Grid* **2023**, *early access*. [CrossRef]
17. Xiang, Z. *Active Distribution Network Planning with Distributed Generation*; Shanghai Jiao Tong University: Shanghai, China, 2014.
18. Wenqing, L. *Distribution Network Planning with Wind Turbine Photovoltaic System, Storage System and Electric Vehicles*; Shanghai Jiao Tong University: Shanghai, China, 2013.
19. Koutsoukis, N.C.; Georgilakis, P.S. A multistage distribution network planning method considering distributed generation active management and demand response. *IET Renew. Power Gener.* **2022**, *16*, 65–76. [CrossRef]
20. Gan, L.; Li, N.; Topcu, U.; Low, S.H. Exact convex relaxation of optimal power flow in radial networks. *IEEE Trans. Autom. Control.* **2015**, *60*, 72–87. [CrossRef]
21. Soliman, M.S.; Belkhier, Y.; Ullah, N.; Achour, A.; Alharbi, Y.M.; Al Alahmadi, A.A.; Abeida, H.; Khraisat, Y.S.H. Supervisory energy management of a hybrid battery/PV/tidal/wind sources integrated in DC-microgrid energy storage system. *Energy Rep.* **2021**, *7*, 7728–7740. [CrossRef]

22. Sahri, Y.; Belkhier, Y.; Tamalouzt, S.; Ullah, N.; Shaw, R.N.; Chowdhury, M.S.; Techato, K. Energy management system for hybrid PV/wind/battery/fuel cell in microgrid-based hydrogen and economical hybrid battery/super capacitor energy storage. *Energies* **2021**, *14*, 5722. [CrossRef]
23. Moradi-Sarvestani, S.; Jooshaki, M.; Fotuhi-Firuzabad, M.; Lehtonen, M. Incorporating direct load control demand response into active distribution system planning. *Appl. Energy* **2023**, *339*, 120897. [CrossRef]
24. Xu, W.; Yu, B.; Song, Q.; Weng, L.; Luo, M.; Zhang, F. Economic and low-carbon-oriented distribution network planning considering the uncertainties of photovoltaic generation and load demand to achieve their reliability. *Energies* **2022**, *15*, 9639. [CrossRef]
25. He, S.; Gao, H.; Liu, J.; Zhang, X.; Chen, Z. Distribution system planning considering peak shaving of energy station. *Appl. Energy* **2022**, *312*, 118692. [CrossRef]
26. Ziegler, D.U.; Prettico, G.; Mateo, C.; Román, T.G.S. Methodology for integrating flexibility into realistic large-scale distribution network planning using Tabu search. *Int. J. Electr. Power Energy Syst.* **2023**, *152*, 109201. [CrossRef]
27. Alsokhiry, F.; Siano, P.; Annuk, A.; Mohamed, M.A. A novel time-of-use pricing based energy management system for smart home appliances: Cost-effective method. *Sustainability* **2022**, *14*, 14556. [CrossRef]
28. Ma, H.; Liu, Z.; Li, M.; Wang, B.; Si, Y.; Yang, Y.; Mohamed, M.A. A two-stage optimal scheduling method for active distribution networks considering uncertainty risk. *Energy Rep.* **2021**, *7*, 4633–4641. [CrossRef]
29. Tan, H.; Ren, Z.; Yan, W.; Wang, Q.; Mohamed, M.A. A wind power accommodation capability assessment method for multi-energy microgrids. *IEEE Trans. Sustain. Energy* **2021**, *12*, 2482–2492. [CrossRef]
30. Selçuk, M.; Ercan, Ş. Literature review of transmission expansion planning problem test systems. detailed analysis of IEEE-24. *Electr. Power Syst. Res.* **2021**, *201*, 107543.

sustainability

Article

A Novel Evolving Framework for Energy Management in Combined Heat and Electricity Systems with Demand Response Programs

Ting Chen [1], Lei Gan [2], Sheeraz Iqbal [3,*], Marek Jasiński [4], Mohammed A. El-Meligy [5], Mohamed Sharaf [5] and Samia G. Ali [6]

1 Department of Electric Power Engineering, Fujian Vocational & Technical College of Water Conservancy & Electric Power, Yong'an 366000, China
2 State Grid Hubei Shiyan Power Supply Company, Shiyan 442000, China
3 Department of Electrical Engineering, University of Azad Jammu and Kashmir, Muzaffarabad 13100, Pakistan
4 WWSIS "Horyzont", 54-239 Wrocław, Poland
5 Department of Industrial Engineering, College of Engineering, King Saud University, Riyadh 11421, Saudi Arabia
6 Department of Electrical Power and Machines, Faculty of Engineering, Kafrelsheikh University, Kafrelsheikh 33516, Egypt
* Correspondence: sheeraz.iqbal@ajku.edu.pk

Abstract: In recent years, demand response programs (DRPs) have become an effective method of encouraging users to participate in energy system operations. The problem of optimal energy flow (OEF) is a complex challenge in multiple power systems. Accordingly, this study aims to propose a novel evolving framework for optimal OEF operation of an electricity, heat, and gas integrating system, taking into account flexible heat and electricity demands. To this end, a switching idea between input energy carriers has been introduced to combine the traditional DRP with demand-side energy supply management. Switching between the feeding energy carriers could change how power is supplied to the end users and thus would affect the total cost of the grid. Operators of integrated systems minimize the operational costs associated with supplying flexible power to users in this study. Considering the high nonlinearity of the problem, a novel optimization algorithm is presented for solving the complex OEF based on the improved teaching–learning-based optimization algorithm (ITLBOA). According to the outcomes, flexible DRP reduces operational prices and smooths power demand curves for power and heating networks.

Keywords: demand response; optimum power flow; improved teaching–learning optimization; heat and electrical demands; combined heat and electricity systems

Citation: Chen, T.; Gan, L.; Iqbal, S.; Jasiński, M.; El-Meligy, M.A.; Sharaf, M.; Ali, S.G. A Novel Evolving Framework for Energy Management in Combined Heat and Electricity Systems with Demand Response Programs. *Sustainability* **2023**, *15*, 10481. https://doi.org/10.3390/su151310481

Academic Editor: Pallav Purohit

Received: 5 May 2023
Revised: 22 June 2023
Accepted: 29 June 2023
Published: 3 July 2023

1. Introduction

Societies and economies develop as a result of energy. As fossil fuels are depleted and pollution increases, technologies such as combined heat and power (CHP), gas turbines (GT), electric boilers (EB), photovoltaics (PV), power-to-gas (P2G), wind turbines (WT), and gas storage (GS) are beneficial in terms of reducing emissions and costs [1]. The multi-energy system (MES) was developed as a solution to the issues of efficient power usage, environmental friendliness, and optimal use of energy from renewable sources [2]. In the MES, a variety of energy carriers are integrated in order to interact with energy on diverse levels, offering great potential for advancement [3].

An MES is able to enhance power performance and provide more advantages to society by combining CHP, WT, PV GS, EB, GT, and P2G [4]. The coordination of electrical and heating systems significantly facilitates renewable energy (RE) utilization. P2G technology enhances the connection between gas and electrical systems, allowing RE to be accommodated. According to [1], wind energy can be a cost-effective way to generate heat in excess conditions; thus, the deployment of EBs may offer a viable option under continuous wind excess conditions. With EBs, wind power could be better exploited; in addition to producing useful heat, EBs could also decrease the operating limitations of CHPs and minimize CO_2 emissions. As well as breaking the CHP thermoelectric coupling, storage facilities and EBs provide indirect interaction with RE, thus improving MES flexibility and economics [2]. The conversion of surplus electricity to different kinds of energy can maximize the use of RE, thereby minimizing the need to curtail wind and solar energy production. CHP units maximize power performance and environmental protection by utilizing energy cascades. P2G has been extensively researched by many experts in the field of MES. Ref. [3] examined the economic impact of P2G technology. A surplus of electricity could be converted into gas with P2G technology, providing a range of advantages to the power system, including decreasing wind power curtailments, utilizing new energy more efficiently, and offering auxiliary services. Ref. [4] indicated that P2G could reduce RE curtailment while easing gas and electrical transmission network limitations.

In order to ensure the MES's security, uncertainties of renewable sources must also be taken into account. As the world moves toward green energy, RE exploitation is becoming increasingly important [5]. RE's unpredictable nature presents serious reliability and protection concerns [6], limiting its expansion and application. As a result, RE production is subject to stochastic fluctuations, reducing the flexibility of supply for the power grid. A scheme is proposed in Ref. [7] that maximizes the social benefit for unpredictable cases by meeting safety requirements through hourly power demand response (DR). Ref. [8] examined a two-step, robust, centralized-optimum dispatch scheme that is resilient to PV production uncertainties. It is difficult to compute two-step, robust optimization schemes since the problem is NP-hard. A study has been conducted on two solution methods for overcoming the difficulty of computing. Approximation algorithms are used in the first method, where two-stage decisions assume easy functions, for example, affine functions, of uncertainty [9]. In the second kind of algorithm, the Benders decomposition technique [10] and column-and-constraint generation (CCG) technique [11] are used for obtaining accurate results. Based on ref. [11], CCG converges in fewer iterations due to its superior computing efficiency.

The concept of DR covers any intentional modification of power usage from users as well as incentive payments aimed at affecting usage time, overall demand, and instantaneous demand. DR can be realized through load regulation and control, as demonstrated in ref. [12]. As discussed in ref. [13], DR has proven crucial for enhancing performance and security. According to [14], DR may enhance the reliability of electrical systems and contribute to the scheduling of PV systems. Temperature variations do not affect the thermal inertia of thermodynamic systems and consumers, and a suitable temperature range can be maintained, thus allowing heat loads to be controlled. According to [15], electricity DR can decrease operational prices by providing reserves for managing predicted wind and solar generation uncertainties. A residential energy management controller that incorporates various types of home devices, such as deferrable, curtailable, and thermal devices, was presented in ref. [16], thereby reducing energy prices and ensuring user satisfaction. In ref. [17], a more feasible and effective household energy management planning scheme was proposed in order to reduce power prices without compromising a predetermined level of satisfaction.

There are a number of issues that should be properly considered according to the available studies: (1) Heat and power are strongly coupled in conventional CHP units. Generally, thermoelectric decoupling is obtained by considering storage. It takes extensive research to fully realize the possibilities of using energy conversion devices and energy storage for accommodating and storing additional renewable production for ensuring energy balance, especially when CHP power exceeds electrical demand due to high thermal loads. (2) In response to uncertainties about RE, the majority of research applies stochastic optimization, which may negatively affect the safety of the system [6,7]. In contrast, robust optimization is being investigated in power systems for ensuring safety, despite uncertainties [10,11]. There have been relatively few studies on the robust planning of MES.

(3) A number of investigations were considered on electricity DR [13–15], but not enough has been done on integrated heat–electric DR.

The DR program (DRP) optimal energy flow (OEF) model influences outcomes in a complicated, non-convex optimization problem involving intricate computations for the integrated gas, electric, and heat system (IGEHS). In the present study, a new optimization algorithm called improved teaching–learning-based optimization algorithm (ITLBOA) is presented to lessen the OEF issue's complexity of multi-carrier energy (MCE).

This study mainly focuses on the following contributions: (i) Developing a novel participating management system for flexible heat and electricity users' demand-side energy (DSE). (ii) Proposing a new modification for the OEF model based on IGEHS, considering flexible electricity and heat requirements. (iii) Providing a new modified method called ITLBOA for solving the proposed non-linear and non-convex OEF problem in the MCE system.

Following are the remaining sections of this study. Part 2 outlines the suggested architecture for the OEF for a composed heating, power, and gas system in relation to the integrated disaster response (IDR) strategy. Part 3 models IGEHS. In addition, sub-network modeling is discussed in detail. Part 4 introduces the suggested ITLBOA. Part 5 examines the performance of the suggested OEF architecture using DR power users. Part 6 provides the conclusion.

2. The MCE System Operator's Decision-Making Issue with Flexible DSE Activity

When reducing the overall system operating price of an MCE system, the OEF problem is typically considered. An MCE system's overall price can be calculated by calculating the fuel costs of each production unit. The energy carriers are natural gas, coal, or biomass. The integrated MES is operated by a specific independent system operator (ISO). According to the ISO, Equation (1) is used to formulate the MCE's OEF problem:

$$Min\ (cost) = Min(\sum_{t=1}^{24}\sum_{q=1}^{N_{fuel}}(U_q \cdot fuel(\tau_{q,fuel}(t))^2 + V_{q,fuel}\tau_{q,fuel}(t) + W_{q,fuel} + \sum_{t=1}^{24}\sum_{r=1}^{N_{hub}}(L^t_{e,out,r} \cdot \exists_e + L^t_{h,out,r} \cdot \exists_h) \quad (1)$$

Integrated system operating costs are minimized through the adjustment of the ISO parameters in the following manner:

$$X(t) = \left[P^{gen}_{E,j}, \phi^{chp}_{H,j}, \phi^{boil}_{H,j}, T^{chp}_{s,j}, T^{boil}_{s,j}, H^{comp}_{j}, L^t_{e,out,r}, L^t_{h,out,r}\right] \quad (2)$$

OEF solves its price-minimizing problem using limitations Equation (21) to Equation (43) as follows:

$$P^{gen}_{min,j} \le P^{gen}_{E,j} \le P^{gen}_{max,j} \quad j = 1, 2, \ldots, N_{gen} \quad (3)$$

$$Q^{gen}_{min,j} \le Q^{gen}_{E,j} \le Q^{gen}_{max,j} \quad j = 1, 2, \ldots, N_{gen} \quad (4)$$

$$V^{min}_{j} \le V_j \le V^{max}_{j} \quad j = 1, 2, \ldots, N^{bus}_{E} \quad (5)$$

$$\pi^{min}_{j} \le \pi_j \le \pi^{max}_{j} \quad j = 1, 2, \ldots, N^{bus}_{G} \quad (6)$$

$$T^{min}_{s,j} \le T_{s,j} \le T^{max}_{s,j} \quad j = 1, 2, \ldots, N^{bus}_{H} \quad (7)$$

$$T^{min}_{r,j} \le T_{r,j} \le T^{max}_{r,j} \quad j = 1, 2, \ldots, N^{bus}_{H} \quad (8)$$

$$H^{comp}_{j,min} \le H^{comp}_{j} \le H^{comp}_{j,max} \quad j = 1, 2, \ldots, N_{comp} \quad (9)$$

$$\phi^{chp,min}_{H,j} \le \phi^{chp}_{H,j} \le \phi^{chp,max}_{H,j} \quad j = 1, 2, \ldots, N_{CHP} \quad (10)$$

$$\phi^{boil}_{min,j} \le \phi^{boil}_{H,j} \le \phi^{boil}_{max,j} \quad j = 1, 2, \ldots, N_{boil} \quad (11)$$

$$T^{chp}_{min,s,j} \le T^{chp}_{s,j} \le T^{chp}_{max,s,j} \quad j = 1,2,\ldots,N_{comp} \tag{12}$$

$$T^{boil}_{min,s,j} \le T^{boil}_{s,j} \le T^{boil}_{max,s,j} \quad j = 1,2,\ldots,N_{boil} \tag{13}$$

$$\left|P^{line}_{E,ij}\right| \le P^{line,max}_{E,ij} \; j = 1,2,\ldots,N^{line}_{E} \tag{14}$$

$$\left|F^{line}_{G,ij}\right| \le F^{line,max}_{G,ij} \; j = 1,2,\ldots,N^{line}_{G} \tag{15}$$

$$\left|\phi^{line}_{H,ij}\right| \le \phi^{line,max}_{H,ij} \; j = 1,2,\ldots,N^{line}_{H} \tag{16}$$

in which $P^{gen}_{min,j}$ represents minimal active power and $P^{gen}_{max,j}$ shows maximal active power of generators and $Q^{gen}_{min,j}$ shows minimal reactive power and $Q^{gen}_{max,j}$ shows maximal reactive power of generators. $\left|P^{line}_{E,\,ij}\right|$ shows the transmission flow of the electrical line, $\left|F^{line}_{G,\,ij}\right|$ represents the transmission flow of the gas line, and $\left|\varnothing^{line}_{H,\,ij}\right|$ shows the transmission flow of the heat line.

The generators' active and reactive powers are defined by the limitations Equations (3) and (4). In addition, the boundaries of magnitudes of voltage for the power networks, nodal pressures for the gas networks, and temperatures for the heat networks are represented by limitations Equations (5)–(8). Equation (9) defines the compressor ratio restrictions. Heat power produced by CHPs and boilers is limited by limitations Equations (10) and (11). A CHP and boiler's temperature boundaries are determined by Equations (12) and (13). Equations (3)–(16) present the transmission line flow restrictions of electricity, gas, and heating systems.

In addition, in order to minimize OEF on any flexible DR hub, it is necessary to satisfy the below formulas for the incentive-driven DRP.

There is no change in the overall demand for heat and electricity over the course of the day, and these demands are shifting from one-period intervals to other periods. E^{day}_{r} and H^{day}_{r} represent the overall amounts for the flexible power and heating requirements per day for the rth hub in the following way:

$$E^{day}_{r} = \sum_{t\in T} L^{t}_{e,\,out} \qquad r = 1,2,\ldots,N_{hub} \tag{17}$$

$$H^{day}_{r} = \sum_{t\in T} L^{t}_{h,\,out} \qquad r = 1,2,\ldots,N_{hub} \tag{18}$$

The demand for electricity and heat of the rth hub fall between the minimal and maximal limitations $L^{t,\,max}_{e,\,out}$, $L^{t,\,max}_{h,\,out}$, $L^{t,\,min}_{e,\,out}$, and $L^{t,\,min}_{h,\,out}$ for t. Thus:

$$L^{t,\,min}_{e,\,out} \le L^{t}_{e,\,out} \le L^{t,\,max}_{e,\,out} \tag{19}$$

$$L^{t,\,min}_{h,\,out} \le L^{t}_{h,\,out} \le L^{t,\,max}_{h,\,out} \tag{20}$$

According to the simultaneous power flow scheme for the IGEHS, the electric voltages, gas pressures, starting and ending temperatures, and any line power flow can be computed to minimize Equation (1). The following part describes the gas, heat, and electrical subsystems in detail.

3. IGEHS Model

3.1. Electrical System

The AC power flow equations were employed in this study for modeling the power network. Every node's reactive and active powers needed to satisfy the formulas below to be power balanced:

$$P_i^{gen} + P_i^{chp} = P_i^{dem} + P_i^{pump} + P_i^{comp} + Re(V_i \sum_{j=1}^{N_E} Y_{ij}^* V_j^*) \tag{21}$$

$$Q_i^{gen} + Q_i^{chp} + Q_i^{ch} = Q_i^{dem} + im(V_i \sum_{k=1}^{N_E} Y_{ij}^* V_j^*) \tag{22}$$

It is important to note that the previous equations accounted for the equal generation of electricity and the equal load demand plus the grid losses.

3.2. Gas System

There are several types of devices that make up an inherent gas grid, such as gas sources, pipelines, compressor stations, storage facilities, and users. Natural gas frameworks for steady-state evaluation can be illustrated with the below formulas. This formula is used to calculate gas flow in pipelines based on the pressures at either end [18]:

$$F_{gk} = sign(F_{gk}) F_{gk} \left(\left| \Pi_g^2 - \Pi_k^2 - H_p^{gk} \right| \right)^{0.5} \tag{23}$$

in which $sign(F_{gk})$ represents a sign function of pressures, in which its amount equals $+1$ when $\Pi_g \geq \Pi_k$ and power flow would transpose from node g to node k and -1; the reverse is also true. H_p^{gk} considers the altitude variation impacts over the gas pipeline and can be determined as follows:

$$H_p^{gk} = \frac{0.0375 g (H_g - H_k) (P_a^{gk})^2}{Z_a T_a^{gk}} \tag{24}$$

in which H_g and H_k represent the altitudes for the gas pipelines for the gth and k^{th} gas nodes. The average pressure Π_a^{gk} considers the nonlinear pressure decrease using distance [18] and can be determined as follows:

$$\Pi_a^{gk} = \frac{2}{3} [(\Pi_g + \Pi_k) - (\frac{\Pi_g \Pi_k}{\Pi_g + \Pi_k})] \tag{25}$$

$\mho_{gk}$ shows one constant for the physical features for all pipelines obtained according to the information about the network and engineers' experiences operating the gas assets and is determined as follows:

$$\mho_{gk} = \frac{1.14 \times 10^{-3} T_0 (D_{gk}^{GL})^{2.5} E_p}{(L^{gk} \gamma_G Z T_a \Theta_{gk}^{GL})^{0.5}} \tag{26}$$

The gas pipeline friction factor (Θ_{gk}^{GL}) is determined according to the Colebrook formula [19]:

$$\frac{1}{\sqrt{\Theta_{gk}^{GL}}} = -2 \log (\frac{\varepsilon_G}{3.71 D_{gk}^{GL}} + \frac{2.51}{R_{gk}^{GL}} \frac{1}{\sqrt{\Theta_{gk}^{GL}}}) \tag{27}$$

The gas pressure is maintained by the compressors in the gas network by consuming horsepower. In kW/h, the compressor consumes the below power:

$$E_{gk}^{comp} = \frac{112.989\,\Pi_O}{\eta_{gk}^{comp}\,T_O}\frac{\lambda_G}{\lambda_G - 1} \times ZT_a F_{gk}((H_{gk}^{comp})^{\frac{\lambda_G}{\lambda_G - 1}} - 1) \tag{28}$$

$$H_{gk}^{comp} = \frac{\Pi_{k\cdot out}^{comp}}{\Pi_{g\cdot in}^{comp}} \tag{29}$$

Electric motors and gas-fired turbines can provide the compressor with energy. A power network's electricity is calculated as follows:

$$P_{E,g}^{comp} = (\frac{745.7 \times 10^{-6}}{3600})E_{gk}^{comp} \tag{30}$$

Natural gas consumption by gas-fired generators is typically determined in the following way:

$$f^{GG} = \frac{1}{GHV}(a^{gen}(P_E^{gen})^2 + b^{gen}P_E^{gen} + c^{gen} + \left|d^{gen}\sin(e^{gen}(P_E^{gen,min} - P_E^{gen}))\right|) \tag{31}$$

For a gas system, the nodal gas flow balance can be represented by Equation (32):

$$F_{g\cdot S} = F_{g\cdot L} + F^{GG} + F_{gk}^{comp} + F_g^{chp} + F_g^{boiler} + \sum_{k=1}^{N_G} F_{gk} \tag{32}$$

3.3. Heating System

Water transports heat from heat sources to users in the heating system. Hot water or steam is circulated from production heat sources to heat load demands via the electric pumps in the system. A leading heat network infrastructure consists of heat sources, circulation pumps, heat users, and pipelines. Node h transfers heat power to node b in the following manner:

$$\phi_{hb} = \dot{m}_{hb}c_p(T_{start,h} - T_{end,b}) \tag{33}$$

$$T_{end} = (T_{start} - T_g)\exp(-\frac{l\Gamma}{\dot{m}_{hb}c_p}) + T_g \tag{34}$$

Based on the Darcy–Weisbach formula [20], friction within pipelines causes pressure changes in mass flows as a result of a loss of pressure. Losses due to pressure can be calculated as follows:

$$v_{hb}^{H,\,L} = K_{hb}^{H,\,L} sign_m(\dot{m}_{hb}) - sign_m(-\dot{m}_{hb})\cdot(-\dot{m}_{hb})^2 \tag{35}$$

$$K_{hb}^{H,\,L} = \frac{8L_{hb}^{H,\,L}\Theta_{hb}^{H,L}}{\rho_W^2\pi^2 g(D_{hb}^{H,\,L})^5} \tag{36}$$

in which $K_{hb}^{H,\,L}$ shows the coefficient of resistance for the heating pipelines, g represents the gravitational acceleration, ρ_W shows the density of water, and $\Theta_{hb}^{H,L}$ shows the friction factor for the heating pipelines and can be calculated via the Colebrook formula in the following manner:

$$\frac{1}{\sqrt{\Theta_{hb}^{H,L}}} = -2\log(\frac{\varepsilon_H}{3.71D_{hb}^{H,L}} + \frac{2.51}{R_{hb}^{H,L}}\frac{1}{\sqrt{\Theta_{hb}^{H,L}}}) \tag{37}$$

$$R_{hb}^{H,L} = \frac{U_{hb}^{H,\,L}D_{hb}^{H,L}}{\mu_W} \tag{38}$$

$$U_{hb}^{H,L} = \frac{4m_{hb}^{\cdot}}{\rho_W \pi \left(D_{hb}^{H,L}\right)^2} \tag{39}$$

in which ε_H shows the pipe roughness and R_{hb}^{HL} represents the Reynolds number. U_{hb}^{HL} shows the flow velocity and μ_W represents the kinematic viscosity of water.

Water pumps consume the following amount of electric power:

$$P_E^{HP} = \frac{4\dot{m}_H^{HP} g H_p}{\eta^{HP}} \tag{40}$$

It is necessary to satisfy four formulas to balance the heat flow in the heating network. The hydraulic heating flow evaluation is based on two formulas, while the thermal heating flow evaluation is based on the other two formulas. As a result of the hydraulic formulas, each node and loop in the heating system is in balance with heat and pressure loss. Following is the formula for heat balance:

$$\phi_{h,load} = \sum_{h=1}^{Nh} c_p(\dot{m_{hb}} + \dot{m}_H^{CHP} + \dot{m}_H^{boiler}) \times (T_{start} - T_{end,load}) \tag{41}$$

In contrast, pressure loss balance is defined as follows:

$$\sum_{h=1}^{Nh} \sum_{b=1}^{Nh} sign_m(\dot{m_{hb}}) \cdot v_{hb}^{H \cdot L} = 0 \tag{42}$$

Each node must satisfy supply and return temperature balance restrictions, excluding slack nodes. The return and supply temperature balances of all heating nodes can be expressed in Equations (43) and (44).

$$\begin{aligned} \dot{m}_H^{CHP}(T_{start,CHP} - T_g) + \dot{m}_H^{boiler}(T_{start,boiler} - T_g) + \sum_{b=1}^{Nh} sign_m(\dot{m_{hb}}) \cdot \left(\dot{m_{hb}}(T_{start} - T_g) \exp\left(-\frac{lU}{\dot{m_{hb}} c_p}\right)\right) \\ = (T_{start} - T_g)(\dot{m}_H^{CHP} + \dot{m}_H^{boiler} + \sum_{b=1}^{Nh} sign_m(\dot{m_{hb}}) \cdot \dot{m_{hb}} \end{aligned} \tag{43}$$

in which indexes h and b are associated with the hth and b^{th} nodes of the heat system. It should be noted that mass flows in the same node may be combined. A node's leaving water temperature is calculated by taking the average of all entering pipeline temperatures and weighting them by mass flows. Therefore, the mixed mass node's temperature depends on the associated node's power conservation. Equation (44) defines the temperature of a mixed node as follows:

$$T_{out} = \sum (\dot{m}_{in} T_{in}) / (\sum_{out}^{\dot{m}}) \tag{44}$$

It is noteworthy that for all energy flows in the OEF issue, the return and supply temperatures in every heat node are determined according to Equations (33) and (44).

4. Solution Methodology

4.1. State Variables Scheme and Decomposing Solution

In order to optimize the OEF problem, control variables must be identified. Using the ITLBOA approach, the overall cost function objective is optimized. Using ITLBOA, control variables are generated for OEF problems based on a predetermined objective cost. Afterward, upon determining the pressures in every gas node, angle, and magnitude of voltages in all electric nodes, and supply and return temperatures in all heating nodes, the decomposed power flow problem for the multi-energy carrier system can be analyzed. It is important to verify that the sub-networks restrictions Equations (5)–(8) and (14)–(16) are met following obtaining the energy flow solution.

A combined system power flow problem decomposes into an evaluation for the whole sub-system power flow, taking into account each device's energy flow. As a result of avoiding creating a large Jacobian matrix to address the simultaneous power flow for a composed system, this suggested approach for determining the power flow for a composed system is quicker and more flexible. In the decomposed method, the power flow for electric and heating systems is solved using a graph method and a holomorphic embedding method, respectively.

4.2. Suggested Algorithm

An ITLBOA would be an effective meta-heuristic method that is greatly affected by the student–teacher learning procedure [21]. In the same way as various meta-heuristic methods, ITLBOA would be a population-driven approach that uses a variety of solutions to obtain the desired outcome. A set of learners constitute the population of the algorithm. There are two phases in the ITLBOA method: (1) the teacher phase and (2) the learner phase. In the teacher phase, a teacher is used to generate a set of potential solutions to the problem. This set is then evaluated using a fitness function to determine the best solution. This best solution is then used as the starting point for the learner phase. In the learner phase, the algorithm uses a set of learners to improve the solution generated in the teacher phase. Each learner is given a set of parameters to modify in order to improve the solution. These parameters are then evaluated using the fitness function to determine if the solution has been improved. If it has, the new solution is used as the starting point for the next iteration. This process is repeated until the best possible solution is found. The TLBO algorithm is an effective optimization technique for solving complex problems. It combines the power of evolutionary optimization with the advantages of teaching and learning to find the best possible solution. It is also relatively easy to implement and has been successfully used in a variety of applications. In this formulation, Xold shows the previous position of the students and Xteacher shows the position of the best student (called teacher). In fact, in each iteration, the most fitting solution (most knowledgeable student) is assigned as Xteacher to guide the class to a higher level of knowledge. Therefore, it must update in each iteration.

The algorithm can be enhanced in many ways. As a result, the notion of the "self-learning capability of the learners" is explored.

Having a self-propelled ability according to one's own abilities is essential for achieving desirable outcomes. In this way, the self-based ability can be added to learners for optimizing the reproduction of populations, leading to a variety of variants. Table 1 summarizes the ITLBOA algorithm in detail.

Table 1. ITLBOA algorithm.

1. Setting up parameters:	The step defines the primary learners $X_{old,i}$, count of the population (NP), teaching factor (T_F), self-learning factor (S_F), and maximal count of iterations (G_{max}).
2. Primary population:	The primary randomly selected population of each learner is created within its bounds, as shown below: $X_{old,i} = X^{min}_{old,i} + rand(0,1)(X^{max}_{old,i} - X^{min}_{old,i})$ $Xold, i = \left[P^{gen}_{E,j}, \phi^{chp}_{H,j}, \phi^{boil}_{H,j}, T^{chp}_{s,j}, T^{boil}_{s,j}, H^{comp}_{j}, L^{t}_{e,out,r}, L^{t}_{h,out,r}\right]$ in which i = 1, …, D; $X^{min}_{old,i}$ is the minimal bound of the ith learner and $X^{max}_{old,i}$ is the maximal bound of the ith learner.

Table 1. *Cont.*

3. Teacher step:	The updated learner vectors are created via achieving the knowledge from the trained teachers within the step. The created updated vector can be defined as follows: $X_{new1,i} = X_{old,i} + Diff - Mean$ in which $Diff - Mean\ _i = rand_i(M_{new\_}\ T_F M_i)$ in which M_i shows the mean knowledge level of the learners at time i, and M_{new} shows the predicted knowledge level of learners; T_F shows the teacher factor deciding the amount of M_i to be 1 or 0 and $rand_i$ shows a randomly selected number between 0 and 1. In addition, T_F can be defined as follows: $T_F = rnd[1 + rand(0,1)]$
4. Learner phase using self-learning capability:	During this step, the newly acquired knowledge level of the learners is improved (i) through interacting with their peers or (ii) their self-learning capability: $X_{new2,i} = \begin{cases} X_{new1,i} + r_1(X_{old,i} - X_{old,j}) + r_2(X_{teacher} - S_F X_{old,i}) \\ \quad if\ f(X_{old,i}) > f(X_{old,j}) \\ X_{new1,i} + r_1(X_{old,j} - X_{old,i}) + r_2(X_{teacher} - S_F X_{old,i}) \\ \quad if\ f(X_{old,j}) > f(X_{old,i}) \end{cases}$ in which r_1 andr_2 are the randomly selected numbers between [0, 1] and S_F, the self-driven learning factor, which can be determined as follows: $S_F = rnd[1 + rand]$
5. Analyze/ choosing:	The newly created learner vector $X_{new2,i}$ at stage 4 competes with its previous individuals $X_{old,i}$ on the basis of the below evaluation criteria: $X_i^1 = \begin{cases} X_{new2,i} & if \quad f(X_{old,i}) \le f(X_{new2,i}) \\ X_{old,i} & otherwise \end{cases}$
6. The end:	The procedure is repeated until the preset G_{max} is reached.

5. Numerical Evaluation

A numerical analysis of the suggested OEF taking into account DRP is presented in the following section. The outcomes examine the involvement of flexible energy users in the MCE's OEF. The following part analyzes the advantages and features of the suggested OEF. The software and hardware needed are as follows: MATLAB (R2015a or later): Simulink and Simscape Power Systems; computer hardware requirements: processor: Intel Core i5 processor or equivalent; RAM: 8 GB or higher; hard disk: 500 GB or higher; graphics card: NVIDIA Quadro K4000 or equivalent; operating System: Windows 10. The proposed algorithm data are as below: primary learners: 10-20; count of population (NP): 50–100; teaching factor (α): 0.1–2; self-learning factor (β): 0.5–2; maximal count of iterations: 50–100.

For evaluating the suggested ITLBOA's optimization capability, this paper compares the suggested algorithm with the outcomes of various non-linear optimization approaches. This paper then presents the outcomes of comparing the optimization algorithms for the MCE's OEF. For traditional benchmark functions with different dimensions, Table 2 compares the suggested ITLBOA to a variety of different optimization algorithms. In this table, famous math benchmarks are deployed to measure the quality of the algorithms:

Schwefel: Schwefel's function is a non-convex optimization benchmark used to test optimization algorithms. It is often used to evaluate the performance of global optimization algorithms as it is multimodal and has many local minima.

Ackley: The Ackley function is another non-convex optimization benchmark used to test the performance of optimization algorithms. It is often used to evaluate the performance of global optimization algorithms as it is multimodal and has many local minima.

Rosenbrock: The Rosenbrock function is a classic optimization benchmark used to test the performance of optimization algorithms. It is a convex optimization problem and is often used to evaluate the performance of local optimization algorithms.

Rastrigin: The Rastrigin function is a non-convex optimization benchmark used to test the performance of optimization algorithms. It is often used to evaluate the performance of global optimization algorithms as it is multimodal and has many local minima.

Table 2. Comparing outcomes of algorithms in benchmark testing functions.

Benchmark Test Function		Algorithms		
		ITLBOA	**TLBO**	**PSO**
Schwefel	Mean	$1.6 \times 10^3 \pm 57$	$3.7 \times 10^3 \pm 284$	$5.2 \times 10^3 \pm 374$
	Time (s)	0.89	0.93	1.04
Ackley	Mean	$7.3 \times 10^{-17} \pm 1 \times 10^{-19}$	$8.5 \times 10^{-10} \pm 2 \times 10^{-9}$	$5.9 \times 10^{-6} \pm 7 \times 10^{-7}$
	Time (s)	0.91	0.93	1.04
Rosenbrock	Mean	28.75 ± 0.4	68.6 ± 3.7	256.8 ± 6.9
	Time (s)	1.31	1.22	1.3
Rastrigin	Mean	0	82.4 ± 4.3	163.7 ± 5.8
	Time (s)	1.16	1.11	1.23
Griewank	Mean	0	$6.7 \pm 16 \times 10^{-2}$	$25.2 \pm 8 \times 10^{-1}$
	Time (s)	1.17	1.14	1.21

Griewank: The Griewank function is a non-convex optimization benchmark used to test the performance of optimization algorithms. It is often used to evaluate the performance of global optimization algorithms as it is multimodal and has many local minima.

In the experimental MCE system, there are 14-bus IEEE buses, 20-node gas networks, and 14-node heat networks [22,23]. In the sample MCE system, there are 2 electrical gas compressors, 5 boilers, 2 gas generators, 4 CHP units, 5 electric pumps, and direct links to gas, electrical power, and heat. Figures 1–3 show a schematic diagram of sub-networks for the electrical network (Figure 1), gas network (Figure 2), and heat system (Figure 3). The integrated testing system considers four industrial IDR users as energy hubs, including flexible electricity and heat loads. Variables have the following minimums and maximums: $V = [0.9,\ 1.1]$ pu, pressures $= [5,\ 80]$ bar, supply temperatures $[130,\ \ 110]$ °C, and return temperatures $[55,\ \ 40]$ °C. Coefficients of heating ratio curves for gas-powered generators are $a^{gen} = 0.008$, $b^{gen} = 4$, $c^{gen} = 150$, $d^{gen} = 15$, and $e^{gen} = 0.5$. Cost factors are $u^{fuel} = 0$, $v^{fuel} = 0.14126$ $/m^3$, and $v^{fuel} = 0.$ of the gas. The information needed to optimize the proficiency of the integrated system is illustrated in Table 3. Table 4 provides the technical features of the three systems of the integrated system. For the ITLBOA algorithm, the size of the population is 80 and the overall number of iterations is 500. Flexible users' shift of energy is limited to 25% of every hub's energy demand. Figure 4 shows electricity and heat demand percentages for the integrated hubs over a 24 h period. Table 5 shows the interdependencies between the three heat, electricity, and gas systems within IDR hubs.

Figure 5 shows the power prices of the DRP per hour. Specifically, every hub's needs are met by its devices or via the transmission sub-network within the MCE systems. When the hubs' production power exceeds their loads, the excess energy is delivered to the associated network. An integrated system's OEF problem poses a computational challenge due to the large-scale nature of gas, electricity, and heat sub-networks. It is, fortunately, possible to significantly minimize this problem with the suggested highly efficient approach. Since the proposed approach utilizes a combination of optimization techniques and machine learning algorithms to identify the optimal energy flows in the MCE system, it is capable of providing an accurate solution to the OEF problem while also reducing the computational complexity of the problem. Furthermore, the use of machine learning algorithms allows the system to adapt to changes in power prices and adjust the optimal energy flows accordingly. In addition, the proposed approach can be used to identify the most cost-effective energy sources for each hub in the MCE system, thus allowing for the most efficient use of energy resources. In the suggested OEF approach, the inverse of a variable Jacobian is not used for all iterations, as in Newton–Raphson approaches [24]. In addition, the holomorphic approach to address electric power flow can achieve a viable solution when all viable solutions exist; holomorphic methods do not require a primary guess of the system [25].

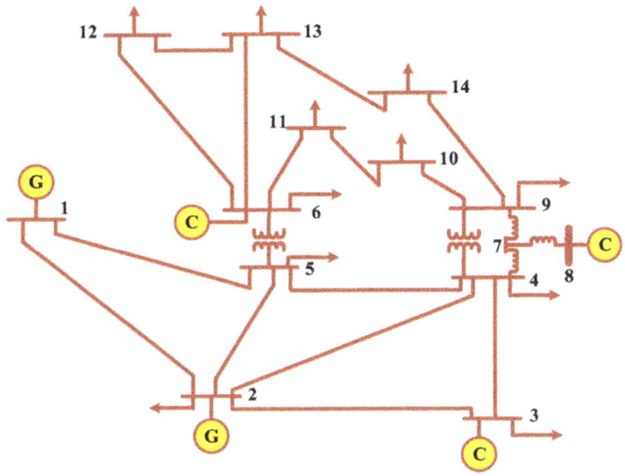

Figure 1. Diagram of the electrical system.

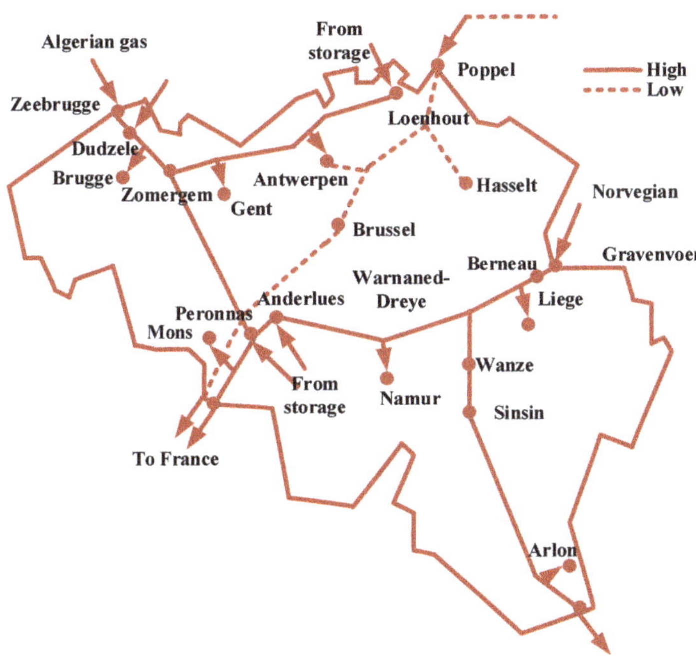

Figure 2. Diagram of the gas system.

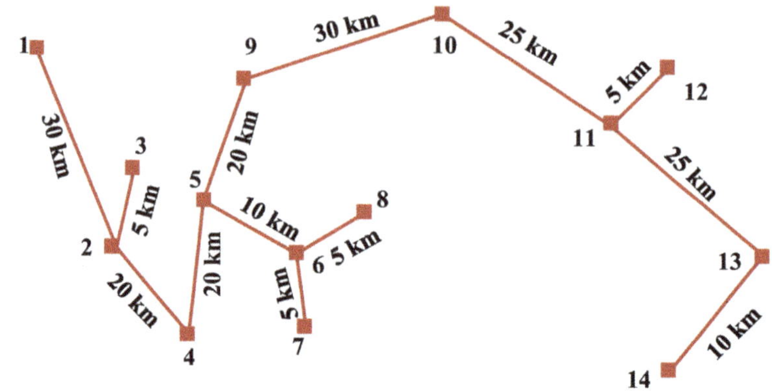

Figure 3. Diagram of the heat system.

Table 3. Technical information about energy sources in the MCE.

Unit	Boiler of hub_4	Boiler of hub_3	Boiler of hub_2	Boiler of hub_1	Slack Boiler
$T^{boil}_{min,s,j}$, C	115	115	115	115	115
$T^{boil}_{max,s,j}$, C	125	125	125	125	125
$\Phi^{boiler,min}_{H,j}$, MW	0	0	0	0	0
$\Phi^{boiler,max}_{H,j}$, MW	5	5	5	5	15
Unit	CHP of hub_4	CHP of hub_3	CHP of hub_2	CHP of hub_1	
$T^{chp}_{min,s,j}$, C	115	115	115	115	
$T^{chp}_{max,s,j}$, C	125	125	125	115	
$\Phi^{chp,min}_{H,j}$, MW	0	0	0	0	
$\Phi^{chp,max}_{H,j}$, MW	30	30	30	30	
Unit	Generator 2			Generator 1	
$Q^{gen}_{max,j}$, j	50			80	
$Q^{gen}_{min,j}$, j	0			0	
$P^{gen}_{min,j}$, MW	10			15	
$P^{gen}_{max,j}$, MW	80			332.4	

Table 4. Information about the sub-networks.

Heat system	$c_p = 4182$ kJ/kg K $U = 0.2$ W/m K $\quad \eta^{chp}_t = 0.88$	$\rho_w = 960$ kg/m^3 $\quad g = 9.81$ kgm/S^2 $\Phi^{boil,max}_H = 5$ MW $\quad b^{boil} = 0.822$	$H_p = 100$ m $a^{boil} = 0.0169$	$\eta_{HP} = 0.65$ $T_g = 10$ °C
Electrical system	$a^{chp} = 0.463$ $d^{chp}_1 = 0.8$	$b^{chp} = -0.0491$ $d^{chp}_2 = 0.6$	$c^{chp} = 4.49$ $e^{chp}_1 = 0.0736$	 $e^{chp}_2 = 0.0845$
Gas system	$GHV = 40.611$ MBTU/m^3 $\pi_0 = 1.0133$ bar	$\varepsilon_G = 0.05$ mm $\quad Z_a = 0.8$ $T_G = 281.15$ k	$\gamma_G = 0.6106$ $T_0 = 273.2$ k	$\lambda_G = 1.309$ $\eta^{GC} = 0.8$

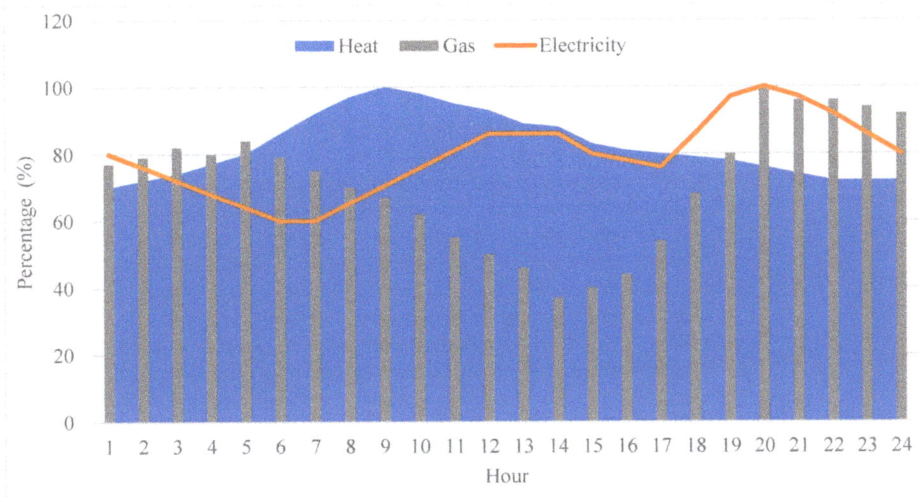

Figure 4. Percentage of heat and electricity requirements of the combined hubs over a day.

Table 5. MCE configuration.

Devices	Generator	Slack Generator	Moto-Compressor	Moto-Compressor	Slack Boiler and Pump
Heat node	-	-	-	-	1
Power bus	2	1	5	3	30
Gas node	19	12	9	18	3
Devices	**IDR *hub*$_4$**	**IDR *hub*$_3$**	**IDR *hub*$_2$**	**IDR *hub*$_1$**	
Heat node	13	10	9	4	
Power bus	7	17	23	14	
Gas node	10	6	7	15	

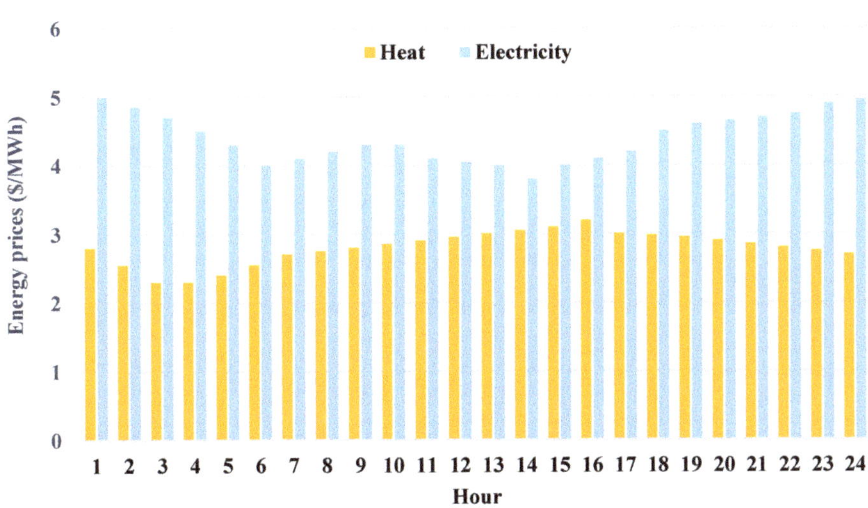

Figure 5. Energy prices of the DRP per hour.

The holomorphic approach to addressing electric power flow in power systems is a mathematical approach that uses holomorphic functions to solve the equations of power flow in a power system. This approach is based on the fact that the power flow equations can be written as a set of holomorphic functions. This approach is advantageous because it allows for the solution of the equations without having to resort to numerical methods, which can be time-consuming and difficult to implement. Additionally, holomorphic functions can be used to analyze system behavior under different operating conditions, providing a more detailed understanding of system dynamics. This approach is also useful for detecting and diagnosing faults in power systems as it allows for the identification of potential issues that may not be visible with other methods. The main advantages of this method over the Newton–Raphson method are as below:

The holomorphic approach provides an exact solution to the power flow equations, while the Newton–Raphson method only provides an approximate solution.

The holomorphic approach is computationally efficient and does not require any iteration, while the Newton–Raphson method requires multiple iterations to reach a solution.

The holomorphic approach is more accurate as it takes into account the non-linearity of the power flow equations, while the Newton–Raphson method is based on linear approximations.

Figure 6 depicts the impact of DRP on the performance of the OEF of the experimental MCE system. According to the figure, the IDR program for the MCE systems reduces operational costs. The figure shows that with the greater flexibility of users' involvement, the operational costs of the system are significantly decreased. A direct control method allows the ISO for managing flexible heat and electrical loads by means of an acceptable level of user involvement [26]. Incentives must be paid by the ISO to flexible users, though.

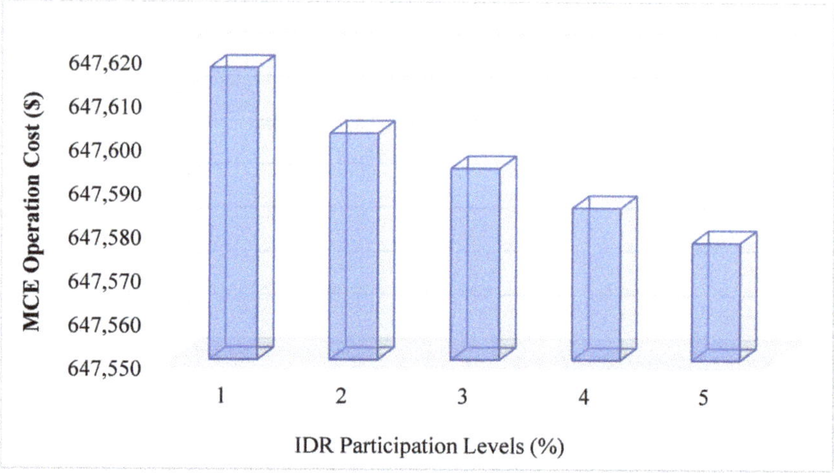

Figure 6. Operating costs of the MCE system in the presence of DR involvement levels.

The load duration curves for the power and heating sub-systems in the presence and sans of the DRP are presented in Figure 7. To meet the heating and electricity needs of IDR users during peak times, the ISO consumes gas through CHP and hub boilers. The overall electrical loads of the system including and excluding the DRP during 24 h are shown in Figure 7a. The figures indicate that the total system electrical loads are decreased within peak durations. As a result of the DRP, the electrical load shape is better balanced as well. Due to the DRP per hour, the heat and electricity peak loads are shifted to times in which the price of providing electricity and heat is lower [27–29]. The DR led to a decrease in peak loads and a smoothing of the load curve of the heating system in Figure 7b.

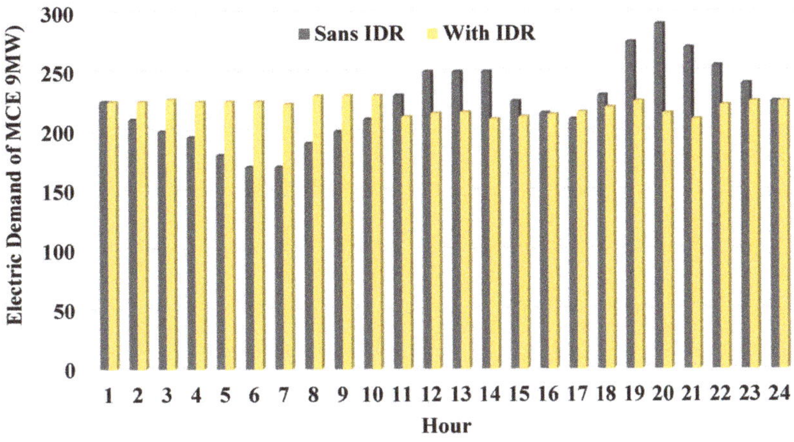

(a)

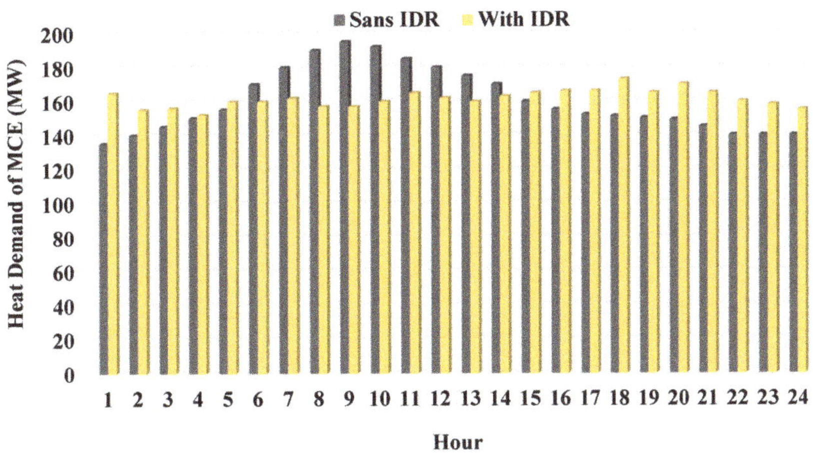

(b)

Figure 7. Load duration curves for the heat and electricity sub-systems including and excluding the DRP. (**a**) Overall power curve for MCE systems; (**b**) overall heating curve for MCE systems.

For the purpose of investigating the DRP, three sub-networks were examined. Figure 8 shows the overall electrical power generated by generators in the MCE system in the presence and absence of DRP. The figure shows the reduction in electrical power generated by generators in the MCE in the presence of DRP.

Figure 9 shows the overall heat produced by the MCE system including and excluding DRP. The integrated heat DRP reduces the overall heat generated outside the energy hubs in the MCE. Heat network efficiency is improved by the DRP.

The overall gas produced by the system from the outside DR hubs in the presence and sans of DRP is depicted in Figure 10. A higher amount of gas is produced by the system in the presence of the DRP. This is because, in the DR hubs, equipment that is interdependent like CHPs and boilers consumes greater quantities of natural gas to provide both electricity and heat. Despite this, the overall operational prices of the multi-energy carrier system decreased because gas has a lower cost compared to various energy sources.

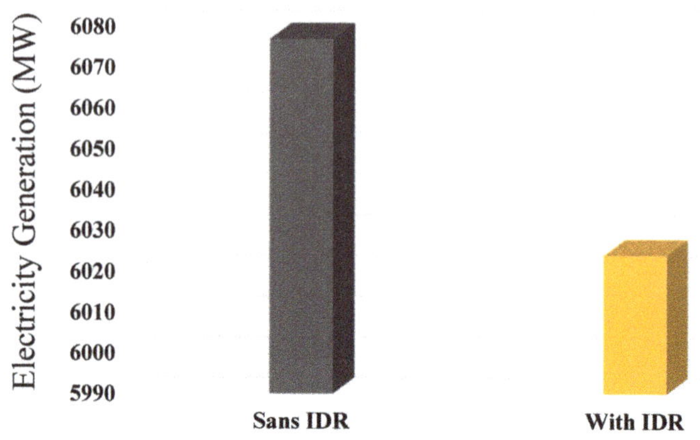

Figure 8. Overall electricity produced via the electrical system.

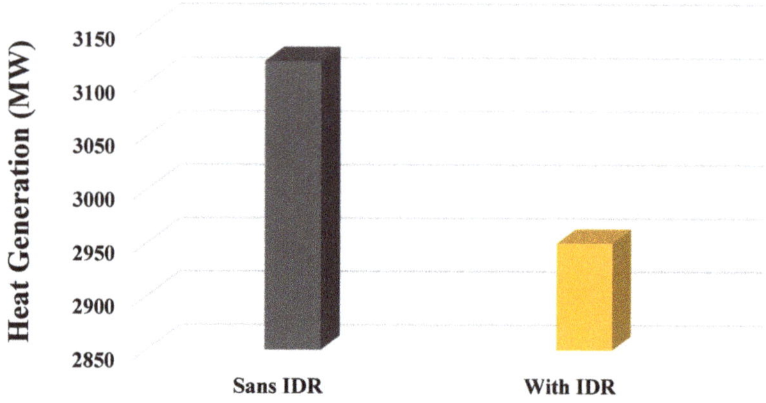

Figure 9. Overall heat produced via the heating system.

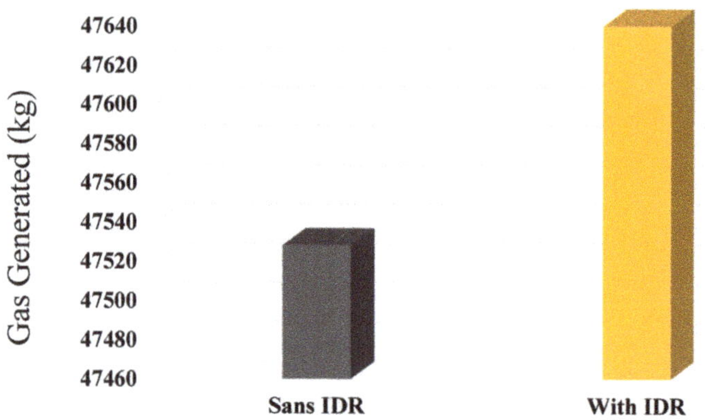

Figure 10. Overall gas produced via the gas system.

The losses of the networks in the presence of DRP and in the absence of DR in the two systems are depicted in Figure 11. It is concluded that the use of DRP diminishes electrical losses. Furthermore, if the DRP is incorporated into an integrated system, the overall heat loss decreases. In view of the assumption that the gas network is zero-loss, no gas loss is considered.

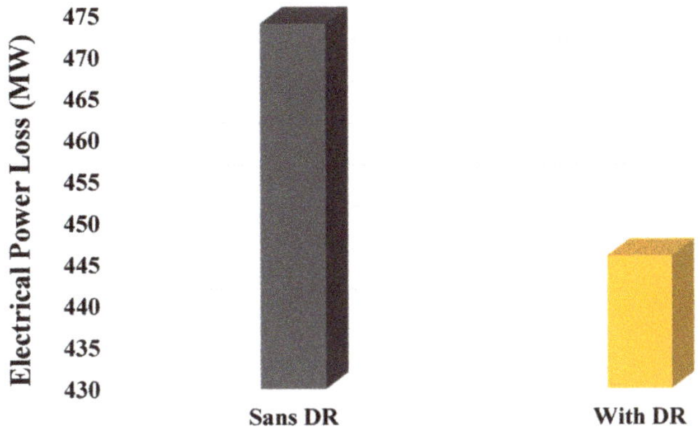

(a)

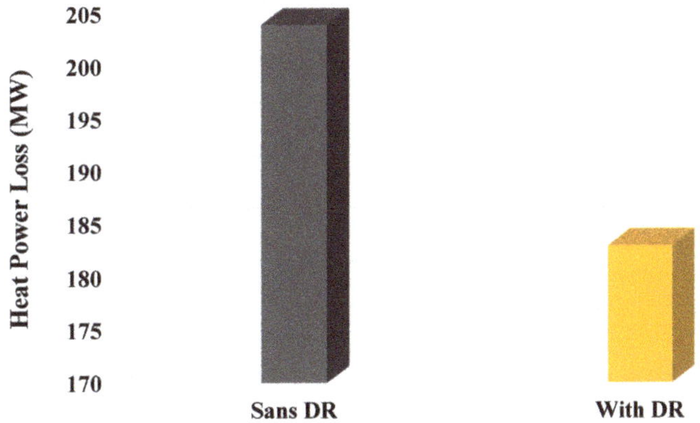

(b)

Figure 11. Altering in loss power in the presence and absence of DRP. (**a**) Power system; (**b**) heating system.

Figure 12 illustrates the change in nodal operations of the integrated network. When DRP is used, nodal voltages on the electrical system tend to be nearer the optimum voltage (1 pu), and such values tend to be near their maximum or minimum levels when not using DRP. Gas network nodal pressures in the absence of DRP outcomes are similar to those obtained in the presence of DRP outcomes. Gas network nodal operation is not affected significantly by flexible electricity and heat demands. According to Figure 12c, the nodal supply temperatures of the heating system in the absence of DRP tend to be closer to the maximum and minimum bounds than when DRP is utilized.

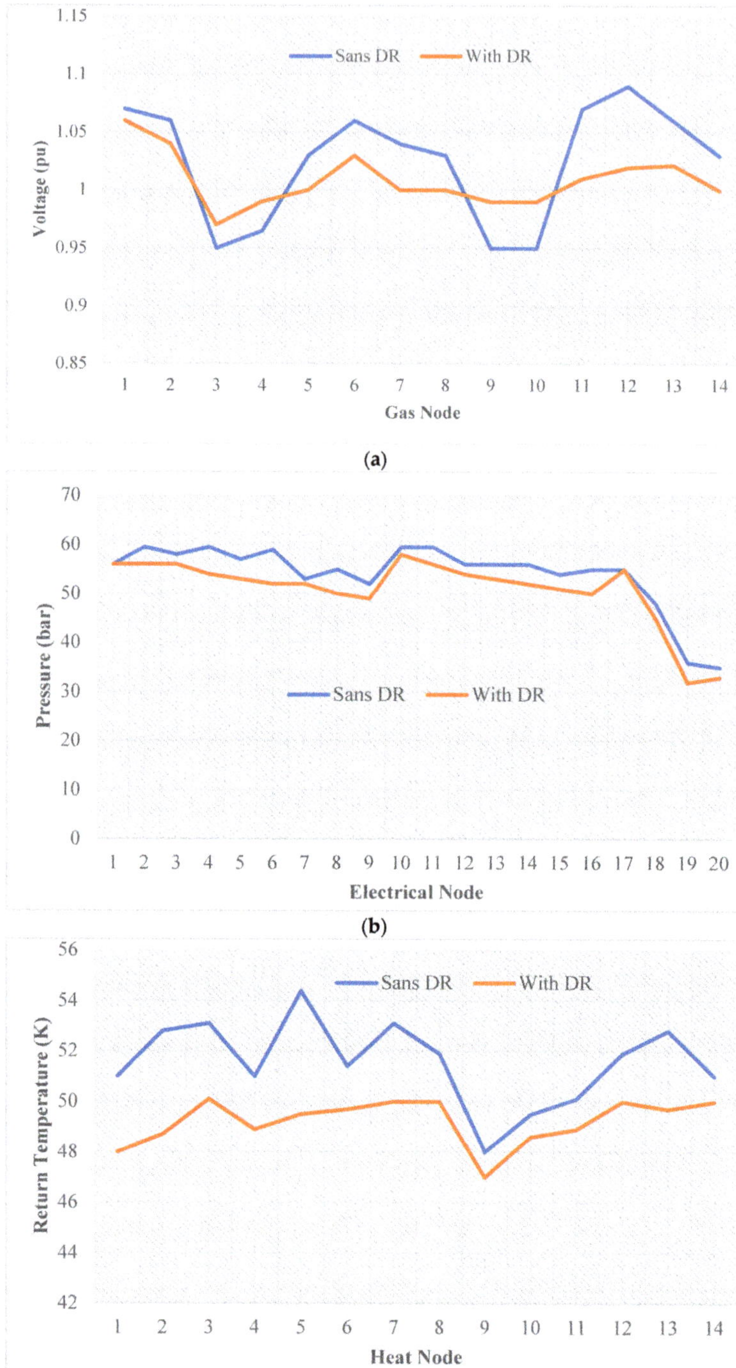

Figure 12. Altering in nodal operation in the presence and absence of DRP. (**a**) Electrical system; (**b**) gas system; (**c**) heat system.

This paper compares the outcomes of the algorithm to the outcomes of other optimizer schemes to evaluate how the ITLBOA affects the OEF of the MCE experimental system. Table 6 illustrates this analysis.

Table 6. Algorithm outcomes of OEF for testing MCE systems.

Method of Solution		ITLBOA	TLBO	PSO
Overall operational costs ($)	Time, s	4268	4311	4401
	SD	0.137	0.314	0.529
	Worst	647,871	647,893	647,912
	Mean	647,859	647,878	647,887
	Optimal	647,858	647,876	647,884

6. Conclusions

The goal of the study was to peruse the involvement of integrated flexible users, a particular type of demand-side management activity, in MCE systems such as electricity, heat, and gas. This will be beneficial for both the energy provider and the customers as it will reduce the cost of energy. In addition, the coordination of multiple energy carriers will also help to reduce the environmental impact of energy production and consumption. This is because the coordination of multiple energy carriers will lead to more efficient use of energy, which will reduce the amount of energy wasted. This, in turn, will reduce the amount of greenhouse gases released into the atmosphere. Finally, the integration of multiple energy carriers will also make the energy system more resilient as it will be better able to cope with unexpected energy demand changes. An OEF scheme of the multi-power carrier system has been used for determining DR hubs' impact. The OEF scheme considers electrical as well as heat loads in real time. The MCE is operated by elastic energy users. With the developed OEF scheme, electric, gas, and heat networks can be coordinated with industrial DR hubs while taking network security into consideration. Energy demand curves are smoothed by flexible energy users based on simulation outcomes. In addition, flexible energy users decrease the overall operational costs for integrated multi-power carrier systems. In addition, as the number of flexible electric and heat users increases, multi-energy carrier system operation prices decrease even further. As the number of flexible users in the system increases, DRP's impact will become more evident.

Author Contributions: Conceptualization, T.C., L.G., S.I., M.J., M.A.E.-M., M.S. and S.G.A.; Methodology, S.I.; Software, S.I.; Validation, L.G., M.J. and M.S.; Formal analysis, T.C., L.G. and M.S.; Investigation, M.J.; Data curation, S.G.A.; Writing—review & editing, T.C., L.G., S.I., M.J., M.A.E.-M., M.S. and S.G.A.; Visualization, M.A.E.-M. All authors have read and agreed to the published version of the manuscript.

Funding: This research was funded by King Saud University through Researchers Supporting Program number (RSPD2023R704).

Institutional Review Board Statement: Not applicable.

Informed Consent Statement: Not applicable.

Data Availability Statement: Not applicable.

Acknowledgments: The authors present their appreciation to King Saud University for funding this research through Researchers Supporting Program number (RSPD2023R704), King Saud University, Riyadh, Saudi Arabia.

Conflicts of Interest: The authors declare no conflict of interest.

Abbreviations

DRP	Demand response program
ITLBOA	Improved teaching–learning-based optimization algorithm
GT	Gas turbine
PV	Photovoltaic
WT	Wind turbine
MES	Multi-energy system
IGEHS	Integrated gas, electric, and heat system
RE	Renewable energy
DSE	Demand-side energy
ISO	Independent system operator
OEF	Optimal energy flow
CHP	Combined heat and power
EB	Electric boilers
P2G	Power-to-gas
GS	Gas storage
DR	Demand response
CCG	Column-and-Constraint Generation
MCE	Multi-carrier energy
IDR	Integrated disaster response

Nomenclature

Variable	Definition
$U_{q,fuel}$, $V_{q,fuel}$, $W_{q,fuel}$	Fuel price's ratio ($/ton)
$\tau_{q,fuel}$	Amount of fuel type consumed
$L^t_{e,out,r}/L^t_{h,out,r}$	Output electrical/heat energy of every hub from the demand side (MW)
$P^{gen}_{E,j}$	Generators' production active power (MW)
$P^{gen}_{min,j}/P^{gen}_{max,j}$	Minimum/maximum generators' production active power (MW)
$Q^{gen}_{E,j}$	Generators' production reactive power (MVAR)
$Q^{gen}_{min,j}/Q^{gen}_{max,j}$	Minimum/maximum generators' production reactive power (MVAR)
V_i, V_j	Voltage of the node in the system (pu)
ϕ^{chp}_H	CHP heat production (MW)
ϕ^{boiler}_H	Boiler heat production (MW)
$T^{chp}_{s,j}$	Initial temperatures of the CHP (K)
$T^{boiler}_{s,j}$	Initial temperatures of the boiler (K)
$\vert P^{line}_{E,\,ij} \vert$	Transmission flow of electrical line
$\vert F^{line}_{G,\,ij} \vert$	Transmission flow of gas line
$\vert \varnothing^{line}_{H,\,ij} \vert$	Transmission flow of heat line
H^{day}_r	Entire amounts of heat energy demand of the hub per day (MW)
E^{day}_r	Entire amounts of electrical energy demand of the hub per day (MW)
$F_{g \cdot k}$	Gas transmission from node g^{th} to k^{th} ($\mathbf{m^3/day}$)
$F_{g \cdot L}$	Gas demand at node ($\mathbf{m^3/day}$)
$F_{g \cdot S}$	Gas injection via node g^{th} ($\mathbf{m^3/day}$)
$F^{CHP}_g/F^{comp}_g/F^{boiler}_g$	CHP/turbo-compressor/boiler consumption gas ($\mathbf{m^3/day}$)
$\dot{m}^{boiler}_H/\dot{m}^{CHP}_H$	Boiler/CHP 's mass flow ratio (**kg/s**)
$\dot{m}_{out}/\dot{m}_{in}$	Output/input mass flow via a pump (**kg/s**)
$\dot{m}_{hb}$	Mass flow via the pipeline among node h^{th} and b^{th} (kg/s)
$\dot{m}^{HP}_H$	Heat pump's mass flow (kg/s)
$\dot{m}^{load}_H$	Heat load's mass flow (kg/s)
T_{end}	End of the pipeline's temperature (K)
T_{start}	Beginning of the pipeline's temperature (**K**)
l	Heat pipeline's length (km)
$\eta^{chp}/\eta^{comp}/\eta^{HP}$	Efficiency of the CHP/compressor/pump

$D_{hb}^{H,L}$	Pipeline's diameter (mm)
$L_{hb}^{H,L}$	Pipeline's length (km)
$\Theta_{hb}^{H,L}$	Pipeline's friction ratio
ρ_W	Water density/**kg/m^3**
T_{in}/T_{out}	Input/output mass flow's temperature in a mixed node (K)
$P_{k,in}^{comp}/P_{k,out}^{comp}$	Pressure of input/output gas compressor
$Q_i^{dem/gen}/P_i^{dem/gen}$	Reactive/active power demand
Q_i^{CHP}/P_i^{CHP}	Reactive/active power production of CHP
Q_i^{SH}	Shunt capacitors reactive production (MVAR)
P_i^{comp}	Consumed electric power via compressor (MVA)
P_i^{pump}	Consumed power via heating pump (MVA)
H_p	Network's pump head (m)
N_E	Number of electrical bus
θ_i	Voltage angle
Y_{ij}	Electrical transmission line admittance
N_h	Number of heat nodes
a^{boil}, a^{boil}	Partial boiler ratios
$\exists_h/\exists_e$	Incentive cost that operator should pay to the flexible Heat/electric customers (**$**)
Γ	Heat transition ratio (W/mK)
T_g	Temperature of ground (**K**)
c_p	Specific heat of the water (**kJ/kgK**)
$\alpha_{gk}^{comp}/\alpha_{gk}^{comp}/\alpha_{gk}^{comp}$	Compressor consumption ratio
γ_G	Gas's gravity ratio
λ_G	Natural gas specific heat proportion
E_p	Gas pipelines' absolute rugosity ratio (**mm**)
L_{gk}	Length of pipeline (km)
Z	Gas's compressibility at the gas flow's temperature
T_a	Gas flow's temperature (K)
T_0	Base temperature, (**K**)
P_0	Base pressure (**kPa**)
P_E^{CHP}	CHP's active power (**MW**)
P_E^{HP}	Heat power's electrical demand (**MW**)
$P_{E,g}^{comp}$	Power demand of compressor (**MW**)
$\phi_{h,load}$	Customer point's heat energy demand (MW)
ϕ_{hb}	Heat energy flow via the heat pipeline (MW).
$a^{gen}, b^{gen}, c^{gen}, d^{gen}, e^{gen}$	Consumption ratios of generator
R_{gk}^{GL}	Reynolds number
$K_{hb}^{H,L}$	Pipe's resistance ratio
D_{gk}^{GL}	Pipeline diameter (**mm**)
$\Pi_a^{\mathbf{gk}}$	Pipeline mean pressure (**kPa**)
$\Pi_{g,in}^{\mathbf{comp}}$	Compressor's input pressure (**kPa**)
$\Pi_{k,out}^{\mathbf{comp}}$	Compressor's output pressure (**kPa**)
$H_p^{\mathbf{gk}}$	Pipeline's slope pipeline correction (kPa^2)
$H_{gk}^{\mathbf{comp}}$	Compressor ratio among g^{th} node and k^{th} node
F^{GG}	Consumption gas via gas-fired power agent (m^3/day)

F_t^{P2G}	Gas generation of P2G agent ($\mathbf{m^3/day}$)
$\vartheta_{hb}^{H,L}$	Flow velocity ($\mathbf{m/s}$)
$\phi_H^{boil,maax}$	Maximum heat production of boiler ($\mathbf{MW}$)
ϕ_h^{dem}	Demand for heat energy at h^{th} node (MW)

References

1. Huang, X.; Xu, Z.; Sun, Y.; Xue, Y.; Wang, Z.; Liu, Z.; Li, Z.; Ni, W. Heat and power load dispatching considering energy storage of district heating system and electric boilers. *J. Mod. Power Syst. Clean Energy* **2018**, *6*, 992–1003. [CrossRef]
2. Wang, H.; Wang, B.; Luo, P.; Ma, F.; Zhou, Y.; Mohamed, M.A. State Evaluation Based on Feature Identification of Measurement Data: For Resilient Power System. *CSEE J. Power Energy Syst.* **2021**, *8*, 983–992.
3. Askari, M.; Dehghani, M.; Razmjoui, P.; GhasemiGarpachi, M.; Tahmasebi, D.; Ghasemi, S. A novel stochastic thermo-solar model for water demand supply using point estimate method. *IET Renew. Power Gener.* **2022**, *16*, 3559–3572. [CrossRef]
4. Belderbos, A.; Valkaert, T.; Bruninx, K.; Delarue, E.; D'haeseleer, W. Facilitating renewables and power-to-gas via integrated electrical power-gas system scheduling. *Appl. Energy* **2020**, *275*, 115082. [CrossRef]
5. Song, Y.; Mu, H.; Li, N.; Shi, X.; Zhao, X.; Chen, C.; Wang, H. Techno-economic analysis of a hybrid energy system for CCHP and hydrogen production based on solar energy. *Int. J. Hydrogen Energy* **2022**, *47*, 24533–24547. [CrossRef]
6. Ma, H.; Liu, Z.; Li, M.; Wang, B.; Si, Y.; Yang, Y.; Mohamed, M.A. A two-stage optimal scheduling method for active distribution networks considering uncertainty risk. *Energy Rep.* **2021**, *7*, 4633–4641. [CrossRef]
7. AlHajri, I.; Ahmadian, A.; Elkamel, A. Stochastic day-ahead unit commitment scheduling of integrated electricity and gas networks with hydrogen energy storage (HES), plug-in electric vehicles (PEVs) and renewable energies. *Sustain. Cities Soc.* **2021**, *67*, 102736. [CrossRef]
8. Gan-yun, L.; Bin, C.; De-xiang, J.; Nan, W.; Jun, L.; Guangyu, C. Optimal scheduling of regional integrated energy system considering integrated demand response. *CSEE J. Power Energy Syst.* **2021**. [CrossRef]
9. Gabrel, V.; Murat, C.; Thiele, A. Recent advances in robust optimization: An overview. *Eur. J. Oper. Res.* **2014**, *235*, 471–483. [CrossRef]
10. Yang, N.; Liu, S.; Deng, Y.; Xing, C. An improved robust SCUC approach considering multiple uncertainty and correlation. *IEEJ Trans. Electr. Electron. Eng.* **2021**, *16*, 21–34. [CrossRef]
11. Wu, G.; Xiang, Y.; Liu, J.; Shen, X.; Cheng, S.; Hong, B.; Jawad, S. Distributed energy-reserve co-optimization of electricity and natural gas systems with multi-type reserve resources. *Energy* **2020**, *207*, 118229. [CrossRef]
12. de Souza Dutra, M.D.; da Conceição Júnior, G.; de Paula Ferreira, W.; Chaves, M.R. A customized transition towards smart homes: A fast framework for economic analyses. *Appl. Energy* **2020**, *262*, 114549. [CrossRef]
13. Zhang, Y.; Ai, X.; Wen, J.; Fang, J.; He, H. Data-adaptive robust optimization method for the economic dispatch of active distribution networks. *IEEE Trans. Smart Grid* **2018**, *10*, 3791–3800. [CrossRef]
14. Qaeini, S.; Nazar, M.S.; Varasteh, F.; Shafie-khah, M.; Catalão, J.P. Combined heat and power units and network expansion planning considering distributed energy resources and demand response programs. *Energy Convers. Manag.* **2020**, *211*, 112776. [CrossRef]
15. Yang, S.; Tan, Z.; Liu, Z.; Lin, H.; Ju, L.; Zhou, F.; Li, J. A multi-objective stochastic optimization model for electricity retailers with energy storage system considering uncertainty and demand response. *J. Clean. Prod.* **2020**, *277*, 124017. [CrossRef]
16. Abdallah, W.J.; Hashmi, K.; Faiz, M.T.; Flah, A.; Channumsin, S.; Mohamed, M.A.; Ustinov, D.A. A Novel Control Method for Active Power Sharing in Renewable-Energy-Based Micro Distribution Networks. *Sustainability* **2023**, *15*, 1579. [CrossRef]
17. Lu, Q.; Guo, Q.; Zeng, W. Optimization scheduling of home appliances in smart home: A model based on a niche technology with sharing mechanism. *Int. J. Electr. Power Energy Syst.* **2022**, *141*, 108126. [CrossRef]
18. Mokhatab, S.; Poe, W.A.; Mak, J.Y. *Handbook of Natural Gas Transmission and Processing: Principles and Practices*; Gulf Professional Publishing: Houston, TX, USA, 2018.
19. Lamri, A.A.; Easa, S.M. Computationally efficient and accurate solution for Colebrook equation based on Lagrange theorem. *J. Fluids Eng.* **2022**, *144*, 014504. [CrossRef]
20. Yang, J.; Zhang, N.; Botterud, A.; Kang, C. On an equivalent representation of the dynamics in district heating networks for combined electricity-heat operation. *IEEE Trans. Power Syst.* **2019**, *35*, 560–570. [CrossRef]
21. Rao, R.V.; Savsani, V.J.; Vakharia, D.P. Teaching–learning-based optimization: An optimization method for continuous non-linear large scale problems. *Inf. Sci.* **2012**, *183*, 1–15. [CrossRef]
22. De Wolf, D.; Smeers, Y. The gas transmission problem solved by an extension of the simplex algorithm. *Manag. Sci.* **2000**, *46*, 1454–1465. [CrossRef]
23. Ceseña, E.A.; Mancarella, P. Energy systems integration in smart districts: Robust optimisation of multi-energy flows in integrated electricity, heat and gas networks. *IEEE Trans. Smart Grid* **2018**, *10*, 1122–1131. [CrossRef]
24. Cui, X.; Liu, Y.; Yuan, D.; Jin, T.; Mohamed, M.A. A Hierarchical Coordinated Control Strategy for Power Quality Improvement in Energy Router Integrated Active Distribution Networks. *Sustainability* **2023**, *15*, 2655. [CrossRef]
25. Shi, Y.; Hou, Y.; Yu, Y.; Jin, Z.; Mohamed, M.A. Robust Power System State Estimation Method Based on Generalized M-Estimation of Optimized Parameters Based on Sampling. *Sustainability* **2023**, *15*, 2550. [CrossRef]

26. Jasim, A.M.; Jasim, B.H.; Flah, A.; Bolshev, V.; Mihet-Popa, L. A new optimized demand management system for smart grid-based residential buildings adopting renewable and storage energies. *Energy Rep.* **2023**, *9*, 4018–4035. [CrossRef]
27. Jasim, A.M.; Jasim, B.H.; Aymen, F.; Kotb, H.; Althobaiti, A. Consensus-based intelligent distributed secondary control for multiagent islanded microgrid. *Int. Trans. Electr. Energy Syst.* **2023**, *2023*, 6812351. [CrossRef]
28. Tan, H.; Yan, W.; Ren, Z.; Wang, Q.; Mohamed, M.A. A robust dispatch model for integrated electricity and heat networks considering price-based integrated demand response. *Energy* **2022**, *239*, 121875. [CrossRef]
29. Duan, Q.; Quynh, N.V.; Abdullah, H.M.; Almalaq, A.; Do, T.D.; Abdelkader, S.M.; Mohamed, M.A. Optimal scheduling and management of a smart city within the safe framework. *IEEE Access* **2020**, *8*, 161847–161861. [CrossRef]

sustainability

Article

Multi-Stage Incentive-Based Demand Response Using a Novel Stackelberg–Particle Swarm Optimization

Suchitra Dayalan [1], Sheikh Suhaib Gul [1], Rajarajeswari Rathinam [1], George Fernandez Savari [1], Shady H. E. Abdel Aleem [2], Mohamed A. Mohamed [3,*] and Ziad M. Ali [4,5]

1 Department of EEE, SRMIST, Kattankulathur 603203, India
2 Department of Electrical Engineering, Valley High Institute of Engineering and Technology, Science Valley Academy, Qalyubia 44971, Egypt
3 Electrical Engineering Department, Faculty of Engineering, Minia University, Minia 61519, Egypt
4 Electrical Engineering Department, College of Engineering, Prince Sattam Bin Abdulaziz University, Wadi Addawaser 11991, Saudi Arabia
5 Electrical Engineering Department, Aswan Faculty of Engineering, Aswan University, Aswan 81542, Egypt
* Correspondence: dr.mohamed.abdelaziz@mu.edu.eg

Abstract: Demand response programs can effectively handle the smart grid's increasing energy demand and power imbalances. In this regard, price-based DR (PBDR) and incentive-based DR (IBDR) are two broad categories of demand response in which incentives for consumers are provided in IBDR to reduce their demand. This work aims to implement the IBDR strategy from the perspective of the service provider and consumers. The relationship between the different entities concerned is modelled. The incentives offered by the service provider (SP) to its consumers and the consumers' reduced demand are optimized using Stackelberg–particle swarm optimization (SPSO) as a bi-level problem. Furthermore, the system with a grid operator, the industrial consumers of the grid operator, the service provider and its consumers are analyzed from the service provider's viewpoint as a tri-level problem. The benefits offered by the service provider to its customers, the incentives provided by the grid operator to its industrial customers, the reduction of customer demand, and the average cost procured by the grid operator are optimized using SPSO and compared with the Stackelberg-distributed algorithm. The problem was analyzed for an hour and 24 h in the MATLAB environment. Besides this, sensitivity analysis and payment analysis were carried out in order to delve into the impact of the demand response program concerning the change in customer parameters.

Keywords: demand response; energy; smart-grid; grid operator; industrial customer; Stackelberg–particle swarm optimization

Citation: Dayalan, S.; Gul, S.S.; Rathinam, R.; Fernandez Savari, G.; Aleem, S.H.E.A.; Mohamed, M.A.; Ali, Z.M. Multi-Stage Incentive-Based Demand Response Using a Novel Stackelberg–Particle Swarm Optimization. *Sustainability* **2022**, *14*, 10985. https://doi.org/10.3390/su141710985

Academic Editor: Alberto-Jesus Perea-Moreno

Received: 3 August 2022
Accepted: 29 August 2022
Published: 2 September 2022

1. Introduction

A smart grid (SG) epitomizes an unprecedented chance to motivate the energy sector into a new age of dependability, availability, and efficacy, contributing to future economic and environmental conditions. As per the strength, weakness, opportunities, and threat analysis report, the increasing per capita electricity consumption could be controlled by implementing the demand response programs (DRPs) that form a focal point of SG [1]. The evolution of the renewable energy systems, in addition to the altered consumption pattern of the consumers, aids in the effective implementation of DRPs. With the evolution of multi-energy systems (MES), demand response (DR) programming has been broadened into integrated DR [2]. DR is a sub-classification of a broader concept called demand side management, which is further classified into price-based DR (PBDR) and incentive-based DR (IBDR). PBDR relies on pricing in different ways, such as real-time and day-ahead pricing, etc. IBDR programs are based on contracts involving various market entities such as the grid operator (GO), retailers, and customers, etc. An IBDR scheme that provides coupon incentives to the customers was designed by several researchers [3,4]. It benefits both the customers and the load-serving entities (LSE). In [5,6], a novel IBDR scheme has been framed considering the various market operators that benefit the customers and the utilities. The SG technology advancement has made DR program implementation easier. The communication infrastructure of the SG technology has made it possible to implement DR programs for all kinds of customers. The modeling of the IBDR program for the residential customers, including the loads of the customer to reduce the power consumption, are carried out in the intelligent environment, thereby reducing the customer bill. The IBDR models for handling DR problems, with optimization techniques such as the Stackelberg game, GAMS, CPLEX, and others, are used in SG to obtain a better result of demand reduction and cost minimization [7–10]. In recent years, game-theory-based algorithms have been implemented to solve the IBDR strategy bidding model from the SP and GO ends [11–13]. IBDR problems with pricing schemes such as day-ahead pricing and intra-day renewable energy sources for procuring electricity at a minimal cost from the market were proposed in [14,15]. A multi-stage algorithm involving uncertain renewable sources, for serving the loads, considering entities such as GO and SP, was proposed in [16–19]. An integrated IBDR model considering the uncertainties coupled with the renewable sources [20,21] was proposed to improve the profit obtained by the customer [22]. An experimental setup based on flexible incentive-based DR to reduce critical demand and maximize consumer profits was proposed by Luo Zhe [23]. The IBDR was programmed using data mining for a virtual power plant. The authors in [24] proposed the IBDR problem for the Danish low-voltage (LV) grid with battery storage devices. Deepan et al. [25] proposed a novel self-reporting baseline estimation and outperformed the method for solving the IBDR program involving aggregators and consumers. The optimization techniques for effectively handling more variables maximization and minimization problems along with parameter selection and tuning are dealt with in [26–29].

Table 1 summarizes the highlights of the surveyed literature based on the objective of the DR problem, the entities considered, adapted pricing schemes, and optimization technology. The effect of inclusion and the interaction of the three market entities together has not been analyzed, although many algorithms were developed to solve IBDR. The influence of the customer parameters on the IBDR programs has also not been adequately shown. Furthermore, the literature has not concentrated on the IBDR work on 24-h timing, including all of the entities.

Table 1. Highlights of the literature survey.

Reference	Entities/Participants	Pricing Schemes	Objective	Methodology and Simulation Tools
[3]	LSE and retail customers	Flat rate pricing	Optimizing social welfare	CPLEX
[4]	Service provider and customer	A day ahead of electricity pricing	Optimal incentives for SPs	Stackelberg game theory-GAMS tool
[6]	Utility and customers	Spot pricing	Maximizing benefits of retailers	MATLAB Yalmip toolbox
[7]	Grid operator, multi-service provider and customer	Incentive based pricing	Resource utilization in minimizing cost and maximizing profit of operators	Stackelberg game approach
[12]	Service provider and end-user	Real-time pricing	Peak demand and electricity bill reduction	MATLAB toolbox for optimization
[16]	Retailer and end-user	A day ahead of electricity pricing	Minimizing peak demand and finding hourly financial incentives for customers	NSGA II
[19]	LSE and ISO	Real-time market	LSE net revenue maximization	CPLEX

Accordingly, the proposed work compares the changes involved in the inclusion of various entities in the IBDR program using a novel SPSO algorithm. The IBDR problem designed concerning Case 1 (SP-customers) and Case 2 (GO-SP-industrial consumers (IC) customers) was optimized using SPSO algorithms, and was compared with the Stackelberg distributed algorithm for an hour and an entire day. A sensitivity analysis was also adapted by altering the customer parameters, namely the discomfort factor, the customer attitude to demand reduction, and the magnitude of demand change required to review the impact of IBDR. Furthermore, payment analysis was made to focus on the benefits obtained by the entities by varying the customer parameters.

The content of this paper is organized as follows: Section 2 elaborates on the problem formulation of IBDR and Case 1. Section 3 focuses on the problem formulation of Case 2 of the IBDR problem, followed by a briefing on the optimization technique adapted to solve the IBDR problem in Section 4. Section 5 discusses the results obtained for the two different cases. Finally, Section 6 presents the conclusions.

2. Problem Formulation of IBDR with One SP and Two Customers: Case 1

This section modelled and formulated an IBDR problem including GO, SP, IC and customers to maximize or minimize the utility function depending on the entities considered and the demand reduction.

An SP sells electricity to its retail customers and procures their capacity through demand reduction. The SP can even sell the procured capacity and gain profit. The SP provides incentives to those customers who agrees to reduce their demand when they are told to do so by the SP. In this work, a system with one SP and two consumers is considered, and the interaction between them is modelled using Stackelberg's game theory.

a. Customer model

Let N be the total number of customers. Here, N is taken to be 2. Every customer i, when provided with incentives from the SP, tries to increase their demand reduction to gain more incentives. Here, time t ranges from 1 to T, where T is taken as 24. The customer will aim to maximize the utility function by using their demand reduction, and the above can be framed as follows [4]:

$$\max_{D_i} U_i = \sum_{t=1}^{T} D_{i,t} \times \pi_t - \mu_i \times \sum_{t=1}^{T} \varphi_{i,t}(D_{i,t}) \tag{1}$$

which is subjected to

$$0 \leq D_{i,t} \leq D_{i,t}^{tar} - D_{i,t}^{min}, \ \forall\, i \in N, \forall\, t \in T \tag{2}$$

Equation (1) represents the customer utility function, which has to be maximized using demand reduction as the variable. The constraint given by Equation (2) restricts the demand reduction from going beyond $D_{i,t}^{tar} - D_{i,t}^{min}$, which represents the available quantity of demand reduction. The first term of Equation (1) means the income gained by the customer by reducing the demand $D_{i,t}$ for the incentive π_t. The second term represents the discomfort of the customer. The weight factor μ_i decides the level of discomfort each customer can accept. The small value of μ_i represents less importance given to discomfort. The dissatisfaction cost ($\varphi_{i,t}$) represents the customer's discomfort involved in demand reduction. The dissatisfaction cost is modelled as follows:

$$\varphi_{i,t}(D_{i,t}) = \frac{\theta_i}{2}\left(D_{(i,t)}^2 + \lambda_i \times D_{i,t}\right)\theta_i > 0,\ \lambda_i > 0 \tag{3}$$

θ_i and λ_i are parameters set based on the customer's attitude towards demand reduction.

b. Service provider model

The SP gains profit by selling the capacity procured from the customers, and can profit by procuring the capacity (i.e., the demand reduction) at a minimal incentive. Thus, the SP aims to minimize the utility function using the incentive offered as the variable. This can be modeled as follows [4]:

$$\min_{\pi} U_{SP} = \sum_{t=1}^{T}\sum_{i=1}^{N} D_{i,t} \times \pi_t - \sum_{t=1}^{T}\sum_{i=1}^{N} D_{i,t} \times P_t \tag{4}$$

which is subjected to

$$\pi_t^{min} \leq \pi_t \leq \pi_t^{max}, \quad \forall\, t \in T \tag{5}$$

$$\sum_{i=1}^{N} D_{i,t} \geq D_t^{req}, \quad \forall\, t \in T \tag{6}$$

The first term of Equation (4) represents the incentive payments given by the SP to the consumers according to their demand reductions, and the second term represents the amount gained by the SP in selling the procured capacity to the market.

c. Stackelberg game formulation and analysis

The interaction between the SP and the customers is framed as a Stackelberg game. Here, the SP acts as a leader and sets incentives for its customers; thus, a one-leader *N*-follower Stackelberg game is framed (here, one leader and two followers). The SP act provides incentives as a motivation for the customers to reduce their demand. The optimization equations are solved using PSO. The SP model is solved to obtain the optimal incentive set by the SP to its customers, and this is utilized in solving the customer model, thereby obtaining the corresponding optimal demand reduction; this iteration process tries to reach the global condition. The customer and SP models are reformulated using the Stackelberg theory. If the incentive value set by the leader (SP) is identified, then the customer's optimal demand reduction can be obtained.

d. Optimal solution for customers

By solving the first-order derivative of the Equation (1), concerning $D_{i,t}$, the optimal value of the demand reduction can be obtained as:

$$D_{i,t} = \frac{\pi_t - \mu_i \times \lambda_i}{\mu_i \times \theta_i} \tag{7}$$

The value of the second-order derivative of Equation (1) for $D_{i,t}$ gives $-\mu_i\theta_i < 0$. This equation is always negative; as such, the customer utility function is strictly concave for the feasible values of $D_{i,t}$. The optimal value of π_t obtained from the SP is substituted in Equation (7) to obtain the optimal demand reduction.

e. Optimal solution for the service provider

Equation (7) is substituted in Equation (4), and the SP utility function becomes

$$\min_{\pi} U_{SP} = \sum_{t=1}^{T}\sum_{i=1}^{N}\left(\frac{\pi_t - \mu_i \times \lambda_i}{\mu_i \times \theta_i}\right) \times \pi_t - \sum_{t=1}^{T}\sum_{i=1}^{N}\left(\frac{\pi_t - \mu_i \times \lambda_i}{\mu_i \times \theta_i}\right) \times P_t \tag{8}$$

which is subjected to

$$\pi_t^{min} \leq \pi_t \leq \pi_t^{max}, \quad \forall\, t \in T \tag{9}$$

The constraints can be modified by arranging equation (7) as

$$\pi_t = D_{i,t}(\mu_i \times \theta_i) + \mu_i \times \lambda_i \tag{10}$$

Constraints (5) and (6) can be modified using the above-obtained equations. Thus, the value of π_t^{min} and π_t^{max}, in (9), can be replaced by Equations (11) and (12).

$$\pi_t^{max} = min\left(\pi_t^{max}, min\left\{\left(\left(D_{i,t}^{tar} - D_{i,t}^{min}\right) \times \mu_i * \theta_i + \mu_i \times \lambda_i\right) \forall\, i \in N\right\}\right) \tag{11}$$

$$\pi_t^{min} = max\left(\pi_t^{min}, max\{\mu_i \times \lambda_i, \forall\, i \in N\}, \frac{D_t^{req} + \sum_{i=1}^{N} \frac{\lambda_i}{\theta_i}}{\sum_{i=1}^{N} \frac{1}{\theta_i}}\right) \tag{12}$$

The optimization equation in (8) is solved using PSO, and the optimal incentive is obtained.

3. Problem Formulation of IBDR with GO, ICs, SPs and Consumers: Case 2

The role of the GO in IBDR is re-studied in this section. Here, IBDR is implemented in a system consisting of GO, ICs, and SPs under the GO, and the customers under each SP. The GO sets the incentive for the ICs and the SPs. The GO procures the capacities from its SPs and ICs to meet the demand deficit. The SPs set the incentive for the customers under them. The SPs procure the capacity from the customers regarding demand reduction and sell it to the GO. The GO meets the required demand deficit of the day either through the generators or by procuring the capacity from the SPs and ICs (i.e., through their demand reductions), and the GO aims to reduce the cost of procuring these capacities. Figure 1 depicts the role of each domain in the smart grid. The mathematical models of the entities are framed and solved using the Stackelberg game theory, where the outcome of the leader is first obtained (i.e., the incentive of GO), and then this outcome is utilized to solve the model of the IC and the customers of the SP for demand reduction.

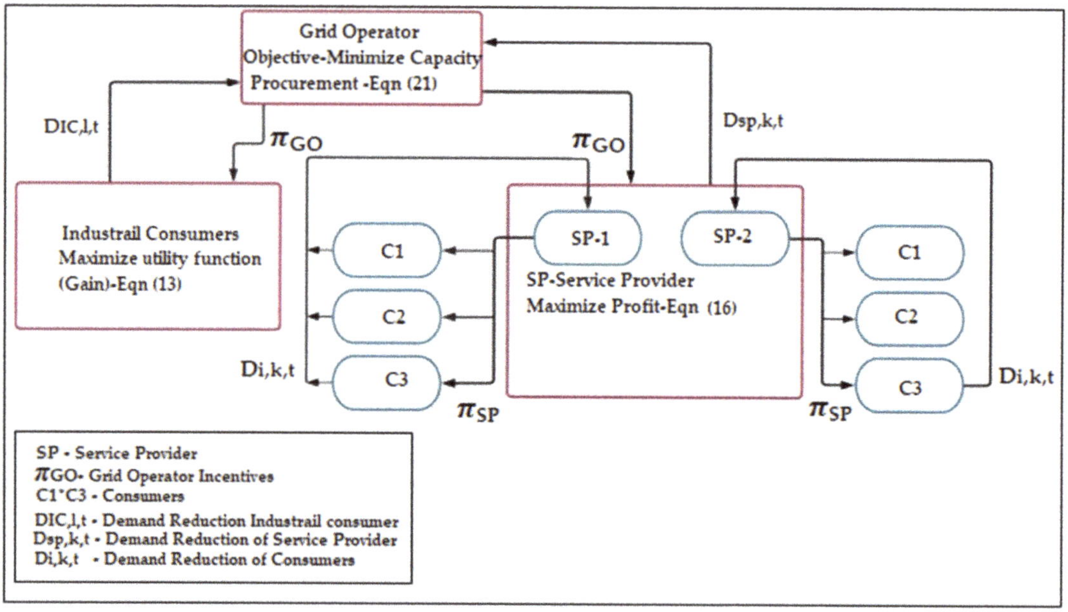

Figure 1. Hierarchy of entities in a smart grid.

a. Industrial consumer model

The IC aims to maximize its utility function, thereby maximizing the gain obtained as an incentive from the GO. The IC's utility function involves the gain obtained by the energy consumed and the incentive profit obtained from the GO for demand reduction. Let L be the total number of ICs. For each IC_l, where $l \in L$, the utility function is given by [15].

$$\max_{D_{IC,l}} U_{IC,l} = \sum_{t=1}^{T} \Psi_l\left(D_{l,t}^{ava} - D_{IC,l,t}\right) + D_{IC,l,t} \times \pi_{GOIC,t} \tag{13}$$

which is subjected to

$$0 \le D_{IC,l,t} \le D_{l,t}^{ava} \quad (14)$$

The constraint in (14) restricts the value of the demand reduction within the range of the available amount of demand reduction, and Ψ_l is the term used to describe the ICs' profit; it is calculated by the following formula:

$$\Psi_l = \begin{cases} \omega_l \times \left(D_l^{ava} - D_{IC,l}\right) - \frac{\sigma_l}{2} \times \left(D_l^{ava} - D_{IC,l}\right)^2 & if\ 0 \le \left(D_l^{ava} - D_{IC,l}\right) \le \frac{\omega_l}{\sigma_l} \\ \frac{\omega_l{}^2}{2\sigma_l} & if\ \left(D_l^{ava} - D_{IC,l}\right) \ge \frac{\omega_l}{\sigma_l} \end{cases} \quad (15)$$

σ_l and ω_l denote the rate and magnitude of the profit change of the IC when the power consumption of the IC is varied. As σ_l increases, the IC gains more by selling its resource to the GO.

b. Service provider model

The SP procures capacity from its customers by giving incentives to them, and it sells the procured capacity to the GO at the incentive rate the GO gives. As such, the utility function of the SP involves these terms. The SP aims to maximize its profit obtained by trading with the GO and its customers. Here, if the SP model is compared with Case 1, the second term of the equation is modified into Equation (16). In Case 1, the second term describes the capacity sold by the SP to the market at the electricity pricing, whereas in (16), it describes the capacity sold by SP to the GO at the incentive rate given by GO. Let K be the total number of SPs, and for each SP, the utility function can be written as follows [15]:

$$\max_{\pi_{SP,k}\ D_{SP,k}} U_{SP,k} = \sum_{t=1}^{T} D_{SP,k,t} \times \pi_{GOSP,t} - D_{SP,k,t} \times \pi_{SP,k,t} \quad (16)$$

which is subjected to

$$D_{SP,k,t} = \sum_{i \in N_k} D_{i,k,t} \quad (17)$$

The customer receives incentives from SP in return for their demand reduction. The utility function of the customers will also involve the dissatisfaction cost. While comparing Equation (18) with Case 1 customer model, it could be seen that the number of SPs considered in Case 2 is more than that in Case 1, and the incentives the SPs give to their customers differ. As such, the customers belonging to different SPs receive different incentives. The utility function of the customers is written as follows [4,7]:

$$\max_{D_{i,k}} U_{i,k,t} = \sum_{t=1}^{T} D_{i,k,t} \times \pi_{SP,k,t} - \mu_{i,k,t} \times \varphi_{i,k,t} \quad (18)$$

which is subjected to

$$0 \le D_{i,k,t} \le D_{i,k,t}^{ava} \quad (19)$$

The dissatisfaction cost function $\varphi_{i,k,t}$ is a function of $D_{i,k}$, and the level of discomfort that a customer might experience due to the reduction of demand is modeled by $\varphi_{i,k}$.

$$\varphi_{i,k} = \frac{\theta_{i,k}}{2}\left(D_{(i,k)}^2 + \lambda_{i,k} \times D_{i,k}\right), \theta_{i,k} > 0,\ \ \lambda_{i,k} > 0 \quad (20)$$

Here, $\theta_{i,k}$ and $\lambda_{i,k}$ are the parameters set based on the customer's attitude towards demand reduction. By increasing the value of $\theta_{i,k}$, the customer is more reluctant towards the demand reduction.

c. Grid operator model

The GO tries to minimize the cost of capacity procurement. Let the expected demand deficit be taken as D^{req}. The GO compensates for the demand deficit by acquiring capacities from SP and IC. The utility function of the GO can be framed as follows [15]:

$$min\ C_{GO} = \sum_{t=1}^{T} C_{gen}(G) + \pi_{GOIC,t} \times \sum_{i \in L} D_{IC,l,t} + \pi_{GOSP,t} \times \sum_{i \in L} D_{SP,k,t} \tag{21}$$

which is subjected to

$$\pi_{GO,t}^{min} \leq \pi_{GOSP,t} \leq \pi_{GOIC,t} \leq \pi_{GO}^{max} \tag{22}$$

$$G = D^{req} - \sum\nolimits_{i \in L} D_{IC,l} - \sum\nolimits_{i \in L} D_{SP,k} \tag{23}$$

The value of the cost of generating the required quantity G is calculated as follows:

$$C_{gen}(G) = a\left(G^2\right) + b(G) + c \tag{24}$$

In Equation (21), the first term represents the cost required for generation, and the second and the third terms represent the cost of procuring the capacities from the IC and SP.

d. Stackelberg game formulation and analysis

The interaction between the various entities involved in IBDR is modelled using Stackelberg's game theory. Including GO and IC in the problem makes it a leader–multi-follower game, increasing the complexity compared to Case 1. The GO model is solved, and the optimal incentive set by the GO for IC and SP is found, and then those optimal incentive values are utilized for the calculation of the optimal demand reductions. The GO fixes the incentives for the IC and SP using the following equations:

$$\pi_{GOSP} = \pi_{GO} \tag{25}$$

$$\pi_{GOIC} = \rho \times \pi_{GO},\ \ 0 \leq \rho \leq 1 \tag{26}$$

e. Optimal solution for the industrial consumer

Using the first-order derivative of the utility function in (13) with respect to $D_{IC,l}$ when equated to zero, we can obtain the equation for $D_{IC,l}$, as follows:

$$D_{IC,l} = D_l^{ava} - \frac{\omega_l}{\sigma_l} + \frac{\rho \times \pi_{GO}}{\sigma_l} \tag{27}$$

The total demand reduction of all of the ICs could be summed as

$$\sum_{i \in L} D_{IC,l} = \sum_{l \in L} D_l^{ava} - \frac{\omega_l}{\sigma_l} + \sum_{l \in L} \frac{\rho \times \pi_{GO}}{\sigma_l} \tag{28}$$

For convenience, the constant terms in the equation are taken as follows:

$$\eta = \sum_{l \in L} D_l^{ava} - \frac{\omega_l}{\sigma_l}, \gamma = \sum_{l \in L} \frac{1}{\sigma_l} > 0 \tag{29}$$

For Equation (29), when it is substituted into (28), we obtain

$$\sum_{i \in L} D_{IC,l} = \eta + \gamma \times \rho \times \pi_{GO} \tag{30}$$

The constraint in (14) can be regulated to make the IC contribute the minimum load by replacing the zero to D^{min}.

$$D^{min} = max\left(0, min\left(D_l^{ava} - \frac{\omega_l}{\sigma_l}\right)\right) \tag{31}$$

The optimal demand reduction of IC can be found by substituting the optimal GO incentive using (30).

f. Optimal solution for the service provider and its customers

The optimal demand reduction of the SP can be calculated as follows:

$$D_{SP,k} = \frac{1}{2}\pi_{GO} \times \sum\nolimits_{i \in N_k} \frac{1}{\mu_{i,k} \times \theta_{i,k}} - \frac{1}{2}\sum\nolimits_{i \in N_k} \frac{\lambda_{i,k}}{\theta_{i,k}} \tag{32}$$

The total demand reduction of all SPs could be found as follows:

$$\sum_{k \in K} D_{SP,k} = \frac{1}{2}\pi_{GO} \times \sum_{k \in K}\sum_{i \in N_k} \frac{1}{\mu_{i,k} \times \theta_{i,k}} - \frac{1}{2}\sum_{k \in K}\sum_{i \in N_k} \frac{\lambda_{i,k}}{\theta_{i,k}} \tag{33}$$

For convenience, the constant terms in the equation are taken as follows:

$$\alpha = \sum\nolimits_{k \in K}\sum\nolimits_{i \in N_k} \frac{1}{\mu_{i,k} \times \theta_{i,k}} > 0, \beta = \sum\nolimits_{k \in K}\sum\nolimits_{i \in N_k} \frac{\lambda_{i,k}}{\theta_{i,k}} > 0 \tag{34}$$

For Equation (34), when substituted in (33), we obtain

$$\sum_{k \in K} D_{SP,k} = \frac{\alpha}{2}(\pi_{GO}) - \frac{\beta}{2} \tag{35}$$

In order to solve the customer equation, the second derivative of Equation (18) with reference to $D_{i,k}$ is taken and equated to zero. The demand reduction of each customer under each SP can be found as follows:

$$D_{i,k} = \frac{\pi_{SP,k} - \mu_{i,k} \times \lambda_{i,k}}{\mu_{i,k} \times \theta_{i,k}} \tag{36}$$

g. Optimal solution for the grid operator

The cost function of the GO in (21) can be rewritten by substituting (24) and (23) into it, as follows:

$$\begin{aligned} min\ C_{GO} = a\left((D^{req} - \sum_{i \in L} D_{IC,l} - \sum_{i \in L} D_{SP,k})^2\right) + b\left(D^{req} - \sum_{i \in L} D_{IC,l} - \sum_{i \in L} D_{SP,k}\right) + \\ c + \pi_{GOIC} \times \sum_{i \in L} D_{IC,l} + \pi_{GOSP} \times \sum_{i \in L} D_{SP,k} \end{aligned} \tag{37}$$

By substituting (29) and (34) into (37) and equating the first derivative of the obtained equation to zero, we can realize the optimal incentive of the GO.

$$\pi^*_{GO} = \frac{2a\,(D^{req} + 0.5\beta - \eta) \times (0.5\alpha + \gamma\rho) + b(0.5\alpha + \gamma\rho) + 0.5\beta - \rho\eta}{2a(0.5\alpha + \gamma\rho)^2 + \alpha + 2\gamma\rho^2} \tag{38}$$

The obtained optimal incentive value of the GO is utilized to obtain the optimal demand reduction of ICs and SPs.

h. Stackelberg distributed algorithm

The parameters of the IC and SP customers must be disclosed to the GO, which is difficult in practice. As such, the distributed algorithm is used [15]. Here, the incentive of the GO is used as the variable, and SP and IC demand reductions are found.

The initial value of the GO incentive π^*_{GO} is initialized to π^{min}_{GO}. A value of ρ is chosen, and the value of π_{GOIC} *and* π_{GOSP} are calculated. Then, the initial values of $D^*_{IC,l}$ and $D^*_{SP,k}$ are calculated using Equations (27) and (32), respectively.

The initial value of the procurement cost of the GO is calculated as

$$C^*_{GO} = \left(a(D^{req} - \sum_{i\in L} D_{IC,l} + \sum_{i\in L} D_{SP,k})^2 + b\left(D^{req} - \sum_{i\in L} D_{IC,l} + \sum_{i\in L} D_{SP,k}\right) + c\right) \\ +\pi_{GOIC} \times \sum_{i\in L} D_{IC,l} + \pi_{GOSP} \times \sum_{i\in L} D_{SP,k} \tag{39}$$

These initial values are then used to calculate the optimal value of C_{GO} with π_{GO} as the variable which is updated for every iteration, as shown below:

Step 1: for the iteration $m = m + 1$;
Step 2: update the value of π^m_{GO} using

$$\pi^{m+1}_{GO} = \pi^m_{GO} + \Delta\pi_{GO} \tag{40}$$

where

$$\Delta\pi_{GO} = \delta\left(e^{\varpi|C^m_{GO} - C^{m-1}_{GO}|} - 1\right) \tag{41}$$

Step 3: the value of $D^{m+1}_{IC,l}$ and $D^{m+1}_{SP,k}$ are updated from Equations (27) and (32), respectively, and the π^{m+1}_{GO} value;
Step 4: with the values found in step 3, the value of C^{m+1}_{GO} is calculated as follows:

$$\begin{aligned} C^{m+1}_{GO} = \; & (a(D^{req} - \sum_{i\in L} D^{m+1}_{IC,l} + \sum_{i\in L} D^{m+1}_{SP,k}\Big)^2 \\ & +b\left(D^{req} - \sum_{i\in L} D^{m+1}_{IC,l} + \sum_{i\in L} D^{m+1}_{SP,k}\right) + c\Big) \\ & +\pi_{GOIC} \times \sum_{i\in L} D^{m+1}_{IC,l} + \pi_{GOSP} \times \sum_{i\in L} D^{m+1}_{SP,k} \end{aligned} \tag{42}$$

Step 5: f $C^{m+1}_{GO} \leq C^*_{GO}$ and $\pi^{m+1}_{GO} \leq \pi^*_{GO}$, then update the values of GO as $\pi^*_{GO} = \pi^{m+1}_{GO}$ and $C^*_{GO} = C^{m+1}_{GO}$;
Step 6: end if;
Step 7: end for;
Step 8: these steps are repeated until the condition shown in (43)

$$\left|C^{m+1}_{GO} - C^m_{GO}\right| \leq \varepsilon \tag{43}$$

Step 9: the equilibrium is reached when the value of C_{GO} does not decrease further.

4. Optimization Technique

Optimization techniques are utilized to find the best results from the set of feasible solutions. In this work, the interaction between the market entities is modelled using Stackelberg game theory, and the minimization and maximization equations of the entities are solved using PSO [30]. In the IBDR problem considered, there is a need for a repeated iteration with either maximization or minimization equations for the leader (GO) and the followers (IC, SP and customers) with two to three variables. As the Stackelberg game theory-based IBDR problem considered for optimization is a complicated iterative problem, the inertia-based PSO is considered for optimization in order to reduce its complexity.

In Case 1, the maximization equation of the customer is solved by setting the demand reduction as a variable, and the minimization equation of the SP is solved by setting the incentive as a variable. In Case 2, the industrial customer's maximization equations are solved using the demand reduction as the variable; the SP is maximized by compromising the GO and customer by varying the demand reduction from customers and the incentives.

Furthermore, the GO tries to minimize the cost of capacity procurement with the demand reduction obtained from IC and SP by varying the incentives.

The PSO procedure to solve the formulated Stackelberg-based problem for all of the entities is given below:

Step 1: initialize a set of particles in the search space;
Step 2: each particle will have a position and velocity. Initially, we generate the positions randomly based on the minimum and maximum limits of the variables shown in Table 2;
Step 3: for each particle, evaluate the incentives and utility function (objective) as per Equations (13), (16), (18) and (21), and in the entities, each particle will have a position and velocity;
Step 4: store the local and global best values (*pbest* and *gbest*);
Step 5: if the utility function for the new particle changes, based on maximization or minimization of the problem, update the *pbest* and *gbest* values;
Step 6: update the inertia weight factor by using $w = w \times wdamp$;
Step 7: update the position and velocity of the particle. The velocity update is given in (44);

$$Velocity = w \times velocity + C_1 \times \text{rand}() \times Pbest + C_2 \times \text{rand}() \times gbest \quad (44)$$

Step 8: velocity clamping is performed to maintain the velocity of the particle within the limit;
Step 9: check the termination condition; if satisfied, stop; else, go to Step 3.

Table 2. Variables and their bounds for Case 2's optimization.

Entities	Variables	Lower and Upper Bounds	Number of Variables
Grid operator	Incentive (π_{igo})	(3, 10)	1
Industrial consumer	Demand reduction of IC_1 (D_{IC1})	(0, 45.4)	3
	Demand reduction of IC_2 (D_{IC2})	(0, 36.2)	
	Demand reduction of IC_3 (D_{IC3})	(0, 56.5)	
Service provider	Incentive of SP_1 (π_{SP1})	(3, 10)	2
	Incentive of SP_2 (π_{SP2})	(3, 10)	
Consumer	Demand reduction of customer 1 ($D_{k,1,t}$)	(0, 11.35)	3
	Demand reduction of customer 2 ($D_{k,2,t}$)	(0, 16.55)	
	Demand reduction of customer 3 ($D_{k,3,t}$)	(0, 12.77)	

The optimization is carried out by considering the maximum iteration count as 1000, the number of particles as 100, and the acceleration coefficients C_1 and C_2 as 1.5 and 2. In this work, the inertia weight is dynamic, with a weight-damping ratio of 0.99.

5. Results and Discussions

The IBDR problem for Cases 1 and 2 is analyzed and optimized with the load data and the customer parameters considered.

a. Results of Case 1 with one SP and two customers

In this case, one SP and two customers are considered. The minimum and target demands of customers 1 and 2 for each hour are illustrated in Figures 2 and 3. The customer-related parameters (μ_i,θ_i,λ_i) for the two customers are shown in Table 3.

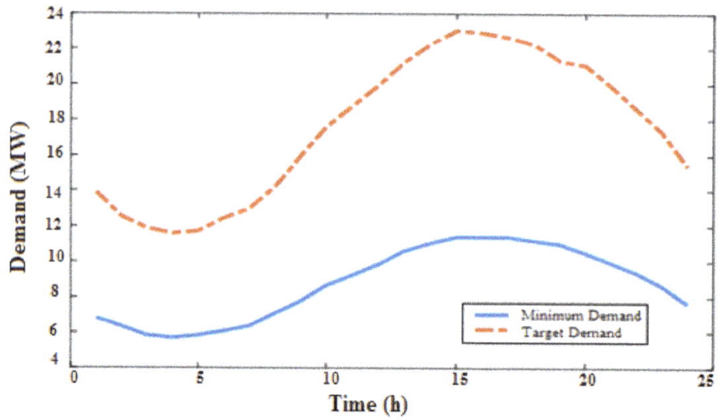

Figure 2. Minimum and target demands of customer 1.

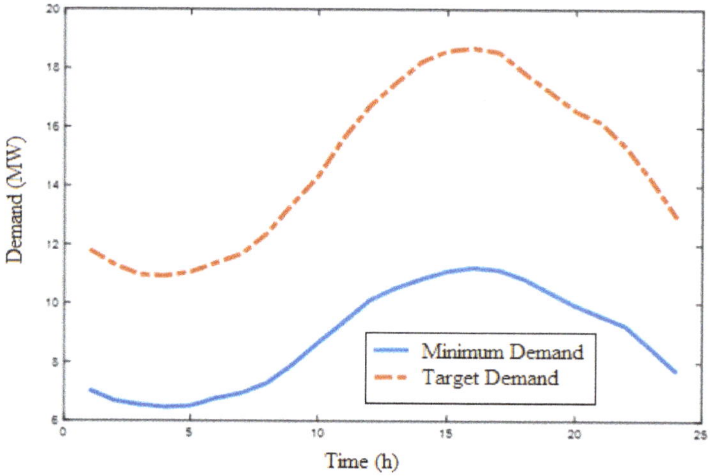

Figure 3. Minimum and target demands of customer 2.

Table 3. Parameters of Case 1, for customers 1 and 2.

Parameters	Customer 1	Customer 2
μ	(0.8, 1)	(0.8, 1)
θ	3.0	4.5
λ	10.0	10.0

The hourly electricity market pricing required for the calculation of the SP's utility function is represented in Figure 4.

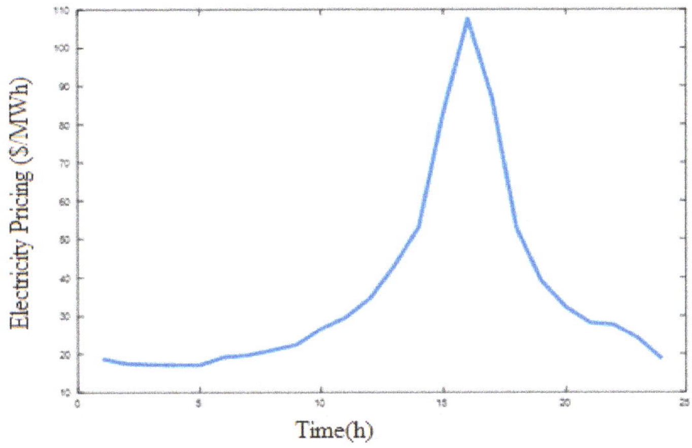

Figure 4. Hourly electricity pricing.

The value of D_t^{req}, i.e., the required demand reduction of the SP and the available demand reduction, are given in Figure 5.

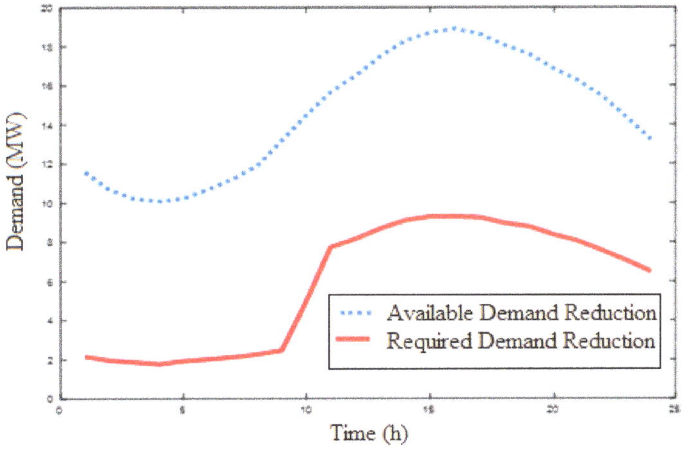

Figure 5. Available and required demand reduction of Case 1.

The optimal incentive of the SP is obtained using PSO for two different values of μ, namely 0.8 and 1, for both customers. The optimal SP incentive and the optimal mean demand reductions of customers 1 and 2 are taken and plotted. The obtained optimal solutions for the corresponding incentives are given in Figures 6 and 7. In order to ensure that the Stackelberg game is implemented correctly, Case 1 is tested with an existing system [4], and the results are obtained. The comparisons of the results for μ = 1.0 and 0.8 are listed in Tables 4 and 5.

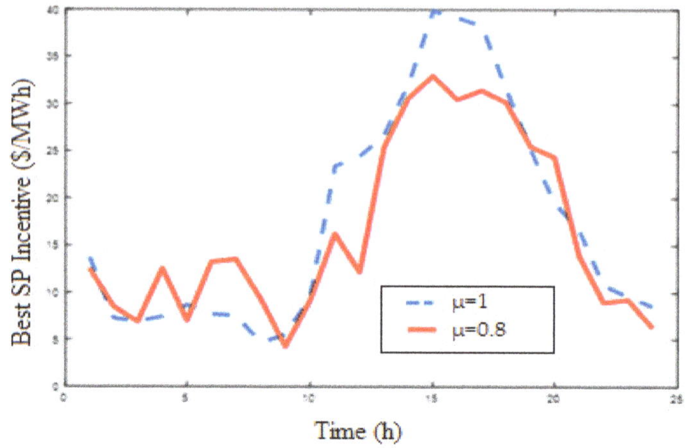

Figure 6. Optimal SP incentive for $\mu_1 = \mu_2 = 0.8$ and $\mu_1 = \mu_2 = 1.0$.

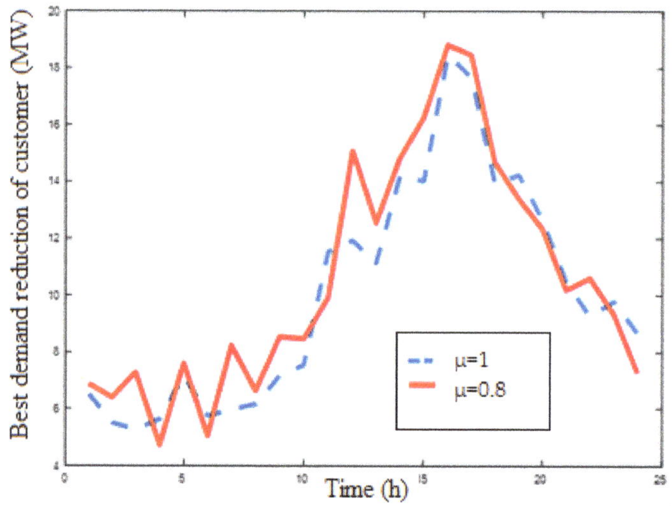

Figure 7. Optimal mean demand reduction for $\mu_1 = \mu_2 = 0.8$ and $\mu_1 = \mu_2 = 1.0$.

Table 4. Comparison with the existing system for $\mu = 1.0$.

Incentive by SP ($)		Demand Reduction (MW)		Incentive by SP ($)		Demand Reduction (MW)	
For one hour (16th hour)				For 24 h			
Stackelberg [4]	SPSO	Stackelberg [4]	SPSO	Stackelberg [4]	SPSO	Stackelberg [4]	SPSO
38	39	18.2	18.3	486.8	425.2	195.4	240.59

Table 5. Comparison with the existing system for $\mu = 0.8$.

Incentive by SP ($)		Demand Reduction (MW)		Incentive by SP ($)		Demand Reduction (MW)	
For one hour (16th hour)				For 24 h			
Stackelberg [4]	SPSO	Stackelberg [4]	SPSO	Stackelberg [4]	SPSO	Stackelberg [4]	SPSO
32	32	18.5	19	439	394.8	230.02	253.44

The tables show that the results obtained are better than the existing system results when optimization is performed using SPSO to solve the IBDR program.

The value of the weight factor represents the importance given by the customer to the discomfort involved in reducing the demand. A small value of μ_i represents less importance given to discomfort. This can be seen in Figure 7. The demand reduction curve for μ = 0.8 is high compared to that for μ = 1.0, which shows that the customers with a smaller μ value impose less weightage on the dissatisfaction cost, i.e., they can afford to have more demand reduction, without giving too much consideration to the discomfort caused. However, the customers with a higher μ value impose more weightage on dissatisfaction, such that they reduce their demand less compared to those with lower values of μ.

b. Sensitivity analysis for Case 1

Here, two conditions for which the value of μ is different for both customers are considered. At first, the values of $\mu_1 = 0.8$ and $\mu_2 = 1.0$ were taken. It could be inferred from the output graphs in Figures 8 and 9 that for the same optimal incentive curve obtained by the SP, the customer with a μ = 0.8 value shows more interest in demand reduction than the customer with μ = 1.0. This also happens when the condition is vice-versa; i.e., when $\mu_1 = 1.0$ and $\mu_2 = 0.8$, the demand reduction is more for customer 2, as depicted in Figures 10 and 11.

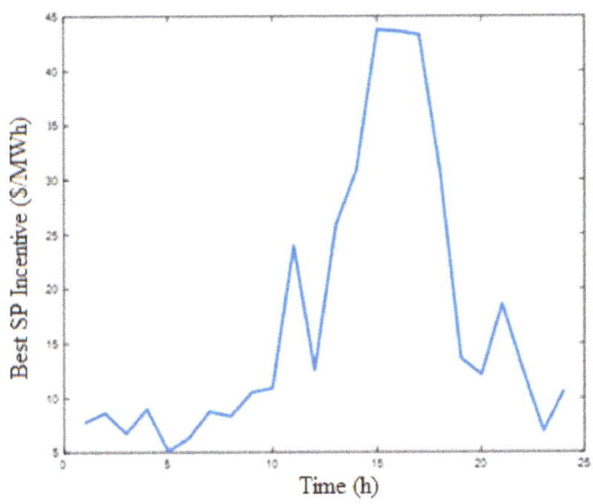

Figure 8. Optimal SP incentive for $\mu_1 = 0.8$ and $\mu_2 = 1.0$.

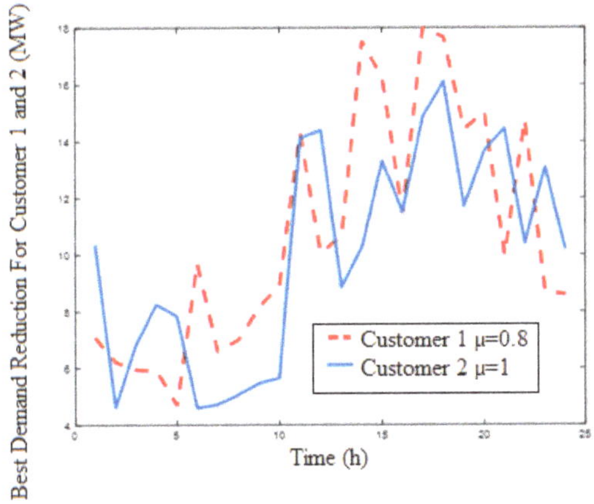

Figure 9. Optimal demand reduction for $\mu_1 = 0.8$ and $\mu_2 = 1.0$.

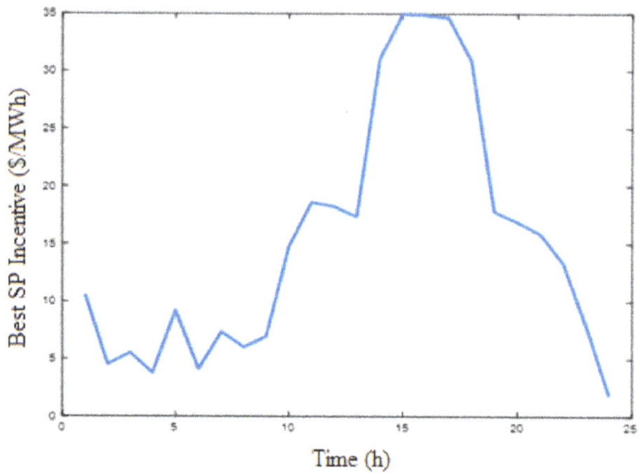

Figure 10. Optimal SP incentive for $\mu_1 = 1.0$ and $\mu_2 = 0.8$.

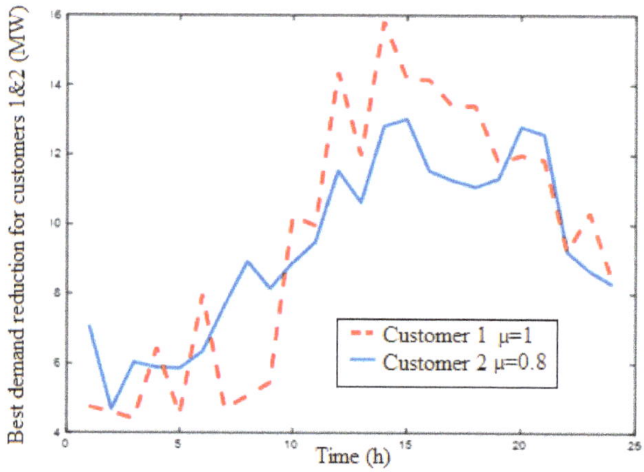

Figure 11. Optimal demand reduction for $\mu_1 = 1.0$ and $\mu_2 = 0.8$.

c. Case 2 with GO, ICS and SPS optimized using Stackelberg-distributed and SPSO algorithms

This case considers GO, IC, SP and its customers. The number of ICs is three and the number of SPs is two, with each SP being connected to three customers. The generator coefficients used are $a = 0.2$, $b = 0$, $c = 0$, and the value of ρ is chosen as 0.6. The value of δ is taken as 2.5, ϖ is taken as 0.13, ε is taken as 10^{-4}, and μ is taken as 1. Initially, the results are tested for SPSO and the Stackelberg-distributed algorithm for an hour using the test data and parameters specified in Tables 6 and 7.

Table 6. Test data used in Case 2 for the ICs [15].

ICs	IC_1	IC_2	IC_3
Load (kW)	45.4	36.2	56.5
σ	0.1	0.12	0.13
ω	8	8	8

Table 7. Test data used in Case 2 for the SPs.

SPs		SP_1			SP_2	
End Users	**User 1**	**User 2**	**User 3**	**User 1**	**User 2**	**User 3**
Load (kW)	11.4	7.5	14.4	5.5	13.7	9.2
θ	3.0	4.5	5.0	4.0	5.5	6.0
λ	2.0	2.0	2.0	3.0	3.0	3.0
μ	1.0	1.0	1.0	1.0	1.0	1.0

Figures 12 and 13 depict the optimal incentive and total cost of GO executed using the Stackelberg PSO. The results show that the SPSO can obtain the optimal converged output obtained by the Stackelberg distributed algorithm [31]. As per [15], the optimal GO incentive should be 7.27 cents/kWh, and that obtained by the distributed algorithm was 7.3, whereas with SPSO, the obtained global optimal solution was exactly 7.273 cents/kWh, as shown in Figure 12. From the results, it can be observed that the optimal incentive for the GO is 7.27 cents/kWh [4]. With the distributed algorithm, the result obtained was 7.29, with an error percentage of 0.275 [15], whereas with SPSO, the result was 7.273, with an error percentage of 0.041.

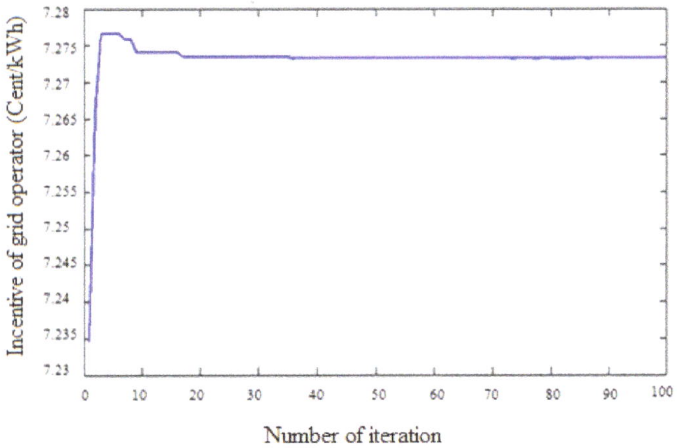

Figure 12. Optimal PSO- GO incentive for one hour.

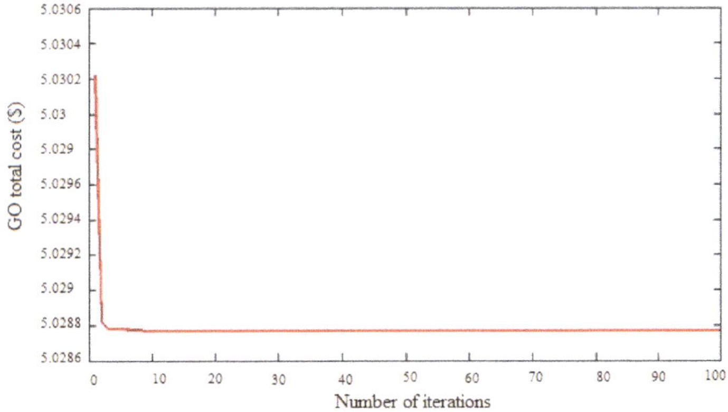

Figure 13. Optimal PSO-total cost of the GO for one hour.

Furthermore, the optimal incentive of the *SPs* obtained is shown in Figure 14, and the demand reduction of the three customers under SP_1 and SP_2 is given in Figures 15 and 16, respectively. The comparison of the obtained optimal incentive for GO, IC, SP, and the total cost of GO with SPSO and the distributed algorithm are tabulated in Table 8. Compared with the Stackelberg distributed algorithm, the SPSO can reach the exact optimal solution. Similarly, Case 2 with GO, IC, SP and its customers is considered for 24 h, and optimization is carried out using both Stackelberg-distributed and SPSO algorithms. Figures 17–19 illustrate the evaluated load data for Case 2 for 24 h of the ICs and the customers under SP_1 and SP_2.

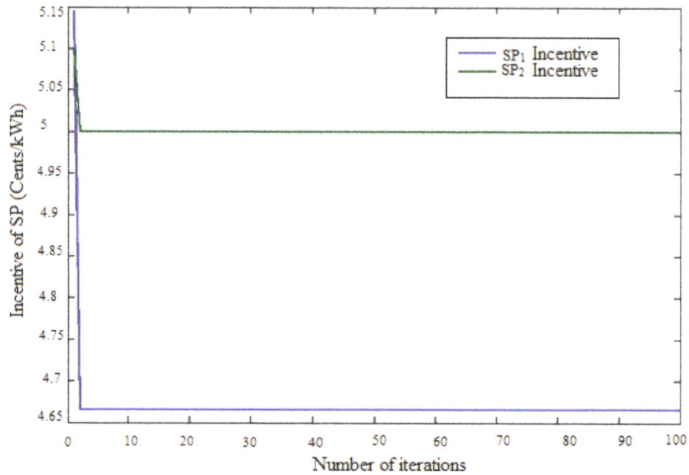

Figure 14. Optimal PSO-SP incentive for one hour.

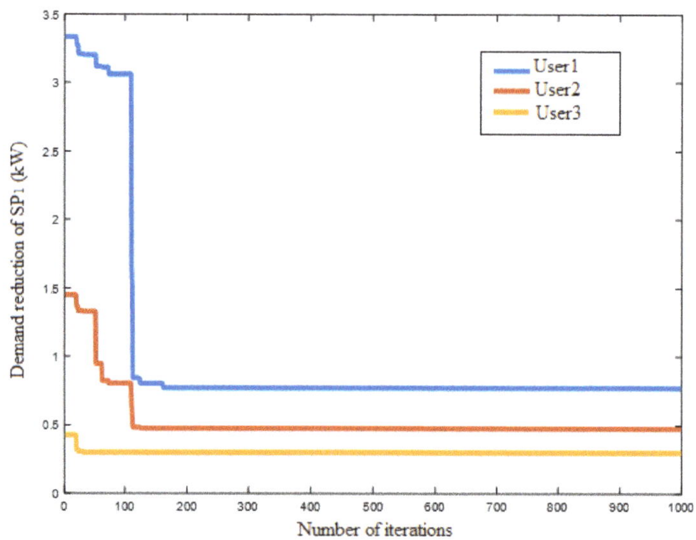

Figure 15. Optimal demand reduction of the three customers of SP_1 for one hour.

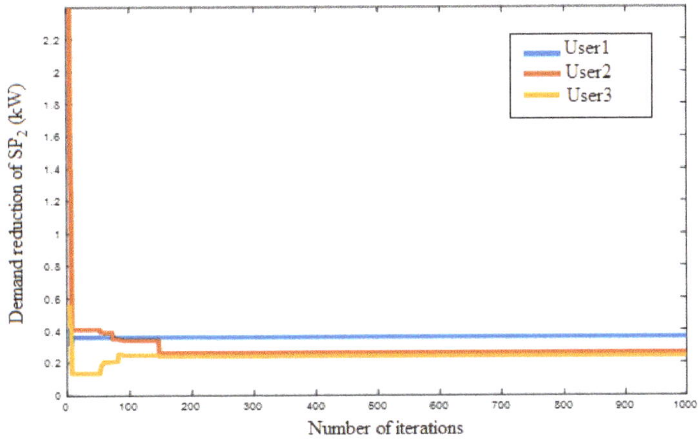

Figure 16. Optimal demand reduction of the three customers of SP_2 for one hour.

Table 8. Result comparison of Case 2's Stackelberg-PSO and distributed algorithm.

Parameters	**Stackelberg**			
	PSO Algorithm		**Distributed Algorithm**	
Grid operator incentive (cents/kWh)	7.273		7.290	
Total cost ($)	5.028		5.060	
IC Incentive (cents/kWh)	4.363		4.38	
SP incentive to users (cents/kWh)	SP_1 4.665	SP_2 5.0	SP_1 4.7	SP_2 5.1

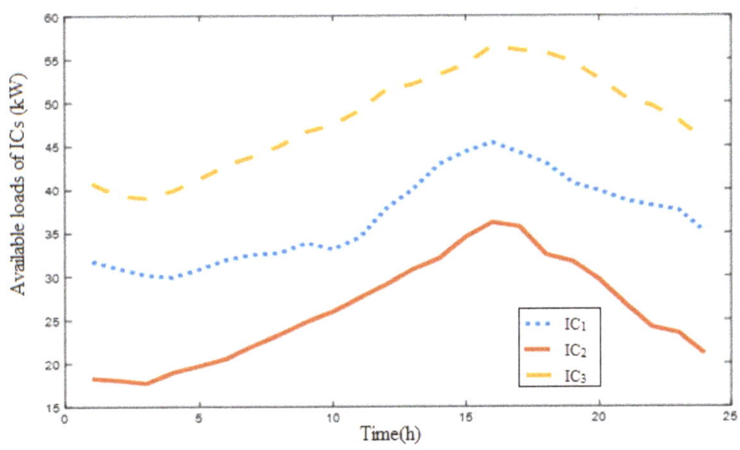

Figure 17. Available load data of the ICs.

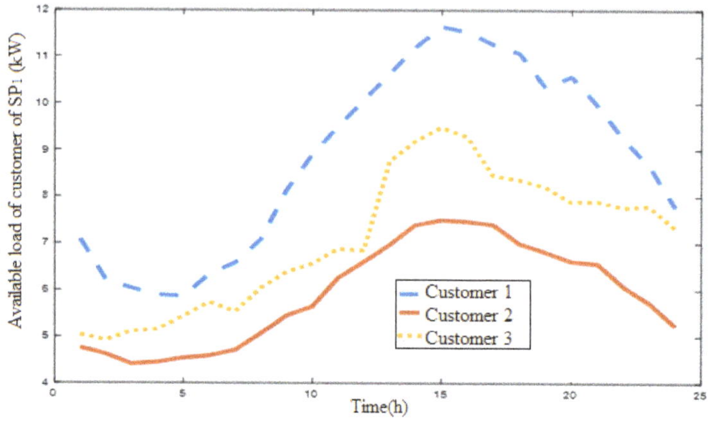

Figure 18. Available load data of the customers belonging to SP_1.

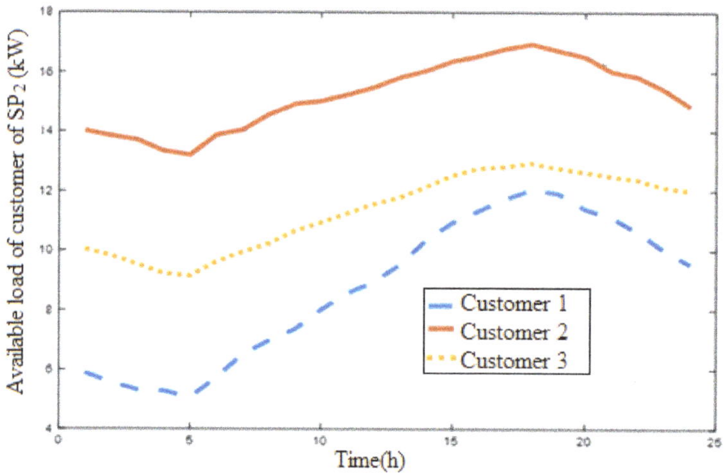

Figure 19. Available load data of the customers belonging to SP_2.

The customer and IC parameters are the same as those in Tables 6 and 7. The load data was taken for the entire day. These parameters remain the same throughout the day. The value of ρ is taken as 0.6. The required demand reduction optimized for the whole day is obtained, as illustrated in Figure 20. This demand reduction was used in the modeling of GO required for the calculation of the quantity of power generated [32].

The optimal incentive of GO is found using the Stackelberg equation which was derived, and the optimized incentives are utilized in the evaluation of the total cost. The total cost is further minimized using SPSO. The optimized incentive and the total cost incurred by the GO in procuring the capacities are depicted in Figure 21. The entire study is conducted for a day; therefore, the incentives and demand reductions vary daily.

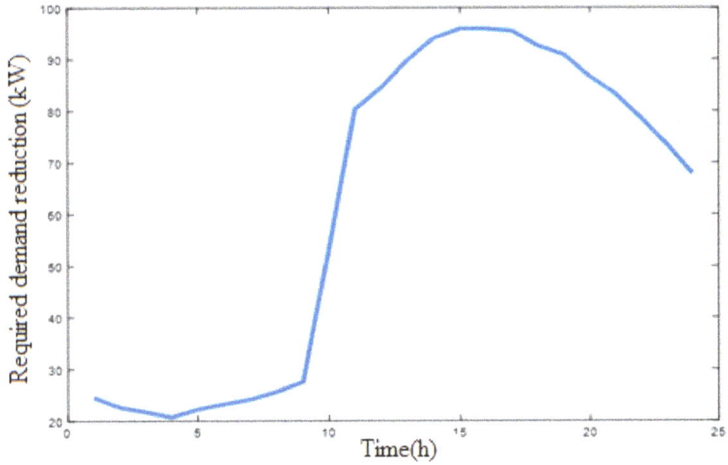

Figure 20. Required demand reduction in Case 2.

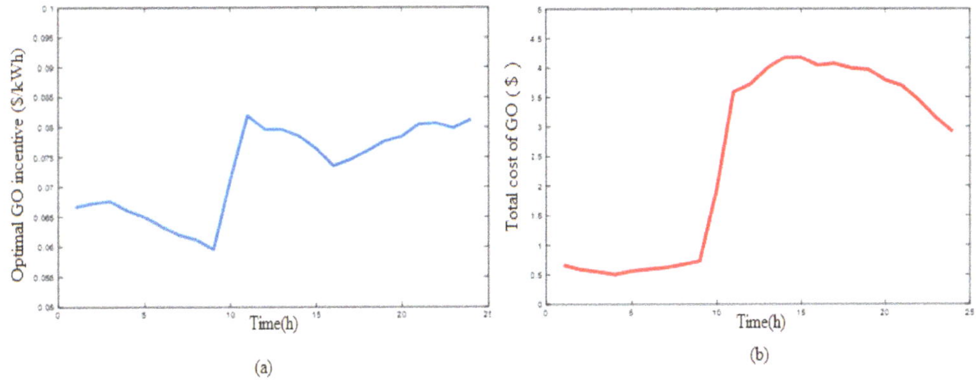

Figure 21. Results obtained: (**a**) the optimal GO incentive, and (**b**) the total cost of the GO.

The demand reductions of each ICs in response to the optimal GO incentive are optimized using PSO, and the results are given in Figure 22a. The customers with a larger value of σ are the ones contributing to more load reduction. It could be observed that customer 3, with a higher σ value, contributes to a greater demand reduction. The optimized GO incentives are utilized, and the incentives for SP_1 and SP_2 are found. The optimized values of the incentives of the SPs are given in Figure 22b. It is seen that SP_2 has a higher incentive compared to SP_1. The corresponding demand reduction of each customer under SP_1 and SP_2 is depicted in Figure 22c,d. These figures show that the demand reduction of customer 1 is high compared to customers 2 and 3.

The effect of various parameters on the results is studied using sensitivity analysis. Before starting with the altering of the parameters, from the results obtained from Case 2, it could be inferred that the demand reduction of customer 1 of SP_1 is higher compared to that of customer 1 of SP_2, which is visible in Figure 22c,d. This result could be attributed to the fact of the θ value. The increase in the θ value means that the customer's interest in demand reduction is low. From the input data of Tables 6 and 7, it is visible that customer 1 of SP_1 has a lower θ value (i.e., 3.0) compared to customer 1 of SP_2 (i.e., 4.0); therefore, the demand reduction is high for customer 1 of SP1. Furthermore, from Figure 22a, it can be said that the ICs with a larger value of σ will reduce more load. It could be seen that customer 3, with a higher σ value, contributes to a greater demand reduction. μ for the SP customers and ω for the ICs are changed to investigate the changes caused by the parameters, and one hour's (hour 16) output was observed and analyzed. Tables 9 and 10 present the input data of the performed analysis. The required demand deficit of the hour considered is 96 kW.

Table 9. Load data of the IC for the hour considered for the sensitivity analysis in Case 2.

ICs	IC_1	IC_2	IC_3
Available load (kW)	45.4	36.2	56.5

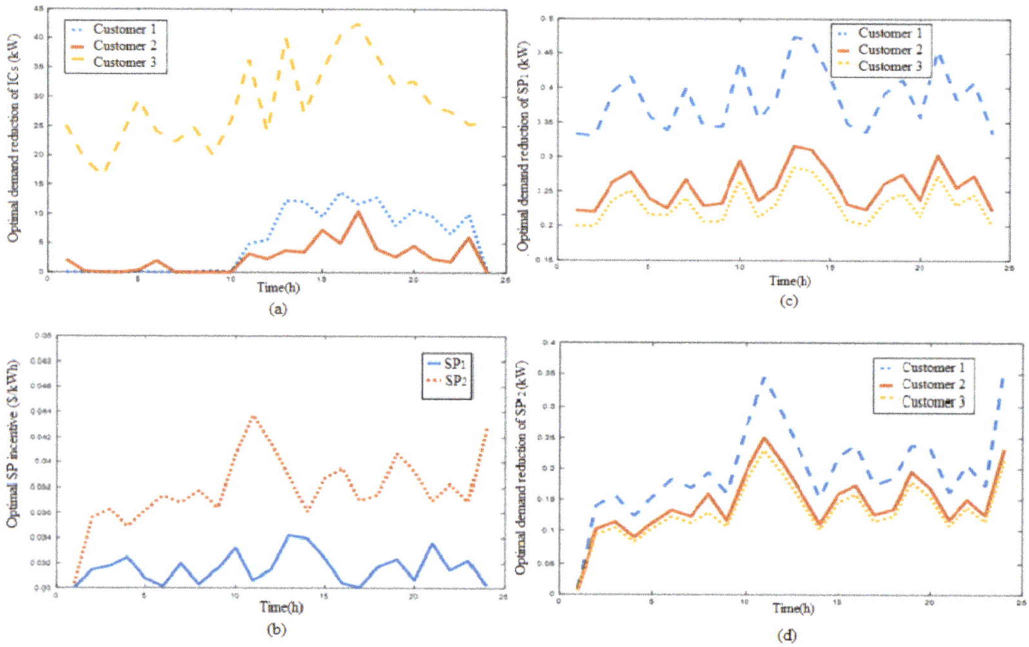

Figure 22. Optimization results: (**a**) the optimal demand reduction of the three ICs, (**b**) the optimal SP incentive, (**c**) the optimal demand reduction of the three customers of SP_1, and (**d**) the optimal demand reduction of the three customers of SP_2.

Table 10. Load data of SP for the hour considered for the sensitivity analysis in Case 2.

SPs		SP_1			SP_2	
Users	User 1	User 2	User 3	User 1	User 2	User 3
Load (kW)	11.54	7.47	9.29	11.35	16.55	12.77

The values of ω of the ICs considered are 8 and 5 for the three ICs. The values of μ for the SP customers considered are 1 and 0.8. These changes in the outputs are also shown in Tables 11 and 12 for IC and SP, respectively. The demand reduction of IC, the total cost of GO, and the incentive of GO are reduced when the value of ω is diminished.

Table 11. Optimized output with changes in the ω value.

Parameters		ω=8			ω=5	
Incentive of GO (cents/kWh)		7.3593			3.3383	
Total cost of GO ($)		4.0397			2.0634	
IC demand reduction (kW)	13.6714	4.9452	40.5349	8.2034	6.3111	34.7609

Table 12. Optimized output for changes in the μ value.

Parameters		μ=1.0			μ=0.8	
Incentive of GO (cents/kWh)		7.3593			9.9254	
Total cost of GO ($)		4.0397			5.2703	
SP incentive (cents/kWh)	3.0415		3.9544	3.1819		3.8468
Demand reduction of SP_1 customers (kW)	0.18	0.12	0.11	0.52	0.35	0.31
Demand reduction of SP_2 customers (kW)	0.11	0.08	0.07	0.42	0.31	0.28

The demand reduction of SP, and the total cost and the incentive of GO increase when the value of μ is reduced. The customer output of both of the SPs shows that as the value of μ is small, the customer shows more interest in demand reduction, and when observed closely, it is noted that the θ value also affects the demand reduction.

As θ increases, the value of the demand reduction decreases, which highlights that the customers show less interest in demand reduction as the value of θ increases.

A payment analysis is used to evaluate the benefits gained by each entity, and the results are tabulated. In Case 1, the demand reductions obtained by this approach for the entire day are much higher—the SP gains by selling these demand reductions at market prices. The incentive payment to be given to the customer is calculated. The incentive given by the SP to its customers is 5263 dollars for $\mu = 1.0$. If the IBDR implementation was not performed, the SP must purchase the required demand reduction at the market price as tabulated in Table 13.

Table 13. Payment analysis in Case 1.

Entity Considered	Incentive Benefits ($)
Incentives given by SP to customers in dollars	5263
SP purchasing D_t^{req} at market price (with no IBDR) in dollars	6357.7

The gain obtained by SP_1 and SP_2 can be found in Table 14 as USD 126.77 cents. The total cost incurred by the GO from Figure 22 is USD 60.1813, which is USD 6018.13 cents; lesser than the cost obtained using generators to serve the entire demand. Therefore, all of the entities involved in the IBDR program is benefited.

Table 14. Payment analysis in Case 2.

Entity Considered	Incentive Benefits (Cents)
Incentives obtained by IC from GO (cents)	3900
Total incentives obtained by SP1 and SP2 from GO (cents)	237.4676
Total incentives given by SP1 and SP2 to its customers (cents)	110.69

6. Conclusions

A novel IBDR is implemented considering the viewpoint of the SP and the GO. It could be said that the introduction of the GO into the system brings many changes in how incentives are provided, and in the demand reductions being evaluated. The interactions between the various entities are modeled using the Stackelberg game, and the contradicting parameters are brought to equilibrium. In order to solve the optimization problems, PSO is used, which provides the best-optimized results. The interaction between the SP and its two customers is a one-leader, two-follower game, but due to the presence of multiple entities with the introduction of GO, the game becomes a little complex as a one-leader multi-follower game. For the first IBDR case implemented (considering only the SP and its two customers), the incentive of the SP and the corresponding demand reductions of its customers were optimized. For the second case, with the involvement of the GO, the IC, and the SP, the incentives of the GO, the total cost of the GO, the corresponding demand reduction of the ICs under GO, the incentives set by the SP to its customers, and the demand reductions of the customers under SP were optimized. A sensitivity analysis was conducted to study the influence of varying customer parameters on the proposed IBDR program. The results proved that various changes in the outcomes are obtained due to the effect of the customer parameters, and a change made to a parameter of one entity affects the results of other entities. In the first case, the discomfort weight factor and the customer's attitude towards demand reduction are modified, and the outputs are monitored. In the second case, the discomfort weight factor and the customer's attitude towards the demand reduction of the SP and the IC parameters are varied, and the results are evaluated. The results show that the IBDR proves itself as a valuable tool to help the SP and GO procure resource capacities, thereby enabling them to solve the demand deficit issue. This work can be extended in the future by increasing the number of customers, SPs and ICs, and by considering the effect of integrating renewable energy sources on the results.

Author Contributions: Conceptualization: S.D.; methodology: S.D. and R.R.; software, S.D.; validation: R.R., S.D., S.H.E.A.A. and G.F.S.; visualization, M.A.M. and Z.M.A.; investigation: R.R., S.D., S.H.E.A.A., G.F.S., M.A.M. and Z.M.A.; data curation: S.D., S.S.G. and R.R., writing—original draft preparation, R.R. and S.D.; writing—review and editing, R.R., S.D., S.H.E.A.A., G.F.S., M.A.M. and Z.M.A.; supervision, S.D. and S.H.E.A.A.; project administration, S.D. and S.H.E.A.A.; funding acquisition, Z.M.A. All authors have read and agreed to the published version of the manuscript.

Funding: This research received no external funding.

Institutional Review Board Statement: Not applicable.

Data Availability Statement: Not applicable.

Conflicts of Interest: The authors declare no conflict of interest.

Glossary

DR	Demand response
IBDR	Incentive-based demand response
PBDR	Price-based demand response
SPSO	Stackelberg–particle swarm optimization
LSE	Load serving entities
SG	Smart grid
MES	Multi-energy systems
GO	Grid operator
SP	Service provider
IC	Industrial customer
PSO	Particle swarm optimization
Sets, parameters and variables	
i	Customer
t	Time
π_t	Hourly incentives
μ_i	Weight factor
$D_{i,t}$	Demand reduction
$\varphi_{i,t}(D_{i,t})$	Dissatisfaction cost

$D_{i,k,t}^{min}$	Minimum demand
$D_{i,k,t}^{tar}$	Target demand
D_t^{req}	Required demand
P_t	Electricity pricing
π_t^{min}, π_t^{max}	Minimum and maximum incentive
π_{GOIC}	Incentive for the IC
$D_{l,t}^{ava}$	Available load for the IC
$D_{IC,l,t}$	Demand reduction of the IC
Ψ_l	Profit for the IC
$D_l^{ava} - D_{IC,l}$	Energy consumed by the the IC
$U_{IC,l}$	Utility function of the IC
σ_l, ω_l	Rate and magnitude of profit of the IC
K	Number of service providers
N_k	Total number of customers under the *k*th SP
$D_{SP,k,t}$	Demand reduction of all customers belonging to the *k*th SP
N_k	Number of customers under the *k*th SP
$\pi_{SP,k,t}$	Incentive offered by the *k*th SP
$\pi_{GOSP,t}$	Incentive offered by the GO to the SP
G	Quantity of power being generated
$C_{gen}(G)$	Cost of generating power
a, b, c	Coefficients of generation
π_{GO}	Incentive of the GO
π_{GOSP}	Incentive set by GO for the SP
π_{GOIC}	Incentive set by GO for the IC
$\pi_{GO}^{max}, \pi_{GO}^{min}$	Maximum and minimum incentive of the GO

References

1. Wang, B.; Ma, H.; Wang, F.; Dampage, U.; Al-Dhaifallah, M.; Ali, Z.M.; Mohamed, M.A. An IoT-Enabled Stochastic Operation Management Framework for Smart Grids. *IEEE Trans. Intell. Transp. Syst.* **2022**, 1–10. [CrossRef]
2. Ahmed, E.M.; Rathinam, R.; Dayalan, S.; Fernandez, G.S.; Ali, Z.M.; Abdel Aleem, S.H.; Omar, A.I. A comprehensive analysis of demand response pricing strategies in a smart grid environment using particle swarm optimization and the strawberry optimization algorithm. *Mathematics* **2021**, *9*, 2338. [CrossRef]
3. Zhong, H.; Xie, L.; Xia, Q. Coupon Incentive-Based Demand Response: Theory and Case Study. *IEEE Trans. Power Syst.* **2013**, *28*, 1266–1276. [CrossRef]
4. Yu, M.; Hong, S.H.; Kim, J.B. Incentive-based demand response approach for aggregated demand side participation. Workshop. In Proceedings of the IEEE Conference on Smart Grid Communication: Smart City Infrastructure and Applications, Sydney, NSW, Australia, 6–9 November 2016; p. 16526293. [CrossRef]
5. Zhong, H.; Xie, L.; Xia, Q.; Kang, C.; Rahman, S. Incentive-based demand response for smart grid with reinforcement learning and deep neural network. In Proceedings of the IEEE Power & Energy Society Innovative Smart Grid Technologies Conference (ISGT), Washington, DC, USA, 18–20 February 2015; pp. 1–5. [CrossRef]
6. Chai, Y.; Xiang, Y.; Liu, J.; Gu, C.; Zhang, W.; Xu, W. Incentive-based demand response model for maximizing benefits of electricity retailers. *J. Mod. Power Syst. Clean Energy* **2019**, *7*, 1644–1650. [CrossRef]
7. Yu, M.; Hong, A.H. Incentive-based demand response considering hierarchical electricity market: A Stackelberg game approach. *Appl. Energy* **2017**, *203*, 267–279. [CrossRef]
8. Sahebi, M.M.; Duki, E.A.; Kia, M.; Soroudi, A.; Ehsan, M. Simultanous emergency demand response programming and unit commitment programming in comparison with interruptible load contracts. *IET Gener. Transm. Distrib.* **2012**, *6*, 605–611. [CrossRef]
9. Chrysikou, V.; Alamaniotis, M.; Tsoukalas, L.H. A Review of Incentive Based Demand Response Methods in Smart Electricity Grids. *Int. J. Monit. Surveill. Technol. Res.* **2015**, *3*, 62–73. [CrossRef]
10. Gadham, K.R.; Ghose, T. Demand response program using incentive and dis-incentive based scheme. *Energy Syst.* **2020**, *11*, 417–442. [CrossRef]
11. Imani, M.H.; Ghadi, M.J.; Ghavidel, S.; Li, L. Demand Response Modeling in Microgrid Operation: A Review and Application for Incentive-Based and Time-Based Programs. *Renew. Sustain. Energy Rev.* **2018**, *94*, 486–499. [CrossRef]
12. Sivanantham, G.; Gopalakrishnan, S. A Stackelberg game theoretical approach for demand response in smart grid. *Pers. Ubiquitous Comput.* **2020**, *24*, 511–518. [CrossRef]
13. Maharjan, S.; Zhu, Q.; Zhang, Y.; Gjessing, S.; Basar, T. Dependable Demand Response Management in the Smart Grid: A Stackelberg Game Approach. *IEEE Trans. Smart Grid* **2013**, *4*, 120–132. [CrossRef]

14. Shahryari, E.; Shayeghi, H.; Mohammadi-Ivatloo, B.; Moradzadeh, M. An Improved Incentive-based Demand Response Program in Day-Ahead and Intra-Day Electricity Markets. *Energy* **2018**, *155*, 205–214. [CrossRef]
15. Yu, M.; Hong, S.H.; Ding, Y.; Ye, X. An Incentive-Based Demand Response (DR) Model Considering Composited DR Resources. *IEEE Trans. Ind. Electron.* **2019**, *66*, 1488–1498. [CrossRef]
16. Ghazvini, M.A.F.; Soares, J.; Horta, N.; Neves, R.; Castro, R.; Vale, Z. A multi-objective model for scheduling of short-term incentive-based demand response programs offered by electricity retailers. *Appl. Energy* **2015**, *151*, 102–118. [CrossRef]
17. Kia, M.; Etemad, R.; Heidari, A.; Lotfi, M.; Catalão, J.P.; Shafie-khah, M.; Osório, G.J. Two-Stage Stochastic Mixed Integer Programming Approach for Optimal SCUC by Economic DR Model. In Proceedings of the IEEE Milan Power Tech, Milan, Italy, 23–27 June 2019; pp. 1–6. [CrossRef]
18. Paterakis, N.G.; Erdinç, O.; Catalão, J.P. An overview of Demand Response: Key-elements and international experience. *Renew. Sustain. Energy Rev.* **2016**, *69*, 871–891. [CrossRef]
19. Fang, X.; Hu, Q.; Li, F.; Wang, B.; Li, Y. Coupon-Based Demand Response Considering Wind Power Uncertainty: A Strategic Bidding Model for Load Serving Entities. *IEEE Trans. Power Syst.* **2016**, *31*, 1025–1037. [CrossRef]
20. RezaeeJordehi, A. Optimization of demand response in electric power systems, a review. *Renew. Sustain. Energy Rev.* **2019**, *103*, 308–319. [CrossRef]
21. O'Connell, N.; Pinson, P.; Madsen, H.; O'Malley, M. Benefits and challenges of electrical demand response: A critical review. *Renew. Sustain. Energy Rev.* **2014**, *39*, 686–699. [CrossRef]
22. Zheng, S.; Sun, Y.; Li, B.; Qi, B.; Zhang, X.; Li, F. Incentive-based integrated demand response for multiple energy carriers under complex uncertainties and double coupling effects. *Appl. Energy* **2020**, *283*, 116254. [CrossRef]
23. Luo, Z.; Hong, S.; Ding, Y. A data mining-driven incentive-based demand response scheme for a virtual power plant. *Appl. Energy* **2019**, *239*, 549–559. [CrossRef]
24. Nainar, K.; Pillai, J.R.; Bak-Jensen, B. Incentive Price-Based Demand Response in Active Distribution Grids. *Appl. Sci.* **2021**, *11*, 180. [CrossRef]
25. Muthirayan, D.; Baeyens, E.; Chakraborty, P.; Poolla, K.; Khargonekar, P.P. A Minimal Incentive-Based Demand Response Program With Self-Reported Baseline Mechanism. *IEEE Trans. Smart Grid* **2020**, *11*, 2195–2207. [CrossRef]
26. Savari, G.F.; Krishnasamy, V.; Sugavanam, V.; Vakesan, K. Optimal Charging Scheduling of Electric Vehicles in Micro Grids Using Priority Algorithms and Particle Swarm Optimization. *Mob. Netw. Appl.* **2019**, *24*, 1835–1847. [CrossRef]
27. Fernandez, G.S.; Krishnasamy, V.; Kuppusamy, S.; Ali, J.S.; Ali, Z.M.; El-Shahat, A.; Abdel Aleem, S.H. Optimal Dynamic Scheduling of Electric Vehicles in a Parking Lot Using Particle Swarm Optimization and Shuffled Frog Leaping Algorithm. *Energies* **2020**, *13*, 6384. [CrossRef]
28. Almalaq, A.; Alqunun, K.; Refaat, M.M.; Farah, A.; Benabdallah, F.; Ali, Z.M.; Aleem, S.H.E.A. Towards Increasing Hosting Capacity of Modern Power Systems through Generation and Transmission Expansion Planning. *Sustainability* **2022**, *14*, 2998. [CrossRef]
29. Ali, Z.M.; Diaaeldin, I.M.; El-Rafei, A.; Hasanien, H.M.; Abdel Aleem, S.H.E.; Abdelaziz, A.Y. A Novel Distributed Generation Planning Algorithm via Graphically-Based Network Reconfiguration and Soft Open Points Placement Using Archimedes Optimization Algorithm. *Ain Shams Eng. J.* **2021**, *12*, 1923–1941. [CrossRef]
30. Mohamed, M.A.; Eltamaly, A.M.; Alolah, A.I. PSO-based smart grid application for sizing and optimization of hybrid renewable energy systems. *PLoS ONE* **2016**, *11*, e0159702.
31. Rizwan, M.; Waseem, M.; Liaqat, R.; Sajjad, I.A.; Dampage, U.; Salmen, S.H.; Obaid, S.A.; Mohamed, M.A.; Annuk, A. SPSO Based Optimal Integration of DGs in Local Distribution Systems under Extreme Load Growth for Smart Cities. *Electronics* **2021**, *10*, 2542.
32. Refaat, M.M.; Aleem, S.H.E.A.; Atia, Y.; Ali, Z.M.; Sayed, M.M. Multi-stage dynamic transmission network expansion planning using lshade-spacma. *Appl. Sci.* **2021**, *11*, 2155. [CrossRef]

sustainability

MDPI

Article

Management Optimization of Electricity System with Sustainability Enhancement

Wei Hou [1,2,*], Rita Yi Man Li [3] and Thanawan Sittihai [1]

1 Chakrabongse Bhuvanarth International Institute for Interdisciplinary Studies, Rajamangala University of Technology Tawan-Ok, Bangkok 10400, Thailand; thanawansittihai@gmai.com
2 State Grid Beijing Electric Power Corporation, Beijing 100031, China
3 Sustainable Real Estate Research Center, Hong Kong Shue Yan University, Hong Kong, China; ritayimanli@gmai.com
* Correspondence: hhbbj1108@163.com

Abstract: Based on new policies and social changes, renewable energies have highly penetrated electrical systems, making the system more vulnerable than before. On the other hand, it leads to congestion and competition within the network. To this end, this paper developed a probabilistic multi-objective-based congestion management approach and applied it to the optimal transmission switching (OTS) strategies, to maximize system suitability and minimize total production costs. A point estimation economic method (PEM) has been applied, as one of the best management and economic tools to handle the uncertainties associated with a wind turbine's power production and load demand (LD). Results demonstrate the effectiveness and merit of the proposed technique, compared to the existing one, which can lead to higher reliability and sustainability for the grids.

Keywords: dynamic thermal rating; probabilistic energy demand; two points approximation scheme; wind power; congestion management; electricity optimization; MPSO multi-objective optimization; suitability

Citation: Hou, W.; Man Li, R.Y.; Sittihai, T. Management Optimization of Electricity System with Sustainability Enhancement. *Sustainability* **2022**, *14*, 6650. https://doi.org/10.3390/su14116650

Academic Editor: Mohamed A. Mohamed

Received: 24 April 2022
Accepted: 18 May 2022
Published: 29 May 2022

1. Introduction

Electric utilities must use renewable energy resources, due to the growing demand for power and the increase in the world population. The popularity of such resources was, also, boosted by growing environmental concerns and reducing investment prices. Renewable energy resources, such as wind, are one of the most widely used resources, with 651 GW of installed capacity globally, in 2019 [1]. A significant challenge for network operators is associated with wind power fluctuations [2]. Wind power resources present a number of challenges in the transmission networks (TNs) operation, including that many lines lack sufficient capacity for transmitting the energy, so dealing in the day-ahead power market leads to congestion [3]. Generally, congestion of transmission happens whenever variations in demand/production cause a transfer of power that reaches or exceeds the transmission network's physical capacity. Additionally, the transmission infrastructure on the system is rapidly deteriorating. Consequently, the amount of congestion in the grid enhances markedly, which is a major factor in the increase in power production costs and the restriction of the use of renewable energy [4]. It is difficult and costly to construct new transmission lines (TLs) to reduce congestion, so utilities are seeking a more cost-efficient technology to make use of the current infrastructure [5]. The flexible AC-TS (FACTS), high voltage DC system, and enhanced conductor system are developing transmission technologies, which are increasing the capacity of the transmission network. Nevertheless, a number of technologies, such as FACTS controllers and large-scale energy storage, require significant investment. Accordingly, novel transmission technologies, designed to utilize the potential of the current transmission infrastructure, would be more appealing and are becoming, increasingly, used in practice. As a result, transmission service providers are searching for alternatives that could be used effectively for their networks. This paper uses DTR and network topology optimization (NTO) as well as, for improving the network's reliability, increasing wind power's penetration rate and reducing congestions of TL.

A network topology optimization method can be used to take advantage of the current transmission infrastructure, to achieve a timely and significant purpose: The transmission network's performance and flexibility can be improved. As early as the 1980s, the concept of changing transmission network topology was offered [6], giving system operators the chance to temporarily remove TLs from service, giving them the flexibility they need to manage transmission network topologies. As a precaution or corrective measure, the TL is turned on/off to decrease voltage violations, line overloads, secure the system, and restore the load when an outage has occurred [7]. Transmission switching is also examined, as a means of harnessing the flexibility of the current transmission infrastructure to reduce the operational costs of the system, in addition to being a control action or as a solution to reduce system losses. Optimal transmission switching (OTS) was, firstly, introduced in [8], as part of the DC optimal power flow. A binary variable shows the on/off status of TLs, within the suggested formulation, while [9] extended the OTS real-time utilization to the AC optimal power flow (ACOPF) context. The two-level frequent structure was presented, using 2th-order cone programming, for generating ideal switching solutions at high levels; afterward, those solutions were screened for achieving AC reliability at low levels. The transmission switching of power systems is proven to lower the investment cost.

Through the development of heuristic solution techniques, transmission switching operations have become more adaptable, by finding solutions to OTS problems more quickly. Stochastic optimal transmission switching (SOTS) requires a suitable stochastic programming formulation. The SOTS must, explicitly, represent and consider the uncertain renewable generation and loads, while determining the OTS decisions. Then, [10] recommended an OTS model with adaptive robust optimization (RO), according to the uncertainty of net LD, and [11] used a linearized OTS model, according to AC OPF, for adapting to the random nature and intermittent nature of wind power for practical applications. Finally, [12] offered a very conservative TL switching solution, using an RO that just took into account the worst-case uncertainty.

Along with optimizing stochastic and robust programming, probable power flow has proven effective at examining the uncertain nature of electric power systems as well, and is used widely. Point estimation methods (PEM) are adapted, by such methods, to a wide range of applications [13]. Generally, Static line ratings (SLRs) are used to plan and operate many TLs. Severe weather conditions limit TS ratings, in this case. It might be difficult to fully utilize the TLs, due to this conservative method [14]. Therefore, transmission service providers are exploring alternatives, in order to optimize the use of the transmission network. As an example, one of those methods uses the DTR of TLs, which is dependent on the actual climate, for creating conditions of actual loading [15]. Taking the benefit of the cold temperatures and high winds that occur during such events will enable grid operators to significantly increase the thermal limit of TLs with DTR [15]. TS operators benefit most from this solution, as by utilizing current assets for TLs, capital investment in this part of the system is much less, in comparison to reconstructing more resilient systems from scratch. Power systems have been the subject of many DTR study programs. DTR's integration into real-time operations improved congestion management (CM) as well as related costs throughout such alterations and events in [16]. Moreover, Ref. [17] discussed the aging impact of transmission overhead lines based on the DTR technique, and its parametric uncertainty towards line failure probability. Despite this, the majority of research exists on the basis of fixed networks. Coordinating DTR and other new transmission-control technologies should be investigated further. The OTS and DTR share a similar benefit, namely that, together, they can add more power over the transmission network, by mitigating system limitations and decreasing operational costs. In summary, the co-optimization of these two technologies in the power system would increase its efficiency effectively. Power system reliability has been enhanced recently, by combining DTR and OTS with other technology. In addition, Ref. [18] discussed DTR progress scheduling with NTO deployment and its reliability evaluation, while Ref. [19] suggested combining OTS and DTR in a unit commitment problem, to increase power distribution uniformity. Moreover, Ref. [20] discussed the effects of OTS and dynamic thermal ratings on decreasing emission of carbon dioxide, particularly in the case of diversifying energy resources. A key aim of this research is to schedule the power system day-ahead, by simultaneously implementing OTS and DTR. Compared to the previous investigations, there are still a limited number of suitable limitations, which are, directly, incorporated into the formulation of OTS problems, for assuring that an OTS problem is truly networked, and that the load and wind power are random and variable.

The probable multi-objective CM, according to the PEM using the DTR and OTS methods, is presented in the present study, which supports utilities in making better use of current transmission infrastructure, by providing better congestion relief efficiencies. In total, two levels of the problem are planned. Firstly, MPSO examines choices from a range of possible switching schemes for TLs, and, then, it sends the results to the next level. In the next step, (2PEM + 1) has been applied, for solving the probable optimum power flow problems that have been utilized to estimate the expected costs of production and the reliability index, related to various strategies.

The paper consists of the following parts: Part 1 provides an overview of the study. Part 2 deals with the modeling of system uncertainties. Part 3 introduces the probabilistic power flow (PPF) estimation process (2PEM + 1). Part 4 discusses the problem formulation. Part 5 presents the MPSO algorithm, whereas Part 6 covers the test system and scenarios. Part 7 concludes the paper.

2. Modeling of Uncertainties

The increasing uncertainty associated with load and renewable energy sources is one of the major characteristics of modern power grids. There are various methods to handling uncertainty in the electrical grid, and they usually have been grouped into RO methods, probable methods, and interval-set analyses. RO models uncertainties in the form of intervals and uses optimization for solving state boundary issues. Interval set decomposition involves demonstrating the uncertainty in the form of a set/interval and using an interval set method for estimating the output boundary. In probabilistic models, uncertain parameters are modeled as random variables, with known probability density functions, and propagations of uncertainty are modeled through analytic methods or Monte Carlo simulation. The present paper uses the probabilistic method, since this can be a widely applied method, in order to handle uncertainties in electrical grids [1] (See Appendix A).

Regarding the distribution types, please note it that it is widely accepted in the literature, [1,21], that normal distribution and the Weibull type can be used for modeling the behavior of the load and wind speed/power. Nevertheless, it should be noted that any other PDF function can be used in the same manner, without loss of generalization. In other words, we can use other PDF types in quite the same way as we used the normal or Weibull functions. Meanwhile since these assumptions are made based on long-term big data, there is no way to check their accuracy in this work. In fact, in practice, we need to, first, make an initial data analysis to find the most-fitting PDF, based on our real-time data, and, then, start working with our model.

2.1. Modeling of Probable Load

There are two parameters that define probability distribution functions (PDFs); the mean (μ) and standard deviation (σ) of the uncertain parameter are described in the following way [1]:

$$\mathrm{PDF(s)} = \frac{1}{\sigma\sqrt{2\pi}} * exp^{\left(\frac{(-(S-\mu)^2)}{2\pi^2}\right)} \quad (1)$$

Here, the apparent power of the load is shown by S, and the assumption is that every bus has a mean (μ) that equals the base load and a standard deviation (σ) of $\pm$ 5% of the base load [1].

2.2. Probabilistic Wind Modeling

Wind turbines (WTs) generate power based on the wind speed (v), which can typically be modeled using the Weibull distribution PDF [22]:

$$\mathrm{PDE(v)} = \left(\frac{k}{c}\right)\left(\frac{v}{c}\right)^{k-1} exp\left[-\left(\frac{v}{c}\right)^{k}\right] \quad (2)$$

Here, the scale factor is shown by c, and the shape factor of the Weibull function is indicated by k.

The PDF of the wind speed can be supposed to be known in every region, and the variation of wind speed to WT output power can be determined via [23]:

$$P(v) = \begin{cases} 0, & v \geq v_o \quad or \quad v \leq v_i \\ P_r\left(\frac{v-v_i}{v_r-v_o}\right) & v_i \leq v \leq v_r \\ P_r & v_i \leq v \leq v_r \end{cases} \quad (3)$$

Here, v_i indicates the cut-in wind speed, v_r shows the rated wind velocity, v_o shows the cut-out wind speed, and P_r represents the rated power.

2.3. Probabilistic Line Rating Modeling

Various approaches exist for implementing DTR, always requiring various input parameters. Those approaches are: [24] (1) predicting DTR based on climate forecasts and system loading; (2) an estimate of DTR based on indirect measurements; and (3) an actual DTR assessment based on real climatic information, as applied in this study. Therefore, the ampacity of the line must be considered as a probabilistic variable, in which the PDF of a thermal limit (MVA) would correspond to a generalized extreme value distribution, as illustrated in Equation (4). [1,21].

$$\mathrm{PDF(t)} = \left(\frac{1}{\sigma_1}\right)\cdot\left(1+\xi\frac{(t-\mu l)^{\frac{-1}{\xi}}}{\sigma_1}\right)\cdot e^{\left(-\left(1+\xi\frac{(t-\mu l)^{\frac{1}{\xi}}}{\sigma_1}\right)\right)} \quad (4)$$

Here, the DTR of the line is shown by t, the location parameter is represented by μ_l, the shape parameter is shown by ξ, and the scale parameter is represented by σ_l. Weather conditions determine these parameters, which are computed according to [25]. Therefore, the DTR mean has been achieved for every period (for example, season) and is applied as a limit to the formulas of power flow [1].

3. Probabilistic Power Flow

This paper has proposed a stochastic framework based on PEM, to handle the uncertainties of the problem, including the renewable sources' output power uncertainties and the load demand uncertainties. Therefore, Section 2 is devoted to describing this section completely. The core idea in PEM is to replace the PDF functions with some appropriate fitting concentration points. In this approach, the PEM solves probabilistic problems via a deterministic process, although it needs less computation [13]. Therefore, the PEM would decompose the stochastic problem into a 2m + 1 equivalent deterministic problem, with different probabilities. PEMs have the advantage of requiring basic information about random variables, in order to model them effectively. The skewness, variance, and kurtosis of the variables are included in the data. This study uses the (2m + 1) scheme, in which the kurtosis of the input random parameters is taken into account, and just one more computation has been performed [26]. More details, including the complete formulations, are provided in this paper, which are highlighted in yellow. The PPF problem is solved using the scheme (2m + 1), in the following way [1]:

$$\xi_{l,k} = \frac{\lambda_{l,3}}{2} + (-1)^{3-k}\sqrt{\lambda_{l,4} - \frac{3}{4}\lambda^2{}_{l,3}} \\ k = 1,2,\ \xi_{l,3} = 0 \tag{5}$$

$$p_{l,k} = \mu_{pl} + \sigma_{pl} k = 1,2,3 \tag{6}$$

$$w_{l,k} = \frac{(-1)^{3-k}}{m\xi_{l,k}(\xi_{l,1} - \xi_{l,2})} \quad k = 1,2 \tag{7}$$

$$w_{l,3} = \frac{1}{m} - \frac{1}{m(\lambda_{l,4} - \lambda^2{}_{l,3})} \tag{8}$$

Here, l,k represents the standard location, μ_{Pl} shows the mean, σ_{Pl} indicates the standard deviation, and $\lambda_{l,j}$ shows the jth standard central moment of the input random parameters p_l. Based on Equation (11), location $l,3 = 0$ gives $p_{l,3} = \mu_{pl}$ and, thus, the locations are the same $\left(\mu_{p1},\ \mu_{p2},\ \ldots,\ \mu_{pl},\ \ldots,\ \mu_{pm}\right)$ point. Therefore, it should be possible to perform a single calculation at this location, given equivalent weight w_0, in the following way:

$$w_0 = 1 - \sum_{l=1}^{m} \frac{1}{m(\lambda_{l,4} - \lambda^2{}_{l,3})} \tag{9}$$

In addition, Equation (11) proves that the scheme yields non-real locations if $\lambda_{l,4} - \frac{3}{4}\lambda^2_{l,3}$ has a negative value. Furthermore, in power system problems, the probability distributions have been typically applied to the normal, uniform, and binomial models; thus, the places can always be actual.

The PPF issue, utilizing the (2m + 1) layout, is solved by modeling the power flow input information as random variables, and calculating the weights and places by applying Equations (11) and (13). A solution can be found in [26]:

$$E\left(Z^j\right) = \sum_{l=1}^{m}\sum_{k=1}^{2} w_{l,k}(Z(l,k))^j + w_o Z_o^j \tag{10}$$

$$(l,\ k) = F(\mu_{p1},\ \mu_{p2}, \ldots, P_{1,k}, \ldots,\ \mu_{pm}) \quad , k = 1,2 \tag{11}$$

$$Z_o = F(\mu_{p1},\ \mu_{p2}, \ldots, \ldots,\ \mu_{pm}) \tag{12}$$

Here, $Z(l,\ k)$ shows the output of the RVs associated with the kth concentration $(\mu_{p1},\ \mu_{p2}, \ldots, P_{1,k}, \ldots,\ \mu_{pm})$ of random parameters, representing the relation between the output and input in the probable power flow (PPF). The gathering scheme determines the dense number of definite PFs needed to be executed. $Z(l,\ k)$ has been used for assessing the raw moments of the yield, and the computation ends once entire centralizations of data RVs have been taken into account. Afterwards, applying Equation (12), the analyzed raw moments of the yield have been used for calculating the necessary statistical data [1].

4. Formulation of the Issue

4.1. Objective Functions

In TNs, the TL switching method reduces congestion. Through optimization, TLs that need to be disconnected would be identified, decreasing congestion. In order to determine the optimal switching plan, in terms of maximum probabilistic reliability and lowest cost of production, a multi-objective-based methodology is proposed in this study. Following are details on the objective functions.

4.1.1. Overall Cost of Power Production

In order to realize this goal, total network savings must be maximized, via a reduction in total power production costs, which are dependent on physical limitations, such as TL flow limitations and voltage of bus restrictions. It should be noted that it is well perceived that the cost function is considered in a quadratic format, since the case study is the transmission system. From a technical point of view, this models the nonlinear opening and closing process of the steam valves, which looks like sinusoidal curves. It is clear that we have to use a linear equation (rather than a quadratic function), if the case study is a distribution system. The paper uses the secondary supply bid price function:

$$min \sum_{i=1}^{n_g} \left(a_i + b_i P_{gi} + c_i P^2{}_{gi} \right) \tag{13}$$

Subject to:

$$P_{gi} + P_{wi} + P_{Di} - V_i \sum_{j=1}^{N_b} V_j \left(G_{ij} \cos \delta_{ij} + B_{ij} \sin \delta_{ij} \right) = 0 \,, \;\; i \in N_b \tag{14}$$

$$Q_{gi} + Q_{wi} + Q_{Di} - V \sum_{j=1}^{N_b} V_{ji} \left(G_{ij} \cos \delta_{ij} - B_{ij} \cos \delta_{ij} \right) = 0 \,, \;\; i \in N_b \tag{15}$$

$$P_{gi}^{min} \leq E\left(P_{gi}\right) \leq P_{gi}^{max} \,, i \in N_G \tag{16}$$

$$Q_{gi}^{min} \leq E\left(Q_{gi}\right) \leq Q_{gi}^{max} \,, i \in N_G \tag{17}$$

$$V_{gi}^{min} \leq E\left(V_{gi}\right) \leq V_{gi}^{max} \,, i \in N_G \tag{18}$$

$$V_{Li}^{min} \leq E(V_{Li}) \leq V_{Li}^{max} \,, i \in N_L \tag{19}$$

$$(S_{li}) \leq S_l^{max} * \gamma_k \;\; , \, i \in N_L \tag{20}$$

$$\delta_i^{min} \leq \delta_i \leq \delta_i^{max} \,, i \in N_b \tag{21}$$

$$\sum_{k=1}^{N_L} (1 - \gamma_k) \leq \varphi \tag{22}$$

Here, Q_{gi} and P_{gi} represent the reactive and active power production of the i generator unit; Q_{wi} and P_{wi} show the reactive and active power generation of wind farm i; Q_{D1} and P_{D1} represent the reactive and active LD at load bus i; P_{gi}^{max}, P_{gi}^{min}, Q_{gi}^{max}, and Q_{gi}^{min} show the maximum and minimum restrictions of the reactive and active power injection of the ith generator agent; V_{gi}^{min} and V_{gi}^{max} represent the minimum and maximum limits of the voltage magnitude at bus i; $V_j \angle \delta_j$ and $V_i \angle \delta_i$ show the termination buses j and i voltages; S_l represents the power flow via the line, and S_l^{max} shows the loading limit; B_{ij} and G_{ij} represent the substance and conductance of the branch linked between bus i and j; φ, N_b, N_g, and N_L indicate the groups of the switching lines number, branches number, generator buses, and buses of load, respectively; γ_k shows the state of line l that can be 0 or 1 (open/close); and a_i, b_i, and c_i represent generator fuel ratios.

4.1.2. Probable Reliability Objective

Maximum probabilistic reliability has been regarded as an additional target to optimize, in the suggested multi-objective optimization problem. OTS strategy can effectively change the structure of the grid and, thus, provide new power flow paths, through which the optimal values of the objective functions would enhance. OTS would, then, minimize the power losses and cost, enhancing the EENS in the system. From the market point of view, this is very useful and can be considered as a powerful ancillary service for the market players. The expected energy not supplied (EENS) factor has been applied as the OTS's goal of OTS, showing the reliability and efficiency of the grid, in the status of structure reconfiguration. The probabilistic analysis state enumeration method is used to calculate the EENS factor of the TS, which can follow each OTS plan [27]. The objective function, which should be minimized, can be described via Equation (23):

$$min(EENS) = \sum_{s \in \Omega} P_t(s) \cdot PC(s) \tag{23}$$

$$P_t(s) = \left(\frac{\lambda_i}{\lambda_i + \mu_i}\right) \prod_{j=1}^{\psi} \left(1 - \frac{\lambda_i}{\lambda_i + \mu_i}\right) \tag{24}$$

Here, $PC(s)$ is the total load curtailment and $P_t(s)$ is the probability of occurrence of grid status s; Ω shows the set for grid status s; λ_i and μ_i represent the blackout and element i maintenance rate; and the accessible parts number is shown by ψ.

Failures of TL are modeled as an independent single blackout $(N - 1)$; therefore, just first-order contingencies are taken into account, and, as far as is related to the cause, that will not be affected by any concurrent failures. Reconstruction state probabilities have been computed in Equation (24), in which the probability of occurrence of every contingency (s) has been obtained, by multiplying the existing part's probability by the failed part's probability. The overall load curtailment, $PC(s)$, of any status s, has been computed via an ACOPF method, as illustrated in the Equations (25)–(29) sub-problem, which can minimize the overall load decrement of every possibility:

$$min \sum_{i \in N_b} PC_i \tag{25}$$

Subject to:

$$P_{gi} + P_{wi} - P_{Di} + PC_i - V_i \sum_{j=1}^{N_b} V_j (G_{ij} \cos \delta_{ij} + B_{ij} \sin \delta_{ij}) = 0\,,\ \ i \in N_b \tag{26}$$

$$Q_{gi} + Q_{wi} - Q_{Di} + QC_i - V_i \sum_{j=1}^{N_b} V_j (G_{ij} \sin \delta_{ij} - B_{ij} \cos \delta_{ij}) = 0\,,\ \ i \in N_b \tag{27}$$

$$0 \leq PC_i \leq P_{Di} i \in N_b \tag{28}$$

$$0 \leq QC_i \leq Q_{Di} i \in N_b \tag{29}$$

It is necessary to determine the reactive and active power at the bus of the generator and the magnitude of the voltage of bus in the inequality restriction, according to Equations (16)–(20).

The EENS index was assessed in networks using the (2PEM + 1) process, and the load's uncertainties and WT's power have been taken into account. Figure 1 shows the steps for calculating the EENS index, using the PEM-based approach.

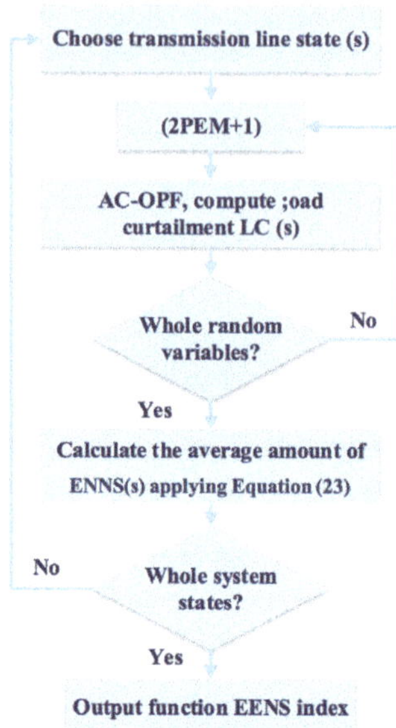

Figure 1. Process for determining the EENS reliability index.

5. Solution Method

5.1. Enhanced Particle Swarm Optimization

PSO is an evolutionary optimization technique, in order to minimize an objective that mimics the behavior of flocks of birds flying overhead or a group of fish. Particle swarm optimizers are comprised of particles and update empirical data about a search space, iteratively. Individuals in the population illustrate potential solutions to problems and can be viewed as particles that move in a ψ-dimensional search space.

Particles in a general PSO algorithm will adjust their position in accordance with their experiences and those of their neighbors, such as their current position, velocity, and the best prior position. By including the worst experience of each particle, MPSO enhances convergent performance of PSO and offers additional exploration capacity to the swarm [28]. By remembering its worst experience, the particle explores the search space more efficiently, to determine the best solution area. The MPSO algorithm updates positions and velocities of particles, in the following way [28]:

$$v_i^d(k+1) = \xi v_i^d(k) + c_1 \times r_1 \times (P^d{}_{best,i} - X_i^d(k) + c_2 \times r_2 \times (G^d{}_{best} \\ -X_i^d(k)) + c_3 \times r_3 \times \left(X_i^d(k) - P^d{}_{worst,i}\right) \quad (30)$$

$$x_i^d(k+1) = x_i^d(k) + \sigma v_i^d(k+1) \quad (31)$$

Here, the current velocity of the ith particle is shown by $v_i^d(k)$, $i = 1, \ldots, P$, where P shows the population size; k indicates the kth iteration; superscript $d = 1, \ldots, \psi$ shows the dimensions of the particle; $P_{best,i}$ represents the best prior location of the ith particle; $P_{worst,i}$ shows the worst prior location of the ith particle; G_{best} represents the best prior location between whole of particles in the swarm; $X_i^d(k)$ shows the current location of the ith particle; c_1, c_2, and c_3 represent the acceleration factors; r_1, r_2, and r_3 show the monotonous random numbers among 0 to 1; and σ shows the learning factor. By using designed inertia weight ζ, the previous updated feature is copied to the next iteration. When the greater ζ has been chosen, the previous $v_i^d(k)$ has a significant influence on $v_i^d(k+1)$. It should be noted that the PSO method in Equation (30) does not include the last section on the right part. The amount of MPSO variables have been presented in Table 1.

Table 1. Amount of MPSO variables.

Variable	P	c_1	c_2	c_3	ψ	σ	ζ	k_{max}
Value	5	0.5	0.5	0.5	1	5×10^{-3}	1	30

5.2. Islanding Prevention

System-wide cascading interruptions and complete blackouts can be avoided, by using power system island detection. When the TL is disconnected, it can lead to bus isolation. Consequently, the optimization process must include measures to avoid islanding. The prior investigations have not included restrictions that could be directly incorporated into the OTS problem formulation, to make sure the network connection is properly enforced. A novel island detection process, which can provide a robust approach for connection matrix detection, is presented in this paper [29]. Figure 2 shows the suggested MPSO algorithm.

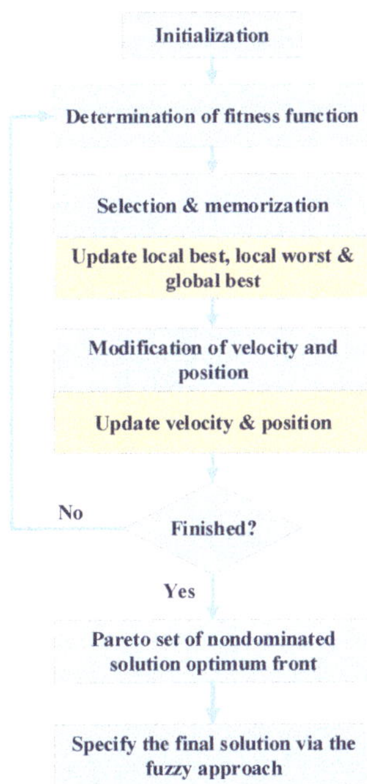

Figure 2. MPSO procedure.

5.3. Suggested Method

Two levels are planned for the complete problem. During the first level, MPSO determines the optimal TL switching strategies from various feasible ones, and the results are transferred to the 2th level. The 2th level involves the use of the (2PEM + 1), for the solution of the issues of the PPF, which can be necessary for determining the production cost and reliability index for different strategies.

6. Test System and Scenarios

An IEEE RTS 96 test system [30] is applied, for illustrating the suggested approach in cases where transmission switching is needed to reduce operating expenses and ease congestion. Interconnections among three identical 24-bus networks make this system possible. This system has an active power load of 8.55×10^3 (MW) and a reactive power of 1.74×10^3 (MVAr). A total of 9.832×10^3 (MW) active power and 2.001×10^3 (MVAr) reactive power have been added to the system. For every bus, the active and reactive power has been expressed as the usual PDF, whose mean (μ) equals the basic load and whose standard deviation (σ) equals 5% of the basic load.

Table 2 gives producing unit cost and kinds of data information. Figure 3 shows a schematic scheme of the RTS. Six wind farm generations, with 285 MW of installation capacity, have been permitted to be linked at bus 10, with 14 in every region of the RTS. There are 190 WTs in each wind farm. The whole optimization runs start with population size $[N = 100]$ and the maximum number of iterations $[GN = 200]$, as primary control parameters for the MO MPSO method. Pareto optimal fronts, with a maximum size of 25 solutions, were chosen. An analysis of the operational and economic advantages of implementing the OTS approach was conducted, using two scenarios.

Table 2. Producing agent kinds and cost data.

Generator Unit	U_{12}	U_{20}	U_{50}	U_{76}	U_{100}	U_{155}	U_{197}	U_{350}	U_{400}
Size (MW)	12.00	20.00	50.00	76.00	100.00	155.00	197.00	350.00	400.00
Fuel	Steam/Oil	CT/Oil	Hydro	Steam/Coal	Steam/Oil	Steam/Coal	Steam/Oil	Steam/Coal	Nuclear
Fuel (**\$/Mbtu**)	8.4	15.17	0	1.78	8.4	1.78	8.4	1.78	0.6
Cost (**\$/Mwh**)	85.5	149.56	0.1	17	67.95	14.71	74.75	14.96	22

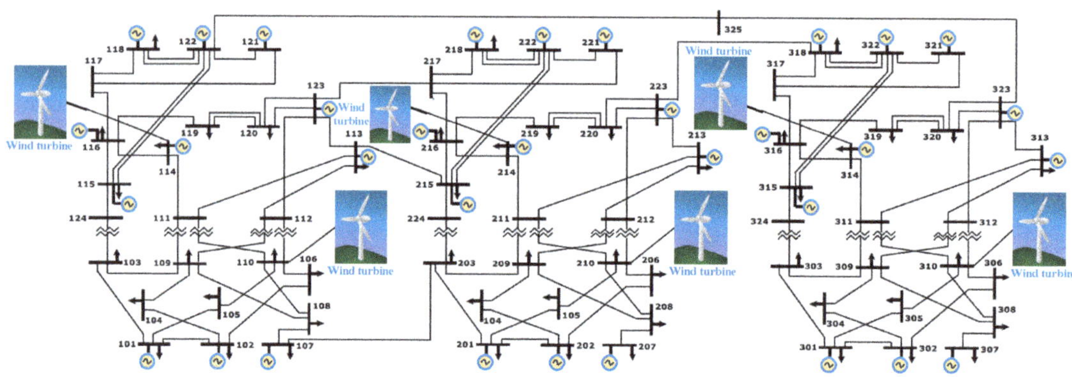

Figure 3. Enhanced IEEE RTS 96-bus network.

6.1. First Case Study: OTS with SLR

Figure 4 shows the Pareto optimal front for two objectives: probabilistic reliability and the production costs for DTR. MPSO has been used to derive the Pareto optimal front, and the fuzzy approach has been used for defining the BCS for diverse wind farm production capacities, according to Tables 3 and 4. In order to reduce grid dispatch prices with OTS, just one line is open ($\phi = 1$). For the optimization issue, the line that links bus 109 to bus 111 is switched off. Eight producing sets in the grid have their output power changed by this reconfiguration. According to Table 5, generator agents U179 and U100 decrease output, whereas generator agents U12, U350, and U67 increase output. Reducing the generation costs of generators, by transferring power from costly devices to low-price devices, without compromising the network's safety, is possible. Continuing the analysis for $[\phi = 2\ to\ 4]$, for every further open line, the cost of the network reduces, but at a decreasing rate. Various wind farm capacities have been investigated in four subcases. Tables 3 and 4 show the outcomes. As wind farm capacity increases, power production costs decrease, if the number of switched TLs is zero. Tables 3 and 4 show that if the number of TLs increases, the mean of the production cost of the network decreases, with an increase in the capacity of the wind farm, due to the greater dispatch of cheaper generators, according to Figure 5. As a consequence, if $N - 1$ security has been kept, *EENS* greatly enhances those probabilistic indicators for the four topologies, which have disconnected lines. Figure 6 indicates that up to four lines can be opened, without compromising the system operating constraints.

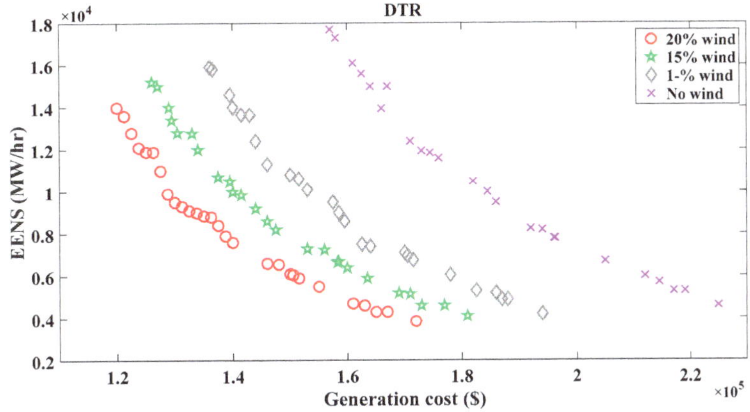

Figure 4. The EENS Pareto optimal front and overall production cost objective applying MOMPSO at ($\phi = 4$) for DTR.

Table 3. The BCS outcomes for OTS methods with SLR (with wind influence 0% and 10%).

φ	0	1	2	3	4
Wind penetration (0%) Lines open	-	[111–114]	[210,211]; [111–114]	[118–121]; [209–111]; [210,211]	[310,311]; [106–110]; [118–121]; [109–111]
μ[*Gen.cost*] ($/h)	189,685	186,439	185,285	184,633	184,125
μ[*EENS*] (Mwh/y)	2.77×10^3	5.831×10^3	6.93×10^3	7.78×10^3	10.273×10^3
Wind penetration (10%) Lines open	-	[111–114]	[210,211]; [109–111]	[110–112]; [109–111]; [215,216]	[312–314]; [106–110]; [118–121]; [109–111]
μ[*Gen.cost*] ($/h)	164,860	161,340	160,215	159,730	159,310
μ[*EENS*] (Mwh/y)	2.49×10^3	5.31×10^3	6.33×10^3	7.08×10^3	9.273×10^3

Table 4. The BCS outcomes for OTS methods with SLR (with wind influence 15% and 20%).

φ	0	1	2	3	4
Wind penetration (15%) Lines open	-	[109–111]	[210,211]; [114–111]	[118–121]; [109–108]; [210–205]	[310–305]; [106–110]; [118–117]; [109–111]
μ[*Gen.cost*] ($/h)	189,685	186,439	185,285	184,633	184,125
μ[*EENS*] (Mwh/y)	2.77×10^3	5.831×10^3	6.93×10^3	7.78×10^3	10.273×10^3
Wind penetration (20%) Lines open	-	[119–114]	[210,211]; [109–103]	[118–121]; [109–108]; [210,211]	[311–314]; [106–110]; [218–222]; [109–104]
μ[*Gen.cost*] ($/h)	164,820	143,620	142,510	140,715	140,030
μ[*EENS*] (Mwh/y)	2.36×10^3	5.01×10^3	5.95×10^3	6.69×10^3	8.83×10^3

Table 5. Alterations in generator output, following switching off the line (109–111).

Generator Unit	U_{12}	U_{20}	U_{50}	U_{76}	U_{100}	U_{155}	U_{197}	U_{350}	U_{400}
Size (MW)	12.00	20.00	50.00	76.00	100.00	155.00	197.00	350.00	400.00
Fuel kind	Oil/Steam	Oil/CT	Hydro	Coal/Steam	Oil/Steam	Coal/Steam	Oil/Steam	Coal/Steam	Nuclear
Fuel ($/**MBtu**)	8.4	15.17	0	1.78	8.4	1.78	8.4	1.78	0.6
Cost ($/**Mwh**)	85.5	149.56	0.1	17	67.95	14.71	74.75	14.96	22
Change in output (Mw)	4.5	0	0	29.74	−17.37	0	−98	81.13	0

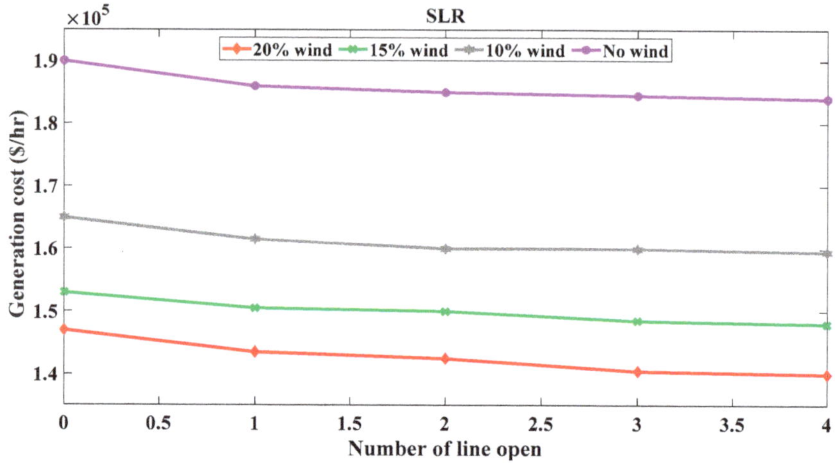

Figure 5. Anticipated producing cost for the OTS methods, for various wind levels for SLR.

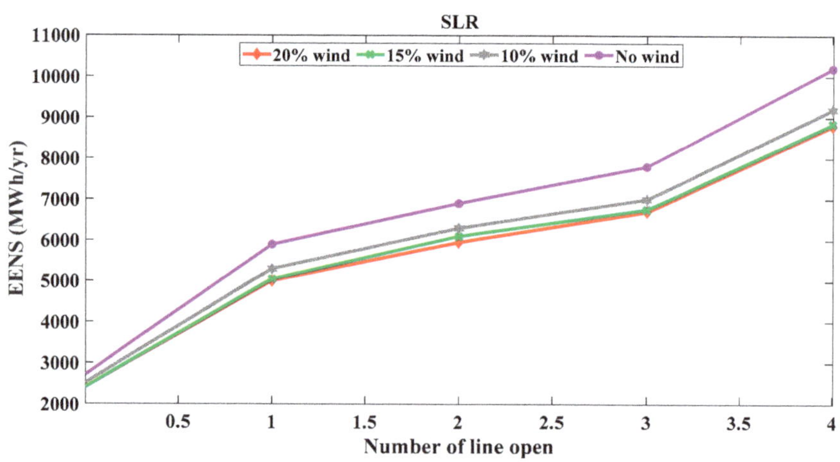

Figure 6. EENS for the OTS methods, for various SLR wind stages.

6.2. Case Study: OTS with DTR

Based on the suggested approach in Part 5, Tables 6 and 7 illustrate the achieved production cost and EENS index. The power, of the great-cost generator agent U197 in the DTR, has been decreased and moved to the small-cost generator agents U155, U76, and U100, compensating for the power loss of U197. Based on Figure 7, it can be concluded that up to four lines can be opened without compromising $N-1$ security. The BCS achieved, for this case, shows that when opening TLs 310–311, 106–110, 118–121, and 109–111, the overall production reduces from 141,520 USD/h to 131,830 USD/h, as a result of the greater dispatch of low-cost generators, and the system EENS of reliability relevant to the optimum OTS method changes from 2.1 × 10 ^{3} MWh/yr to 7.53 × 10 ^{3} MWh/yr, according to Figure 8. The dispatch cost-saving amount with OTS and DTR is 7.15%.

Table 6. The BCS outcomes for OTS methods with DTR (with wind influence 0% and 10%).

φ	0	1	2	3	4
Wind penetration (0%) Lines open	-	[109–111]	[210,211]; [109–111]	[118–121]; [109–111]; [210,211]	[310,311]; [106–110]; [118–121]; [109–111]
$\mu[Gen.cost]$ ($/h)	189,625	186,412	185,235	184,587	184,098
$\mu[EENS]$ (Mwh/y)	2.56 × 10^3	5.42 × 10^3	6.44 × 10^3	7.23 × 10^3	9.53 × 10^3
Wind penetration (10%) Lines open	-	[119–111]	[210,211]; [109–111]	[118–121]; [109–111]; [210,211]	[310,311]; [106–110]; [118–121]; [109–111]
$\mu[Gen.cost]$ ($/h)	164,710	161,240	160,215	159,510	159,205
$\mu[EENS]$ (Mwh/y)	2.31 × 10^3	4.91 × 10^3	5.85 × 10^3	6.54 × 10^3	8.57 × 10^3

Table 7. The BCS outcomes for OTS methods with DTR (with wind influence 15% and 20%).

φ	0	1	2	3	4
Wind penetration (15%) Lines open	-	[109–111]	[210,211]; [114–111]	[118–121]; [109–108]; [210–205]	[310–305]; [106–110]; [118–117]; [109–111]
$\mu[Gen.cost]$ ($/h)	153,101	150,05	149,520	148,490	147,850
$\mu[EENS]$ (Mwh/y)	2.41 × 10^3	5.01 × 10^3	6.12 × 10^3	6.77 × 10^3	8.93 × 10^3
Wind penetration (20%) Lines open	-	[119–114]	[210,211]; [109–103]	[118–121]; [109–108]; [210,211]	[311–314]; [106–110]; [218–222]; [109–104]
$\mu[Gen.cost]$ ($/h)	141,520	140,320	137,210	134,415	131,830
$\mu[EENS]$ (Mwh/y)	2.36 × 10^3	5.01 × 10^3	5.95 × 10^3	6.69 × 10^3	8.83 × 10^3

DTR provides benefits, in order to solve the grid limitations, particularly considering the growing production and consumption of renewable energy resources. When OTS and DTR are coordinated, system efficiency can be improved, in comparison with the SLR scenario. Table 8 shows the exact cost of dispatch and the reliability of the grid criteria for both SLR and DTR case studies, for wind farms with an installation capacity of 20%. As a conclusion, DTR has a lower total system cost compared to SLR. Furthermore, as opposed to the optimal approach, which is to open four lines, the execution of DTR allows the dispatching of the grid generator sets to be less expensive than SLR, where the production cost decreased 2.03% to 6.78%. Figure 9 shows the comparison between DTR and SLR. According to Figure 10, DTR-OTS has the lowest increase in EENS, 7.71% to 8.03%, in comparison with SLR.

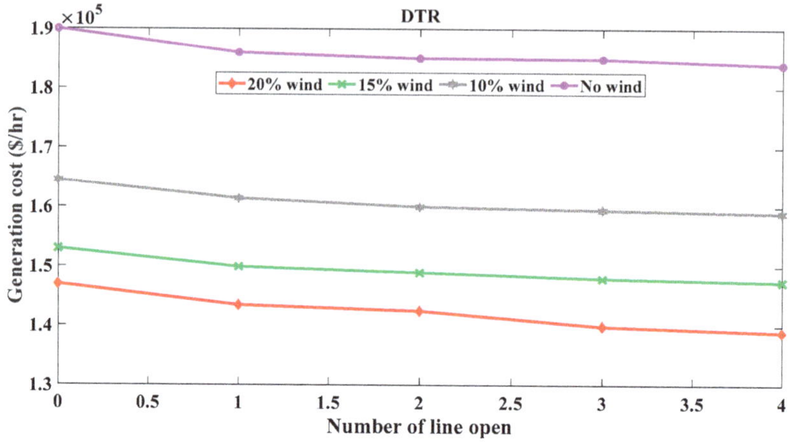

Figure 7. Expected producing cost with various DTR wind stages.

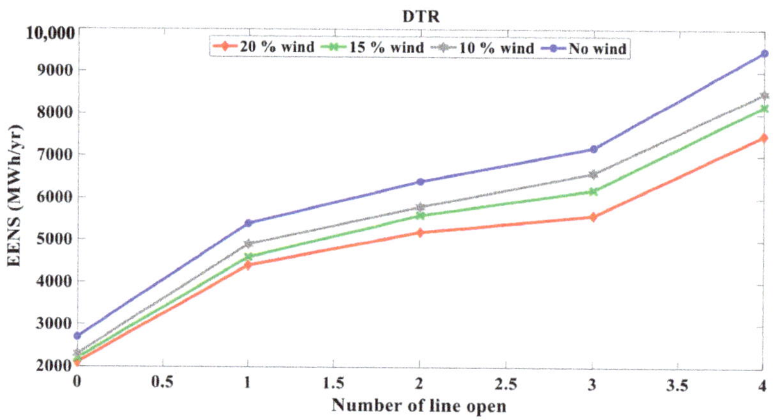

Figure 8. EENS for the OTS methods, for various DTR wind stages.

Table 8. Comparisons of BCS for OTS methods with DTR and SLR (with wind influence 20%).

φ	0	1	2	3	4
SLR Lines open	-	[109–111]	[210,211]; [114–111]	[118–121]; [109–108]; [210–205]	[310–305]; [106–110]; [118–117]; [109–111]
$\mu[Gen.cost]$ ($/h)	146,820	143,620	142,510	140,715	140,030
$\mu[EENS]$ (Mwh/y)	2.41×10^3	5.01×10^3	6.12×10^3	6.77×10^3	8.93×10^3
DTR Lines open	-	[109–111]	[210,211]; [109–111]	[118–121]; [109–111]; [210,211]	[310,311]; [106–110]; [118–121]; [109–111]
$\mu[Gen.cost]$ ($/h)	141,520	140,320	137,210	134,415	131,830
$\mu[EENS]$ (Mwh/y)	2.10×10^3	4.41×10^3	5.23×10^3	5.64×10^3	7.53×10^3

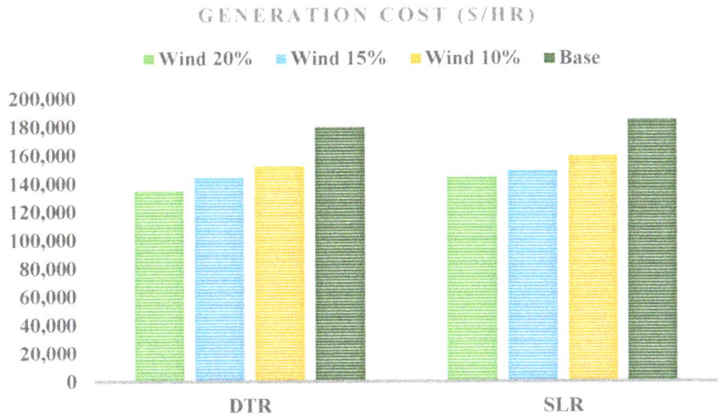

Figure 9. Comparing the production costs of DTR and SLR at various wind influence stages.

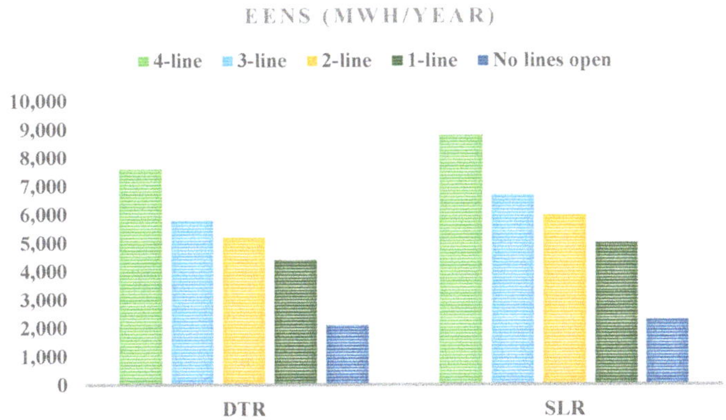

Figure 10. Comparing EENS for SLR and DTR for various OTS methods, with wind penetration of 20%.

7. Conclusions

This paper uses TS capacity as the primary source for flexibility. The technologies employed in this paper are DTR and OTS, which have both been anticipated to contribute to improved congestion mitigation and enhanced network utilization. As a result, the paper proposes a simple and efficient method to solve the CM problem, by applying the OTS approach, which considers the total cost of production as well as the system reliability index, as multiple objectives with the greatest influence rate on wind production. A modified IEEE RTS-96 system was used to evaluate the validity of the suggested method, and (MOMPSO + PEM) was used to solve it. Based on the simulation outcomes, applying the OTS approach improves the performance of the TS, by regarding the feasibility of altering its topology. Through integration of DTR, the system can achieve a number of benefits, including increased grid capacity. Due to simplified implementation procedures and a lack of major infrastructure or line rebuilding, the DTR method is remarkably appealing in systems that are, presently, suffering from a crescent status of overload. According to the outcomes, OTS could make better use of transmission capacity by using NTO, and DTR can improve reliability and resilience by improving grid operators' awareness of individual assets, thus enhancing flexibility.

Author Contributions: Data curation, W.H. and R.Y.M.L.; formal analysis, T.S.; investigation, W.H.; methodology, W.H. and T.S.; project administration, T.S.; software, R.Y.M.L.; supervision, W.H.; validation, R.Y.M.L. All authors have read and agreed to the published version of the manuscript.

Funding: This research received no external funding.

Conflicts of Interest: The authors declare no conflict of interest.

Abbreviation

ACOPF	AC optimal power flow
CM	Congestion management
EENS	Expected energy not supplied
FACTs	Flexible AC-TS
LD	Load demand
MPSO	Modified particle swarm optimization
MOMPSO	Multi-objective MPSO
MCS	Monte Carlo simulation
NTO	Network topology optimization
OTS	Optimal transmission switching
PEM	points estimation method
PPF	Probabilistic power flow
PDFs	probability distribution functions
RO	Robust optimization
SOTS	Stochastic optimal transmission switching
SLRs	Static line ratings
SLR	Static line ratings
TL	Transmission lines
TN	Transmission network
WT	Wind turbine

Appendix A

This section provides all the input data used for modeling the uncertainty effects due to the renewable sources and load demand. There are 190 WTs in each wind farm with the specifications as follows:

NEG Micon 1500/64 WT,

Scale parameter c = 8.549 m/s;
Shape parameter k = 1.98;
V_i = 5 m/s;
V_r = 15 m/s;
V_o = 25 m/s;
and P_r = 1.5 MW.

Regarding the PDF functions, all load buses are modeled with the normal density function of the mean value of the base value (active power/reactive power value) and a 7% standard deviation of the base value. For the wind turbine, it is modeled by the Weibull distribution function, with the scale parameter of the base value and the shape parameter of 5% of the base value. It is clear that any other appropriate PDF might be applied, based on the real dataset.

References

1. Alnowibet, K.; Annuk, A.; Dampage, U.; Mohamed, M.A. Effective Energy Management via False Data Detection Scheme for the Interconnected Smart Energy Hub–Microgrid System under Stochastic Framework. *Sustainability* **2021**, *13*, 11836. [CrossRef]
2. Mohamed, M.A.; Mirjalili, S.; Dampage, U.; Salmen, S.H.; Al Obaid, S.; Annuk, A. A Cost-Efficient-Based Cooperative Allocation of Mining Devices and Renewable Resources Enhancing Blockchain Architecture. *Sustainability* **2021**, *13*, 10382. [CrossRef]
3. Abhinav, R.; Pindoriya, N.M. Grid integration of wind turbine and battery energy storage system: Review and key challenges. In Proceedings of the 2016 IEEE 6th International Conference on Power Systems (ICPS), New Delhi, India, 4–6 March 2016; pp. 1–6.
4. Dabbaghjamanesh, M.; Wang, B.; Kavousi-Fard, A.; Hatziargyriou, N.D.; Zhang, J. Blockchain-Based Stochastic Energy Management of Interconnected Microgrids Considering Incentive Price. *IEEE Trans. Control Netw. Syst.* **2021**, *8*, 1201–1211. [CrossRef]
5. Aghajan-Eshkevari, S.; Sasan, A.; Morteza, N.; Mohammad, T.A.; Somayeh, A. "Charging and Discharging of Electric Vehicles in Power Systems: An Updated and Detailed Review of Methods", Control Structures, Objectives, and Optimization Methodologies. *Sustainability* **2022**, *14*, 2137. [CrossRef]

6. Shojaei, F.; Rastegar, M.; Dabbaghjamanesh, M. Simultaneous placement of tie-lines and distributed generations to optimize distribution system post-outage operations and minimize energy losses. *CSEE J. Power Energy Syst.* **2020**, *7*, 318–328.
7. Dabbaghjamanesh, M.; Kavousi-Fard, A.; Dong, Z.Y. A Novel Distributed Cloud-Fog Based Framework for Energy Management of Networked Microgrids. *IEEE Trans. Power Syst.* **2020**, *35*, 2847–2862. [CrossRef]
8. Putranto, L.M.; Irnawan, R.; Priyanto, A.; Isnandar, S.; Savitri, I. Transmission Expansion Planning for the Optimization of Renewable Energy Integration in the Sulawesi Electricity System. *Sustainability* **2021**, *13*, 10477. [CrossRef]
9. Reusser, C.A.; Pérez, J.R. Evaluation of the emission impact of cold-ironing power systems, using a bi-directional power flow control strategy. *Sustainability* **2020**, *13*, 334. [CrossRef]
10. Abdulwahid, A.H.; Wang, S. A Novel Method of Protection to Prevent Reverse Power Flow Based on Neuro-Fuzzy Networks for Smart Grid. *Sustainability* **2018**, *10*, 1059. [CrossRef]
11. Salimi, A.A.; Karimi, A.; Noorizadeh, Y. Simultaneous operation of wind and pumped storage hydropower plants in a linearized security-constrained unit commitment model for high wind energy penetration. *J. Energy Storage* **2019**, *22*, 318–330. [CrossRef]
12. Dehbozorgi, M.R.; Rastegar, M.; Dabbaghjamanesh, M. Decision tree-based classifiers for root-cause detection of equipment-related distribution power system outages. *IET Gener. Transm. Distrib.* **2020**, *14*, 5809–5815. [CrossRef]
13. Shepero, M.; van der Meer, D.; Munkhammar, J.; Widén, J. Residential probabilistic load forecasting: A method using Gaussian process designed for electric load data. *Appl. Energy* **2018**, *218*, 159–172. [CrossRef]
14. Teh, J.; Lai, C.-M. Reliability impacts of the dynamic thermal rating and battery energy storage systems on wind-integrated power networks. *Sustain. Energy Grids Netw.* **2019**, *20*, 100268. [CrossRef]
15. Douglass, D.A.; Gentle, J.; Nguyen, H.; Chisholm, W.; Xu, C.; Goodwin, T.; Chen, H.; Nuthalapati, S.; Hurst, N.; Grant, I.; et al. A review of dynamic thermal line rating methods with forecasting. *IEEE Trans. Power Deliv.* **2019**, *34*, 2100–2109. [CrossRef]
16. Zeng, L.; Xia, T.; Elsayed, S.; Ahmed, M.; Rezaei, M.; Jermsittiparsert, K.; Dampage, U.; Mohamed, M. A Novel Machine Learning-Based Framework for Optimal and Secure Operation of Static VAR Compensators in EAFs. *Sustainability* **2021**, *13*, 5777. [CrossRef]
17. Lai, C.M.; Teh, J.; Cheng, Y.H. Fuzzy evaluation of transmission line end-of-life reliability model. In Proceedings of the 2019 International Automatic Control Conference (CACS), Keelung, Taiwan, 13–16 November 2019; pp. 1–4.
18. Li, Y.; Hu, B.; Xie, K.; Wang, L.; Xiang, Y.; Xiao, R.; Kong, D. Day-Ahead Scheduling of Power System Incorporating Network Topology Optimization and Dynamic Thermal Rating. *IEEE Access* **2019**, *7*, 35287–35301. [CrossRef]
19. Mohamed, M.A.; Awwad, E.M.; El-Sherbeeny, A.M.; Nasr, E.A.; Ali, Z.M. Optimal scheduling of reconfigurable grids considering dynamic line rating constraint. *IET Gener. Transm. Distrib.* **2020**, *14*, 1862–1871. [CrossRef]
20. Kazemi, B.; Kavousi-Fard, A.; Dabbaghjamanesh, M.; Karimi, M. IoT-Enabled Operation of Multi Energy Hubs Considering Electric Vehicles and Demand Response. 2022. IEEE Transactions on Intelligent Transportation Systems. Available online: https://osuva.uwasa.fi/handle/10024/13474 (accessed on 24 April 2022).
21. Karimi, S.; Musilek, P.; Knight, A.M. Dynamic thermal rating of transmission lines: A review. *Renew. Sustain. Energy Rev.* **2018**, *91*, 600–612. [CrossRef]
22. Rizwan, M.; Waseem, M.; Liaqat, R.; Sajjad, I.A.; Dampage, U.; Salmen, S.H.; Al Obaid, S.; Mohamed, M.A.; Annuk, A. SPSO Based Optimal Integration of DGs in Local Distribution Systems under Extreme Load Growth for Smart Cities. *Electronics* **2021**, *10*, 2542. [CrossRef]
23. Luo, L.; Abdulkareem, S.S.; Rezvani, A.; Miveh, M.R.; Samad, S.; Aljojo, N.; Pazhoohesh, M. Optimal scheduling of a renewable based microgrid considering photovoltaic system and battery energy storage under uncertainty. *J. Energy Storage* **2020**, *28*, 101306. [CrossRef]
24. Razmjouei, P.; Kavousi-Fard, A.; Dabbaghjamanesh, M.; Jin, T.; Su, W. Ultra-Lightweight Mutual Authentication in the Vehicle Based on Smart Contract Blockchain: Case of MITM Attack. *IEEE Sens. J.* **2020**, *21*, 15839–15848. [CrossRef]
25. Zhang, L.; Shaby, B. Uniqueness and global optimality of the maximum likelihood estimator for the generalized extreme value distribution. *arXiv* **2020**, arXiv:2008.06400. [CrossRef]
26. Tajalli, S.Z.; Mardaneh, M.; Taherian-Fard, E.; Izadian, A.; Kavousi-Fard, A.; Dabbaghjamanesh, M.; Niknam, T. DoS-resilient distributed optimal scheduling in a fog supporting IIoT-based smart microgrid. *IEEE Trans. Ind. Appl.* **2020**, *56*, 2968–2977. [CrossRef]
27. Ashkaboosi, M.; Ashkaboosi, F.; Nourani, S.M. *The Interaction of Cybernetics and Contemporary Economic Graphic Art as. "Interactive Graphics"*; University Library of Munich: Munich, Germany, 2016.
28. Dabbaghjamanesh, M.; Zhang, J. Deep learning-based real-time switching of reconfigurable microgrids. In Proceedings of the 2020 IEEE Power & Energy Society Innovative Smart Grid Technologies Conference (ISGT), Washington, DC, USA, 17–20 February 2020; pp. 1–5.
29. Ghaffari, S.; Ashkaboosi, M. Applying Hidden Markov Model Baby Cry Signal Recognition Based on Cybernetic Theory. *IJEIR* **2016**, *5*, 243–247.
30. Dabbaghjamanesh, M.; Wang, B.; Kavousi-Fard, A.; Mehraeen, S.; Hatziargyriou, N.D.; Trakas, D.N.; Ferdowsi, F. A Novel Two-Stage Multi-Layer Constrained Spectral Clustering Strategy for Intentional Islanding of Power Grids. *IEEE Trans. Power Deliv.* **2019**, *35*, 560–570. [CrossRef]

sustainability MDPI

Article

EnergyAuction: IoT-Blockchain Architecture for Local Peer-to-Peer Energy Trading in a Microgrid

Felipe Condon [1], Patricia Franco [1], José M. Martínez [1], Ali M. Eltamaly [2,3], Young-Chon Kim [4,*] and Mohamed A. Ahmed [1,*]

1 Department of Electronic Engineering, Universidad Técnica Federico Santa María, Valparaíso 2390123, Chile; felipe.condon.13@sansano.usm.cl (F.C.); patricia.franco@usm.cl (P.F.); jose.martinez@usm.cl (J.M.M.)
2 Sustainable Energy Technologies Center, King Saud University, Riyadh 11421, Saudi Arabia; eltamaly@ksu.edu.sa
3 Electrical Engineering Department, Faculty of Engineering, Mansoura University, Mansoura 35516, Egypt
4 Department of Computer Engineering, Jeonbuk National University, Jeonju 561-756, Republic of Korea
* Correspondence: yckim@jbnu.ac.kr (Y.-C.K.); mohamed.abdelhamid@usm.cl (M.A.A.)

Abstract: The widespread adoption of distributed energy resources (DERs) and the progress made in internet of things (IoT) and cloud computing technologies have enabled and facilitated the development of various smart grid applications and services. This study aims to develop and implement a peer-to-peer (P2P) energy trading platform that allows local energy trading between consumers and prosumers within a microgrid which combines IoT and blockchain technologies. The proposed platform comprises an IoT-cloud home energy management system (HEMS) responsible for gathering and storing energy consumption data and incorporates a blockchain framework that ensures secure and transparent energy trading. The proposed IoT–blockchain architecture utilizes a Chainlink oracle network and a private Ethereum blockchain. Through the use of smart contracts, consumers and prosumers can participate in an open auction to trade energy, while the settlement process involves acquiring external energy data from an API through the oracle network. The performance of the platform is evaluated through a testbed scenario using real-world energy data from a real house in Valparaiso, Chile, while storing those measurements in AWS cloud, validating the feasibility of the proposed architecture in enabling local energy trading. This work contributes to the development of energy management systems by providing a real-world implementation of an IoT–blockchain architecture for local energy trading. The integration of these technologies will allow for a more efficient and secure energy trading system that can benefit prosumers, consumers, and utilities.

Keywords: blockchain; local energy trading; microgrids; oracle networks; peer-to-peer

check for updates

Citation: Condon, F.; Franco, P.; Martinez, J.M.; Eltamaly, A.M.; Kim, Y.-C.; Ahmed, M.A. EnergyAuction: IoT-Blockchain Architecture for Local Peer-to-Peer Energy Trading in a Microgrid. *Sustainability* **2023**, *15*, 13203. https://doi.org/10.3390/su151713203

Academic Editor: Yang (Jack) Lu

Received: 10 July 2023
Revised: 23 August 2023
Accepted: 30 August 2023
Published: 2 September 2023

1. Introduction

Decades ago, traditional power grids were designed with the primary purpose of delivering electricity from large centralized power plants to individual residences. However, significant transformations have occurred in the electricity sector due to the integration of various technologies, including distributed energy resources (DERs), electric vehicles (EVs), advanced metering infrastructures (AMIs), and home energy management systems (HEMSs) [1]. DERs have enabled many end-users to transition from being energy consumers to becoming energy prosumers. This transformation allows prosumers to produce and consume energy simultaneously. In the near future, energy prosumers and consumers will be able to connect with one another and trade energy locally through a marketplace during instances of mismatch between supply and demand [2].

During the period when prosumers have surplus electricity, they can store it using energy storage systems (ESS), export it back to the power grid, or sell it to other customers. The direct energy trading among prosumers and consumers is called peer-to-peer (P2P) energy trading [3]. The implementation of this novel P2P energy trading structure allows the local sharing of surplus energy from small-scale DERs within a microgrid, bringing advantages to prosumers, consumers, and the distribution system operator (DSO). P2P energy trading presents several benefits to all parties involved, as it facilitates local energy balance and improves the overall energy exchange process [4].

Internet of things (IoT) technologies are widely used to enable different smart grid applications. IoT technologies can be used for monitoring and controlling by using various sensors, actuators, and metering devices [5–8]. Such integrations can be found in HEMSs, which facilitates users' operation and management of household appliances and DER. To enable a microgrid application, such as trading within a microgrid, it is necessary to collect the energy consumption and/or generation data of houses. The use of these data serves as a key element to enable the implementation of various applications [9,10].

Blockchain technology has been gaining attention in non-cryptocurrency-related research, such as the integration of heterogeneous IoT systems with the energy market, with features such as data privacy, cybersecurity, and reliability [11]. In general, four layers compose the P2P energy trading system: the power grid layer, information and communication technology (ICT) layer, control layer, and business layer [12]. There have been several projects that have been carried out as energy trading pilots. Among these are the following projects: Piclo in the United Kingdom (see https://www.piclo.energy/ accessed on 2 November 2022), Transactive Grid and Brooklyn Microgrid in the USA [13], Peer Energy Cloud in Germany, and Powerledger in Australia [14,15]. The combination of IoT technology and blockchain technology has been covered in [16], where digitalization and decentralization are stated as the key enablers for a transactive energy internet (EI). A schematic diagram is presented in Figure 1 to provide a visual representation of the interaction between microgrids, distributed peer-to-peer networks, and cloud-based services. This diagram illustrates the main components of this research work, including the relationships and interactions between different entities. The microgrid is defined as a modern and innovative energy system that involves various parties, including electric vehicles (EVs), photovoltaic systems (PVs), energy storage systems (ESSs), prosumers, consumers, and utilities, which interact by exchanging both energy and data. In our previous work, Ref. [17], we presented the basic concept for the development of "EnergyAuction", a P2P energy trading platform that focuses on a proof of concept for the integration of oracles into smart contracts.

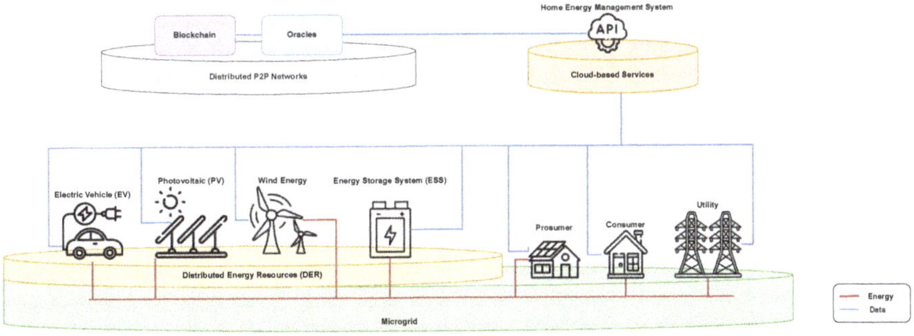

Figure 1. Schematic diagram of microgrid with distributed P2P networks and cloud-based services.

This work proposes the design and implementation of a P2P energy trading platform in a microgrid combining two key components to enable secure, transparent, and decentralized energy trading among participants: a HEMS and blockchain technologies. The HEMS facilitates real-time monitoring and control of energy consumption and generation and supports two different implementations based on edge devices and cloud-based services, respectively. The blockchain platform uses a private Ethereum network and Chainlink oracle network to facilitate energy trading between participants, storing energy transactions through a distributed ledger. Cloud-based services, such as the application programming interface (API), provide access to valuable energy consumption and generation data. To retrieve data from cloud-based services, oracle networks serve as a gateway, such as using available APIs, and route data to blockchain-based services.

Although a lot of research work has been conducted for peer-to-peer energy trading, there is no available information for the detailed technical implementation of such systems nor detailed aspects of such IoT–blockchain integration. The main contributions of this paper are given below:

- A thorough analysis of blockchain network design and market types/attributes for local peer-to-peer energy trading.
- Proposal of IoT–blockchain architecture for local P2P energy trading in a microgrid which consists of two layers: physical (IoT layer) and virtual (off-chain layer and blockchain layer).
- Development, deploying, and testing of four smart contracts for local peer-to-peer energy trading: data access contract, energy trading smart contract, Link Token contract, and oracle client contract.
- A testbed implementation (hardware/software) is carried out in a laboratory environment for one prosumer, one consumer, and a microgrid control center, where the off-chain layer of the HEMS is implemented on AWS platform and the blockchain layer is implemented using an AIO computer.

The remainder of this paper is organized as follows: Section 2 presents a review of the most relevant work related to blockchain technology, smart contracts, oracle networks, and the implementation of P2P energy trading. Section 3 describes necessary concepts for the design of P2P energy trading platforms, while Section 4 details the main components of the proposed architecture. Section 5 explains the process of setting up a market for P2P energy trading in a microgrid, including the definition of market type, archetypal market design, price formation mechanism, market participants, market value proposition, market scale, and market operation. Section 6 shows the implementation of the proposed platform, and in Section 7, the obtained results are analyzed. In addition, Section 8 provides the conclusions and future work.

2. Related Work

The energy internet presents a virtuous relationship among various entities, such as energy producers and end consumers. When these entities trade with one another, they are defined as a local energy market (LEM), which aims to support energy exchange between them in a competitive market and balance supply and demand locally [18]. The two key enablers for energy trading in a LEM are digitalization and decentralization [16]. Both decentralization and energy markets are presented below.

2.1. Decentralization: Blockchain Technology

The accelerated growth of renewable energy resources and the growing adoption of distributed power generation are transforming the energy sector. Peer-to-peer (P2P) energy trading has surfaced as an innovative solution, empowering prosumers and consumers to engage in energy trading in a more adaptable, efficient, and decentralized manner [19,20]. Blockchain technology, characterized by decentralization, transparency, and immutability, offers an ideal foundation for building P2P energy trading platforms [21].

Blockchain technology facilitates a secure and reliable environment for executing business logic and allowing participants in energy trading to interact. It creates transparent and tamper-proof records of energy transactions, ensuring that both producers and consumers can trust the system without relying on intermediaries [22]. This decentralized trust model, along with the ability to automate processes through smart contracts, has the potential to revolutionize energy trading and distribution [23]. Blockchain is also being introduced in the industrial internet of things (IIoT). For example, in [24], the authors used blockchain technology to enhance the security and privacy of a multi-camera multi-tracking using edge cameras. The authors in [25] presented an algorithm that improves the scalability and reduces latency. This improves the performance of blockchain for IIoT solutions. Numerous studies and projects have emphasized the potential of blockchain technology in facilitating P2P energy trading. For instance, the authors in [26] proposed a decentralized framework for coordinating energy supply and demand in a local market using blockchain and smart contracts. The work demonstrated that blockchain technology is a natural facilitator for implementing business logic and bringing energy trading participants together for interaction. As the energy sector continues to progress, the adoption of blockchain-based solutions is expected to rise, laying the foundation for a more decentralized and efficient energy landscape. General elements in the blockchain-enabled system consider network selection and configuration, consensus mechanism, privacy and network access, smart contracts, and oracle networks. These attributes are shown in Figure 2 and discussed in Section 2.2.

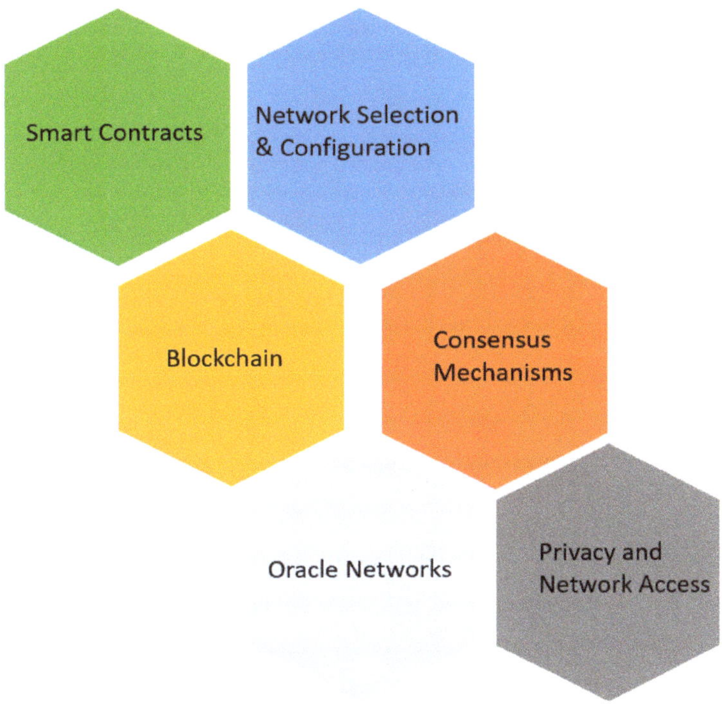

Figure 2. The blockchain system and its enabling elements.

2.2. *Blockchain Network Design*

2.2.1. Network Selection

Selecting the right blockchain network is essential for the success of an energy trading platform. There are several networks available, each with its unique features and characteristics. Ethereum (ETH), Hyperledger Fabric, Avalanche, and Cardano are some of the prominent options. The choice should be based on criteria such as the developer community, ecosystem maturity, and available tools. Ethereum is a popular choice due to its large developer base, robust ecosystem, and the backing of the Enterprise Ethereum Alliance. However, it is crucial to consider the specific requirements of the energy trading platform when selecting a blockchain network.

2.2.2. Consensus Mechanisms

A crucial aspect of the configuration of a blockchain network is choosing the appropriate consensus mechanism which ensures the reliability and trustworthiness of the system.

Popular consensus mechanisms include Proof of Stake (PoS), Proof of Work (PoW), Practical Byzantine Fault Tolerance (PBFT), and Delegated Proof of Stake (DPoS) [27]. The choice of consensus mechanism should be based on factors such as security, energy efficiency, and transaction throughput.

2.2.3. Privacy and Network Access

Several privacy and network access configurations can be considered based on the requirements of the energy trading platform. As an example, a public blockchain allows any participant to interact with the platform and join the network, whereas a permissioned or private blockchain restricts access to a selected group of users [27]. Privacy-preserving techniques, such as zero-knowledge proofs or secure multiparty computation, can be implemented to protect sensitive user data [28]. The selection of privacy and network access configuration should be aligned with the platform's objectives and the desired level of decentralization and security.

2.3. *Smart Contracts*

Smart contracts are a critical component of blockchain-based energy trading platforms, as they allow the automation of transactions and agreements between parties [29]. These self-executing contracts, coded with specific rules and conditions, eliminate the need for intermediaries and enhance the efficiency and transparency of the trading process. Designing and implementing robust smart contracts is essential to ensure the reliability, security, and performance of the platform. Several research papers and projects have demonstrated the use of smart contracts for energy trading. For example, the authors in [13] presented a study of a local energy market based on the Ethereum blockchain, where smart contracts were used to manage energy transactions between prosumers and consumers. The authors in [22] proposed a consortium blockchain-based energy trading framework that leverages smart contracts for trading and settling energy transactions between microgrids. This framework aimed to improve the efficiency and security of energy trading while reducing transaction costs. These examples highlighted the versatility of smart contracts in facilitating energy trading and their potential to transform the energy sector.

2.4. *oracle Networks*

Significant advancements have been made in the domain of blockchain-based networks. However, one notable limitation of smart contracts is their ability to interact with real-world data (also known as off-chain data). To overcome this challenge, the integration of blockchain oracles has become crucial. By incorporating blockchain oracles, the issue of securely and reliably incorporating external data sources into smart contracts can be effectively addressed. The authors in [30] conducted a survey aimed to identify researchers and institutions involved in blockchain oracle research. The survey findings revealed that a total of 162 papers were identified from different databases during the last seven years. This indicates that, despite the significance of the topic, there is still a need for greater efforts to establish a widely accepted oracle taxonomy. The results emphasize the importance of further research and collaboration in order to enhance the understanding and promote standardization in the field of blockchain oracles. In [31], different types of oracles were introduced to facilitate the selection process for specific blockchain-based applications. The authors in [32] introduced a decentralized energy flexibility market based on blockchain, which leverages oracles to fetch results from an off-chain computation process for determining flexibility matching. In this implementation, oracles were utilized to address the limited computational capabilities of smart contracts. In [33], the authors presented a decentralized access control system for IoT data that incorporates blockchain technology and trusted oracles. oracles were employed as intermediaries between entities responsible for hosting the blockchain network and IoT data.

2.5. Pilots and Implementations

Various pilots and testing platforms have been conducted to evaluate the potential of blockchain technology in energy trading. Table 1 summarizes the literature review of energy trading-related surveys, implementations, and testbeds. One notable example is the Brooklyn Microgrid project, which demonstrated the feasibility of a blockchain-based P2P energy trading platform within a microgrid [13]. Other examples include the Energy Web Foundation's EW-DOS platform, and the Power Ledger's trials in Australia [34]. These initiatives presented the potential benefits of blockchain technology in enhancing efficiency, transparency, and decentralization in energy trading while also highlighting the challenges that must be addressed for widespread adoption. In [21], the authors provided a comprehensive overview of blockchain's involvement in the energy sector, discussing its potential and limitations. The authors examined various scenarios where blockchain has been applied to energy trading, including mitigating the impact of electric vehicle charging during peak periods. The authors in [35], presented an energy trading implementation using IOTA, a "Tangle"-based framework, which is considered a third-generation DLT that uses a directed acyclic graph (DAG). The authors provided guidelines for deploying end-to-end transactions for the energy section on a dynamic set of scenarios.

This work differentiates itself from other related works by providing an implementation of a P2P energy trading platform using IoT and blockchain. The proposed implementation is validated in a testbed using real data consumption. Our work provides insight on how to implement and carry out a testbed scenario using the given architecture. Furthermore, a functional and cost analysis was also performed.

Table 1. Literature review for blockchain in smart grids. "✓" represents "included", a cross mark "×" represents "not included" in the reviewed articles.

Reference	Survey	Implementation	Testbed	Energy Trading
[13]	×	✓	×	✓
[21]	✓	×	×	✓
[34]	✓	✓	×	✓
[28]	✓	✓	×	✓
[35]	×	✓	✓	✓
[36]	✓	✓	×	✓
This work	×	✓	✓	✓

3. Local P2P Energy Trading

In order to design and implement an energy trading platform, it is necessary to define the design of the energy trading market. The authors in [37] conducted a systematic literature review of three local energy market (LEM) models: P2P markets, collective self-consumption (CSC) or community, and transactive energy (TE) markets. The work provided detailed information on the market design and aspects of LEM transactions. These models were compared in terms of six categories representing attributes found in surveyed literature, archetypal market design, price formation mechanism, market participants, market value proposition, market scale, and market operation. The relationship between the three energy market models and the aforementioned attributes is shown in Figure 3.

Figure 3. Market types and attributes for energy trading.

3.1. Market Type

Three types of LEMs have emerged [20,37–39]:

- *P2P Markets*: Facilitate trading between users without the need for an intermediary. The primary objective of this market is to encourage active participation by energy users, providing them with an incentive to participate in the energy markets.
- *Collective Source Connected (CSC) or Collaborative*: Users tend to be co-located energy prosumers that trade their surplus energy in a market arrangement.
- *Transactive Energy Markets*: Maintain supply–demand equilibrium through autonomous management of DER using price signals, which ensures system stability. These markets offer services to the electricity system by optimizing the use of resources.

3.2. Archetypal Market Design

Six archetypal market designs have been identified [37,40–42]:

- *Futures market* is the most common market design, in which all the trading happens before the settlement period. When the settlement period arrives, the participants stick to their traded position. Any variation from the trading position is dealt with by the settlement.
- *Real time market* is where all the trading occurs at the settlement period, where participants contract with other participants to buy or sell the energy they are in the process of consuming or producing. This approach usually considers the usage of forecasted consumption or generation curves. One consideration found in this market design is that it is assumed that they are connected to larger traditional energy systems, which are able to provide or absorb an infinite excess supply or demand.
- *Mixed decentralized/centralized market* is a two-phase market design where initial decentralized trading is carried between participants looking to clear the market as far as possible, without intervention from a market operator. The second phase is performed by a centralized entity, which clears the remainder of the market.

- *Mixed futures/real time market* consists of two phases: An initial phase before the settlement, where participants trade based on predicted supply and demand for energy. During the settlement period, participants can adjust their position based on actual supply and demand curves in a real-time energy market.
- *Multi-layer market* starts from a local or bottom level for clearing the market. An aggregator is used for passing the uncleared supply or demand for energy to the next market level.
- *Settlement after the fact* is an uncommon scheme where no trading was carried out before the end of the settlement. Here, participants are charged or paid for the energy bought or sold after the settlement. A fixed price for buying or selling energy is defined by the system. Participants can purchase or sell the amount of energy they require.

3.3. Price Formation Mechanism

Price formation refers to the mechanism by which prices are discovered. There are five main categories [37,43]:

- *Single auction*: Only one party of the market is able to submit bids.
- *Double auction*: Allows sellers and buyers to submit bids.
- *System-determined mechanisms*: Usually defined by the system operator. Some of these mechanisms include uniform or fixed prices or cooperative game theory.
- *Negotiation-based mechanisms*: Automated transactions considered usually through the use of artificial intelligence (AI). Negotiating autonomous agents is one kind of negotiation-based mechanism.
- *Equilibrium-based mechanisms*: Use game theory to establish market equilibrium, which utilizes the bids and offers from the participants.

3.4. Market Participants

There are different market participants such as prosumers, consumers, and grid operators. Based on [2,37], the following are the main market participants:

- *Prosumers* are participants that generate and consume energy at the same time using small-scale distributed energy resources (DERs) such as wind turbines and solar panels.
- *Pure consumer* are participants that only consume energy without any DERs.
- *Pure generator* are participants that only generate energy.
- *Retailer* is an entity that connects to other large markets.
- *Market operator* is a single agent that runs the market operation.
- *Grid operator* is an entity that operates the electricity network and interacts with the market.
- *Aggregator* is an actor that represents a small group of participants.

3.5. Market Value Proposition

Market commodity and the benefits to participants are described in [37,44] as follows:

- *Market commodity* corresponds to the traded commodity in the market. Electrical energy was the most found commodity traded. Other commodities include flexibility alongside the sale of energy or a flexibility-only market. Other forms of energy are also found, such as combined heat and power, combined power and gas, or combined heat, power, and gas.
- *Benefits to market participants* are mainly financial, like increasing the profit from the sale of energy. Other objectives can also be considered, such as increasing the total welfare, which implies multiple financial benefits to all market participants, reducing the grid imbalance, or reducing the electricity cost.

3.6. Market Scale

Market scale considers the number of nodes involved in sizing the market [2,37]. For the participant scale, four different scales of participants are found:

- *Small scale*: Considers residential or individual energy users.
- *Building scale*: Multiple buildings trade with each other.
- *Microgrid/community scale*: Energy users operate as a community, rather than as individual participants.
- *Grid scale*: The trading is at a distribution or transmission network level.

3.7. Market Operation

Market operation refers to the type of data that is shared by the participants with the system, as well as the user preferences. Also, the settlement period and gate closures are discussed [20].

- *Data shared*: Price, volume, price and volume, demand and supply curve, controllable loads, the flexibility available, battery state of charge (SoC), distribution line distance, and discomfort level are some of the data parameters that can be shared for the energy trading.
- *Settlement period and gate closure*: The settlement period is a period of time in which the supply and demand of market participants are balanced. Gate closure is the period of time before the settlement, where bidding and offering can take place.

4. How Does Blockchain Enable Energy Trading?

Modern blockchain networks implement smart contracts which allow automatizing transactions [29,45]. Programming languages can be used to write scripts that implement energy trading business logic. This added to the benefits of an immutable, trustworthy, and decentralized network that provides the foundations to support an energy trading platform. As shown in Figure 4, a simplified diagram illustrates the interaction between the participants including prosumers, consumers, IoT devices, and the blockchain network. Data collected from the cloud-IoT HEMS is used to improve the implemented business logic in smart contracts. Information such as the available energy from an ESS, or the last hourly consumption from a householder, is used to enhance validations in the smart contract, resulting in viable trading scenarios. One challenge for using these data is the restriction for blockchain networks to fetch data from outside the blockchain. oracle networks work as a middleware between the blockchain networks and the data provider, as shown in Figure 5. oracle's primary job is to collect and provide data feeds to smart contracts [46].

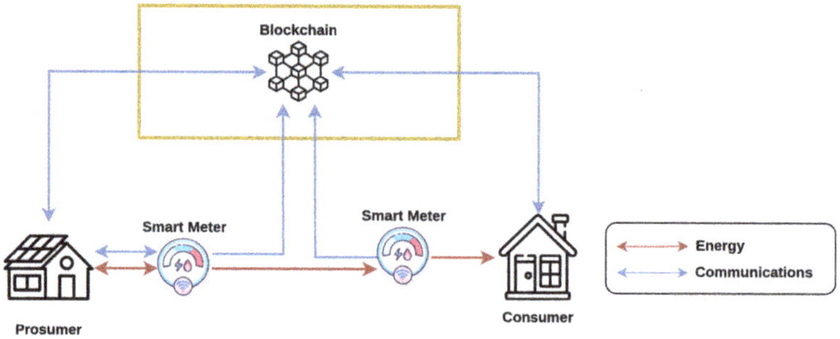

Figure 4. Simplified diagram for interactions among prosumers, consumers, IoT devices, and the blockchain network.

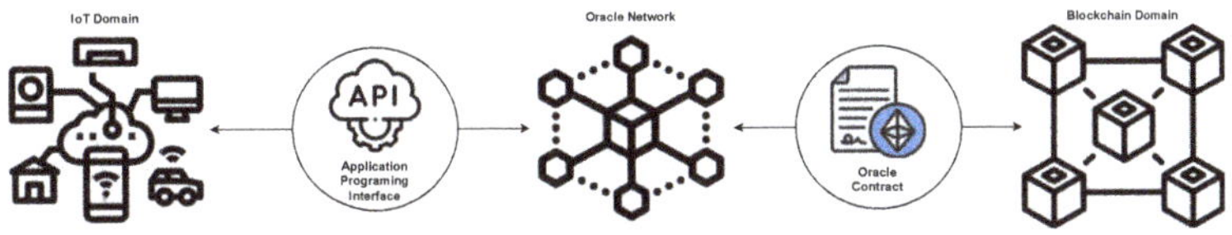

Figure 5. IoT and blockchain interaction using oracles.

The proposed IoT–blockchain architecture for P2P energy trading in a microgrid is shown in Figure 6. The system is composed of multiple entities that are divided into two layers: the physical layer and the virtual layer. The main components of the system are the microgrid, cloud-based services, and distributed P2P networks. The microgrid consists of prosumers, consumers, battery energy storage systems, and photovoltaic systems, which exchange energy and data. Data related to energy consumption and generation are available through cloud-based services via APIs. Transactions of energy are recorded on a distributed ledger, such as a blockchain, through distributed P2P networks. oracle networks, another form of distributed P2P network, act as a connection between cloud-based services and blockchain-based services by retrieving data from APIs and routing them to the blockchain. The data that are routed back to the blockchain are then utilized by smart contracts.

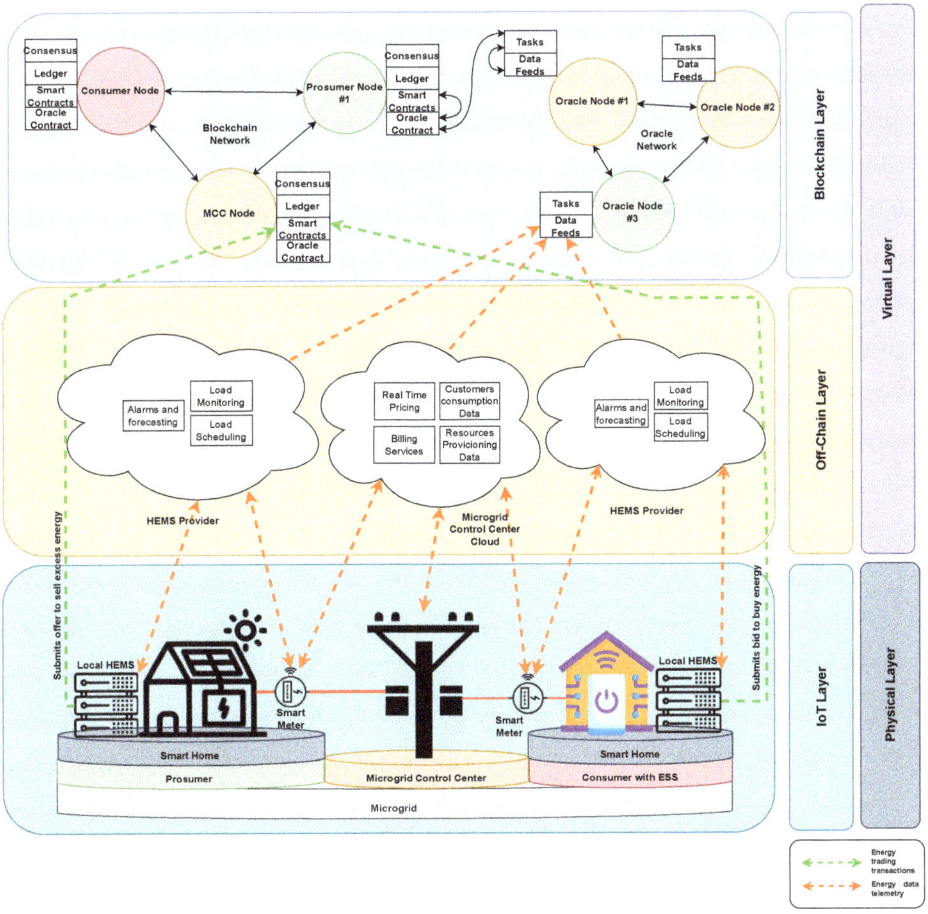

Figure 6. IoT–blockchain architecture for P2P energy trading in a microgrid.

4.1. Physical Layer: IoT Layer

The physical layer considers the physical infrastructure that a microgrid requires to enable the energy trading feature. The microgrid is composed of three main actors: the prosumer, the consumer, and the microgrid control center (MCC). A prosumer is a consumer who has the capability of generating energy through photovoltaic systems or wind turbines. The energy stored in a battery energy storage system can be used as a power supply to satisfy its own consumption or to sell energy to the rest of the microgrid. Consumers consume the energy provided by the MCC, but they can also employ a BESS to store energy bought from the grid. The MCC manages the energy present in the microgrid. It also provides the power infrastructure for energy flow. This energy is bought or sold by households of the microgrid. Energy can be from decentralized energy resources or from the utility service provider. This will define the operation mode of the microgrid, by island mode or grid-connected mode, respectively.

The physical layer generates plenty of energy-related data, which are required to be stored and/or transmitted to the cloud layer in order to enable several smart grid applications. This challenge is addressed by IoT devices, which enable the collection of the intended energy measurements while also communicating the data to cloud-based services on the internet. The devices in the context of microgrids are smart meters. Smart meters are digital hardware devices that allow the reading of energy consumption from households on a regular basis. Other IoT-related devices are HEMSs, which aggregate energy-related data such as state of charge of BESS, appliance loads consumption, or the main household consumption.

4.2. Virtual Layer: Off-Chain Layer

The virtual layer of the IoT–blockchain architecture for P2P energy trading in a microgrid encompasses both energy data and energy trading transactions generated by the physical layer's IoT devices. This virtual layer is comprised of two sub-layers: the off-chain layer and the blockchain layer. The off-chain layer functions as a mediator between the real-world data captured by the IoT devices and the blockchain layer. The data are stored in cloud-based systems belonging to the HEMS provider and used for various purposes, such as load scheduling, load monitoring, alarm triggers, and forecasting. These features are available through cloud-based services offered by various HEMS providers. In addition, the MCC acts as a microgrid energy manager, collecting energy data from smart meters. This function enables the MCC to manage different processes and services on its own cloud servers. Both HEMSs and MCC cloud services may be self-hosted on-premise or through public cloud platforms such as Amazon Web Services (AWS), Google Cloud Platform (GCP), or Microsoft Azure. As these services manage energy storage data that are not stored on the blockchain, they are considered part of the off-chain layer.

4.3. Virtual Layer: Blockchain Layer

The blockchain layer plays a critical role in enabling secure and transparent energy trading in the microgrid. It provides a decentralized and trusted platform for energy transactions. The use of distributed ledger technology (DLT) ensures that all energy trading transactions are recorded in a tamper-proof and transparent manner. Furthermore, it enables all participants to have access to the same information and reduces the risk of fraudulent or malicious activities. The use of consensus mechanisms, such PoA, PoW, or PoS, enables the platform to validate and confirm energy trading transactions in a decentralized manner. This ensures that all parties involved in the energy trading platform can trust the outcome of the transaction and reduces the risk of fraud or errors.

5. Market Setup

This section provides a description of the market setup process for P2P energy trading in the microgrid. The following are the definitions of main elements, including the P2P market type, market design, price mechanism, types of market participants, market value proposition, market scale, and market operation. In order to establish a functional and efficient market, these key elements are crucial for P2P energy trading.

- **Market Type:** It is set up as a peer-to-peer, meaning that the participants are able to exchange energy directly among each other without intermediaries. This allows for greater autonomy and more flexibility in energy trading.
- **Market Design:** It is selected as a future market design, which means that the participants trade for future delivery of energy. Such a design allows for greater predictability and flexibility in energy trading.
- **Price Formation Mechanism:** The market price is determined by the intersection of the energy offered by the prosumer and the energy demanded by the consumer. This mechanism ensures efficiency and fairness in the energy trading process.
- **Market Participants:** The participants in the market are defined as prosumers, pure consumers, and the microgrid control center (MCC). The MCC is responsible for the overall operation of the microgrid and acts as a combination of both the market operator and the grid operator. The decentralization of the system through the use of blockchain technology ensures more efficient management of the microgrid.
- **Market Value Proposition:** The commodity traded in the market is energy, and the main benefit of the model is to put prosumers at the forefront of energy production and supply to consumers. This allows for a more sustainable and decentralized energy system.
- **Market Scale:** The market is defined as small-scale, with participants confined to a single microgrid. This allows for more efficient and localized energy trading.
- **Market Operation:** The market operation involves two categories: data related, including state of charge (SoC) data from the battery energy storage system (BESS), energy consumption and generation data from IoT devices, and trade-related, including price and quantity. The settlement and gate closure procedures are defined in detail in a subsequent figure, promoting a transparent and fair energy trading system.

Settlement and gate closure policy is depicted in Figure 7. The energy trading window and the energy trading sequence diagram are defined as part of the market setup. The trading window is composed of four sequential phases which make up the trading process:

1. **Bidding and Offering Period**: This is a 10 min phase in which consumers and prosumers can submit bidding and offer requests, respectively.
2. **Matching Period**: This is a 2 min phase in which the bids and offers are processed to make a match between them.
3. **Energy Transfer Period**: This is a 60 min phase during which energy is transferred from prosumers to consumers.
4. **Settlement Period**: This is a 3 min phase in which the settlement of funds takes place. The settlement period verifies how much consumers should pay to prosumers based on the actual energy received.

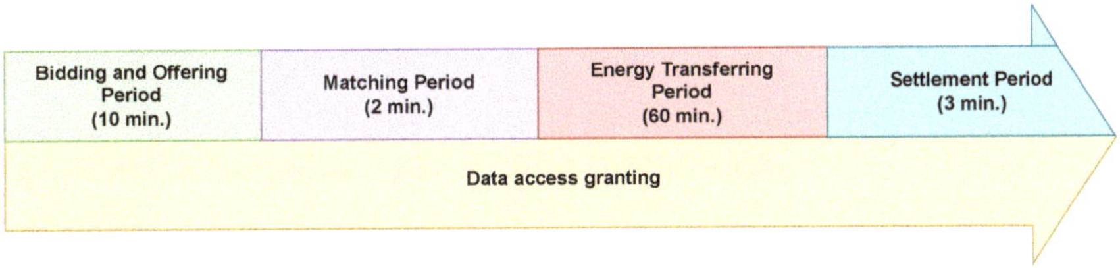

Figure 7. Trading window.

The four phases are modeled in the trading smart contract as valid states that a trading window can take. The transition between the states is executed autonomously by the microgrid control center using the smart contract when the expiration time elapses. Regarding the market operation, Figure 8 presents the interaction in which participants manage to bid and offer energy. Before any trading operation is executed by the households, every householder has to grant permission to the trading smart contract to request energy data from its data sources, which is depicted as the data access granting phase from the trading window.

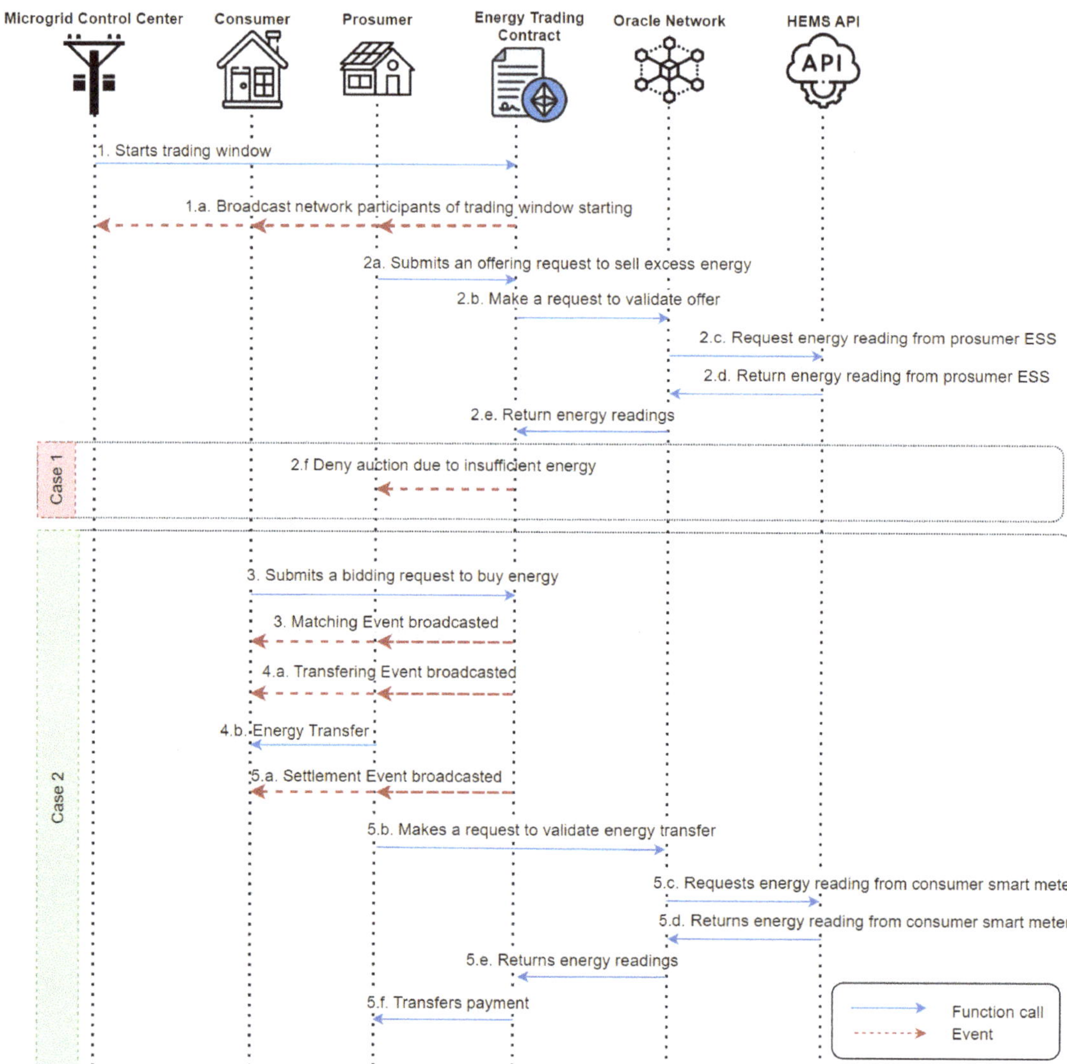

Figure 8. Test case sequence diagram for energy trading in a microgrid.

1. The MCC starts a new energy trading window, which begins with the bidding and offering phase. This is broadcasted as an event in the blockchain network.
2. The prosumer submits an offer to sell excess energy. The energy trading smart contract validates the offer request by obtaining energy readings from the HEMS through an oracle network. If the prosumer is unable to provide the offered energy, the smart contract emits an event to notify the prosumer that their offer will not be considered, and this is depicted in "Case 1". This ensures fairness and transparency by preventing

false offers. If the prosumer has enough energy, the smart contract records the energy offer as valid, and this is depicted in "Case 2".

3. A consumer submits a bid to buy energy, which is recorded in the record of bids. After 10 min, the smart contract matches available energy offerings with the record of bids, providing a match between prosumers and consumers for energy transfer.
4. In the energy-transferring phase, which lasts for 60 min, the prosumer becomes the energy provider of the consumer who charges its ESS with the energy provided. A maximum of 10 kWh is set for this phase due to technological restrictions of a common residential system to transfer more than 10 kW over an hour to the battery.
5. In the settlement phase, the energy trading smart contract obtains energy readings from the consumer smart meter on the HEMS API through an oracle network. If the consumer received less energy than they requested, funds are returned to cover the energy mismatch. The payment is then transferred to the prosumer based on the revenue generated from the sale of energy, which is determined by market demand through the use of bid pricing. The mathematical representation of this process can be represented as $R = E * B$, where R represents the revenue, E represents the energy received by the consumer, and B represents the bid price.

6. Implementation

The testbed was carried out in the research laboratory B110 at Universidad Tecnica Federico Santa Maria in Valparaiso, Chile. In the laboratory, two Raspberry Pi 4 Model B units were used to represent the prosumer and consumer participants. An all-in-one (AIO) computer was used to set the blockchain-related elements and for the MCC operations. Additional services were implemented in Amazon Web Services (AWS). As seen in Figure 9, the implementation is divided into three main components:

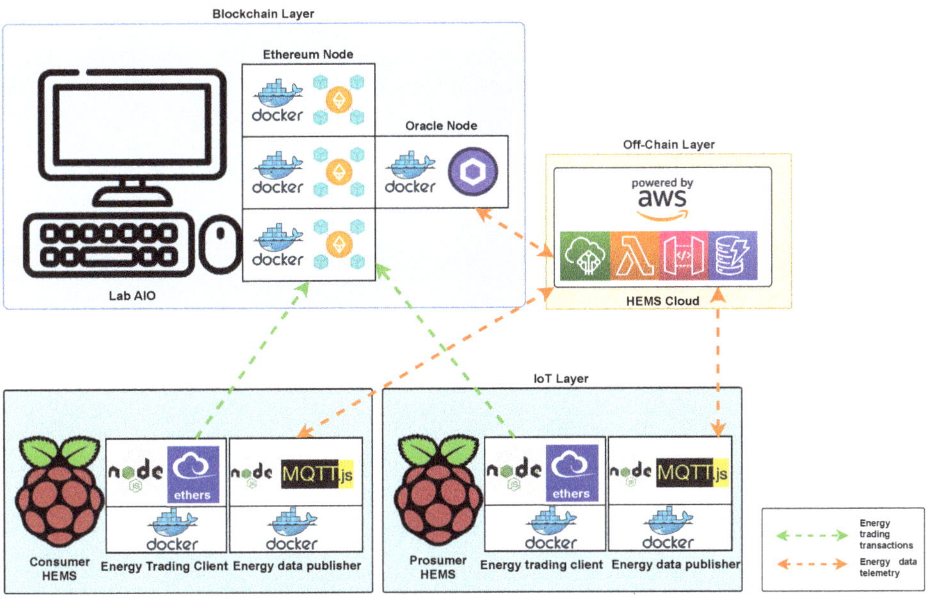

Figure 9. Testbed diagram.

6.1. Off-Chain Layer

The off-chain layer of the home energy management system (HEMS) was implemented on the AWS platform. AWS IoT Core was utilized to provide an MQTT broker for messaging between the energy data publisher microservice available in both the consumer and prosumer HEMS. This broker allows energy consumption and generation data to be sent to the cloud, where it is stored and made available through an API. Energy data for each household can be requested using a unique identifier assigned to each smart meter.

The two HTTP get endpoints implemented are:

- /battery-soc/id: This endpoint returns an object with the readings from the last measurement of the state of charge (SoC) of a battery energy storage system (BESS). The readings are randomly generated each hour, and the id represents the identifier of the battery.
- /smartmeter/id: This endpoint returns the last hour energy consumption in kWh from a user's smart meter. The id represents the identifier of the smart meter.

The implementation of the off-chain layer allows the collection and storage of energy data from both the consumer and prosumer HEMS. The use of AWS and the energy data publisher microservice ensures the feasibility of the IoT–blockchain energy trading system in microgrids. The source code for this implementation is provided in a GitHub repository. (https://github.com/pipegreyback/EViG-Server accessed on 3 November 2022).

6.2. Blockchain Layer

The blockchain layer of the IoT–blockchain energy trading system was implemented in a laboratory using an all-in-one (AIO) computer, which featured an AMD Ryzen 7 5300U CPU, 16 GB RAM, and a 1 TB hard drive. The Hyperledger Besu, an open-source Ethereum client, was used to implement a private Ethereum network. To speed up the network setup, Hyperledger Besu Docker quickstart was used, which allowed for the deployment of four validator nodes and three client nodes. The validator nodes were configured to operate using IBFT 2.0 consensus [47], which was chosen due to its high throughput and finality compared to other consensus mechanisms. This is also the consensus mechanism set by default in Hyperledger Besu. This consensus mechanism uses a dynamic mechanism for validator selection and requires at least a 2/3 validator signature in order to confirm a block. In case no consensus is reached, further validation rounds are required until the consensus is reached. Additional tools, such as a block explorer, were also provided to simplify the development process. Finally, to develop smart contracts, Remix, a popular online Solidity compiler, and a debugger were used.

6.2.1. Smart Contracts

Four smart contracts were deployed to enable energy trading. Figure 10 shows a summary of each smart contract by category.

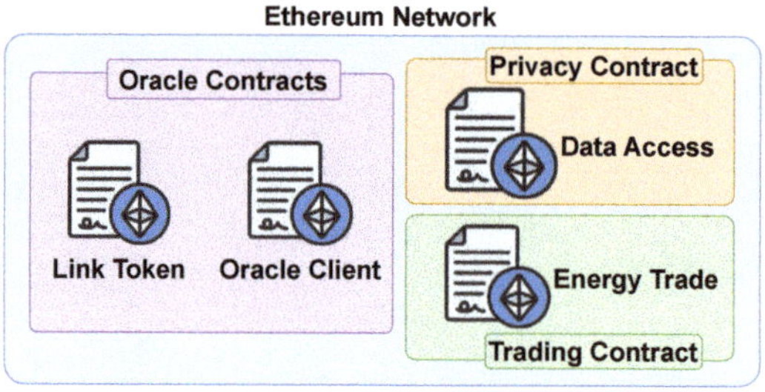

Figure 10. Deployed smart contracts by category.

- In the privacy contract category, the data access contract was implemented to manage an access control list containing entities authorized to request energy data from a user's HEMS. This permission is granted by the user. This smart contract helps ensure that the energy data are only accessible to authorized parties, which helps to protect the privacy and security of the energy data.

- In the trading contract category, the energy trade smart contract was developed, which contains the main business logic of the energy trading system. This smart contract implements a trading window consisting of four states: a bidding and offering state, a matching state, an energy-transferring state, and a settlement state. During the initial phase, the bidding and offering period is initialized when a new trading window is requested in the trading smart contract. Additionally, the trading window allows granting data access at any time, which is required to participate in a trading window. A class diagram is presented in Figure 11, where the solidity structures are presented as classes, that possess attributes and functions.
- In the oracle contracts category, two contracts were developed. The Link Token contract issues the ERC-20 Link Token, which is required to pay for data requests through the oracle network. The oracle client contract is used to request energy data using the HEMS APIs, and this request is handled by the oracle network, which processes the request and provides a callback with the results of the request. Both the Link Token contract and the oracle client contract can be found in the Chainlink official repository (https://github.com/smartcontractkit/chainlink/tree/develop/contracts/src accessed on 3 November 2022).

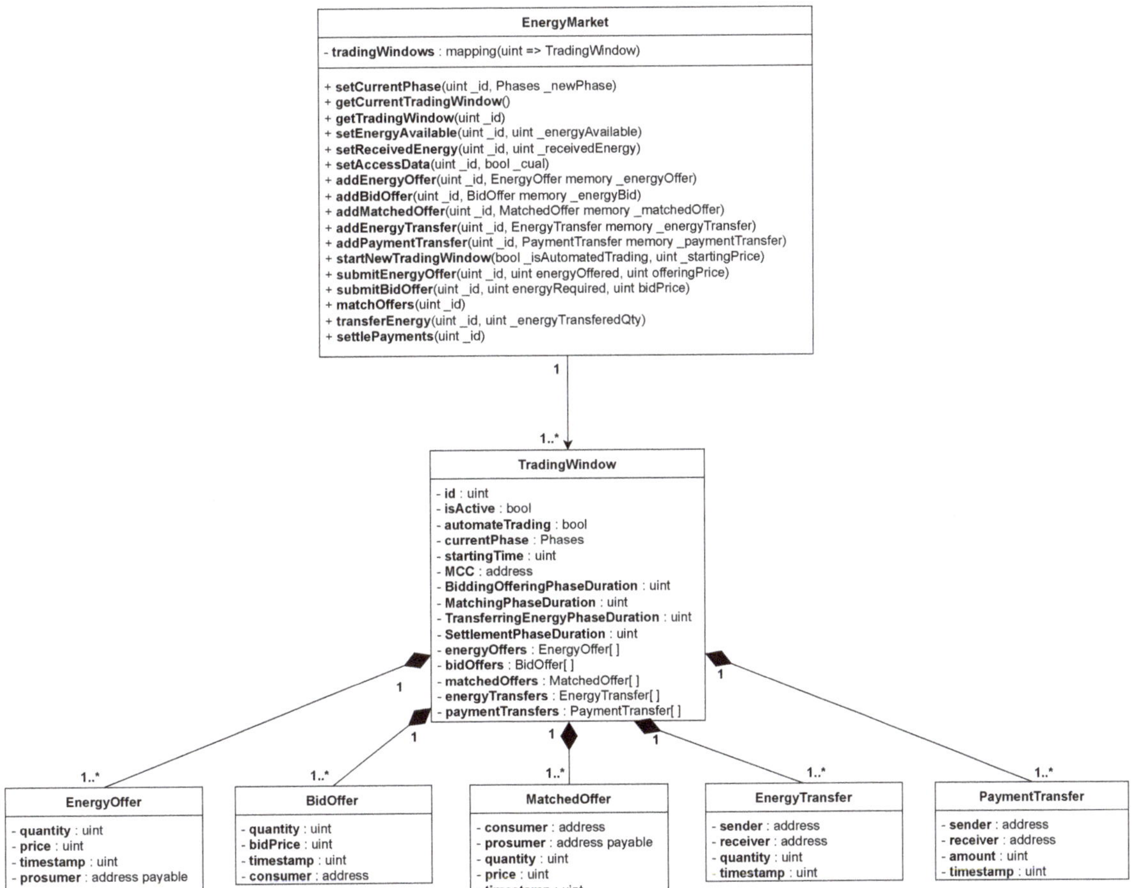

Figure 11. Class diagram of energy trading contract.

6.2.2. oracle Nodes

In order to enable smart contracts to utilize the available API, the deployment of a Chainlink oracle network using docker containers was necessary. Additionally, a PostgreSQL database was deployed using docker containers to support the Chainlink oracle network. Configuration between the Ethereum network and the Chainlink oracle node was established to facilitate interaction between smart contracts and the oracle within the Ethereum network. The primary parameters utilized for communication between the two networks include the RPC URL and the chainId. Figure 12 shows the interaction diagram for the data feed request between smart contracts and oracle nodes.

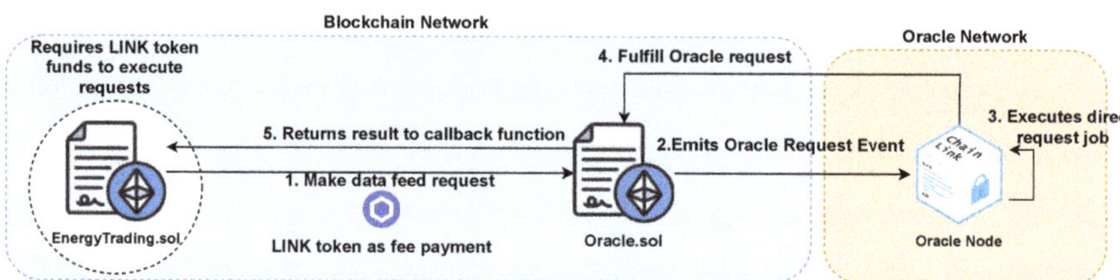

Figure 12. oracle smart contract deployment and invocation diagram.

In order to access the endpoints provided by the HEMS API, it is necessary to create and deploy a job within the oracle node. A job comprises a sequence of steps, called tasks, that fulfill the data feed requirement of a smart contract. In this specific case, a direct request job was created. This job involved requesting data from the HEMS API, parsing the response, encoding the data into a format compatible with Solidity, and ultimately submitting the transaction back to the requesting smart contract. Figure A1 shows the direct request job configured.

6.3. IoT Layer

For the IoT layer, the two Raspberry Pi units were used. Docker was employed to run containers of the services required to enable the energy data client and the energy trading client.

6.3.1. Energy Data Client

The implementation of the IoT layer was done using the AWS platform and the energy data publisher microservice was developed using Node.js and MQTT.js. This microservice sends energy data to the HEMS cloud service using MQTT, allowing the collection and storage of energy data from each household. The microservice was packaged into a docker container for easy deployment and running on any operating system that supports docker containers.

6.3.2. Energy Trading Client

Both the consumer and the prosumer were configured with an energy trading client, which allows for submitting bids and offers to the blockchain to participate in the energy trading within the microgrid. The energy trading client was developed using Node.js and the ethers.js library, which is an Ethereum client that allows the execution of operations within a blockchain network. The energy trading client was also packaged into a docker container for ease of deployment. The trading client subscribes to the four events that the smart contract emits to the blockchain network. The event subscription and execution of smart contract functions are managed by ether.js. Table 2 describes the operations that any participant in the microgrid can execute within the energy trading smart contract. The client source code is provided in the repository, and it considers three possible configurations: consumer, prosumer, and MCC.

Table 2. Available operations for the energy trading smart contract.

Operation	Description
Constructor	Initialize the contract's state variables when the smart contract is deployed.
Phase Transition	Update the current phase of the trading process based on a predetermined timeline or other criteria.
Start Trading Window	Mark the beginning of the trading phase and set the initial trading parameters, such as the duration of the trading window and the maximum amount of energy that can be traded.
Submit Energy Offer	Allow sellers to submit offers to sell their energy at a certain price and quantity. These offers are stored in the contract's state variables until they are matched with a corresponding bid.
Submit Energy Bid	Allow buyers to submit bids to buy energy at a certain price and quantity. These bids are stored in the contract's state variables until they are matched with a corresponding offer.
Matching Phase	Match the submitted offers and bids based on their price and quantity, and assign the matched pairs to a transfer list for the transferring phase.
Transferring Phase	Transfer the energy between the matched parties based on the agreed-upon price and quantity. This phase ensures that the energy is delivered securely and efficiently.
Settlement Phase	Transfer the payment between the matched parties based on the agreed-upon price and quantity, and update the state variables to reflect the completed trade. This phase ensures that the payment is made accurately and securely.

6.4. Use Case and Validation

The testbed consists of one prosumer, one consumer, and a microgrid control center. We assumed that a prosumer owns a PV system and BESS, which will offer excess energy, the consumer with BESS will bid to fulfill his energy consumption requirements, and the microgrid control center will provide the power infrastructure for all households in the microgrid. The smart contract for energy trading was deployed by the lab all-in-one (AIO) computer acting as the MCC. Both the prosumer and consumer grant data access permission to the energy trading smart contract for making energy requests to the HEMS API. Figure A2 shows the event log where the prosumer executes the grant data access function passing the energy trading smart contract address and the smart meter id as the parameters in the invocation.

The energy trading window starts with an energy price of 60,000 gwei per kWh, which was set by the microgrid control center (MCC). This price is equivalent to around CLP 102 per kWh, which is the rate provided by energy utility Chilquinta in Chile (https://www.chilquinta.cl/informacion-de-interes/tarifas-vigentes accessed on 3 November 2022) for the cost of energy per kWh, and the cost of transmission per kWh, for the BT1A tariff. The trading window has now entered the offering and bidding phase, allowing consumers and prosumers to submit their offers to sell or bids to buy energy at the specified price. The energy starting window event is broadcasted to all participants in the trading system, enabling them to react to the event on the clients by event subscription and execute their own state machines. This enhances the overall efficiency and transparency of the trading process, allowing participants to respond quickly and accurately to changes in the trading window and the energy price. The log message shown in Figure A3 confirms the current phase of the trading window and the energy price broadcasted in the event. This information is essential for understanding the initial pricing conditions and the current state of the energy trading system.

Later, a prosumer submits an offer to sell 10 kWh. This offer request triggers the oracle smart contract for making a request to the HEMS API for checking the prosumer available energy. The event of an energy offering being denied due to insufficient energy is illustrated in Figure A4. An event with an error message indicating the reason for the denial, and in this case, "insufficient energy" is provided. In the event that a participant has not granted the energy trading smart contract the required permission to access the energy data, an error is returned by the contract validations, as seen in Figure A5.

In the case of valid energy offers and energy bids, when the matching phase finishes, a result event is broadcasted to the matched participants, as seen in Figure A6. This event indicates that the energy trading smart contract has reached the matching state, where a consumer and prosumer have agreed on a trade for 10 units of energy at a price of 54,000 gwei per unit. The smart contract has successfully matched the consumer's energy demand with the prosumer's energy supply at the agreed price, and the settlement phase can proceed.

Regarding the received energy by the consumer, Figure A7 shows the log from the oracle request. This log was emitted by the energy trading smart contract during the energy-transferring phase, indicating that the consumer received 8 kWh of energy from the prosumer, as measured by the smart meter. It is important to note that the energy data are managed by the Chainlink oracle, and the energy trading smart contract interacts with the oracle contracts to retrieve the data. The discrepancy between the initially agreed 10 kWh, and the 8 kWh received by the consumer may be due to a delay in the initial transfer from the prosumer, or external factors such as network congestion or power outages. However, the flexibility provided by the smart contract allows for mismatches between agreed energy and received energy to be addressed during the settlement phase.

The settlement phase, which follows the energy-transferring phase, is responsible for determining the proportional payment to be made to the prosumer based on the amount of energy received by the consumer. Figure A8 shows an example of the proportional payment calculation for the scenario where the consumer received 8 kWh of energy, which was less than the initially agreed 10 kWh. The implementation of the energy trading smart contract provides a degree of flexibility by allowing for adjustments to be made during the settlement phase in case of mismatches between agreed and received energy.

6.5. Cost Analysis Results

Table 3 shows a cost comparison of different energy trading smart contract operations performed in our energy trading platform and in various speed-based configurations on the public Ethereum network. As our implementation uses a private Ethereum network, there were no transaction fees incurred during contract deployment or execution. Therefore, the gas values listed in the table represent the resources used to perform the operations. In comparison, public Ethereum networks charge transaction fees for contract execution, and gas prices can vary based on the user's preference for transaction speed. This is because the public Ethereum network tends to become congested, where validators prioritize transactions with a higher gas price to process and receive a reward. As a result, users may choose to pay a higher gas price to increase the chances of their transaction being processed sooner. The public Ethereum network offers suggested gas prices for different transaction speeds, namely slow, average, or fast. For the public Ethereum network, each transaction speed has a corresponding gas price, which on the date of this study were 20 gwei, 22 gwei, and 24 gwei. To simplify the comparison, the transaction costs in the table are presented in Chilean pesos (CLP) using the conversion rate from ETH to CLP on 10 April 2023.

Block time in the Ethereum network is dynamic and reflects the network congestion and performance. In the case of Hyperledger Besu, the block time is a critical parameter that determines the block frequency. This time has a direct impact on both throughput and latency. If the block time is very short (e.g., 1 s), blocks will be generated frequently, which results in low latency and high throughput. However, in the case of a block period being very long, the performance will be reduced because the transactions will need to wait a long time to be added to the next block [48].

Table 3. Cost comparison of different energy trading smart contract operations based on the network execution.

Contract Operation	Gas Utilization	Cost in CLP in Our System	Cost in CLP in Slow Public Ethereum	Cost in CLP in Average Public Ethereum	Cost in CLP in Fast Public Ethereum
Constructor	2,619,703	$0	$81.010	$89.112	$97.213
Phase Transition	161,524	$0	$4.995	$5.494	$5.994
Start Trading Window	286,403	$0	$8.857	$9.742	$10.628
Submit Energy Offer	129,613	$0	$4.008	$4.409	$4.810
Submit Energy Bid	130,354	$0	$4.031	$4.434	$4.837
Matching Phase	153,400	$0	$4.744	$5.218	$5.692
Transferring Phase	134,179	$0	$4.149	$4.564	$4.979
Settlement Phase	144,247	$0	$4.461	$4.907	$5.353

In our implementation of the energy trading smart contract, the created Ethereum network provided a 5 s validation block time. In comparison, the last average block time of the public Ethereum network is around 12 s (https://etherscan.io/chart/blocktime accessed on 10 January 2023), as shown in Figure 13. The difference in block times could be due to various factors, such as network congestion or the consensus mechanism used by the network. Understanding the impact of block time on the energy trading smart contract's performance, throughput, and gas usage is critical for future optimizations and improvements.

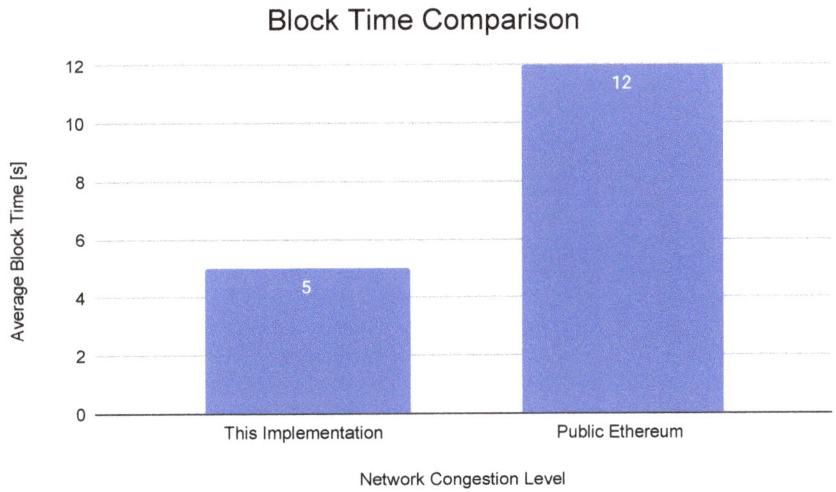

Figure 13. Block time comparison between EnergyAuction network and public Ethereum.

The maximum block size in gas units for a regular block in our implementation and in public Ethereum networks is 30 million gas units. Figure 14 shows a bar chart where the eight operations of the energy trading smart contract are presented in terms of the percentage of block usage. The constructor is the most gas-consuming operation and utilizes 8.73% of the available gas in a block. The remaining operations present lower gas consumption, between 0.43% and 0.95% of gas usage for the available gas in a block. An analysis of the gas usage for the energy trading smart contract revealed that no more than 12 constructors or 200 trading operations could be performed within a single block. This indicates that the platform may face limitations in terms of scalability and network congestion with an increasing number of participants. However, given the 5 s block time of the private Ethereum network used in this work, and the number of participants for the use case validation scenario, this limitation was not explored in detail. Future research could investigate ways to optimize the gas usage of the smart contract or explore alternative solutions to address scalability and congestion issues.

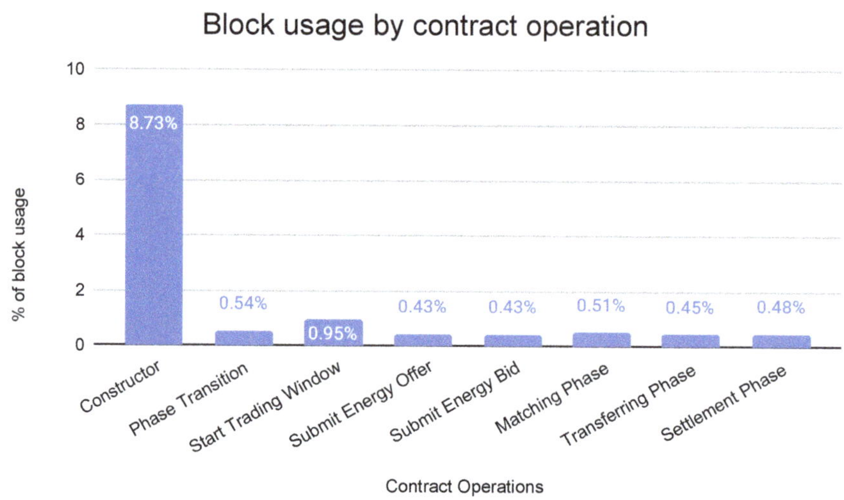

Figure 14. Percentage of block usage per smart contract operation.

7. Discussion and Limitations

The implemented energy trading platform successfully demonstrates the feasibility of a permissioned blockchain network for energy trading. The platform was designed to support energy trading between prosumers and consumers, with the support of an MCC. While the use of an access control list in conjunction with oracles allows the energy trading smart contract to access energy information, the relationship is established between the consumer or prosumer and the contract, instead of using IoT devices as a separate entity in the interaction. Recent developments in the field propose the use of Non-Fungible Tokens (NFTs) as a digital representation or "digital twin" of IoT devices. These tokens, standardized by Ethereum under the ERC-721 and ERC-1155 interfaces, enable the establishment of ownership relationships between the householder, consumer, or prosumer and their energy data-collecting devices. This feature includes the data access functionality, which can simplify the addition of further implementations, such as the one presented here.

In terms of cost analysis, the results indicate that a private implementation of the energy trading platform is more cost-effective than implementing the same business logic in the public Ethereum network. This is mainly due to the public network's high transactional fee and longer validation block times, caused by network congestion and a different consensus mechanism. Therefore, it is recommended to implement such platforms in private Ethereum networks or shared private networks, such as consortiums involving multiple entities related to smart grid applications. If by any means there is a requirement to deploy such a platform in a public network, it is recommended to make a prospection into layer two (L2) solutions, which present lower transactional fees and faster validation block time.

The privacy of the energy trading system is achieved through the use of a permissioned blockchain network, which restricts access to only authorized nodes. However, it is important to note that while the access control list and the oracles provide privacy for off-chain data access, the transactions that occur within the energy trading smart contract are still recorded on the blockchain. This includes the funds transferred between parties during the settlement phase and the readings from the off-chain data that are accessed via the oracles. Therefore, further research is needed to ensure the privacy of on-chain transactions in the energy trading system. For example, a technology that could be used to enhance the privacy and security of energy trading platforms is the zero-knowledge proof (Zk-proofs). Zk-proofs enable the verification of data without revealing the data itself. This could be useful in scenarios where the privacy of the energy trading data is of utmost importance, such as trading between competitors or sensitive business transactions.

The analysis of the gas usage of the smart contract revealed that the constructor had an average gas usage of 9%, while the average gas usage for each function call was 0.5%. Although the implementation was successful in supporting the energy trading use case, the gas usage results suggest that there is room for improvement in the smart contract design. Future work could explore the use of more advanced smart contract technologies, such as verifiable computing (VC) or zero-knowledge proof (ZKP), to optimize gas usage and enable the implementation of more complex business logic.

In this work, the old transactions are still kept on the blockchain. However, the concept of offloading can be applied, but a different approach is needed regarding the blockchain network configuration, IoT data, transactions executed by devices, and energy trading participants. For example, in [49], the authors introduced using different types of nodes including full nodes, semi-full nodes, and light nodes, while in [50], the authors developed a hierarchical blockchain storage structure for industrial IoT architectures.

Regarding the law and regulations, in Chile, the current regulation allows prosumers with PV systems to only inject surplus energy to the power grid; however, consumers and prosumers are not allowed to conduct direct energy trading with one another. To enable such opportunities, new regulations should be developed that will bring benefits not only to utility companies but also all participants, including prosumers, consumers, and utilities.

8. Conclusions

This paper presents the development and deployment of a microgrid-based local peer-to-peer energy trading system in a laboratory environment. It thoroughly examined and discussed the conceptual architecture of a distributed peer-to-peer network. To establish a blockchain-based system in a test environment, a Chainlink oracle and a private Ethereum blockchain network were employed as foundational components. The research focused on exploring the capabilities of oracles as intermediaries between off-chain data sources, such as smart meter energy readings, and the business logic implemented within smart contracts, which typically lack the ability to directly request off-chain data. Overall, this work highlights the potential of IoT and blockchain technologies to enable sustainable and efficient energy management practices. Despite the promising results, the proposed platform has some limitations and challenges that require further research and development. The following are the main directions to extend the current work:

- To support a large number of participants, there is a need to optimize the platform performance and scalability. This includes optimizing the smart contract code, data storage, and processing algorithms.
- To support real-time energy trading, different business models and trading schemes should be developed where participants trade energy in real time based on the current market conditions.
- Cybersecurity and privacy are important factors to be considered. This includes the data exchange among different participants as well as the sensors, smart meters, and networking infrastructure.

Author Contributions: Conceptualization, M.A.A., J.M.M. and F.C.; methodology, M.A.A., J.M.M., F.C., P.F., A.M.E. and Y.-C.K.; software, F.C.; validation, M.A.A., J.M.M. and F.C.; formal analysis, M.A.A., J.M.M., F.C., P.F., A.M.E. and Y.-C.K.; investigation, M.A.A., J.M.M., F.C., P.F., A.M.E. and Y.-C.K.; writing—original draft preparation, F.C. and M.A.A.; writing—review and editing, M.A.A., J.M.M., F.C., P.F., A.M.E. and Y.-C.K.; funding acquisition, M.A.A. and Y.-C.K. All authors have read and agreed to the published version of the manuscript.

Funding: This work was supported in part by the Agencia Nacional de Investigación y Desarrollo (ANID) through the Proyecto Fondecyt de Iniciación en Investigación 2020 under Project ID11200178, DGIIP-UTFSM Chile, and in part by the National Research Foundation of Korea (NSF) Grant through the Korean Government under Grant (2021R1I1A305872911).

Institutional Review Board Statement: Not applicable.

Informed Consent Statement: Not applicable.

Data Availability Statement: Not applicable.

Acknowledgments: The authors would like to acknowledge Proyecto Fondecyt de Iniciacion en Investigacion 2020 under Project ID11200178, DGIIP-UTFSM Chile, and National Research Foundation of Korea (NSF) Grant through the Korean Government under Grant (2021R1I1A305872911).

Conflicts of Interest: The authors declare no conflict of interest.

Appendix A

Figure A1. Job definition in the oracle node operator.

```
[
    {
        "from": "0x9209a142C1DFa4C746B19c4bf3e91BbC96737dC7",
        "topic": "0xb4c6779ceb4a20f448e76a0e11f39bd183cff9c9dbac53df6bfcc202e2eb32f1",
        "event": "AccessGranted",
        "args": {
            "0": "0x1e553735F5DA2B5A68bB37F90fbFb9551c229c8B",
            "1": "2474156",
            "_contract": "0x1e553735F5DA2B5A68bB37F90fbFb9551c229c8B",
            "_id": "2474156"
        }
    }
]
```


Figure A2. Log message from granting permission operation.


```
[
    {
        "from": "0x46cb39166e52d7Ea8077CAa6a764C5507e813FA6",
        "topic": "0x9569a998635142ff1e15cd543af337ca0d78b279ed2ffdecf5568e75625d121f",
        "event": "phaseTransition",
        "args": {
            "0": 0,
            "1": 0,
            "2": "This trading window is transitioning to the bidding and offering phase",
            "currentPhase": 0,
            "newPhase": 0,
            "message": "This trading window is transitioning to the bidding and offering phase"
        }
    },
    {
        "from": "0x46cb39166e52d7Ea8077CAa6a764C5507e813FA6",
        "topic": "0x8c2c6508c93167c2281369e462ef65[illegible]683fa9108d9c74d5b84649b00520162ef",
        "event": "BiddingStarting",
        "args": {
            "0": "60000",
            "1": "Bidding/Offering",
            "price": "60000",
            "message": "Bidding/Offering"
        }
    }
]
```


Figure A3. Log message from a new energy trading window initialization.


```
[
    {
        "from": "0xf9565D6c300882b79800C8Ab16Fb9189Fe518Af7",
        "topic": "0xe13f19f70b29c7eaad932e468eb44464bcba87a95c33cf47888b63d1983b5893",
        "event": "OfferDenied",
        "args": {
            "0": "0x9625900fa6E5418B327E05997F546CAac473BB16",
            "1": "15",
            "2": "Insufficient energy",
            "prosumer": "0x9625900fa6E5418B327E05997F546CAac473BB16",
            "energyOffered": "15",
            "reason": "Insufficient energy"
        }
    }
]
```


Figure A4. Error log when prosumer is offering more energy than available.

```
[
    {
        "from": "0xEB231F11704c3d6e7E7969eE2dc0Ff80ef46697C",
        "topic": "0x903e4d60b59021f144ee487f61a98a583f3a2762dddcd38603816527688de4db",
        "event": "noPermission",
        "args": {
            "0": "0xEB231F11704c3d6e7E7969eE2dc0Ff80ef46697C",
            "1": "This contract does not have permission to access the smart meter data.",
            "contract": "0xEB231F11704c3d6e7E7969eE2dc0Ff80ef46697C",
            "message": "This contract does not have permission to access the smart meter data."
        }
    }
]
```


Figure A5. Error log when the energy trading smart contract does not have permission to request energy data from smart meters.


```
[
    {
        "from": "0x46cb39166e52d7Ea8077CAa6a764C5507e813FA6",
        "topic": "0x62830dd5377daa0dd3d791f401e84cbc08818150be0d78afd850cbf3db239afe",
        "event": "MatchingState",
        "args": {
            "0": "0xD8229d99515faB9159A9b33BA4d868eDeD92Dadc",
            "1": "0xb45CFB124677b22553e88b2B6189bD1055aC7bE7",
            "2": "10",
            "3": "54000",
            "consumer": "0xD8229d99515faB9159A9b33BA4d868eDeD92Dadc",
            "prosumer": "0xb45CFB124677b22553e88b2B6189bD1055aC7bE7",
            "quantity": "10",
            "price": "54000"
        }
    }
]
```


Figure A6. Log from the results of the matching phase.


```
[
    {
        "from": "0x46cb39166e52d7Ea8077CAa6a764C5507e813FA6",
        "topic": "0xb496c71060c8c1de109c75b6a2dacb7d8a9d1cbc9b6f573913b811cf37a0701e",
        "event": "EnergyRecieved",
        "args": {
            "0": "0xb496c71060c8c1de109c75b6a2dacb7d8a9d1cb",
            "1": "0x46cb39166e52d7Ea8077CAa6a764C5507e813FA6",
            "2": "8",
            "oracleAggregator": "0xb496c71060c8c1de109c75b6a2dacb7d8a9d1cb",
            "energyTradingContract": "0x46cb39166e52d7Ea8077CAa6a764C5507e813FA6",
            "energyReceived": "8"
        }
    }
]
```


Figure A7. Log from the oracle request for obtaining the received energy by the smart meter of the consumer.

```
[
    {
        "from": "0x46cb39166e52d7Ea8077CAa6a764C5507e813FA6",
        "topic": "0x7bdd298fee8520ef5fb09d30ad01cd090f764b284bf6ee61c6fa184132705f6b",
        "event": "SettlementState",
        "args": {
            "0": "0xb45CFB124677b22553e88b2B6189bD1055aC7bE7",
            "1": "0x46cb39166e52d7Ea8077CAa6a764C5507e813FA6",
            "2": "432000000000000",
            "prosumer": "0xb45CFB124677b22553e88b2B6189bD1055aC7bE7",
            "lockedFundsContract": "0x46cb39166e52d7Ea8077CAa6a764C5507e813FA6",
            "payment": "432000000000000"
        }
    }
]
```


Figure A8. Log from the settlement phase, where the prosumer is paid proportionally to what we provided.

References

1. Bouhafs, F.; Mackay, M.; Merabti, M. Links to the Future: Communication Requirements and Challenges in the Smart Grid. *IEEE Power Energy Mag.* **2012**, *10*, 24–32. [CrossRef]
2. Bayram, I.S.; Shakir, M.Z.; Abdallah, M.; Qaraqe, K. A survey on energy trading in smart grid. In Proceedings of the 2014 IEEE Global Conference on Signal and Information Processing (GlobalSIP), Atlanta, GA, USA, 3–5 December 2014. [CrossRef]
3. Hamari, J.; Sjölint, M.; Ukkonen, A. The sharing economy: Why people participate in collaborative consumption. *J. Assoc. Inf. Sci. Technol.* **2015**, *67*, 2047–2059. [CrossRef]
4. Jamil, F.; Iqbal, N.; Ahmad, S.; Kim, D. Peer-to-Peer Energy Trading Mechanism Based on Blockchain and Machine Learning for Sustainable Electrical Power Supply in Smart Grid. *IEEE Access* **2021**, *9*, 39193–39217. [CrossRef]
5. Kabalci, Y.; Kabalci, E.; Padmanaban, S.; Holm-Nielsen, J.B.; Blaabjerg, F. Internet of Things Applications as Energy Internet in Smart Grids and Smart Environments. *Electronics* **2019**, *8*, 972. [CrossRef]
6. Tan, S.; Wu, Y.; Xie, P.; Guerrero, J.M.; Vasquez, J.C.; Abusorrah, A. New Challenges in the Design of Microgrid Systems: Communication Networks, Cyberattacks, and Resilience. *IEEE Electrif. Mag.* **2020**, *8*, 98–106. [CrossRef]
7. Wu, Y.; Wu, Y.; Guerrero, J.M.; Vasquez, J.C.; Palacios-Garcia, E.J.; Li, J. Convergence and Interoperability for the Energy Internet: From Ubiquitous Connection to Distributed Automation. *IEEE Ind. Electron. Mag.* **2020**, *14*, 91–105. [CrossRef]
8. Leitao, J.; Gil, P.; Ribeiro, B.; Cardoso, A. A Survey on Home Energy Management. *IEEE Access* **2020**, *8*, 5699–5722. [CrossRef]
9. Zafar, U.; Bayhan, S.; Sanfilippo, A. Home Energy Management System Concepts, Configurations, and Technologies for the Smart Grid. *IEEE Access* **2020**, *8*, 119271–119286. [CrossRef]
10. Llaria, A.; Dos Santos, J.; Terrasson, G.; Boussaada, Z.; Merlo, C.; Curea, O. Intelligent Buildings in Smart Grids: A Survey on Security and Privacy Issues Related to Energy Management. *Energies* **2021**, *14*, 2733. [CrossRef]
11. Chander, B.; Kumaravelan, G. Internet of Things: Foundation. In *Principles of Internet of Things (IoT) Ecosystem: Insight Paradigm*; Peng, S.L., Pal, S., Huang, L., Eds.; Springer International Publishing: Cham, Switzerland, 2020; pp. 3–33. [CrossRef]
12. Zhang, C.; Wu, J.; Cheng, M.; Zhou, Y.; Long, C. A Bidding System for Peer-to-Peer Energy Trading in a Grid-connected Microgrid. *Energy Procedia* **2016**, *103*, 147–152. [CrossRef]
13. Mengelkamp, E.; Gärttner, J.; Rock, K.; Kessler, S.; Orsini, L.; Weinhardt, C. Designing microgrid energy markets: A case study: The Brooklyn Microgrid. *Appl. Energy* **2018**, *210*, 870–880. [CrossRef]
14. Zhou, Y.; Wu, J.; Long, C.; Ming, W. State-of-the-Art Analysis and Perspectives for Peer-to-Peer Energy Trading. *Engineering* **2020**, *6*, 739–753. [CrossRef]
15. Soto, E.A.; Bosman, L.B.; Wollega, E.; Leon-Salas, W.D. Peer-to-peer energy trading: A review of the literature. *Appl. Energy* **2021**, *283*, 116268. [CrossRef]
16. Wu, Y.; Wu, Y.; Guerrero, J.M.; Vasquez, J.C. Digitalization and decentralization driving transactive energy Internet: Key technologies and infrastructures. *Int. J. Electr. Power Energy Syst.* **2021**, *126*, 106593. [CrossRef]
17. Condon, F.; Martínez, J.M.; Kim, Y.C.; Ahmed, M.A. EnergyAuction: oracle Blockchain-Based Energy Trading System for Microgrids. In Proceedings of the 2023 IEEE Conference on Power Electronics and Renewable Energy (CPERE), Luxor, Egypt, 15–17 February 2023; pp. 1–6.
18. Mengelkamp, E.; Notheisen, B.; Beer, C.; Dauer, D.; Weinhardt, C. A blockchain-based smart grid: Towards sustainable local energy markets. *Comput. Sci. Res. Dev.* **2018**, *33*, 207–214. [CrossRef]
19. Luo, Y.; Itaya, S.; Nakamura, S.; Davis, P. Autonomous cooperative energy trading between prosumers for microgrid systems. In Proceedings of the 39th Annual IEEE Conference on Local Computer Networks Workshops, Edmonton, AB, Canada, 8–11 September 2014. [CrossRef]

20. Zhang, C.; Wu, J.; Zhou, Y.; Cheng, M.; Long, C. Peer-to-Peer energy trading in a Microgrid. *Appl. Energy* **2018**, *220*, 1–12. . [CrossRef]
21. Andoni, M.; Robu, V.; Flynn, D.; Abram, S.; Geach, D.; Jenkins, D.; McCallum, P.; Peacock, A. Blockchain technology in the energy sector: A systematic review of challenges and opportunities. *Renew. Sustain. Energy Rev.* **2019**, *100*, 143–174. [CrossRef]
22. Li, Z.; Kang, J.; Yu, R.; Ye, D.; Deng, Q.; Zhang, Y. Consortium Blockchain for Secure Energy Trading in Industrial Internet of Things. *IEEE Trans. Ind. Inform.* **2018**, *14*, 3690–3700. [CrossRef]
23. Aitzhan, N.Z.; Svetinovic, D. Security and Privacy in Decentralized Energy Trading Through Multi-Signatures, Blockchain and Anonymous Messaging Streams. *IEEE Trans. Dependable Secur. Comput.* **2018**, *15*, 840–852. [CrossRef]
24. Wang, S.; Sheng, H.; Zhang, Y.; Yang, D.; Shen, J.; Chen, R. Blockchain-Empowered Distributed Multi-Camera Multi-Target Tracking in Edge Computing. *IEEE Trans. Ind. Inform.* **2023**, 1–10. [CrossRef]
25. Cao, B.; Wang, X.; Zhang, W.; Song, H.; Lv, Z. A Many-Objective Optimization Model of Industrial Internet of Things Based on Private Blockchain. *IEEE Netw.* **2020**, *34*, 78–83. [CrossRef]
26. Mihaylov, M.; Jurado, S.; Avellana, N.; Van Moffaert, K.; de Abril, I.M.; Nowé, A. NRGcoin: Virtual currency for trading of renewable energy in smart grids. In Proceedings of the 11th International Conference on the European Energy Market (EEM14), Krakow, Poland, 28–30 May 2014; pp. 1–6. [CrossRef]
27. Wang, W.; Hoang, D.T.; Hu, P.; Xiong, Z.; Niyato, D.; Wang, P.; Wen, Y.; Kim, D.I. A Survey on Consensus Mechanisms and Mining Strategy Management in Blockchain Networks. *IEEE Access* **2019**, *7*, 22328–22370. [CrossRef]
28. Liu, C.; Zhang, X.; Chai, K.K.; Loo, J.; Chen, Y. A survey on blockchain-enabled smart grids: Advances, applications and challenges. *IET Smart Cities* **2021**, *3*, 56–78.
29. Szabo, N. Formalizing and securing relationships on public networks. *First Monday* **1997**, *2*. [CrossRef]
30. Caldarelli, G. Overview of Blockchain oracle Research. *Future Internet* **2022**, *14*, 175. [CrossRef]
31. Xu, X.; Pautasso, C.; Zhu, L.; Lu, Q.; Weber, I. A Pattern Collection for Blockchain-Based Applications. In Proceedings of the 23rd European Conference on Pattern Languages of Programs, New York, NY, USA, 19 February 2018.
32. Antal, C.; Cioara, T.; Antal, M.; Mihailescu, V.; Mitrea, D.; Anghel, I.; Salomie, I.; Raveduto, G.; Bertoncini, M.; Croce, V.; et al. Blockchain based decentralized local energy flexibility market. *Energy Rep.* **2021**, *7*, 5269–5288. [CrossRef]
33. Al Breiki, H.; Al Qassem, L.; Salah, K.; Habib Ur Rehman, M.; Sevtinovic, D. Decentralized Access Control for IoT Data Using Blockchain and Trusted oracles. In Proceedings of the 2019 IEEE International Conference on Industrial Internet (ICII), Orlando, FL, USA, 11–12 November 2019; pp. 248–257. [CrossRef]
34. Wu, Y.; Wu, Y.; Cimen, H.; Vasquez, J.C.; Guerrero, J.M. Towards collective energy Community: Potential roles of microgrid and blockchain to go beyond P2P energy trading. *Appl. Energy* **2022**, *314*, 119003. [CrossRef]
35. Pradhan, N.R.; Singh, A.P.; Verma, S.; Kavita; Wozniak, M.; Shafi, J.; Ijaz, M.F. A blockchain based lightweight peer-to-peer energy trading framework for secured high throughput micro-transactions. *Sci. Rep.* **2022**, *12*, 14523. [CrossRef]
36. Appasani, B.; Mishra, S.K.; Jha, A.V.; Mishra, S.K.; Enescu, F.M.; Sorlei, I.S.; Bîrleanu, F.G.; Takorabet, N.; Thounthong, P.; Bizon, N. Blockchain-Enabled Smart Grid Applications: Architecture, Challenges, and Solutions. *Sustainability* **2022**, *14*, 8801. [CrossRef]
37. Capper, T.; Gorbatcheva, A.; Mustafa, M.A.; Bahloul, M.; Schwidtal, J.M.; Chitchyan, R.; Andoni, M.; Robu, V.; Montakhabi, M.; Scott, I.J.; et al. Peer-to-peer, community self-consumption, and transactive energy: A systematic literature review of local energy market models. *Renew. Sustain. Energy Rev.* **2022**, *162*, 112403. [CrossRef]
38. Frieden, D.; Tuerk, A.; Neumann, C.; d'Herbemont, S.; Roberts, J. Collective self-consumption and energy communities: Trends and challenges in the transposition of the EU framework. *Compil. Integr. Community Power Energy Isl.* **2020**. [CrossRef]
39. Chen, S.; Liu, C.C. From demand response to transactive energy: State of the art. *J. Mod. Power Syst. Clean Energy* **2017**, *5*, 10–19. [CrossRef]
40. Parag, Y.; Sovacool, B.K. Electricity market design for the prosumer era. *Nat. Energy* **2016**, *1*, 16032. [CrossRef]
41. Wang, Q.; Zhang, C.; Ding, Y.; Xydis, G.; Wang, J.; Østergaard, J. Review of real-time electricity markets for integrating Distributed Energy Resources and Demand Response. *Appl. Energy* **2015**, *138*, 695–706. [CrossRef]
42. Koirala, B.P.; Koliou, E.; Friege, J.; Hakvoort, R.A.; Herder, P.M. Energetic communities for community energy: A review of key issues and trends shaping integrated community energy systems. *Renew. Sustain. Energy Rev.* **2016**, *56*, 722–744. [CrossRef]
43. Saad, W.; Han, Z.; Poor, H.V.; Basar, T. Game-Theoretic Methods for the Smart Grid: An Overview of Microgrid Systems, Demand-Side Management, and Smart Grid Communications. *IEEE Signal Process. Mag.* **2012**, *29*, 86–105. [CrossRef]
44. Abdella, J.; Shuaib, K. Peer to Peer Distributed Energy Trading in Smart Grids: A Survey. *Energies* **2018**, *11*, 1560. [CrossRef]
45. Mohanta, B.K.; Panda, S.S.; Jena, D. An Overview of Smart Contract and Use Cases in Blockchain Technology. In Proceedings of the 2018 9th International Conference on Computing, Communication and Networking Technologies (ICCCNT), Bengaluru, India, 10–12 July 2018; pp. 1–4. [CrossRef]
46. Al-Breiki, H.; Rehman, M.H.U.; Salah, K.; Svetinovic, D. Trustworthy Blockchain oracles: Review, Comparison, and Open Research Challenges. *IEEE Access* **2020**, *8*, 85675–85685. [CrossRef]
47. Saltini, R.; Hyland-Wood, D. Ibft 2.0: A safe and live variation of the ibft blockchain consensus protocol for eventually synchronous networks. *arXiv* **2019**, arXiv:1909.10194.
48. Fan, C.; Lin, C.; Khazaei, H.; Musilek, P. Performance Analysis of Hyperledger Besu in Private Blockchain. In Proceedings of the 2022 IEEE International Conference on Decentralized Applications and Infrastructures (DAPPS), San Francisco, CA, USA, 15–18 August 2022; pp. 64–73. [CrossRef]

49. Zhao, P.; Cheng, H.; Fang, Y.; Wang, X. A Secure Storage Strategy for Blockchain Based on MCMC Algorithm. *IEEE Access* **2020**, *8*, 160815–160824. [CrossRef]
50. Wang, G.; Shi, Z.; Nixon, M.; Han, S. ChainSplitter: Towards Blockchain-Based Industrial IoT Architecture for Supporting Hierarchical Storage. In Proceedings of the 2019 IEEE International Conference on Blockchain (Blockchain), Atlanta, GA, USA, 14–17 July 2019; pp. 166–175. [CrossRef]

sustainability

Article

An Evaluation of ANN Algorithm Performance for MPPT Energy Harvesting in Solar PV Systems

Md Tahmid Hussain [1], Adil Sarwar [1], Mohd Tariq [1], Shabana Urooj [2,*], Amal BaQais [3] and Md. Alamgir Hossain [4,*]

1 Department of Electrical Engineering, ZHCET, Aligarh Muslim University, Aligarh 202002, India; gk3910@myamu.ac.in (M.T.H.); adil.sarwar@zhcet.ac.in (A.S.); tariq.ee@zhcet.ac.in (M.T.)
2 Department of Electrical Engineering, College of Engineering, Princess Nourah bint Abdulrahman University, P.O. Box 84428, Riyadh 11671, Saudi Arabia
3 Department of Chemistry, College of Science, Princess Nourah bint Abdulrahman University, P.O. Box 84428, Riyadh 11671, Saudi Arabia; aabaqeis@pnu.edu.sa
4 Queensland Micro and Nanotechnology Centre, Griffith University, Nathan, QLD 4111, Australia
* Correspondence: smurooj@pnu.edu.sa (S.U.); mdalamgir.hossain@griffith.edu.au (M.A.H.)

Abstract: In this paper, the Levenberg–Marquardt (LM), Bayesian regularization (BR), resilient backpropagation (RP), gradient descent momentum (GDM), Broyden–Fletcher–Goldfarb–Shanno (BFGS), and scaled conjugate gradient (SCG) algorithms constructed using artificial neural networks (ANN) are applied to the problem of MPPT energy harvesting in solar photovoltaic (PV) systems for the purpose of creating a comparative evaluation of the performance of the six distinct algorithms. The goal of this analysis is to determine which of the six algorithms has the best overall performance. In the study, the performance of managing the training dataset is compared across the algorithms. The maximum power point tracking energy harvesting system is created using the environment of MATLAB or Simulink, and the produced model is examined using the artificial neural network toolkit. A total of 1000 datasets of solar irradiance, temperature, and voltage were used to train the suggested model. The data are split into three categories: training, validation, and testing. Eighty percent of the total data is used for training the model, and the remaining twenty percent is divided equally for testing and validation. According to the results, the regression values of LM, RP, BR, and BFGS are 1, whereas the regression values for SCG and GDM are less than 1. The gradient values for LM, RP, BFGS, SCG, BR, and GDM are 7.983×10^{-6}, 0.033415, 1.0211×10^{-7}, 0.14161, 0.00010493, and 11.485, respectively. Similarly, the performance values for these algorithms are 2.0816×10^{-10}, 2.8668×10^{-6}, 9.98×10^{-17}, 0.052985, 1.583×10^{-7}, and 0.15378. Overall, the results demonstrate that the LM and BFGS algorithms exhibit superior performance in terms of gradient and overall performance. The RP and BR algorithms also perform well across various metrics, while the SCG and GDM algorithms show comparatively less effectiveness in addressing the proposed problem. These findings provide valuable insights into the relative performance of the six evaluated algorithms for MPPT energy harvesting in solar PV systems.

Keywords: artificial neural network (ANN); solar photovoltaic (PV); maximum power point tracking; Levenberg–Marquardt (LM); Bayesian regularization (BR); scaled conjugate gradient (SCG); resilient backpropagation (RP)

Citation: Hussain, M.T.; Sarwar, A.; Tariq, M.; Urooj, S.; BaQais, A.; Hossain, M.A. An Evaluation of ANN Algorithm Performance for MPPT Energy Harvesting in Solar PV Systems. *Sustainability* **2023**, *15*, 11144. https://doi.org/10.3390/su151411144

Academic Editor: Mohamed A. Mohamed

Received: 30 April 2023
Revised: 1 June 2023
Accepted: 12 June 2023
Published: 17 July 2023

1. Introduction

Our energy usage includes solar PV electricity, which is also a crucial element of renewable energy networks. The cost of PV modules is falling as technology advances quickly, and PV panels are becoming more reliable. National economies are investing heavily in off-grid and grid-connected PV networks [1,2]. PV electricity is unstable and dependent on solar radiation as well as other meteorological conditions, such as humidity, wind speed direction, cloud cover temperature, and precipitation. This makes it different from typical energy production methods [3]. Power networks have had significant issues because of the advent of large-scale grid-connected solar PV facilities, including a lack of device flexibility, energy balance, and efficiency [4]. For PV networks to have a consistent electricity supply, it is essential to anticipate the solar output energy. The accuracy of the predictive models boosts the device dependability, decreases the cost of extra equipment maintenance, and limits the impact of solar PV performance [5]. Irradiation and temperature are features of a PV module's I-V properties. Solar cell arrays are followed by MPPT controllers for the best utilization performance. A thorough list of 40 distinct MPPT approaches and their categorization was developed in [6]. Several MPPT algorithms and designs are covered in a number of publications in the literature to improve the performance of PV device [7]. The most efficient and widely used methods are perturb and observe (P&O) [8], incremental conductance (INC) [9], fuzzy logic controller (FLC) [10], a P&O technique based on particle swarm optimization (PSO) [11], and ANN [12]. For each of these options, there are differences in the pace of convergence, oscillation across the absolute maximum power point (MPP) complexity, stability, cost, and necessary electrical equipment [13].

After a quick irradiance change that causes a distortion in the algorithm of P&O and the working parameters of PV systems, the controller struggles at first to surpass the MPP [14]. The controller nonetheless mitigates the mistakes of the algorithm, which, with some lag time, follows the MPP once again. Furthermore, the MPP's terminal voltage fluctuates as a consequence amid a power outage. The smallest possible disruption phase size is used to counteract these oscillations. The minor phase again minimizes the transient startup and changes the system sensitivity to the weather. Inc. algorithm; it works well to use a controller such as proportional integral (PI) to deal with quick changes in irradiance and corresponding declines in the oscillation of the MPP during rips. Hence, the response and, according to the INC method, swing pace would still be balanced, but often breaks away from the MPP with sudden irradiance shifts. Authors in [15] claim that these algorithms cannot quickly and accurately detect the entire power because there are oscillations at the highest point. The biological neural networks of the human brain are the motivation for the ANN system development. It is used to train and assess the PV system I-V and P-V nonlinearity relationship. ANN retrieves inputs such as the input voltage, input current, temperature, irradiance, and metrological information and continuously learns to modify how the solar power system behaves for the highest impact [16]. The design of an FLC model is possible using ANN for a more accurate and simple converter actualization [17]. The dataset comes from simply using a simulation or hardware configuration and also introducing solar irradiances, temperatures, and/or the voltage or current of a solar power system to ANN to determine the necessary maximum power (Pmax) or maximum output voltage (Vmax). This information is converted into training data that is provided to the system to teach the desired ANN how to operate. Test datasets are used after training to evaluate the built-in ANN's performance, and for further correction, errors are sent to the ANN [18], which has the capability of predicting MPP by state estimation and SMC filtering (also known as sequential Monte Carlo). The framework of the incremental conductance maximum power point tracking approach (IC MPPT) might be expanded to include a model of state space estimations for successive maximum power points (MPP). The voltage or current and irradiance statistics are used to predict the global MPP (GMPP) in the ANN model to enhance the SMC estimation [19]. Among the advantages of ANN are exceptional precision in modelling and the ability to resolve nonlinearity issues without prior knowledge or models [20]. To speed up and improve tracking through solar power system modelling and forecasting, ANN may be used [21]. It has been demonstrated to have a quicker response time and less oscillation than MPP [22]. In actual operational circumstances, MPPT based on ANN can monitor MPP with little effort. Low ripple and transient time [23]. The square error method is used in the error calculation as a feedback correction [24]. A proper, precise, and systematic training set of data, however, is an important constraint to function well with the ANN without much training error [25]. Yet, the variations in instruction and operation when creating an ANN model and solar system settings make the training technique challenging. The authors in [26] thus recommended using a particle swarm optimization (PSO) model in MATLAB and Simulink to determine the best starting weights for ANN models by selecting the best topology to improve the accuracy of the ANN model. Hence, with the conflict processing speed and the most accurate regression after resolution of the ANN model, the minimization of the mean squared error occurs. The results show that using the real-world approach of the improved feedforward ANN depends on data from the PSO method; the peak power is accurately predicted with average hourly efficiencies of over 99.67% on bright and 99.30% on overcast days, respectively. An ANN-based MPPT controller demonstrates reduced steady-state error and a quicker response to abrupt changes in solar temperature and irradiance in contrast to both P&O and IC [27]. Nevertheless, an improved algorithm of P&O with variable step size aims to improve the tracking speed and reduce the steady-state fluctuation or oscillation under abrupt changes in irradiance or partial shading conditions (PSC). Integrating FLC and ANN with more established MPPT methods such as IC and P&O is a good fit. The ANN method calculates the MPP even when there is no shade or temperature information available from the panel; nevertheless, the hill-climbing (HC) methodology further enhances the outcome. IC-ANN and P&O-ANN are two more hybrid MPPTs that are linked with the stacked autoencoder (SAE) controller via the use of building blocks and deep learning (DL) training. It is recommended to use a greedy layer-wise method to harvest the maximum amount of energy possible from the solar energy system. Backpropagation and supervised learning are then used to fine-tune the deep neural network using traditional MPPT-IC and P&O. This allows for the greatest amount of power to be extracted [28].

In the literature study, different artificial intelligence (AI) algorithms for energy harvesting with the PV system are highlighted; however, only a few studies have made use of LM algorithms, BR algorithms, SCG algorithms, etc. There is reason to be optimistic about the ability of AI systems to anticipate optimum power with low error under a variety of meteorological situations. Processing vast volumes of data is achieved much more speedily and effectively via the use of neural networks [29]. Even though there has been some study on contrasting various MPPT topologies for solar PV systems, there is still a significant research gap when it comes to evaluating the performance of different ANN algorithm-based MPPT methods for solar energy harvesting. In addition to evaluating the performance of ANN algorithms for MPPT energy harvesting in solar PV systems, the study proposed in [30] also aims to contribute to the literature by introducing a new model predictive control method for buck–boost inverter-based photovoltaic systems. By incorporating this study into the proposed study, we expand the scope and relevance of our research. The proposed model predictive control method offers a novel approach to optimize the operation of buck–boost inverters in photovoltaic systems, thereby enhancing the energy harvesting efficiency. This integration of the new control method into the evaluation of ANN algorithms provides a comprehensive analysis of advanced techniques for improving the performance of solar PV systems. Similarly, the study presented in [31] aims to contribute to the literature by considering the topic of optimal control of an energy-storage system in a microgrid for reducing wind-power fluctuations. By incorporating this work into the study, we can also broaden the scope of our research and address the challenges associated with integrating wind power into microgrid systems. The optimal control of an energy-storage system plays a crucial role in mitigating the intermittent nature of wind power generation and ensuring a stable and reliable energy supply. This integration of the optimal control strategy into the evaluation of ANN algorithms provides a comprehensive analysis of advanced techniques for improving the stability and efficiency of renewable energy systems. By considering both solar and wind power aspects, our study offers insights into the integration of multiple renewable energy sources and their control mechanisms, thereby contributing to the overall understanding of sustainable energy systems.

It is also crucial to investigate alternate methods for training neuro-fuzzy systems in addition to analyzing the effectiveness of different ANN algorithms for MPPT energy harvesting in solar PV systems. A notable topic in this context is the training of neuro-fuzzy systems using meta-heuristic algorithms presented in [32]. These algorithms offer a promising approach to optimize the parameters of neuro-fuzzy models and enhance their MPPT capabilities. By incorporating meta-heuristic algorithms into the training process, such as genetic algorithms, particle swarm optimization, or simulated annealing, the neuro-fuzzy models can effectively learn the mapping between input data and optimal power outputs. This integration of meta-heuristic algorithms with neuro-fuzzy systems has the potential to improve the accuracy and efficiency of MPPT algorithms, ultimately leading to enhanced energy harvesting in solar PV systems. Therefore, investigating the training of neuro-fuzzy systems using meta-heuristic algorithms presents an intriguing avenue for further research and advancement in the field of MPPT energy harvesting. Furthermore, recent advancements in cooperative optimization techniques for assessing the performance of various ANN algorithms for MPPT energy harvesting in solar PV systems should also be considered. One such notable topic is the improved cooperative artificial neural network–particle swarm optimization (ANN-PSO) approach for solar photovoltaic systems with maximum power point tracking (MPPT) [33]. This approach combines the power of artificial neural networks and particle swarm optimization to enhance the efficiency and accuracy of MPPT algorithms in solar PV systems. By leveraging the cooperative nature of ANN-PSO, the system can benefit from the collective intelligence of multiple agents working together to find the optimal power output. This cooperative approach offers a promising solution for addressing the challenges of MPPT in solar PV systems, including non-linearity, partial shading, and varying environmental conditions. Therefore, incorporating the improved cooperative ANN-PSO approach in the evaluation of ANN algorithms for MPPT energy harvesting would provide valuable insights into its effectiveness and potential as an advanced optimization technique in solar PV systems.

In [34], a comparative analysis of three algorithm is proposed for energy harvesting of solar PV. The proposed work is compared in terms of the performance of handling the trained dataset, and the authors have described the algorithms in a clear and detailed way.

However, this work provides a detailed performance comparison of six different ANN-based algorithms (LM, BR, RP, GDM, BFGS, and SCG) for MPPT solar energy harvesting and also provides an explanation of each algorithm. Recent research on ANN-based MPPT has solely concentrated on fewer approaches. The created model gives a good grasp of how practical and applicable these algorithms are. It also contributes significantly to the existing literature in several ways. Firstly, it provides a comprehensive evaluation of six different ANN algorithms, specifically in the context of MPPT for solar PV systems. By comparing the performance of these algorithms, the study offers valuable insights into their effectiveness, convergence properties, and accuracy. This comparative analysis helps researchers and practitioners make informed decisions when selecting the most suitable algorithm for MPPT in solar PV systems. Additionally, the study incorporates real-time data on the solar irradiance, panel temperature, and generated voltage for training the ANN algorithms. This aspect enhances the practical relevance of the research findings and increases their applicability to real-world scenarios. The methodology and dataset used in this study can serve as a valuable resource for further research and algorithm development in the field. Furthermore, the study introduces and evaluates various performance metrics, such as regression, error at the middle bin, gradient, performance, momentum parameter, and epochs. These metrics provide a comprehensive framework for assessing the efficacy of ANN algorithms in MPPT energy harvesting. By establishing these metrics, the study contributes to the existing literature by providing a standardized approach to evaluate and compare the performance of different algorithms. Overall, this study advances the understanding of ANN algorithm performance for MPPT energy harvesting in solar PV systems and provides valuable insights and guidance for researchers, engineers, and practitioners working in this field. It sets a foundation for further research and development of advanced algorithms and methodologies for optimizing energy harvesting efficiency in solar PV systems. The performance of the six ANN algorithms is evaluated using a thorough method, which includes training, validation, and testing using generated data of the solar irradiance, temperature, and produced voltage. This method is used to determine how well the algorithm functions. The performance of these algorithms is analyzed via the use of a simulated model that is deployed on MATLAB/Simulink. This offers clear knowledge of the application of ANN algorithms for MPPT in solar PV systems. The (ANN)-based MPPT algorithm is trained using data taken from the actual world, proving both the usefulness and efficacy of this technology.

The article's remaining sections are organized as follows: the status of ANN and MPPT technology is discussed in Section 2. Section 3 describes the modelling of ANN-based MPPT for solar PV system. The results and comments for six training algorithms are shown and discussed in Section 4. Lastly, the work is concluded in Section 5.

2. The State of the Art of ANNs and MPPTs

Artificial neural networks, often known as ANNs, are a specific form of machine learning algorithm that attempt to replicate the function and structure of the biological neural networks found in the human brain [35]. They are designed to connect various parameters to particular data points without the need for mathematical equations or complex mathematical bases [36]. In order to train ANNs, a technique known as supervised learning is applied. In this approach, datasets consisting of input–output parameter values are utilized in order to train the network. The datasets are often divided into two groups: a training dataset and a validation dataset. The training dataset is used to train the network, and the validation dataset is used to assess how well the trained network performed. An artificial neural network (ANN) is made up of many different neurons, all of which are linked together by a fractional number that is referred to as weight [37]. In order to accurately forecast the results of the process, the weights are changed while it is in the training phase. The weights become constant once the error falls below a permissible value. The most common form of ANN is the two-layer model, which is shown in Figure 1.

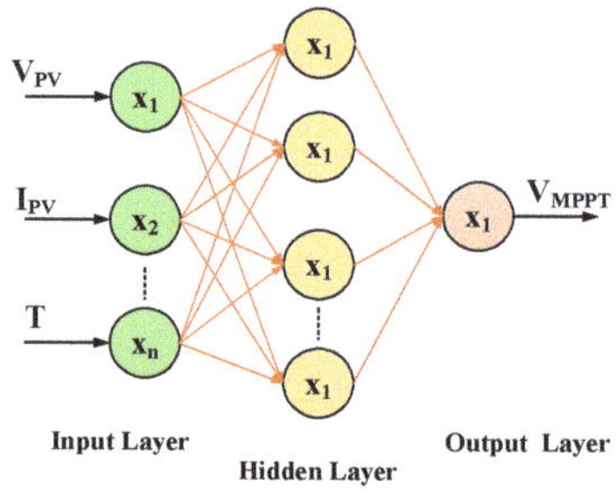

Figure 1. The fundamental design of a two-layer ANN.

The network's inputs are incorporated into the network at various times. The training dataset is used to create the neural network, and the validation dataset is used to assess how well it performed. Once the network has been trained and the error is within permissible limits [38], the validation dataset parameters of input are imported, and the associated values of output parameter are predicted [39]. This process helps to ensure that the ANN is accurate and reliable.

By contrasting the anticipated output parameter values from the validation dataset with the corresponding actual values, the trained ANN's performance is assessed. The trained ANN may be regarded as the best prediction model if the difference between the predicted and actual values is less than the allowable maximum. For the selected training method and number of training iterations, the ANN can forecast the matching input values and output parameters.

The trained ANN along with the training procedure is selected as the optimized model if the error magnitude is smaller than the allowed value. However, if the error is still high, different training algorithms or more training iterations may be tried before an allowable error is obtained [40]. The results obtained from the validation dataset using the optimal ANN model validate the generalization of the trained ANN [41].

MPPT in Solar PV systems is one of the many power electronics applications where ANNs have found widespread usage. By changing the operating point to the MPP under various environmental circumstances, MPPT aims to maximize the power output from a PV system. The MPP of a PV system is a nonlinear and time-varying function that is difficult to model analytically. ANNs have been shown to be effective tools for modeling this nonlinear relationship, making them a popular choice for MPPT in PV systems.

For MPPT in PV systems, a variety of ANN algorithms, such as radial basis function (RBF) and multi-layer perceptron (MLP) networks, may be utilized. A sort of feedforward neural network called an MLP network is made up of layers of linked neurons. They are commonly used for supervised learning tasks, such as function approximation and regression. One of the most common ANN-based MPPT techniques is the Perturb and Observe (P&O) method with an MLP network. In this method, the PV system operating point is perturbed and the change in power is observed. The change in power and the operating point are used as inputs to the ANN, which is trained to predict the MPP. The ANN output is then used to adjust the operating point to the MPP. This process is repeated in real time to track the MPP as it changes due to varying environmental conditions. Another popular ANN-based MPPT technique is the incremental conductance (IC) method with an RBF network. In this method, the PV system operating point is adjusted incrementally, and the change in power is used as an input to the ANN. The ANN is trained to predict the direction of the MPP, and the operating point is adjusted in the direction predicted by the ANN. This process is also repeated in real time to track the MPP as it changes. There are also ANN-based MPPT techniques that combine multiple algorithms, such as the hybrid MPPT algorithm that uses both the P&O and IC methods. In this method, the P&O method is used to quickly find the initial MPP, and the IC method is used to track the MPP as it changes. One of the main advantages of ANN-based MPPT techniques is that they can effectively model the nonlinear relationship between the PV system output power and the operating conditions. They can also adapt to changing environmental conditions in real time, which is essential for efficient MPPT in PV systems. Additionally, ANNs are relatively easy to implement and can be used with a variety of PV systems, from small-scale systems to large-scale power plants. The fact that ANN-based MPPT approaches need a lot of data to train the network properly is one of their key drawbacks. This can be a significant obstacle for some applications, particularly those that are remote or have limited data collection capabilities. Additionally, the training procedure may also be time-consuming and computationally demanding, which can be problematic for real-time applications.

A variety of maximum power point tracking (MPPT) algorithms for photovoltaic (PV) systems were examined and classified in a study that was referenced in [42]. The classifications were based on the number of control variables and the types of control strategies. In order to examine the dynamic response of the PV voltage ripple, the authors made use of MATLAB/Simulink and the dSPACE framework. They provided a hands-on assessment for commonly used MPPT algorithms [43] and compared them with a PI controller. The research modified environmental variables, including changing irradiance and increasing temperatures, to emphasize the benefits and drawbacks of the P&O and INC algorithms in simulated findings. To improve the P&O algorithm in the context of unforeseen irradiance variations, they put up a novel method [44], which includes two algorithms: an original disturbance algorithm and an adaptive control algorithm. Experimental findings were also compared with the proposed algorithm and traditional algorithms. In another study [45], the authors proposed an updated P&O algorithm to address the root cause of the drift phenomenon and compared the results of experiments and simulations using conventional P&O methods using adaptive measures.

The INC MPPT algorithm has been evaluated in [45]. The authors tested the INC algorithm using an isolated PV pumping approach, which impacted service speeds and altered the reference voltage. Here, the influence of prior disturbances on the phase size and disturbance size has been made clear, which often reveals the algorithm's uncertainty as a result of sudden changes in irradiance. The accomplishment also has something to do with the recommended algorithm.

To monitor the PV properties of MPP, a comparison study is performed between the P&O method and the INC algorithm. In contrast, the transient solution for the INC job ratio disturbance and reference voltage disturbance is more quickly implemented. It was shown that the INC MPPT technique is less susceptible to device dynamics and noise. Greater stability at rapidly changing irradiance has been shown by this phenomenon in the suggested device. The authors of [26] have written up experimental studies with high INC algorithm perturbation rates. At greater disruption speeds, INC is shown to provide a quicker transitional reaction. The method also offers a speedier MPP recovery when the MPP is affected by noise or brightness fluctuations. There have been several MPPT algorithms using PI controllers used so far. However, the most uncomfortable and unusual idea for implementing PV systems was using ANN as the MPPT controllers.

The ANN MPPT was characterized as having a number of off-line preparation characteristics, including nonlinear mapping, a faster response time, and less computing effort in [46]. A novel neural network (NN) MPPT controller for PV systems was proposed by the authors in [47]. Using MATLAB and Simulink, data were extracted from the P&O system to be used in the training and testing of the NN model. The simulation results showed that using the suggested NN controllers for swiftly moving insulation will increase monitoring accuracy, response time, and control. In order to accomplish the electronic power grids' optimal design using AI, the research is important because it integrates configuration parameters with reliability measurements [48]. This article explains how switching the frequency and voltage series may be used to determine a device's stability, efficiency, and cost. Additionally, it provides a thorough examination of the data extraction and training for artificial neural networks. The findings of another ANN MPPT [49] are superior to those of the climbing algorithms.

To locate the global peak using the MATLAB NN-Tool, the authors created a feedback network using the LM backpropagation method. The simulation findings demonstrate that the suggested model is effective and has a smaller root mean square error (RMSE) than the climbing method. The fundamental nature and purpose of diverse MPPTs are achieved using modified methodologies. It is recommended that a technique employ a NN controller instead of a PI controller when utilizing conventional MPPT algorithms. PI-dependent algorithms may thus provide better dynamic stability with abrupt changes in the environment.

3. Modelling of ANN-Based MPPT for Solar PV System

3.1. Solar Photovoltaic (PV) System

A solar photovoltaic (PV) cell, also known as a solar cell, is a device that uses a process known as the photovoltaic effect to transform light energy into electrical energy. The photovoltaic effect occurs when photons from sunlight knock electrons into a higher state of energy, allowing them to flow as an electrical current. Solar PV cells are made of semiconductor materials, typically silicon, and are designed to capture the energy from sunlight and convert it into usable electrical power.

Solar PV cells are connected together to form a solar panel, which can then be connected in series to form a solar array. The efficiency of solar PV cells, or the percentage of sunlight energy that is transformed into electrical energy, varies, but most commercial solar cells have an efficiency of around 15–20%. The performance of a solar cell is also influenced by temperature, shading, and other environmental factors.

The equivalent circuit of a solar cell is a mathematical model that represents the behavior of a solar cell as an electrical circuit. This model is used to analyze and understand the performance of a solar cell, including its voltage, current, and power output.

A current source, a series resistance (denoted by R_s), a shunt resistance (denoted by R_{sh}), and a diode are the standard components that make up the equivalent circuit of a solar cell. The current source is symbolic of the photocurrent that is produced by the solar cell, and the R_s is symbolic of the resistance that is posed by the material and connections that are included inside the cell. The R_{sh} represents the resistance to current flow through pathways other than the intended path, such as cracks or defects in the cell. The diode represents the non-linear behavior of the solar cell, including it short-circuit current and open-circuit voltage.

By analyzing the equivalent circuit, one can better understand the factors that influence the performance of a solar cell and make design improvements to increase its efficiency. For example, reducing the values of R_s and R_{sh} can increase the overall performance of the solar cell. Additionally, the equivalent circuit can be used to determine the MPP of a solar cell, which is the point at which the solar cell produces the maximum amount of power.

There are several different types of equivalent circuits used to model the behavior of solar cells, each with its own advantages and limitations. The most common types of equivalent circuits are:

1. One-diode model (Scheme 1): The one-diode model is the simplest type of equivalent circuit and is often used as a first approximation of the behavior of a solar cell. It consists of only a current source and a diode and is relatively easy to analyze and understand. The main advantage of the one-diode model is its simplicity, which makes it suitable for many applications where a quick estimate of the performance of a solar cell is needed.
2. Two-diode model (Scheme 2): The two-diode model is a more sophisticated equivalent circuit that includes two diodes and a series resistance. This model is used to represent the behavior of a solar cell more accurately and is particularly useful for analyzing the performance of cells under varying light and temperature conditions. The main advantage of the two-diode model is its improved accuracy compared to the one-diode model, which makes it suitable for applications where a more detailed understanding of the behavior of a solar cell is needed.
3. Circular model: The circular model is an advanced equivalent circuit that includes a series resistance, a shunt resistance, and a diode. This model represents the behavior of a solar cell more accurately than the one-diode or two-diode models and is particularly useful for analyzing the behavior of cells under complex environmental conditions. The main advantage of the circular model is its improved accuracy and the ability to model the effects of shading and other environmental factors on the performance of a solar cell.
4. Three-diode model (Scheme 3): It is a more sophisticated equivalent circuit that consists of three diodes and a series resistance. This model is used to represent the behavior of a solar cell more accurately, particularly under conditions of high light intensity and high temperature. The three-diode model takes into account the non-linear behavior of a solar cell under these conditions and provides a more accurate representation of the behavior of a solar cell than the circular model or the one- or two-diode models.

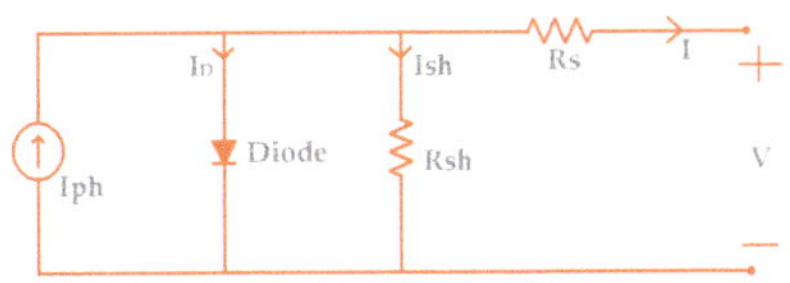

Scheme 1. Equivalent circuit of one diode model.

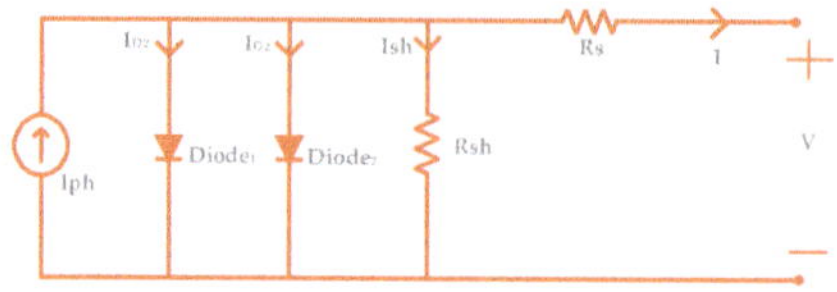

Scheme 2. Equivalent circuit of two diode model.

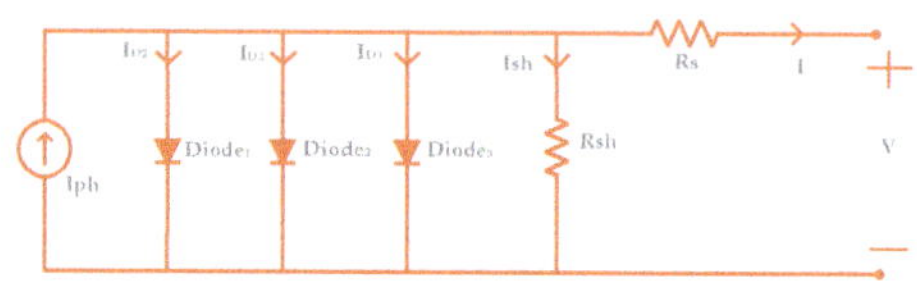

Scheme 3. Equivalent circuit of three diode model.

A simulation model for a photovoltaic (PV) cell is created to optimize its power conversion efficiency. This is achieved by taking into account the impact of light intensity and temperature on the cell's production capacity. The analogous electrical circuit of the PV cell, as displayed in Figure 2, is used as the basis for this simulation model. This permits the measurement of the maximum power point of the solar panels and provides a more accurate forecast of the performance of the cell under various environmental circumstances. Additionally, this makes it possible to determine the maximum power point.

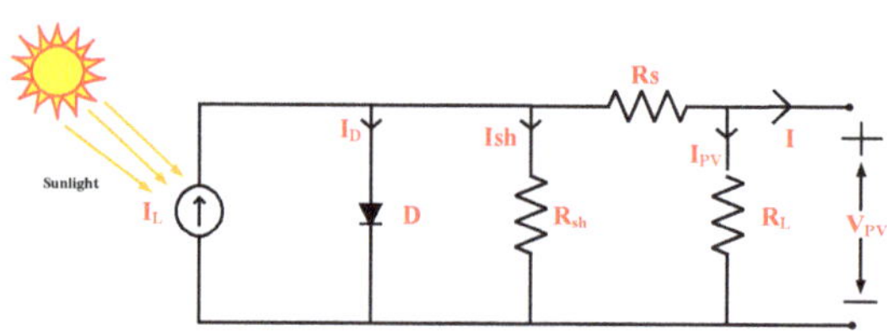

Figure 2. Analogous circuit of a solar PV cell.

In Figure 3a, the current and voltage for a photovoltaic cell are shown in response to various levels of irradiance, including (200 W/m^2, 400 W/m^2, 600 W/m^2, 800 W/m^2, and 1000 W/m^2) at 25 °C. The voltage and current of the cell fluctuate as a consequence of variations in the sun irradiation.

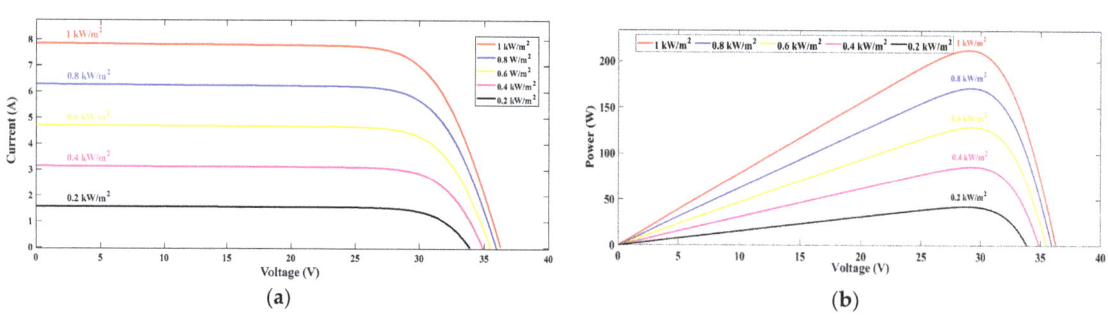

Figure 3. Characteristic curve of module 1Soltech ISTH-215-P: (**a**) I-V curve; (**b**) P-V curve.

Figure 3b demonstrates the impact of light intensity on the production capacity of the photovoltaic cell. It shows that even with the same temperature, the optimal energy point for the cell occurs at a constant level of light intensity.

The solar cell current can be expressed using the variables such as short circuit current (I_{sc}), saturation current (I_o), diode ideality constant (a), number of series-connected cells

(N_s), temperature of the cell (T), Boltzmann constant (K), charge of an electron (q), series resistance (R_S), and shunt resistance (R_{sh}) of the array. These variables are used in Equations (1) and (2) to represent the solar cell current [50].

$$I = I_{sc} - I_o(e^{\frac{V+IR_S}{\frac{N_S KT}{q}a}} - 1) - \frac{V + IR_s}{R_{sh}} \tag{1}$$

$$I = I_{sc} - I_o(e^{\frac{V+IR_S}{V_T a}} - 1) - \frac{V + IR_s}{R_{sh}} \tag{2}$$

where the value of K and q are 1.38×10^{-23} J/K and 1.6×10^{-19} C, respectively. Equation (1) can be rewritten as Equation (2), if the array thermal voltage is replaced by $V_T = \frac{N_s KT}{q}$.

3.2. ANN-Based MPPT for Solar PV System

The MATLAB/Simulink—based simulated model of the solar PV system, illustrated in Figure 4, has been developed. It utilizes the 1Soltech ISTH-215-P solar panel, with the electrical specifications provided in Table 1. The simulated model consists of two primary subsystems: the ANN_MPPT and Switching block. The proposed system incorporates an ANN algorithm, which is represented by an ANN block as depicted in Figure 5.

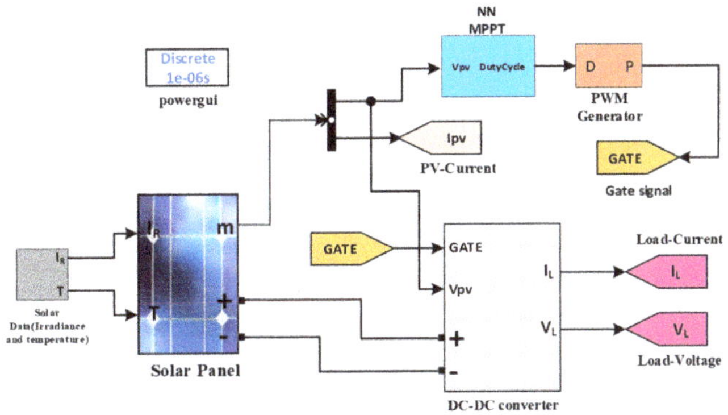

Figure 4. ANN–Based MPPT Solar PV Model.

Table 1. Specification of 1Soltech ISTH–215–P Solar Module.

Specification	Value
Power at STC (W)	215
Power at PTC (W)	189.4
V_{mp}: Voltage at Max Power (V)	29.0
I_{mp}: Current at Max power (A)	7.35
V_{oc}: Open Circuit Voltage (V)	36.3
I_{sc}: Short Circuit Current (A)	7.84
Nominal Operating Cell Temperature (°C)	47.4
Open Circuit Voltage Temp Coefficient (%/°C)	−0.361
Short Circuit Current Temp Coefficient (%/°C)	0.102
Max power Temp Coefficient (%/°C)	−0.495
Power Density at STC (W/m^2)	136.943
Power Density at PTC (W/m^2)	120.637

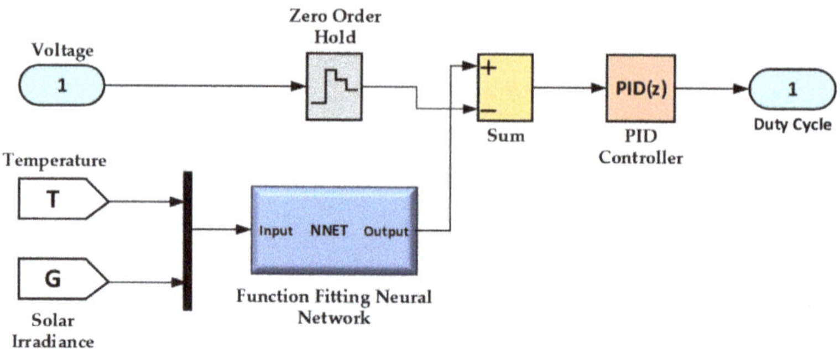

Figure 5. Proposed Block Diagram of ANN Algorithm.

Within the ANN_MPPT subsystem, a comparator compares the output voltage, V1, of the ANN with the voltage generated by the PV array. This generated voltage serves as the reference for the comparator. A Proportional Integral Derivative (PID) controller generates a duty cycle signal based on the difference between V and V1. The Switching block incorporates a boost converter where the insulated-gate bipolar transistor (IGBT) is activated by the gate signal produced by the pulse width modulated (PWM) generator. The duty cycle of PWM is controlled by the voltage difference observed by the comparator. The ANN algorithm ensures a consistent duty cycle for PWM through the ideal correlation between the target and trained values, leading to smooth IGBT switching operations. Additionally, the solar data subsystem sequentially provides input data (irradiance and array temperature) to the PV array, ensuring that the simulation time corresponds to the time taken for transferring the input data.

Solar radiation intensity and panel temperature both have an impact on how much power can be generated by solar panels. Hence, the input data for the artificial neural network (ANN) model is computed depends on the solar irradiance and temperature using Formulas (3) and (4).

Irradiance Calculation, G (W/m^2):

$$G = [(G_{max} - G_{min}) \times rand\,] + G_{min} \tag{3}$$

Temperature Calculation, T (°C):

$$T = [(T_{max} - T_{min}) \times rand\,] + T_{min} \tag{4}$$

Maximum Voltage (V_{MP}), at given G and T

$$V_{MP} = V_{OC} + (beta \times (T - T_S)) \tag{5}$$

where V_{OC} is open circuit voltage of panel and T_S is standard temperature.

Maximum Current (I_{MP}) at given G and T

$$I_{MP} = I_M \times (\frac{G}{G_S}) \times (1 + (alpha \times (T - T_S))) \tag{6}$$

Maximum Power (P_{MP}) at given G and T

$$P_{MP} = I_{MP} \times V_{MP} \tag{7}$$

The MPPT technology that is used for the solar panel system is built with the help of data on solar irradiation, the maximum voltage produced and temperature. The solar PV array receives specific data about the amount of solar irradiation as well as the temperature from the solar data system. Equations (5)–(7) are used in order to compute the solar panel maximum produced power (P_{MP}), maximum generated voltage (V_{MP}), and maximum generated current (I_{MP}) for a variety of irradiance and temperature conditions. The data that are considered to be the output of the neural network are the voltage that is produced by the solar panel.

The rated current and temperature coefficient and rated voltage of the solar panel (1Soltech ISTH-215-P) were used as the basis for the input and output datasets that were used in the design process for the MPPT technology. Both the standard irradiance (G_s) of the sun and the standard temperature (T_s) are assumed to be 1000 W/m^2 and 25 degrees Celsius, respectively. The irradiance levels that are regarded as the highest, or G_{max}, and the lowest, or G_{min}, are, respectively, 1000 W/m^2 and 0 W/m^2. The highest temperature, denoted by the notation T_{max}, is determined to be 35 degrees Celsius, while the lowest temperature, denoted by T_{min}, is determined to be 15 degrees Celsius.

In Figure 6, the flowchart of proposed ANN algorithm is shown in order to implement the MPPT technology. In MATLAB/Simulink, the neural network toolbox function is used to construct the ANN algorithm. In order to collect data, develop and train a network, and assess the network performance, the fitting application of the ANN toolbox is utilized.

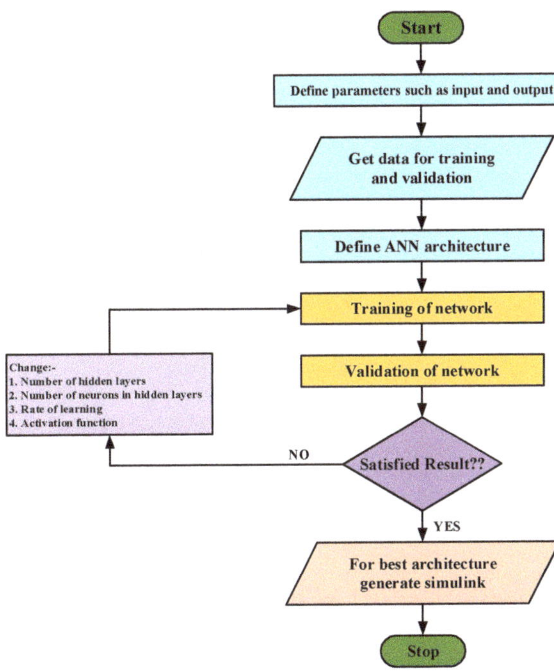

Figure 6. Flowchart of ANN Algorithm.

The created ANN is made up of sigmoid hidden neurons and linear output neurons, and it is built as a two-layer feedforward network. This kind of network is well-suited for solving multi-dimensional mapping issues. In order to train the neural network, one thousand datasets, including information on the temperature, irradiance and produced voltage of the chosen 1Soltech ISTH-215-P solar panel, are used. The data are then arbitrarily split into thirds: 10 percent for validation, 80 percent for training, and 10 percent for testing.

During the process of building the feedforward network, the number of neurons for the hidden layer is estimated to be 20. For training ANN datasets, a number of different training algorithms, including RP, GDM, LM, BR, SCG, and BFGS, may be used. The mean squared error, or MSE, is defined as the average squared difference between outputs and objectives and is used by the LM algorithm, which is sometimes referred to as the damped least-squares approach.

In the LM strategy, "steepest descent" and "Gauss–Newton" are combined as two optimization techniques. When the estimated values are near to the final solution, the LM technique shifts to the Gauss–Newton method, which is initially more resistant to beginning values. This allows for a faster convergence rate. The transition from the steepest descent to the Gauss–Newton method is controlled by the "damping factor" parameter, ensuring the efficient use of the LM approach.

The parameters in question are adjusted at each iteration according to the Equation (8) [51]. The equation involves modifying the parameters based on the Jacobian matrix (J), the damping factor (λ), the identity matrix (I), the discrepancy between the network's intended and actual performance (ε), the number of iterations (k), and the Hessian matrix ($J'J + \lambda_k I$).

$$\theta_{k+1} = \theta_k - (\frac{J'\varepsilon}{J'J + \lambda_k I})_{\theta=\theta_k} \tag{8}$$

The damping factor must be positive for the algorithm to converge. The LM algorithm is fast and has a built-in solution in MATLAB, making it efficient to use in that environment. The MSE is used to quantify the performance of the network, with a lower number signifying greater performance. The number 1 represents a perfect correlation, the value 0 represents a random association, and the regression coefficient R determines the degree to which the outputs and targets are correlated with one another.

Since they need fewer cross-validation steps, Bayesian regularized artificial neural networks (BRANNs) are a more reliable substitute for conventional backpropagation nets. BRANNs use mathematical methods to make ridge regression, such as nonlinear regression, a well-posed statistical problem. With this technique, only one iteration is needed to generate the model that is the "most generalizable", but it must reach a local minimum instead of a global minimum. In contrast to the hundreds or thousands of repetitions required for unregularized ANNs, testing has shown that repeating the approach five times is sufficient to prevent any aberrant behavior. The mathematical model and all aspects of BRANN are described in detail in [52,53].

When it comes to training, the SCG strategy is one that is often used for a number of different kinds of issues. It employs knowledge of the second order rather than the line-search approach, which enables the amount of memory that is utilized to be decreased. Previous research [54,55], provides a full grasp of the approach by offering a detailed description of the final algorithm for the SCG. Similarly, when it comes to handle noisy and ill-conditioned data, resilient backpropagation (RP) is a robust optimization algorithm. RP uses only the sign of the gradient to update the weights and biases, which makes it computationally efficient and suitable for large-scale problems [56]. RP does not require any learning rate parameter to be tuned, which simplifies the training process and makes it less sensitive to hyperparameter tuning.

Broyden–Fletcher–Goldfarb–Shanno (BFGS) is a quasi-Newton optimization algorithm that approximates the inverse Hessian matrix of the loss function, which makes it converge faster than first-order optimization algorithms such as gradient descent [57]. Similar to the RP algorithm, BFGS does not require any learning rate parameter to be tuned, which simplifies the training process and makes it less sensitive to hyperparameter tuning. BFGS is relatively robust to noisy and ill-conditioned data and can handle non-convex optimization problems. Furthermore, for gradient descent momentum, the momentum parameter in GDM allows the algorithm to overcome local minima and converge faster to the global minimum of the loss function [58]. GDM reduces the oscillations and noise in the update direction and therefore provides a smoother convergence path. GDM is easy to implement and computationally efficient compared to other optimization algorithms.

4. Results and Discussion

The suggested model for solar energy collection involves simulating 1000 s of data transfer from a photovoltaic (PV) array to assess optimal analysis. This simulation uses a discrete approach instead of a continuous one. The accuracy of the artificial neural network (ANN) depends on the amount of training data and the selected training algorithm. Generally, larger training datasets result in less error from the ANN. The input data for the solar panel, consisting of solar irradiance and panel temperature, are supplied using a lookup table and a clock for synchronization.

The suitability of six algorithms for solar energy harvesting is also compared in this article. To assess the efficacy and performance of each method, regression, gradient, mean square error, Mu, and validation check are all employed. Regression measures the output's ability to predict the inputs, and error is determined by deducting the goal from the output. The neural network uses three main types of samples: training, validation, and testing. Data are trained through training, and the network is then modified in response to error. By terminating training if faults are found, validation checks the network's generalizability. Contrarily, testing does not alter the training data and offers a neutral evaluation of the network's performance after training.

When referring to the training dataset, an "epoch" is a single cycle. A neural network has to be trained across a number of epochs. The iteration, or quantity of partitioned training data batches or steps required to complete one epoch, is connected to the epoch. Heuristically, the network has the opportunity to view the prior data and revise the parameters of the model. The model is impartial towards the most recent few data points during training. To adjust the ANN's parameters and keep the output divergence to a minimum, a gradient is a numerical computation. Each matrix or vector representation of the network parameters is employed as a reference point in this multivariable derivative of the loss function. The ANN sometimes encounters the local minimum issue and fails to converge; thus, the Mu is added to the weight update phrase to prevent this issue. As a result, its value, which ranges from 0 to 1, directly influences the error of convergence during dataset training. The validation check is represented in training data as error minimization.

One or more selected error metrics are used to evaluate and validate a neural network (ANN) prediction model. An approximation of a function is achieved by the ANN algorithm via the use of a continuous error matrix, such as mean square error (MSE), mean absolute error (MAE), or root mean square error (RMSE).

After tallying up the mistakes across all of the inputs and outputs of the validation set, the results are then normalized based on the size of the set. Each data instance receives an application of a loss function that has been squared and then averaged across the entire dataset in order to maximize the predictive model's operation as a whole. By using error minimization, commonly known as "backpropagation," the ANN modifies its anticipated output in relation to its actual output.

4.1. Levenberg–Marquardt (LM)

Levenberg–Marquardt (LM) is an optimization algorithm widely used for training ANNs. It is particularly effective in solving non-linear regression problems and finding the optimal set of weights to minimize the difference between the network's predicted outputs and the target outputs. LM is known for its fast convergence and robustness, making it a popular choice in the field of machine learning. The LM algorithm combines the benefits of both the Gauss–Newton and steepest descent methods. It iteratively updates the weights of the network by considering both local and global search directions. The primary objective of LM is to minimize a given error function, often represented by the mean squared error (MSE), which quantifies the discrepancy between the predicted and target outputs. The core idea behind LM is to adaptively adjust the step size of weight updates based on the local curvature of the error surface. It achieves this by introducing a damping parameter that controls the trade-off between the local and global search directions. In regions where the error surface is steep and narrow, the damping parameter reduces the step size to avoid overshooting the optimal solution. Conversely, in regions where the error surface is flat, the damping parameter increases the step size to speed up convergence. The LM algorithm starts with an initial set of weights and computes the Jacobian matrix, which represents the sensitivity of the network's outputs to changes in the weights. The Jacobian is used to approximate the Hessian matrix, which describes the curvature of the error surface. The Hessian matrix is modified by adding a damping term to ensure its positive definiteness, which guarantees the convergence of the algorithm. During each iteration, the LM algorithm updates the weights by solving a system of linear equations derived from the modified Hessian matrix and the gradient of the error function. This update step is performed iteratively until the error function reaches a minimum or a convergence criterion is met. The convergence criterion is often based on the change in the error function between iterations. One of the key advantages of LM is its ability to handle non-linear regression problems effectively. Unlike other gradient-based methods, the LM does not require explicit computation of the Hessian matrix, which can be computationally expensive for large networks. Instead, it approximates the Hessian using the Jacobian matrix and adapts the damping parameter to ensure stable convergence.

Another advantage of LM is its robustness to local minima. The combination of local and global search directions allows LM to escape from shallow local minima and find better solutions. This property makes LM particularly useful in situations where the error surface is complex and contains multiple local minima. Implementing LM for ANN training requires careful initialization of the weights and tuning of the damping parameter. In practice, the initial weights can be randomly assigned or set based on prior knowledge of the problem domain. The damping parameter is typically adjusted dynamically during the training process based on the convergence behavior of the algorithm. In conclusion, the Levenberg–Marquardt (LM) algorithm is a powerful optimization method for training artificial neural networks (ANNs). Its combination of local and global search directions, adaptive step size adjustment, and robustness to local minima make it an effective tool for solving non-linear regression problems. LM's fast convergence and stability contribute to its widespread use in the field of machine learning. With its ability to handle complex error surfaces and find optimal solutions, LM plays a significant role in the successful training of ANNs and the advancement of the field.

The plot in Figure 7 demonstrates the accuracy of the ANN's prediction of output in relation to input, indicated by the regression (R = 1) measurement. Error is defined as the variance between the solar panel's produced voltage output and its intended generated voltage, which is determined by deducting the output from the target. The regression plot illustrates that the LM algorithm has effectively trained the data with minimal error, as the output follows the desired value quite closely.

This approach for ANN is further validated by Figure 8, which shows zero error in the data matching training phase, validation phase, and test phases. The bins indicate how many vertical bars there are in the error histogram in Figure 8, where the total ANN error varies from −0.0004 (the leftmost bin) to 0.0000512 (the rightmost bin). It is shown how many samples from the chosen dataset fit into each of the 20 smaller bins that make up the error range. For 100 samples from the validation dataset, the bin with the error value −0.0000015 is in the center of the error histogram. The use of ANN for MPPT shows convergence at 20 bins with 0% error in the error histogram.

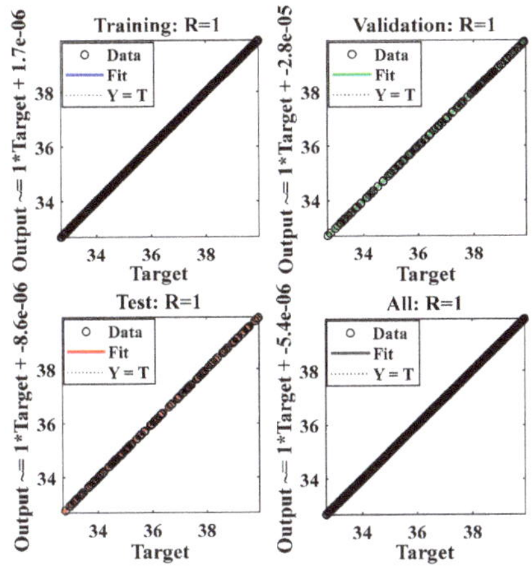

Figure 7. Regression plot of LM algorithm.

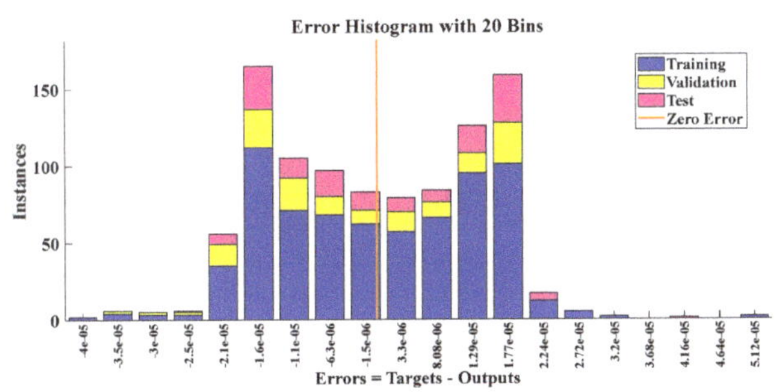

Figure 8. Error histogram plot of LM algorithm.

Figures 9 and 10 illustrate the training stage and performance phase of the ANN for the purpose of processing the selected dataset, respectively. The validation check, as well as the gradient and momentum parameter (Mu), for the training dataset are shown in Figure 9 at 1000 epochs. The simulation indicates that the gradient is 7.983×10^{-6} at the 1000 epoch, which represents the insignificant variances from the training data with a small loss function. The simulation's findings indicate that a choice to produce zero outputs and the mean for each input vector make up the cumulative error. The LM algorithm suitability for MPPT is justified by the extremely low value of gradient, Mu and validation tests of the training dataset.

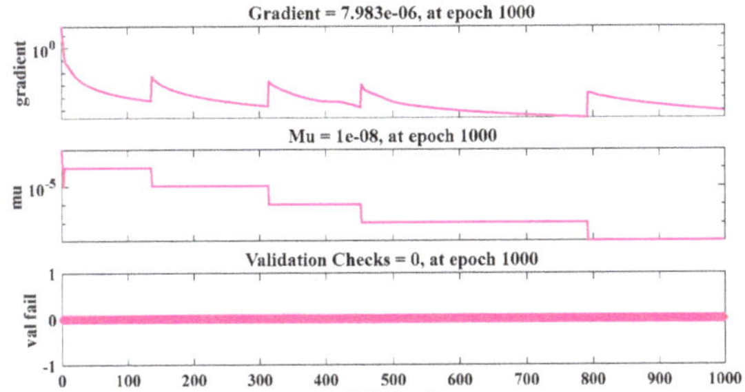

Figure 9. Training test plot of LM algorithm.

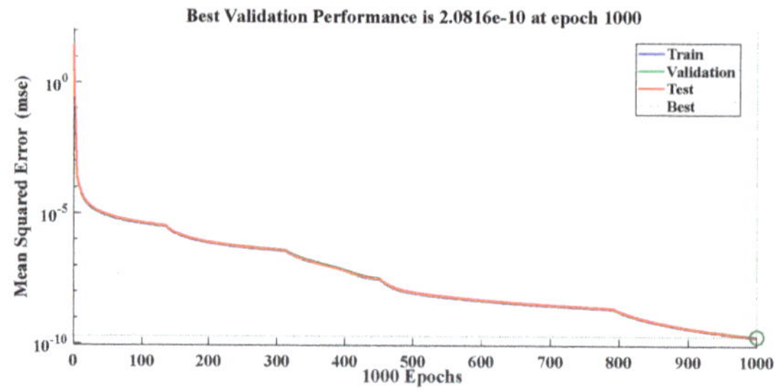

Figure 10. Performance plot of LM algorithm.

Figure 10 depicts the MSE for the trained dataset samples at different epochs, with the best training result being achieved at 1000 epochs. As a consequence of this, the trained dataset displays the highest level of validation performance after 1000 epochs have passed. In accordance with the findings of the simulation, the validation performance of 2.0816×10^{-10}, which is reached at the 1000 epoch, is the best possible value. The estimation of MPPT provided by the LM method had a nearly zero validation performance, which means that its error was minimal.

4.2. Bayesian Regularization (BR)

BR is a mathematical method like ridge regression that is used to solve nonlinear regression problems in a well-posed way. In this method, the nonlinear regression is turned into a statistical problem, and a conjugate gradient descent or a similar minimizer is used to find a solution. The advantage of using BR over traditional backpropagation methods is that it is more stable and eliminates the importance of thorough cross-validation. The method only requires a single iteration to create the model that is "most generalizable", although repeating the process several times may be necessary to ensure a local minimum is reached instead of a global minimum. BRANN is a type of neural network that uses BR for training and is successful in solving a broad range of problems. It uses second-order knowledge instead of line-search methods, which uses less memory. The BR is a powerful technique used for training ANNs that addresses the challenges of overfitting and model complexity. By incorporating Bayesian principles into the training process, BR provides a probabilistic framework that allows for more robust and stable learning. The primary objective of BR is to find the optimal balance between fitting the training data well and avoiding overfitting. Overfitting occurs when the model becomes too complex and starts to memorize the training data instead of generalizing well to unseen data. BR tackles this issue by introducing a regularization term into the training objective, which encourages simpler and more robust models. The core idea behind BR is to impose a prior distribution over the weights of the network and update this distribution during the training process. The prior distribution reflects our prior beliefs about the values of the weights before observing the data. By incorporating prior knowledge, BR provides a regularization mechanism that constrains the model's complexity and prevents it from overfitting.

During training, BR aims to find the posterior distribution of the weights given the observed data. This is achieved by maximizing the posterior probability using the training data and the prior distribution. The posterior distribution represents the updated beliefs about the weights after observing the data. The optimization process involves finding the weight values that maximize the posterior probability, which can be approached using various methods such as Markov chain Monte Carlo (MCMC) or variational inference. In conclusion, Bayesian regularization (BR) is a powerful technique for training artificial neural networks (ANNs) that addresses the challenges of overfitting and model complexity. By incorporating Bayesian principles, BR provides a probabilistic framework that allows for more robust and stable learning. Its ability to automatically determine the regularization strength and provide uncertainty estimates for predictions makes it a valuable tool in the field of machine learning. With its adaptability and robustness, BR contributes to the advancement of ANN training and its successful application in various domains.

From the regression plot in Figure 11, the trained dataset does not need to go through a validation step. In the regression diagram, the best correlation between output and desired generated voltage is shown by the value R = 1, which indicates that the data for the solar PV system were appropriately trained.

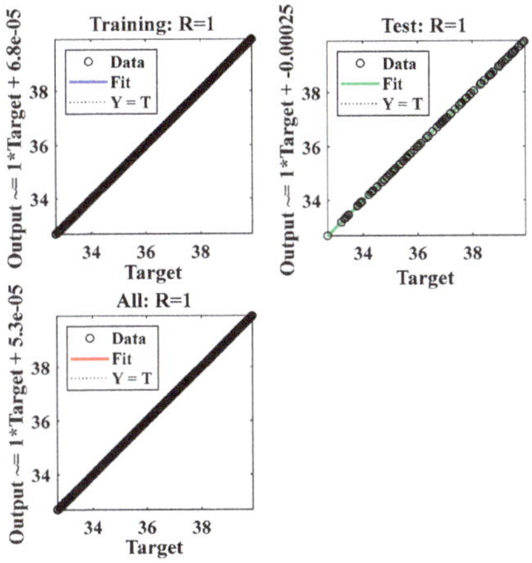

Figure 11. Regression plot of BR algorithm.

Figure 12 depicts the trained dataset, which contains zero errors in both the training and testing phases and total errors that range from −0.00111 (the bin on the left) to 0.00116 (the bin on the right) (the rightmost bin). The gradient and Mu in the training state phase are, at the 1000th epoch, 0.00010493 and 5000, respectively. The error histogram plot of the BR algorithm is shown in Figure 13; there are 20 smaller bins, and the central bin has a near-zero error of 0.000037 for 100 samples, which is larger than that of the LM approach.

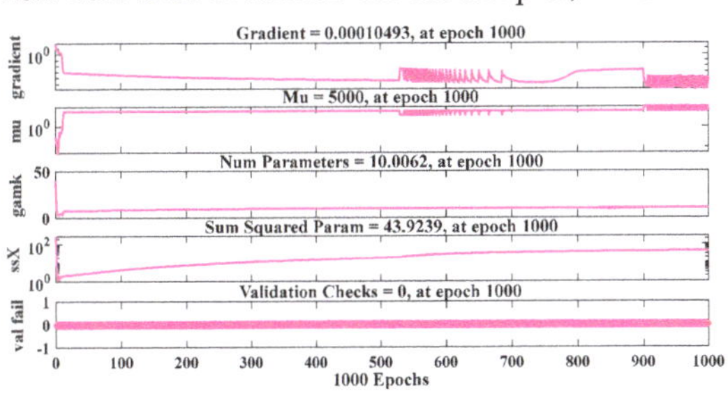

Figure 12. Training test plot of BR.

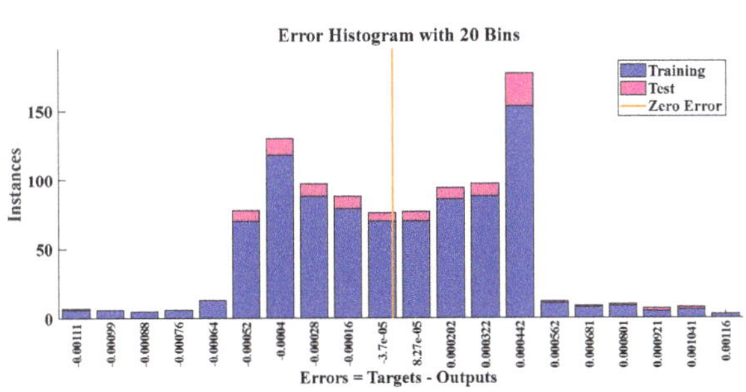

Figure 13. Error histogram plot of BR.

The actual number of parameters at the 1000th epoch is 10.0062; however, the total squared parameters at that time amount to 43.9239. This is shown by the LM method's slower backpropagation capacity for the training dataset and by the high value of Mu as well as an efficient number of parameters and absence of any validation tests.

Compared to the BR approach, which starts with an objective function for the prediction and adds the residual sum of squares and sum of squared weights for reducing the prediction error, the LM algorithm gives quicker convergence in predicting training data with near to zero error. As a consequence, the LM approach generally performs training dataset processing faster than the BR technique. Figure 14, which shows convergence of trained data with the optimum training outcome after 1000 iterations, shows the mean squared error at various epochs. The robustness of the BR method is shown by the best training performance of 1.582×10^{-7} at the 1000 epoch with no validation phase.

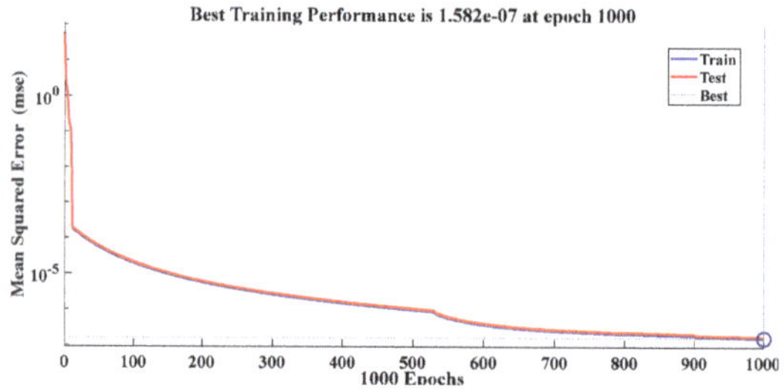

Figure 14. Performance plot of BR.

4.3. Scaled Conjugate Gradient (SCG)

The scaled conjugate gradient (SCG) algorithm is a powerful optimization technique commonly used for training artificial neural networks (ANNs). It offers an efficient and effective approach to updating the network's weights, facilitating convergence and improving the network's performance. ANNs are computational models inspired by the structure and functioning of the human brain. They are composed of interconnected nodes, known as neurons, which work together to process and transmit information. ANNs have gained significant attention in various fields due to their ability to learn from data and make accurate predictions. Training an ANN involves adjusting the weights of its connections to minimize the difference between the predicted outputs and the target outputs. The optimization algorithm used for weight adjustment plays a crucial role in determining the network's performance. The SCG algorithm is a popular choice for this task due to its desirable properties.

The SCG algorithm is based on the conjugate gradient (CG) method, a well-known optimization technique. The CG method aims to find the minimum of a function by iteratively updating the weight values. However, the SCG algorithm introduces a scaling factor that ensures the gradients have the same magnitude, which accelerates convergence. This scaling property allows the algorithm to adaptively adjust the learning rate for each weight update, leading to efficient weight adjustments and improved convergence speed. One of the key advantages of the SCG algorithm is that it eliminates the need for an explicit line search procedure, which is often required in traditional gradient-based optimization algorithms [59]. The line search procedure is responsible for determining an appropriate learning rate at each iteration. By incorporating second-order information and using a scaling approach, the SCG algorithm estimates the learning rate without the need for repetitive line search calculations. This feature significantly reduces the computational burden and makes the algorithm more efficient. The SCG algorithm exhibits excellent performance in handling non-linear optimization problems, making it particularly suitable for training ANNs with complex architectures. It has been successfully applied in various domains, including pattern recognition, data mining, and control systems. Researchers and practitioners often rely on the SCG algorithm to optimize the performance of their neural network models and achieve accurate predictions in real-world applications. One notable advantage of the SCG algorithm is its robustness to noise and ill-conditioned problems. Ill-conditioned problems refer to situations where slight changes in input data or initial weights can have a significant impact on the optimization process. The SCG algorithm's adaptive learning rate and scaling properties help mitigate the effects of such issues, making it more stable and reliable. Another significant benefit of the SCG algorithm is its good generalization capability. Generalization refers to the ability of a trained network to perform well on unseen data. The SCG algorithm's efficient weight adjustments and convergence properties contribute to improved generalization, allowing the network to make accurate predictions on new and unseen instances. To summarize, the SCG algorithm offers several advantages for training ANNs. Its efficient weight adjustment scheme, adaptive learning rate estimation, and scaling properties contribute to faster convergence and improved generalization capabilities. The algorithm's robustness to noise and ill-conditioned problems makes it a reliable choice for various applications. Researchers and practitioners rely on the SCG algorithm to optimize the performance of their neural network models and achieve accurate predictions in diverse fields.

In comparison to the LM and BR algorithms, the R is somewhat less than 1, as seen by the regression plot in Figure 15. A regression value less than 1 for the SCG algorithm implies that there might be some level of inconsistency or error in the predictions made by the algorithm. The deviation from a perfect correlation suggests that there is room for improvement in the algorithm's ability to accurately predict the output based on the given input. In comparison, when the regression value is 1, it indicates a perfect correlation between the input and output data, suggesting that the algorithm is able to accurately capture the relationship between the variables.

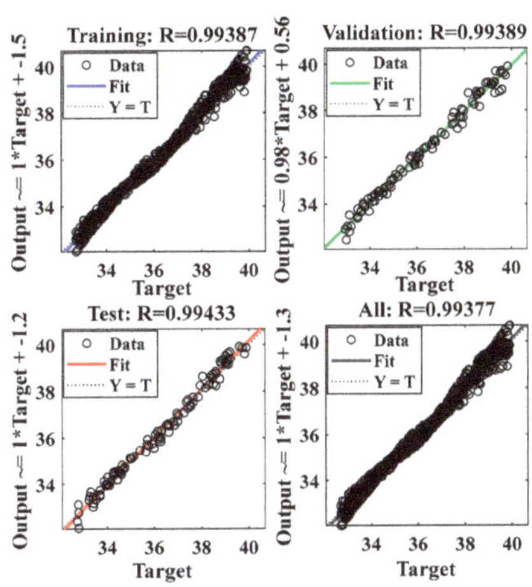

Figure 15. Regression plot of SCG algorithm.

As can be seen in Figure 16, the total error for the trained dataset varies from −0.7406 (the leftmost bin) to 0.8365 (the rightmost bin) when there is no mistake at any point throughout the training phase, validation phase, or test phase. The error histogram shows that the center bin has a value that is −0.006443 (100 samples) higher than the value produced by the LM and BR algorithms. After 75 iterations, Figure 17 displays a gradient with a value of 0.14161, while the validation tests have a value of 6. When data training is terminated after 75 epochs, the SCG's performance in terms of achieving the target objective suffers as a direct result.

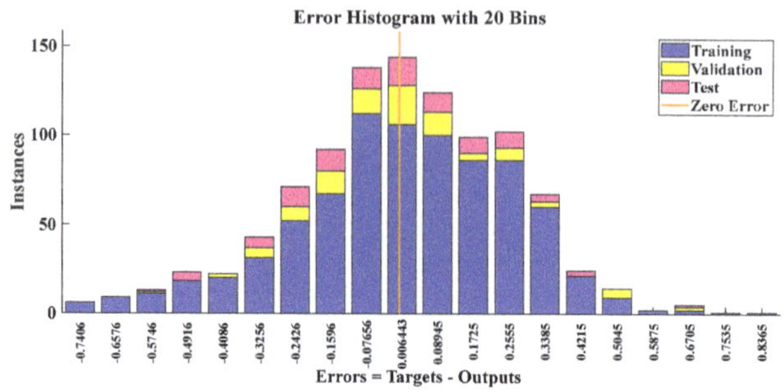

Figure 16. Error histogram of SCG algorithm.

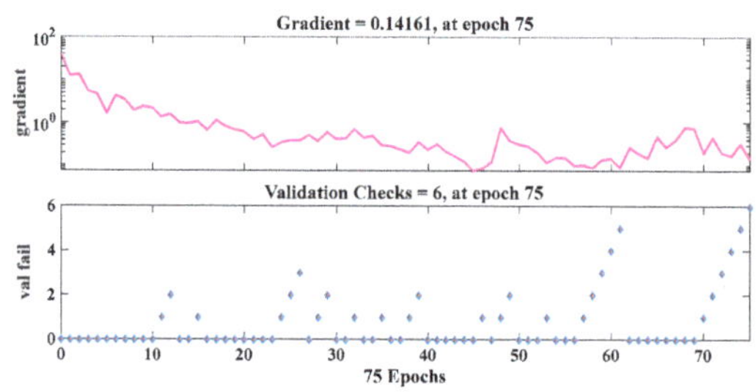

Figure 17. Training test plot of SCG.

The performance gradient for validation is higher than the maximum time for failure and lower than the lowest gradient. Figure 18, which represents the convergence of training data with the highest validation performance of 0.052985 at 75 epochs, displays the mean squared error at various epochs. The learned dataset for the solar PV system do not suit the SCG algorithm, despite having greater validation performance and a lower total training performance than the LM and BR algorithms. Similar to this, the BR method is less suitable than the LM algorithm due to the high processing time for prediction and the large momentum parameter.

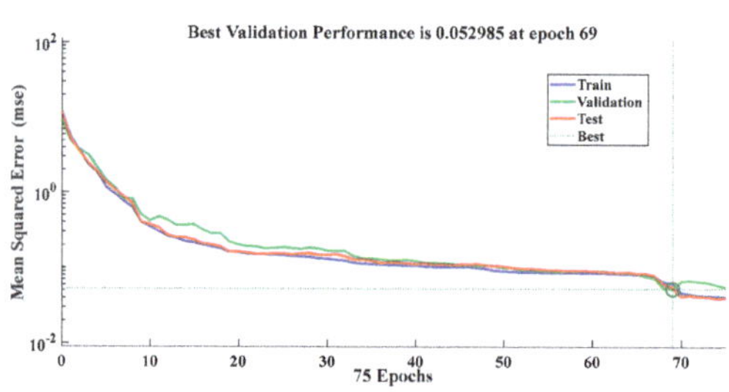

Figure 18. Performance plot of SCG.

4.4. *Resilient Backpropagation (RP)*

Resilient backpropagation (RP) is a gradient-based optimization algorithm commonly used in training artificial neural networks (ANNs). It is a variation of the traditional backpropagation algorithm that uses a different update rule for adjusting the weights of the neural network during training.

The key feature of the RP algorithm is its use of a dynamic learning rate for each weight in the network. The learning rate determines the step size used to update the weights during the backpropagation process and can have a significant impact on the convergence and stability of the training process. In traditional backpropagation, a fixed learning rate is used for all weights, which can lead to slow convergence or even divergence if the learning rate is set too high or too low.

The RP algorithm overcomes this problem by using two different update rules based on the sign of the gradient of the error function with respect to the weight. If the gradient changes sign from one iteration to the next, indicating that the optimization process is moving in the wrong direction, the learning rate for that weight is reduced by a factor (e.g., 0.5). If the gradient has the same sign as the previous iteration, indicating that the optimization process is moving in the right direction, the learning rate is increased by a factor (e.g., 1.2). If the gradient is zero, indicating that the weight has reached a stationary point, the learning rate is unchanged.

This approach is known as a "resilient" update rule, as it allows the algorithm to recover quickly from bad updates and adapt to changing conditions during the optimization process. By adjusting the learning rate dynamically for each weight, the RP algorithm can achieve faster convergence and better stability compared to traditional backpropagation. Another advantage of the RP algorithm is its ability to handle noisy or ill-conditioned data, which can cause traditional optimization algorithms to become stuck in local minima. By adjusting the learning rate dynamically, the RP algorithm is able to navigate complex, high-dimensional search spaces more effectively and avoid becoming trapped in local minima. Hence, the RP algorithm is a robust and efficient optimization algorithm that is well-suited for training ANNs. Its ability to adapt to changing conditions and handle noisy data make it a popular choice for many machine learning applications.

From the regression plot in Figure 19, the trained dataset does not need to go through a validation step. In the regression diagram, the best correlation between output and desired generated voltage is shown by the value R = 1, which indicates that the data for the solar PV system were appropriately trained.

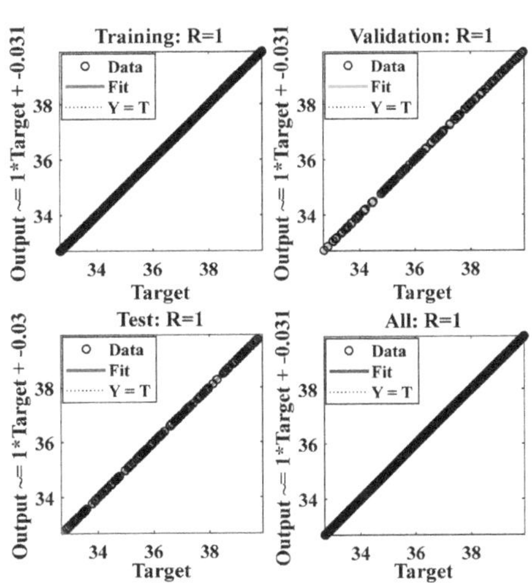

Figure 19. Regression plot of RP algorithm.

After 21 iterations, Figure 20 displays a gradient with a value of 0.033415, while the validation tests have a value of 6 at 21 epochs. A gradient value of 0.033415 indicates the rate of change or steepness of the objective function with respect to the algorithm's parameters. A low gradient value indicates that the algorithm is making progress towards the optimal solution, and the estimate is close to the true minimum. This value suggests that the RP algorithm is making significant progress towards optimizing the objective function and reaching the desired solution. The validation tests having a value of 6 at 21 epochs implies that the algorithm's predictions during the validation phase have a relatively larger deviation from the true values compared to desired accuracy. This indicates that the algorithm may not be performing optimally in terms of accurately predicting the target values during the validation phase.

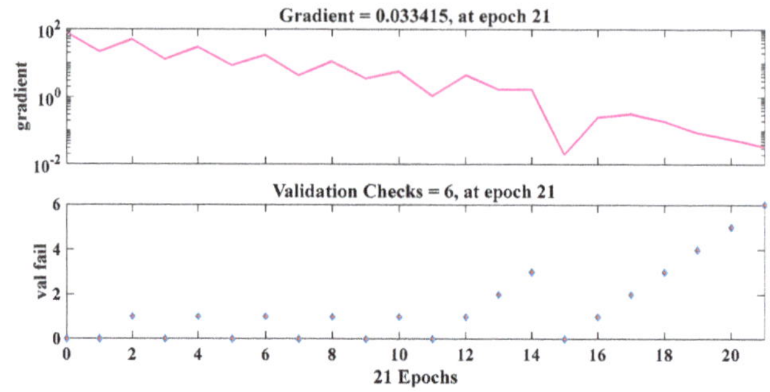

Figure 20. Training plot of RP algorithm.

Figure 21 displays the mean squared error at a number of different epochs. This figure illustrates the convergence of trained data with the optimal training result after 15 iterations. The greatest training performance of 2.8668×10^{-6} was achieved at the 15 epoch even though there was no validation phase, which demonstrates the resilience of the RP algorithm.

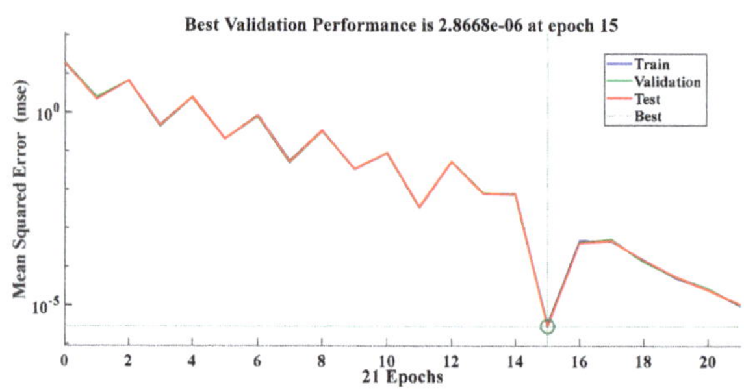

Figure 21. Performance test plot of Rp algorithm.

4.5. Gradient Descent Momentum (GDM)

Gradient descent momentum (GDM) is a popular optimization algorithm for training artificial neural networks (ANNs). It is a variation of the traditional gradient descent algorithm that incorporates a momentum term to help speed up the convergence of the optimization process. The momentum term allows the algorithm to "remember" the direction it has been moving in the past and use that information to help accelerate the optimization process.

The basic idea behind gradient descent with momentum is to add a "velocity" term to the weight updates that takes into account the direction and magnitude of previous weight updates. Specifically, the velocity for weight 'w' at iteration 't' is given by:

$$V_t(w) = \alpha \times V_{t-1}(w) - \eta \times \nabla \in (w_t) \quad (9)$$

where α is the momentum coefficient (typically set to a value between 0 and 1), η is the learning rate, $\in (w_t)$ is the error function at iteration t, and $\nabla \in (w_t)$ is the gradient of the error function with respect to weight w at iteration t.

The weight update at iteration t is then given by:

$$w_{t+1} = w_t + V_t(w) \quad (10)$$

The momentum term allows the algorithm to "smooth out" the weight updates over time, reducing the impact of small, noisy changes in the gradient and helping to avoid getting stuck in local minima. The momentum term can also help the algorithm to accelerate when the gradient is pointing consistently in the same direction, allowing it to move more quickly towards the global minimum. One of the key advantages of GDM is its ability to handle noisy or ill-conditioned data, which can cause traditional optimization algorithms to become stuck in local minima. By incorporating information about the direction and magnitude of previous weight updates, the momentum term allows the algorithm to navigate complex, high-dimensional search spaces more effectively and avoid becoming trapped in local minima. Another advantage of gradient descent with momentum is its ability to handle non-convex objective functions. Non-convex objective functions can be difficult to optimize using traditional gradient descent, as the optimization process can become stuck in local minima. By incorporating information about the direction and magnitude of previous weight updates, the momentum term allows the algorithm to explore more of the search space and avoid becoming trapped in local minima. Hence, gradient descent with momentum is a powerful and efficient optimization algorithm that is well-suited for training ANNs. Its ability to handle noisy or ill-conditioned data and to navigate complex, high-dimensional search spaces make it a popular choice for many machine learning applications.

In comparison to the LM, RP, BFGS, and BR algorithms, the R is somewhat less than 1, as seen by the regression plot in Figure 22 for GDM algorithm.

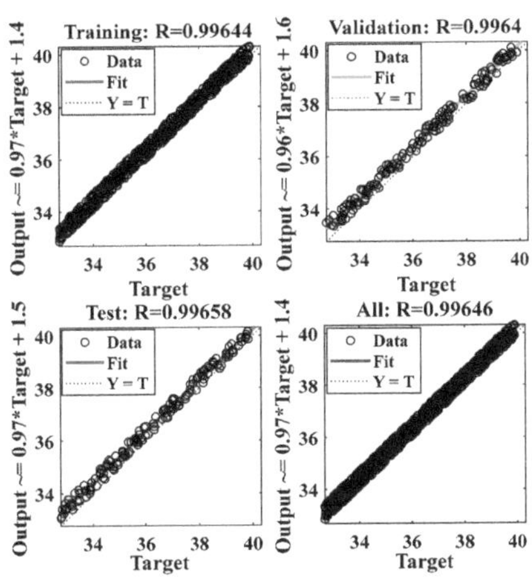

Figure 22. Regression plot of GDM algorithm.

A regression value of 0.996 indicates a strong positive correlation between the predicted values and the actual values in the training dataset, while a regression value of 1 indicates a perfect positive correlation between the predicted values and the actual values.

Figure 23 displays a gradient with a value of 11.485, while the validation tests have a value of 6 at 8 epochs. The higher value of gradient for GDM algorithm indicates that the current estimate is far from the true minimum, and the algorithm needs to take large steps to get closer to it. This could result in overshooting the minimum or becoming stuck in a local minimum instead of the global minimum, which can lead to poor performance.

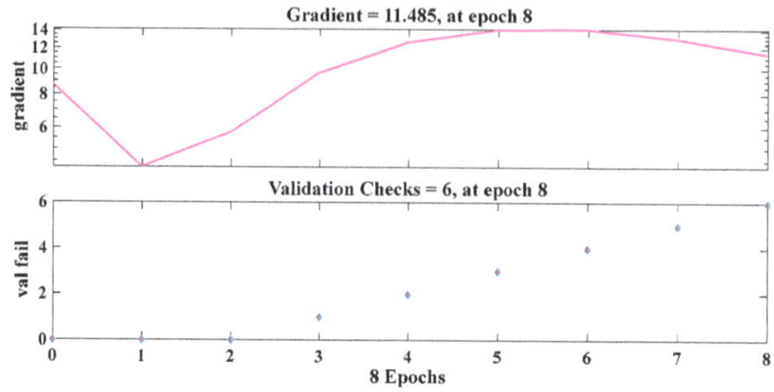

Figure 23. Training plot of GDM algorithm.

Figure 24 displays the performance plot of GDM algorithm. The plot shows that the algorithm achieves its best validation performance of 0.15378 at 2 epochs. A validation performance of 0.15378 indicates that the algorithm's predictions have a relatively moderate level of deviation from the true values of the validation dataset. Although not as low as achieving near-zero validation performance, this value still demonstrates a reasonable level of accuracy in estimating the desired output within a very short training time.

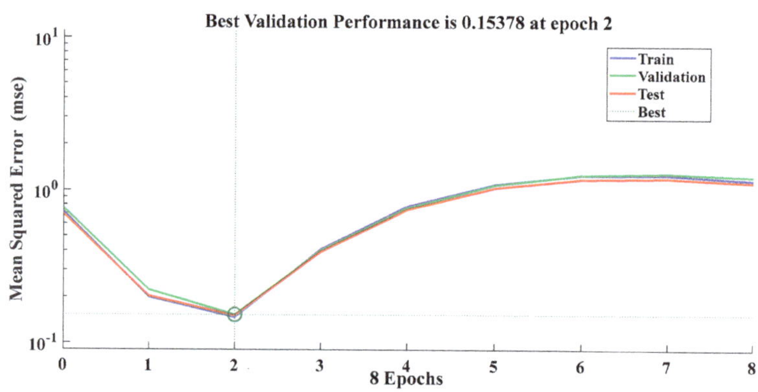

Figure 24. Performance plot of GDM algorithm.

4.6. Broyden–Fletcher–Goldfarb–Shanno (BFGS) Quasi Newton

The Broyden–Fletcher–Goldfarb–Shanno (BFGS) algorithm is a popular quasi-Newton method for unconstrained optimization, commonly used in training artificial neural networks (ANNs). The basic idea behind the BFGS algorithm is to approximate the inverse Hessian matrix of the objective function, which is a measure of how the gradient of the function changes as the weights of the network are updated. By using this approximation, the algorithm is able to estimate the direction and magnitude of weight updates that are likely to lead to a decrease in the objective function.

The BFGS algorithm works by iteratively updating an approximation of the inverse Hessian matrix, denoted by H_k, at each iteration k. The weight update at iteration k is then given by:

$$W_{k+1} = W_k - \alpha_k \times H_k \times \nabla \in (W_k) \tag{11}$$

where α_k is a step size parameter, $\nabla \in (W_k)$ is the gradient of the objective function with respect to the weights at iteration k, and H_k is the approximate inverse Hessian matrix.

The BFGS algorithm uses a rank-2 update formula to update the approximate inverse Hessian matrix at each iteration. Specifically, the update formula is given by:

$$H_{k+1} = (1 - P_k \times S_k \times Y_{kT}) \times H_k(1 - P_k \times Y_k \times S_{kT}) + P_k \times S_k \times S_{kT} \tag{12}$$

where $S_k = W_{k+1} - W_k$, $Y_k = \nabla \in (W_{k+1}) - \nabla \in (W_k)$, and $P_k = \frac{1}{(Y_{kT} \times S_k)}$. This update formula is designed to approximate the true inverse Hessian matrix, while also maintaining the positive definiteness of the approximation.

One of the key advantages of the BFGS algorithm is its ability to handle non-convex objective functions, which are commonly encountered in ANNs. Non-convex objective functions can be difficult to optimize using traditional gradient descent methods, as the optimization process can become stuck in local minima. The BFGS algorithm is able to overcome this problem by using an approximation of the inverse Hessian matrix, which allows it to navigate complex, high-dimensional search spaces more effectively.

Another advantage of the BFGS algorithm is its ability to handle ill-conditioned or noisy data. Traditional optimization algorithms can be sensitive to noisy or ill-conditioned data, which can cause them to become stuck in local minima. The BFGS algorithm is able to overcome this problem by using an approximation of the inverse Hessian matrix, which allows it to smooth out noisy or ill-conditioned data and navigate the search space more effectively. Hence, the BFGS algorithm is a powerful and efficient optimization algorithm that is well-suited for training ANNs. Its ability to handle non-convex objective functions and ill-conditioned or noisy data makes it a popular choice for many machine learning applications. However, like all optimization algorithms, it may require careful tuning of the step size parameter and other hyperparameters to achieve optimal performance on a given problem.

Figure 25 is a figure that indicates how accurate the ANN's prediction of output in relation to input is. This accuracy is demonstrated by the regression measurement of R = 1, which is shown in the plot. The regression plot demonstrates that the BFGS method has successfully trained the data with a small amount of error by displaying the output in a manner that closely reflects the intended value. In other words, the BFGS algorithm accurately captures the relationship between the input variables (such as solar irradiance and temperature) and the corresponding output variable (such as voltage and power output in energy harvesting applications). This high regression value signifies that the BFGS algorithm is effectively modeling and predicting the behavior of the energy harvesting system, making it a reliable and precise algorithm for optimizing energy extraction from solar photovoltaic systems.

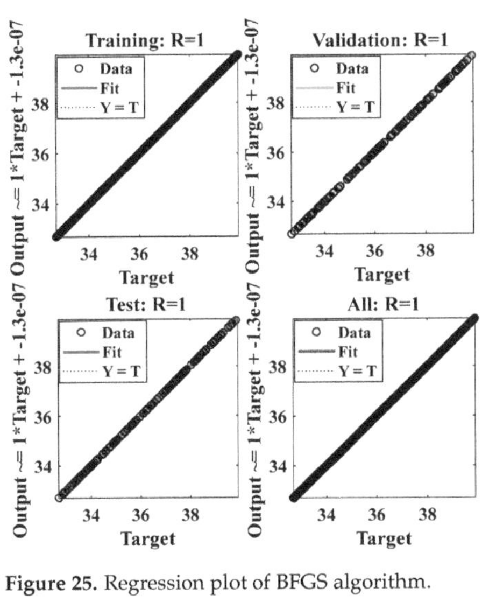

Figure 25. Regression plot of BFGS algorithm.

In Figure 26, a training plot of the BFGS algorithm is shown. The validation check as well as the gradient and reset values for the training dataset are shown at 10 epochs.

The algorithm attained a gradient with a value of 1.0211 × 10^{-7}, while the validation tests and resets had no value at 10 epochs. A gradient value of 1.0211 × 10^{-7} indicates that the algorithm's estimate of the optimal solution is very close to the true minimum. In general, a lower gradient value suggests that the algorithm is converging well and making small, steady steps towards the optimal solution. This indicates that the algorithm is effectively adjusting its parameters to minimize the difference between predicted and actual values, resulting in improved accuracy. The fact that the validation tests and resets have no value at 10 epochs suggests that the algorithm has successfully learned and generalized the underlying patterns in the training data. This means that the algorithm's predictions align well with the validation dataset, and there is no need for further adjustments or resets at this point. Overall, these findings indicate that the algorithm has achieved a high level of accuracy and convergence, making it a reliable choice for MPPT (maximum-power point tracking) energy harvesting in solar PV systems.

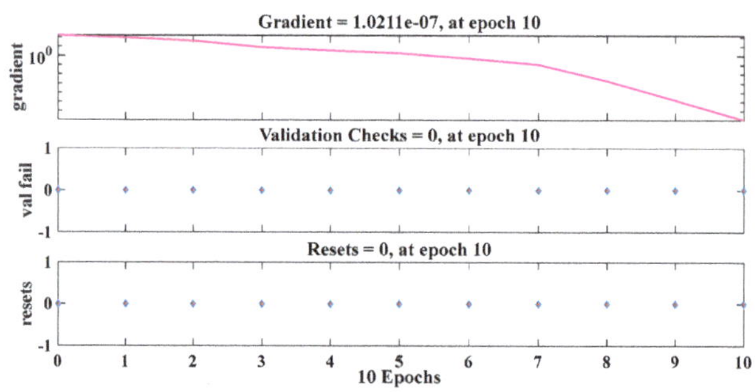

Figure 26. Training plot of BFGS algorithm.

In Figure 27, The convergence of training data with the highest performance value of the BFGS algorithm is displayed. The algorithm attained the best validation performance of 9.98 × 10^{-17} at 10 epochs. A validation performance of 9.98 × 10^{-17} indicates that the algorithm's predictions are extremely close to the true values of the validation dataset. This near-zero validation performance suggests that the algorithm has successfully learned the underlying patterns and relationships in the training data, and it can generalize its predictions accurately to unseen data. It also suggests that the algorithm is capable of accurately capturing the complex relationships between input variables and the desired output, enabling it to make precise predictions for MPPT (maximum-power point tracking) energy harvesting in solar PV systems.

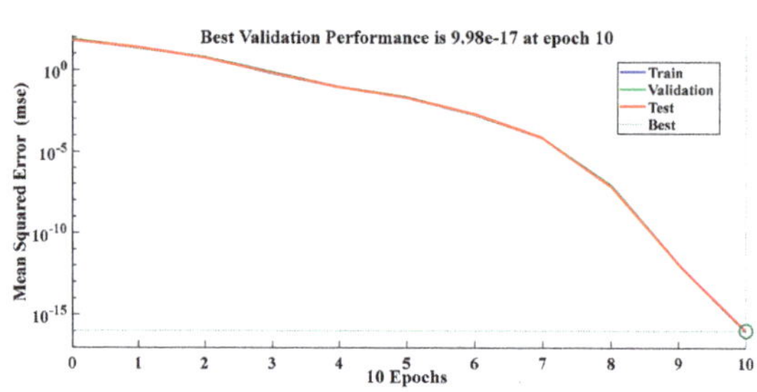

Figure 27. Performance plot of BFGS algorithm.

In relation to solar radiation and array temperature, respectively, Figure 28a–d show the output power and load power of the PV array. According to the MPPT topology of ANN, both powers are claimed to follow the solar irradiation and the temperature of the PV array. Based on the amount of simulation time that has passed, the PV array's produced power, load power, irradiance, and temperature are shown here. The MPPT is satisfied when the array temperature is 15 degrees Celsius and the sun irradiance is 200 watts per square meter. This results in the lowest power output of 100 watts. With an array temperature of 35 degrees Celsius and a solar irradiation of 1000 watts per square meter, the power output reaches its maximum of 450 watts. The ripples in the output power are not smoothed out by this model since it does not have filter component. As a direct consequence of this, the power waveforms that are created and loaded have discernible ripples.

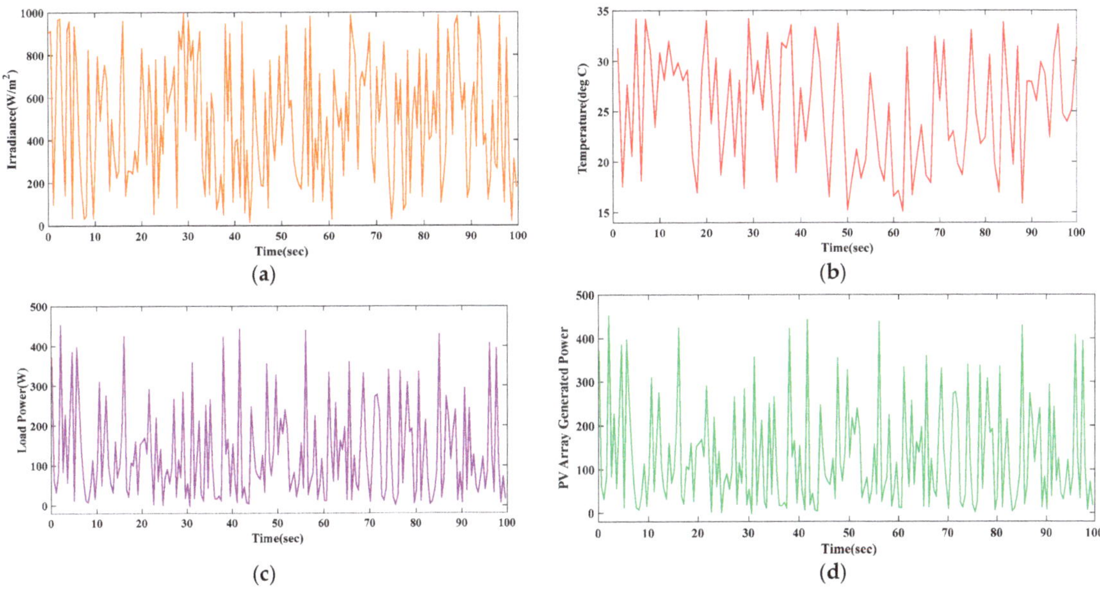

Figure 28. (**a**) Irradiance (W/m^2) versus time (s) plot of the solar panel, (**b**) temperature (°C) versus time (s) plot of the solar panel, (**c**) load power (W) versus time (s) plot of the solar panel, (**d**) and power generated (W) versus time (s) of the panel.

Table 2 presents a dataset consisting of matching values for solar irradiance (G) in watts per square meter (W/m^2), temperature (T) in degrees Celsius (°C), and maximum voltage (Vmax) in volts (V). The table contains 120 rows, each representing a specific observation or data point. These 120 data points are sample data among 1000 data points. Using Equations (3)–(5) of solar irradiance, temperature, and maximum voltage, the data in Table 2 were generated. These equations are used to calculate the solar irradiance (G), temperature (T), and maximum voltage (V_{MP}) based on specific parameters and random values. Furthermore, the data were used to train the neural network.

Table 2. ANN databases of matching values of solar irradiance, G (W/m^2), temperature, T (°C), and maximum voltage, Vmax (V).

S. No.	Solar Irradiance (W/m^2)	Temperature (°C)	Maximum Voltage (V_{max})
1	905.7919371	31.29447373	34.02775793
2	913.3758561	17.53973633	38.99308058
3	97.540405	27.64718492	35.34439271
4	546.8815192	20.56996438	37.89919856

Table 2. *Cont.*

S. No.	Solar Irradiance (W/m^2)	Temperature (°C)	Maximum Voltage (V_{max})
5	964.8885352	34.15013671	32.99689215
6	970.5927818	18.15226163	38.77196507
7	485.3756487	34.14333896	32.99934607
8	141.8863386	31.00560938	34.13203507
9	915.7355252	23.43522565	36.86486789
10	959.4924264	30.84414659	34.19032152
11	35.71167857	28.11481398	35.1755833
12	933.9932478	31.98258612	33.77935624
13	757.7401306	28.5747031	35.00956793
14	392.2270195	29.86264936	34.54463221
15	171.1866878	28.1095578	35.17748073
16	31.83284638	29.12092176	34.81238845
17	46.17139063	20.5384597	37.91057143
18	823.4578283	16.94263562	39.20862797
19	317.0994801	28.89657246	34.89337631
20	34.4460805	34.00444098	33.04948685
21	381.5584571	23.77488719	36.74225347
22	795.1999011	30.31033576	34.38302189
23	489.7643958	18.73745209	38.56071717
24	646.3130101	23.91172401	36.69285675
25	754.686682	29.18729662	34.78842779
26	679.7026769	20.52050154	37.91705415
27	162.6117352	28.10196008	35.18022343
28	498.364052	17.37995363	39.05076054
29	340.3857267	34.19487917	32.98074057
30	223.8119395	26.70535502	35.68438389
31	255.0951155	30.02534119	34.48590209
32	699.0767227	25.11914103	36.25699128
33	959.2914252	32.81806505	33.4777567
34	138.6244428	25.9443106	35.95911332
35	257.5082541	17.98588011	38.83202714
36	254.282179	31.81434512	33.84008956
37	243.5249687	31.28569652	34.03092641
38	349.983766	33.58527246	33.20080249
39	251.083858	18.93190501	38.49052161
40	473.2888489	27.32089352	35.46218065
41	830.8286279	22.03319014	37.37098869
42	549.7236083	26.70528182	35.68441031
43	285.8390188	33.34387328	33.28794519
44	753.7290943	30.14400458	34.44306579

Table 2. *Cont.*

S. No.	Solar Irradiance (W/m^2)	Temperature (°C)	Maximum Voltage (V_{max})
45	567.8216407	22.60891694	37.16315707
46	53.95011867	16.51708579	39.3622472
47	779.1672301	25.61595106	36.07764783
48	129.9062085	33.68021368	33.16652966
49	469.3906411	26.37647322	35.80310693
50	337.1226444	15.23804139	39.82396944
51	794.2845407	18.24364616	38.73897617
52	528.5331355	21.22430084	37.66298964
53	601.9819414	18.31297459	38.7139493
54	654.0790985	20.25942569	38.01129992
55	748.1515928	28.78429006	34.93390913
56	83.821378	24.01083197	36.65707977
57	913.3373615	19.57953937	38.25673208
58	825.8169775	18.04756038	38.80976118
59	996.1347166	25.76684871	36.02317529
60	442.6782698	16.56351058	39.34548832
61	961.8980809	17.1330554	39.13988833
62	774.9104647	15.09268448	39.87644183
63	868.6947054	31.34606441	34.00913421
64	399.7826491	16.68871691	39.30029008
65	800.0684802	20.19740806	38.03368767
66	910.6475944	23.62827655	36.79517845
67	263.8029165	18.63694057	38.59700083
68	136.0685587	17.91077961	38.85913767
69	579.7045874	32.38584415	33.63378412
70	144.9547982	25.99720404	35.94001931
71	622.0551315	32.06062235	33.75118594
72	513.2495399	22.01904762	37.376094
73	75.96669169	23.03616068	37.00892636
74	123.3189348	19.79832307	38.17775335
75	239.9525257	18.67815577	38.58212255
76	49.65443033	23.34534138	36.89731521
77	944.7871897	33.0543222	33.39247023
78	489.2526384	24.81728185	36.36595943
79	900.0538464	21.7543882	37.4716334
80	111.2027553	22.38493562	37.24401209
81	389.738837	30.60504137	34.27663612
82	403.9121456	19.83382572	38.16493725
83	131.9732926	16.9290905	39.21351762
84	956.1345402	33.84101182	33.10848314

Table 2. *Cont.*

S. No.	Solar Irradiance (W/m^2)	Temperature (°C)	Maximum Voltage (V_{max})
85	59.77954295	26.5041719	35.75700899
86	353.1585712	19.69559827	38.21483598
87	15.40343765	31.4238808	33.98104327
88	168.9900295	15.86047603	39.59927676
89	731.7223857	27.9823095	35.22341609
90	450.9237064	27.95491926	35.2333037
91	296.3208056	25.94017785	35.9606052
92	188.955015	29.89385614	34.53336687
93	183.5111557	28.73550867	34.95151873
94	625.6185607	22.36969193	37.24951491
95	81.12576887	30.6045487	34.27681396
96	775.7126786	33.58771942	33.19991917
97	435.8585886	24.73583265	36.39536177
98	306.349472	23.93567499	36.68421069
99	510.7715642	25.17017311	36.23856921
100	794.8314169	31.35255417	34.00679147
101	378.6093827	27.8863626	35.25805196
102	532.8255888	31.23160917	34.05045141
103	939.001562	22.01454207	37.37772046
104	550.1563429	32.51885623	33.58576809
105	587.0447045	27.44950172	35.41575437
106	301.2463303	19.15484585	38.41004219
107	230.4881602	24.41846697	36.50992761
108	194.7642896	31.88617585	33.81415938
109	170.7080471	19.51843562	38.27878993
110	435.6986841	19.55328596	38.2662093
111	923.3796421	21.22204573	37.66380371
112	184.8163201	23.60414783	36.80388868
113	979.7483784	33.09761937	33.37684038
114	111.1192234	23.77739946	36.74134657
115	408.7198461	20.16129392	38.04672451
116	262.2117478	26.89792148	35.61486932
117	711.2157804	27.05686179	35.55749346
118	117.4176509	19.43493468	38.30893293
119	318.7783019	20.93351746	37.76795953
120	507.8582847	23.48333519	36.84750083

Table 3 compares the efficacy of six different ANN algorithms—the LM algorithm, BR algorithm, SCG algorithm, RP algorithm, GDM algorithm, and BFGS algorithm—in this regard.

Table 3. Comparison of the ANN algorithms' performance.

Parameters	Algorithm					
	LM	**RP**	**SCG**	**BR**	**GDM**	**BFGS**
Regression	1	1	0.993	1	0.996	1
Error at middle bin	−0.0000015	-	−0.006443	−0.000037	-	-
Gradient	7.983×10^{-6}	0.033415	0.14161	0.00010493	11.485	1.0211×10^{-7}
Performance	2.0816×10^{-10}	2.8668×10^{-6}	0.052985	1.583×10^{-7}	0.15378	9.98×10^{-17}
Momentum parameter	1.00×10^{-8}	-	-	5000	-	-
Epochs	1000	15	69	1000	2	10

The table shows various performance parameters for each algorithm, including regression, error at the middle bin, gradient, performance, momentum parameter, and epochs. The regression value for the LM, BR, RP, and BFGS algorithms is 1, indicating a perfect correlation between the input and output data, while the SCG and GDM algorithms have a slightly lower regression value of 0.993 and 0.996, respectively. The error at the middle bin is negative for the LM algorithm (−0.0000015), indicating that the algorithm is performing better than the other algorithms. The gradient value for the BFGS algorithm (1.0211×10^{-7}) is much lower than that for the BR, RP, SCG, LM, and GDM algorithms, which are 0.00010493, 0.033415, 0.14161, 7.983×10^{-6}, and 11.485, respectively, which indicates that the BFGS algorithm has better convergence properties. The lower gradient value of BFGS indicates that the algorithm is making progress towards the optimal solution, and the estimate is close to the true minimum. This suggests that the algorithm is converging well and making small, steady steps towards the optimal solution. Similarly, the higher gradient value of the GDM algorithm indicates that the current estimate is far from the true minimum, and the algorithm needs to take large steps to get closer to it. This could result in overshooting the minimum or becoming stuck at a local minimum instead of the global minimum, which can lead to poor performance.

The performance value for the BFGS algorithm (9.98×10^{-17}) is significantly better than that for the BR algorithm (1.583×10^{-7}), SCG algorithm (0.052985), RP algorithm (2.8668×10^{-6}), GDM algorithm (0.15378), and LM algorithm (2.0816×10^{-10}), indicating that the BFGS algorithm provides the most accurate predictions. The LM algorithm secures second place in terms of performance and accuracy of predictions. The momentum parameter is only specified for the LM and BR algorithms, with the LM algorithm having a value of 0.00000001 and the BR algorithm having a value of 5000. The momentum parameter was not applicable for RP, SCG, GDM, and BFGS algorithms, as they do not use momentum. Finally, looking at the number of epochs required for convergence, the RP and BR algorithms required the fewest epochs at 15 and 2, respectively. The LM, GDM, and BFGS algorithms required moderate numbers of epochs, while the SCG algorithm required a relatively large number of epochs at 69. Overall, the results illustrate that the LM algorithm outperforms the other two algorithms in terms of accuracy, convergence, and performance, making it the best choice for MPPT energy harvesting in a solar PV system.

Overall, the results suggest that the LM and BFGS algorithms are the best performers in terms of gradient and performance, while the RP and BR algorithms also perform well across the various metrics. The SCG and GDM algorithms appear to be less effective for this particular problem.

5. Conclusions

In conclusion, this paper presented a novel approach for evaluating the efficacy of six ANN algorithms (LM, BR, RP, GDM, BFGS, and SCG) in the context of MPPT for solar PV systems. The authors utilized a two-layer feedforward neural network from the ANN toolbox and trained it with generated data on solar irradiance, panel temperature, and generated voltage. Based on the outcomes, the regression values of the LM, RP, BR, and BFGS algorithms are all equal to 1, whereas the regression values for the SCG and GDM algorithms are less than 1. The gradient values for the LM, RP, BFGS, SCG, BR, and GDM algorithms are 7.983×10^{-6}, 0.033415, 1.0211×10^{-7}, 0.14161, 0.00010493, and 11.485, respectively. Likewise, the performance values for these algorithms are 2.0816×10^{-10}, 2.8668×10^{-6}, 9.98×10^{-17}, 0.052985, 1.583×10^{-7}, and 0.15378. Overall, the findings demonstrated that the LM and BFGS algorithms performed exceptionally well in terms of performance and gradient, with LM showing almost no error in the middle epoch. At 1000 epochs, the validation efficiency was highest, indicating close-to-zero values for the momentum parameter, validation, and gradient checks. The LM and BFGS algorithms exhibit superior performance in terms of regression, gradient, and overall performance. Additionally, the RP and BR algorithms also demonstrate good performance across various metrics, while the SCG and GDM algorithms show relatively less effectiveness in addressing the proposed problem. These findings provide valuable insights into the relative performance of the six evaluated algorithms for MPPT energy harvesting in solar PV systems.

The proposed ANN-based MPPT energy harvesting concept holds potential for solving problems on a large scale and can be incorporated into multilayer neural networks, making it highly versatile. Furthermore, the study revealed a perfect correlation between input and output data, exemplified by the consistency between generated power and load power after MPPT. This ANN-based MPPT energy harvesting model can find applications in standalone and grid-connected solar PV systems, as well as military equipment, telecommunications, and space satellites. The suggested ANN-based MPPT energy harvesting model exhibits applicability across various domains and can be integrated into numerous technologies. Additionally, the model can be utilized for solar radiation and temperature prediction, energy estimation, energy management in smart homes and cities, and forecasting solar radiation and temperature. Looking ahead, the future scope of solar PV MPPT based on ANN appears promising, offering several areas for research and development. One potential direction is exploring the use of reinforcement learning algorithms to further optimize MPPT control in solar PV systems. Additionally, integrating AI-based control systems with power electronics devices, such as DC–DC converters and inverters, could enhance overall system efficiency. Advancements in sensors and data acquisition systems hold the potential to develop more sophisticated and accurate ANN models for MPPT control. Furthermore, investigating hybrid control algorithms that combine multiple AI-based techniques could lead to improved performance in solar PV MPPT. The proposed work offers valuable insights but has limitations. Firstly, it focuses only on six specific ANN algorithms, potentially limiting the understanding of the overall algorithmic landscape. Secondly, the evaluation is based on specific performance parameters, which may not capture all aspects of algorithm performance. Additionally, the study focuses solely on solar PV systems, excluding other renewable energy sources. This limits the applicability of the findings to a broader range of energy harvesting systems. Overall, this research contributes valuable insights into the evaluation of ANN algorithm performance for MPPT in solar PV systems and sets the stage for future advancements in the field.

Author Contributions: Conceptualization, A.S. and M.T.; formal analysis, M.T.H., A.S., M.T., S.U. and M.A.H.; funding acquisition, S.U. and A.B.; investigation, M.T.H., A.S., M.T. and S.U.; methodology, M.T.H., A.S., M.T. and M.A.H.; project administration, M.T., S.U. and A.B.; resources, M.T.; supervision, A.S. and M.T.; validation, M.T.H., A.S., M.T., A.B. and M.A.H.; writing—original draft, M.T.H., A.S. and M.T.; writing—review and editing, S.U., A.B. and M.A.H. All authors have read and agreed to the published version of the manuscript.

Funding: Princess Nourah bint Abdulrahman University (43-PRFA-P-38), Princess Nourah bint Abdulrahman University, Riyadh, Saudi Arabia.

Institutional Review Board Statement: Not applicable.

Informed Consent Statement: Not applicable.

Data Availability Statement: Not applicable.

Acknowledgments: This research project was funded by the Deanship of Scientific Research, Princess Nourah bint Abdulrahman University, through the Program of Research Project Funding After Publication, grant No (43-PRFA-P-38).

Conflicts of Interest: The authors declare no conflict of interest.

References

1. Kouro, S.; Leon, J.I.; Vinnikov, D.; Franquelo, L.G. Grid-Connected Photovoltaic Systems: An Overview of Recent Research and Emerging PV Converter Technology. *IEEE Ind. Electron. Mag.* **2015**, *9*, 47–61. [CrossRef]
2. Tarroja, B.; Shaffer, B.; Samuelsen, S. The importance of grid integration for achievable greenhouse gas emissions reductions from alternative vehicle technologies. *Energy* **2015**, *87*, 504–519. [CrossRef]
3. Rokonuzzaman; Mishu, M.K.; Amin, N.; Nadarajah, M.; Roy, R.B.; Rahman, K.S.; Buhari, A.M.; Binzaid, S.; Shakeri, M.; Pasupuleti, J. Self-Sustained Autonomous Wireless Sensor Network with Integrated Solar Photovoltaic System for Internet of Smart Home-Building (IoSHB) Applications. *Micromachines* **2021**, *12*, 653. [CrossRef]
4. Koohi-Kamali, S.; Rahim, N.; Mokhlis, H.; Tyagi, V. Photovoltaic electricity generator dynamic modeling methods for smart grid applications: A review. *Renew. Sustain. Energy Rev.* **2016**, *57*, 131–172. [CrossRef]
5. De Giorgi, M.G.; Congedo, P.M.; Malvoni, M. Photovoltaic power forecasting using statistical methods: Impact of weather data. *IET Sci. Meas. Technol.* **2014**, *8*, 90–97. [CrossRef]
6. Karami, N.; Moubayed, N.; Outbib, R. General review and classification of different MPPT Techniques. *Renew. Sustain. Energy Rev.* **2017**, *68*, 1–18. [CrossRef]
7. Mishu, M.K.; Rokonuzzaman, M.; Pasupuleti, J.; Shakeri, M.; Rahman, K.S.; Hamid, F.A.; Tiong, S.K.; Amin, N. Prospective Efficient Ambient Energy Harvesting Sources for IoT-Equipped Sensor Applications. *Electronics* **2020**, *9*, 1345. [CrossRef]
8. Rokonuzzaman; Shakeri, M.; Hamid, F.A.; Mishu, M.K.; Pasupuleti, J.; Rahman, K.S.; Tiong, S.K.; Amin, N. IoT-Enabled High Efficiency Smart Solar Charge Controller with Maximum Power Point Tracking—Design, Hardware Implementation and Performance Testing. *Electronics* **2020**, *9*, 1267. [CrossRef]
9. Sivakumar, P.; Kader, A.A.; Kaliavaradhan, Y.; Arutchelvi, M. Analysis and enhancement of PV efficiency with incremental conductance MPPT technique under non-linear loading conditions. *Renew. Energy* **2015**, *81*, 543–550. [CrossRef]
10. Roy, R.B.; Basher, E.; Yasmin, R.; Rokonuzzaman, M. Fuzzy logic based MPPT approach in a grid connected photovoltaic system. In Proceedings of the the 8th International Conference on Software, Knowledge, Information Management and Applications (SKIMA 2014), Dhaka, Bangladesh, 18–20 December 2014; pp. 1–6. [CrossRef]
11. Bhatti, A.R.; Salam, Z.; Sultana, B.; Rasheed, N.; Awan, A.B.; Sultana, U.; Younas, M. Optimized sizing of photovoltaic grid-connected electric vehicle charging system using particle swarm optimization. *Int. J. Energy Res.* **2018**, *43*, 500–522. [CrossRef]
12. Dixit, T.V.; Yadav, A.; Gupta, S. Experimental assessment of maximum power extraction from solar panel with different converter topologies. *Int. Trans. Electr. Energy Syst.* **2018**, *29*, e2712. [CrossRef]
13. Rokonuzzaman, H.-E.-H. Design and implementation of maximum power point tracking solar charge controller. In Proceedings of the 2016 3rd International Conference on Electrical Engineering and Information Communication Technology (ICEEICT), Dhaka, Bangladesh, 22–24 September 2016; pp. 1–5. [CrossRef]
14. Mishu, M.K.; Rokonuzzaman; Pasupuleti, J.; Shakeri, M.; Rahman, K.S.; Binzaid, S.; Tiong, S.K.; Amin, N. An Adaptive TE-PV Hybrid Energy Harvesting System for Self-Powered IoT Sensor Applications. *Sensors* **2021**, *21*, 2604. [CrossRef] [PubMed]
15. Esram, T.; Chapman, P.L. Comparison of Photovoltaic Array Maximum Power Point Tracking Techniques. *IEEE Trans. Energy Convers.* **2007**, *22*, 439–449. [CrossRef]
16. Basha, C.H.; Rani, C. Different Conventional and Soft Computing MPPT Techniques for Solar PV Systems with High Step-Up Boost Converters: A Comprehensive Analysis. *Energies* **2020**, *13*, 371. [CrossRef]
17. Lopez-Guede, J.M.; Ramos-Hernanz, J.; Altın, N.; Ozdemir, S.; Kurt, E.; Azkune, G. Neural Modeling of Fuzzy Controllers for Maximum Power Point Tracking in Photovoltaic Energy Systems. *J. Electron. Mater.* **2018**, *47*, 4519–4532. [CrossRef]
18. Khanam, J.J.F.; Simon, Y. Modeling of a photovoltaic array in MATLAB simulink and maximum power point tracking using neural network. *J. Electr. Electron. Syst.* **2018**, *7*, 40–46.
19. Chen, L.; Wang, X. Enhanced MPPT method based on ANN-assisted sequential Monte–Carlo and quickest change detection. *IET Smart Grid* **2019**, *2*, 635–644. [CrossRef]
20. Chtouki, I.; Wira, P.; Zazi, M. ICIT—Comparison of several neural network perturb and observe MPPT methods for photovoltaic applications. In Proceedings of the 2018 IEEE International Conference on Industrial Technology (ICIT), Lyon, France, 20–22 February 2018; pp. 909–914. [CrossRef]
21. Bouakkaz, M.S.; Boukadoum, A.; Boudebbouz, O.; Bouraiou, A.; Attoui, I. ANN based MPPT Algorithm Design using Real Operating Climatic Condition. In Proceedings of the 2020 2nd International Conference on Mathematics and Information Technology (ICMIT), Adrar, Algeria, 18–19 February 2020. [CrossRef]
22. Algarín, C.R.; Hernández, D.S.; Leal, D.R. A Low-Cost Maximum Power Point Tracking System Based on Neural Network Inverse Model Controller. *Electronics* **2018**, *7*, 4. [CrossRef]

23. Divyasharon, R.; Banu, R.N.; Devaraj, D. Artificial Neural Network based MPPT with CUK Converter Topology for PV Systems Under Varying Climatic Conditions. In Proceedings of the 2019 IEEE International Conference on Intelligent Techniques in Control, Optimization and Signal Processing (INCOS), Tamilnadu, India, 11–13 April 2019; pp. 1–6. [CrossRef]
24. Fatima, K.; Alam, M.A.; Minai, A.F. Optimization of Solar Energy Using ANN Techniques. In Proceedings of the 2019 2nd International Conference on Power Energy, Environment and Intelligent Control (PEEIC), Greater Noida, India, 18–19 October 2019. [CrossRef]
25. Al-Majidi, S.D.; Abbod, M.F.; Al-Raweshidy, H.S. Design of an intelligent MPPT based on ANN using a real photovoltaic system data. In Proceedings of the 2019 54th International Universities Power Engineering Conference (UPEC), Bucharest, Romania, 3–6 September 2019. [CrossRef]
26. Elgendy, M.A.; Atkinson, D.J.; Zahawi, B. Experimental investigation of the incremental conductance maximum power point tracking algorithm at high perturbation rates. *IET Renew. Power Gener.* **2016**, *10*, 133–139. [CrossRef]
27. Jyothy, L.P.; Sindhu, M.R. An Artificial Neural Network based MPPT Algorithm For Solar PV System. In Proceedings of the 2018 4th International Conference on Electrical Energy Systems (ICEES), Chennai, India, 7–9 February 2018. [CrossRef]
28. Vimalarani, C.; Kamaraj, N. Improved method of maximum power point tracking of photovoltaic (PV) array using hybrid intelligent controller. *Optik* **2018**, *168*, 403–415. [CrossRef]
29. Roy, R.B.; Cros, J.; Nandi, A.; Ahmed, T. Maximum Power Tracking by Neural Network. In Proceedings of the 2020 8th International Conference on Reliability, Infocom Technologies and Optimization (Trends and Future Directions) (ICRITO), Noida, India, 4–5 June 2020. [CrossRef]
30. Danyali, S.; Aghaei, O.; Shirkhani, M.; Aazami, R.; Tavoosi, J.; Mohammadzadeh, A.; Mosavi, A. A New Model Predictive Control Method for Buck-Boost Inverter-Based Photovoltaic Systems. *Sustainability* **2022**, *14*, 11731. [CrossRef]
31. Aazami, R.; Heydari, O.; Tavoosi, J.; Shirkhani, M.; Mohammadzadeh, A.; Mosavi, A. Optimal Control of an Energy-Storage System in a Microgrid for Reducing Wind-Power Fluctuations. *Sustainability* **2022**, *14*, 6183. [CrossRef]
32. Kaya, C.B.; Kaya, E.; Gökkuş, G. Training Neuro-Fuzzy by Using Meta-Heuristic Algorithms for MPPT. *Comput. Syst. Sci. Eng.* **2023**, *45*, 69–84. [CrossRef]
33. Ibnelouad, A.; El Kari, A.; Ayad, H.; Mjahed, M. Improved cooperative artificial neural network-particle swarm optimization approach for solar photovoltaic systems using maximum power point tracking. *Int. Trans. Electr. Energy Syst.* **2020**, *30*, 12439. [CrossRef]
34. Roy, R.B.; Rokonuzzaman, M.; Amin, N.; Mishu, M.K.; Alahakoon, S.; Rahman, S.; Pasupuleti, J. A Comparative Performance Analysis of ANN Algorithms for MPPT Energy Harvesting in Solar PV System. *IEEE Access* **2021**, *9*, 102137–102152. [CrossRef]
35. Li, Z.; Rahman, S.M.; Vega, R.; Dong, B. A Hierarchical Approach Using Machine Learning Methods in Solar Photovoltaic Energy Production Forecasting. *Energies* **2016**, *9*, 55. [CrossRef]
36. Xia, L.; Ma, Z.; Kokogiannakis, G.; Wang, Z.; Wang, S. A model-based design optimization strategy for ground source heat pump systems with integrated photovoltaic thermal collectors. *Appl. Energy* **2018**, *214*, 178–190. [CrossRef]
37. Ding, M.; Wang, L.; Bi, R. An ANN-based Approach for Forecasting the Power Output of Photovoltaic System. *Procedia Environ. Sci.* **2011**, *11*, 1308–1315. [CrossRef]
38. Porrazzo, R.; Cipollina, A.; Galluzzo, M.; Micale, G. A neural network-based optimizing control system for a seawater-desalination solar-powered membrane distillation unit. *Comput. Chem. Eng.* **2013**, *54*, 79–96. [CrossRef]
39. Chine, W.; Mellit, A.; Lughi, V.; Malek, A.; Sulligoi, G.; Pavan, A.M. A novel fault diagnosis technique for photovoltaic systems based on artificial neural networks. *Renew. Energy* **2016**, *90*, 501–512. [CrossRef]
40. Vaz, A.; Elsinga, B.; van Sark, W.; Brito, M. An artificial neural network to assess the impact of neighbouring photovoltaic systems in power forecasting in Utrecht, the Netherlands. *Renew. Energy* **2016**, *85*, 631–641. [CrossRef]
41. Elsheikh, A.H.; Sharshir, S.W.; Abd Elaziz, M.; Kabeel, A.E.; Guilan, W.; Haiou, Z. Modeling of solar energy systems using artificial neural network: A comprehensive review. *Sol. Energy* **2019**, *180*, 622–639. [CrossRef]
42. Subudhi, B.; Pradhan, R. A Comparative Study on Maximum Power Point Tracking Techniques for Photovoltaic Power Systems. *IEEE Trans. Sustain. Energy* **2013**, *4*, 89–98. [CrossRef]
43. De Brito, M.A.G.; Galotto, L.; Sampaio, L.P.; Melo, G.D.A.E.; Canesin, C.A. Evaluation of the Main MPPT Techniques for Photovoltaic Applications. *IEEE Trans. Ind. Electron.* **2013**, *60*, 1156–1167. [CrossRef]
44. Kollimalla, S.K.; Mishra, M.K. A Novel Adaptive P&O MPPT Algorithm Considering Sudden Changes in the Irradiance. *IEEE Trans. Energy Convers.* **2014**, *29*, 602–610. [CrossRef]
45. Elgendy, M.A.; Zahawi, B.; Atkinson, D.J. Assessment of the Incremental Conductance Maximum Power Point Tracking Algorithm. *IEEE Trans. Sustain. Energy* **2013**, *4*, 108–117. [CrossRef]
46. Elobaid, L.M.; Abdelsalam, A.K.; Zakzouk, E.E. Artificial neural network-based photovoltaic maximum power point tracking techniques: A survey. *IET Renew. Power Gener.* **2015**, *9*, 1043–1063. [CrossRef]
47. Mohammed, K.K.; Buyamin, S.; Shams, I.; Mekhilef, S. Maximum power point tracking based on adaptive neuro-fuzzy inference systems for a photovoltaic system with fast varying load conditions. *Int. Trans. Electr. Energy Syst.* **2021**, *31*, e12904. [CrossRef]
48. Dragicevic, T.; Wheeler, P.; Blaabjerg, F. Artificial Intelligence Aided Automated Design for Reliability of Power Electronic Systems. *IEEE Trans. Power Electron.* **2019**, *34*, 7161–7171. [CrossRef]

49. Ramana, V.V.; Jena, D. Maximum power point tracking of PV array under non-uniform irradiance using artificial neural network. In Proceedings of the 2015 IEEE International Conference on Signal Processing, Informatics, Communication and Energy Systems (SPICES), Kozhikode, India, 19–21 February 2015; pp. 1–5. [CrossRef]
50. Ali, F.; Sarwar, A.; Bakhsh, F.I.; Ahmad, S.; Shah, A.A.; Ahmed, H. Parameter extraction of photovoltaic models using atomic orbital search algorithm on a decent basis for novel accurate RMSE calculation. *Energy Convers. Manag.* **2023**, *277*, 116613. [CrossRef]
51. Dkhichi, F.; Oukarfi, B.; Fakkar, A.; Belbounaguia, N. Parameter identification of solar cell model using Levenberg–Marquardt algorithm combined with simulated annealing. *Sol. Energy* **2014**, *110*, 781–788. [CrossRef]
52. Gimazov, R.; Shidlovskiy, S. Simulation Modeling of Intelligent Control Algorithms for Constructing Autonomous Power Supply Systems with Improved Energy Efficiency. In *MATEC Web of Conferences*; EDP Sciences: Les Ulis, France, 2018; Volume 155, p. 01032. [CrossRef]
53. Macaulay, J.; Zhou, Z. A Fuzzy Logical-Based Variable Step Size P&O MPPT Algorithm for Photovoltaic System. *Energies* **2018**, *11*, 1340. [CrossRef]
54. Cho, T.-H.; Hwang, H.-R.; Lee, J.-H.; Lee, I.-S. Comparison of Intelligent Methods of SOC Estimation for Battery of Photovoltaic System. *Int. J. Adv. Comput. Sci. Appl.* **2018**, *9*, 90907. [CrossRef]
55. Ul-Haq, A.; Sindi, H.F.; Gul, S.; Jalal, M. Modeling and Fault Categorization in Thin-Film and Crystalline PV Arrays Through Multilayer Neural Network Algorithm. *IEEE Access* **2020**, *8*, 102235–102255. [CrossRef]
56. Abd Ellah, A.R.; Essai, M.H.; Yahya, A. Comparison of different backpropagation training algorithms using robust M-estimators performance functions. In Proceedings of the 2015 Tenth International Conference on Computer Engineering & Systems (ICCES), Cairo, Egypt, 23–24 December 2015; pp. 384–388. [CrossRef]
57. Dong, C.L.; Nocedal, J. On the limited memory BFGS method for large scale optimization. *Math. Program* **1989**, *45*, 503–528. Available online: https://link.springer.com/content/pdf/10.1007%2FBF01589116.pdf (accessed on 11 June 2023).
58. Luo, X.; Qin, W.; Dong, A.; Sedraoui, K.; Zhou, M. Efficient and High-quality Recommendations via Momentum-incorporated Parallel Stochastic Gradient Descent-Based Learning. *IEEE/CAA J. Autom. Sin.* **2020**, *8*, 402–411. [CrossRef]
59. Møller, M.F. A scaled conjugate gradient algorithm for fast supervised learning. *Neural Netw.* **1993**, *6*, 525–533. [CrossRef]

sustainability

MDPI

Article

Tourism Service Scheduling in Smart City Based on Hybrid Genetic Algorithm Simulated Annealing Algorithm

Pannee Suanpang [1,*], Pitchaya Jamjuntr [2], Kittisak Jermsittiparsert [3,4,5,6,7] and Phuripoj Kaewyong [1]

1 Faculty of Science & Technology, Suan Dusit University, Bangkok 10300, Thailand
2 Faculty of Engineering, King Mongkut's University of Technology Thonburi, Bangkok 10140, Thailand
3 Faculty of Education, University of City Island, Famagusta 9945, Cyprus
4 Faculty of Social and Political Sciences, Universitas Muhammadiyah Sinjai, Kabupaten Sinjai 92615, Sulawesi Selatan, Indonesia
5 Faculty of Social and Political Sciences, Universitas Muhammadiyah Makassar, Kota Makassar 90221, Sulawesi Selatan, Indonesia
6 Publication Research Institute and Community Service, Universitas Muhammadiyah Sidenreng Rappang, Sidenreng Rappang Regency 91651, South Sulawesi, Indonesia
7 Sekolah Tinggi Ilmu Administrasi Abdul Haris, Kota Makassar 90000, Sulawesi Selatan, Indonesia
* Correspondence: pannee_sua@dusit.ac.th

Abstract: The disruptions in this era have caused a leap forward in information technology being applied in organizations to create a competitive advantage. In particular, we see this in tourism services, as they provide the best solution and prompt responses to create value in experiences and enhance the sustainability of tourism. Since scheduling is required in tourism service applications, it is regarded as a crucial topic in production management and combinatorial optimization. Since workshop scheduling difficulties are regarded as extremely difficult and complex, efforts to discover optimal or near-ideal solutions are vital. The aim of this study was to develop a hybrid genetic algorithm by combining a genetic algorithm and a simulated annealing algorithm with a gradient search method to the optimize complex processes involved in solving tourism service problems, as well as to compare the traditional genetic algorithms employed in smart city case studies in Thailand. A hybrid genetic algorithm was developed, and the results could assist in solving scheduling issues related to the sustainability of the tourism industry with the goal of lowering production requirements. An operation-based representation was employed to create workable schedules that can more effectively handle the given challenge. Additionally, a new knowledge-based operator was created within the context of function evaluation, which focuses on the features of the problem to utilize machine downtime to enhance the quality of the solution. To produce the offspring, a machine-based crossover with order-based precedence preservation was suggested. Additionally, a neighborhood search strategy based on simulated annealing was utilized to enhance the algorithm's capacity for local exploitation, and to broaden its usability. Numerous examples were gathered from the Thailand Tourism Department to demonstrate the effectiveness and efficiency of the proposed approach. The proposed hybrid genetic algorithm's computational results show good performance. We found that the hybrid genetic algorithm can effectively generate a satisfactory tourism service, and its performance is better than that of the genetic algorithm.

Keywords: service scheduling; hybrid genetic algorithms; simulated annealing algorithms; tourism services; sustainability tourism

Citation: Suanpang, P.; Jamjuntr, P.; Jermsittiparsert, K.; Kaewyong, P. Tourism Service Scheduling in Smart City Based on Hybrid Genetic Algorithm Simulated Annealing Algorithm. *Sustainability* **2022**, *14*, 16293. https://doi.org/10.3390/su142316293

Academic Editor: Mohamed A. Mohamed

Received: 27 October 2022
Accepted: 2 December 2022
Published: 6 December 2022

1. Introduction

In this era of disruption, people around the world are utilizing information technology to create competitive advantages, especially in the tourism industry. This approach should provide the best solutions and prompt responses that will improve experiences and enhance suitability tourism. Since scheduling is involved in tourism service applications, it is regarded as a crucial topic in production management and combinatorial optimization. One of the most significant subjects with a wide range of applications in tourism services is scheduling. When assigning resources to a collection of tasks, scheduling must take time, capability, and capacity limits into consideration [1]. To increase profit, the fundamental goal is to increase production efficiency and resource use. Many scheduling issues in the industrial sector are regarded as being extremely complicated, making it challenging to resolve them using precise techniques and standard algorithms. Since 1950, scheduling issues have drawn the attention of several scholars, and significant research efforts have been made in a variety of engineering and scientific domains, including operations research, computer science, industrial engineering, mathematics, and management science.

Scheduling problems are well-known, significant, and difficult issues in the domains of combinatorial optimization and production management. The complexity of scheduling difficulties can be estimated in relation to all potential schedules, and as the size of the problem increases, this complexity increases dramatically.

Scheduling issues are among the hardest problems, and Garey et al. [2] and Ullman [3] demonstrated that this means that polynomial time methods cannot be used to solve them. In cases of scheduling issues have primarily been addressed using traditional approaches and exact methods, such as Lagrangian relaxation, branch and bound, heuristic rules, and shifting bottleneck (e.g., Carlier and Pinson [4], Adams et al. [5], Vancheeswaran and Townsend [6], Brucker et al. [7], and Lageweg et al. [8]). Various techniques have drawn inspiration from nature, biology, and physical processing over the last ten years. These methods have been effectively used to solve a variety of optimization issues, including scheduling issues. Genetic algorithms [9–11], ant colony optimization [12], imperialist competitive algorithms [13,14], tabu search [15], simulated annealing (SA) [16,17], particle swarm optimization [18], and immune systems [19] are a few of these metaheuristic approaches. For further information, see the thorough studies of scheduling problem solutions performed by Jain and Meeran [20] and Aliş and Bulkan [21].

A potent search method that mimics biological evolution and natural selection is the genetic algorithm (GA). Holland [22] was the first to come up with the idea for GA, and David Goldberg expanded on it. In contrast to other algorithms, this metaheuristic approach is frequently used to locate optimal or nearly optimal solutions to a variety of optimization problems. Davis [23] was the one who first used GA in the scheduling problem, and several GA-based methods for solving scheduling problems have since been introduced. A GA for solving scheduling problems with an encoding technique based on preference rules was proposed by Croce et al. in [11]. A GA with a novel representation technique based on operation completion time was proposed by Yamada and Nakano [24], and its crossover can provide active schedules. For the scheduling problems, Lee et al. [25] presented a GA with an operation-based representation and an order-based crossover that preserves precedence. For scheduling problems, Sun et al. [26] created a modified GA with clonal selection and a life span strategy. Out of 23 key instances, the generated algorithm was able to discover the 21 best solutions. The literature makes it clear that GA is a potent search approach with strong global search capabilities; yet, this metaheuristic algorithm suffers from early convergence and poor local search capabilities.

The major goal of the hybridization strategy is to improve the efficiency and potency of GA by addressing its limitations in terms of local search capability and premature convergence. In their hybrid optimization technique, Wang and Zheng [27] combined local search and GA. The hybrid framework of this method lowers the parameter dependence of both algorithms and reduces the likelihood of GA becoming stuck on local optima. In order to solve the scheduling problems using an adaptive mutation operator and avoid premature convergence, Zhou et al. [28] created a hybrid heuristic GA. In this approach, heuristics were utilized to identify the remaining operations in the constrained solution space, while GA was applied to the machines' first operations.

In order to further boost the quality of the hybrid heuristic GA-derived solutions, a neighborhood search strategy was used. A deadlock-free local search GA with an operation-based representation and UOX crossover was presented by Ombuki and Ventresca [29], and this was able to produce workable solutions. In this technique, a local search operator was used for local exploitation, while the GA was used to perform global search. They also created a second hybrid genetic algorithm that uses tabu search; according to computational results, the hybrid GA with tabu search performed better than the local search GA. Gonçalves et al. [30] combined GA, a schedule builder, and a local search operator to create a hybrid GA for addressing scheduling problems. In this approach, a priority rule was utilized to produce schedules, and GA was employed to establish priorities. The quality of the solutions was then further enhanced using a local search operator. A hybrid approach using the novel representation technique known as random keys encoding was created by Lin and Yugeng [31]. In this approach, a neighborhood search was added to perform local exploitation and improve the quality of the solution acquired by GA, after the GA was used to obtain an optimal schedule. A hybrid GA was suggested by Zhou et al. [32] to reduce weighted tardiness in the scheduling problem. The first operations of this algorithm were determined using GA, and the remaining operations were assigned using a heuristic. This method was utilized to produce optimal solutions. The hybrid framework outperformed GA and heuristic used alone, according to the results. Agents were employed to implement the parallel GA that Asadzadeh and Zamanifar [33] proposed, as well as to establish the initial populations. A hybrid SA immune system algorithm was presented by Zhang and Wu [17] for reducing the overall weighted tardiness of workshop scheduling. A hybrid micro-GA created by Yusof et al. [34,35] was used in scheduling problems. This method combined an autonomous immigration GA with subpopulations with an asynchronous colony GA, which had colonies with low populations. This algorithm is suitable for use in the complex optimization problem, and can also be applied in tourism service scheduling.

The problem addressed in this research is to find effective algorithms for a tourism recommendation system in Thailand. For the purpose of solving non-preemptive scheduling problems, an efficient hybrid GA is described in this study. An operation-based representation is utilized to more efficiently address the given problem. Additionally, the context of function evaluation involves the invention and use of a new knowledge-based operator. In order to determine the operations of machines during their downtime, this knowledge-based operator mimics the features of the scheduling problem. In order to create offspring and mutants during the reproduction phase of the GA, two different mutation operators and a machine-based, precedence-preserving and order-based crossover are devised. Additionally, SA is used to some extent to increase the population variety in GA, as well as to improve the solution quality of the schedules obtained from GA. After highlighting its key components, the suggested approach is designed to reduce time of schedules. Finally, the computational outcomes of the given issues are given to demonstrate the effectiveness of the suggested algorithm.

To bridge the gap in knowledge regarding the research problem, the aim of this study was to develop a hybrid genetic algorithm by combining a genetic algorithm and a simulated annealing algorithm using a gradient search method to optimize the complex processes used to deal with the tourism service problem, and to compare this to the traditional genetic algorithms used in Chiang Mai smart city case studies in Thailand. The remainder of this paper is structured as follows: The scheduling problems are described in Section 2, while Section 3 contains the suggested hybrid GA. In Section 4, the computational results of exemplar instances are shown, followed by the discussion. Section 5 provides the reasons for choosing the GA and SAA algorithms as effective approaches, then we give our recommendations for tourism systems, ideas, conclusions and proposals for further research.

2. Materials and Methods

2.1. Tourism Pacagke Tours and Service Scheduling

Package tours, according to Enoch [36], are a useful and more economical way to see tourist locations, but they lack flexibility because of their predetermined itinerary. The advantages of package vacations frequently exceed their drawbacks for travel agents and customers. Travelers can save money, for instance, when travel agents offer them a discount due to bulk purchasing. Additionally, through the tour operator, the travel firms handle every aspect of travel, transfers, and lodging. By making just one call to confirm their package trip, passengers can save a lot of time compared to having to communicate with several service providers separately. Due to the benefits they provide, package tours are frequently chosen by customers as their means of departure.

Tour operators cater to particular customer groups and provide a variety of excursions, including fully escorted, adventure, special interest, city or regional, and group trips. GPTs are frequently used by citizens of Asian nations and regions, including Taiwan, Japan, South Korea, Hong Kong, and China, as one of their preferred outbound modes of transportation [37]. These trips are now crucial for the growth of individual nations' economies, as well as the global tourist industry [38].

With a wealth of literature on GPT services, previous studies on GPT management have focused on issues with passenger happiness, consumer behavior, and customer service [39]. The practice of combining airline seats and beds in hotels (or other forms of accommodation), in a manner that will make the purchase price acceptable to potential vacationers [39], is termed a package holiday. Seddighi and Theocharous [40] established a framework for analyzing the nature, shape, and character of vacation decision-making, as well as a micro-econometric approach to individual travel. The topic of vertical integration in the European tourism sector was investigated by [41]. Jin et al. [41] looked into how customer decision behavior in similar package tours was affected by the selection framework for upgrading and downgrading. From the viewpoint of the asymmetry in tourism information, Ref. [42] addressed the causes of low tourism quality.

Alao and Batabyal [43] indicated that the characteristics associated with the problem of how to sell package tours to tourists include asymmetric information, uncertain demands, and differentiated customers. Taking these three factors into account, they applied the first contract–theoretic analysis in order to provide the sharing of valuable information between package tour-selling firms and tourists. Aguiar-Quinatna et al. [44] analyzed customers' motivations and perceptions, and proposed strategies to improve the competitiveness of traditional travel agencies in Spain.

Seddighi and Theocharous [45] developed a framework for analyzing the nature, shape, and personality of vacation decision-making, as well as a micro-econometric approach to individual travel. The subject of vertical integration in the European tourism sector was examined by Theuvsen [46]. Jin et al. [47] looked into how client decision-making in similar package trips is impacted by the choice framework for upgrading and downgrading. From the standpoint of information asymmetry in the tourism business, Chen et al. [48] investigated the causes of low tourist quality.

Batabyal and Yoo [49] used probabilistic analysis to determine the long-run loss in demand for guided tours during the low season by modeling the pattern of client arrival as a stochastic process. In their investigation of the UK package tour market in 2007, Davies and Downward examined the dynamics of profitability and market share, as well as the function of pricing and non-pricing decisions, in order to identify the causal mechanisms behind both pricing and competition. In order to examine how Norwegians' income levels affect their demand for inclusive tour charters during a specific time period, Jrgensen and Solvoll [50] created an econometric model. Some tour providers are also focused on offering package holidays or single-destination included tours (Tepelus) [51]. However, to the best of our knowledge, no research has looked at tour selection and bus planning. A number of specific studies have also been conducted on themes relating to package tour management. Portfolio management, facility location selection, supplier selection, and investment portfolio management are examples of issues that are somewhat linked to our research (e.g., Arabani and Farahani [52]).

There is no requirement to take into account whether a preset minimum demand or load is met when assessing a reward or service in relation to these issues. Revenue is counted once a product has been purchased—for instance, in the product portfolio management problem. When an investment is sold off in the portfolio investment problem, a profit or loss is realized. If a facility can deliver a service to a geographic area within a given distance, the facility location-selection problem's requirement is satisfied. The issue with staff assignments includes any service provided to a customer by an allocated employee. These observations help us understand how the proposed selection differs from other selection-related problems.

The proposed selection problem relates to the minimal number of passengers, whereas other challenges do not. This is the first distinction between the tour portfolio and other portfolio selection problems. The second difference is that while lost sales in most industries are caused by excessive demand, they can also result from insufficient demand in the tourism sector. In other words, new reservations are only turned down once the number of customers who have made reservations has reached capacity.

However, if there are fewer than the required number of consumers for a tour, all reserved clients are refused the service product. This issue is crucial for the tourism sector, because bad proposed selection and bus planning can result in financial losses and a negative business image. There is a research gap in the analysis of proposed selection and bus planning as a result.

Our research offers a mathematical model to look at the particular issues so as to close this gap in the literature. Through the use of a mathematical model, travel companies can profit from quicker and more accurate decision-making by analyzing the costs and benefits via data from the actual world. Travel agencies can utilize the selection method to arrange their package tours and their sightseeing buses, which can provide insights into how to do so. Because they are computationally challenging, the majority of portfolio selection problems that fall under the category of combinatorial optimization problems and traditional mathematical techniques frequently fall short of producing ideal or promising solutions. Due to their high computational efficiency, meta-heuristics-like genetic algorithms have been effectively used to solve real-world combinatorial optimization issues (e.g., Luan et al. [53]; Niknamfar and Niaki [54]). This article also suggests a hybrid heuristic based on using a genetic algorithm to solve the suggested mathematical model in order to accelerate computation.

2.2. Tourism Service Scheduling in Smart City

A smart city is a city with physical characteristics, and complex social and digital innovation and information technology. The concept of a smart city refers to a city that is as physically and socially connected as it is digitally [55]. Since the beginning of smart city development, the smart tourism city has emerged as a new concept for utilizing information technology to support the tourists and also the tourism industry in a city [55,56].

In Thailand, Chiang Mai is one of the smart cities that was developed in 2017 to build smart agriculture, smart economy, smart safety, smart health and smart tourism. The policy of smart tourism, which implements ICT infrastructure and technology to serve tourists, aims to support good living and the provision of useful information to people in the area, and also support the tourism industry [57].

In order to improve the competitiveness of the tourism industry, many scholars have addressed the problem of tourist trip design and tour route planning. An attempt to schedule tour guides for the tourism and leisure industry was carried out by Chen and Wang [58]. Their findings demonstrate that the Genetic Algorithm is useful for tourism scheduling service centers. Furthermore, Thumrongvut, Sethanan, Pitakaso, Jamrus, and Golinska-Dawson [59] created a mobile application using a mixed-integer linear programming approach for small-size problems, to help the suppliers of tourism services arrange trip and route planning for visitors, in order to improve the timeliness of visitor services.

2.3. Hybrid Genetic Algorithm

There is still much to learn about the extent of the flow shop scheduling challenge established in earlier investigations. Initializing the population will enable us to begin the process in GA. Numerous methods have been employed due to the significance of this process, including opposition-based (Li and Yin, [60]), quasi-oppositional, space transformation searches, and linear regression (Hassanat et al.) [61].

Due to a flow's increased complexity, metaheuristics must be used instead of the traditional heuristic approach (Lee and Loong) [62]. Instead of a memetic or hybrid scheme, a single algorithm dominates the majority of the previous studies (Tosun et al.) [63]. The first metaheuristic used to solve shop flow was simulated annealing (Osman and Potts). GA (Rahman et al.) and the well-known local search optimization technique TS were the next two (Ben Cheikh-Graiet et al.) [64]. Integrating the GA and TS is an alternative technique of solving shop flow, since the hybridization concept provides a more reliable method by combining two or more algorithms. The reason for choosing the GA and SAA algorithms is that they are effective in verifying our recommendations for the tourism system.

Some hybridized algorithms, particularly metaheuristic algorithms, have been used to tackle optimization issues, such as GA, which has already been merged with TS via a scatter search mechanism (Glover et al., 1995). There is the general algorithm (GA) that has been hybridized with the insertion of the search cut and repair algorithm (GA-ISCR) (Tseng and Lin, 2010), the co-evolutionary quantum genetic algorithm (CQGA) (Deng et al. [65]), and the general algorithm combined with simulated annealing (HGSA) (Wei et al.) [66].

In addition to GA, a number of different metaheuristics have been proposed to decrease the life span of shop flows, including those by Ding et al. (2015), who developed a local search approach using variable neighborhood search, called the tabu-mechanism with improved iterated greedy algorithm (TMIIG); there is also the iterated greedy with jumping probability (IG-JP), developed by Tasgetiren et al. [67], and Davendra and Bialic-Davendra (2013) developed DSOMA. Additionally, the aquila optimizer, which is based on how an aquila behaves when hunting its prey, has been applied in general optimization problems in recent years (Abualigah et al.) [68]. CFD analysis and the optimization of the design of a centrifugal pump use an effective, artificially intelligent algorithm that optimizes the compliant gripper mechanism's design by employing an effective bialgorithm comprising fuzzy logic and ANFIS, which is an efficient hybrid approach to the finite element method. The artificial neural network-based multi-objective genetic algorithm has also been used for the computational optimization of a linear compliant mechanism of a nanoindentation tester.

Wang et al. (2022) [69] proposed a cutting-edge strategy to raise the efficiency, power, and head of centrifugal pumps. First, an ideal numerical model that took pump efficiency and head into consideration was explored in order to set restrictions. Before conducting testing, the pump was built, and an artificial intelligence algorithmic technique was applied to the pump. By choosing the parameters for the centrifugal pump's casing section area, the impeller interference, the volute tongue length, and the volute tongue angle were investigated in a number of models. On the optimization indices, the weights of the safety and displacement factors were estimated. The weights matrix for the ideal process ranged from less than 38% to more than 62%. This method ensures that the optimization approach is strong and multi-objective. The outcomes demonstrate an improvement in centrifugal pump performance.

2.4. Simulated Annealing Algorithm

The original single-objective simulated annealing algorithm was slightly modified by Serafini (1994) by changing the method of calculating the probability of accepting new candidate solutions to narrow the search. This was one of the first works to use a simulated annealing algorithm to solve multi-objective problems. In addition to the well-known traveling salesman issue, a smaller problem with two objectives was also solved using the approach. The technique performs well for minor issues, but with large problems, only limited parts of the Pareto front can be reached, as claimed by Czyzak and Jaszkiewicz [70].

In order to find a good approximation of the entire Pareto front in somewhat large multi-objective knapsack problems with two, three, and four objectives, they invented Pareto-simulated annealing (PSA). In an effort to ensure the dispersion of the nondominated solutions discovered throughout the procedure, PSA employs objective weights that are changed after each iteration. In order to solve the two-objective knapsack problem, Ulungu et al. [71] proposed the UMOSA (Ulungu Multi-Objective Simulated Annealing) approach and investigated the impact of various neighborhood architectures (for candidate solution generation) and aggregation functions. All of these older MOSA techniques combine the goals using a variety of aggregation functions and objective weights, and the aggregated values are then minimized, much like in the case of a single simulated annealing process. Although it is necessary to create objective weights in advance, doing so can be difficult because, as Das and Dennis [72] found when using a number of instances, inadequate weight values render some areas of the Pareto front unreachable.

Following these initial attempts, several academics began utilizing the dominance idea to direct the algorithm. SMOSA (Suppapitnarm multi-objective simulated annealing), a technique developed by Suppapitnarm et al. in 2000, uses nondominated solutions that are archived in a solution list. The nondominated solutions discovered during the procedure had to be saved in an archive because SMOSA only produces one solution in each iteration (as opposed to MOEAs algorithms, which deal with populations of solutions). It employs a cutting-edge "return to base" tactic. In order to repeat the search and attempt to reach a new area of the front, the return to base approach occasionally chooses a random solution from the archive. Sankararao and Yoo [73] claim that even with this revised formulation, few nondominated solutions are found, and the convergence velocity is slow due to the search method employed. The acceptance determination process for this algorithm is primarily based on direct comparisons between the candidate and existing solution, which means that several excellent solutions may be overlooked while searching.

In his 2003 study, Suman [33] compared the performance of the UMOSA, SMOSA, and PSA algorithms, as well as two new ones: the PDMOSA (Pareto dominance multi-objective simulated annealing), which uses the Pareto dominance concept to direct the search, and the WMOSA (weighted multi-objective simulated annealing), which uses a weight vector to handle the constraints that are not verified. Certain comparisons aim to demonstrate how effective these algorithms are at addressing multi-objective optimization problems.

All of the approaches reached nondominated solutions, according to the comparisons, although the PDMOSA results are subpar compared to those of the other methods, and in some cases, PSA performs best. To assure a large collection of nondominated solutions, the authors contend that all five techniques need little processing effort and should be applied simultaneously. This means that if each of these methods is utilized independently, a larger number of front-end solutions cannot be guaranteed.

Suman and Kumar [74] reviewed SA-based algorithms and suggested new lines of inquiry to enhance MOSA algorithm performance. Suman et al. [75] proposed the orthogonal simulated annealing (OSA) technique in order to avoid utilizing conventional approaches that could inadvertently choose a new candidate solution at random, rendering the algorithm ineffective in exploring the solution space in complicated optimization problems. To increase the algorithm's convergence, they suggested employing methods to first search the space of potential candidate solutions in the vicinity of the current solution, and then pick the best candidate solution. However, when compared to other algorithms, OSA's performance was problem-dependent; while there are some situations in which OSA succeeds, there are other instances in which it fails. It is essential to develop an algorithm that performs well in a wide range of circumstances.

Bandyopadhyay et al. [76], who pioneered the archive multi-objective simulated annealing (AMOSA) approach, made a significant advancement in this subject. To calculate trade-offs between solutions, this algorithm makes use of an archive that holds the nondominated solutions discovered during the optimization process. Additionally, they devised a process that uses the notion of the quantity of domination and the domination status amongst solutions to direct the algorithm search.

AMOSA's results were superior to those provided by NSGA-II (Deb et al.) [77] and PAES (Knowles and Corne) [78], in that a greater number of nondominated solutions were obtained, according to the results of various tests. The archive taken into account by AMOSA has two separate size restrictions, with an upper limit on the number of nondominated solutions that can be stored. The number of answers is then reduced to a smaller limit using a clustering strategy, which involves controlling and separating a huge amount of data into groups or subsets with comparable characteristics. This lowers the algorithm's computing requirements.

No constraints should be taken into account, however, to arrive at a more varied collection of nondominated solutions. Furthermore, as Sankararao and Yoo [79] did not seek a process to obtain a uniform Pareto front, it was discovered that the large number of solutions obtained by AMOSA were concentrated in specific regions of the front. Sankararao and Yoo [80] developed the multi-objective simulated annealing technique (rMOSA). This is a process to select the "most uncrowded solution" in the archive to generate a new candidate solution and obtain a well-distributed number of solutions in the algorithm's final results. This process seeks to select a random solution from the archive to generate a new candidate solution, with the aim of increasing the convergence speed of the algorithm.

The optimization process begins with the simultaneous employment of these search techniques, and they are applied periodically during the iterations. On the basis of comparisons with the existing solution, rMOSA employs a straightforward process for accepting candidate solutions. This can be ineffective, since it is important to consider the nondominated solutions that have already been found when accepting new candidate solutions (Cao and Tang) [81]. Additionally, the result derived when utilizing both processes concurrently from the start ignores the algorithm's evolution.

In an effort to improve the intensification and diversity aspects of the UMOSA algorithm, Rincón-Garca et al. [82] created a variation of the algorithm. They proposed various aggregation strategies for meeting the aims. The fundamental drawback of this form of weight-based algorithm is still, however, the establishment of weights allocated to targets in advance.

To address issues in air traffic control systems, Mateos and Jiménez-Martn [83] suggested a better AMOSA. To calculate the likelihood of adopting inferior solutions in comparison to the original AMOSA scheme, a new "element" was proposed. By only accepting solutions that increase the variety of the front, this element is utilized to speed up the algorithm's convergence rate. The authors showed that the algorithm performs well, but they contend that in order to demonstrate its superiority, it should be compared to other algorithms that have been applied to the same problem. It is important to carefully consider throwing out some answers that would increase the convergence velocity.

Sengupta and Saha [84] introduced a novel unconstrained many-objective optimization technique that was evaluated on academic optimization issues—the reference point-based simulated annealing (RSA) method. In contrast to the earlier "point-to-point" MOSA algorithms, the authors suggested an "archive-to-archive" procedure, which greatly improves the outcomes. The authors demonstrated that RSA performs effectively in many types of problems, based on comparisons with the outcomes of other methods.

Similar to the AMOSA algorithm, RSA uses clustering to shrink the archive, and as previously mentioned, this should be carefully considered to maintain the variety in the final set of answers. The AMOSA method is also used by Saadatpour et al. [85] to tackle a multi objective waste load allocation problem in rivers. In this instance, the archive is designed to hold just a small number of nondominated solutions. As a result, the finished front's quality may suffer.

Our work introduces a novel MOSA-GR technique, which was motivated by the encouraging outcomes from a wide range of applications of MOSA algorithms. The new method was developed by using the finest ideas from the MOSA algorithms already mentioned above, and avoiding their shortcomings, notably with regard to the convergence and spreading of solutions along the front. The procedure for creating new candidate solutions and the reannealing of the algorithm are the two key areas where we suggest new enhancements to a previous version that Marques et al. [86] already provided. By utilizing a special tool that excels in a wide range of situations, this novel method seeks to identify a nearly optimal and sizable collection of nondominated, well-dispersed solutions.

Furthermore, there are two studies related to our research that apply an optimization technique in other areas of industry; for example, Zhang et al. (2018) [87] evaluated trade-offs between the main factors in green manufacturing (energy, noise and cost) through cutting parameter optimization, which introduces the new optimization technique to the manufacturing industry. Moreover, Zhang et al. (2020) [88] studied a multi-objective cellular genetic algorithm for the energy-oriented balancing and sequencing problem in a mixed-model assembly line in the energy industry. Finally, many algorithms have been designed to solve multi-objective optimization issues in the literature. However, the creation of new algorithms continues to be crucial. New algorithms are required to solve these problems in an acceptable amount of time, because the complexity of the problems is continuously increasing.

3. Methodology

3.1. Simulated Annealing Algorithm

Problem Definition

The JSSP can be expressed as follows: JOB SHOP SCHEDULING PROBLEM. Each of the "n" jobs consists of a number of procedures that must be carried out on "m" different machines. A single machine is employed for a set amount of time for each operation. Each operation can only be processed by one machine at a time, and processing must continue uninterrupted until completion once processing on a given machine begins. A specific job's processes must be completed in a specific sequence. The fundamental goal of the challenge is to plan the activities or operations of the tasks on the machines so that the overall time to complete all of the jobs, or the makespan (Cmax), is as short as possible. The phrase "makespan" describes the total amount of time needed to finish all operations for all tasks on all machines. It is a measurement of the amount of time elapsed between the beginning of the first operation and its conclusion. Sometimes, there may be several solutions with the shortest makespan, but the objective is to identify any one of them; finding all feasible optimum solutions is not always necessary.

The scheduling problem is a broad kind of the following classical scheduling problems:

given tourists n, $N = \{j1, j2, \ldots, jn\}$;

given tourists consists of s operations $S = \{Oj1, Oj2, \ldots, Oj,s\}$.

Each tourist gives rise to tasks that need to be completed on services in a specific technological order: $S = \{Oj1, Oj2, \ldots, Oj,s\}$. The notation indicates the third operation of a project that must be completed on one of the services and has a known processing time. In this setting, each computer can only handle one operation at a time, and a single tourist's operation cannot be handled by two services concurrently. Operation Oj,s must be processed on one of the designated machines until it is finished, without being interrupted. Additionally, a job can only visit one machine once; subsequent visits to the same machine are not permitted. Additionally, it is believed that the machines are always available and that operation travel times are minimal. In contrast to flow shop scheduling, scheduling problems assign each job a certain planned route. The order of all activities on all machines is planned to reduce Cmax, or the maximum time for a project to be completed (max $C1, C2, \ldots, Cj$).

A scheduling problem uses a mathematical model with the goal of reducing the makespan. In this model, B is taken to be a large integer, where tj,s represents the start of operation Oj,s, Smi,l represents the start of machine I in the priority of l, and hi,j,s is given a value of 1 if operation Oj,s is carried out on machine I, and a value of 0 if otherwise:

$$\min Z = \max\{C1, C2, \ldots, Cj\}\ j = 1, 2, \ldots, n \tag{1}$$

$$S.t.\ tj,s + Pj,s \leq tj,s + 1\ j = 1, 2, \ldots, n;\ s = 1, 2, \ldots, sj - 1 \tag{2}$$

$$Smi,l + Pj,sXi,j,s,l \leq Smi,l + 1\ i = 1, 2, \ldots, m;\ j = 1, 2, \ldots, n;\ l = 1, 2, \ldots, li - 1;\ s = 1, 2, \ldots, sj \tag{3}$$

$$Smi,l \leq tj,s + (1 - Xi,j,s,l)\ B\ i = 1, 2, \ldots, m;\ j = 1, 2, \ldots, n;\ l = 1, 2, \ldots, li;\ s = 1, 2, \ldots, sj \tag{4}$$

$$Xi,j,s,l \leq hi,j,s\ i = 1, 2, \ldots, m;\ j = 1, 2, \ldots, n;\ l = 1, 2, \ldots, li;\ s = 1, 2, \ldots, sj \tag{5}$$

$$\sum j \sum s\ Xi,j,s,l = 1\ i = 1, 2, \ldots, m;\ l = 1, 2, \ldots, li \tag{6}$$

$$\sum i \sum l\ Xi,j,s,l = 1\ j = 1, 2, \ldots, n;\ s = 1, 2, \ldots, sj \tag{7}$$

$$tj,s \geq 0\ j = 1, 2, \ldots, n;\ s = 1, 2, \ldots, sj \tag{8}$$

$$Xi,j,s,l \in \{0, 1\}\ i = 1, 2, \ldots, m;\ j = 1, 2, \ldots, n;\ l = 1, 2, \ldots, li;\ s = 1, 2, \ldots, sj. \tag{9}$$

Constraint (2) in this model deals with the sequences of operations that must occur in a particular order. The third constraint forbids machine overlap and mandates that no machine process more than one operation at once. Constraint (4) forbids operations from running simultaneously; therefore, an operation is assigned to a specific idle machine under the condition that its preceding operation is completed. Additionally, a machine is selected in a constraint for each of the operations (5). Operations are assigned to the machines and are performed in the order specified by constraint (6). The number of operations that can be carried out on a single machine is constrained by constraint (7), based on the priority of the machine.

3.2. The Proposed Hybrid Genetic Algorithm-Simulated Annealing Algorithm

Even though the proposed hybrid genetic algorithm–simulated annealing algorithm is a traditional algorithm, it still works effectively for our recommendation system. The hybrid algorithm that is being suggested combines the GA and SA algorithms. The benefits of both algorithms are combined in the suggested framework to get the best answers to the scheduling problems. A flowchart for the suggested hybrid genetic algorithm (HGA) is shown in Figure 1. The random initialization of the population is where it all begins. A new knowledge-based operator is used in this stage to enhance the individual's solution quality, as the formed population is assessed using the fitness function. This knowledge-based operator also integrates with the function assessment stage and operates during machine downtime.

A selection operator is used during the reproduction phase to choose the mating pool's parents, followed by a crossover to create the progeny, when the operator is used. To create the mutants, a mutation operator is also applied to a group of people chosen at random. The evaluation of the produced offspring and mutants is followed by consideration of the termination circumstances, which directs the algorithm in the appropriate direction. Section 3.7 details the termination requirements.

However, the simulated annealing algorithm will converge to a fixed point in simulated annealing; the acceptance probability for a new state in step k is traditionally defined as

$$(accept\ new) = \begin{cases} exp\left(-\frac{\Delta}{T_k}\right) & , if \Delta \geq 0 \\ 1 & , x < 0 \end{cases}$$

where $\Delta = f(new) - f(old)$ is the change in objective function f, which is to be minimized, and T_k is a strictly decreasing positive sequence with $\lim_{k \to \infty} T_k = 0$.

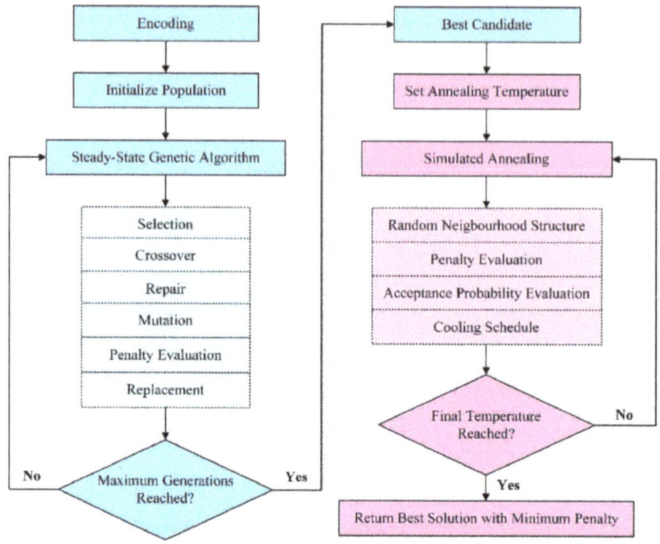

Figure 1. The proposed hybrid genetic algorithm–simulated annealing algorithm.

When the algorithm runs through the termination condition, which is when GA's generations come to an end, SA will begin with the top GA performer. In SA, a neighborhood search structure is used with three separate operators as the proposal mechanism: swapping, insertion, and reversion. Additionally, as each SA cooling criterion is met, a pool of people with the beta percent of accepted solutions is maintained.

In Section 3.6, the SA algorithm is covered in detail. Three conditions are set in this phase of the algorithm to either stop it or proceed with fresh parameters. The parameters of SA are reset and fresh GA parameters are applied if the most recent condition, or condition 4, which is the termination of SA's outer loop, is reached. Additionally, from the migration pool of SA, zeta percent of unique persons are relocated to GA.

3.3. *Encoding and Decoding*

Finding the right encoding and decoding techniques to describe the problem is the first and most crucial stage in any algorithm. The permutation of operations for various occupations is represented in this paper using an operation-based representation [34,37]. If the technology limitations are met, the timetable could be built using this representational approach. According to this method, a chromosome is made up of nm genes, each of which encodes the order of operations to be carried out on the machines.

Each operation is indicated by a positive integer number ranging from 1 to n (see in Figure 2). Each of the integer values has an equal number of occurrences and operations. In other words, the kth instance of an integer value in a chromosome corresponds to the kth technological procedure of the task. Each tourist on this chromosome consists of three processes, and as a result, every job appears three times along the length of the chromosome. For instance, this chromosome's sixth and ninth genes, respectively, reflect the second and third operations of jobs one and two. Each chromosome also has other data that are related to it, such as the machine number, processing time, start time, and finish time.

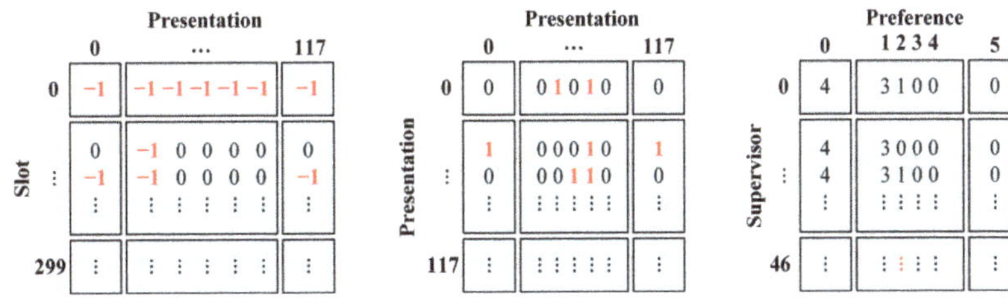

Figure 2. From the left, presentation-by-presentation matrix, supervisor-by-preference matrix, and slot-by-presentation matrix.

3.4. The Knowledge-Based Operator and the Fitness Function

The likelihood that a solution can be passed on to the following generation in an optimization problem is often determined by the fitness function. In other words, by using this operator, the solution quality is assessed, and the better-fitting chromosomes will have a higher chance of survival; nevertheless, the less well-fitting chromosomes must be eliminated from the population. There are numerous possible performance evaluators for defining the fitness function in scheduling problems. In this study, the fitness function used to assess each chromosome is either makespan or Cmax. A new knowledge-based operator is created in this algorithm and in the context of the fitness function based on the characteristics of the challenge. Based on the machine idle times present in job shop settings, this operator was created. Additionally, in order to shorten the algorithm's computation time and cover all the chromosomes that need to be evaluated, this knowledge-based operator is used during the function evaluation phase. The next stages are used to develop this operator more effectively.

The first step is to locate each machine's idle positions. The idle start time, idle finish time, and idle time must then be determined for each of these inactive points.

Step 2. Taking into account the length of idle time as well as the processing time of the candidate operation, a candidate operation is selected from the right side of the machine sequence list based on the position of the idle point on the machine sequence list.

Step 3: If the amount of idle time is greater than or equal to the candidate operation's processing time, it will be conditionally accepted. If not, it will be turned down.

Step 4. If the candidate operation is not accepted for transfer, the operator returns to step 2 and selects the next operation.

Step 5: To decide whether to accept or reject the shift for the conditionally accepted candidate operation, all processing limitations must be taken into account. For instance, the candidate operation's previous operation must be finished.

Step 6: The candidate operation will be moved to its new location if all of the constraints are met. If not, the candidate operation will remain in its current place.

Step 7. Steps 2 through 6 should be repeated for each machine until the very last action on the machine sequence list.

3.5. Encoding Selection Operator

An effective selection method can improve GA's performance by enabling it to find the best solutions more quickly. The most popular operator, the Roulette Wheel selection technique, is used in this study to choose the parents [34]. Additionally, the elitism strategy is used in this selection process to keep the fittest chromosomes for the following generation and stop the problem from getting worse over time.

We determined the likelihood of each chromosome in Roulette Wheel selection using Boltzmann Probability $P(x) = \exp(f(x))/f(x)$. P(x) in this equation stands for the likelihood of each chromosome, indicating selection pressure, while f(x) stands for each individual's fitness, and f(x) is the fitness of the generation's lowest performer. It should be noted that in order to make the selection pressure independent of the problem scale, f(x) was added to the original Boltzmann Probability equation. Additionally, each chosen person's normalized probability is stated as P(x)norm = P(x)/max.pop x = 1 P(x).

3.6. Crossover

In GA, crossover is the most crucial operator compared to other operators, and is actually the algorithm's foundation. By merging the data from the first and second parents, the crossover operator creates offspring with traits from both parents that may be better or worse than their parents. Additionally, the major objective of this operator is to create viable offspring using parental knowledge that is better.

In this study, a realistic offspring generation method called a machine-based precedence-preserving order-based crossover (POX) is proposed [25]. The POX operator is implemented using the specific steps listed below. Figure 3 show the process of chromosome which in the red frame show the presentation of process by the following step.

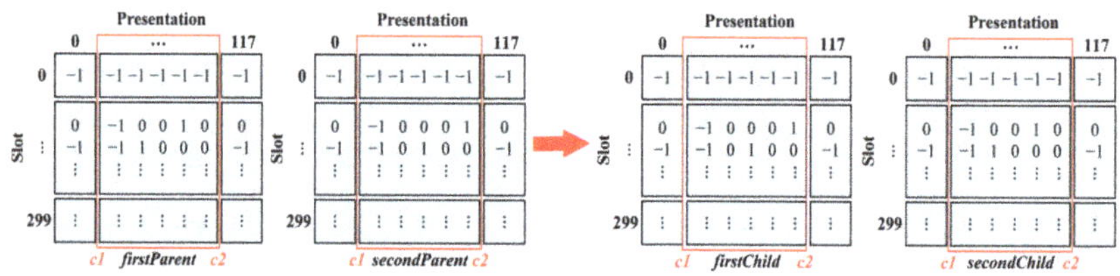

Figure 3. Process of a chromosome.

Step 1: Using Roulette Wheel selection, two people are first chosen to serve as parents (P1, P2).

Step 2: Next, two groups of tourists are chosen and given the names sj1 and sj2. One of the tourists is chosen at random from the remaining n jobs, and one is chosen from the bottleneck machines.

Step 3 copies the components of the first tourists (sj1) from the first parent (P1) to the precise locations; in other words, the elements of the second tourists (sj2) are replicated exactly from the second parent (P2) to the alleles in the second child (O2), and the same is true for the second tourists (sj2).

Step 4: In the first tourists (sj1) and the second tourists (sj2), all of the alleles are deleted in the second parent (P2) and, in the case of the second tourists (sj2), all of the alleles are deleted in the first parent (P1).

Step 5: From the furthest left to the furthest right, the second and first children (O2, O1) receive the remaining alleles from the first and second parents (P1, P2), respectively.

3.7. Mutation

Mutation is the second method of examining the solution space during the reproduction phase. The mutation operator can make the algorithm faster at finding better solutions and prevent it from getting stuck in the local optima. Additionally, it might alter the chromosome to broaden the diversity of the population.

Two different kinds of mutation operators, namely, swapping and insertion, are used in this approach. The insertion operator could perform a thorough search in addition to increasing the population's variety through mutation. It should be emphasized that one of these mutation operators, which are listed below, should be chosen at random in order to produce an offspring. Figure 4 show the process of chromosome mutation which the red frame shows the result of the process of the following operator.

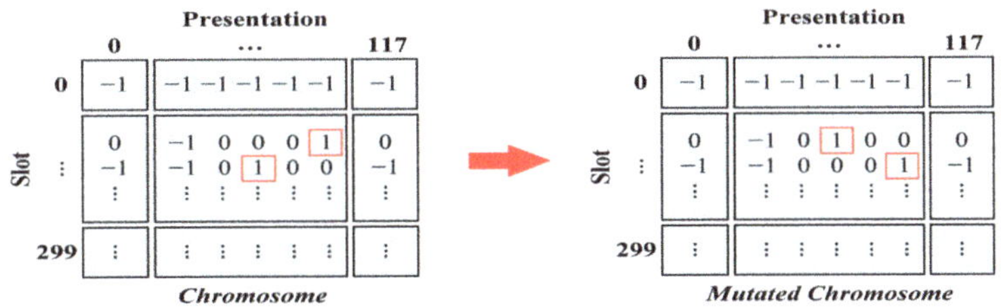

Figure 4. Process of chromosome mutation.

(1) Swapping operator: Before using the swapping operator, two random values (R = 2, 5, for example) are created that represent the locations of two alleles on the chromosome. Except for the randomly chosen alleles that must be switched or exchanged in the offspring, all of the parental information is then copied to the exact places in the child. Consider the parent who was chosen at random from the population.

(2) Insertion operator: To use the insertion operator, two random values are first created to represent the locations of two alleles on the chromosome. The offspring's genetic makeup is then identically duplicated from the parent, with the exception of the randomly chosen alleles. The other randomly chosen allele with a greater job-value is positioned to the left of the smaller randomly chosen allele.

Figure 5 show the simulated annealing process to evaluate the accuracy of this algorithm via convergence, the plot of cost vs. iterations makes it simple to see how the SAA is convergent. Other elements with a similar viewpoint include round-off errors and local extrema. In Figure 6, the cost function is the minimum, beginning at generation 1125.

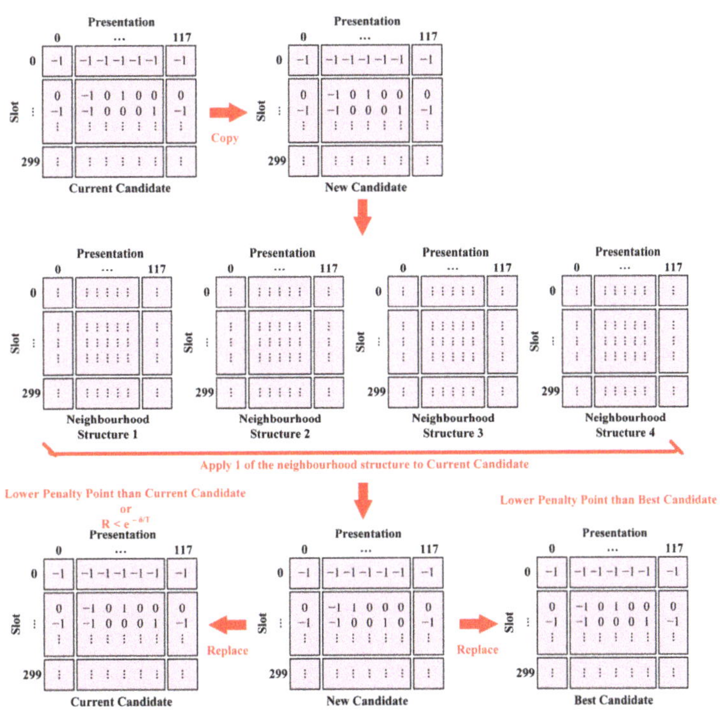

Figure 5. Simulated annealing process.

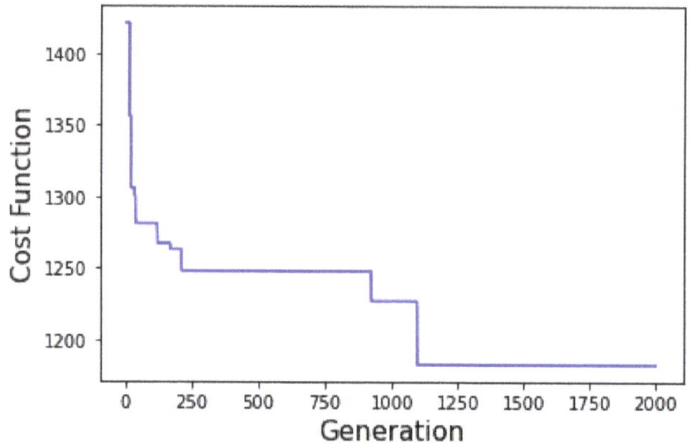

Figure 6. Cost function.

4. Results

4.1. Simulated Annealing

The SA technique, which is categorized as a stochastic local search approach, was motivated by the physical annealing process of solid materials [35]. In SA, the search operator is periodically allowed to take bad turns, which allows the algorithm to divert from local solutions and move toward global ones. In SA, this trait could be acquired by probabilistically opting for the less desirable options.

In the suggested HGA, SA begins with the top GA solution. Then, using the current solution (S), a proposal mechanism made up of the three operators (swapping, insertion, and reversal) is used to create a new neighborhood solution (S). The freshly created solution (S) is subjected to the new knowledge-based operator, after which it is assessed using the objective function. The newly evaluated neighborhood solution will be accepted (f(S)) f(S)) if it is on par with or superior to the current solution. If not, the algorithm will carry on searching for a solution (S or S) determined by a probabilistic acceptance function. Acceptance of a solution is also determined by the value of the individual's goal function and the algorithm's current temperature (T) (e [(f(S)f(S))/T]). An acceptable answer is maintained among a fresh group of people in each inner loop of SA. The starting value of the temperature should be changed in accordance with the annealing schedule as the algorithm's inner loops are ended. Additionally, in each outer loop of SA, the remaining individuals from the new pool of individuals are discarded, while a beta percent of the unique individuals are maintained. Additionally, 25% of unique individuals are retained once the outer loop is closed for migration-related purposes, while the remaining remainders are deleted. It should be emphasized that a low migration rate is required in order to somewhat increase the variety of the GA population. Three operators (swapping, insertion, and reversion) are utilized in SA's proposal method, and one of them is arbitrarily chosen and used to produce the neighborhood solution.

4.2. Ending Conditions

Four distinct conditions are offered in the proposed HGA in order to fully or partially terminate the algorithm. The achievement of the best-known solution serves as the first termination condition, and the maximum number of generations in the HGA main loop serves as the third termination condition. The entire algorithm will be terminated at once if it meets either the first or third criteria. The maximum number of generations in GA and the maximum number of outer loops in SA are the second and fourth partial conditions, respectively.

4.3. Computational Experiments and Discussion

The following table provides a comparison between the planned HGA and the traditional GA. The same parameters are used to implement GA and HGA in a Python 3 environment. The fitness and execution time associated with the quantity of convergence iterations serves as the comparison criterion. For the HGA, all parameters are set for the best result after a specific number of iterations (in Table 1).

Table 1. Parameters.

Parameters	Values
Total number of iterations	2000
Number of individuals per generation	50
Chromosomes' size	8
Mutation probability	0.2
Mutant gene per individual	1
Crossover probability	1
Number of individuals chosen during the selection	1
Number of iterations for the LS	6
Number of neighbors per individual	4

The optimization results are compiled in the table. One thousand repetitions later, the ideal value had been reached. By providing a solid resolution, this also demonstrates that all genetic processes function effectively in general. In actuality, the program consistently moves closer to the ideal outcome. HGA, on the other hand, offers a reduction in the number of iterations required to converge from 15 s to 11 s. As a result, HGA can perform better than regular GA on an optimization problem, which demonstrates the hybridization mechanism's ability to improve convergence (in Table 2).

Table 2. Genetic algorithm and hybrid genetic algorithm performance.

	Convergence (Iterations/Time CPU)
Genetic Algorithm	15/60 s
Hybrid Genetic Algorithm	11/46 s

The proposed HGA was implemented using Python, and the technique was tested on a machine running Windows 10, an Intel Core i5 running at 2.5 GHz, with 8 GB of RAM. The results of the computational experiments were derived using the 10 well-researched tourism services with the aforementioned tweaked parameters and 10 replications for each benchmarked instance (Figure 7). The dimension of the problem is tourism services, the best-known solution, and the outcomes of our proposed HGA, which are presented in columns and include the best replication-level solution, relative deviation (RD), average solution, and worst algorithmic solution over ten runs. These are the most crucial indexes to compare in order to evaluate the algorithm's efficiency and consistency.

The notion of the "smart city" has improved the effectiveness and procedure of city management while integrating connection in many different domains. It requires strategic and innovative thinking to implement the smart city program, because it is not simple to do so. The first steps in this strategic program include problem identification, problem grouping, an abstraction process, solution selection, selection of efficient techniques, and implementation planning. The community and organizations can benefit from the innovative and valuable works produced by creativity. It is possible to research a variety of organizational traits, including human resource management and leadership planning. One tactic for raising public service companies' effectiveness is innovation. The limitations in the implementation of smart cities have not stopped efforts to advance numerous applications and other technologies. Chiang Mai has a plan to become a smart city, aiming to improve services to the community by utilizing local knowledge and maximizing the application of technology. One of the applications is called the Chiang Mai Smart City For Smart Mobility, which stores all the route data in Chiang Mai such that our proposed method can be applied effectively. A smart city is a form of the application of digital technology or information and communication to enhance the efficiency and quality of travel services, and help reduce costs and population consumption. Increasing the capacity of people to achieve a better quality of life, the smart city is a project that many cities around the world have sought to develop, in order to keep up with the new era by using technology. Technology is integrated with the life of the people, whether in terms of transportation power consumption or infrastructure, to make the city more comfortable. Chiang Mai is ready to drive smart city developments via a strategy to improve performance in a systematic way, and thus facilitate the development of a pattern that will drive Chiang Mai to truly become a city of life and wealth.

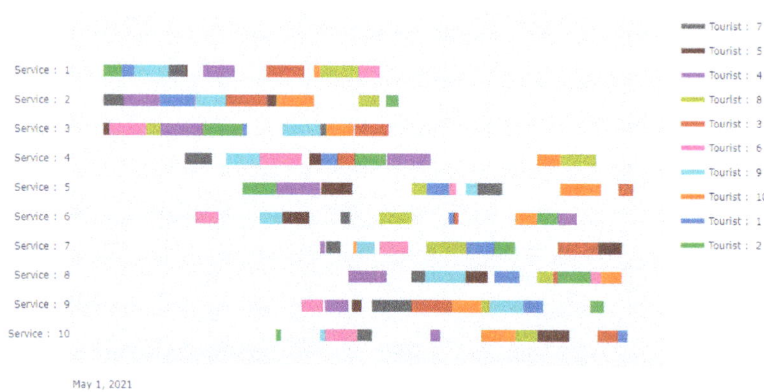

Figure 7. Tourism service scheduling.

5. Conclusions and Discussion

The algorithm's result was encoded in an operation-based form. The problem features were used to develop a new knowledge-based operator, which was able to improve the schedule solutions' quality. In order to boost population variety and sharpen the search, a machine-based precedence-preserving order-based crossover and two types of mutation operators, namely, swapping and insertion, were used to produce the offspring and mutants. Additionally, the SA method was used to further enhance the GA solution's quality due to its neighborhood search functionality.

The research library's well-studied benchmarked cases were used to test the proposed HGA, and the results were compared to those of other methods.

The usefulness and efficiency of the suggested technique are demonstrated by the computational results, which generally reveal that the proposed algorithm has a lower average relative deviation than the compared algorithms.

This paper adopts a hybrid genetic algorithm–simulated annealing algorithm to create schedules for tourism services, which is meets the interests of active tourists by collecting data from Thailand's tourism industry on the preferences of tourists. Our proposed algorithm is able to create schedules for tourism services and develop recommendations. The system employs minimum costs to create schedules for tourism services by the using genetic algorithm approach. Our proposed hybrid recommendation algorithm focuses on the tourists in Thailand. In future research, this can be extended to compare with other algorithms and improve the recommendation system. The result of this study is a recommended system, related to the study of Duan et al. (2020) [89], for the optimal scheduling and management of a smart city within a safe framework. Moreover, the results show that the hybrid genetic algorithm can effectively generate a satisfaction tourism service that saves energy which related the study of Alsokhiry et al. (2022) [90]—and thus develops a framework using pricing-based energy management.

A limitation of this study is that the method is time-consuming, which can cause the amount of services to vary if this method is applied in a large area, and will result in a very long time for the result to be output. Moreover, the accuracy of this work will be developed in future research.

We advise future researchers to consider the greenness issue in scheduling, because it is a new field that is expanding in manufacturing sectors. Additionally, recently created algorithms, such as the imperialist competitive algorithm, can be tested using the proposed framework and knowledge-based operator.

Additionally, one might think about creating new operators to further broaden the algorithm's population diversity, and even creating an operator to quantify population diversity.

Author Contributions: Conceptualization, P.S. and P.J.; methodology, P.S. and P.J; software, P.S. and P.J.; validation, P.S. and P.J.; formal analysis P.S. and P.J.; investigation, P.S. and P.J, resources, P.S. and P.J.; data curation, P.S. and P.J.; writing—original draft preparation, P.S. and P.J.; writing—review and editing, P.S., P.J., and P.K.; visualization, P.S. and P.J.; supervision, K.J.; project administration, P.S.; funding acquisition, P.S. All authors have read and agreed to the published version of the manuscript.

Funding: This research was funded by Suan Dusit University under the Ministry of Higher Education, Science, Research and Innovation, Thailand, grant number 65-FF-003 "Innovation of Smart Tourism to Promote Tourism in Suphan Buri Province".

Institutional Review Board Statement: The study was conducted in accordance with the ethical and approved by the Ethics Committee of Suan Dusit University (SDU-RDI-SHS 2022-030, 1 June 2022) for studies involving humans.

Informed Consent Statement: Not applicable.

Data Availability Statement: Not applicable.

Acknowledgments: The research team would like to thank Suan Dusit University for the support with funding. Also, we would like to thank AVIC Research Center, Chulalongkorn University for consulting this paper. Finally, we would like to thank the Chiang Mai Municipality for all their cooperation and for providing the necessary information for the research.

Conflicts of Interest: The authors declare no conflict of interest.

References

1. AuPinedo, M.L. Introduction. In *Scheduling: Theory, Algorithms, and Systems*; Springer: New York, NY, USA, 2012; pp. 1–10.
2. Garey, M.R.; Johnson, D.S.; Sethi, R. The complexity of flowshop and jobshop scheduling. *Math. Oper. Res.* **1976**, *1*, 117–129. [CrossRef]
3. Ullman, J.D. NP-complete scheduling problems. *J. Comput. Syst. Sci.* **1975**, *10*, 384–393. [CrossRef]
4. Carlier, J.; Pinson, E. An algorithm for solving the job-shop problem. *Manag. Sci.* **1989**, *35*, 164–176. [CrossRef]
5. Adams, J.; Balas, E.; Zawack, D. The shifting bottleneck procedure for job shop scheduling. *Manag. Sci.* **1988**, *34*, 391–401. [CrossRef]
6. Vancheeswaran, R.; Townsend, M.A. Two-stage heuristic procedure for scheduling job shops. *J. Manuf. Syst.* **1993**, *12*, 315–325. [CrossRef]
7. Brucker, P.; Jurisch, B.; Sievers, B. A branch and bound algorithm for the job-shop scheduling problem. *Discret. Appl. Math.* **1994**, *49*, 107–127. [CrossRef]
8. Lageweg, B.J.; Lenstra, J.K.; Rinnooy Kan, A.H. Job-shop scheduling by implicit enumeration. *Manag. Sci.* **1977**, *24*, 441–450. [CrossRef]
9. Fisher, H.; Thompson, G.L. *Probabilistic Learning Combinations of Local Job-Shop Scheduling Rules*; Prentice-Hall: Englewood Cliffs, NJ, USA, 1963.
10. Qing-Dao-Er-Ji, R.; Wang, Y.; Wang, X. Inventory based two-objective job shop scheduling model and its hybrid genetic algorithm. *Appl. Soft Comput.* **2013**, *13*, 1400–1406. [CrossRef]
11. Croce, F.D.; Tadei, R.; Volta, G. A genetic algorithm for the job shop problem. *Comput. Oper. Res.* **1995**, *22*, 15–24. [CrossRef]
12. Zhang, J.; Hu, X.; Tan, X.; Zhong, J.H.; Huang, Q. Implementation of an ant colony optimization technique for job shop scheduling problem. *Trans. Inst. Meas. Control* **2006**, *28*, 93–108. [CrossRef]
13. Zhang, J.; Zhang, P.; Yang, J.; Huang, Y. Solving the Job Shop Scheduling Problem using the imperialist competitive algorithm. *Adv. Mater. Res.* **2012**, *845*, 737–740. [CrossRef]

14. Piroozfard, H.; Wong, K.Y. An imperialist competitive algorithm for the job shop scheduling problems. In Proceedings of the IEEE International Conference on Industrial Engineering and Engineering Management (IEEM '14), Selangor, Malaysia, 9–12 December 2014; pp. 69–73.
15. Armentano, V.A.; Scrich, C.R. Tabu search for minimizing total tardiness in a job shop. *Int. J. Prod. Econ.* **2000**, *63*, 131–140. [CrossRef]
16. Rakkiannan, T.; Palanisamy, B. Hybridization of genetic algorithm with parallel implementation of simulated annealing for job shop scheduling. *Am. J. Appl. Sci.* **2012**, *9*, 1694–1705.
17. Zhang, R.; Wu, C. A hybrid immune simulated annealing algorithm for the job shop scheduling problem. *Appl. Soft Comput. J.* **2010**, *10*, 79–89. [CrossRef]
18. Lei, D. A Pareto archive particle swarm optimization for multi-objective job shop scheduling. *Comput. Ind. Eng.* **2008**, *54*, 960–971. [CrossRef]
19. Coello, C.A.C.; Rivera, D.C.; Cortés, N. Use of an artificial immune system for job shop scheduling. In *Artificial Immune Systems;* Timmis, J., Bentley, P., Hart, E., Eds.; Springer: Berlin/Heidelberg, Germany, 2003; pp. 1–10.
20. Jain, A.S.; Meeran, S. Deterministic job-shop scheduling: Past, present and future. *Eur. J. Oper. Res.* **1999**, *113*, 390–434.
21. Çaliş, B.; Bulkan, S. A research survey: Review of AI solution strategies of job shop scheduling problem. *J. Intell. Manuf.* **2015**, *26*, 961–973. [CrossRef]
22. Holland, J.H. Genetic algorithms. *Sci. Am.* **1992**, *267*, 66–72. [CrossRef]
23. Davis, L. Job shop scheduling with genetic algorithms. In Proceedings of the 1st International Conference on Genetic Algorithms, Hillsdale, NJ, USA, 1 July 1985; pp. 136–140.
24. Yamada, T.; Nakano, R. A genetic algorithm applicable to large-scale job-shop problems. In Proceedings of the Parallel Problem Solving from Nature (PPSN-II '92), Brussels, Belgium, 28–30 September 1992; Elsevier Science: Amsterdam, The Netherlands, 1992; pp. 281–290.
25. Lee, K.M.; Yamakawa, T.; Lee, K.M. Genetic algorithm for general machine scheduling problems. In Proceedings of the 1998 Second International Conference. Knowledge-Based Intelligent Electronic Systems. Proceedings KES'98 (Cat. No.98EX111), Adelaide, SA, Australia, 21–23 April 1998; Volume 2, pp. 60–66. [CrossRef]
26. Sun, L.; Cheng, X.; Liang, Y. Solving job shop scheduling problem using genetic algorithm with penalty function. *Int. J. Intell. Inf. Process.* **2010**, *1*, 65–77.
27. Wang, L.; Zheng, D.Z. An effective hybrid optimization strategy for job-shop scheduling problems. *Comput. Oper. Res.* **2001**, *28*, 585–596. [CrossRef]
28. Zhou, H.; Feng, Y.; Han, L. The hybrid heuristic genetic algorithm for job shop scheduling. *Comput. Ind. Eng.* **2001**, *40*, 191–200. [CrossRef]
29. Ombuki, B.M.; Ventresca, M. Local search genetic algorithms for the job shop scheduling problem. *Appl. Intell.* **2004**, *21*, 99–109. [CrossRef]
30. Gonçalves, J.F.; Mendes, J.J.D.M.; Resende, M.G.C. A hybrid genetic algorithm for the job shop scheduling problem. *Eur. J. Oper. Res.* **2005**, *167*, 77–95. [CrossRef]
31. Lin, L.; Yugeng, X. A hybrid genetic algorithm for job shop scheduling problem to minimize makespan. In Proceedings of the 2006 6th World Congress on Intelligent Control and Automation, Dalian, China, 21–23 June 2006; pp. 3709–3713. [CrossRef]
32. Zhou, H.; Cheung, W.; Leung, L.C. Minimizing weighted tardiness of job-shop scheduling using a hybrid genetic algorithm. *Eur. J. Oper. Res.* **2009**, *194*, 637–649. [CrossRef]
33. Asadzadeh, L.; Zamanifar, K. An agent-based parallel approach for the job shop scheduling problem with genetic algorithms. *Math. Comput. Model.* **2010**, *52*, 1957–1965. [CrossRef]
34. Yusof, R.; Khalid, M.; Hui, G.T.; Md Yusof, S.; Othman, M.F. Solving job shop scheduling problem using a hybrid parallel micro genetic algorithm. *Appl. Soft Comput. J.* **2011**, *11*, 5782–5792. [CrossRef]
35. Kirkpatrick, S.; Gelatt, C.D., Jr.; Vecchi, M.P. Optimization by simulated annealing. *Science* **1983**, *220*, 671–680. [CrossRef]
36. Enoch, Y. Contents of tour packages: A cross-cultural comparison. *Ann. Tour. Res.* **1996**, *23*, 599–616. [CrossRef]
37. March, R. The Japanese travel life cycle. *J. Travel Tour. Mark.* **2000**, *9*, 185–200. [CrossRef]
38. Seddighi, H.R.; Theocharous, A.L. A model of tourism destination choice: A theoretical and empirical analysis. *Tour. Manag.* **2002**, *23*, 475–487. [CrossRef]
39. Wang, K.C.; Hsieh, A.T.; Chou, S.H.; Lin, Y.S. GPTCCC: An instrument for measuring group package tour service. *Tour. Manag.* **2007**, *28*, 361–376. [CrossRef]
40. Seddighi, H. Does cultural background of tourists influence the destination choice? an empirical study with special reference to political instability. *Tour. Manag.* **2001**, *22*, 181–191. [CrossRef]
41. Theuvsen, L. Vertical integration in the European package tour business. *Ann. Tour. Res.* **2004**, *31*, 475–478. [CrossRef]
42. Chen, Y.; Mak, B.; Li, Z. Quality deterioration in package tours: The interplay of asymmetric information. *Tour. Manag.* **2013**, *38*, 43–54. [CrossRef]
43. Alao, O.; Batabyal, A.A. Selling package tours to tourists: A contract theory perspective. *Ann. Tour. Res.* **2013**, *42*, 439–442. [CrossRef]
44. Aguiar-Quinatna, T.; Moreno-Gil, S.; Patricia-Peral, P. How could traditional travel agencies improve their competitiveness and survive? A qualitative study in Spain. *Tour. Manag.* **2016**, *20*, 98–108.

45. Reay, P.; Seddighi., H. An empirical evaluation of management and operational capabilities for innovation via co-creation. European. *J. Innov. Manag.* **2012**, *15*, 259–275. [CrossRef]
46. Nuray, A.; Theuvsen, L. Organic agriculture in Turkey: Status, achievements, and shortcomings. *Organ. Agric.* **2021**, *11*, 501–507. [CrossRef]
47. Jin, L.; He, Y.; Song, H. Service customization: To upgrade or to downgrade? An investigation of how option framing affects tourists' choice of package-tour service. *Tour. Manag.* **2012**, *33*, 266–275. [CrossRef]
48. Chen, H.; Weiler, B.; Young, M. Examining service shortfalls: A case study of Chinese group package tours to Australia. *J. Vacat. Market.* **2018**, *24*, 371–386. [CrossRef]
49. Batabyal, A.A.; Yoo, S.J. A probabilistic analysis of guided tours for tourists during the slack season. *Tour. Manag.* **2010**, *31*, 482–485. [CrossRef]
50. Jørgensen, F.; Solvoll, G. Demand models for inclusive tour charter: The Norwegian case. *Tour. Manag.* **1996**, *17*, 17–24. [CrossRef]
51. Tepelus, C. Aiming for sustainability in the tour operating business. *J. Clean. Prod.* **2005**, *13*, 99–107. [CrossRef]
52. Arabani, A.B.; Farahani, R.Z. Facility location dynamics: An overview of classifications and applications. *Comput. Ind. Eng.* **2012**, *62*, 408–420. [CrossRef]
53. Luan, J.; Yao, Z.; Zhao, F.; Song, X. A novel method to solve supplier selection problem: Hybrid algorithm of genetic algorithm and ant colony optimization. *Math. Comput. Simul.* **2019**, *76*, 294–309. [CrossRef]
54. Niknamfar, A.H.; Niaki, S.T.A. A binary-continuous invasive weed optimization algorithm for a vendor selection problem. *Knowl. Based Syst.* **2018**, *140*, 158–172. [CrossRef]
55. Tyan, I.; Yagüe, M.I.; Guevara-Plaza, A. Blockchain Technology for Smart Tourism Destinations. *Sustainability* **2021**, *12*, 9715. [CrossRef]
56. Koo, C.; Park, J.; Lee, J.N. Smart tourism: Traveler, business, and organizational perspectives. *Inf. Manag.* **2017**, *54*, 683–686. [CrossRef]
57. Taweesaengsakulthai, S.; Laochankham, S.; Kamnuansilpa, P.; Wongthanavasu, S. Thailand Smart Cities: What is the Path to Success? *Asian Politics Policy* **2019**, *11*, 144–156. [CrossRef]
58. Chen, R.C.; Wang, H.L. Application of Genetic Algorithm to Scheduling of Tour Guides for Tourism and Leisure Industry. *INFOS Cairo-Egypt* **2008**, *43*, 4083–4101. [CrossRef]
59. Thumrongvut, P.; Sethanan, K.; Pitakaso, R.; Jamrus, T.; Golinska-Dawson, P. Application of Industry 3.5 approach for planning of more sustainable supply chain operations for tourism service providers. *Int. J. Logist. Res. Appl.* **2022**, 1–24. [CrossRef]
60. Li, X.; Yin, M. An opposition-based differential evolution algorithm for permutation flow shop scheduling based on diversity measure. *Adv. Eng. Softw.* **2013**, *55*, 10–31. [CrossRef]
61. Hassanat, A.B.; Prasath, V.B.S.; Abbadi, M.A.; Abu-Qdari, S.A.; Faris, H. An improved genetic algorithm with a new initialization mechanism based on Regression techniques. *Information* **2018**, *9*, 167. [CrossRef]
62. Lee, T.S.; Loong, Y.T. A review of scheduling problem and resolution methods in flexible flow shop. *Int. J. Ind. Eng. Comput.* **2019**, *10*, 67–88. [CrossRef]
63. Tosun, Ö.; Marichelvam, M.K.; Tosun, N. A literature review on hybrid flow shop scheduling. *Int. J. Adv. Oper. Manag.* **2020**, *12*, 156–194. [CrossRef]
64. Ben Cheikh-Graiet, S.; Dotoli, M.; Hammadi, S. A tabu search based metaheuristic for dynamic carpooling optimization. *Comput. Ind. Eng.* **2020**, *140*, 106217. [CrossRef]
65. Deng, G.; Wei, M.; Su, Q.; Zhao, M. An effective co-evolutionary quantum genetic algorithm for the no-wait flow shop scheduling problem. *Adv. Mech. Eng.* **2015**, *7*, 1687814015622900. [CrossRef]
66. Wei, H.; Li, S.; Jiang, H.; Hu, J.; Hu, J.J. Hybrid genetic simulated annealing algorithm for improved flow shop scheduling with makespan criterion. *Appl. Sci.* **2018**, *8*, 2621. [CrossRef]
67. Tasgetiren, M.F.; Kizilay, D.; Pan, Q.K.; Suganthan, P.N. Iterated greedy algorithms for the blocking flowshop scheduling problem with makespan criterion. *Comput. Oper. Res.* **2017**, *77*, 111–126. [CrossRef]
68. Wang, C.-N.; Yang, F.-C.; Nguyen, V.T.T.; Vo, N.T.M. CFD Analysis and Optimum Design for a Centrifugal Pump Using an Effectively Artificial Intelligent Algorithm. *Micromachines* **2022**, *13*, 1208. [CrossRef]
69. Abualigah, L.; Yousri, D.; Abd Elaziz, M.; Ewees, A.A.; Al-qaness, M.A.A.; Gandomi, A.H. Aquila Optimizer: A novel meta-heuristic optimization algorithm. *Comput. Ind. Eng.* **2021**, *157*, 107250. [CrossRef]
70. Czyżżak, P.; Jaszkiewicz, A. Pareto simulated annealing—A metaheuristic technique for multiple-objective combinatorial optimization. *J. Multi-Criteria Decis. Anal.* **1998**, *7*, 34–47. [CrossRef]
71. Ulungu, E.L.; Teghem, J.; Fortemps, P.H.; Tuyttens, D. MOSA method: A tool for solving multiobjective combinatorial optimization problems. *J. Multi-Criteria Decis. Anal.* **1999**, *8*, 221–236. [CrossRef]
72. Das, I.; Dennis, J.E. A closer look at drawbacks of minimizing weighted sums of objectives for Pareto set generation in multicriteria optimization problems. *Struct. Optim.* **1997**, *14*, 63–69. [CrossRef]
73. Sankararao, B.; Yoo, C.K. Development of a robust multiobjective simulated annealing algorithm for solving multiobjective optimization problems. *Ind. Eng. Chem. Res.* **2011**, *50*, 6728–6742. [CrossRef]
74. Suman, B. Simulated annealing-based multiobjective algorithms and their application for system reliability. *Eng. Optim.* **2003**, *35*, 391–416. [CrossRef]

75. Suman, B.; Kumar, P. A survey of simulated annealing as a tool for single and multiobjective optimization. *J. Oper. Res. Soc.* **2006**, *57*, 1143–1160. [CrossRef]
76. Suman, B.; Hoda, N.; Jha, S. Orthogonal simulated annealing for multiobjective optimization. *Comput. Chem. Eng.* **2010**, *34*, 1618–1631. [CrossRef]
77. Bandyopadhyay, S.; Saha, S.; Maulik, U.; Deb, K. A simulated annealing-based multiobjective optimization algorithm: AMOSA. *IEEE Trans. Evol. Comput.* **2008**, *12*, 269–283. [CrossRef]
78. Deb, K.; Pratap, A.; Agarwal, S.; Meyarivan, T. A fast and elitist multiobjective genetic algorithm: NSGA-II. *IEEE Trans. Evol. Comput.* **2002**, *6*, 182–197. [CrossRef]
79. Knowles, J.; Corne, D. The Pareto archived evolution strategy: A new baseline algorithm for Pareto multiobjective optimization. In Proceedings of the 1999 Congress on Evolutionary Computation-CEC99 (Cat. No. 99TH8406), Washington, DC, USA, 6–9 July 1999; Volume 1, pp. 98–105. [CrossRef]
80. Liu, H.; Kim, M.; Kang, O. Sensor Validation for Monitoring Indoor Air Quality in a Subway Station. *Indoor Built Environ.* **2012**, *21*, 205–221. [CrossRef]
81. Cao, P.; Tang, J. A reinforcement learning hyper-heuristic in multi-objective single point search with application to structural fault identification. *arXiv* **2018**, arXiv:1812.07958.
82. Rincón-García, E.A.; Gutiérrez-Andrade, M.A.; De-Los-Cobos-Silva, S.G.; Lara-Velázquez, P.; Ponsich, A.S.; Mora-Gutiérrez, R.A. A multiobjective algorithm for redistricting. *J. Appl. Res. Technol.* **2013**, *11*, 324–330. [CrossRef]
83. Mateos, A.; Jiménez-Martín, A. Multiobjective simulated annealing for collision avoidance in ATM accounting for three admissible maneuvers. *Math. Probl. Eng.* **2016**, *2016*, 8738014. [CrossRef]
84. Sengupta, R.; Saha, S. Reference point based archived many objective simulated annealing. *Inf. Sci.* **2018**, *467*, 725–749. [CrossRef]
85. Saadatpour, M.; Afshar, A.; Khoshkam, H. Multi-objective multi-pollutant waste load allocation model for rivers using coupled archived simulated annealing algorithm with QUAL2Kw. *J. Hydroinformatics* **2019**, *21*, 397–410. [CrossRef]
86. Marques, J.; Cunha, M.; Savić, D. Many-objective optimization model for the flexible design of water distribution networks. *J. Environ. Manag.* **2018**, *226*, 308–319. [CrossRef] [PubMed]
87. Zhang, L.; Zhang, B.; Bao, H.; Huang, H. Optimization of Cutting Parameters for Minimizing Environmental Impact: Considering Energy Efficiency, Noise Emission and Economic Dimension. *Int. J. Precis. Eng. Manuf.* **2018**, *19*, 613–624. [CrossRef]
88. Zhang, B.; Xu, L.; Zhang, J. A multi-objective cellular genetic algorithm for energy-oriented balancing and sequencing problem of mixed-model assembly line. *J. Clean. Prod.* **2020**, *244*, 118845. [CrossRef]
89. Duan, Q.; Nguyen, V.Q.; Heba, A.; Abdulaziz, A. Optimal scheduling and management of a smart city within the safe framework. *IEEE Access* **2020**, *8*, 161847–161861. [CrossRef]
90. Alsokhiry, F.; Siano, P.; Annuk, A.; Mohamed, M.A. A Novel Time-of-Use Pricing Based Energy Management System for Smart Home Appliances: Cost-Effective Method. *Sustainability* **2022**, *14*, 14556. [CrossRef]

sustainability

MDPI

Article

GIS Fault Prediction Approach Based on IPSO-LSSVM Algorithm

Hengyang Zhao [1,*], Guobao Zhang [2] and Xi Yang [3]

1 State Grid Anhui Electric Power Company Limited, Hefei 230022, China
2 Electric Power Research Institute, State Grid Anhui Electric Power Co., Ltd., Hefei 230601, China
3 School of Electrical Engineering and Automation, Hefei University of Technology, Hefei 230009, China
* Correspondence: jianyanghubei@gmail.com

Abstract: With the improvement of industrialization, the importance of equipment failure prediction is increasing day by day. Accurate failure prediction of gas-insulated switchgear (GIS) in advance can reduce the economic loss caused by the failure of the power system to operate normally. Therefore, a GIS fault prediction approach based on Improved Particle Swarm Optimization Algorithm (IPSO)-least squares support vector machine (LSSVM) is proposed in this paper. Firstly, the future gas conditions of the GIS to determine the characteristic data of SF6 decomposition gas are analyzed; Secondly, a GIS fault prediction model based on LSSVM is established, and the IPSO algorithm is used to normalize the parameters LSSVM. The parameters of c and radial basis kernel function σ^2 are optimized, which can meet the needs of later search accuracy while ensuring the global search capability in the early stage. Finally, the effectiveness of the proposed method is verified by the fault data of gas-insulated switch. Simulation results shows that, compared with the prediction methods based on IGA-LSSVM and PSO-LSSVM, the prediction accuracy rate of the proposed method reached 92.1%, which has the smallest prediction absolute error, higher accuracy and stronger prediction ability.

Keywords: gas-insulated switchgear; failure prediction; parameter optimization; improved particle swarm optimization algorithm

Citation: Zhao, H.; Zhang, G.; Yang, X. GIS Fault Prediction Approach Based on IPSO-LSSVM Algorithm. *Sustainability* **2023**, *15*, 235. https://doi.org/10.3390/su15010235

Academic Editor: Mohamed A. Mohamed

Received: 27 November 2022
Revised: 12 December 2022
Accepted: 15 December 2022
Published: 23 December 2022

1. Introduction

Gas-insulated switchgear (GIS) has the advantages of a small footprint, high operational reliability, and a long maintenance cycle [1–3]. So, it has been widely used in power systems, especially in high-voltage power grids [4,5]. Although GIS equipment has the advantage of high operational reliability, insulation and mechanical defects and failures still occur from time to time during its operation [6–8]. In order to meet the high efficiency and reliability of equipment operation, fault prediction can process the information measured by each sensor through certain theories and algorithms to infer the trend of fault development, so as to check the equipment to reduce the loss caused by the fault [9]. Therefore, to propose a GIS fault prediction method for accurate fault prediction in advance is of great significance to ensure the stable operation of the power system [10].

Because the fault prediction is carried out before the fault occurs, it often has the characteristics of small fault amplitude, inconspicuous fault characteristics, and easy to be covered by noise. Therefore, the signal processing and analysis before the fault occurs are the focus of fault prediction technology research [11–13]. The authors in [14] used the combination of deep neural network (DNN) and principal component analysis (PCA) to realize the early diagnosis and life prediction of small and slowly changing faults. In [15], in order to realize the active early warning of familial defect faults of power grid equipment, a familial defect fault database is established, and a probability prediction model between a single influencing factor and the defect decision data is established by the

Copula function. In order to solve the problem of large error and low accuracy in predicting equipment failure rate with a single model, a method for equipment failure rate prediction based on Autoregressive Moving Average Model-Back Propagation (ARMA-BP) combined model was proposed in [16]. The prediction results show that compared with the single prediction model, the equipment failure rate prediction result of the ARMA-BP combined model is more accurate. Reference [17] predicts the next batch of statistic values in GIS by establishing an autoregressive model, and compares the calculated value with the control limit, thereby realizing the prediction of S fault, but this method assumes that the trend of the fault is linear, and prediction results of the quantity are independent, so it is difficult to predict the nonlinear and multi-coupling data.

SF_6 gas-insulated metal-enclosed combined electrical appliances have become the main component of modern power systems because of their small footprint and high reliability [18]. Reference [19] proposed a fault detection method for gas-insulated switchgear based on a support vector machine and nuclear principal component analysis, and verified the reliability and accuracy of the algorithm through the actual sample test. Reference [20] established a GIS internal insulation electric field degradation calculation model, and used the central limit theorem to estimate the failure probability of the most harmful metal tip. In the literature [21], the online monitoring of the content of SO_2, SOF_2, H_2S, CO and other components in the SF_6 gas decomposition products can quickly find the location of the gas chamber where the fault point is located, but no research has been conducted on the types of faults that cause gas composition changes. Reference [22] proposes a GIS state assessment method combining the subjective weighting method and the objective weighting method to judge the operating status of GIS equipment. This method can systematically and comprehensively evaluate the health status of GIS, but it cannot provide a reference for more detailed troubleshooting, which is difficult to apply in practical engineering. The applicability of the model to the data characteristics is not highly consistent, and the prediction accuracy cannot meet the requirements.

In summary, how to accurately diagnose GIS faults according to operating parameters has become a key technology that needs to be overcome. For this reason, a GIS fault prediction approach based on IPSO-LSSVM is proposed in this paper. The contributions are as follows:

(1) A GIS fault prediction model based on IPSO-LSSVM is established.
(2) The PSO algorithm is improved through weight nonlinear adjustment and iterative speed optimization, so as to improve the optimization ability and convergence speed of the algorithm.
(3) The normalization parameter c of LSSVM and the parameter of radial basis kernel function σ^2 are optimized by the improved PSO algorithm, so as to improve the accuracy of the prediction model.

The organization of this paper is as follows: Section 2 analyzes the GIS fault characteristics; Section 3 introduces the GIS fault prediction model based on the improved PSO-LSSVM; Section 4 introduces the improvement strategy of the PSO algorithm; the Section 4 is the simulation verification; Section 5 is the conclusion.

2. GIS Fault Feature Analysis

The main faults of SF_6 electrical equipment are discharge and overheating. The discharge mainly includes three types: arc discharge, spark discharge and corona discharge [23]. Both discharge and overheating will generate energy to promote the decomposition of SF_6 gas. The main components decomposed under the action of discharge are SF_6 and metal oxides of electrodes or containers. In the presence of water vapor and oxygen, SF_4 reacts with them and finally generates SO_2, HF, SO_2F_2 and SOF_2 and other compounds.

In the discharge state, the energy generated by different discharge types is different, which promotes the decomposition of SF_6 gas into different stable substances under different energy conditions. Studies have found that the ratio of SO_2F_2/SO_2 has certain regularity [24]. During corona discharge, its ratio is often in the range of 4.0~6.0 or even higher, and spark discharges are mostly concentrated in 2.0~3.5. During arc discharge, the ratio is the smallest, mostly 0.1~0.3, which decreases with the increase in intensity of discharge. SO_2 and SO_2F_2 are substances that are produced in large quantities during discharge faults. Considering that they have good stability at the same time, monitoring will be more convenient and accurate. Based on the observation data of 220 kV voltage level GIS in a substation, some data of SF_6 decomposition gas characteristics are shown in Table 1.

Table 1. Characteristic data of SF_6 decomposition gas.

Fault State	Sample Eigenvalues		
	$SO_2/(\mu L \cdot L^{-1})$	$SO_2F_2/(\mu L \cdot L^{-1})$	SO_2F_2/SO_2
Arc discharge	19.47	4.14	0.213
Arc discharge	16.41	3.70	0.225
Arc discharge	12.73	3.29	0.258
Spark discharge	18.74	55.95	0.986
Spark discharge	21.11	44.87	2.126
Spark discharge	17.55	45.35	2.584
Corona discharge	26.77	151.91	5.675
Corona discharge	20.99	84.50	4.026
Corona discharge	23.12	121.07	5.237

When predicting the future failure condition of GIS, it is necessary to analyze the future gas condition of GIS. In this paper, based on the historical data of decomposed gas content in GIS gas chamber, the GIS fault prediction method based on IPSO-LSSVM is used to predict the decomposition gas of GIS gas chamber in the future.

3. GIS Fault Prediction Approach Based on Improved PSO-LSSVM

3.1. Least Squares Support Vector Machine

The least squares support vector machine is a new type of support vector machine developed on the basis of the standard support vector machine. It replaces the inequality constraints of the traditional support vector machine (SVM) with equality constraints, and uses the sum of squares of errors as the empirical loss of the training set [25]. By transforming the quadratic programming problem into a system of linear equations, the solution speed and convergence accuracy are improved.

For the training sample set $Tr = \{(x_1, y_1), (x_2, y_2), \cdots, (x_n, y_n)\}$ $x_i \in R^d$, $y_i \in \{1, 2, \cdots, k\}, i = 1, 2, \cdots, n$. The sample is mapped from the input space R^d to the feature space, the nonlinear mapping is $\varphi(\cdot)$, namely:

$$\psi(x) = (\varphi(x_1), \varphi(x_2), \cdots, \varphi(x_n)) \tag{1}$$

construct an optimal decision function in the mapped feature space as follows:

$$y = w^T \cdot \varphi(x) + b \tag{2}$$

The decision function parameters w and b are determined based on the principle of structural risk minimization, and the solution process can be equivalent to solving the following optimization problems:

$$\begin{cases} \min R = c \cdot \sum\limits_{i=1}^{n} \xi_i^2 + \frac{1}{2} \cdot \|w\|^2 \\ s.t.\ y_i = w^T \cdot \varphi(x_i) + b\ i = 1, 2, \cdots, n \end{cases} \tag{3}$$

The solution using the Lagrange method can be written as:

$$L(w,b,\xi_i,\alpha) = c\cdot\sum_{i=1}^{n}\xi_i^2 + \frac{1}{2}\|w\|^2 - \sum_{i=1}^{n}\left\{\alpha_i\cdot\left[w^T\cdot\varphi(x_i)+b-y_i\right]\right\} \tag{4}$$

where $\alpha = [\alpha_1, \alpha_1, \cdots \alpha_n]$ is the Lagrange multiplier.

According to the optimization conditions $\frac{\partial L}{\partial w} = 0, \frac{\partial L}{\partial b} = 0, \frac{\partial L}{\partial \xi_i} = 0, \frac{\partial L}{\partial \alpha_i} = 0$, we can get:

$$w = \sum_{i=1}^{n}\alpha_i\cdot\varphi(x_i) \tag{5}$$

$$\sum_{i=1}^{n}\alpha_i = 0 \tag{6}$$

$$2c\xi_i = \alpha_i \tag{7}$$

$$y_i = w^T\cdot\varphi(x_i)+b+\xi_i \tag{8}$$

Substituting Equations (5) and (7) into Equation (8), we can get:

$$y_i = \sum_{j=1}^{n}\left(\alpha_i\cdot\langle\varphi(x_j),\varphi(x_i)\rangle\right) + b + \frac{1}{2c}\alpha_i \tag{9}$$

Assuming that the kernel function $K(x_j, x_i) = \langle\varphi(x_j),\varphi(x_i)\rangle$ is defined, then

$$y_i = \sum_{j=1}^{n}\left(\alpha_i\cdot K(x_i,x_j)\right) + b + \frac{1}{2c}\alpha_i \tag{10}$$

Combining Equations (6) and (10) into a system of linear equations is as follows:

$$\begin{bmatrix} 0 & 1 & 1 & \cdots & 1 \\ 1 & K(x_1,x_2)+\frac{1}{2c} & K(x_1,x_2) & \cdots & K(x_1,x_n) \\ 1 & K(x_2,x_1) & K(x_2,x_2)+\frac{1}{2c} & \cdots & K(x_2,x_n) \\ \vdots & \vdots & \vdots & \ddots & \vdots \\ 1 & K(x_n,x_1) & K(x_n,x_2) & \cdots & K(x_n,x_n)+\frac{1}{2c} \end{bmatrix} \cdot \begin{bmatrix} b \\ \alpha_1 \\ \alpha_2 \\ \vdots \\ \alpha_n \end{bmatrix} = \begin{bmatrix} 0 \\ y_1 \\ y_2 \\ \vdots \\ y_n \end{bmatrix} \tag{11}$$

Based on the previously set training sample set $(x_i, y_i), i = 1, 2, \cdots n, x \in R^d, y \in R$, the model parameters can be obtained by solving the linear equation system (11). Then the decision function $\begin{bmatrix} b & \alpha_1 & \alpha_2 & \cdots & \alpha_n \end{bmatrix}$ can be determined:

$$f(x) = \sum_{i=1}^{n}\alpha_i\cdot K(x,x_i)+b \tag{12}$$

In the coefficient α of the support vector machine, if the element α_i is not equal to zero, the so-called support vector α_i is the corresponding sample (x_i, y_i).

On the basis of LSSVM binary classification, a one-to-many classification algorithm is used to establish a multi-class classifier. For the k ($k > 2$) class classification problem, take all the training samples of $y = i$ and $y \neq i$, and construct k training subsets Tr_i. When constructing the m-th classifier among the k classifiers ($m \in y_i$), the m-class training samples are used as the first class, the class number is $y_i^m = \pm 1$, and the other m-1 classes are used as one class, the class number is $y_i^m = -1$.

When classifying the test sample, the test sample is first input to the classifier 1. If the output of the discriminant function is 1, the judgment category is the first type of fault; otherwise, it is automatically input to the next classifier 2, and so on, until the classifier k. If the output of the judgment function is 1, the judgment category is the k-th fault; if the output is -1, the test sample is not in the k categories and belongs to other categories.

3.2. Kernel Function and Parameter Selection

LSSVM transforms the optimization problem into solving a system of linear equations, which makes its model have better generalization ability, and the parameters in LSSVM have a great influence on the performance of the algorithm. In this paper, the Gaussian radial basis kernel function is selected as the kernel function. The normalization parameter c and the parameters of the radial basis kernel function σ^2 play a key role in the learning ability, generalization ability and training calculation amount of the algorithm [26]. The size of is σ^2 closely related to the sparsity of the coefficient matrix in formula (11). Another thing that needs to be determined is the regularization parameter c (also called penalty factor) of the model. Based on the principle of minimizing structural risk, its size determines the proportion of empirical risk in structural risk. The smaller c is the weaker the proportion of empirical risk will be, so the model will be simpler, but the accuracy will be reduced. If c is larger, the proportion of empirical risk will be larger, which can improve the accuracy of the model, but the cost is the model will be more complicated. Therefore, in practical applications, an optimal choice should be made between the accuracy and complexity of the model. Take a section of c and σ^2, and form a two-dimensional plane with the two sections, and perform continuous search based on the accuracy rate. It can be determined that the only one (c, σ^2) has the highest accuracy rate. It should be noted that this (c, σ^2) may be not the optimal solution of above, but it is also an acceptable satisfactory solution. Therefore, finding the most parameters and applying them to the state estimation model of LSSVM will improve the accuracy of state estimation.

This paper has utilized the improved particle swarm optimization algorithm to find the optimal parameters of LSSVM.

4. Improved LSSVM Parameter Optimization for PSO

In the prediction method proposed in this paper, the parameters c and σ^2 of the LSSVM are determined by the improved particle swarm optimization algorithm, and the behavior of a group of particles with randomly selected initial values is modeled by the particle swarm optimization algorithm. The position and velocity of each particle over k iterations in the search space are described by X_k^i and V_k^i, respectively, and each particle records its best local position P_{lbest}^i. Then, at (k + 1) iterations, the velocity of particle i is obtained by the following equation:

$$V_{k+1}^i = \omega \cdot V_k^i + C_1 \cdot R_1 \left(P_{\text{lbest}}^i - X_k^i \right) + C_2 \cdot R_2 \left(P_{\text{global}}^i - X_k^i \right) \tag{13}$$

where R_1 and R_2 are random functions that generate random numbers between 0 and 1; C_1 and C_2 are training coefficients; ω is inertia weight factors, which linearly decrease from 0.9 to 0.4, and it can be obtained by the following formula:

$$\omega = \omega_{\text{max}} - ((\omega_{\text{max}} - \omega_{\text{min}})/k_{\text{max}}) \times k \tag{14}$$

where k_{max} is the maximum number of iteration. At the end of each iteration, the new position of each particle is obtained by the sum of its old position and new velocity:

$$X_{k+1}^i = X_k^i + V_{k+1}^i \tag{15}$$

In addition to improving the accuracy of the GIS fault prediction, the training time and convergence speed are also particularly important. Therefore, we use the particle swarm algorithm to optimize the parameters of the LSSVM model in this section, the particle swarm algorithm is mainly improved in terms of optimization effect and calculation time.

4.1. PSO Weight Nonlinear Adjustment

The inertia weight factor ω has a great influence on the position update of the PSO algorithm. When the inertia weight factor ω is relatively large, the particle can quickly converge to the optimal position, but the volatility is large; when the inertia weight factor is relatively small, the particle convergence speed is slow, and it is easy to fall into the local optimum when faced with uncertain bionic optimization. Equation (14) is only processed by linearly decreasing the weight, and there is still room for improvement in the actual optimization process. Therefore, in this paper, the formula (14) is improved, and the weight is decreased by a nonlinear method. The improved inertia weight factor ω is:

$$\omega = \omega_{\min}\left(\frac{\omega_{\max}}{\omega_{\min}}\right)^{\frac{1}{1+\sqrt{k/k_{\max}}}} \tag{16}$$

The nonlinear decreasing change of the inertia weight factor can be realized through the above formula, which not only ensures the global search ability in the early stage, but also meets the needs of the later search accuracy.

4.2. PSO Iteration Speed Optimization

After each iteration, the new position of each particle will also change, so the search state will also be different. To adapt to this situation, this chapter optimizes and improves the iteration speed of Equation (15), namely:

$$X_{k+1}^{i} = X_{k}^{i} + \xi V_{k+1}^{i} \tag{17}$$

where ξ is the speed coefficient.

The parameter optimization process is shown in the Figure 1.

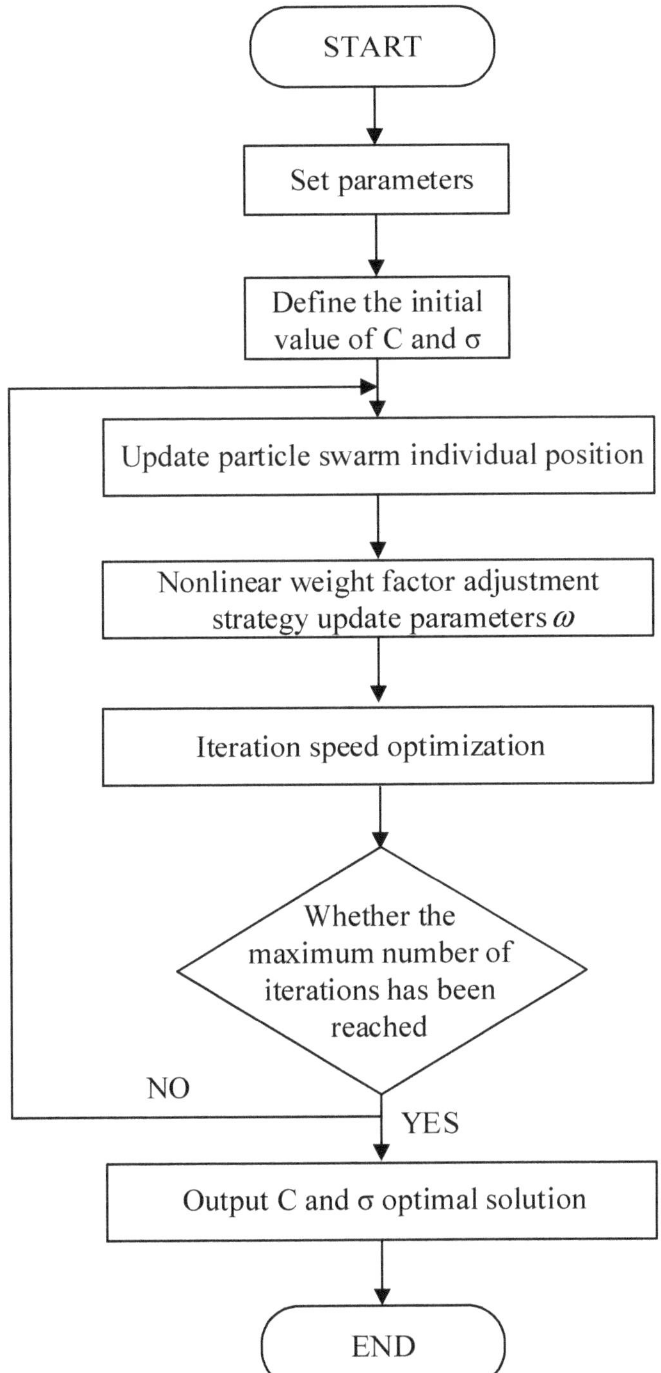

Figure 1. LSSVM parameter optimization process.

5. Simulation Results

Taking the data collected by the SF_6 online monitoring device of GIS in a distribution network as an example. A total of 400 sets of data from 1 March to 5 December 2021 are selected as the training set, and a total of 50 sets of samples of fault type data are selected as the test set. Code c and σ^2 according to the real number coding mode, and the optimization interval is (0, 200), (0, 15), respectively. The basic parameters of the PSO algorithm are shown in Table 2. The IPSO iteration curve is shown in Figure 2.

Table 2. Related parameters of PSO.

Parameter	Numerical Value
ω_{max}	0.8
ω_{min}	0.5
k_{max}	25
C_1, C_2	1.5
number of particles	100

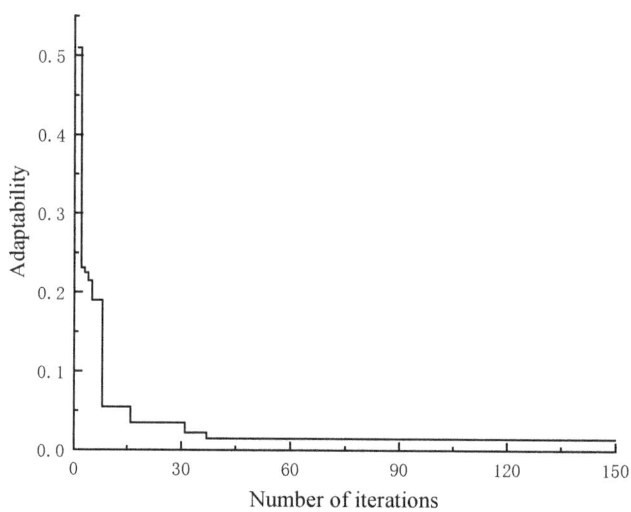

Figure 2. IPSO iteration curve.

It can be seen from Figure 2 that IPSO reaches the minimum fitness value after 37 iterations, and after that the value no longer changed. In addition, the corresponding optimal solutions c and σ^2 are 64.35 and 2.47, respectively. In order to compare the optimization effect of IPSO, PSO and IGA are used to optimize c and σ^2, respectively, and the maximum number of iterations is set to 300. The calculation results of the three optimization algorithms are shown in Table 3.

Table 3. Comparison of optimization algorithms.

Optimization Algorithm	IGA	PSO	IPSO
Number of iterations at convergence	154	126	52
Minimum fitness value	0.0918	0.0642	0.0223
Convergence time/ms	7.49	8.66	4.24

It can be seen from Table 3 that IPSO is better than PSO and IGA in terms of the number of iterations, optimal value and convergence time. It can be seen that the improvement strategy of the improved PSO algorithm proposed in this paper is successful. IPSO can effectively reduce the number of iterations and accelerate the convergence of the algorithm.

The ratio of SO_2F_2/SO_2 was calculated to find the PD fault types, in which the number of samples of arc discharge fault types was 14, the classification error was 2 times, and the model prediction accuracy was 85.7%. There are 10 spark discharge fault type samples, one classification error, and the prediction accuracy rate is 90.3%. There are 18 normal samples, one prediction error, and the prediction accuracy rate is 94.4%. The overall accuracy of the model reached 92.1%. Prediction results and accuracy comparisons of different algorithms are shown in Tables 4 and 5.

Table 4. Analysis data of GIA gas prediction results.

Method of Prediction	SO_2	HF	SO_2F_2	SOF_2
Actual value	11.84	0.67	0.15	0.8
IGA-LSSVM predictive value	11.21	0.69	0.175	0.72
Absolute error	−0.63	0.02	0.025	0.08
PSO-LSSVM predictive value	11.19	0.64	0.159	0.75
Absolute error	0.65	−0.03	0.009	−0.05
IPSO-LSSVM predictive value	11.79	0.68	0.143	0.82
Absolute error	−0.05	0.01	0.007	0.02

Table 5. Comparison of prediction results of different algorithms.

Fault Type	Number of Samples	Accuracy		
		IGA-LSSVM	PSO-LSSVM	IPSO-LSSVM
1	15	86.4	77.6	85.7
2	10	81.7	89.5	90.1
3	8	77.4	87.4	99.8
4	20	83.2	85.2	95.6

It can be seen from Tables 4 and 5, compared with the prediction methods based on IGA-LSSVM and PSO-LSSVM, the method proposed in this paper has the smallest prediction absolute error, higher accuracy and stronger prediction ability.

6. Conclusions

A GIS fault prediction approach based on IPSO-LSSVM is proposed in this paper. The PSO algorithm is improved by nonlinear weight adjustment and iterative speed optimization, so as to optimize the normalization parameters of LSSVM. Simulation results show that: on the one hand, the improved PSO algorithm is superior to PSO and IGA in terms of the number of iterations, optimal value and convergence time, which can effectively reduce the number of iterations and speed up the algorithm convergence; on the other hand, compared with the traditional algorithm, prediction results of the proposed method are more accurate, so that the failure trend can be detected before the failure occurs, which can provide reference for the relevant staff.

In the aspect of power transformer fault diagnosis, with the development of electronic technology and sensor technology, the means of monitoring power transformers are increasing day by day. How to make full use of multi-dimensional information to comprehensively diagnose power transformers is a problem that needs further research in the field of power transformer fault diagnosis.

Author Contributions: All authors contributed equal. All authors have read and agreed to the published version of the manuscript.

Funding: This research received no external funding.

Institutional Review Board Statement: Not applicable.

Informed Consent Statement: Not applicable.

Data Availability Statement: Not applicable.

Conflicts of Interest: The authors declare no conflict of interest.

References

1. Liu, L.; Wang, B.; Ma, F.; Zheng, Q.; Yao, L.; Zhang, C.; Mohamed, M.A. A concurrent fault diagnosis method of transformer based on graph convolutional network and knowledge graph. *Front. Energy Res.* **2022**, *10*, 127. [CrossRef]
2. Zhuang, Y.; Hu, X.; Tang, B.; Wang, S.; Cui, A.; Hou, K.; He, Y.; Zhu, L.; Li, W.; Chu, J. Effects of SF6 decomposition components and concentrations on the discharge faults and insulation defects in GIS equipment. *Sci. Rep.* **2020**, *10*, 15039–15048. [CrossRef] [PubMed]
3. Orrù, P.F.; Zoccheddu, A.; Sassu, L.; Mattia, C.; Cozza, R.; Arena, S. Machine learning approach using MLP and SVM algorithms for the fault prediction of a centrifugal pump in the oil and gas industry. *Sustainability* **2020**, *12*, 4776. [CrossRef]
4. Wang, Q.; Xiao, Y.; Dampage, U.; Alkuhayli, A.; Alhelou, H.H.; Annuk, A. An effective fault section location method based three-line defense scheme considering distribution systems resilience. *Energy Rep.* **2022**, *8*, 10937–10949. [CrossRef]
5. Wang, Q.; Jin, T.; Mohamed, M.A. An innovative minimum hitting set algorithm for model-based fault diagnosis in power distribution network. *IEEE Access* **2019**, *7*, 30683–30692. [CrossRef]
6. Faraz, B.; Alesheikh, A.A.; Sharif, M.; Farnaghi, M. FLCSS: A fuzzy-based longest common subsequence method for uncertainty management in trajectory similarity measures. *Trans. GIS* **2022**, *26*, 2244–2262.
7. Wen, T.; Zhang, Q.; Ma, J.; Wu, Z.; Shimomura, N.; Chen, W. A new method to evaluate the effectiveness of impulse voltage for detecting insulation defects in GIS equipment. *IEEE Trans. Dielectr. Electr. Insul.* **2019**, *26*, 1301–1307. [CrossRef]
8. Thacker, H.; Shah, Y.; Borah, A.J.; Jadeja, Y.; Thakkar, M.; Bhimani, S.; Chauhan, G. Assessment of groundwater potential zones across Katrol hill fault, Kachchh, Western India: A remote sensing and GIS approach. *Open J. Geol.* **2022**, *12*, 25–33. [CrossRef]
9. Wang, Q.; Jin, T.; Mohamed, M.A.; Deb, D. A novel linear optimization method for section location of single-phase ground faults in neutral noneffectively grounded systems. *IEEE Trans. Instrum. Meas.* **2021**, *70*, 1–10. [CrossRef]
10. Chen, Z.; He, G.; Li, J.; Liao, Y.; Gryllias, K.; Li, W. Domain adversarial transfer network for cross-domain fault diagnosis of rotary machinery. *IEEE Trans. Instrum. Meas.* **2020**, *23*, 61–73. [CrossRef]
11. Wu, Y.; Zhao, R.; Jin, W.; He, T.; Ma, S.; Shi, M. Intelligent fault diagnosis of rolling bearings using a semi-supervised convolutional neural network. *Appl. Intell.* **2020**, *5*, 21–29. [CrossRef]
12. Wu, Z.; Zhang, Q.; Ma, J.; Li, X.; Wen, T. Effectiveness of on-site dielectric test of GIS equipment. *IEEE Trans. Dielectr. Electr. Insul.* **2018**, *25*, 1454–1460. [CrossRef]
13. Zhang, L.; He, C.; Guo, R.; Yuan, W.; Li, J. Research on effectiveness of lightning impulses with different parameters for detecting protrusion defects in GIS. *IEEE Trans. Dielectr. Electr. Insul.* **2020**, *27*, 1354–1362. [CrossRef]
14. Wen, T.; Zhang, Q.; Ma, J.; Liu, X.; Wu, Z.; Zhang, L.; Zhao, J.; Shimomura, N.; Chen, W. Research on the detecting effectiveness of on-site lightning impulse test for GIS equipment with insulation defects. *IEEE Trans. Dielectr. Electr. Insul.* **2018**, *25*, 551–558. [CrossRef]
15. Deng, X.; Zhang, G. Family defect early warning of multi-factor GIS equipment based on ARMA and copula theory. *High Volt. Electr. Appl.* **2022**, *58*, 9–16.
16. Xu, D.; Zhou, C.; Guan, C.; Wang, X. Equipment failure rate prediction method based on ARMA-BP combined model. *Firepower Command. Control.* **2021**, *46*, 83–87.
17. Ji, H.; Ma, G.; Li, C.; Pang, Z.; Zheng, S. Influence of voltage waveforms on partial discharge characteristics of protrusion defect in GIS. *IEEE Trans. Dielectr. Electr. Insul.* **2016**, *23*, 1058–1067. [CrossRef]
18. Han, X.; Li, J.; Zhang, L.; Liu, Z. Partial discharge characteristics of metallic protrusion in GIS under different lightning impulse voltage waveforms based on UHF method. *IEEE Trans. Dielectr. Electr. Insul.* **2017**, *24*, 3722–3729. [CrossRef]
19. He, S.; Zhang, T.; Zhou, L.; Peng, X.; He, L. GIS fault diagnosis strategy based on feature classification algorithm. *Autom. Instrum.* **2019**, *232*, 197–200.
20. Wu, X.; Li, Y. Pang, W. Research on the electric field degradation characteristics and fault probability prediction of GIS bus latent faults. *Power Grid Technol.* **2014**, *12*, 1–12.
21. Wu, Z.; Lyu, B.; Zhang, Q.; Liu, L.; Zhao, J. Phase-space joint resolved PD characteristics of defects on insulator surface in GIS. *IEEE Trans. Dielectr. Electr. Insul.* **2020**, *27*, 156–163. [CrossRef]
22. Yuan, Y.; Ma, S.; Wu, J.; Jia, B.; Li, W.; Luo, X. Frequency feature learning from vibration information of GIS for mechanical fault detection. *Sensors* **2019**, *19*, 1949. [CrossRef] [PubMed]
23. Yang, Y.; Ma, S.; Wu, J.; Jia, B.; Li, W.; Luo, X. Fault diagnosis in gas insulated switchgear based on genetic algorithm and density-based spatial clustering of applications with noise. *IEEE Sens. J.* **2019**, *21*, 965–973. [CrossRef]
24. Ma, S.; Chen, M.; Wu, J.; Wang, Y.; Jia, B.; Jiang, Y. Intelligent fault diagnosis of HVCB with feature space optimization-based random forest. *Sensors* **2018**, *18*, 1221. [CrossRef] [PubMed]

25. He, L.; Yang, J.; Zhang, Z.; Li, Z.; Ding, D.; Yuan, M.; Li, R.; Chen, M. Research on mechanical defect detection and diagnosis method for GIS equipment based on vibration signal. *Energies* **2021**, *14*, 5507. [CrossRef]
26. Zhang, C.; Dong, M.; Ren, M.; Huang, W.; Zhou, J.; Gao, X.; Albarracín, R. Partial discharge monitoring on metal-enclosed switchgear with distributed non-contact sensors. *Sensors* **2018**, *18*, 551. [CrossRef]

sustainability

MDPI

Article

An Efficient MPPT Technique-Based Single-Stage Incremental Conductance for Integrated PV Systems Considering Flyback Central-Type PV Inverter

Ahmed Ismail M. Ali [1,*], Zuhair Muhammed Alaas [2], Mahmoud A. Sayed [1], Abdulaziz Almalaq [3], Anouar Farah [3] and Mohamed A. Mohamed [4,*]

1 Electrical Engineering Department, South Valley University, Qena 83523, Egypt
2 Electrical Engineering Department, Jazan University, Jazan 45142, Saudi Arabia
3 Department of Electrical Engineering, Engineering College, University of Ha'il, Ha'il 55476, Saudi Arabia
4 Electrical Engineering Department, Faculty of Engineering, Minia University, Minia 61519, Egypt
* Correspondence: a.ismail@eng.svu.edu.eg (A.I.M.A.); dr.mohamed.abdelaziz@mu.edu.eg (M.A.M.)

Abstract: Central-type photovoltaic (PV) inverters are used in most large-scale standalone and grid-tied PV applications due to the inverter's high efficiency and low-cost per kW generated. The perturbation and observation (P&O) and incremental conductance (IncCond) have become the most common techniques for maximum power point tracking (MPPT) strategies of PV/wind generation systems. Typically, the MPPT technique is applied in a two-stage operation; the first stage tracks the MPP and boosts the PV voltage to a certain level that complies with grid voltage, whereas the second stage represents the inversion stage that ties the PV system to the grid. Therefore, these common configurations increase the system size and cost as well as reduce its overall footprint. As a result, this paper applies two IncCond MPPT techniques on a proposed single-stage three-phase differential-flyback inverter (DFI). In addition, the three-phase DFI is analyzed for grid current negative-sequence harmonic compensation (NSHC). The proposed system efficiently provides a MPPT of the PV system and voltage boosting property of the DC-AC inverter in a single-stage operation. Moreover, the MPPT technique has been applied through the DFI using the conventional and modified IncCond tracking strategies. Furthermore, the system is validated for the grid-tied operation with the negative-sequence harmonic compensation strategy using computer-based simulation and is tested under uniform, step-change, as well as fast-changing irradiance profiles. The average efficiencies of the proposed system, considering the conventional and modified IncCond MPPT techniques, are 94.16% and 96.4% with tracking responses of 0.062 and 0.035 s and maximum overshoot of 46.15% and 15.38%, respectively.

Keywords: MPPT; incremental conductance (IncCond); differential inverter; differential flyback inverter (DFI); high-frequency transformer (HFT); continuous modulation scheme (CMS); harmonic compensation

Citation: Ali, A.I.M.; Alaas, Z.M.; Sayed, M.A.; Almalaq, A.; Farah, A.; Mohamed, M.A. An Efficient MPPT Technique-Based Single-Stage Incremental Conductance for Integrated PV Systems Considering Flyback Central-Type PV Inverter. *Sustainability* **2022**, *14*, 12105. https://doi.org/10.3390/su141912105

Academic Editor: Alberto-Jesus Perea-Moreno

Received: 10 August 2022
Accepted: 21 September 2022
Published: 25 September 2022

1. Introduction

Photovoltaic (PV) energy supplies the utility grid with electricity for public utilization purposes. PV energy generation offers many advantages compared with conventional fossil fuels, such as inexhaustible energy sources, environmentally-friendly nature of energy resources, and reduced cost per kWh [1,2]. Moreover, the modular operation characteristic of the PV generation system allows for the construction of solar energy systems at different power levels and load ratings. However, PV generation systems suffer from the drawback of low conversion efficiency and high installation cost [3]. In addition, the level of generated power from the PV system depends on many aspects, such as solar irradiance level and ambient temperature. However, the recent advancement in the interfaced power electronic technologies provides different solutions for solar energy utilization with enhanced operational efficiency [4]. Considering different irradiance levels, there is only a unique point where the PV system generates the maximum power, as depicted in Figure 1. Therefore, it is very important for solar energy applications to track the point of peak power for better solar energy utilization.

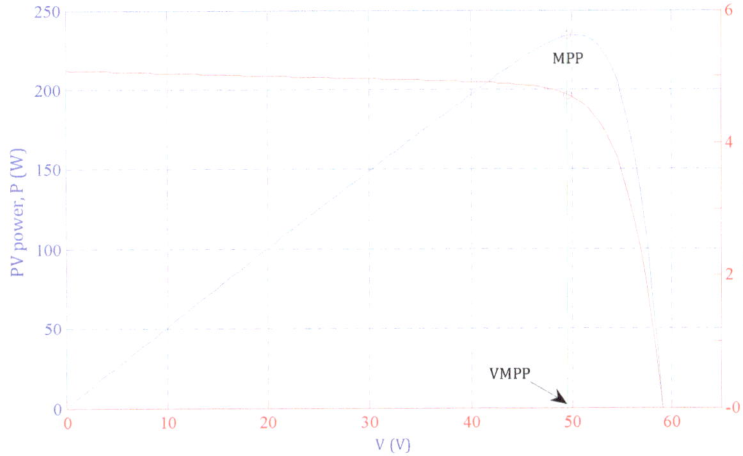

Figure 1. Characteristic curves of PV module.

To track the optimal power point, numerous maximum power point tracking (MPPT) techniques have been presented through interfaced power electronic circuits [5–7]. These MPPT algorithms vary in many aspects, such as tracking efficiency, system dynamic response (system settling time and overshoot), cost, and the required implementation equipment. An analysis of more than 30 different MPPT topologies is illustrated in [7,8]. Among the different MPPT topologies, perturbation and observation (P&O) [9–15] and incremental conductance (IncCond) [16–19] MPPT techniques are the most common topologies due to their simplicity, low implementation cost, and direct operation as a result of their independency on PV array parameters. In addition, different MPPT techniques have been proposed based on the parameters of the array, such as the fractional short-circuit current that predicts the location of optimal power from the short-circuit current [14,20], fractional open-circuit voltage that estimates the MPP from the array open-circuit voltage [21], fuzzy logic control-based tracker [22,23], neural network control-based algorithm [24], model-predictive control-based algorithm [25], and the sliding mode control-based MPPT techniques [26,27]. These tracking topologies are iterative MPPT topologies, which track the MPP with high precision under uniform irradiance conditions. However, these topologies suffer from slow response under load change and fast-changing weather conditions. In addition, the MPP spends a large execution time for MPPT control decision, which slows the system tracking response. In [28], an enhanced P&O technique has been presented to improve the PV system tracking operation under fast-changing irradiance conditions. In this control technique, the I-V curve of PV array has been used to differentiate between the irradiance change and operating point perturbation. However, this topology exhibits a slow system dynamic response with high contained steady-state oscillations. Moreover, an improved IncCond MPPT algorithm has been proposed in [16], which applies an allowable error level to improve the dynamic response of the conventional MPPT strategy under fast-changing irradiance conditions. However, it executes complex calculations that slow the system response. In [29], a modified IncCond strategy has been presented that efficiently tracks the optimal power point under fast-changing irradiance conditions. The presented topology improves the steady-state oscillations. However, the algorithm implements this function through complex computations that slows the system dynamic response. In [10], a modified P&O MPPT technique has been recommended to improve the tracking efficiency. However, the proposed controller slows the system response to the long computational execution time. In [30], the basic IncCond MPPT technique has been implemented through Cuk converter, which directly perturbs the duty-cycle of the Cuk converter to track the MPP until the optimal power point location is reached. However, it uses the conventional fixed step-size voltage perturbation that slows the system dynamic response and decreases the system tracking efficiency. Therefore, many research efforts are exerted to improve the tracking efficiency as well as system efficiency by applying a few variable step-size perturbation control algorithms [31,32]. However, the fixed and variable step-size MPPT techniques are applied to improve the tracking efficiency and dynamic behavior of the proposed controller through the DC-DC converter or two-stage DC-AC conversion stage, which increases the system size and cost, and controls complexity. Recently, different MPPT control parameters and topics are raised for the optimal operation of MPPT techniques for modern distribution networks with multi-types of intermittent renewable energy resources [33–35].

1.1. Motivation

Stimulated by the aforementioned literature survey, this work aims to apply a modified IncCond MPPT algorithm via a single-stage isolated DC-AC grid-connected differential flyback inverter (DFI). Both MPPT techniques are direct MPPT controllers, which can operate directly with the PV without any pre-stored data requirements [1,2]. In addition, the conventional and modified IncCond MPPT algorithms are compared by a grid-connected PV system through the three-phase isolated DFI. The two MPPT algorithms are applied to the grid-tied PV system under uniform, step-changing, and fast-changing irradiance profiles. The modified MPPT algorithm improves the system dynamic performance by decreasing the system overshoot by 30% and the settling time by 0.03 s compared with the conventional fixed-step MPPT algorithm. Ultimately, the average tracking efficiency of the isolated three-phase grid-integrated PV system is increased from 94.17% to 96.42%.

1.2. Contribution

The significant contributions of this paper are summarized as follows:

1. The study provides a simple single-stage three-phase isolated grid-tied PV system through a differential flyback inverter.
2. The proposed system is applicable under conventional as well as modified MPPT algorithms and at different irradiance conditions.
3. The single-stage operation of the PV system exhibits improved average efficiencies of 94.17% and 96.42% for the conventional and modified algorithms, respectively, with grid isolation for protection purposes.
4. The modified MPPT algorithm enhances the system overshoot from 46.45% to 15.38% and the system settling time from 0.062 to 0.035 s.
5. The paper exhibits a simple, low-cost, and efficient single-stage three-phase isolated grid-tied PV system through DFI with a 1:2 voltage boosting property.

1.3. Organization

This paper is organized into the following sections. Section 2 illustrates the PV array model with a detailed analysis of PV array specifications. Section 3 analyzes the single-stage three-phase DFI circuit structure and operation, whereas the single-stage conventional and modified IncCond MPPT techniques are explained in Section 4. The system specifications and results considering the conventional and modified single-stage IncCond MPPT algorithms are deeply analyzed in Section 5. Finally, the conclusions are provided in Section 6.

2. Solar Cell Modeling

In general, different mathematical modeling topologies are presented for the solar cell, which form the basic constructional element in the PV array [3]. A single-diode model has been used for simplicity purposes as depicted in Figure 2. The solar cell voltage-current mathematical relationship formula is as follows [1]:

$$I = I_{Ph} - I_{Sat} \cdot \left[exp\left\{ q \cdot \frac{V_{PV} + I_{PV} \cdot R_S}{A \cdot K \cdot T} \right\} - 1 \right] - \frac{V_{PV} + I_{PV} \cdot R_s}{R_{Sh}} \quad (1)$$

where

I_{Ph}: light-produced current/photocurrent (A).
I_{Sat}: reverse saturation-current (A).
q: electron charge (C).
V_{PV}: solar-cell output-voltage (V).
I_{PV}: solar-cell output-current (A).
R_S and R_{Sh}: series and shunt resistors (Ω), respectively.
A: P-N junction ideality-factor.
K: Boltzmann's constant (1.38×10^{-23} J/K)
T: solar-cell temperature (K)

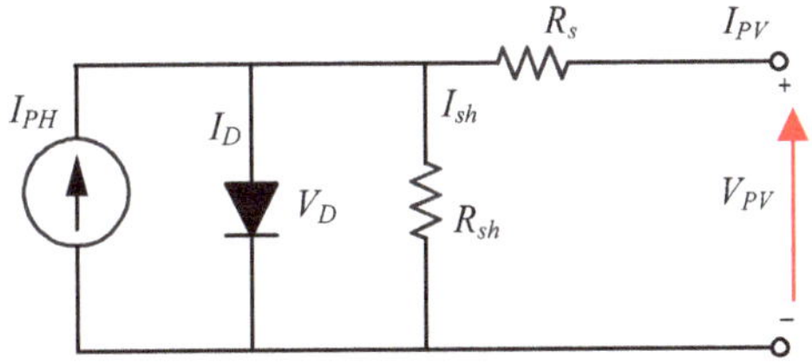

Figure 2. Solar cell model.

In addition, the produced photo-current depends on the level of irradiance (G) and the cell ambient temperature (T) as follows [2]:

$$I_{Ph} = \{I_{sc}^* + K_i(T - T^*)\} \cdot \frac{G}{G^*} \quad (2)$$

where I_{sc}* is the short-circuit current of PV array at standard temperature and irradiation conditions (STICs) (T^* = 298 K, G^* = 1000 W/m^2) and K_i is the PV array temperature co-efficient. In addition, the reverse saturation current is related to the cell temperature as follows [2]:

$$I_{Sat} = \frac{I_{sc}^* + K_i(T - T^*)}{exp\left[\frac{V_{oc}^* + K_v(T - T^*)}{V_t}\right] - 1} \quad (3)$$

where V_{oc}*, K_v, and V_t are the cell open-circuit voltage at STICs, open-circuit voltage co-efficient, and the thermal voltage ($V_t = K \cdot T/q$), respectively.

For a PV array combining series-connected cells of Ns per string and parallel-connected strings of N_p, the PV array generated output-current can be formulated as follows [36]:

$$I = N_p \cdot I_{Ph} - N_p \cdot I_{Sat} \cdot \left[exp\left\{ q \cdot \frac{\frac{V_{PV}}{N_s} + I_{PV} \cdot \frac{R_S}{N_p}}{A \cdot K \cdot T} \right\} - 1 \right] - \frac{\frac{N_p}{N_s} \cdot V_{PV} + I_{PV} \cdot R_S}{R_{Sh}} \quad (4)$$

Moreover, Canadian solar module CSP-220M PV is utilized in the proposed single-stage grid-tied system, where its characteristics under a uniform irradiance profile (1000 W/m^2) are depicted in Table 1. In addition, the system characteristics are directly related to the applied irradiance level as depicted in Figure 3.

Table 1. Canadian solar module characteristics CSP-220M PV.

Maximum power, PMPP	220 (W)
MPP Voltage, VMPP	48.3159 (V)
MPP Current, IMPP	4.54758 (A)
Open-circuit voltage, VOC	59.2618 (V)
Short-circuit current, ISC	5.09261 (A)
Temperature, STC	25 °C

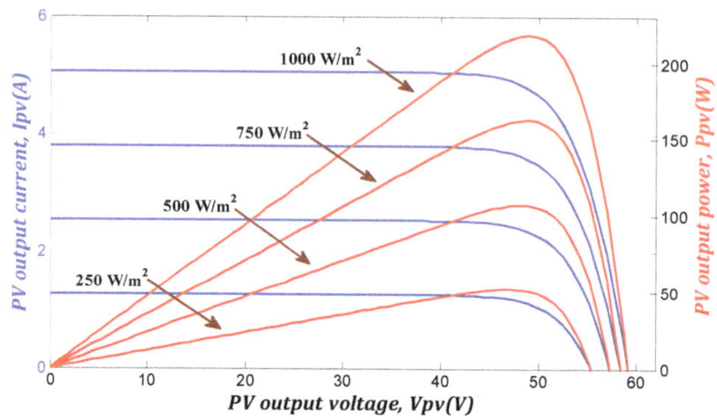

Figure 3. Output characteristics of the solar cell under different irradiance levels [1].

3. Three-Phase Single-Stage DFI

Figure 4 shows the three-phase DFI circuit structure, which consists of three parallel-connected input and differential-connected output flyback converter modules. Therefore, each module utilizes two switches (S_{Mx}, S_{Rx}), high frequency transformer (T_{rx}), and output capacitor (C_{ox}), where $x = a$, b, or c. Each flyback module temporarily transfers the input power to the secondary side with voltage bucking/boosting property as depicted in the operational modes in Figure 5. In addition, a variable duty-cycle is applied to control each flyback module under a wide range of duty-cycle changes to form DC output-voltage over module-output capacitors, which have a sinusoidal envelope and 120° phase-shift between the different modules (see Figure 4). Consequently, the differential connection between the flyback modules cancels the DC-voltage component and forms sinusoidal voltage waveforms at the grid terminals. Moreover, the success of the single-stage DFI depends on the flyback HFT design. As the DFI transfers the power temporarily to the grid, the HFT performs two main operational functions: (a) Storage element for the input power over the HFT magnetizing inductance, L_M, (b) Galvanic isolation for grid integration applications. In addition, the HFT offers further voltage bucking/boosting properties.

Different modulation schemes are used to control variable duty-cycle inverter topologies as reported in [37,38]. The grid-tied DFI synthesizes a sinusoidal grid current deceived with low-order harmonics due to the voltage boosting property with a mismatch between the converter modules and the input-to-output nonlinear relationship [39]. Therefore, the continuous modulation scheme (CMS) combined with the static linearization method (SLM) is used for low-frequency odd harmonics elimination in the grid-injected currents [40]. However, a feedback separate compensation loop is required for NS low-frequency even harmonics compensation. In addition, the CMS-based control scheme improves the input current ripple for PV applications as well as grid current THD to follow the IEEE and IEC standard harmonic limitations [41,42].

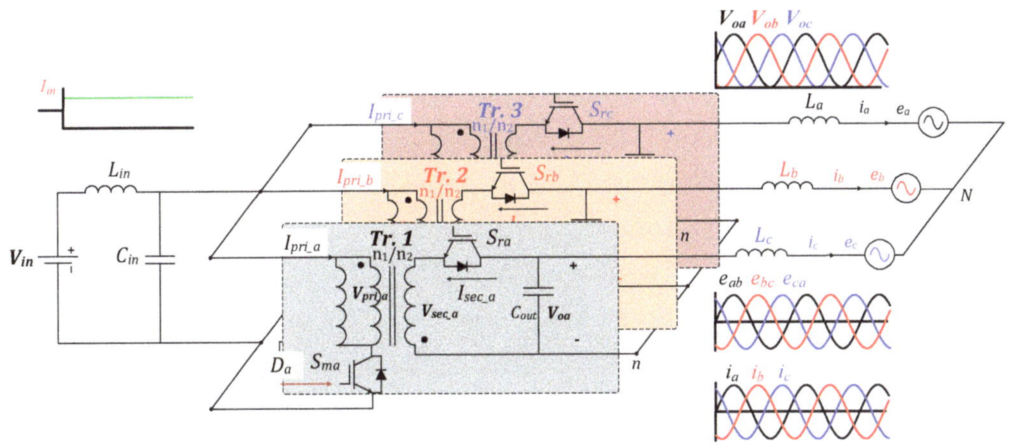

Figure 4. Differential-based three-phase flyback inverter.

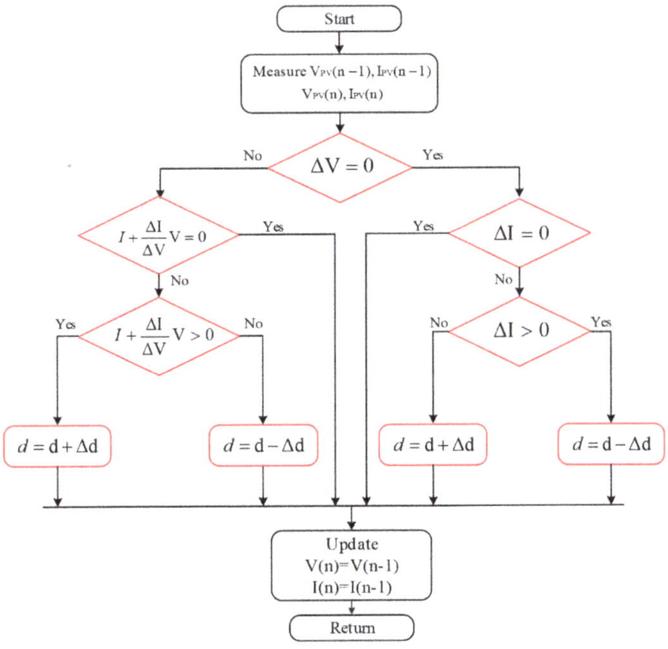

Figure 5. Conventional IncCond MPPT algorithm flowchart.

4. Incremental Conductance MPPT

The incremental conductance is a common MPPT technique that tracks the optimal PV array power directly without pre-defined information about the PV system [1]. It tracks the MPP by comparing the addition of the instantaneous conductance (I/V) and the incremental conductance ($\Delta i/\Delta v$) with zero to decide the location of the MPP ($I/V + \Delta i/\Delta v$). The slope of the solar array P-V curve is positive on the left side of MPP, zero at the MPP, and negative on the right side of MPP. Consequently, the MPPT algorithm adjusts the PV array voltage to control the operating point location. Therefore, when the P-V curve slope is negative, the controller moves the operating point to the left side toward the MPP locations by decreasing the PV array voltage. Similarly, when the slope is positive, it increases the

PV array voltage to move the operating point to the right side toward the MPP. Finally, as a null slope is reached, the voltage adjustment stops and the MPP is reached as depicted in the control algorithm in Figure 5. This simple and conventional MPPT algorithm tracks the MPP effectively at a uniform irradiance condition. However, it causes oscillations around the steady-state location and can easily divert the MPPT under fast-changing irradiance conditions. This conventional algorithm has the former handicap since it cannot differentiate between the fast-changing irradiance effects and the MPP location change.

Considering the former drawbacks, a modified IncCond MPPT technique is applied to track the optimal operating point location for efficient MPPT as shown in the algorithm flowchart in Figure 6 [3]. The modified algorithm tracks the MPP efficiently by checking the current variation sign with voltage variations. If the voltage perturbation results in current variations of the same sign, a sudden irradiance change is applied to the solar array (see Figure 6). Otherwise, the conventional IncCond technique is adequate to track the optimal MPP efficiently. In addition, the modified IncCond MPPT technique uses different step sizes in the PV array perturbation according to the location of the MPP on the P-V curve. Therefore, the closer the operating point to the MPP, the smaller the step size of voltage perturbation as depicted in the modified technique flowchart in Figure 6.

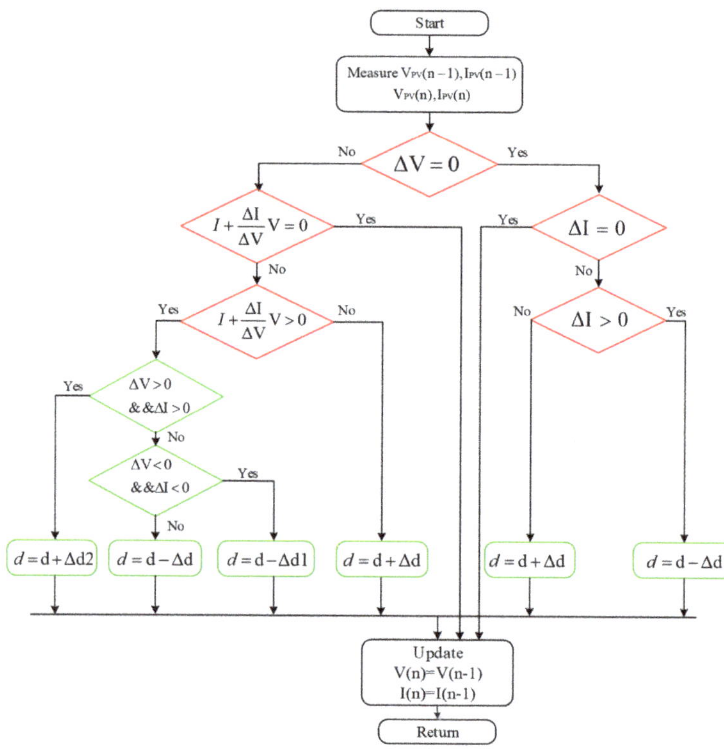

Figure 6. Modified IncCond MPPT algorithm flowchart.

5. System Results and Discussion

5.1. System Specifications

Based on the aforementioned analysis and modeling of different parts of the proposed system, the PSIM software is used to simulate the single-stage grid-tied PV system combined with the MPPT algorithm. In addition, the paper did not consider the experimental system verifications due to the unavailability of the system prototype at this moment, which will be considered in the future with low power capacity. This section illustrates the validation of the three-phase DFI using CMS control techniques through 200 V, 5 kW, and 50 kHz switching frequencies as the down-scale system for central-type PV inverters. The DFI closed-loop control scheme with the main and secondary control-loops is described in Figure 7. The parameters of the grid-tied system are listed in Table 2, which consists of the input PV array-based Canadian solar module with the parameters in Table 1, the DFI power-stage, and the grid-tied current filter (grid-inductance). Moreover, a stringent designed snubber-circuit is used to mitigate the high voltage spikes over the power switches.

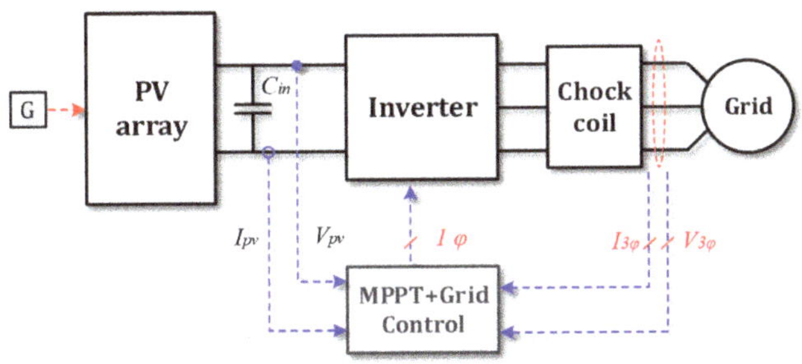

Figure 7. Proposed DFI control scheme.

Table 2. Overall DFI system parameters.

Inverter rating, P	5 kW
Input DC-voltage, V_{in}	100 V
Input-filter, L_{in}, C_{in}	150 μH, 10 μF
Input filter-resistance, r_{in}	4 Ω
Grid-voltage (L.L.), E, ω	200 V, $2 \times \pi \times 60$ rad/s
HFT magnetizing-inductance, L_m	100 μH
HFT primary-resistance, r_m	2 mΩ
Output-capacitor, C_o	10 μF
HFT leakage-inductance, $L_{Leakage}$	2.25 μH
HFT turns-ratio, n	1:1
Grid-inductance, L_g	4 mH
Grid inductor-resistance, r_g	5 mΩ
Switching-frequency, F_{SW}	50 kHz
MPPT update frequency	0.025 ms [43]
PI controller-gains, K_P, K_I	0.097 A/V, 280 rad·s^{-1}
PV array specifications	Canadian solar module
Δd, $\Delta d1$, $\Delta d2$	0.0002, 0.0005, 0.00006 s

In addition, this section aims to exhibit the performance of the conventional and proposed IncCond MPPT techniques through a compact three-phase DFI under a uniform irradiance profile and constant temperature fixed at 25 °C. Moreover, the fast-changing irradiance profile is applied to the proposed grid-tied system to investigate the MPPT response under stringent irradiance conditions. Furthermore, the DFI connects the PV system to the utility grid with a voltage boosting factor of (2), as shown in the input and grid voltage specifications listed in Table 2. The system circuit diagram has been illustrated in Figure 4 and the duty-cycle increments are listed in Table 2.

5.2. System Results

First, the proposed single-stage grid-tied PV system is tested under the conventional and modified IncCond MPPT techniques using a step-change uniform irradiance profile, as depicted in Figure 8. The solar irradiance profile is depicted in Figure 8a. The solar irradiance is uniform at 1000 W/m^2 for 1 s of the operating period, which is step-changed to 500 W/m^2 for another 1 s. Figure 8b shows the PV output-voltage using the conventional and modified IncCond MPPT techniques compared with the reference PV voltage, which is plotted in red color as depicted in the legend of each waveform. In addition, the MPP reference voltage variations are small compared to the variations in the actual waveforms. The results of the two tracking techniques are matched. However, the modified IncCond algorithm improves the system settling time and overshoot from (0.062 s and 46.15%) for the conventional algorithm to (0.035 s and 15.38%), respectively, as shown in the two zoomed regions. In addition, the d-axis current of the conventional and modified MPPT techniques are compared with the reference value as depicted in Figure 8c,d, respectively. Therefore, it proves the single-stage IncCond-based MPPT operation of the proposed grid-tied PV system under the two tracking algorithms using a step-changing irradiance profile. In addition, the settling time and overshoot of the d-axis current is improved using the modified IncCond from (0.062 s and 46.15%) for the conventional technique to (0.035 s and 15.38%), respectively.

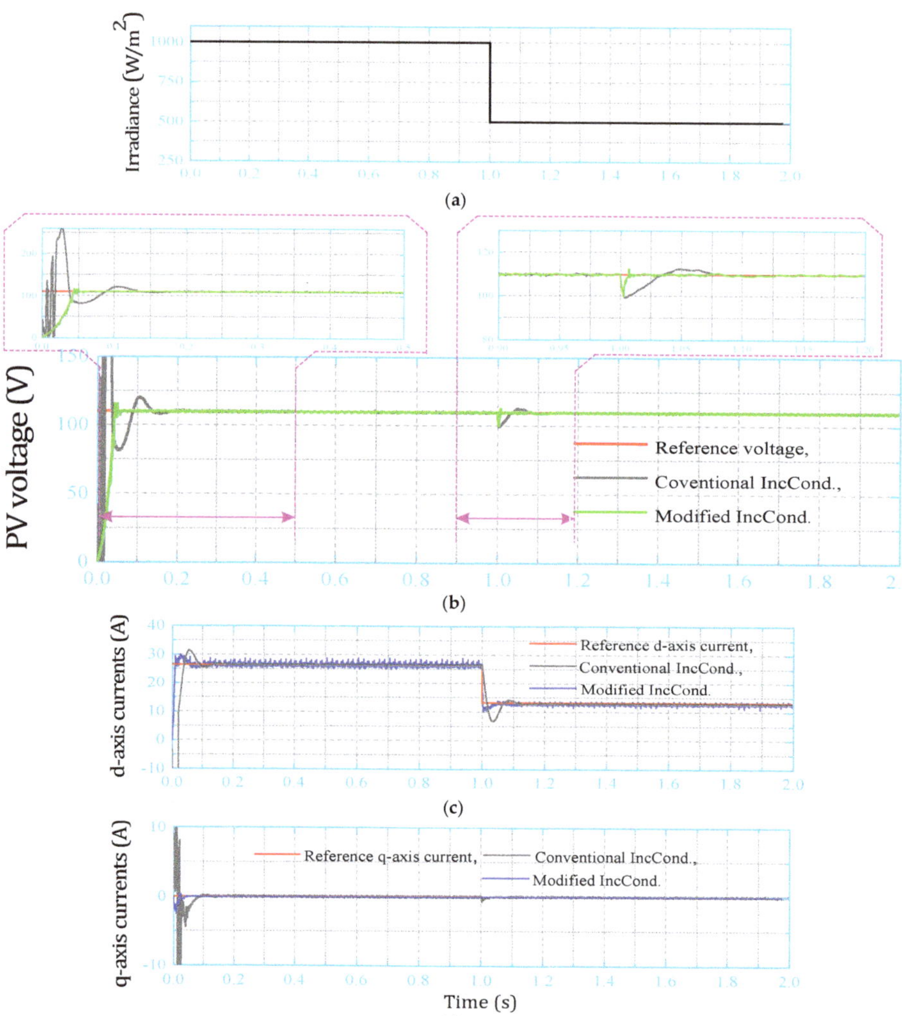

Figure 8. Grid-tied system results under the step-change uniform irradiance profile. (**a**) Step-change irradiance profile; (**b**) PV output-voltage under the two IncCond MPPT techniques; (**c**) grid-injected d-axis current; (**d**) grid-injected q-axis current.

Furthermore, the fast-changing irradiance profile is applied to the single-stage grid-connected PV system to reveal the tracking response of the conventional and modified IncCond MPPT techniques under these hard and fast-changing irradiance conditions. The fast-changing irradiance profile is depicted in Figure 9a,b. As can be seen, the PV output-voltage under the conventional and modified MPPT techniques is compared with the reference value. The system output-voltage using the modified IncCond MPPT technique shows an improved response compared with the conventional one. The d-axis current of the conventional and modified MPPT algorithms is compared with the reference value in Figure 9a,c comparison of the q-axis current is illustrated in Figure 9d.

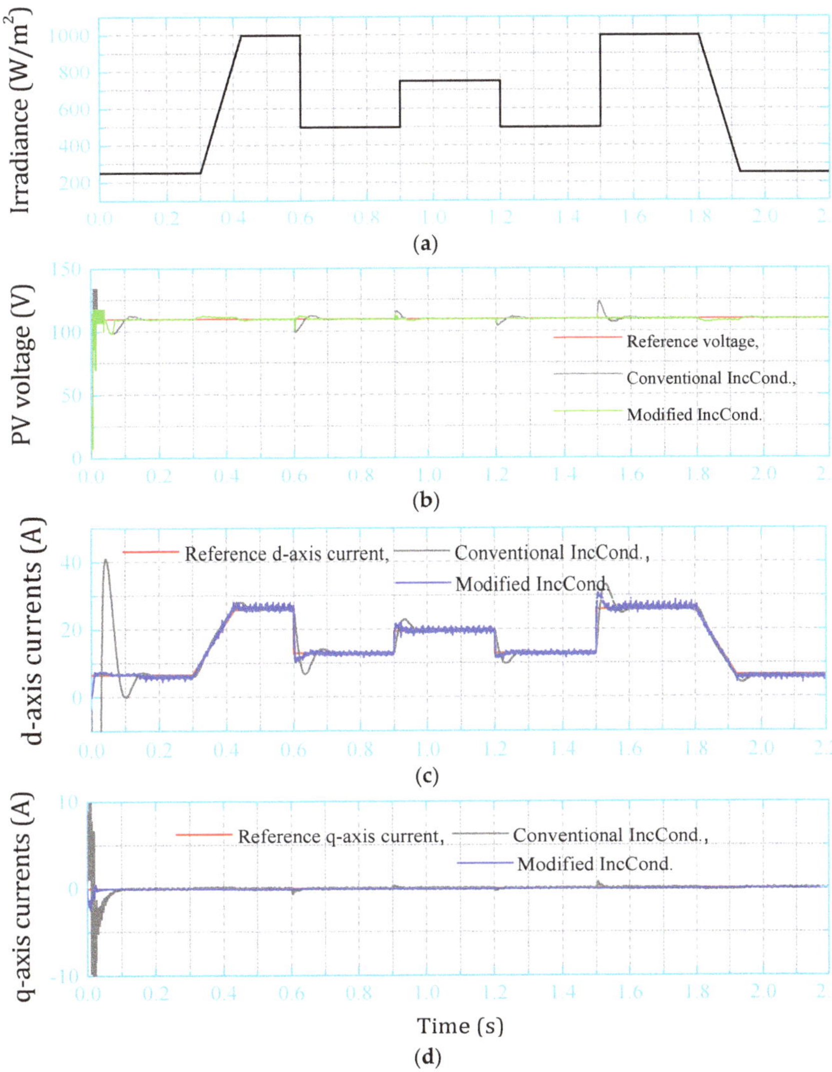

Figure 9. Grid-tied system results under the fast-changing irradiance profile. (**a**) Step-change irradiance profile; (**b**) PV output-voltage under the two IncCond MPPT techniques; (**c**) grid-injected d-axis current; (**d**) grid-injected q-axis current.

It is clear that the PV-side and grid-side results guaranteed the single-stage DC-AC PV system under the conventional and modified IncCond MPPT techniques. Moreover, the modified algorithm exhibits a better performance and dynamic response than the conventional algorithm's poor response. It improves the classical tracking behavior by decreasing the system settling time and dynamic overshoot. In addition, Figure 10 shows the FFT harmonic spectrum of the grid-injected current. The grid current THD is (3.95%), which follows the IEC-61000 and IEEE-519 harmonic standard limits.

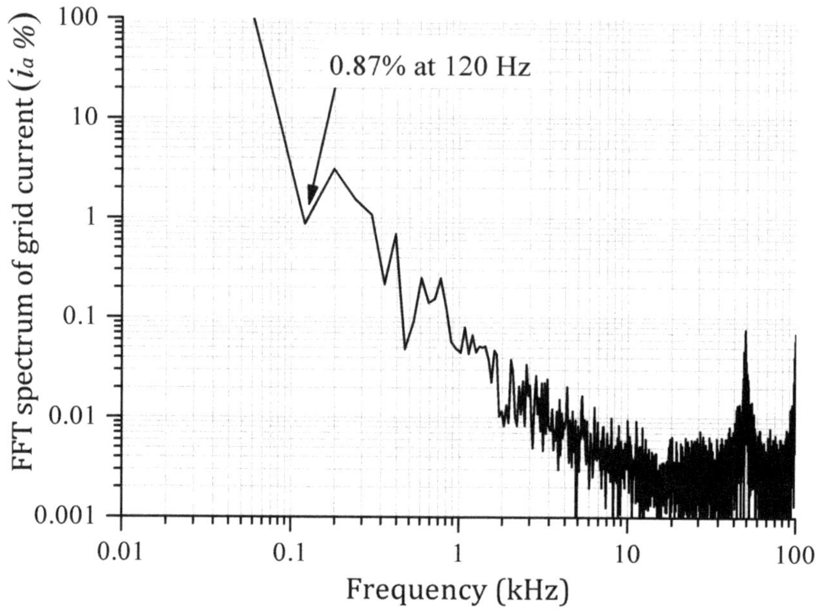

Figure 10. Grid current FFT harmonic spectrum (3.95%).

In addition, Figure 11 shows the instantaneous grid-injected powers and instantaneous tracking efficiency of the proposed single-stage PV system based on the following formula [1]:

$$\eta_T = \frac{P_G}{P_{th}} \cdot 100[\%] \tag{5}$$

where η_T is the instantaneous tracking efficiency, P_G is the grid-injected power, and P_{th} refers to the theoretical grid-injected power. In addition, the average tracking efficiency can be calculated from the following formula [2]:

$$\eta_{avg} = \langle \eta_T \rangle = \frac{\int_0^t P_G}{\int_0^t P_{th}} \cdot 100[\%] \tag{6}$$

The grid-injected power of the conventional and modified MPPT techniques is compared with the theoretical grid-injected power in Figure 11a. Moreover, Figure 11b shows the instantaneous tracking efficiency of the two IncCond-based MPPT algorithms based on Equation (5). Therefore, the modified algorithm improved the system average efficiency from 94.46% to 96.41%, which improved the system efficiency by 2.25%. Finally, the aforementioned results guaranteed the single-stage IncCond MPPT algorithm operation for the grid-tied PV system through the three-phase DFI.

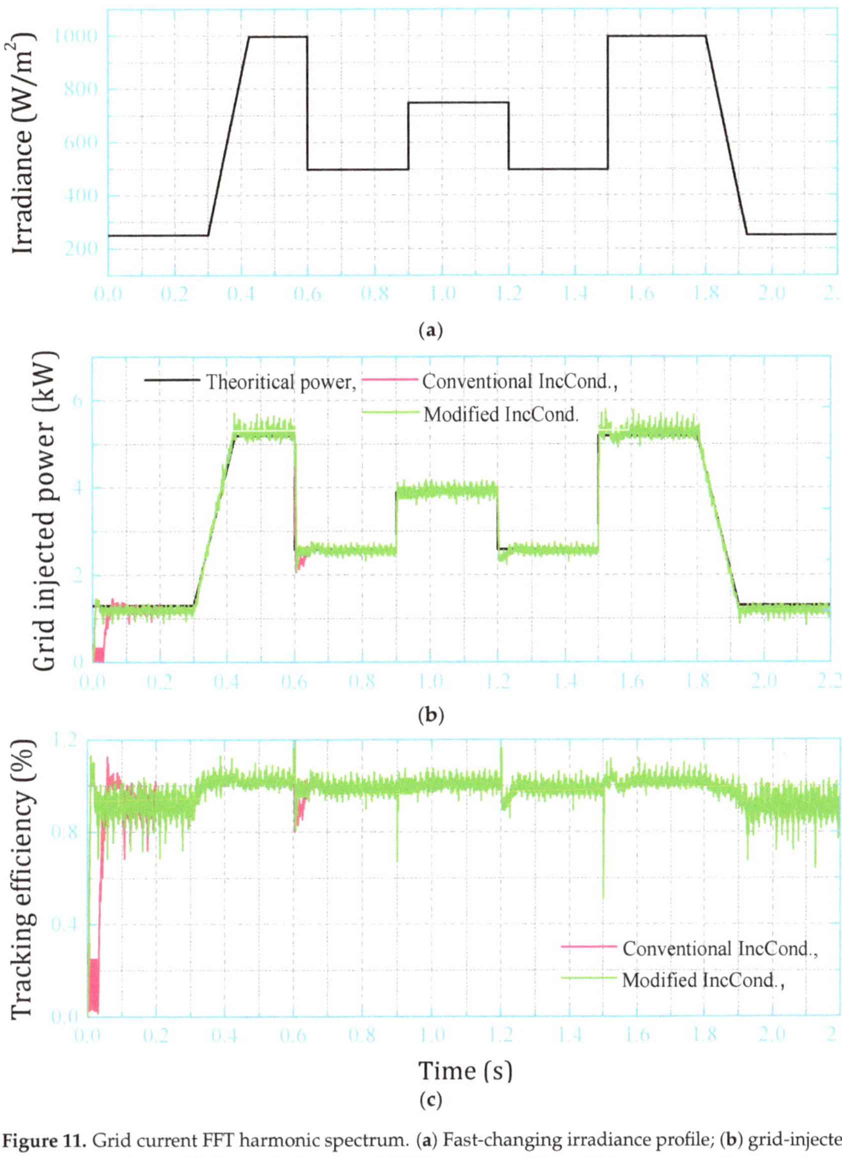

Figure 11. Grid current FFT harmonic spectrum. (**a**) Fast-changing irradiance profile; (**b**) grid-injected power comparison with theoretical power; (**c**) PV system tracking efficiency comparison.

6. Conclusions

This paper proposes a single-stage incremental conductance-based MPPT technique for the grid-tied PV system through the three-phase DC-AC isolated DFI. The proposed system decreases the required number of operational stages, components, and footprints. In addition, the system operation has been investigated by considering the conventional and modified IncCond MPPT algorithms. Moreover, the two tracking techniques are applied to the PV system under the step-change uniform irradiance as well as fast-changing irradiance profile. The system results guaranteed the single-stage operation of the proposed grid-tied PV system under the two MPPT techniques. Furthermore, the proposed modified IncCond algorithm exhibits an improved dynamic response compared with the conventional technique. It improves the system overshoot and settling time by (0.027 s and 30.77%), respectively. Ultimately, the tracking efficiency of the proposed single-stage grid-tied PV system has been improved from (94.16%) for the conventional IncCond algorithm to (96.41%) for the modified algorithm, as listed in Table 3. As a result, this paper achieved the single-stage MPPT operation using two techniques with the efficiency enhancement and minimized inverter footprint, which is applicable for industrial central-type PV tracking systems. Finally, the experimental verification of the proposed system will be investigated in future work considering a low-scale system capacity and standalone operation.

Table 3. Conventional vs. modified IncCond MPPT technique comparison at 1000 W/m^2 and 25 °C.

MPPT Algorithm	Settling Time (s)	Max. Overshoot (%)	η_T (%)
Conventional IncCond	0.062	46.15	94.1648
Modified IncCond	0.035	15.38	96.4168

Author Contributions: Conceptualization, A.I.M.A., Z.M.A., M.A.S. and M.A.M.; methodology, A.I.M.A.; software, A.I.M.A.; validation, A.I.M.A., Z.M.A., M.A.S., A.A., A.F. and M.A.M.; formal analysis, A.I.M.A., Z.M.A., M.A.S., A.A., A.F. and M.A.M.; investigation, A.I.M.A., Z.M.A., M.A.S., A.A., A.F. and M.A.M.; resources, A.I.M.A., Z.M.A., M.A.S., A.A., A.F. and M.A.M.; data curation, A.I.M.A., Z.M.A., M.A.S., A.A., A.F. and M.A.M.; writing—original draft preparation, A.I.M.A.; writing—review and editing, A.I.M.A., Z.M.A., M.A.S., A.A., A.F. and M.A.M.; visualization, A.I.M.A., Z.M.A., M.A.S., A.A., A.F. and M.A.M.; supervision, A.I.M.A., Z.M.A., M.A.S., A.A., A.F. and M.A.M. All authors have read and agreed to the published version of the manuscript.

Funding: This research has been funded by Scientific Research Deanship at the University of Ha'il—Saudi Arabia through project number RG-21079.

Institutional Review Board Statement: Not applicable.

Informed Consent Statement: Not applicable.

Data Availability Statement: Not applicable.

Conflicts of Interest: The authors declare no conflict of interest.

References

1. Ali, A.I.; Sayed, M.A.; Mohamed, E.E. Modified efficient perturb and observe maximum power point tracking technique for grid-tied PV system. *Int. J. Electr. Power Energy Syst.* **2018**, *99*, 192–202. [CrossRef]
2. Ali, A.I.M.; Mohamed, H.R.A. Improved P&O MPPT algorithm with efficient open-circuit voltage estimation for two-stage grid-integrated PV system under realistic solar radiation. *Int. J. Electr. Power Energy Syst.* **2022**, *137*, 107805.
3. Eltamaly, A.M.; Mohamed, M.A.; Abo-Khalil, A.G. Design and Comprehensive Analysis of Maximum Power Point Tracking Techniques in Photovoltaic Systems. In *Advanced Technologies for Solar Photovoltaics Energy Systems*; Springer: Cham, Switzerland, 2021; pp. 253–284.
4. Chen, W.; Liu, B.; Nazir, M.S.; Abdalla, A.N.; Mohamed, M.A.; Ding, Z.; Bhutta, M.S.; Gul, M. An Energy Storage Assessment: Using Frequency Modulation Approach to Capture Optimal Coordination. *Sustainability* **2022**, *14*, 8510. [CrossRef]
5. Rao, C.; Hajjiah, A.; El-Meligy, M.A.; Sharaf, M.; Soliman, A.T.; Mohamed, M.A. A novel high-gain soft-switching DC-DC converter with improved P&O MPPT for photovoltaic applications. *IEEE Access* **2021**, *9*, 58790–58806.
6. Subudhi, B.; Pradhan, R. A comparative study on maximum power point tracking techniques for photovoltaic power systems. *IEEE Trans. Sustain. Energy* **2012**, *4*, 89–98. [CrossRef]
7. Ali, M.N.; Mahmoud, K.; Lehtonen, M.; Darwish, M.M. An efficient fuzzy-logic based variable-step incremental conductance MPPT method for grid-connected PV systems. *IEEE Access* **2021**, *9*, 26420–26430. [CrossRef]
8. Hanzaei, S.H.; Gorji, S.A.; Ektesabi, M. A scheme-based review of MPPT techniques with respect to input variables including solar irradiance and PV arrays' temperature. *IEEE Access* **2020**, *8*, 182229–182239. [CrossRef]
9. Bianconi, E.; Calvente, J.; Giral, R.; Mamarelis, E.; Petrone, G.; Ramos-Paja, C.A.; Spagnuolo, G.; Vitelli, M. Perturb and observe MPPT algorithm with a current controller based on the sliding mode. *Int. J. Electr. Power Energy Syst.* **2013**, *44*, 346–356. [CrossRef]
10. Ahmed, J.; Salam, Z. An improved perturb and observe (P&O) maximum power point tracking (MPPT) algorithm for higher efficiency. *Appl. Energy* **2015**, *150*, 97–108.
11. Tafticht, T.; Agbossou, K.; Doumbia, M.; Cheriti, A. An improved maximum power point tracking method for photovoltaic systems. *Renew. Energy* **2008**, *33*, 1508–1516. [CrossRef]

12. Ishaque, K.; Salam, Z.; Lauss, G. The performance of perturb and observe and incremental conductance maximum power point tracking method under dynamic weather conditions. *Appl. Energy* **2014**, *119*, 228–236. [CrossRef]
13. Femia, N.; Petrone, G.; Spagnuolo, G.; Vitelli, M. Optimization of perturb and observe maximum power point tracking method. *IEEE Trans. Power Electron.* **2005**, *20*, 963–973. [CrossRef]
14. Kollimalla, S.K.; Mishra, M.K. A new adaptive P&O MPPT algorithm based on FSCC method for photovoltaic system. In Proceedings of the 2013 International Conference on Circuits, Power and Computing Technologies (ICCPCT), Nagercoil, India, 21–22 March 2013; pp. 406–411.
15. Kamran, M.; Mudassar, M.; Fazal, M.R.; Asghar, M.U.; Bilal, M.; Asghar, R. Implementation of improved Perturb & Observe MPPT technique with confined search space for standalone photovoltaic system. *J. King Saud Univ. Eng. Sci.* **2020**, *32*, 432–441.
16. Elgendy, M.A.; Zahawi, B.; Atkinson, D.J. Assessment of the incremental conductance maximum power point tracking algorithm. *IEEE Trans. Sustain. Energy* **2012**, *4*, 108–117. [CrossRef]
17. Sayed, M.A.; Mohamed, E.; Ali, A. Maximum power point tracking technique for grid tie PV system. In Proceedings of the 7th International Middle-East Power System Conference, (MEPCON'15), Mansoura, Egypt, 15–17 December 2015; Mansoura University: Mansoura, Egypt, 2015.
18. Lin, C.-H.; Huang, C.-H.; Du, Y.-C.; Chen, J.-L. Maximum photovoltaic power tracking for the PV array using the fractional-order incremental conductance method. *Appl. Energy* **2011**, *88*, 4840–4847. [CrossRef]
19. Radjai, T.; Rahmani, L.; Mekhilef, S.; Gaubert, J.P. Implementation of a modified incremental conductance MPPT algorithm with direct control based on a fuzzy duty cycle change estimator using dSPACE. *Sol. Energy* **2014**, *110*, 325–337. [CrossRef]
20. Hiyama, T.; Kouzuma, S.; Imakubo, T. Identification of optimal operating point of PV modules using neural network for real time maximum power tracking control. *IEEE Trans. Energy Convers.* **1995**, *10*, 360–367. [CrossRef]
21. Ahmad, J. A fractional open circuit voltage based maximum power point tracker for photovoltaic arrays. In Proceedings of the 2010 2nd International Conference on Software Technology and Engineering, San Juan, PR, USA, 3–5 October 2010; pp. V1-247–V1-250.
22. Alajmi, B.N.; Ahmed, K.H.; Finney, S.J.; Williams, B.W. Fuzzy-logic-control approach of a modified hill-climbing method for maximum power point in microgrid standalone photovoltaic system. *IEEE Trans. Power Electron.* **2010**, *26*, 1022–1030. [CrossRef]
23. Salam, Z.; Ahmed, J.; Merugu, B.S. The application of soft computing methods for MPPT of PV system: A technological and status review. *Appl. Energy* **2013**, *107*, 135–148. [CrossRef]
24. Rai, A.K.; Kaushika, N.; Singh, B.; Agarwal, N. Simulation model of ANN based maximum power point tracking controller for solar PV system. *Sol. Energy Mater. Sol. Cells* **2011**, *95*, 773–778. [CrossRef]
25. Ali, A.I.; Mohamed, E.E.; Youssef, A.-R. MPPT algorithm for grid-connected photovoltaic generation systems via model predictive controller. In Proceedings of the 2017 Nineteenth International Middle East Power Systems Conference (MEPCON), Cairo, Egypt, 19–21 December 2017; pp. 895–900.
26. Farhat, M.; Barambones, O.; Sbita, L. Real-time efficiency boosting for PV systems using MPPT based on sliding mode. *Energy Procedia* **2015**, *75*, 361–366. [CrossRef]
27. Farhat, M.; Barambones, O.; Sbita, L. A new maximum power point method based on a sliding mode approach for solar energy harvesting. *Appl. Energy* **2017**, *185*, 1185–1198. [CrossRef]
28. Ghassami, A.A.; Sadeghzadeh, S.M.; Soleimani, A. A high performance maximum power point tracker for PV systems. *Int. J. Electr. Power Energy Syst.* **2013**, *53*, 237–243. [CrossRef]
29. Tey, K.S.; Mekhilef, S. Modified incremental conductance MPPT algorithm to mitigate inaccurate responses under fast-changing solar irradiation level. *Sol. Energy* **2014**, *101*, 333–342. [CrossRef]
30. Safari, A.; Mekhilef, S. Implementation of incremental conductance method with direct control. In Proceedings of the TENCON 2011—2011 IEEE Region 10 Conference, Bali, Indonesia, 21–24 November 2011; pp. 944–948.
31. Liu, F.; Duan, S.; Liu, F.; Liu, B.; Kang, Y. A variable step size INC MPPT method for PV systems. *IEEE Trans. Ind. Electron.* **2008**, *55*, 2622–2628.
32. Yau, H.-T.; Liang, Q.-C.; Hsieh, C.-T. Maximum power point tracking and optimal Li-ion battery charging control for photovoltaic charging system. *Comput. Math. Appl.* **2012**, *64*, 822–832. [CrossRef]
33. Zhou, S.; Han, Y.; Yang, P.; Mahmoud, K.; Lehtonen, M.; Darwish, M.M.; Zalhaf, A.S. An optimal network constraint-based joint expansion planning model for modern distribution networks with multi-types intermittent RERs. *Renew. Energy* **2022**, *194*, 137–151. [CrossRef]
34. Al-Gabalawy, M.; Mahmoud, K.; Darwish, M.M.; Dawson, J.A.; Lehtonen, M.; Hosny, N.S. Reliable and robust observer for simultaneously estimating state-of-charge and state-of-health of lifepo4 batteries. *Appl. Sci.* **2021**, *11*, 3609. [CrossRef]
35. Ghoneim, S.S.; Mahmoud, K.; Darwish, M.M. Effective Transmission Congestion Management via Optimal DG Capacity Using Hybrid Swarm Optimization for Contemporary Power System Operations. *IEEE Access* **2022**, *10*, 71091–71106.
36. Ali, A.I.; Sayed, M.A.; Mohamed, E.E. Maximum Power Point Tracking technique applied on partial shaded grid connected PV system. In Proceedings of the 2016 Eighteenth International Middle East Power Systems Conference (MEPCON), Cairo, Egypt, 27–29 December 2016; pp. 656–663.
37. Diab, M.; Elserougi, A.; Abdel-Khalik, A.; Massoud, A.; Ahmed, S. Modified modulation scheme for photovoltaic fed grid-connected three-phase boost inverter. In Proceedings of the IECON 2013-39th Annual Conference of the IEEE Industrial Electronics Society, Vienna, Austria, 10–13 November 2013; pp. 1735–1740.

38. Ali, A.I.M.; Takeshita, T.; Sayed, M.A. In depth mathematical-analysis and experimentation of high-power SiC-FET based single-stage three-phase differential-based flyback inverter with practical design issues for grid-tied applications. *Int. J. Electr. Power Energy Syst.* **2022**, *140*, 108041. [CrossRef]
39. Mehrnami, S.; Mazumder, S.K.; Soni, H. Modulation scheme for three-phase differential-mode Ćuk inverter. *IEEE Trans. Power Electron.* **2015**, *31*, 2654–2668. [CrossRef]
40. Darwish, A.; Massoud, A.M.; Holliday, D.; Ahmed, S.; Williams, B. Single-stage three-phase differential-mode buck-boost inverters with continuous input current for PV applications. *IEEE Trans. Power Electron.* **2016**, *31*, 8218–8236. [CrossRef]
41. Ali, A.I.M.; Tuan, C.A.; Takeshita, T.; Sayed, M.A.; Alaas, Z.M. Modular Single-Stage Three-Phase Flyback Differential Inverter for Medium/High-Power Grid Integrated Applications. *Sensors* **2022**, *22*, 2064. [CrossRef] [PubMed]
42. Hammami, M.; Grandi, G. A single-phase multilevel PV generation system with an improved ripple correlation control MPPT algorithm. *Energies* **2017**, *10*, 2037. [CrossRef]
43. Zurbriggen, I.G.; Ordonez, M. PV energy harvesting under extremely fast changing irradiance: State-plane direct MPPT. *IEEE Trans. Ind. Electron.* **2018**, *66*, 1852–1861. [CrossRef]

sustainability

MDPI

Article

Investigation on New Metaheuristic Algorithms for Solving Dynamic Combined Economic Environmental Dispatch Problems

Benyekhlef Larouci [1,*], Ahmed Nour El Islam Ayad [1], Hisham Alharbi [2], Turki E. A. Alharbi [2], Houari Boudjella [1], Abdelkader Si Tayeb [3], Sherif S. M. Ghoneim [2,*] and Saad A. Mohamed Abdelwahab [4,5]

1 Department of Electrical Engineering, Kasdi Merbah University, Ghardaia Road, P.O. Box 511, Ouargla 30000, Algeria; ayad.ahmed@univ-ouargla.dz (A.N.E.I.A.); boudjella.houari@univ-ouargla.dz (H.B.)
2 Department of Electrical Engineering, College of Engineering, Taif University, P.O. Box 11099, Taif 21944, Saudi Arabia; h.alharbi@tu.edu.sa (H.A.); turki.alharbi@tu.edu.sa (T.E.A.A.)
3 Applied Research Unit for Renewable Energies "URAER Ghardaia", Renewable Energy Development Center (CDER), Ghardaïa 47133, Algeria; si_tayeb12@yahoo.fr
4 Electrical Department, Faculty of Technology and Education, Suez University, Suez 43533, Egypt; saad.abdelwahab@suezuniv.edu.eg
5 Department of Computers and Systems Engineering, High Institute of Electronic Engineering, Ministry of Higher Education, Bilbis-Sharqiya 44621, Egypt
* Correspondence: larouci.benyekhlef@univ-ouargla.dz (B.L.); s.ghoneim@tu.edu.sa (S.S.M.G.)

Abstract: In this paper, the dynamic combined economic environmental dispatch problems (DCEED) with variable real transmission losses are tackled using four metaheuristics techniques. Due to the consideration of the valve-point loading effects (VPE), DCEED have become a non-smooth and more complex optimization problem. The seagull optimization algorithm (SOA), crow search algorithm (CSA), tunicate swarm algorithm (TSA), and firefly algorithm (FFA), as both nature and biologic phenomena-based algorithms, are investigated to solve DCEED problems. Our proposed algorithms, SOA, TSA, and FFA, were evaluated and applied on the IEEE five-unit test system, and the effectiveness of the proposed CSA approach was applied on two-unit, five-unit, and ten-unit systems by considering VPE. We defined CSA for different objective functions, such as cost of production, emission, and CEED, by considering VPE. The obtained results reveal the efficiency and robustness of the CSA compared to SOA, TSA, FFA, and to other optimization algorithms reported recently in the literature. In addition, Matlab simulation results show the advantages of the proposed approaches for solving DCEED problems.

Keywords: dynamic economic environmental dispatch; cow search algorithm; seagull optimization algorithm; tunicate swarm algorithm; firefly algorithm; optimal scheduling; optimization methods

Citation: Larouci, B.; Ayad, A.N.E.I.; Alharbi, H.; Alharbi, T.E.A.; Boudjella, H.; Tayeb, A.S.; Ghoneim, S.S.M.; Abdelwahab, S.A.M. Investigation on New Metaheuristic Algorithms for Solving Dynamic Combined Economic Environmental Dispatch Problems. *Sustainability* **2022**, *14*, 5554. https://doi.org/10.3390/su14095554

Academic Editors: Mohamed A. Mohamed and Elena Cristina Rada

Received: 13 March 2022
Accepted: 28 April 2022
Published: 5 May 2022

1. Introduction

A competitive electricity generation market necessitates high concentration to solve the economic load dispatch (ELD) problem of power generation. The essential objective of the ELD is to distribute the production of the available production units in such a way that the load demand is covered with a minimum total cost of fuel, taking into account the physical and operational limitations of the electrical system [1].

Due to these physical and operational limitations, the ELD problem is a non-linear optimization problem [1]. As a result, the conventional mathematical methods fall somewhat short in solving such a problem. On the other hand, with the growing concern about polluting emissions and the total cost of fuel, finding a solution to emissions has become a critical issue for the power generation systems around the world. However, the functions of pollutant emission and operating cost are incompatible; reducing emissions increases operating costs and vice versa [2]. Therefore, a grid operator allocates generation units to the tradeoff between the total operation cost reduction and/or reducing emissions under the newest generations, such as prohibited operating zones [3] and ramp rate limits [4], which have a high order of nonlinearities, discontinuities, dimensions, and valve loading points that determine the characteristics of the ELD problem [5]. Therefore, considering more non-convex and multiple local minima increases the complexity of obtaining the optimal fuel cost function. In contrast, metaheuristic optimization techniques have become common to find the best solution to the EED problem. Such algorithms include a colony optimizer [6], multi-verse optimizer [7], particle swarm optimization [8], gray wolf optimizer [9], biogeography-based optimization [10], enhanced exploratory whale optimization algorithm (EEWOA) [7], and hybrid bat–crow search algorithm (HBACSA); they are used exceptionally, in a unique, improved, or hybrid form with others approaches.

Thermal power plants release toxic gases into the atmosphere, endangering human health. Environmental pollution caused by these chemicals has the potential to harm not only humans but also animals and birds. It also has a negative impact on visibility, material quality, and contributes to global warming. DEED provides a solution to these problems by scheduling a renewable energy source and backup power generation based on the predicted load demand to decrease the operational generator's cost and emissions [2].

2. Literature Review

According to the importance of DEED, researchers have adapted and proposed these new algorithms to address the DEED problem. In [11], the authors proposed a method based on improved sailfish algorithm to solve the hybrid dynamic economic emission dispatch (HDEED). In addition, the paper in [12] presented a novel multi-objective optimization based on enhanced moth-flame optimization approach to locate the optimal solution of the hybrid DEED, including renewable energy generation. Ref. [13] proposed an improved tunicate swarm method to explore the search space for DEED. The proposed ITSA in [13] was applied to the 5, 10, and 15-unit systems, respectively. Ref. [14] proposed a novel approach, namely a multi-objective differential evolution algorithm, to deal with the constraints in DEED problems by considering the difference in the power generation range of the units. Authors in [15] suggested a multi-objective virus colony search algorithm (MOVCS) for solving the DEED problem in the power system integrated electric vehicle and wind units over a 24 h period. Ref. [16] formulated DEED as a third-order polynomial fitting curve method to balance between emissions gas and fuel cost by utilizing renewable energy and PEVs. In [17], the authors proposed an enhanced exploratory whale optimization algorithm (EEWOA) to solve the dynamic economic dispatch (DED), which coordinates the behavior of whales, random exploration, and local random search. The feasibility of EEWOA is validated on 5, 10, and 15 units considering VPE and power loss constraints.

In the last two years, various research works have focused on solving the DEED problems by using novels metaheuristic techniques. The research in [18] developed a new model of the multi-objective DCEED using a hybridized flower pollination algorithm (FPA) with sequential quadratic programming (SQP). The FPA–SQP is tested on 5 and 10-unit systems for a 24 h period. Ref. [19] suggested a novel multi-objective hybrid optimization-algorithm-based equilibrium optimizer (EO) and differential evolution (DE) to solve the DEED. The proposed algorithm is validated and verified on the test system containing ten thermal power generators and one wind farm. The study showed that the hybrid EO–DE method with a constraint management system is able to counter balance the tasks of exploitation and exploration. The results obtained in [20] show that the new chaotic artificial bee colony (IABC) rides the local trap and improves the convergence of the solution in terms of solving the CEED with different constraints. A 10-unit system without and with a wind farm and a 40-unit system were investigated to prove the effectiveness and accuracy of the suggested IABC for tackling the CEED problem. In [21], the improved slime mold algorithm (ISMA) is developed and applied to optimize the single-and bi-objective economic emission dispatch (EED) problems considering VPE. Five test systems

(6-unit, 10-unit, 11-unit, 40-unit, and 110-unit) are used to validate the proposed ISMA. Additionally, ref. [22] presented an optimal allocation of the power system for the combined economic emission dispatch problem using a moth swarm algorithm (MSA). The method was tested on two different systems. The first system is a combination of six thermal generators and thirteen solar plants, and the second test system comprises three thermal units combined with thirteen photovoltaic plants.

This paper proposes, presents, and applies more effective methods, such as, TSA, SOA, CSA, and FFA, for solving the DCEED problem, including VPE. The most important contributions of this article can be summarized as follows:

- Improvement of some optimization methods, such as TSA, SOA, CSA, and FFA.
- Application of the proposed algorithms to solve single-and bi-objective DEED problems.
- The four techniques are validated and tested by applying them on the IEEE standard five-unit test system to demonstrate their robustness and accuracy.

The rest of the article is organized as follows. The DEED problem with the mathematical model is presented in Section 3. Section 4 demonstrates the four metaheuristic approaches: TSA, SOA, CSA, and FA. The investigation of the simulation results and discussion are shown in Section 5. Finally, we conclude this paper in Section 6.

3. DCEED Problem Formulation Including VPE

Due to the dynamic behavior of the electrical network and the prodigious variations in load demand on the consumer side, the DCEED problem can be described as a multi-objective mathematical optimization problem, which is non-linear and dynamic. DCEED is a constraint optimization problem that minimizes simultaneously the fuel cost and emission effects in order to meet a power system's load demand over some appropriate periods while meeting certain equality and inequality constraints [23].

3.1. Objective Function

DCEELDP's objective is to minimize total fuel costs while also reducing the level of emissions emitted by generating units. Thus, the objective function is mathematically defined as the weighted summation of the production cost of generating units and emissions caused by fossil fuel thermal plants, which is shown below [16]:

$$\min F = \sum_{t=1}^{T}\sum_{i=1}^{Ng} C_{i,t}(P_{i,t}) + ppf_i \sum_{t=1}^{T}\sum_{i=1}^{Ng} E_{i,t}(P_{i,t}) \tag{1}$$

where F indicates the single objective to be minimized; $C_{i,t}(P_{i,t})$ denotes the fuel cost of Ng generators in the tth (t =1, 2, ... , T) time interval in USD/h; $E_{i,t}(P_{i,t})$ stands for the emissions generated by the generation stations over T dispatch intervals in kg/h; $P_{i,t}$ denotes the dynamic dispatch power in MW. ppf_i is the price penalty factor determined by the ratio of $C_i(P^{max})$ and $E_i(P^{min})$ in USD/kg.

3.1.1. Dynamic Economic Load Dispatch Model (DED)

The objective of the DED problem is to minimize the overall economic cost of fossil fuel during a 24-h period. In some large generators, their cost functions are also non-linear, due to the effect of the valve opening [24]. Consequently, the valve dynamics increase several local minimum points in the cost function, hence complicating the problem. The DED problem involved with VPE is expressed as minimization of the production cost of power dispatch in the following way [25]:

$$C_{i,t}(P_{i,t}) = a_i + b_i\, P_{i,t} + c_i\, P_{i,t}^2 + \left| e_i \times sin\left(f_i \times \left(\, P_i^{min} - P_i\right)\right)\right| \tag{2}$$

where a_i, b_i, andc_i are the coefficients of the fuel cost corresponding to the generator i; e_i, and f_i stand for the fuel cost coefficients of the ith generator due to VPE; and P_i^{min} denotes the minimum real power of the $i^{th}(i = 1, 2, \ldots, Ng)$ generating unit.

3.1.2. Dynamic Environmental Dispatch Model (DEnD)

Global warming and increased movements to protect the environment have forced producers to reduce gas emissions caused by the combustion of fossil fuels in various power plants mainly due to sulfur dioxide (SO_2) and nitrate oxide (NO_X) [26,27].

Each thermal power plant will produce its power according to a dynamic non-smooth emission function given by the following quadratic form [28]:

$$E_{i,t}(P_{i,t}) = \alpha_i + \beta_i\, P_{i,t} + \gamma_i\, P_{i,t}^2 + \eta_i \times exp(\delta_i P_{i,t}) \quad (3)$$

where α_i, β_i, γ_i, η_i, and δ_i are the emission curve coefficients.

3.2. *Constraints Functions*

The minimization of the DCEED problem is subject to the following constraints and limits:

3.2.1. Power Balance Constraint

The sum of total power generated by all generators at each time interval t should be matched with the load demand PD and the total transmission losses P_L in the corresponding time period, which is given as follows [29]:

$$\sum_{i=1}^{Ng} P_{i,t} = P_{D,t} + P_{L,t} \quad (4)$$

The power losses incurred in the transmission lines can be computed by using Kron's loss coefficients formula given below [30]:

$$P_{L,t} = \sum_{i=1}^{Ng}\sum_{j=1}^{Ng} P_{i,t}\, B_{i,j} P_{j,t} \quad (5)$$

where $B_{i,j}$ denotes the transmission loss coefficients. Wood et al.in [31] provide detailed procedures for calculating the B coefficients.

3.2.2. Power Output Limits

The dispatch active power outputs of each generator must be between the capacities of each specific generating unit at each time interval t [32]:

$$P_i^{min} \leq P_{i,t} \geq P_i^{max} \quad (6)$$

where P_i^{min} and P_i^{max} indicate, respectively, the minimum and the maximum power limits of $P_{i,t}$.

4. Metaheuristic Approaches Applied to DEED

4.1. *Seagull Optimization Algorithm (SOA)*

The seagull optimization algorithm (SOA) [33] is a recently proposed metaheuristic technique inspired by the natural behaviors of seagulls. The seagull optimization algorithm is a metaheuristic algorithm that mimics gull behavior. Seagulls live in colonies. Seagulls are omnivorous, eating reptiles, earthworms, insects, fish, and other animals.

Seagulls are highly intelligent birds. This aids the seagull in its hunt for prey. The migratory and hunting habits of seagulls are well known. Seagulls migrate in search of a new plentiful food source. After arriving at a new home, seagulls attack their prey (Figure 1 [33]).

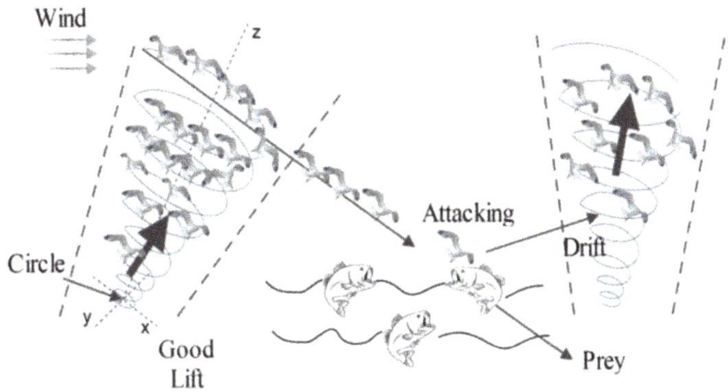

Figure 1. Migration and attacking behaviors of seagulls.

The most important thing about seagulls is their migrating and attacking behavior [34]. Thus, the SOA focuses on these two natural behaviors. This behavior is clarified as follows:

4.1.1. Migration (Exploration)

During migration, the members of a seagull swarm should avoid colliding with each other (see Figure 2a). To achieve this purpose, an additional variable A is employed [35]:

$$\vec{C_s} = A \times \vec{P_s}(j) \tag{7}$$

where $\vec{C_s}$ indicates the position of each agent, $\vec{P_s}(j)$ denotes the current position of seagulls in the *jth* iteration (j = 1, 2, . . . , $\text{Max}_{\text{iteration}}$), and A depicts the movement behavior of seagulls.

$$A = a - \left(j \times (a/Max_{iteration}) \right) \tag{8}$$

where a is a constant and responsible for controlling the frequency of employing variable A, which linearly decreases from a to 0.

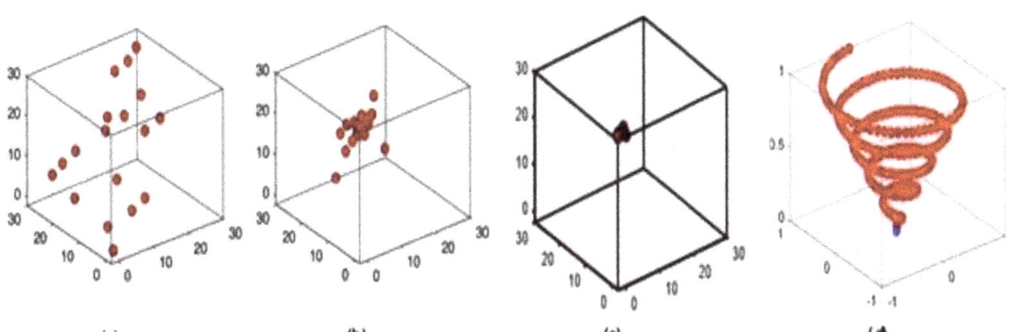

Figure 2. The basic comportment of seagull (**a**) Avoiding collisions between search agents, (**b**) Displacement of search agents toward the best seagull, (**c**) Convergence in the direction of the best seagull, and (**d**) Seagulls' natural aggressive behavior.

To find the richest food resources, seagulls move toward the best search agent as shown in Figure 2b.

$$\vec{M_s} = B \times \left(\vec{P_{bs}}(j) - \vec{P_s}(j) \right) \quad (9)$$

where $\vec{M_s}$ indicates seagull position $\vec{P_s}(j)$ toward the best search agent $\vec{P_{bs}}(j)$. The coefficient B is a random value responsible for making a tradeoff between exploitation and exploration and is expressed as follows:

$$B = 2 \times A^2 \times rd \quad (10)$$

where rd is a random number that lies in the interval (0,1).

As seagulls move toward the fittest search agent, they might remain close to each other (see Figure 2c). Thus, seagulls can update their position according to the following rule [33]:

$$\vec{D_s} = |\vec{C_s} + \vec{M_s}| \quad (11)$$

where $\vec{D_s}$ stands for the distance between seagulls and the best search agent.

4.1.2. Attacking (Exploitation)

Seagulls attack their prey in a spiral shape after arriving at a new place, as presented in Figure 2d. Their attacking behavior in the x, y, and z planes is explained as follows [33]:

$$\begin{cases} x' = r \times cos(k) \\ y' = r \times sin(k) \\ z' = r \times k \end{cases} \quad (12)$$

$$r = u \times exp(kv) \quad (13)$$

where r denotes the radius of each turn of the spiral; k is a random number with $[0 \leq k \leq 2\pi]$; u and v are constants to define the spiral shape [33,36].

The updated position of the search agent $\vec{P_s}(j)$ is determined by using Equations (11)–(13) [36].

$$\vec{P_s}(j) = (\vec{D_s} \times x' \times y' \times z') + \vec{P_{bs}}(j) \quad (14)$$

4.2. Crow Search Algorithm (CSA)

Crow search algorithm (CSA) is a nature-inspired algorithm suggested by A. Askarzadeh in 2016 [37]. This evolutionary computation technique based-population algorithm imitates crow birds' conduct and social interaction [38].

One of the best indications of their cleverness is hiding food and remembering its location. Moreover, the mechanism of exploration and exploitation of CSA can be learned from Figure 3 [39]. As shown in Figure 3a, small values of *fl* (<1) result in a solution to an optimum local; otherwise, if fl of more than 1is selected, the optimal solution leads in the global search, as shown in Figure 3b. Overall, the pseudo code of CSA can be explained as shown in Algorithm 1.

The CSA process is discussed in this subsection [39]:

Step 1: Initializing crow swarm in the d-dimension randomly.

Step 2: A fitness function is used to evaluate each crow, and its value is put as an initial memory value. Each crow stores its hiding place in its memory variable P_i.

Step 3: The crow updates its position by selecting another random crow, (i.e, *Mj*) and generating a random value. If this value is greater than the awareness probability *"AP"*, then crow *Pi* will follow *Pj* to know *Mj*.

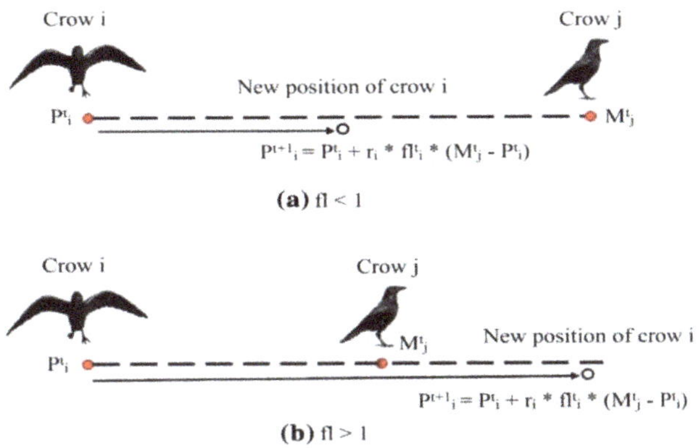

Figure 3. Searching mechanism by the crow in both states.

Step 4: The crow updates its position by selecting another random crow (i.e., *Pj*) and following it to know *Mj*. Then new *Pj* is calculated as follows:

$$P_i^{t+1} = \begin{cases} P_i^t + r_i \times fl_i^t \times \left(M_j^t - P_i^t\right) r_j \geq AP_i^t \\ a \quad random\ position \qquad otherwise \end{cases} \tag{15}$$

where t refers to iteration number; r_i and r_j refer to random numbers; fl_i^t is the crow i flight length to denote crow j memory.

Step 5: Updating memory

The crows' memory is updated as follows:

$$M_i^{t+1} = \begin{cases} P_i^{t+1} & f(P_i^{t+1}) < f(M_i^t) \\ M_i^t & otherwise \end{cases} \tag{16}$$

Step 6: Check termination criterion. Steps 3–5 are repeated until *tmax* is reached.

Algorithm 1. Crow Search Algorithm

Input: N number of crows in the population, and Maximum number of iteration (tmax).
Output: Optimal crow position
Initialize position of crows, and crows' memory
while t < t_{max} do
    for i= 1:N (all N crows of the flock)
  Choose a random crow (i.e. Mj), and determine a value of an awareness probability AP
      if $r_j \geq AP_i^t$
        $P_i^{t+1} = P_i^t + r_i \times fl_i^t \times \left(M_j^t - P_i^t\right)$
      else
        $P_i^{t+1} = a$ random position of search space
      end if
  end for
Check solution boundaries.
Calculate the fitness of each crow
    Update Crows' memory
end while

4.3. *Tunicate Swarm Algorithm (TSA)*

The standard tunicate swarm algorithm is a very simple bio-inspired metaheuristic optimization technique, which was first proposed by S. Kaur et al. in 2020 [40]. Its inspiration and performance were proven over the seventy-four benchmark problems compared to several other optimization approaches. Its efficacy and unpretentious structure draw the attention to employ and improve this algorithm for the considered problem. The swarm behavior of TSA is given in Figure 4 [40]. TSA main limitates the swarming behaviors of the marine tunicates and their jet propulsions during its navigation and foraging procedure [41].

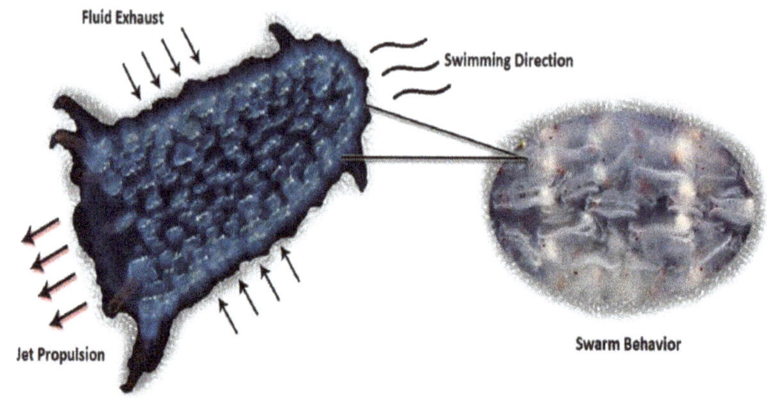

Figure 4. Inspiration of TSA.

In TSA, a population of tunicates (*PT*) is swarming to search for the best source of food (*SF*), representing the fitness function. In this swarming, the tunicates update their positions related to the first best tunicates stored and upgraded in each iteration. The TSA begins where the tunicate population is initialized randomly, considering the permissible bounds of the control variables. The dimension of the control variables composes each tunicate (*T*), which can be initially created as [40]

$$T_n(m) = T^n_{min} + r \times (T^n_{max} - T^n_{min}) \quad \forall m \in PT_{size} \; \& \; n \in Dim \tag{17}$$

where $T(m)$ stands for the position of each tunicate (m); n refers to each control variable in dimension Dim; r is a random number within the range (0:1); and PT_{size} indicates the number of tunicates in the population.

The update process of the tunicates position is executed by the following formula [41]:

$$T_n(m) = \frac{T^*_n(m) - T^*_n(m-1)}{2 + c_1} \tag{18}$$

where T^* denotes the updated position of the mth tunicate based on Equation (18); $T(m-1)$ refers to the neighbor tunicate; c_1 is a random number, uniformly distributed between 0 and 1.

$$T^*_n(m) = \begin{cases} SF + A \times |SF - rand \times T_n(n)| & if\ rand \geq 0.5 \\ SF - A \times |SF - rand \times T_n(n)| & if\ rand < 0.5 \end{cases} \tag{19}$$

where SF is the source of food, which is represented by the best tunicate position in the whole population; A is a randomized vector to avoid any conflicts between tunicates and each other, which is expressed as: [40]

$$A = \frac{c_2 + c_3 - 2c_1}{VT_{min} + c_1(VT_{max} - VT_{min})} \tag{20}$$

where c_2 and c_3 are random numbers within the range (0:1); VT_{min} and VT_{max} represent the initial and subordinate speeds to produce social interaction.

The TSA method's key steps can be described as [42]:

Step 1: Create the initial tunicate population.
Step 2: Determine the control units of TSA and stopping criteria.
Step 3: Compute the fitness values of the initial population.
Step 4: Select the position of the tunicate with the best fitness value.
Step 5: Create the new position for each tunicate by using Equation (18).
Step 6: Update the position of the tunicates that are out of the search space.
Step 7: Compute the fitness values for the new positions of tunicates.
Step 8: Until stopping criteria is satisfied, repeat steps 5–8.
Step 9: After stopping criteria is satisfied, save the best tunicate position.

4.4. *Overview of the Firefly Algorithm (FFA)*

The firefly algorithm developed by X-S. Yang in 2007 [43] is one of the swarm-based optimization algorithms, which works based on the sparkling performance of all the fireflies. Fireflies use their flashing property to communicate. There are three rules in the firefly algorithm, which are based on the idealized behavior of the flashing characteristics of fireflies [44].

4.4.1. Attractiveness

All fireflies are unisex, and they will move toward more attractive and brighter ones regardless of their sex. In the firefly algorithm, the form of attractiveness function of a firefly is the following monotonically decreasing function [44]:

$$\beta(r) = \beta_0 \times exp(-\gamma\, r^m), \quad m \geq 1 \tag{21}$$

where r is the distance between any two fireflies β_0 is the initial attractiveness at $r = 0$, and γ is an absorption coefficient which controls the decrease in light intensity.

4.4.2. Distance

The degree of attractiveness of a firefly is proportional to its brightness, which decreases as the distance from the other firefly increases because the air absorbs light. If there is no brighter or more attractive firefly than a particular one, it will move randomly [45].

The distance between any two fireflies i and j, at positions xi and xj, respectively, can be defined as a Cartesian or Euclidean distance as follows [46]:

$$r_{ij} = \|x_i - x_j\| = \sqrt{\sum_{k=1}^{d} \left(x_{i,k} - x_{j,k}\right)^2} \tag{22}$$

where $x_{i,k}$ is the kth component of the spatial coordinate x_i of the ith firefly.

In the dimensions 2-d, we have:

$$r_{ij} = \sqrt{\left(x_i - x_j\right)^2 + \left(y_i - y_j\right)^2} \tag{23}$$

4.4.3. Movement

The movement of a firefly i, which is attracted by a more attractive (brighter) firefly j, is calculated by [46]

$$x_i = x_i + \beta_0 \times exp\left(-\gamma\, r_{ij}^2\right)\left(x_j - x_i\right) + \alpha\left(rand - \frac{1}{2}\right) \tag{24}$$

where the second term is due to the attraction, while the third term is randomization with α being the randomization parameter, and rand is a random number within the range (0:1).

5. Simulation Results and Discussion

In order to solve the dynamic combined economic environmental dispatch problem, we developed and executed the CSA, SOA, TSA, and FFA algorithms in MATLAB 2017, and they were run on a personal computer with an Intel Core(TM) i5 with a processor of 2.60 GHz and a Ram of 8.0 GB under MS Windows 8.1. In the first part, the four proposed techniques were tested on the five-unit test system by considering VPE for three case studies, and in the second part, the CSA method was applied on the IEEE ten-unit system, including VPE, for three cases. The constraints involved in all cases were power balance limit with consideration of transmission losses and generator operating limits constraints.

The obtained results were compared with the optimization approaches recently published in the literature. Table 1 stands for the parameter values of CSA, SOA, TSA, and FFA algorithms, for all cases studies.

Table 1. Parameters of CSA, SOA, TSA, and FFA algorithms for DEED problem.

Algorithm	Parameters
SOA	$N = 50$, $t_{max} = 1000$, $u = 1$, $v = 0.011$
CSA	$N = 50$, itermax= 1000, $fl = 2$, $AP = 0.1$
TSA	m = 50, $iter_{max} = 1000$, $VT_{min} = 1$, $VT_{max} = 4$
FFA	$N = 100$, itermax= 1000, $\beta_0 = 1$, $\gamma = 1$, $\alpha = 0.1$,

5.1. IEEE Five-Unit Test System

The five-unit test system was derived from [47,48]. Tables 2 and 3 include the generation data. In this subsection, the dynamic load dispatch (DLD) cases through the application of the four metaheuristics techniques include:

- Case 1: Dynamic economic dispatch DED;
- Case 2: Dynamic environmental dispatch DEnD;
- Case 3: Dynamic economic emission dispatch DEED.

Table 2. Cost and emission coefficients of generators for IEEE five-unit power system.

Unit	a_i \$/h	b_i \$/MWh	c_i \$/MW^{2}h	e_i USD/h	f_i rad/MW	α_i Ib/h *	β_i Ib/MWh	γ_i Ib/MW^{2}h	η_i Ib/h	δ_i 1/MW	P_i^{min} MW	P_i^{max} MW	ppf_i \$/Ib
P_1	25	2.0	0.0080	100	0.042	80	−0.805	0.0180	0.6550	0.02846	10	75	1.8201
P_2	60	1.8	0.0030	140	0.040	50	−0.555	0.0150	0.5773	0.02446	20	125	1.5436
P_3	100	2.1	0.0012	160	0.038	60	−1.355	0.0105	0.4968	0.02270	30	175	3.4911
P_4	120	2.0	0.0010	180	0.037	45	−0.600	0.0080	0.4860	0.01948	40	250	1.7278
P_5	40	1.8	0.0015	200	0.035	30	−0.555	0.0120	0.5035	0.02075	50	300	0.7578

* Pound per hour (lb/h) = 0.453 59237 kilogram per hour (kg/h).

Table 3. Load demands.

Time (h)	Load (MW)	Time (h)	Load (MW)	Time (h)	Load (MW)
1	410	9	690	17	558
2	435	10	704	18	608
3	475	11	720	19	654
4	530	12	740	20	704
5	558	13	704	21	680
6	608	14	690	22	605
7	626	15	654	23	527
8	654	16	580	24	463

The B matrix of the test system we use is given by Equation (25) [48]:

$$B = 10^{-4}{}_{i} \times \begin{bmatrix} 0.49 & 0.14 & 0.15 & 0.15 & 0.20 \\ 0.14 & 0.45 & 0.16 & 0.20 & 0.18 \\ 0.15 & 0.16 & 0.39 & 0.10 & 0.12 \\ 0.15 & 0.20 & 0.10 & 0.40 & 0.14 \\ 0.20 & 0.18 & 0.12 & 0.14 & 0.35 \end{bmatrix} \quad (25)$$

5.1.1. Case 1

The dynamic economic dispatch (DED) considering VPE with variables power losses is studied in this subsection (for total load = 14,577 MW).

The optimal schedule for a five-thermal-generator test system considering VPE and transmission losses, using the crow search algorithm (CSA), is presented in Table 4. The best results are highlighted in bold font.

Table 4. Simulation results of best solutions of CSA for Case 1.

Time h	P_1 MW	P_2 MW	P_3 MW	P_4 MW	P_5 MW	P_D MW	P_L MW	Fuel Cost USD/h	Emission Ib/h
1	84.804	98.539	50.652	40.000	139.759	410	3.756	**1363.640**	546.604
2	48.075	98.539	112.673	40.000	139.759	435	4.048	**1380.410**	510.742
3	88.876	98.539	112.673	40.000	139.759	475	4.849	**1423.790**	584.122
4	59.954	98.539	112.673	124.908	139.759	530	5.835	**1584.710**	590.865
5	84.139	98.539	112.673	40.000	229.519	558	6.872	**1604.790**	969.866
6	55.079	98.539	112.673	209.816	139.759	608	7.868	**1777.140**	784.062
7	73.529	98.539	112.673	209.816	139.759	626	8.317	**1783.770**	814.087
8	97.449	98.539	112.673	124.908	229.519	654	9.090	**1882.450**	1071.515
9	49.619	98.539	112.673	209.816	229.519	690	10.168	**1977.660**	1175.437
10	64.011	98.539	112.673	209.816	229.519	704	10.559	**1996.590**	1194.648
11	80.483	98.539	112.673	209.815	229.519	720	11.032	**1989.970**	1226.653
12	101.112	98.539	112.673	209.816	229.519	740	11.662	**2106.450**	1282.648
13	64.0110	98.539	112.673	209.816	229.519	704	10.559	**1996.590**	1194.648
14	49.619	98.539	112.673	209.816	229.519	690	10.168	**1977.660**	1175.437
15	97.449	98.539	112.673	124.908	229.519	654	9.090	**1882.450**	1071.515
16	84.800	40.058	112.673	209.816	139.759	580	7.108	**1746.070**	745.131
17	84.139	98.539	112.673	40.000	229.519	558	6.872	**1604.790**	969.866
18	55.079	98.539	112.673	209.816	139.759	608	7.868	**1777.140**	784.062
19	97.449	98.539	112.673	124.908	229.519	654	9.090	**1882.450**	1071.515
20	64.011	98.539	112.673	209.816	229.519	704	10.559	**1996.590**	1194.648
21	39.353	98.539	112.673	209.816	229.519	680	9.902	**1944.590**	1166.578
22	52.007	98.539	112.673	209.816	139.759	605	7.796	**1771.650**	780.351
23	56.889	98.539	112.673	124.908	139.759	527	5.771	**1581.460**	586.584
24	81.432	98.539	112.673	124.908	50.000	463	4.553	**1392.530**	468.968

The comparison results of the optimization techniques (CSA, SOA, TSA, and FFA) and the statistical analysis of the optimal results of Case 1 are given in Table 5. The best values, the robustness and effectiveness of the proposed CSA in finding optimal solutions to the DED problem with a reasonable number of iterations are compared and the constraints verified.

The results shown in Table 4 present the total cost found by the CSA algorithm, which is equal to 42,425.455 USD/h, is lower compared to that found by the SOA, TSA, and FFA algorithms, which is estimated at 48,609.77 USD/h, 46,672.4787 USD/h and 45,474.198 USD/h respectively.

As seen in Table 5, we infer that the CSA obtained the best fuel cost compared to PSO [47], PSOGSA [48], NEHS [49], HS-NPSA [49], and DE-SQP [50].

The CSA algorithm, which offers the best production cost, on the other hand, offers a relatively higher amount of transmission losses compared to FFA, i.e., 193.393 MW for the CSA algorithm, 204.66 MW for SOA, 198.278 MW for TSA, and 191.298 MW for FFA.

Table 5. Comparison of CSA with previous algorithms for DED.

Methods	Total Loss MW	Total Fuel Cost USD/h	Total Emission Ib/h
CSA	193.393	42,425.455	21,960.553
SOA	204.660	48,609.770	32,652.860
TSA	198.278	46,672.479	27,641.230
FFA	191.298	45,474.198	24,862.338
PSOGSA [48]	NA	42,853.339	22,087.887
NEHS [49]	NA	43,066.073	NA
MHS [49]	NA	45,497.740	NA
HS-NPSA [49]	NA	43,927.305	NA
DE-SQP [50]	NA	45,590.000	23,567.000
PSO [47]	NA	47,852.000	22,405.000

NA means not available in the corresponding literature.

It is also worth mentioning that in Table 5, the total emissions found by the CSA algorithm, which amount to 21,960.553 lb/h are more reduced compared to those found by the SOA, TSA, and FFA algorithms, which are estimated at 32,652.86 lb/h, 27,641.23 lb/h, and 24,862.338 lb/h respectively.

Figure 5, shows the graphical depiction of the convergence of the DED problem with time (iterations) obtained by applying our proposed methods corresponding to the peak hour (P_D = 740 MW).

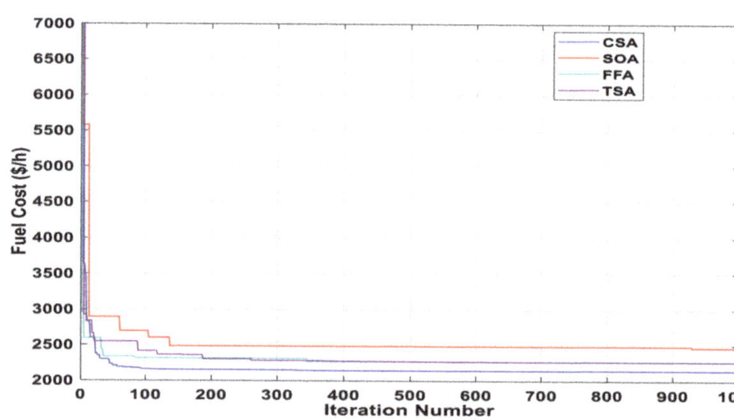

Figure 5. ° Convergence of the ED problem with time (iterations) for P_D = 740 MW.

In comparison to SOA, TSA, and FFA, we can see that CSA has an incredibly fast convergence ability. In addition, CSA was trapped into the local optimum at around 230 iterations.

5.1.2. Case 2

The dynamic environmental dispatch (DEnD) considering the valve-point loading effect and transmission losses as given in Equation (3) are discussed in this case.

Table 6 shows the results of the optimum power dispatch for the five-unit system over 24 h obtained by CSA.

Table 6. Simulation results of best solutions of CSA for Case 2.

Time h	P_1 MW	P_2 MW	P_3 MW	P_4 MW	P_5 MW	P_D MW	P_L MW	Fuel Cost USD/h	Emission Ib/h
1	54.679	58.236	116.571	110.598	73.364	410	3.448	352.453	**1723.627**
2	58.067	62.383	121.851	117.982	78.601	435	3.885	385.960	**1784.169**
3	63.526	69.080	130.221	129.751	87.063	475	4.641	446.642	**1912.114**
4	71.120	78.439	141.552	145.801	98.890	530	5.797	544.648	**2135.506**
5	75.032	83.262	147.232	153.895	105.008	558	6.431	601.208	**2203.616**
6	82.107	92.034	157.218	168.176	116.119	608	7.654	713.801	**2241.188**
7	84.684	95.241	160.760	173.252	120.183	626	8.121	758.078	**2229.487**
8	88.729	100.286	166.212	181.072	126.577	654	8.876	831.019	**2240.807**
9	93.996	106.880	173.119	190.970	134.934	690	9.898	932.292	**2270.804**
10	96.065	109.478	175.772	194.768	138.227	704	10.311	974.021	**2274.187**
11	98.446	112.472	178.783	199.072	142.020	720	10.794	1023.365	**2305.246**
12	101.444	116.253	182.514	204.393	146.809	740	11.414	1087577	**2365.552**
13	96.065	109.478	175.772	194.768	138.227	704	10.311	974.021	**2274.190**
14	93.996	106.879	173.119	190.970	134.934	690	9.898	932.292	**2270.804**
15	88.729	100.286	166.212	181.072	126.577	654	8.876	831.019	**2240.804**
16	78.130	87.098	151.652	160.208	109.866	580	6.955	648.892	**2233.356**
17	75.032	83.262	147.233	153.895	105.008	558	6.431	601.208	**2203.615**
18	82.107	92.034	157.218	168.176	116.119	608	7.654	713.801	**2241.189**
19	88.729	100.286	166.212	181.072	126.577	654	8.876	831.019	**2240.807**
20	96.066	109.478	175.772	194.768	138.227	704	10.311	974.021	**2274.187**
21	92.525	105.036	171.212	188.239	132.596	680	9.609	903.298	**2265.782**
22	81.678	91.502	156.625	167.326	115.445	605	7.577	706.617	**2241.906**
23	70.703	77.915	140.940	144.931	98.239	527	5.727	538.858	**2126.253**
24	61.883	67.063	127.721	126.227	84.514	463	4.407	427.514	**1850.281**

The optimal results of the pollutant power generation, fuel cost, and active power loss obtained by the CSA are depicted in Table 6.

The total losses, total fuel cost, and total emissions computed by our proposed approaches are illustrated in Table 7. We compared them with those determined by other algorithms recently applied in the literature.

Table 7. Comparison of CSA with previous algorithms for DEnD.

Methods	Total Loss MW	Total Fuel Cost USD/h	Total Emission Ib/h
CSA	187.901	51,149.500	17,733.600
SOA	189.766	51,385.300	18,743.100
TSA	188.825	51,878.700	18,602.800
FFA	187.070	51,263.500	19,840.200
NEHS [49]	NA	NA	17,853.003
MHS [49]	NA	NA	17,937.408
HS-NPSA [49]	NA	NA	17,872.348
PSOGSA [48]	NA	51,953.905	17,852.979
DE-SQP [50]	NA	52,611.000	18,955.000
PSO [47]	NA	53,086.000	19,094.000

By analyzing the results given in Table 7, we infer that the total emissions found by the CSA algorithm, which amount to 17,733.6 lb/h, are lower, compared to those found by the SOA TSA, and FFA algorithms, which are estimated at 18,743.1 lb/h, 18,602.8 lb/h, and 19,840.2 lb/h, respectively.

From the results shown in Table 7, we notice that the total cost determined by the CSA algorithm, which amounts to 51,149.5 USD/h, is lower compared to that found by the SOA, TSA, and FFA algorithms, which is estimated at 51,385.3 USD/h, 51,878.7 USD/h, and 51,263.5 USD/h, respectively.

In addition, when compared to CSA, SOA, and TSA, FFA has a relatively low amount of active power losses. The CS algorithm, which offers the minimum pollutant power generation, in return offers relatively higher transmission losses than FFA, and lower than SOA and TSA, i.e., 187.901 MW, 204.66MW, 188.825 MW, and, 187.07 MW for CSA, SOA, TSA, and FFA, respectively.

Figure 6 represents the convergence curves of our proposed approaches for the DEnD problem. It describes the stable convergence of the objective function of the problem given in Equation (3). The convergence characteristics shown in Figure 6 prove that CSA has better qualities than SOA, TSA, and FFA. We conclude that CSA is favorable for large-scale power systems.

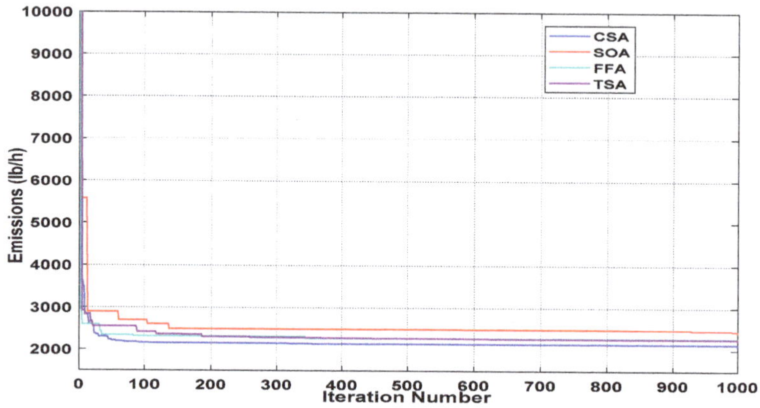

Figure 6. Convergence characteristics of the EnD with iterations for P_D = 740 MW.

5.1.3. Case 3

In this case, we deal with the dynamic combined economic environmental dispatching (DCEED) with the variable losses by applying the price penalty factor under the impact of the valve-point. Electrical losses are variable depending on the power generated.

Table 8 summarizes the best solutions achieved for DCEED for a given load. We can deduce from the same table that CSA and our three competitive algorithms adjust the value of the output power ($P_1 - P_5$) such that load demand and the constraint limits are satisfied, and the best fuel cost and pollutants emission results are achieved.

Table 8. The best solutions obtained by CSA for the DCEED of the IEEE five-unit test system.

Time h	P_1 MW	P_2 MW	P_3 MW	P_4 MW	P_5 MW	P_D MW	P_L MW	Fuel Cost USD/h	Emission Ib/h	CEED USD/h
1	23.221	95.066	118.179	125.507	51.646	410.00	3.619	1343.493	400.897	**1814.930**
2	42.464	84.193	112.348	124.800	75.142	435.00	3.946	1607.846	399.951	**2078.248**
3	46.363	94.690	111.484	87.447	139.713	475.00	4.697	1670.469	519.509	**2254.467**
4	26.838	99.530	114.280	126.507	168.745	530.00	5.899	1745.194	669.218	**2475.641**
5	33.564	92.891	123.185	175.142	139.753	558.00	6.536	1902.752	665.000	**2649.852**
6	73.336	98.983	157.530	146.016	139.760	608.00	7.625	2031.153	728.731	**2893.762**
7	57.750	99.767	123.604	210.658	142.515	626.00	8.294	1914.203	816.010	**2838.928**
8	51.884	88.674	103.432	204.515	214.625	654.00	9.131	2131.606	1040.647	**3248.435**
9	53.142	97.776	114.207	209.557	225.462	690.00	10.144	2019.883	1153.740	**3260.318**
10	54.142	109.352	112.662	209.810	228.632	704.00	10.598	2075.861	1204.781	**3370.372**
11	73.636	100.102	117.640	210.105	229.527	720.00	11.010	2052.002	1223.114	**3381.405**

Table 8. *Cont.*

Time h	P_1 MW	P_2 MW	P_3 MW	P_4 MW	P_5 MW	P_D MW	P_L MW	Fuel Cost USD/h	Emission Ib/h	CEED USD/h
12	72.069	116.167	123.841	210.056	229.502	740.00	11.635	2224.386	1274.535	**3613.856**
13	66.908	98.680	173.744	145.451	229.517	704.00	10.300	2236.929	1154.978	**3523.693**
14	61.029	98.536	102.648	209.815	228.176	690.00	10.204	2036.995	1171.356	**3293.254**
15	66.581	117.798	115.698	213.160	149.886	654.00	9.123	2078.304	903.466	**3098.415**
16	22.005	101.970	112.492	124.576	226.155	580.00	7.200	1701.781	948.532	**2700.991**
17	24.788	95.120	109.230	195.916	139.614	558.00	6.667	1740.035	706.614	**2521.975**
18	66.807	83.120	156.707	124.906	184.080	608.00	7.620	2199.492	803.606	**3116.141**
19	61.025	110.762	116.453	209.863	164.984	654.00	9.087	2122.958	915.351	**3145.552**
20	61.190	96.517	118.054	209.294	229.473	704.00	10.528	2049.371	1189.705	**3333.294**
21	61.279	77.640	112.693	208.694	229.519	680.00	9.825	2055.372	1140.486	**3280.867**
22	66.280	38.471	112.702	209.818	185.459	605.00	7.729	2051.944	873.120	**3013.830**
23	47.683	93.663	126.697	124.929	139.771	527.00	5.741	1693.775	582.538	**2357.800**
24	41.562	70.450	112.609	124.880	117.903	463.00	4.404	1696.097	452.880	**2212.678**

Table 9 shows the values of the active power losses, the best cost, the best emission, and the total cost objectives against 0.5 weight for the test system. The results displayed in the mentioned table are obtained from NEHS, MHS, HS-NPSA, PSO-GSA, DE-SQO, PSO, and our proposed approaches.

Table 9. The DCEED comparison results of different algorithms of the five-unit test system.

Methods	Total Loss MW	Total Fuel Cost USD/h	Total Emission Ib/h	Total Cost (0.5 Weight) USD/h	Change % w.r.t CSA
CSA	192.386	46,381.900	20,938.800	33,660.350	//
SOA	196.961	48,500.800	21,130.100	34,815.450	3.43
TSA	194.204	45,816.300	22,424.800	33,942.950	0.84
FFA	192.451	47,030.700	22,069.600	34,550.150	2.64
NEHS [49]	NA	45,398.016	18,392.337	31,895.177	−5.24
MHS [49]	NA	47,390.956	18,423.776	32,907.366	−2.24
MOHDESAT [51]	NA	48,214.000	18,011.000	33,112.500	−1.63
WOA [52]	NA	46,475.090	18,827.980	32,651.530	−2.99
PSOGSA [48]	NA	45,702.6001	18,267.179	**31,984.985**	**−5.24**
DE-SQP [50]	NA	46,625.000	20,527.000	33,576.000	−0.002
PSO [47]	NA	50,893.000	20,163.000	35,528.000	5.55
MONNDE [53]	NA	49,135.000	18,233.000	33,684.240	0.06
EP [54]	NA	48,628.000	21,154.000	34,891.000	3.66
SA [51]	NA	48,621.000	21,188.000	33,904.500	0.73
PS [55]	NA	47,911.000	18,927.000	33,419.000	−0.72
MODE [51]	NA	47,330.000	18,116.000	32,723.000	−2.78

Column five of Table 9 depicts the value of the single-objective function of the problem, as specified in Equation (1). The effectiveness of the suggested CSA is demonstrated by the percentage change of some algorithms listed in the same table.

As seen in Table 9, the results reveal that CSA has a good performance for solving DCEED at less than SOA (3.43%), than TSA (0.84%), than FFA (2.64%), than PSO (5.55%), and then EP (3.66%).

According to Table 9, we infer that minimizing the cost of generation and minimizing emissions are contradictory goals. Emissions are highest when the cost of production is minimized.

Figure 7 shows the convergence curves of the total cost with iterations, obtained by applying CSA, SOA, TSA, and FFA methods corresponding to the power demand equal to

740 MW. We infer that CSA and FFA reach the optimum solution more quickly than SOA and TSA, which is shown in Figure 7.

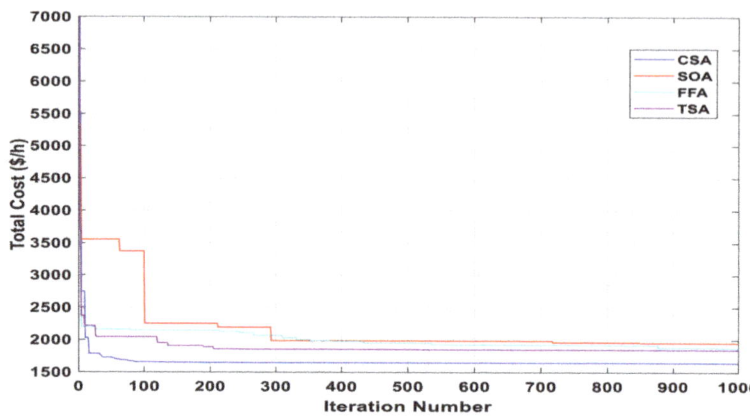

Figure 7. Convergence curves of the CEED with time (iterations) for P_D = 740 MW.

Figure 8 presents a graphical comparison of the total fuel cost, the total emissions, and the total cost of our proposed methods and other algorithms for a five-unit test system. This clearly demonstrates that CSA has the lowest total cost when compared to other algorithms.

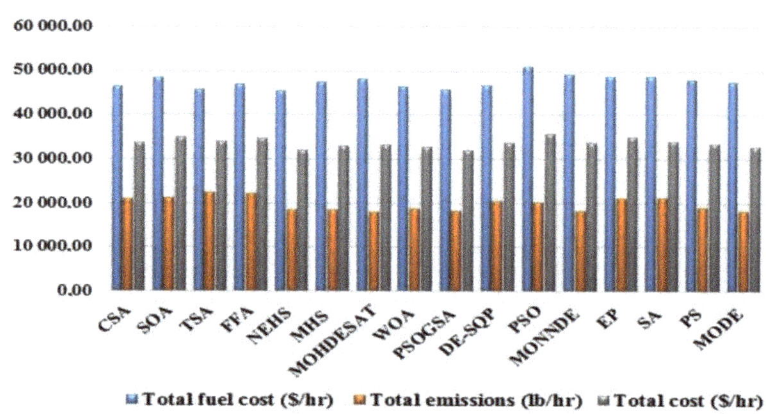

Figure 8. Total fuel cost, total emissions, and total cost obtained by CSA, SOA, TSA, and FFA vs. other algorithms applied on the IEEE five-unit test system.

Percentage enhancement of the suggested CSA over other algorithms is shown in Figure 9. From this figure, the positive bars show the largest values of the total cost compared to those obtained by CSA, and negative bars represent smaller values of the total cost compared to those obtained by CSA.

The statistical summary of 50 runs for each study case performed (i.e., Case 1 to Case 3) using CSA, SOA, TSA, and FFA approaches is depicted in Table 10.

The columns display the min, average, worst, and standard deviation values for the objective function in each case. According to Table 10, it is clear the CSA method delivers the best fitness, best mean, and best standard deviation values in all the cases.

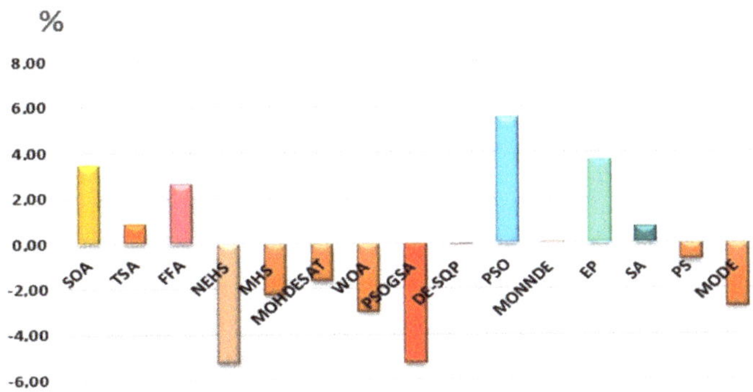

Figure 9. Percentage enhancement of CSA vs. other algorithms for the IEEE five-unit test system.

Table 10. Statistical summary of case studies using CSA, SOA, TSA, and FFA algorithms for load demand of 740 MW for a IEEE five-unit system.

Case 1	Economic Load Dispatch ELD (USD/h)			
Algorithm	**Min**	**Average**	**Worst**	**Std Dev**
CSA	**2142.3181**	2164.8135	2593.5470	**55.4145**
SOA	2456.3468	2537.9394	5577.4323	290.6773
TSA	2269.2981	2293.2777	3635.3608	113.8379
FFA	2265.5019	2310.7733	2926.1582	100.0485
Case 2	**Environmental Dispatch EnD (lb/h)**			
CSA	**1090.3380**	1111.2373	1421.8938	**58.9938**
SOA	1160.6740	1757.2506	3536.7399	499.4472
TSA	1295.4652	1506.9932	1728.0597	83.5214
FFA	1241.5852	1323.0107	5121.1719	331.6223
Case 3	**Combined Economic Emission Dispatch CEED (USD/h)**			
CSA	**1646.9098**	1652.9305	1784.2619	**22.1540**
SOA	1955.3266	2153.2196	3553.8243	418.7256
TSA	1879.5231	1992.3606	2203.5179	99.1567
FFA	1848.8820	1878.6942	2216.4237	67.0644

Figure 10 obtained values (min, mean, and worst values) for 50 runs of the CSA approach for the case of reducing the cost of production (Case 1).

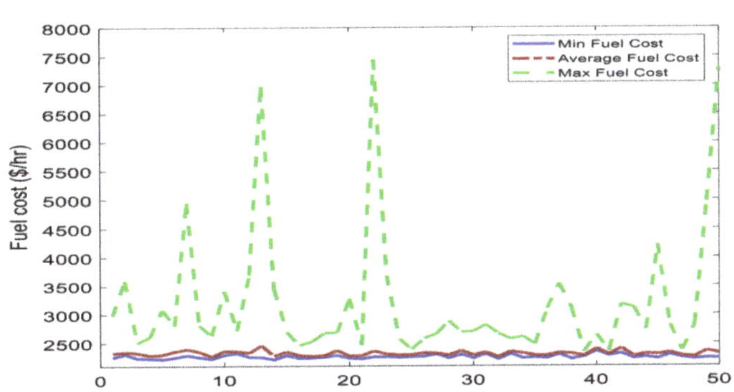

Figure 10. Min, Mean and Max Fuel cost obtained in 50 runs using CSA method.

5.2. IEEE 10-Unit Test System

In this case study, the then-unit generating system is utilized to evaluate the performance of the proposed CSA by considering the valve-point loading effect. In this experiment, a 24-h scheduling period with 1-h intervals is used. The original data of the 10-unit system are shown in Table 11, which are available in the literature [53]. Tables 12 and 13 are the 10-unit generator loads demand and the B matrix, respectively, which are derived from [53]. In this research, the weighting function approach (w.f.a) is used to converts multi-objective functions into a single problem [56]. Hence, by the usage of the w. f. a, Equation (1) can be reformulated as:

$$\min F = w_1 \sum_{t=1}^{T} \sum_{i=1}^{Ng} C_{i,t}(P_{i,t}) + w_2 \sum_{t=1}^{T} \sum_{i=1}^{Ng} E_{i,t}(P_{i,t}) \tag{26}$$

where $\sum_{i=1}^{n} w_i = 1$ (n: number of problems).

Table 11. IEEE 10-unit generator coefficients.

Unit	a_i \$/h	b_i \$/MWh	c_i \$/MW^{2}h	e_i \$/h	f_i rad/MW	α_i Ib/h*	β_i Ib/MWh	γ_i Ib/MW^{2}h	η_i Ib/h	δ_i 1/MW	P_i^{min} MW	P_i^{max} MW
P_1	786.7988	38.5397	0.1524	450	0.0410	103.3908	−2.4444	0.0312	0.5035	0.0207	150	470
P_2	451.3251	46.1591	0.1058	600	0.0360	103.3908	−2.4444	0.0312	0.5035	0.0207	135	470
P_3	1049.9977	40.3965	0.0280	320	0.0280	300.3910	−4.0695	0.0509	0.4968	0.0202	73	340
P_4	1243.5311	38.3055	0.0354	260	0.0520	300.3910	−4.0695	0.0509	0.4968	0.0202	60	300
P_5	1658.5696	36.3278	0.0211	280	0.0630	320.0006	−3.8132	0.0344	0.4972	0.0200	73	243
P_6	1356.6592	38.2704	0.0179	310	0.0480	320.0006	−3.8132	0.0344	0.4972	0.0200	57	160
P_7	1450.7045	36.5104	0.0121	300	0.0860	330.0056	−3.9023	0.0465	0.5163	0.0214	20	130
P_8	1450.7045	36.5104	0.0121	340	0.0820	330.0056	−3.9023	0.0465	0.5163	0.0214	47	120
P_9	1455.6056	39.5804	0.1090	270	0.0980	350.0056	−3.9524	0.0465	0.5475	0.0234	20	80
P_{10}	1469.4026	40.5407	0.1295	380	0.0940	360.0012	−3.9864	0.0470	0.5475	0.0234	10	55

Table 12. IEEE 10-unit B matrix ($\times 10^{-4}$).

0.49	0.14	0.15	0.15	0.16	0.17	0.17	0.18	0.19	0.20
0.14	0.45	0.16	0.16	0.17	0.15	0.15	0.16	0.18	0.18
0.15	0.16	0.39	0.10	0.12	0.14	0.14	0.16	0.16	0.16
0.15	0.16	0.10	0.40	0.14	0.10	0.11	0.12	0.14	0.15
0.16	0.17	0.12	0.14	0.35	0.11	0.13	0.13	0.15	0.16
0.17	0.15	0.12	0.10	0.11	0.36	0.12	0.12	0.14	0.15
0.17	0.15	0.14	0.11	0.13	0.12	0.38	0.16	0.16	0.18
0.18	0.16	0.14	0.12	0.13	0.12	0.16	0.40	0.15	0.16
0.19	0.18	0.16	0.14	0.15	0.14	0.16	0.15	0.42	0.19
0.20	0.18	0.16	0.15	0.16	0.15	0.18	0.16	0.19	0.44

Table 13. Load demands.

Time (h)	Load (MW)	Time (h)	Load (MW)	Time (h)	Load (MW)
1	1036	9	1924	17	1480
2	1110	10	2022	18	1628
3	1258	11	2106	19	1776
4	1406	12	2150	20	1972
5	1480	13	2072	21	1924
6	1628	14	1924	22	1628
7	1702	15	1776	23	1332
8	1776	16	1554	24	1184

Dynamic load dispatch (DLD) cases in this subsection include:

- Case 4: Dynamic economic dispatch DED ($w_1 = 1\ ,\ \ w_2 = 0$);
- Case 5: Dynamic environmental dispatch DEnD ($w_1 = 0\ ,\ \ w_2 = 1$);
- Case 6: Dynamic economic emission dispatch DEED ($w_1 = 0.5,\ \ w_2 = 0.5$);

5.2.1. Case 4

The dynamic economic dispatch (DED) considering VPE with variables power losses is studied in this subsection (for total load = 39.848 MW).

Hourly generation (MW) schedule obtained from DED using CSA for a 10-unit system by considering VPE and transmission losses are presented in Table 14. The best results are highlighted in bold font.

Table 14. Best solutions obtained from CSA for test Case 4.

Time h	P_1MW	P_2MW	P_3MW	P_4MW	P_5MW	P_6MW	P_7MW	P_8MW	P_9MW	P_{10}MW	P_LMW	Fuel Cost USD/h	Emission Ib/h
1	150.510	135.104	80.705	62.302	169.338	122.675	125.946	119.882	54.877	34.491	19.832	**61,600.519**	3803.894
2	150.006	135.034	146.567	120.465	175.863	122.983	128.759	118.303	20.993	13.373	22.336	**64,906.264**	4689.769
3	150.279	135.165	181.085	182.838	221.095	138.049	129.652	89.063	49.058	10.227	28.514	**72,304.859**	6228.359
4	150.009	135.007	224.237	191.995	242.419	159.944	129.963	119.998	44.613	43.418	35.606	**80,271.190**	7718.284
5	150.034	135.001	290.188	242.097	222.606	124.793	129.457	117.073	61.531	47.082	39.862	**84,408.221**	9763.984
6	150.032	144.110	339.873	298.516	242.647	159.853	98.936	118.058	79.997	44.509	48.532	**94,297.297**	13,307.920
7	222.445	135.032	330.607	299.754	239.465	159.924	129.960	116.990	70.059	51.333	53.563	**100,946.056**	13,761.749
8	224.611	213.686	339.936	291.807	242.827	139.143	129.115	119.997	79.959	54.035	59.116	**107,884.313**	14,564.055
9	272.168	323.499	339.776	299.530	242.880	159.038	129.950	120.000	65.817	42.489	71.142	**124,932.384**	17,534.742
10	299.063	395.517	339.995	293.040	242.928	159.991	122.587	114.212	79.889	55.000	80.219	**137,626.028**	20,656.966
11	341.347	449.088	339.950	296.452	243.000	159.982	129.965	110.813	79.997	44.020	88.609	**150,876.695**	26,836.868
12	386.873	450.590	339.134	299.972	242.996	156.337	129.852	119.998	79.960	37.402	93.114	**157,522.576**	29,006.003
13	312.707	441.906	332.353	299.267	242.973	144.923	128.829	119.947	79.461	54.994	85.359	**145,630.761**	24,855.409
14	272.168	323.499	339.776	299.530	242.880	159.038	129.950	120.000	65.817	42.489	71.142	**124,932.384**	17,534.742
15	224.611	213.686	339.936	291.807	242.827	139.143	129.115	119.997	79.959	54.035	59.116	**107,884.313**	14,564.055
16	152.831	135.356	293.140	299.998	237.653	158.625	129.971	93.751	52.366	44.230	43.915	**89,095.065**	11,635.439
17	150.034	135.001	290.188	242.097	222.606	124.793	129.457	117.073	61.531	47.082	39.862	**84,408.221**	9763.984
18	150.032	144.110	339.873	298.516	242.647	159.853	98.936	118.058	79.997	44.509	48.532	**94,297.297**	13,307.920
19	224.611	213.686	339.936	291.807	242.827	139.143	129.115	119.997	79.959	54.035	59.116	**107,884.313**	14,564.055
20	247.097	405.611	338.935	300.000	241.615	159.957	130.000	116.665	64.768	43.158	75.807	**131,752.041**	20,566.632
21	272.168	323.499	339.776	299.530	242.880	159.038	129.950	120.000	65.817	42.489	71.142	**124,932.384**	17,534.742
22	150.032	144.110	339.873	298.516	242.647	159.853	98.936	118.058	79.997	44.509	48.532	**94,297.297**	13,307.920
23	150.015	135.032	215.063	180.844	222.867	143.145	129.793	119.964	56.899	10.332	31.955	**76,082.212**	7006.619
24	150.001	135.024	184.626	131.224	222.612	123.171	129.588	86.119	24.598	22.486	25.447	**68,742.483**	5589.037

Total fuel cost and total emission results obtained from the CSA algorithm are compared with other methods in Table 15. As can be seen from Table 15, the proposed CSA algorithm produces fewer fuel cost value by 2.487512×10^6 USD than other algorithms. CSA was able to achieve an enhanced in total fuel cost in rang 10^6 of 0.01659 USD, 0.013315 USD and 0.029288 USD, respectively as compared to MHS [49], SA [57], and NSGAII [58]. IBFA [59], GCABC [60], and TLBO [61] gives a best solutions (less than this obtained by our proposed approach of 5812 USD, 13,039 USD, and 15,396 USD, respectively).

To elaborate the transition of CSA for different phases of the search convergence characteristics for the 10-unit test system under a fixed hour (P_D = 2150 MW) for 1000 iterations is shown in Figure 11. From the figure, we can be seen that CSA has an incredibly fast convergence ability it has been quickly trapped the optimal values.

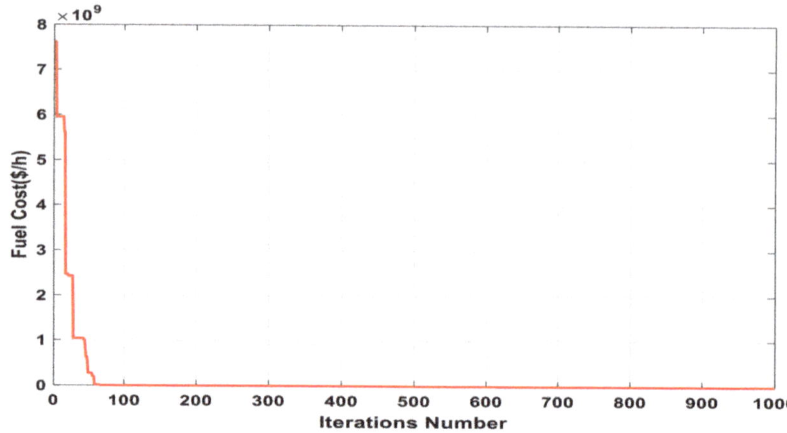

Figure 11. Convergence curve for10-unit test system for a fixed hour (P_D = 2150 MW).

Table 15. Comparison of the 10-unit test system with previous algorithms for best minimum cost solution (Case 4).

Methods	Total Fuel Cost ($\times 10^6$) USD	Total Emission ($\times 10^5$) Ib
CSA	2.487512	3.38103
MHS [49]	2.504106	NA
SA [57]	2.537200	NA
NSGAII [58]	2.516800	3.17400
IBFA [59]	2.481700	3.27500
GCABC [60]	2.474473	NA
TLBO [61]	2.472116	3.30411

5.2.2. Case 5

The dynamic environmental dispatch (DEnD) considering the valve-point loading effect and transmission losses as given in Equation (3) is discussed in this case.

Table 16 shows the results of the optimum power dispatch for the 10-unit system over 24 h obtained by CSA. The optimal results of the pollutant power generation and fuel cost obtained by the CSA are depicted in Table 16. The total fuel cost and total emissions computed by the CSA method are illustrated in Table 17. We compared the obtained solutions with those determined by other algorithms recently published in the literature.

By analyzing the results given in Table 17, we infer that the total emissions found by the CSA algorithm, which amount to 2.9354 $\times$ 10^5lb, are lower, compared to that found by IHS [49], MHS in [49], NSGA-II [58], TLBO [61], CRO [62], and HCRO [62], which are estimated at 296,044 lb, 302,093 lb, 304,120 lb, 294,153 lb, 317,400 lb, and 327,500 lb, respectively.

Method GCABC in [60] obtains the less total emission values of 293,416 lb compared with those values obtained by the methods cited in Table 17.

According to Table 17, we notice that the total cost determined by the CSA algorithm which is equal to 2.62594 $\times$ 10^6 USD is lower compared to that found by the NSGA-II [58] algorithm, which is estimated at 2.6563 $\times$ 10^6 USD.

Table 16. Best solutions obtained from CSA for test Case 5.

Time h	P_1MW	P_2MW	P_3MW	P_4MW	P_5MW	P_6MW	P_7MW	P_8MW	P_9MW	P_{10}MW	P_LMW	Fuel Cost USD/h	Emission Ib/h
1	150.002	135.718	90.566	91.469	118.016	136.066	109.538	94.773	74.663	54.913	19.725	63,282.134	**3479.107**
2	157.856	153.661	110.735	93.173	142.569	139.932	110.889	89.120	79.790	54.989	22.712	68,271.277	**3945.886**
3	160.271	167.053	120.413	132.252	178.038	155.820	122.997	119.962	75.131	54.997	28.936	75,457.550	**5060.796**
4	204.728	221.786	134.213	140.423	197.409	159.938	129.528	119.976	79.999	55.000	37.000	86,669.943	**6532.429**
5	188.357	229.602	151.075	171.912	239.342	158.088	129.982	119.990	77.798	54.576	40.723	89,962.573	**7507.212**
6	277.420	270.540	161.914	183.738	242.939	159.996	130.000	117.692	79.999	54.583	50.822	105,456.159	**9689.460**
7	303.679	253.234	215.080	207.820	234.626	159.789	129.996	118.557	80.000	54.731	55.512	110,376.988	**11,046.974**
8	284.213	276.631	242.805	257.079	232.163	159.480	129.906	119.925	79.154	54.847	60.204	114,622.220	**12,591.579**
9	329.264	351.175	263.618	277.043	237.413	153.536	129.956	119.969	79.838	54.610	72.424	131,422.312	**16,492.225**
10	391.022	365.874	298.133	265.150	242.871	159.900	129.864	119.824	75.641	54.994	81.275	143,558.138	**20,215.772**
11	403.610	398.763	323.390	299.975	233.988	159.970	129.997	112.664	79.995	52.485	88.838	152,258.869	**24,291.715**
12	415.007	433.955	340.000	295.831	242.986	159.987	129.954	98.248	79.986	47.515	93.474	159,891.196	**28,568.055**
13	403.752	376.588	323.914	265.781	242.984	159.996	129.998	119.990	79.991	54.733	85.728	148,855.883	**22,349.281**
14	329.264	351.175	263.618	277.043	237.413	153.536	129.956	119.969	79.838	54.610	72.424	131,422.312	**16,492.225**
15	284.213	276.631	242.805	257.079	232.163	159.480	129.906	119.925	79.154	54.847	60.204	114,622.220	**12,591.579**
16	223.433	240.864	174.031	189.920	227.097	159.983	129.867	119.965	79.171	54.950	45.282	95,730.049	**8464.578**
17	188.357	229.602	151.075	171.912	239.342	158.088	129.982	119.990	77.798	54.576	40.723	89,962.573	**7507.212**
18	277.420	270.540	161.914	183.738	242.939	159.996	130.000	117.692	79.999	54.583	50.822	105,456.159	**9689.460**
19	284.213	276.631	242.805	257.079	232.163	159.480	129.906	119.925	79.154	54.847	60.204	114,622.220	**12,591.579**
20	349.961	338.292	287.234	299.742	242.953	157.864	129.890	119.878	68.079	54.236	76.132	134,994.987	**18,062.886**
21	329.264	351.175	263.618	277.043	237.413	153.536	129.956	119.969	79.838	54.610	72.424	131,422.312	**16,492.225**
22	277.420	270.540	161.914	183.738	242.939	159.996	130.000	117.692	79.999	54.583	50.822	105,456.159	**9689.460**
23	167.768	180.865	146.464	142.876	189.040	159.016	129.864	113.622	79.999	54.997	32.511	80,208.210	**5752.844**
24	166.459	153.534	105.409	106.892	152.447	159.990	110.254	119.831	79.968	54.989	25.774	71,963.076	**4438.329**

Table 17. Comparison of the 10-unit test system with previous algorithms for best minimum cost solution (Case 5).

Methods	Total Fuel Cost ($\times 10^6$) USD	Total Emission ($\times 10^5$) Ib
CSA	2.625940	**2.93540**
GCABC [60]	NA	2.93416
TLBO [61]	2.594148	2.94153
IHS [49]	NA	2.96044
MHS [49]	NA	3.02093
NSGA-II [58]	2.656300	3.04120
CRO [62]	NA	3.17400
HCRO [62]	NA	3.27500

Figure 12 represents the convergence curves of our proposed approach for the DEnD problem. It describes the stable convergence of the objective function of the problem given in (3) (Case 5). The convergence characteristics shown in Figure 12 prove that CSA has good qualities. We conclude that CSA is favorable for large-scale power systems.

5.2.3. Case 6

In this case, we deal with the dynamic combined economic environmental dispatching (DCEED) by considering the power losses and the valve point-effect loading. Table 18 summarizes the best compromise solutions achieved for DCEED for a given load.

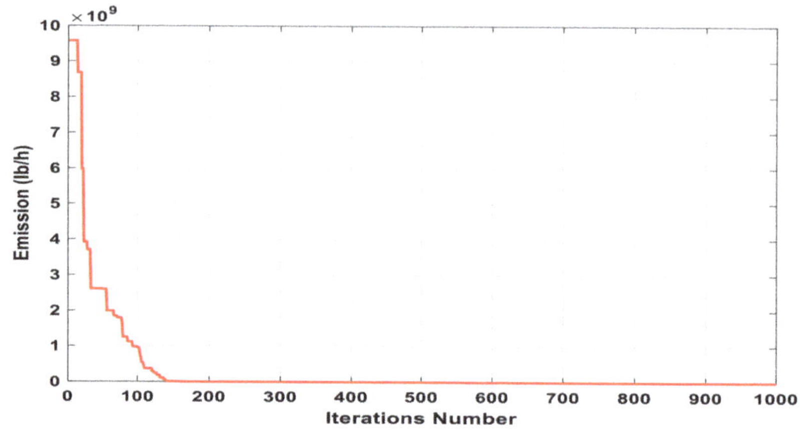

Figure 12. Convergence characteristics of the EnD with iterations for a fixed hour (P_D = 2150 MW).

Table 18. The best solutions obtained by CSA for the DCEED of the IEEE 10-unit test system.

Time h	P_1MW	P_2MW	P_3MW	P_4MW	P_5MW	P_6MW	P_7MW	P_8MW	P_9MW	P_{10}MW	P_LMW	FC USD/h	E Ib/h	TC USD/h
1	150.406	135.246	85.444	94.209	181.135	122.376	129.280	94.366	20.605	42.622	19.691	61,700.233	3934.478	**32,955.958**
2	150.059	135.192	148.162	120.431	172.562	122.461	93.065	119.995	22.348	48.148	22.423	65,198.802	4454.610	**34,966.155**
3	150.093	135.208	101.472	182.259	221.659	149.384	129.872	119.939	55.572	41.141	28.607	72,288.444	5650.903	**39,113.119**
4	150.060	135.019	190.752	204.190	224.454	159.847	124.653	119.999	79.982	52.633	35.589	80,461.980	7112.270	**43,928.263**
5	150.231	135.171	241.644	242.567	241.338	159.980	129.383	119.991	52.892	46.297	39.493	84,292.688	9023.037	**46,796.853**
6	152.020	164.386	320.567	277.526	222.190	159.921	129.914	119.986	80.000	49.903	48.411	94,530.118	12,104.35	**53,457.173**
7	161.851	223.048	310.663	299.756	235.796	141.568	129.795	119.940	78.244	55.000	53.662	100,886.75	13,127.88	**57,150.064**
8	150.069	269.756	339.843	299.405	242.988	158.986	124.745	119.937	75.404	53.775	58.903	107,645.16	14,941.13	**61,436.135**
9	289.083	313.982	339.999	280.201	234.134	156.640	129.595	120.000	78.241	53.458	71.328	125,468.68	16,952.94	**71,353.584**
10	364.296	316.930	339.330	299.170	242.913	159.675	129.700	119.025	76.855	54.122	80.023	137,757.86	19,776.86	**78,914.408**
11	391.758	400.840	336.936	293.570	242.999	159.970	129.998	119.999	63.537	54.997	88.603	151,135.50	24,136.97	**87,778.694**
12	442.035	396.653	337.897	299.108	242.982	157.061	127.818	119.996	79.999	39.772	93.323	158,774.96	28,337.51	**93,695.444**
13	359.815	391.269	339.912	299.467	242.964	151.758	126.070	118.443	79.987	47.449	85.138	145,326.69	22,318.79	**83,964.399**
14	289.083	313.982	339.999	280.201	234.134	156.640	129.595	120.000	78.241	53.458	71.328	125,468.68	16,952.94	**71,353.584**
15	150.069	269.756	339.843	299.405	242.988	158.986	124.745	119.937	75.404	53.775	58.903	107,645.16	14,941.13	**61,436.135**
16	150.156	135.036	276.528	283.184	240.900	159.073	129.657	120.000	51.168	52.036	43.739	88,890.958	10,861.13	**50,016.892**
17	150.231	135.171	241.644	242.567	241.338	159.980	129.383	119.991	52.892	46.297	39.493	84,292.688	9023.037	**46,796.853**
18	152.020	164.386	320.567	277.526	222.190	159.921	129.914	119.986	80.000	49.903	48.411	94,530.118	12,104.35	**53,457.173**
19	150.069	269.756	339.843	299.405	242.988	158.986	124.745	119.937	75.404	53.775	58.903	107,645.16	14,941.13	**61,436.135**
20	326.074	347.521	309.290	299.994	242.997	160.000	129.970	119.999	67.957	44.017	75.820	133,249.35	18,331.48	**75,929.053**
21	289.083	313.982	339.999	280.201	234.134	156.640	129.595	120.000	78.241	53.458	71.328	125,468.68	16,952.94	**71,353.584**
22	152.020	164.386	320.567	277.526	222.190	159.921	129.914	119.986	80.000	49.903	48.411	94,530.118	12,104.35	**53,457.173**
23	150.173	137.534	180.818	163.649	238.569	149.328	129.739	117.812	51.515	44.869	32.009	76,243.211	6439.102	**41,480.912**
24	150.015	135.021	126.085	104.167	220.872	159.474	129.251	119.979	21.132	43.459	25.457	68,625.467	5068.830	**36,985.679**

Note: FC: Fuel cost, E: Emission and TC: Total cost (FC+E).

According to Table 18, we can deduce that CSA adjust the value of the output power ($P_1 - P_{10}$) such that load demand and the constraint limits are satisfied, and the best fuel cost and pollutants emission results are achieved. However, the statistical comparative results of CSA method along with various other methods for both cost and emission objective functions are provided in Table 19. The minimum cost obtained by the proposed CSA approach is 2,492,050 USD among 50 trials in the economic dispatch problem and considered as the ever best solution. It is also clear from the simulation results that the proposed approach provides better statistical results as compared with many other available methods given in Table 19.

Table 19. The DCEED comparison results of different algorithms of the 10-unit test system.

Methods	Total Loss MW	Total Fuel Cost ($\times 10^6$) USD	Total Emission ($\times 10^5$) Ib	Total Cost ($\times 10^6$) USD
CSA	**1298.9956**	**2.492050**	**3.19592**	**2.843985**
MONNDE [53]	1307.8000	2.557900	2.95220	2.769100
NSGA-II [58]	NA	2.521000	3.12460	2.823600
IBFA [59]	1299.8760	2.517117	2.99037	2.753743
TLBO [61]	1301.1900	2.472116	2.94153	2.776644
HCRO [62]	1299.8723	2.517076	2.99065	2.753863
CRO [62]	1298.4666	2.517821	3.01941	2.768615
NSPSO [63]	NA	2.474472	2.93416	2.704316
PSO-CSC [64]	1303.1000	2.524700	3.05240	2.788550

Figure 13 shows the convergence curves of the CEED with time (iterations) obtained by applying the CSA method corresponding to a fixed hour with a power demand of 2150 MW. It is observed from the convergence curve that a better reduction in combined cost and emission can be obtained with less number of iterations using the proposed CSA.

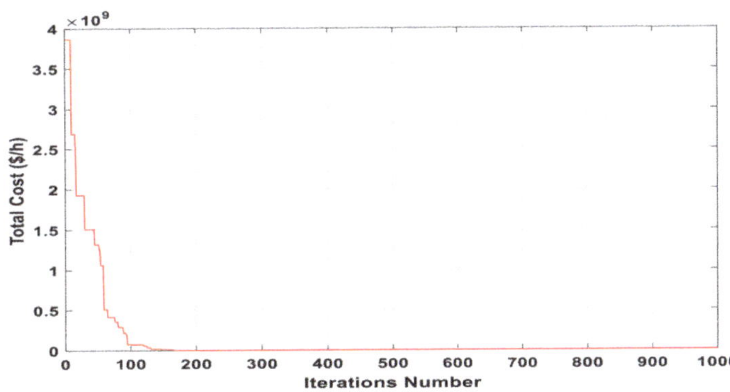

Figure 13. Convergence curves of the CEED with time (iterations) for a fixed hour (P_D = 2150 MW).

Figure 14 displays the obtained values (best, average, and max values) among 50 runs of the CSA technique for the case of minimizing the fuel cost of production (Case 4).

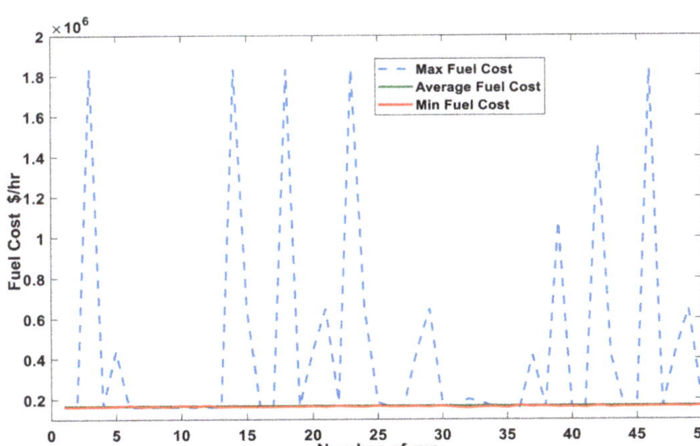

Figure 14. Min, Mean, and Max Fuel cost obtained in 50 runs using CSA method for 10-unit system.

6. Conclusions and Future Work

In this paper, new optimization approaches were proposed, presented and applied to solve the problem of the dynamic economic emissions dispatch of generator units considering the valve-point loading effects. The main inspiration of these optimization techniques are the fact that metaheuristic algorithms are easy to implement and can be used for a variety of other problems.

The proposed strategies are validated by simulating MATLAB and testing on the two standard IEEE power systems, 5-unit and 10-unit systems. The numerical results of this system are presented to show the capabilities of the proposal algorithms to establish an optimal solution of the dynamics problem of combined economic emissions dispatch in several passages. From Tables 4, 6, 8, 14, 16 and 18, it is obviously clear that the optimal generation schedule of a 5-units and 10-unit system obtained by CSA satisfy power balance constraint with considering power losses and generator operating limits constraint. The proposed CSA gives better performance compared to methods cited in the literature.

In all cases, our proposed algorithms can reach the optimum solution more quickly which. In future works, we intend to combine CSA with TSA, and introduce it to other kinds of optimization issues, such as large-scale economic load dispatch problems, integrated renewable energy sources, multi-objective ED problems with many complex constraints.

Author Contributions: Conceptualization, B.L., A.N.E.I.A., H.B. and A.S.T.; Data curation, B.L.; Formal analysis, B.L., A.N.E.I.A. and H.B.; Funding acquisition, H.A., T.E.A.A., S.S.M.G. and S.A.M.A.; Investigation, H.A., T.E.A.A., A.S.T. and S.S.M.G.; Methodology, B.L., A.N.E.I.A. and H.B.; Project administration, B.L. and S.S.M.G.; Resources, A.N.E.I.A., H.A., T.E.A.A., S.S.M.G. and S.A.M.A.; Software, B.L., A.N.E.I.A., H.B. and A.S.T.; Supervision, B.L. and S.S.M.G.; Validation, B.L., A.N.E.I.A. and A.S.T.; Visualization, H.A., T.E.A.A., S.S.M.G. and S.A.M.A.; Writing—original draft, B.L., A.N.E.I.A., H.B. and A.S.T.; Writing—review and editing, H.A., T.E.A.A., S.S.M.G. and S.A.M.A. All authors have read and agreed to the published version of the manuscript.

Funding: This work was supported by Taif University Researchers Supporting Project number (TURSP-2020/34), Taif University, Taif, Saudi Arabia.

Institutional Review Board Statement: Not applicable.

Informed Consent Statement: Not applicable.

Data Availability Statement: Not applicable.

Acknowledgments: The authors appreciate Taif University Researchers Supporting Project number (TURSP-2020/34), Taif University, Taif, Saudi Arabia, and the General Directorate of Scientific Research and Technological Development (DGRSDT) of Algeria.

Conflicts of Interest: The authors declare no conflict of interest.

Abbreviations

CEED	Combined economic environmental dispatch
DED	Dynamic economic dispatch
DEnD	Dynamic environmental dispatch
DEED	Dynamic economic emission dispatch
DCEELDP	Dynamic combined economic emission load dispatch problem
EED	Economic emission dispatch
FC	Fuel cost
a_i, b_i, c_i	Coefficients of the fuel cost corresponding of generator i
e_i, and f_i	Fuel cost coefficients of ith generator due to VPE
$\alpha_i, \beta_i, \gamma_i, \eta_i, and\ \delta_i$	Emission curve coefficients
$P_{L,t}$	Power losses
w_i	Weighting factor

References

1. Larouci, B.; Benasla, L.; Belmadani, A.; Rahli, M. Cuckoo Search Algorithm for Solving Economic Power Dispatch Problem with Consideration of Facts Devices. *UPB Sci. Bull. Ser. C* **2017**, *79*, 43–54.
2. Ryu, H.-S.; Kim, M.-K. Combined Economic Emission Dispatch with Environment-Based Demand Response Using WU-ABC Algorithm. *Energies* **2020**, *13*, 6450. [CrossRef]
3. Razali, N.S.; Mansor, M.H.; Musirin, I.; Othman, M.M.; Akoury, M. Embedded Immune-Evolutionary Programming for Economic Dispatch of Generators with Prohibited Operating Zones. *Int. J. Eng. Technol.* **2018**, *7*, 183–186. [CrossRef]
4. Vahidi, M.; Vahdani, S.; Rahimian, A.; Jamshidi, M.-N.; Kanee, A.-T. Evolutionary-base finite element model updating and damage detection using modal testing results. *Struct. Eng. Mech.* **2019**, *70*, 339–350.
5. Al-Betar, M.A.; Awadallah, M.A.; Abu Doush, I.; Alsukhni, E.; Alkhraisat, H. A Non-convex Economic Dispatch Problem with Valve Loading Effect Using a New Modified β -Hill Climbing Local Search Algorithm. *Arab. J. Sci. Eng.* **2018**, *43*, 7439–7456. [CrossRef]
6. Gao, Y.; Li, X.; Dong, M.; Li, H.-P. An enhanced artificial bee colony optimizer and its application to multi-level threshold image segmentation. *J. Cent. South Univ.* **2018**, *25*, 107–120. [CrossRef]
7. Mirjalili, S.; Mirjalili, S.-M.; Hatamlou, A. Multi-verse optimizer: A nature-inspired algorithm for global optimization. *Neural Comput. Appl.* **2016**, *27*, 495–513. [CrossRef]
8. Wang, F.; Zhang, H.; Zhou, A. A particle swarm optimization algorithm for mixed-variable optimization problems. *Swarm Evol. Comput.* **2021**, *60*, 100808. [CrossRef]
9. Nadimi-Shahraki, M.H.; Taghian, S.; Mirjalili, S. An improved grey wolf optimizer for solving engineering problems. *Expert Syst. Appl.* **2021**, *166*, 113917. [CrossRef]
10. Albashish, D.; Hammouri, A.I.; Braik, M.; Atwan, J.; Sahran, S. Binary biogeography-based optimization based SVM-RFE for feature selection. *Appl. Soft Comput.* **2021**, *101*, 107026. [CrossRef]
11. Li, L.-L.; Shen, Q.; Tseng, M.-L.; Luo, S. Power system hybrid dynamic economic emission dispatch with wind energy based on improved sailfish algorithm. *J. Clean. Prod.* **2021**, *316*, 128318. [CrossRef]
12. Liu, Z.-F.; Li, L.-L.; Liu, Y.-W.; Liu, J.-Q.; Li, H.-Y.; Shen, Q. Dynamic economic emission dispatch considering renewable energy generation: A novel multi-objective optimization approach. *Energy* **2021**, *235*, 121407. [CrossRef]
13. Li, L.-L.; Liu, Z.-F.; Tseng, M.-L.; Zheng, S.-J.; Lim, M.K. Improved tunicate swarm algorithm: Solving the dynamic economic emission dispatch problems. *Appl. Soft Comput.* **2021**, *108*, 107504. [CrossRef]
14. Qiao, B.; Liu, J.; Hao, X. A multi-objective differential evolution algorithm and a constraint handling mechanism based on variables proportion for dynamic economic emission dispatch problems. *Appl. Soft Comput.* **2021**, *108*, 107419. [CrossRef]
15. Zou, Y.; Zhao, J.; Ding, D.; Miao, F.; Sobhani, B.F. Solving dynamic economic and emission dispatch in power system integrated electric vehicle and wind turbine using multi-objective virus colony search algorithm. *Sustain. Cities Soc.* **2021**, *67*, 102722. [CrossRef]
16. Yao, L.; Li, J.; Liang, H. Dynamic Economic/Emission Dispatch Considering Renewable Energy and PEVs. In Proceedings of the 2021 40th Chinese Control Conference (CCC), Shanghai, China, 26–28 July 2021. [CrossRef]
17. Yang, W.; Peng, Z.; Yang, Z.; Guo, Y.; Chen, X. An enhanced exploratory whale optimization algorithm for dynamic economic dispatch. *Energy Rep.* **2021**, *7*, 7015–7029. [CrossRef]
18. Gul, R.N.; Ahmad, A.; Fayyaz, S.; Sattar, M.K.; Haq, S.S.U. Hybrid Flower Pollination Algorithm with Sequential Quadratic Programming Technique for Solving Dynamic Combined Economic Emission Dispatch Problem. *Mehran Univ. Res. J. Eng. Technol.* **2021**, *40*, 371–382. [CrossRef]
19. Xia, A.; Wu, X.; Bai, Y. A new multi-objective hybrid optimization algorithm for wind-thermal dynamic economic emission power dispatch. *Int. Trans. Electr. Energy Syst.* **2021**, *31*, e12966. [CrossRef]
20. Alshammari, M.; Ramli, M.; Mehedi, I. A New Chaotic Artificial Bee Colony for the Risk-Constrained Economic Emission Dispatch Problem Incorporating Wind Power. *Energies* **2021**, *14*, 4014. [CrossRef]
21. Hassan, M.H.; Kamel, S.; Abualigah, L.; Eid, A. Development and application of slime mould algorithm for optimal economic emission dispatch. *Expert Syst. Appl.* **2021**, *182*, 115205. [CrossRef]
22. Ajayi, O.; Heymann, R. Day-ahead combined economic and emission dispatch with spinning reserve consideration using moth swarm algorithm for a data centre load. *Heliyon* **2021**, *7*, e08054. [CrossRef] [PubMed]
23. Fayyaz, S.; Sattar, M.K.; Waseem, M.; Ashraf, M.U.; Ahmad, A.; Hussain, H.A.; Alsubhi, K. Solution of Combined Economic Emission Dispatch Problem Using Improved and Chaotic Population-Based Polar Bear Optimization Algorithm. *IEEE Access* **2021**, *9*, 56152–56167. [CrossRef]
24. Larouci, B.; Si Tayeb, A.; Boudjella, H.; Ayad, A.N.E.I. Cuckoo search algorithm to solve the problem of economic emission dispatch with the incorporation of facts devices under the valve-point loading effect. *Facta Univ. Ser. Electron. Energ.* **2021**, *34*, 569–588. [CrossRef]
25. Si Tayeb, A.; Larouci, B.; Rezzak, D.; Houam, Y.; Bouzeboudja, H.; Bouchakour, A. Application of a new hybridization to solve economic dispatch problem on an algerian power system without or with connection to a renewable energy. *Diagnostyka* **2021**, *22*, 101–112. [CrossRef]

26. Kuk, J.N.; Gonçalves, R.A.; Pavelski, L.M.; Venske, S.M.G.S.; de Almeida, C.P.; Pozo, A.T.R. An empirical analysis of constraint handling on evolutionary multi-objective algorithms for the Environmental/Economic Load Dispatch problem. *Expert Syst. Appl.* **2021**, *165*, 113774. [CrossRef]
27. Prasad, R.S.; Sud, R. The pivotal role of UNFCCC in the international climate policy landscape: A developing country perspective. *Glob. Aff.* **2021**, *7*, 67–78. [CrossRef]
28. Van Hong, T.P.; Ngoc, D.V.; Tuan, K.D. Environmental Economic Dispatch Using Stochastic Fractal Search Algorithm. In Proceedings of the 2021 International Symposium on Electrical and Electronics Engineering (ISEE), Ho Chi Minh, Vietnam, 15–16 April 2021; pp. 214–219.
29. Chakrabarti, S.; Panja, A.K.; Mukherjee, A.; Bar, A.K. Application of Multi-Objective Particle Swarm Optimization Technique for Analytical Solution of Economic and Environmental Dispatch. In *Intelligent Electrical Systems: A Step towards Smarter Earth*; CRC Press: Boca Raton, FL, USA, 2021; pp. 313–319.
30. Boudjella, H.; Laouer, M.; Bouzeboudja, H.; Ayad, A.N.E.I.; Benhamida, F.; Saad, A. Solution of Economic Load Dispatch Problems Using Novel Improved Harmony Search Algorithm. *Int. J. Electr. Eng. Inform.* **2021**, *13*, 218–241. [CrossRef]
31. Wood, A.J.; Wollenberg, B.F.; Sheblé, G.B. *Power Generation, Operation, and Control*; John Wiley & Sons: Hoboken, NJ, USA, 2013.
32. Rezaee Jordehi, A. Dynamic environmental-economic load dispatch in grid-connected microgrids with demand response programs considering the uncertainties of demand, renewable generation and market price. *Int. J. Numer. Model. Electron. Netw. Devices Fields* **2021**, *34*, e2798. [CrossRef]
33. Dhiman, G.; Kumar, V. Seagull optimization algorithm: Theory and its applications for large-scale industrial engineering problems. *Knowl. Based Syst.* **2019**, *165*, 169–196. [CrossRef]
34. Kumar, V.; Kumar, D.; Kaur, M.; Singh, D.; Idris, S.A.; Alshazly, H. A Novel Binary Seagull Optimizer and its Application to Feature Selection Problem. *IEEE Access* **2021**, *9*, 103481–103496. [CrossRef]
35. Sharifi, M.R.; Akbarifard, S.; Qaderi, K.; Madadi, M.R. Comparative analysis of some evolutionary-based models in optimization of dam reservoirs operation. *Sci. Rep.* **2021**, *11*, 15611. [CrossRef] [PubMed]
36. Jia, H.; Xing, Z.; Song, W. A New Hybrid Seagull Optimization Algorithm for Feature Selection. *IEEE Access* **2019**, *7*, 49614–49631. [CrossRef]
37. Askarzadeh, A. A novel metaheuristic method for solving constrained engineering optimization problems: Crow search algorithm. *Comput. Struct.* **2016**, *169*, 1–12. [CrossRef]
38. Han, X.; Xu, Q.; Yue, L.; Dong, Y.; Xie, G.; Xu, X. An Improved Crow Search Algorithm Based on Spiral Search Mechanism for Solving Numerical and Engineering Optimization Problems. *IEEE Access* **2020**, *8*, 92363–92382. [CrossRef]
39. Meraihi, Y.; Gabis, A.B.; Ramdane-Cherif, A.; Acheli, D. A comprehensive survey of Crow Search Algorithm and its applications. *Artif. Intell. Rev.* **2021**, *54*, 2669–2716. [CrossRef]
40. Kaur, S.; Awasthi, L.K.; Sangal, A.; Dhiman, G. Tunicate Swarm Algorithm: A new bio-inspired based metaheuristic paradigm for global optimization. *Eng. Appl. Artif. Intell.* **2020**, *90*, 103541. [CrossRef]
41. Fetouh, T.; Elsayed, A.M. Optimal Control and Operation of Fully Automated Distribution Networks Using Improved Tunicate Swarm Intelligent Algorithm. *IEEE Access* **2020**, *8*, 129689–129708. [CrossRef]
42. Sharma, A.; Sharma, A.; Dasgotra, A.; Jately, V.; Ram, M.; Rajput, S.; Azzopardi, B.; Averbukh, M. Opposition-Based Tunicate Swarm Algorithm for Parameter Optimization of Solar Cells. *IEEE Access* **2021**, *9*, 125590–125602. [CrossRef]
43. Yang, X.S. Firefly algorithms for multimodal optimization. In Proceedings of the 5th International Symposium on Stochastic Algorithms, Sapporo, Japan, 26–28 October 2009; pp. 169–178.
44. Strumberger, I.; Bacanin, N.; Tuba, M. Enhanced firefly algorithm for constrained numerical optimization. In Proceedings of the 2017 IEEE Congress on Evolutionary Computation (CEC), Donostia, Spain, 5–8 June 2017; pp. 2120–2127. [CrossRef]
45. Chahnasir, E.S.; Zandi, Y.; Shariati, M.; Dehghani, E.; Toghroli, A.; Mohamad, E.T.; Shariati, A.; Safa, M.; Wakil, K.; Khorami, M. Application of support vector machine with firefly algorithm for investigation of the factors affecting the shear strength of angle shear connectors. *Smart Struct. Syst.* **2018**, *22*, 413–424.
46. Wang, C.-F.; Song, W.-X. A novel firefly algorithm based on gender difference and its convergence. *Appl. Soft Comput.* **2019**, *80*, 107–124. [CrossRef]
47. Basu, M. Particle Swarm Optimization Based Goal-Attainment Method for Dynamic Economic Emission Dispatch. *Electr. Power Components Syst.* **2006**, *34*, 1015–1025. [CrossRef]
48. Hardiansyah, H. Hybrid PSOGSA technique for solving dynamic economic emission dispatch problem. *Eng. Rev.* **2020**, *40*, 96–104. [CrossRef]
49. Li, Z.; Zou, D.; Kong, Z. A harmony search variant and a useful constraint handling method for the dynamic economic emission dispatch problems considering transmission loss. *Eng. Appl. Artif. Intell.* **2019**, *84*, 18–40. [CrossRef]
50. Shehata, A.M.; Elaiw, A.M. Hybrid DE-SOP for solving dynamic economic emission dispatch with prohibited operating. *Int. J. Sci. Eng. Res.* **2015**, *6*, 1136–1141.
51. Zhang, H.; Yue, D.; Xie, X.; Hu, S.; Weng, S. Multi-elite guide hybrid differential evolution with simulated annealing technique for dynamic economic emission dispatch. *Appl. Soft Comput.* **2015**, *34*, 312–323. [CrossRef]
52. Mehdi, M.F.; Ahmad, A.; Haq, S.S.U.; Saqib, M.; Ullah, M.F. Dynamic economic emission dispatch using whale optimization algorithm for multi-objective function. *Electr. Eng. Electromech.* **2021**, *2*, 64–69. [CrossRef]

53. Mason, K.; Duggan, J.; Howley, E. A multi-objective neural network trained with differential evolution for dynamic economic emission dispatch. *Int. J. Electr. Power Energy Syst.* **2018**, *100*, 201–221. [CrossRef]
54. Basu, M. Dynamic Economic Emission Dispatch Using Evolutionary Programming and Fuzzy Satisfying Method. *Int. J. Emerg. Electr. Power Syst.* **2007**, *8*. [CrossRef]
55. Alsumait, J.S.; Qasem, M.; Sykulski, J.; Al-Othman, A.K. An improved Pattern Search based algorithm to solve the Dynamic Economic Dispatch problem with valve-point effect. *Energy Convers. Manag.* **2010**, *51*, 2062–2067. [CrossRef]
56. Kothari, D.P.; Dhillon, J.S. *Power System Optimization*, 2nd ed.; PHI Learning Private Ltd.: New Delhi, India, 2011.
57. Pattanaik, J.K.; Basu, M.; Dash, D.P. Dynamic economic dispatch: A comparative study for differential evolution, particle swarm optimization, evolutionary programming, genetic algorithm, and simulated annealing. *J. Electr. Syst. Inf. Technol.* **2019**, *6*, 1–8. [CrossRef]
58. Basu, M. Dynamic economic emission dispatch using nondominated sorting genetic algorithm-II. *Int. J. Electr. Power Energy Syst.* **2008**, *30*, 140–149. [CrossRef]
59. Pandit, N.; Tripathi, A.; Tapaswi, S.; Pandit, M. An improved bacterial foraging algorithm for combined static/dynamic environmental economic dispatch. *Appl. Soft Comput.* **2012**, *12*, 3500–3513. [CrossRef]
60. Marouani, I.; Boudjemline, A.; Guesmi, T.; Abdallah, H.H. A Modified Artificial Bee Colony for the Non-Smooth Dynamic Economic/Environmental Dispatch. *Eng. Technol. Appl. Sci. Res.* **2018**, *8*, 3321–3328. [CrossRef]
61. Alshammari, B.M. Teaching-Learning-Based Optimization Algorithm for the Combined Dynamic Economic Environmental Dispatch Problem. *Eng. Technol. Appl. Sci. Res.* **2020**, *10*, 6432–6437. [CrossRef]
62. Roy, P.K.; Bhui, S. A multi-objective hybrid evolutionary algorithm for dynamic economic emission load dispatch. *Int. Trans. Electr. Energy Syst.* **2016**, *26*, 49–78. [CrossRef]
63. Alshammari, M.E.; Ramli, M.A.; Mehedi, I.M. An Elitist Multi-Objective Particle Swarm Optimization Algorithm for Sustainable Dynamic Economic Emission Dispatch Integrating Wind Farms. *Sustainability* **2020**, *12*, 7253. [CrossRef]
64. Qian, S.; Wu, H.; Xu, G. An improved particle swarm optimization with clone selection principle for dynamic economic emission dispatch. *Soft Comput.* **2020**, *24*, 15249–15271. [CrossRef]

sustainability

MDPI

Article

Investigating the Impact of Wind Power Integration on Damping Characteristics of Low Frequency Oscillations in Power Systems

Jian Chen [1,*], Tao Jin [2], Mohamed A. Mohamed [3,*], Andres Annuk [4] and Udaya Dampage [5]

1 Robotics School of Fuzhou Polytechnic, Fuzhou 350108, China
2 Department of Electrical Engineering, Fuzhou University, Fuzhou 350116, China; jintly@fzu.edu.cn
3 Electrical Engineering Department, Faculty of Engineering, Minia University, Minia 61519, Egypt
4 Institute of Forestry and Engineering, Estonian University of Life Sciences, 51006 Tartu, Estonia; andres.annuk@emu.ee
5 Faculty of Engineering, Kotelawla Defence University, Kandawala Estate, Ratmalana 10390, Sri Lanka; dampage@kdu.ac.lk
* Correspondence: chenjian@fvti.edu.cn (J.C.); dr.mohamed.abdelaziz@mu.edu.eg (M.A.M.)

Abstract: This paper investigates the impact of doubly-fed induction generator (DFIG) wind farms on system stability in multi-generator power systems with low-frequency oscillations (LFOs). To this end, this paper establishes the interconnection model of the equivalent generators and derives the system state equation. On this basis, an updated system state equation of the new power system with integrated wind power is further derived. Then, according to the updated system state equation, the impact factors that cause changes in the system damping characteristics are presented. The IEEE two-area four-machine power system is used as a simulation model in which the LFOs occur. The simulation results demonstrate that the connection of wind power to the power feeding area (PFA) increases the damping ratio of the dominant mode of inter-area oscillation from −0.0263 to −0.0107, which obviously improves the system stability. Furthermore, the wind power integration into PFA, as the connection distance of the wind power increases, gradually decreases the damping ratio of the dominant mode of the inter-area oscillation to −0.0236, approaching that of no wind power in the system. Meanwhile, with the increase in the wind power output capacity, the damping ratio of the dominant mode of the intra-area and inter-area oscillation increases, and the maximum damping ratios reach 0.1337 and 0.0233, respectively.

Keywords: wind power integration; low-frequency oscillations (LFOs); doubly-fed induction generators (DFIGs); power system stability; damping characteristics

Citation: Chen, J.; Jin, T.; Mohamed, M.A.; Annuk, A.; Dampage, U. Investigating the Impact of Wind Power Integration on Damping Characteristics of Low Frequency Oscillations in Power Systems. *Sustainability* **2022**, *14*, 3841. https://doi.org/10.3390/su14073841

Academic Editor: Marc A. Rosen

Received: 14 February 2022
Accepted: 20 March 2022
Published: 24 March 2022

1. Introduction

The expansion of the integration of clean and efficient renewable energy sources such as wind and solar energy has exacerbated the gradual expansion of the modern power grid interconnection and caused an unprecedented complexity of the energy system [1–3]. Hence, the security and stability of the power system should be taken seriously [4]. A typical hazard in a power system is the interaction among generators or regional power grids, which results in the damping characteristic changes of the power systems and thus may produce low-frequency oscillations (LFOs) [5]. The synchronous generator angles start oscillating with each other in low frequencies of 0.1–2.5 Hz following grid disturbances. These oscillations are usually poorly damped and can lead to system collapse [6]. A famous example of a grid failure is the blackout that occurred at Western Systems Council Coordinated (WSCC) in August 1996 [7]. A series of system failures occurred in eastern North America, which eventually led to system instability and resulted in the blackout disaster on August 14, 2003 [8]. There was a major grid disturbance in the northern region at 02:33 on 30 July 2012. The northern regional grid load was about 36,000 MW at the time of the disturbance. Subsequently, there was another grid disturbance at 13:00 on 31 July

2012, resulting in the collapse of the northern, eastern, and north-eastern regional grids. A total load of about 48,000 MW was affected in this blackout [9]. The power oscillations at different stages and degrees were observed during these big blackouts. LFO seriously affects the power transmission capacity and threatens the stable operation of the power grid and even causes power failure accidents due to grid instability [10].

Due to the increase in power demand and to reduce global warming, the power grid is moving more and more to renewable energy integration, in which wind power is the most important renewable source of energy all over the world. The impact of integrated wind power on the power system stability is also a long-term problem which in turn arouses the interest of many researchers around the world [11]. The authors in [12] proposed the stability problems of the power systems early, including large-scale wind power in which the impact of intra-area and inter-area oscillations is studied under the condition of wind power integrated into the power systems. It is found that the damping effect of the constant speed wind turbine (WT) on the power systems is more effective than that of the variable speed WT because of the effects of the induction generator. In [13], the authors have analyzed the power system oscillation modes of the two-area four-machine system with small disturbance and studied the changes of the oscillation modes under the situation of a wind farm replacing a synchronous generator in the system. Aiming at the integration of the wind power into Norwegian power grids, E. Hagstrøm et al. [14] studied the impact of the integrated wind power on the interval oscillation damping of northern European power grids and tested the impact of different types of generators on the inter-area oscillation damping of Norway and Sweden. In [15], the generic dynamic model of WT with a doubly-fed induction generator (DFIG) is presented, and the small-signal stability mathematical model is derived. G. Tsourakis et al. [16] studied the response modes of DFIG in wind power plants and their impact on the electromechanical oscillation of interconnected power systems. L. Yang et al. [17] carried out a mathematical model for DFIG. Through eigenvalue sensitivity analysis, the most sensitive parameters of the Hopf bifurcation of DFIG power generation system can be determined. Thus, the Hopf bifurcation boundaries of the key parameters are analyzed to facilitate the parameterization of DFIG WT system and ensure system stability. H. Li et al. [18] proposed a fuzzy control strategy to suppress the LFOs of the power system after the doubly-fed induction generators (DFIGs) are integrated into the power grid. M. Singh et al. [19] studied the potential damping interval oscillation mode of the wind farm. The result shows that the intra-area oscillation mode is a phenomenon of oscillation in a single or a group of generators with another set of generators on a weak transmission link. R. Effatnejad et al. [20] studied the LFO suppression strategy of power systems in the case of the integrated wind power. The angular velocity difference signal of the two generators is used as the input signal of the damping controller. The conventional damping controller is replaced by the fuzzy controller, and the controller parameters are optimized by the intelligent optimization algorithm. In [21], a method was proposed for the separate examination of the two impact factors, the change of load flow and dynamic interaction brought about by the DFIG on the electromechanical oscillation modes of the power systems. J. G. Slootweg and W. L. Kling [22] established a 7-order WT mathematical model for doubly-fed wind turbines and revealed the impact of shafting and the wind speed on the system's forced LFOs in the condition of the wind power integrated into the power systems through the small-signal linear analysis method. In order to solve the dynamic stability problem of the large wind farms, X. Zhang et al. [23] proposed a power grid operation method based on variable inertia control. A variable inertia control strategy for WT was proposed to improve the damping and frequency stability of power oscillation between the areas. In [24], a line modal potential energy (LMPE) method was proposed for the inter-area oscillation analysis and damping control of DFIGs. The study attempts to design the damping control loop using the LMPE approach from the network viewpoint. L. Simon et al. [25] described an optimization enabled wide-area damping control (WADC) for DFIG to mitigate both the intra-area and the inter-area oscillations. The main purpose in [26] was to improve the synchrony generator's transient and dynamic performance using local DFIGs. The impact of wind power integrated into the existing power system is multi-dimensional. There are still many uncertain results that cannot be ignored in the coordinated development of wind power and the large power grid to maintain system stability in the future power grid. Based on the above investigation and explanations, the main contributions of this research are summarized as follows:

- The two-area interconnection system and its mathematical model are established and analyzed, and the system state equation is deduced before and after the wind power integration.
- Three impact factors of the integrated wind power on the LFOs of the power system are proposed and the eigenvalues of the intra-area and inter-area oscillation of the power system are identified by the total least squares-estimation of signal parameters via rotational invariance techniques (TLS-ESPRIT) algorithm.
- By comparing the changes of the damping characteristics of the system before and after wind power integration, the impact on the LFOs of the system is obtained, which provides a reference for the design of the wind power plants, as well as the stability of the power system with small disturbance.

The remaining sections of this paper are organized as follows: Section 2 establishes the model of the two-area interconnection power system and derives the system state equation. The damping characteristics of the system integrated with the wind power consisting of the DFIGs are analyzed. The impact factors of wind power connection on the LFOs are mainly related to the area option of the integrated wind power, including a power feeding area (PFA) or a power receiving area (PRA), the connection distance, and the output capacity of the integrated wind power. Section 3 shows the example simulation and result analysis based on the simulation of the IEEE two-area four-machine power system. Firstly, the eigenvalues of the intra-area and inter-area oscillation modes of the system are calculated. Secondly, considering the variation of each impact factor, the changes in the damping characteristics of the simulation model are reflected by the eigenvalues of the modes of the intra-area and inter-area oscillation. Finally, through the quantitative analysis of the modal eigenvalues and the changes of the damping ratios, the effect of the different impact factors of the integrated wind power on the LFOs of the power system is obtained and discussed. Section 4 addresses the conclusion of the paper.

2. Analysis of Damping Characteristics

The LFOs of large-scale power systems usually include intra-area and inter-area oscillation. The intra-area oscillation is mainly the relative sway between different generators in the area. The inter-area oscillation is the relative swing between multiple generator groups in one area and multiple generator groups in other areas. According to the equivalent principle of the center of inertia (COI) [27], multiple generator groups in the area are taken to be COI equivalent, which can simplify the damping analysis of the system. Hence, the interconnection between two groups of generators can be studied by using a simplified system consisting of two equivalent generators.

2.1. Analysis of Damping Characteristics of the Two-Area Interconnection Power Systems

The equivalent model of the two-area interconnection system is shown in Figure 1. The inter-area oscillation is equivalent to the oscillation between the synchronous generators G_{eq1} and G_{eq2}. Taking the active power transmitted from area A1 to area A2 as an example, it can be said that area A1 is PFA, while area A2 is PRA. Then, the equivalent rotor angular velocity increment of G_{eq1} and G_{eq2} will show the same frequency reverse characteristic, as well as the equivalent power angle increment.

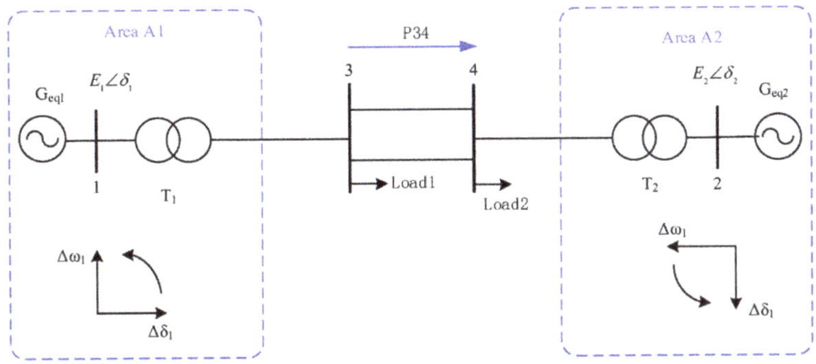

Figure 1. Equivalent model of the two-area interconnection system.

The classical second-order equations are adopted as the rotor motion mathematical model of G_{eq1} and G_{eq2}. The mechanical power of the prime movers is constant, and the excitation system dynamics are ignored while the load is treated as constant impedance. The rotor motion equation of the COI equivalent synchronous generator is represented as follows:

$$\begin{cases} \Delta\dot{\delta}_j = \Delta\omega_j \\ \Delta\dot{\omega}_j = \frac{1}{T_{Jj}}(-\Delta P_{je} - D_j\Delta\omega_j) \end{cases} \tag{1}$$

Using the Heffron–Phillips model, rewriting (1) in the matrix form, it is represented as follows:

$$\begin{bmatrix} \Delta\dot{\delta}_1 \\ \Delta\dot{\delta}_2 \\ \Delta\dot{\omega}_1 \\ \Delta\dot{\omega}_2 \end{bmatrix} = \begin{bmatrix} 0 & 0 & 1 & 0 \\ 0 & 0 & 0 & 1 \\ -\frac{K_{11}}{T_{J1}} & -\frac{K_{11}}{T_{J1}} & -\frac{D_1}{T_{J1}} & 0 \\ -\frac{K_{21}}{T_{J2}} & -\frac{K_{21}}{T_{J2}} & 0 & -\frac{D_2}{T_{J2}} \end{bmatrix} \begin{bmatrix} \Delta\delta_1 \\ \Delta\delta_2 \\ \Delta\omega_1 \\ \Delta\omega_2 \end{bmatrix} = \boldsymbol{A}_{eq} \begin{bmatrix} \Delta\delta_1 \\ \Delta\delta_2 \\ \Delta\omega_1 \\ \Delta\omega_2 \end{bmatrix} \tag{2}$$

According to the system's circuit analysis, based on the superposition theorem of the model shown in Figure 1, the output electromagnetic power P_{1e} and P_{2e} are calculated as represented in (3).

$$\begin{cases} P_{1e} = \frac{E_1^2}{|Z_{11}|}\sin\alpha_{11} + \frac{E_1E_2}{|Z_{12}|}\sin(\delta_{12} - \alpha_{12}) \\ P_{2e} = \frac{E_2^2}{|Z_{22}|}\sin\alpha_{22} + \frac{E_1E_2}{|Z_{21}|}\sin(\delta_{21} - \alpha_{12}) \end{cases} \tag{3}$$

where

$$Z_{12} = Z_{21},\ Z_{11} = |Z_{11}|\angle\varphi_{11},\ Z_{22} = |Z_{22}|\angle\varphi_{22},\ Z_{12} = |Z_{12}|\angle\varphi_{12},\ \alpha_{11} = 90^\circ - \varphi_{11},\ \alpha_{22} = 90^\circ - \varphi_{22},\ \alpha_{12} = 90^\circ - \varphi_{12}.$$

At a certain operating point, the rotor angle difference of G_{eq1} and G_{eq2} has the relation $\delta_{120} = -\delta_{210}$. According to (3), the synchronous torque coefficients K_{jk} can be calculated as follows:

$$\begin{cases} K_{11} = -K_{12} = \frac{E_1E_2}{|Z_{12}|}\cos(\delta_{120} - \alpha_{12}) \\ K_{22} = -K_{21} = \frac{E_1E_2}{|Z_{21}|}\cos(\delta_{210} - \alpha_{12}) \end{cases} \tag{4}$$

The state matrix $\boldsymbol{A}_{eq}$ can be extracted from (2), and its characteristic equation is represented by:

$$\det(\boldsymbol{A}_{eq} - \lambda\boldsymbol{I}) = 0 \tag{5}$$

Through matrix transformation and the Schur theorem, Equation (5) can be transformed into a fourth-degree equation with one variable with respect to λ, as represented in (6).

$$\lambda^4+\left(\frac{D_1}{T_{J1}}+\frac{D_2}{T_{J2}}\right)\lambda^3+\left(\frac{K_{11}}{T_{J1}}+\frac{K_{22}}{T_{J2}}+\frac{D_1}{T_{J1}}\frac{D_2}{T_{J2}}\right)\lambda^2+\frac{1}{T_{J1}T_{J2}}(D_1K_{22}+D_2K_{11})\lambda=0 \quad (6)$$

Idealizing the damping moment coefficient and the same inertia time constant of the equivalent generator results in $D_1 = D_2, T_{J1} = T_{J2}$, or $\frac{D_1}{T_{J1}} = \frac{D_2}{T_{J2}}$. Therefore, (6) can be simplified as follows:

$$\lambda(\lambda+\gamma)(\lambda^2+\gamma\lambda+\frac{K_{11}}{T_{J1}}+\frac{K_{22}}{T_{J2}})=0 \quad (7)$$

The eigenvalues are obtained by solving the equation as follows:

$$\begin{cases} \lambda_1 = 0 \\ \lambda_1 = -\gamma = -\frac{D_1}{T_{J1}} \\ \lambda_{3,4} = \sigma \pm j\omega = -\frac{D_1}{2T_{J1}} \pm \frac{j\sqrt{\left(-\frac{D_1}{T_{J1}}\right)^2+4\left(\frac{K_{11}}{T_{J1}}+\frac{K_{22}}{T_{J2}}\right)}}{2} \end{cases} \quad (8)$$

In general, K_{11}, K_{22}, D_1, D_2 are greater than zero in normal operation. According to the Routh system stability criterion, Equation (8) shows that the real parts of the three non-zero eigenvalues are all negative, and the conjugate eigenvalues λ_3 and λ_4 represent the electromechanical oscillation modes of the interconnected system. If the system is abnormal, such as $D_1 \leq 0$ or $D_2 \leq 0$, that is $-D_1/T_{J1} \geq 0$ or $-D_2/T_{J2} \geq 0$, the real part of eigenvalue will be located on the imaginary axis or the right half-space of the complex plane. The corresponding consequences are that the system damping is zero or negative, which results in continuous or growth oscillation of the power systems. If it is not intervened in, the power systems under the conditions of instability will be out-of-step, splitting finally.

2.2. Analysis of Damping Characteristics of the Two-Area Interconnection Power Systems Integrated with Wind Power

The wind farm marked by ellipse dotted lines in Figure 2 is incorporated into the two-area interconnection system. The wind farm can be represented by a single machine of a large-capacity DFIG [25]. The schematic diagram of a DFIG integrating the power grid is shown in Figure 3.

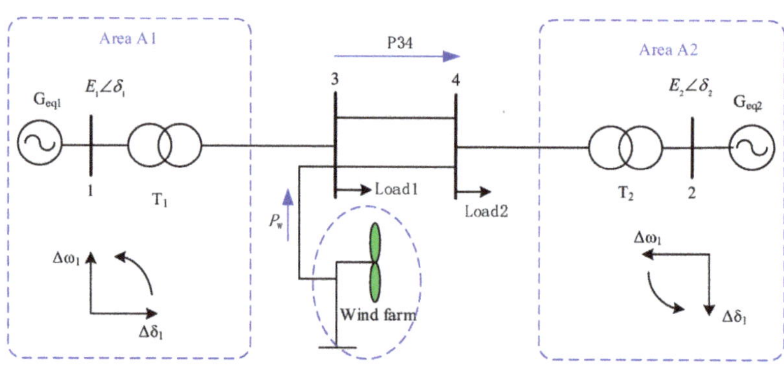

Figure 2. Two-area interconnection system integrated with wind farm.

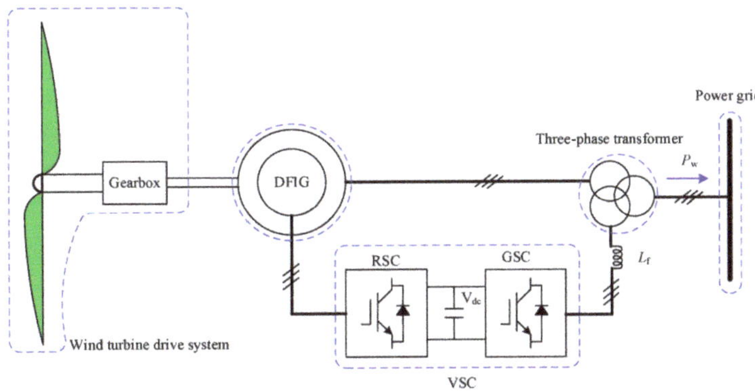

Figure 3. Schematic diagram of DFIG integrated into power grid.

Neglecting the system line loss, the electromagnetic power increments ΔP_{1e} and ΔP_{2e} of G_{eq1} and G_{eq2} mentioned in (1) are rewritten as (9).

$$\begin{cases} \Delta P_{1e} = \frac{E_1 U_w}{X_{13}} \cos(\delta_{10} + \delta_{w0})(\Delta\delta_1 + \Delta\delta_w) \\ \Delta P_{2e} = \frac{E_2 U_w}{X_{23}} \cos(\delta_{20} + \delta_{w0})(\Delta\delta_2 + \Delta\delta_w) \end{cases} \quad (9)$$

Labeling $k_1 = \frac{E_1 U_w}{X_{13}} \cos(\delta_{10} + \delta_{w0})$, $k_2 = \frac{E_2 U_w}{X_{23}} \cos(\delta_{20} + \delta_{w0})$, Equation (9) can be simplified as follows:

$$\begin{cases} \Delta P_{1e} = k_1(\Delta\delta_1 + \Delta\delta_w) \\ \Delta P_{2e} = k_2(\Delta\delta_2 + \Delta\delta_w) \end{cases} \quad (10)$$

As the system structure is stable and the load remains unchanged, the active power of the PFA and PRA is increased or decreased by the same amount. Hence, the active power increment of the interconnected two-area systems in Figure 2 remains in the balance after the integration of the wind farm to the grid, and the mathematical relationship is represented as follows:

$$\Delta P_{1e} + \Delta P_w = \Delta P_{2e} \quad (11)$$

ΔP_w is not a state variable. When LFOs occur in the system, the changing trend of the power increment at any node is consistent with that of the frequency increment at the same point of the system. The power and frequency for any point have the same oscillation mode. Therefore, it can be considered that their increments have a proportional relationship, which can be simplified as follows:

$$\Delta P_w = g_1 \cdot \Delta f_B \quad (12)$$

Similarly, there is a linear relationship between the frequency increment of the system and the rotor angular velocity of the system generator. In the case that the regional structure is not completely symmetrical, the proportional relation can be represented as follows:

$$\Delta f_B = g_2 \cdot \Delta\omega_1 = g_3 \cdot \Delta\omega_2 \quad (13)$$

The linear relationships between (12) and (13) with regard to the state variables $\Delta\omega_1$ and $\Delta\omega_2$ are represented in (14).

$$\begin{cases} \Delta P_w = k_3\Delta\omega_1 = k_4\Delta\omega_2 \\ k_3 = g_1 g_2 \\ k_4 = g_1 g_3 \end{cases} \quad (14)$$

Uniting (10)–(14), the state variables can represent ΔP_{1e} and ΔP_{2e}. Combining with rotor motion (1) derived from the above second-order classical model, the wind farm is incorporated into the PFA corresponding to area A1, and the state equation is represented in (15). When the wind power is incorporated into the PRA corresponding to area A2, the state equation is represented in (16).

$$\begin{bmatrix} \Delta\dot{\delta}_1 \\ \Delta\dot{\delta}_2 \\ \Delta\dot{\omega}_1 \\ \Delta\dot{\omega}_2 \end{bmatrix} = \begin{bmatrix} 0 & 0 & 1 & 0 \\ 0 & 0 & 0 & 1 \\ -\frac{k_1k_2}{T_{J1}(k_1+k_2)} & \frac{k_1k_2}{T_{J1}(k_1+k_2)} & -\frac{D_1}{T_{J1}} + \frac{k_1k_3}{T_{J1}(k_1+k_2)} & 0 \\ -\frac{k_1k_2}{T_{J2}(k_1+k_2)} & \frac{k_1k_2}{T_{J2}(k_1+k_2)} & \frac{k_1k_3}{T_{J2}(k_1+k_2)} & -\frac{D_2}{T_{J2}} \end{bmatrix} \begin{bmatrix} \Delta\delta_1 \\ \Delta\delta_2 \\ \Delta\omega_1 \\ \Delta\omega_2 \end{bmatrix} \tag{15}$$

$$\begin{bmatrix} \Delta\dot{\delta}_1 \\ \Delta\dot{\delta}_2 \\ \Delta\dot{\omega}_1 \\ \Delta\dot{\omega}_2 \end{bmatrix} = \begin{bmatrix} 0 & 0 & 1 & 0 \\ 0 & 0 & 0 & 1 \\ -\frac{k_1k_2}{T_{J1}(k_1+k_2)} & \frac{k_1k_2}{T_{J1}(k_1+k_2)} & -\frac{D_1}{T_{J1}} & \frac{k_1k_4}{T_{J2}(k_1+k_2)} \\ -\frac{k_1k_2}{T_{J2}(k_1+k_2)} & \frac{k_1k_2}{T_{J2}(k_1+k_2)} & 0 & -\frac{D_2}{T_{J2}} + \frac{k_1k_4}{T_{J2}(k_1+k_2)} \end{bmatrix} \begin{bmatrix} \Delta\delta_1 \\ \Delta\delta_2 \\ \Delta\omega_1 \\ \Delta\omega_2 \end{bmatrix} \tag{16}$$

The impact of integrated wind power on the power system damping is mainly investigated through the quantitative changes of the system damping characteristics. Before and after the integration, the change of damping characteristics can be discussed according to (2), (15), and (16).

Keeping the system structure unchanged, the damping sum of all the modes of the system is constant and equals the sum of all the characteristic roots of the system [28]. Furthermore, the sum of diagonal elements $\boldsymbol{A}_{eq}$ in (2) is always equal. When wind power is incorporated into the PFA, it can be seen from (15) that the damping sum of the system is $-\frac{D_1}{T_{J1}} - \frac{D_2}{T_{J2}} + \frac{k_1k_3}{T_{J1}(k_1+k_2)}$. The damping increment is $\frac{k_1k_3}{T_{J1}(k_1+k_2)}$, which is the relation of k_1, k_2, k_3. $k_1 = \frac{E_1U_w}{X_{13}}\cos(\delta_{10}+\delta_{w0})$, $k_2 = \frac{E_2U_w}{X_{23}}\cos(\delta_{20}+\delta_{w0})$, and $k_3 = g_1g_2$ are a relation of E_1, E_2, U_w, X_{13}, and X_{23}, which are fixed. However, δ_{20}, δ_{w0}, g_1, and g_2 may result in changes of k_1, k_2, and k_3, which cause the change of $\frac{k_1k_3}{T_{J1}(k_1+k_2)}$. When the wind power is incorporated into the PRA, it can be observed from (16) that the damping sum of the system is $-\frac{D_1}{T_{J1}} - \frac{D_2}{T_{J2}} + \frac{k_2k_4}{T_{J2}(k_1+k_2)}$. The damping increment is $\frac{k_2k_4}{T_{J2}(k_1+k_2)}$, in which there is $k_4 = g_2g_3$. Similarly, δ_{20}, δ_{w0}, g_2, and g_3 result in the changes of k_1, k_2, and k_4, which cause the change of $\frac{k_1k_4}{T_{J2}(k_1+k_2)}$.

Based on the mathematical derivation and analysis above, the system damping increment may be positive or negative in the impact of the integrated wind power on the power systems. The effect depends on the various parameters of the integrated wind power, such as how far it is connected to the grid, how large-scale the capacity is, and in which area the integrated location is, PFA or PRA. The wind power impact is related to the actual changes of system damping, according to the practical structure of the power systems containing the wind farm.

3. Simulation Example and Result Analysis

3.1. The Impact of the Different Areas of Wind Power on the System LFOs

In this paper, the wind farm is connected to the double loop line of the IEEE two-area four-machine power systems [27]. Before the connection of the wind farm, area A1 is mainly composed of generators G1 and G2, and area A2 is mainly composed of generators G3 and G4. Area A1 is set to transmit 413 MW of active power to area A2 through the tie line. Hence, area A1 is the PFA, and area A2 is the PRA. According to the literature [25], the wind farm consisting of DFIGs can be simulated by an equivalent DFIG denoted with G_w, which is connected to node 7 in the PFA and node 9 in the PRA, as shown in Figures 4 and 5, respectively. The green ellipse dotted line frame in Figures 4 and 5 is the integrated wind farm. The wind farm is equivalent, practically, to the WT of 25 DFIGs, which has the capacity of 2 MW for each one. Hence, the capacity of G_W equals 50 MW.

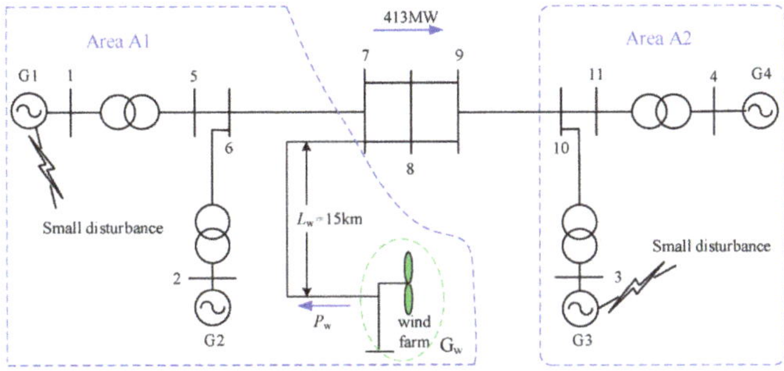

Figure 4. The wind power integrated into the PFA.

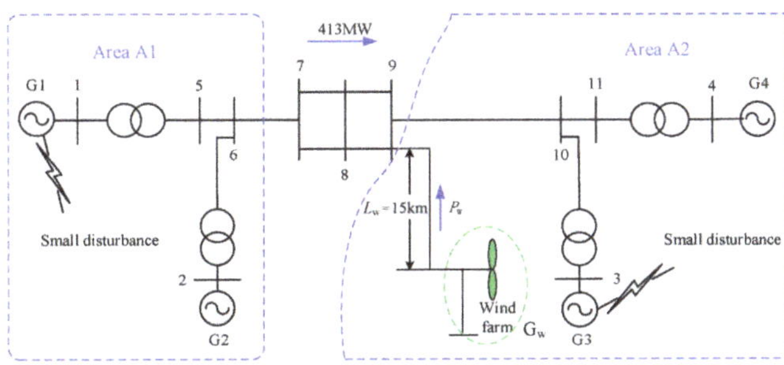

Figure 5. The wind power integrated into the PRA.

Before the DFIG reaches a stable operating state, it needs a period of 10–15 s from the start to the stable state. Therefore, the small disturbance is set at t = 20 s after the DFIG starts. The square wave pulse signals, with 5% of the stable transmission voltage, which is set as the small disturbance with a duration of 0.1 s, are added to G1 and G3 at the same time. Under the perturbation excitation, the intra-area oscillation of the rotor angular velocity of the synchronous generator and the inter-area oscillation of the active power of the tie line between node 7 and node 9 are observed.

In the paper, three operating conditions are studied under the condition of the no-power system stabilizer (PSS) in the system. First, the IEEE two-area four-machine power systems suffer small disturbances without integrated wind power. Second, the power system suffers small disturbance with the wind power integrated into the PFA, as shown in Figure 4. Third, the power system suffers small disturbance with the wind power integrated into the PRA, as shown in Figure 5. According to the data of the oscillation modes in Tables 1–3, the mode distributions of the intra-area oscillation in G1–G4 are drawn out in Figure 6a–d and that of the intra-area oscillation in the tie line are illustrated in Figure 6e in three operating conditions. The variables σ and ω denote the real part and imaginary part of the eigenvalue, respectively. The eigenvalues of the oscillation modes are identified by the TLS-ESPRIT algorithm [29] in the paper. The variable ξ is the damping ratio of the corresponding mode, which is equal to $\frac{-\sigma}{\sqrt{\sigma^2+\omega^2}}$. The intra-area oscillation modes are characterized with the oscillation mode parameters of the rotor angular velocity of the four generators (G1–G4). The inter-area oscillation modes are characterized with the oscillation mode parameters of the active power in the tie line [30]. The conjugate eigenvalues are symmetrically distributed about the real axis in the complex plane; so, the imaginary part is not depicted in Figure 6.

Table 1. Oscillation modes of the system without wind power.

Object	Mode	σ	ω (rad/s)	ξ
G1	Mode #1	−0.5809	±7.0282	0.0824
	Mode #2	0.0871	±4.0228	−0.0217
	Mode #3	−1.0705	±1.1010	0.6971
G2	Mode #1	−0.6319	±7.0404	0.0894
	Mode #2	0.0620	±4.0274	−0.0154
	Mode #3	−0.1028	±0.8767	0.1165
G3	Mode #1	−0.5637	±7.2388	0.0776
	Mode #2	0.0927	±4.0292	−0.0230
	Mode #3	−0.8445	±1.0654	0.6212
G4	Mode #1	−0.6301	±7.2762	0.0863
	Mode #2	0.1092	±4.0185	−0.0272
	Mode #3	−0.0681	±0.8490	0.0800
P_L	Mode #1	−0.2247	±6.9636	0.0323
	Mode #2	0.1072	±4.0257	−0.0266
	Mode #3	−0.0695	±0.7260	0.0953

Table 2. Oscillation modes of the system with wind power integrated into PFA.

Object	Mode	σ	ω (rad/s)	ξ
G1	Mode#1	−0.5868	±7.0125	0.0834
	Mode #2	0.0463	±4.0702	−0.0114
	Mode #3	−1.3740	±1.7076	0.6269
G2	Mode #1	−0.5559	±6.9174	0.0801
	Mode #2	0.0438	±4.0686	−0.0108
	Mode #3	−0.0484	±0.6640	0.0727
G3	Mode #1	−0.5835	±7.2156	0.0806
	Mode #2	0.0429	±4.0650	−0.0106
	Mode #3	−0.1245	±0.7685	0.1599
G4	Mode #1	−0.6608	±7.2553	0.0907
	Mode #2	0.0451	±4.0676	−0.0111
	Mode #3	−0.0244	±0.7069	0.0345
P_L	Mode #1	0.0436	±4.0634	−0.0107
	Mode #2	0.0303	±1.1072	−0.0274

Table 3. Oscillation modes of the system with wind power integrated into PRA.

Object	Mode	σ	ω (rad/s)	ξ
G1	Mode #1	−0.5953	±7.0269	0.0844
	Mode #2	0.1081	±4.0544	−0.0267
	Mode #3	−0.9478	±1.0427	0.6726
G2	Mode #1	−0.5597	±6.9155	0.0807
	Mode #2	0.1108	±4.0523	−0.0273
	Mode #3	−0.0468	±0.5066	0.0920
G3	Mode #1	−0.5480	±7.2324	0.0756
	Mode #2	0.1084	±4.0557	−0.0267
	Mode #3	0.1571	±0.6128	−0.2483
G4	Mode #1	−0.5347	±7.1724	0.0743
	Mode #2	0.1089	±4.0558	−0.0268
	Mode #3	−0.0682	±0.4766	0.1417
P_L	Mode #1	0.1092	±4.0525	−0.0269
	Mode #2	0.2013	±0.7349	−0.2642

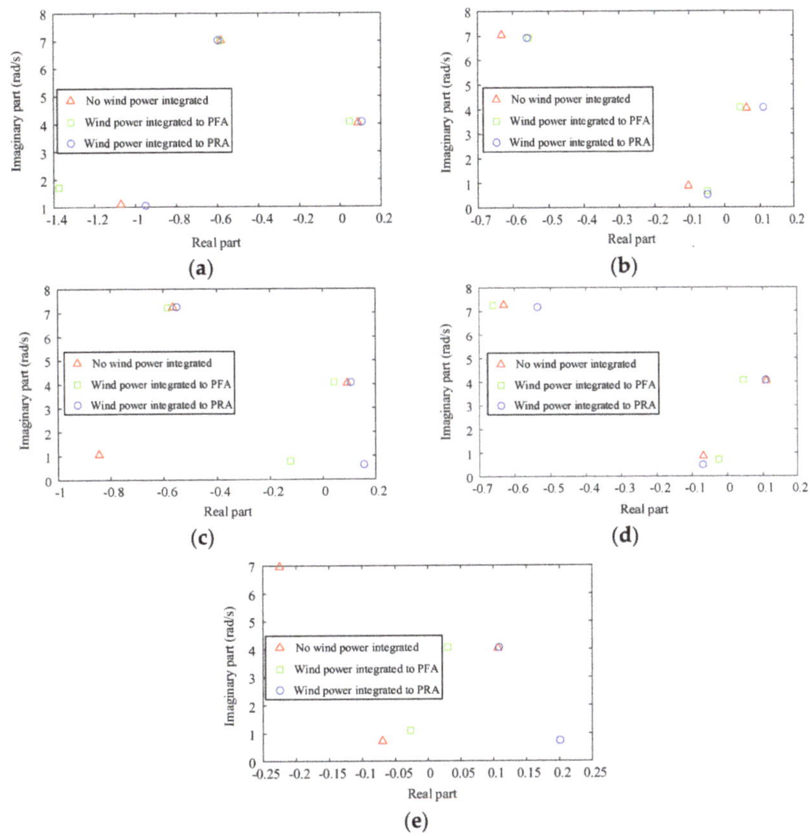

Figure 6. The modes distributions of intra-area oscillation and inter-area oscillation. (**a**) G1; (**b**) G2; (**c**) G3; (**d**) G4; and (**e**) tie line.

The following are the discussions on Figure 6:

(1) The green squares corresponding to the eigenvalues of each mode under the three conditions are mainly located on the left of the red angle. The eigenvalues on the corresponding complex plane generally shift to the left under the condition of wind power integrated into the PFA, compared with no wind power in the system. The damping characteristics are improved and the LFOs are suppressed.

(2) The blue circles corresponding to the eigenvalues of each mode in the three operating conditions are mainly located on the right of the red angle. The eigenvalues on the corresponding complex plane generally shift to the right when the wind power is integrated into the PRA, compared with the condition of no wind power in the system. When the wind power is integrated into the PRA, the damping characteristics become worse and the LFOs are aggravated slightly.

(3) The wind power integration has an impact on the dominant mode of the rotor angular velocity oscillation of the four generators G1~G4. The variation range of the eigenvalues is small, and the impact is not significant. However, the intra-area oscillation of the active power of the tie line is obviously suppressed, which can be observed from the existence of the red angle but no blue circle or green square nearby in Figure 6e. In addition, the corresponding eigenvalue of the blue circle in the lower right corner of Figure 6e indicates that wind power integrated into the PRA increases the oscillation modes and weakens the damping characteristics of the system.

Figure 7 represents the fact that the damping ratio of Mode#1 of the dominant mode of the intra-area oscillation fluctuates in the range of 0.07 to 0.09, which shows that the wind power has a slight impact. However, with the wind power integration into the PFA, the damping ratio of the dominant mode of the inter-area oscillation changes from −0.0266 to −0.0107, which in turn improves the damping characteristics of the system. In contrast, the damping ratio of the wind power integration into PRA is −0.0269, which is a slight decrease. The minimum damping ratio of the intra-area oscillation is reduced to −0.2483, corresponding to Mode#3 on G3, and that of the inter-area oscillation in Mode#2 is reduced to −0.2642.

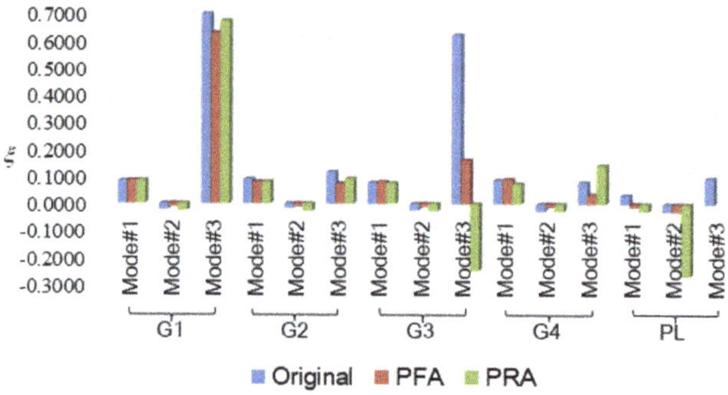

Figure 7. Changes of damping ratio before and after wind power integration.

3.2. *The Impact of Connection Distance of the Wind Power on the System LFOs*

The schematic diagram of a single line of the IEEE two-area four-machine power systems, including the integrated wind power whose connection distance can be adjusted, is shown in Figure 8. Area A1 consists of the generators G1 and G2 and the wind farm. The capacity of the wind farm is set at 50 MW, which is equivalently provided by 25 DFIGs of 2 MW in the simulation. The thick red line marked with l_w represents the connection distance of the wind power and is used as an adjustable object. The area A2 mainly consists of G3 and G4. As area A1 is set to transmit 413 MW of active power to area A2 through the tie line, A1 and A2 represent the PFA and the PRA, respectively. The small disturbance, including the disturbance amplitude and the disturbance period, is set the same as Section 3.1. l_w is set as 10 km, 50 km, 100 km, 200 km, and 300 km, respectively. With the different connection distances of the integrated wind power, the oscillation modes of the rotor angular velocity of G1–G4 and the active power of the tie line are tested and recorded in Tables 4–8.

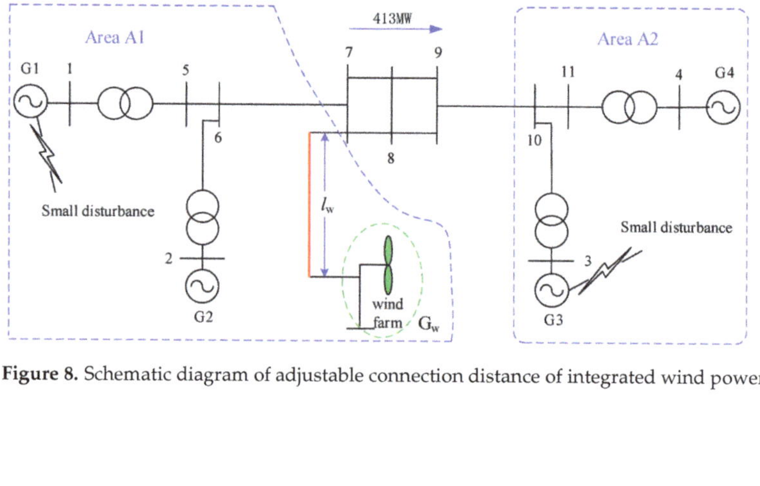

Figure 8. Schematic diagram of adjustable connection distance of integrated wind power.

Table 4. l_w = 10 km.

Object	Mode	σ	ω (rad/s)	ξ
G1	Mode #1	−0.5869	±7.0114	0.0834
	Mode #2	0.0514	±4.1136	−0.0125
	Mode #3	−1.5415	±2.0295	0.6049
G2	Mode #1	−0.5544	±6.9159	0.0799
	Mode #2	0.0471	±4.1054	−0.0115
	Mode #3	−0.0512	±0.6694	0.0763
G3	Mode #1	−0.5832	±7.2138	0.0806
	Mode #2	0.0429	±4.1124	−0.0104
	Mode #3	−0.1772	±0.7630	0.2262
G4	Mode #1	−0.6592	±7.2562	0.0905
	Mode #2	−4.9058	±3.2432	0.8342
	Mode #3	0.0481	±4.1112	−0.0117
	Mode #4	−0.0398	±0.7003	0.0567
P_L	Mode #1	0.0468	±4.1099	−0.0114
	Mode #2	−0.2904	±6.2922	0.0461
	Mode #3	0.0964	±0.6716	−0.1421

Table 5. l_w = 50 km.

Object	Mode	σ	ω (rad/s)	ξ
G1	Mode #1	−0.5897	±7.0044	0.0839
	Mode #2	0.0721	±3.9967	−0.0180
	Mode #3	−0.2326	±0.7583	0.2933
G2	Mode #1	−0.5646	±6.9093	0.0814
	Mode #2	0.0699	±3.9986	−0.0175
	Mode #3	−0.0710	±0.6669	0.1059
G3	Mode #1	−0.5054	±7.2594	0.0695
	Mode #2	0.0683	±3.9903	−0.0171
	Mode #3	−0.0708	±0.9064	0.0779
G4	Mode #1	−0.5606	±7.1206	0.0785
	Mode #2	0.0732	±3.9895	−0.0183
	Mode #3	−0.0461	±0.6661	0.0690
P_L	Mode #1	0.0696	±3.9922	−0.0174
	Mode #2	0.2988	±0.6986	−0.3933

Table 6. l_w = 100 km.

Object	Mode	σ	ω (rad/s)	ξ
G1	Mode #1	−0.5768	±7.0187	0.0819
	Mode #2	0.0801	±3.9932	−0.0201
	Mode #3	−0.8717	±0.8444	0.7183
G2	Mode #1	−0.5540	±6.8997	0.0800
	Mode #2	0.0788	±3.9940	−0.0197
	Mode #3	−0.1318	±0.6801	0.1903
G3	Mode #1	−0.5027	±7.2622	0.0691
	Mode #2	0.0766	±3.9886	−0.0192
	Mode #3	0.0419	±0.8395	−0.0498
G4	Mode #1	−0.5194	±7.1183	0.0728
	Mode #2	0.0800	±3.9868	−0.0201
	Mode #3	−0.1182	±0.7211	0.1618
P_L	Mode #1	0.0775	±3.9902	−0.0194
	Mode #2	0.3224	±0.6805	−0.4281

Table 7. l_w = 200 km.

Object	Mode	σ	ω (rad/s)	ξ
G1	Mode #1	0.0891	±3.9848	−0.0224
	Mode #2	−0.4344	±7.0341	0.0616
	Mode #3	0.0302	±1.0444	−0.0289
G2	Mode #1	−0.5224	±6.8893	0.0756
	Mode #2	0.0840	±3.9819	−0.0211
	Mode #3	−0.1453	±0.9354	0.1535
G3	Mode #1	0.0837	±3.9824	−0.0210
	Mode #2	−0.5783	±7.2186	0.0799
	Mode #3	0.1729	±1.0368	−0.1645
G4	Mode #1	0.0868	±3.9804	−0.0218
	Mode #2	−0.3566	±7.0398	0.0506
	Mode #3	0.0629	±1.0646	−0.0590
P_L	Mode #1	0.0857	±3.9860	−0.0215
	Mode #2	0.3259	±0.6785	−0.4330

Table 8. l_w = 300 km.

Object	Mode	σ	ω (rad/s)	ξ
G1	Mode #1	−0.3745	±7.0774	0.0528
	Mode #2	0.0915	±3.9652	−0.0231
	Mode #3	0.1439	±1.3245	−0.1080
G2	Mode #1	0.2916	±7.5212	−0.0387
	Mode #2	−0.4981	±6.8604	0.0724
	Mode #3	0.0912	±3.9643	−0.0230
	Mode #4	0.3169	±1.3077	−0.2355
G3	Mode #1	−0.6302	±7.0251	0.0893
	Mode #2	0.0913	±3.9695	−0.0230
	Mode #3	0.3995	±1.3254	−0.2886
G4	Mode #1	−0.2189	±6.9959	0.0313
	Mode #2	0.0937	±3.9657	−0.0236
	Mode #3	0.3436	±1.3607	−0.2448
P_L	Mode #1	0.1813	±7.2515	−0.0250
	Mode #2	0.0935	±3.9681	−0.0236
	Mode #3	0.2956	±0.7389	−0.3714

The following are discussions on Figures 9–13:

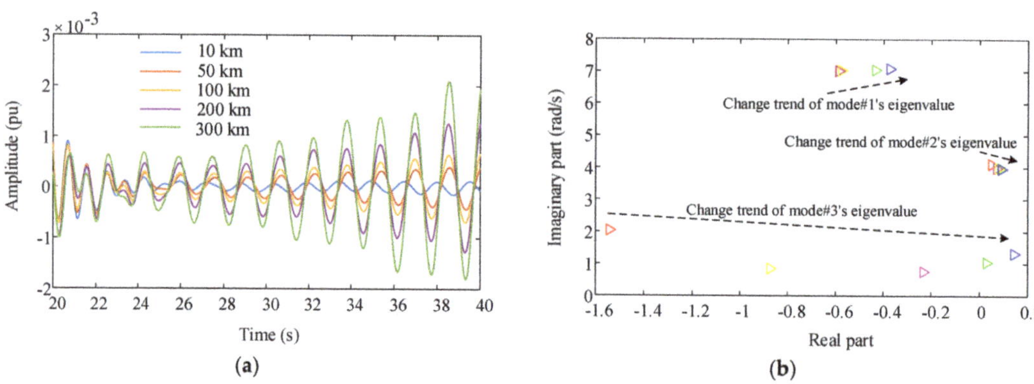

Figure 9. Rotor angular velocity oscillation of G1: (**a**) Oscillation output; (**b**) eigenvalues distribution.

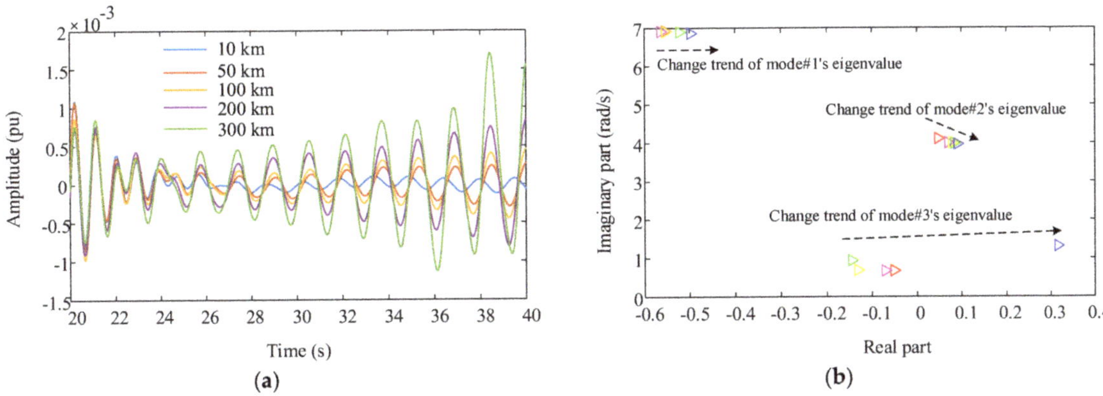

Figure 10. Rotor angular velocity oscillation of G2: (**a**) Oscillation output; (**b**) eigenvalues distribution.

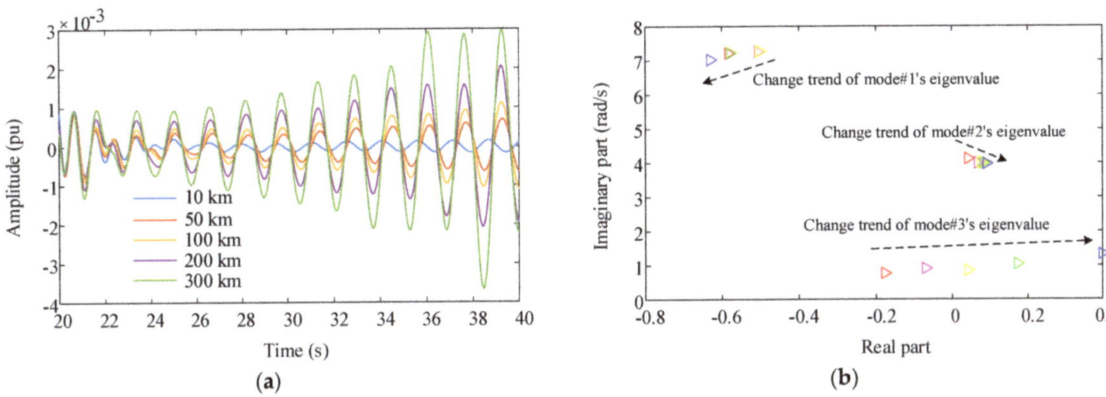

Figure 11. Rotor angular velocity oscillation of G3: (**a**) Oscillation output; (**b**) eigenvalues distribution.

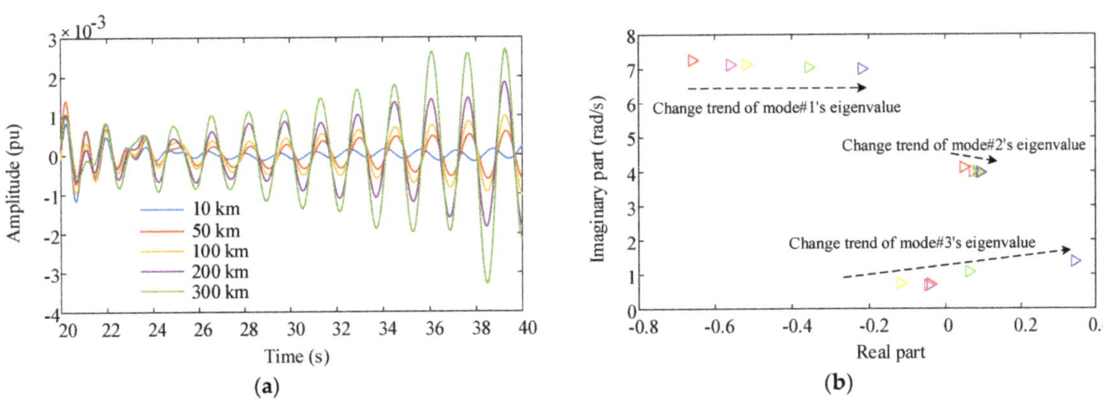

Figure 12. Rotor angular velocity oscillation of G4: (**a**) Oscillation output; (**b**) eigenvalues distribution.

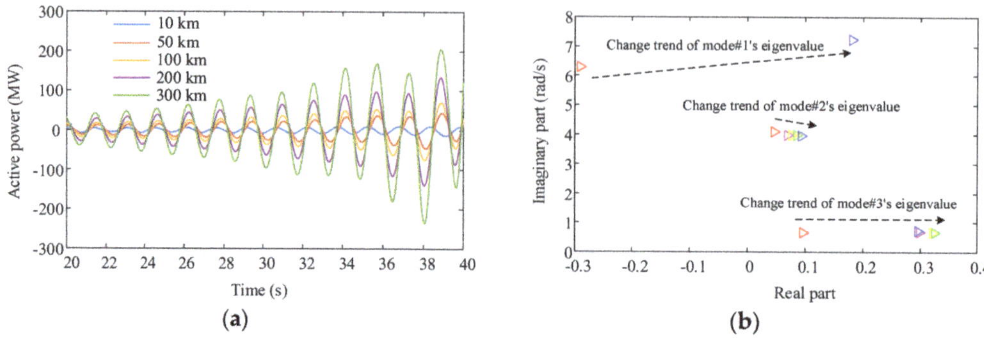

Figure 13. Active power oscillation of the tie line: (**a**) Oscillation output; (**b**) eigenvalues distribution.

According to Figures 9a, 10a, 11a, 12a and 13a, as the connection distance of the wind power increases from 10 km to 300 km; the vibration amplitudes of both intra-area and inter-area oscillations present an increasing trend. In the period of 20 s after the small disturbance occurs, the overall damping ratio of the system decreases. Therefore, the LFO problem worsens, and the system stability weakens.

Figure 9b, Figure 10b, Figure 11b, and Figure 12b describe the eigenvalue distributions of the oscillation modes of the rotor angular velocity of G1–G4 related to the different connection distances of the wind power. Figure 13b describes the eigenvalue distributions of the active power oscillation modes of the tie line with the connection distance changes. It is illustrated that the eigenvalues of each mode generally move to the right in the complex plane with the connection distance increased, which makes the overall system stability worse. Although mode#1′s eigenvalue of G3 moves to the left in the complex plane, the weak and negative damping characteristics of the other two dominant modes determine the increasing amplitude of the intra-area LFOs. On the other hand, it is shown that the inter-area LFOs increase significantly with the connection distance increased in Figure 13.

In addition, it can be seen from Figure 14 that the damping ratio of the dominant mode of the intra-area oscillation is little changed with the increase in distance. With the change of distance from 10 km to 300 km, the damping ratios of the active power of the tie line are −0.0114, −0.0174, −0.0194, −0.0215, and −0.0236, successively, showing an obvious decrease. The damping characteristic of the inter-area oscillation obviously deteriorates, and the system stability becomes worse.

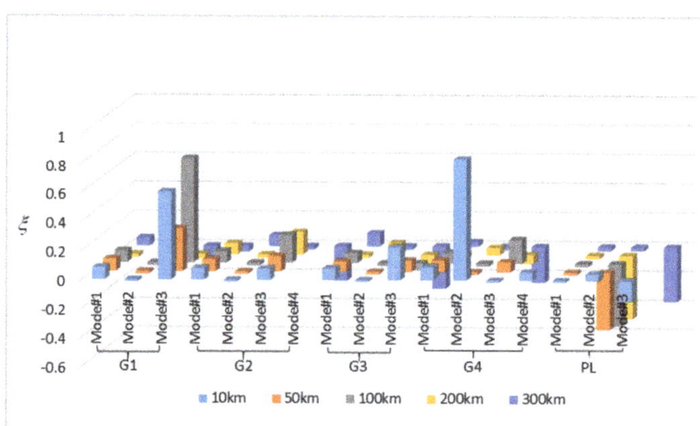

Figure 14. Changes of damping ratio for a different connection distance.

3.3. The Impact of the Capacity of the Wind Power on the System LFOs

The wind farm with adjusted capacity is integrated into the IEEE two-area four-machine power systems, as shown in Figure 15. The wind farm is equivalent to a DFIG whose capacity can be set at a different value. The equivalent DFIG is connected to bus node 7 in area A1 of the power systems. The connection distance l_W is fixed at 15 km, which benefits the wind power transmission. In order to study the impact of the capacity of the integrated wind power on the LFOs of the power systems, the capacity of the wind power is set at different values. The wind power output P_W is adjusted by the number of DFIGs of 2 MW, and the other parameters of the system were the same as those in Figure 6 in Section 3.2.

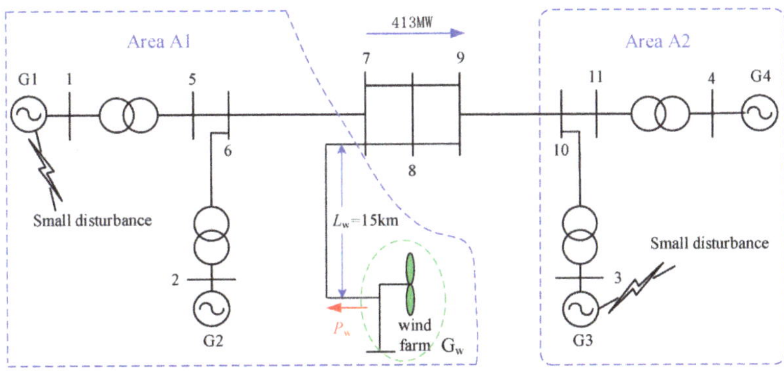

Figure 15. Schematic diagram of the adjustable capacity of integrated wind power.

The capacity of the integrated wind power is set as 10 MW, 50 MW, 100 MW, 300 MW, and 500 MW, respectively. The impact of different capacities on the oscillation modes of the intra-area and inter-area LFOs is tested. The oscillation modes of the rotor angular velocity of G1–G4 and the active power of the tie line are tested and shown in Tables 9–13. Figures 16a, 17a, 18a, 19a and 20a show that the amplitude of both the intra-area oscillations and the inter-area oscillations presents a decreasing trend with the capacity of the wind power increased from 10 MW to 500 MW. Figures 16b, 17b, 18b, 19b and 20b describe the eigenvalues of the oscillation modes of the rotor angular velocity of G1–G4 and the active power of the tie line with the changes of the capacity of the integrated wind power.

Table 9. P_W = 10 MW.

Object	Mode	σ	ω (rad/s)	ξ
G1	Mode #1	−0.5686	±7.0010	0.0810
	Mode #2	0.0968	±4.1159	−0.0235
	Mode #3	0.0681	±0.6361	−0.1065
G2	Mode #1	−0.5362	±6.9257	0.0772
	Mode #2	0.0978	±4.1147	−0.0238
	Mode #3	−0.0478	±0.5757	0.0827
G3	Mode #1	−0.5566	±7.2391	0.0767
	Mode #2	0.0991	±4.1140	−0.0241
	Mode #3	−0.0799	±0.5212	0.1515
G4	Mode #1	−0.5426	±7.1518	0.0757
	Mode #2	0.1018	±4.1155	−0.0247
	Mode #3	−0.0372	±0.4784	0.0775
P_L	Mode #1	0.0988	±4.1140	−0.0240
	Mode #2	0.1435	±0.6649	−0.2110

Table 10. P_W = 50 MW.

Object	Mode	σ	ω (rad/s)	ξ
G1	Mode #1	−0.5867	±7.0125	0.0834
	Mode #2	0.0464	±4.0702	−0.0114
	Mode #3	−0.0257	±0.7032	0.0365
G2	Mode #1	−0.5559	±6.9174	0.0801
	Mode #2	0.0438	±4.0686	−0.0108
	Mode #3	−0.0484	±0.6640	0.0727
G3	Mode #1	−0.5834	±7.2156	0.0806
	Mode #2	0.0430	±4.0650	−0.0106
	Mode #3	−0.1244	±0.7685	0.1598
G4	Mode #1	−0.6607	±7.2553	0.0907
	Mode #2	0.0451	±4.0676	−0.0111
	Mode #3	−0.0244	±0.7069	0.0345
P_L	Mode #1	0.0436	±4.0635	−0.0107
	Mode #2	0.0303	±1.1071	−0.0274

Table 11. P_W = 100 MW.

Object	Mode	σ	ω (rad/s)	ξ
G1	Mode #1	−0.6124	±6.9808	0.0874
	Mode #2	0.0349	±3.9510	−0.0088
	Mode #3	−0.0466	±0.7909	0.0588
G2	Mode #1	−0.5914	±6.8905	0.0855
	Mode #2	0.0320	±3.9483	−0.0081
	Mode #3	−0.0318	±0.6593	0.0482
G3	Mode #1	−0.6149	±7.1886	0.0852
	Mode #2	0.0341	±3.9480	−0.0086
	Mode #3	−0.0616	±0.6992	0.0878
G4	Mode #1	−0.5889	±7.0858	0.0828
	Mode #2	0.0399	±3.9481	−0.0101
	Mode #3	−0.0265	±0.6084	0.0435
P_L	Mode #1	0.0339	±3.9422	−0.0086
	Mode #2	0.0674	±1.2951	−0.0520

Table 12. P_W = 300 MW.

Object	Mode	σ	ω (rad/s)	ξ
G1	Mode #1	−0.7655	±6.8725	0.1107
	Mode #2	−0.0346	±3.6981	0.0094
	Mode #3	−0.0061	±0.6995	0.0087
G2	Mode #1	−0.7323	±6.7197	0.1083
	Mode #2	−0.0253	±3.6977	0.0068
	Mode #3	−0.0282	±0.6665	0.0423
G3	Mode #1	−0.7612	±7.0388	0.1075
	Mode #2	−0.0295	±3.7020	0.0080
	Mode #3	−0.1315	±0.7283	0.1777
G4	Mode #1	−0.7511	±6.8930	0.1083
	Mode #2	−0.0321	±3.7044	0.0087
	Mode #3	−0.0316	±0.7136	0.0442
P_L	Mode #1	−0.0250	±3.6985	0.0068
	Mode #2	−0.0864	±1.5045	0.0573

The following are the discussions on Figures 16–20:

According to Figures 16a, 17a, 18a, 19a and 20a, in the period of 20 s after the small disturbance occurs, the overall damping ratio of the system increases. Therefore, the LFOs have been alleviated, and the power system stability is enhanced. In addition, from the point of view of the eigenvalue change shown in Figures 16b, 17b, 18b, 19b and 20b, the eigenvalue of each mode generally moves to the left in the complex plane. It means that the system's stability is enhanced. The imaginary part of the modal eigenvalues shows a downward trend as a whole. Therefore, the dominant mode frequency of the local oscillation and the interval oscillation decreases with the increase in the wind power output. It is worthy of attention that the damping of the dominant modes of the active power of the tie line increases significantly with the increase in the wind power output capacity. The larger capacity of the integrated wind power suppresses the LFOs better; the damping characteristics of the system are improved, and the system stability is obviously enhanced.

Table 13. P_w = 500 MW.

Object	Mode	σ	ω (rad/s)	ξ
G1	Mode #1	−0.9017	±6.6822	0.1337
	Mode #2	−0.0750	±3.3351	0.0225
	Mode #3	−0.0097	±0.6907	0.0140
G2	Mode #1	−0.8103	±6.4869	0.1240
	Mode #2	−0.0950	±3.3221	0.0286
	Mode #3	−0.0361	±0.6752	0.0534
G3	Mode #1	−0.7154	±6.9459	0.1025
	Mode #2	−0.0820	±3.3208	0.0247
	Mode #3	−0.0320	±0.7344	0.0435
G4	Mode #1	−0.8395	±6.6771	0.1247
	Mode #2	−0.0685	±3.3253	0.0206
	Mode #3	−0.0496	±0.6883	0.0719
P_L	Mode #1	−0.0776	±3.3239	0.0233
	Mode #2	−0.1143	±0.7413	0.1524

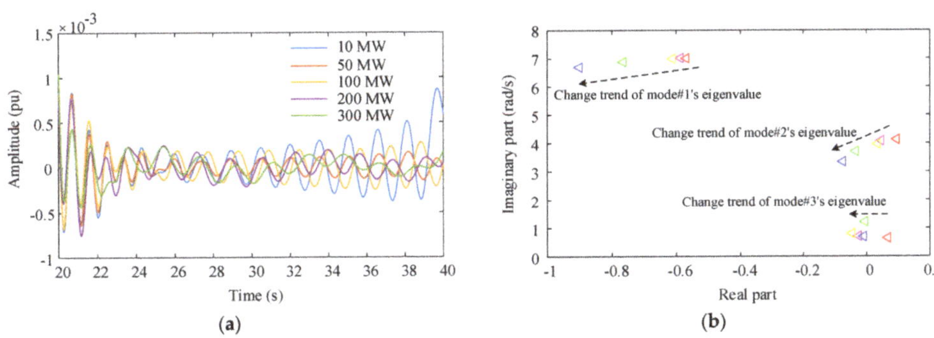

Figure 16. Rotor angular velocity oscillation of G1: (**a**) Oscillation output; (**b**) eigenvalues distribution.

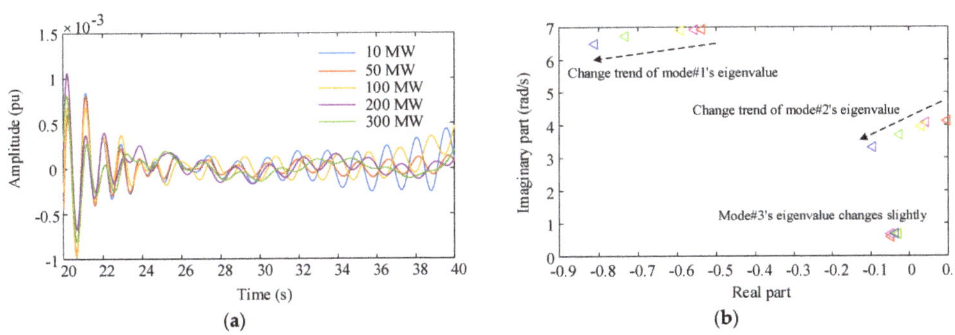

Figure 17. Rotor angular velocity oscillation of G2: (**a**) Oscillation output; (**b**) eigenvalues distribution.

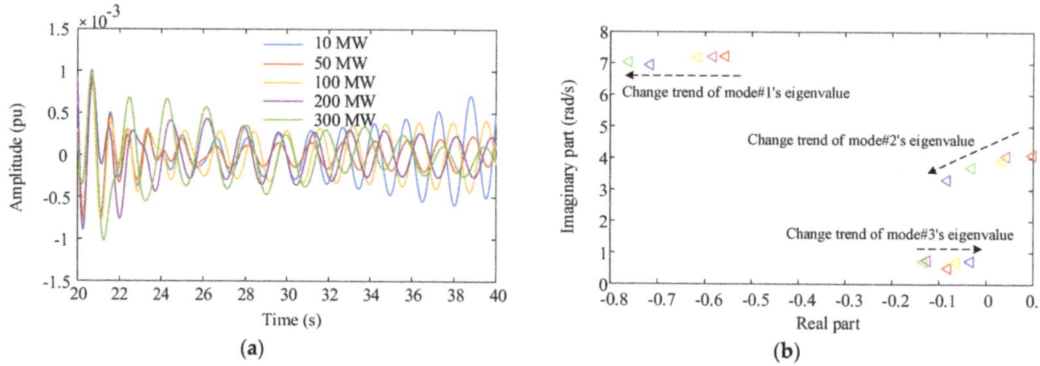

Figure 18. Rotor angular velocity oscillation of G3: (**a**) Oscillation output; (**b**) eigenvalues distribution.

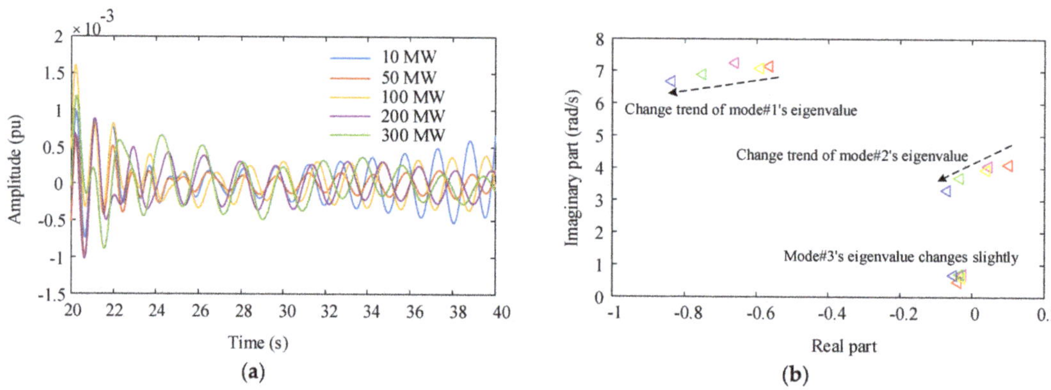

Figure 19. Rotor angular velocity oscillation of G4: (**a**) Oscillation output; (**b**) eigenvalues distribution.

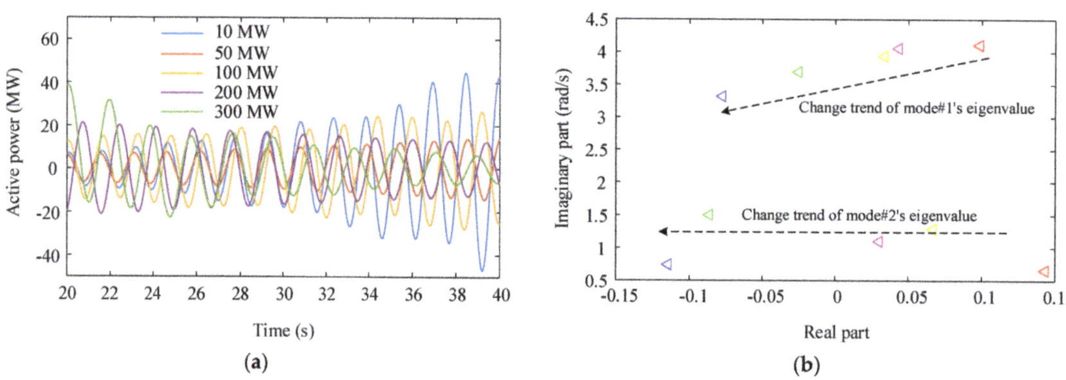

Figure 20. Active power oscillation of the tie line: (**a**) Oscillation output; (**b**) eigenvalues distribution.

As shown in Figure 21, the damping ratio of the dominant mode of the intra-area oscillation (Mode#1) shows an increasing trend, and the damping ratio of the dominant mode of the regional oscillation also shows an increasing trend. The maximum damping ratio will reach 0.1337 and 0.0233, respectively, and the stability of the system will be improved.

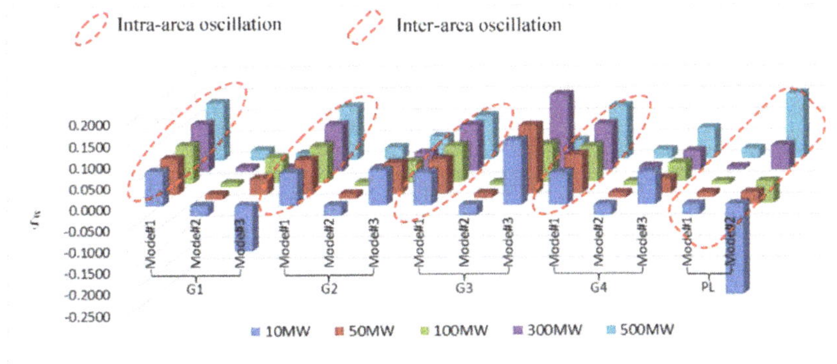

Figure 21. Changes of damping ratio for different output capacities.

4. Conclusions

This paper conducts a detailed theoretical derivation and analysis of the damping characteristics of the two-area interconnected power system. The mathematical relations are used to analyze the changes in the damping characteristics caused by wind power integration into the two-area interconnected system. The LFO model of the IEEE two-area four-machine power system is selected as an example to analyze the impact of wind power integration on the damping characteristics of the power system. The results are summarized as follows:

(1) The wind power consisted of DFIGs which are integrated into the PFA to suppress the system's LFOs. In contrast, the integration to the PRA increases the LFOs slightly.

(2) With the increase in the connection distance of the wind power, the amplitudes of both the intra-area and the inter-area oscillations tend to increase. The system's LFOs increase, and its stability weakens.

(3) As the capacity of the integrated wind power increases, the amplitudes of the intra-area and inter-area oscillations tend to decrease, and the damping ratio of the system increases as a whole. The stability of the system is enhanced, and the LFO problem is alleviated to some extent.

As is known from the analysis of the above results, wind power integration not only changes the structure of the power systems, but also alters the damping characteristics of the large-scale interconnected systems. Not only should the design and construction of the wind farms be optimized, but the appropriate grid parameters should also be constructed to improve the stability of the power system.

Author Contributions: Conceptualization, J.C., T.J., M.A.M., A.A. and U.D.; methodology, J.C., T.J. and M.A.M.; software, J.C., T.J. and M.A.M.; validation, J.C., T.J., M.A.M., A.A. and U.D.; formal analysis, J.C., T.J. and M.A.M.; investigation, J.C., T.J., M.A.M., A.A. and U.D.; methodology, J.C., T.J. and M.A.M.; resources, J.C., T.J. and M.A.M.; data curation, J.C., T.J., M.A.M., A.A. and U.D.; writing—original draft preparation, J.C., T.J. and M.A.M.;writing—review and editing, J.C., T.J., M.A.M., A.A. and U.D.; visualization, J.C., T.J. and M.A.M.d; supervision, T.J. and M.A.M.; project administration, A.A. and U.D.; funding acquisition, A.A. and U.D.; All authors have read and agreed to the published version of the manuscript.

Funding: This research received no external funding.

Institutional Review Board Statement: Not applicable.

Informed Consent Statement: Not applicable.

Data Availability Statement: The data supporting the reported results are available in the manuscript.

Acknowledgments: The authors would like to thank the Estonian Centre of Excellence in Zero Energy and Resource Efficient Smart Buildings and Districts, ZEBE, grant TK146, funded by the European Regional Development Fund to support this research. Furthermore, this work was supported by the scientific research project of Fuzhou Polytechnic under Grant FZYKJRCQD202101.

Conflicts of Interest: The authors declare no conflict of interest.

Nomenclature

Subscripts, Superscripts, and Sets

G_{eq1}, G_{eq2}	Equivalent synchrony generator.
j, k	Serial number of the equivalent synchrony generator.
γ	Mark of $\frac{D_1}{T_{J1}}$.
G1, G2, G3, G4	Synchrony generators of the IEEE two-area four-machinepower systems.
Mode#1, Mode#2, Mode#3, Mode#4	Oscillation modes of the rotor angular velocity of the synchrony generators G1–G4 and the active power oscillation modes of the tie line from node 7 to node 9 in the IEEE two-area four-machine power systems.

Variables and Constants

$\Delta\delta_j$	Power angle increment of the jth synchrony generator.
$\Delta\omega_j$	Rotor angular velocity increment of the jth synchrony generator.
T_{Jj}	Inertia time constant of the synchrony generator j.
ΔP_{je}	Electromagnetic power increment of the synchrony generator j.
D_j	Damping torque coefficient of the synchrony generator j.
K_{jk}	Synchronous torque coefficient between the synchrony generator j and k, $K_{jk} = \partial P_{je}/\partial\delta_k$.
A_{eq}	State matrix of the interconnected systems with two equivalent synchrony generators.
E_1, E_2	No-load electromotive force of the synchrony generator G_{eq1} and G_{eq2}.
U_w	Stable voltage on the integrated node of the wind power, which is assumed to be constant.
Z_{11}, Z_{22}	System impedance of G_{eq1} and G_{eq2} acting as source, respectively.
Z_{12}, Z_{21}	Transfer impedance between G_{eq1} and G_{eq2}.
φ_{11}, φ_{12}, φ_{21}, φ_{22}	Impedance angle of Z_{11}, Z_{12}, Z_{21}, Z_{22}, respectively.
α_{11}, α_{12}, α_{21}, α_{22}	Impedance complementary angle of Z_{11}, Z_{12}, Z_{21}, Z_{22}, respectively.
δ_{10}, δ_{20}	Rotor angle of G_{eq1} and G_{eq2}, respectively.
δ_{12}, δ_{21}	Rotor angle difference between G_{eq1} and G_{eq2}, $\delta_{12} = -\delta_{21}$.
δ_{120}, δ_{210}	Rotor angle difference between G_{eq1} and G_{eq2} at some operating point, $\delta_{120} = -\delta_{210}$.
δ_{w0}	Power angle of the wind farm as an equivalent DFIG.
$\Delta\delta_w$	Power angle increment of the wind farm as an equivalent DFIG.
X_{13}, X_{23}	Reactance between G_{eq1} and the integrated node of wind farm. Reactance between G_{eq2} and the integrated node of wind farm.
λ	The eigenvalue of the equation $\lvert A_{eq} - \lambda I\rvert = 0$.
I	Identity matrix.
ΔP_w	Increment of active power provided by the wind farm.
Δf_B	Frequency increment of the integrated node of the wind farm.
g_1	System characteristic coefficient of the active power increment of the wind farm with respect to Δf_B, that is, $\Delta P_w = g_1 \cdot \Delta f_B$.
g_2	Proportional coefficient between Δf_B and $\Delta\omega_1$.
g_3	Proportional coefficient between Δf_B and $\Delta\omega_2$.
k_1, k_2	$k_1 = \frac{E_1 U_w}{X_{13}}\cos(\delta_{10} - \delta_{w0})$, $k_2 = \frac{E_2 U_w}{X_{23}}\cos(\delta_{20} - \delta_{w0})$.
k_3, k_4	$k_3 = g_1 \cdot g_2$, $k_4 = g_1 \cdot g_3$.
σ	Real part of a mode eigenvalue or damping factor.
ω	Imaginary part of a mode eigenvalue or oscillation angle frequency.
ξ	Damping ratio, $\xi = \frac{-\sigma}{\sqrt{\sigma^2+\omega^2}}$.

References

1. Tan, H.; Yan, W.; Ren, Z.; Wang, Q.; Mohamed, M.A. A robust dispatch model for integrated electricity and heat networks considering price-based integrated demand response. *Energy* **2022**, *239*, 121875.
2. Chen, J.; Alnowibet, K.; Annuk, A.; Mohamed, M.A. An effective distributed approach based machine learning for energy negotiation in networked microgrids. *Energy Strategy Rev.* **2021**, *38*, 100760.
3. Alnowibet, K.; Annuk, A.; Dampage, U.; Mohamed, M.A. Effective Energy Management via False Data Detection Scheme for the Interconnected Smart Energy Hub–Microgrid System under Stochastic Framework. *Sustainability* **2021**, *13*, 11836.
4. Mohamed, M.A.; Mirjalili, S.; Dampage, U.; Salmen, S.H.; Obaid, S.A.; Annuk, A. A cost-efficient-based cooperative allocation of mining devices and renewable resources enhancing blockchain architecture. *Sustainability* **2021**, *13*, 10382.
5. Du, W.; Bi, J.; Wang, H.; Yi, J. Inter-area low-frequency power system oscillations caused by open-loop modal resonance. *IET Gener. Transm. Distrib.* **2018**, *12*, 4249–4259.
6. Shen, C.; An, Z.; Dai, X.; Wei, W.; Ding, L. Measurement-based solution for low frequency oscillation analysis. *J. Mod. Power Syst. Clean Energy* **2016**, *4*, 406–413.
7. Kosterev, D.N.; Taylor, C.W.; Mittelstadt, W.A. Model validation for the August 10, 1996 WSCC system outage. *IEEE Trans. Power Syst.* **1999**, *14*, 967–979.
8. Fairley, P. The unruly power grid. *IEEE Spectr.* **2004**, *41*, 22–27.
9. Report of the Enquiry Committeeon Grid Disturbance in Northern Region on 30th July 2012 India Northern, Eastern & North-Eastern Region on 31 July 2012. New Dehli, India, 2012. Available online: http://nrpc.gov.in/wp-content/uploads/2017/11/grid_disturbance_report.pdf (accessed on 13 February 2022).
10. Chen, J.; Mohamed, M.A.; Dampage, U.; Rezaei, M.; Salmen, S.H.; Obaid, S.A.; Annuk, A. A multi-layer security scheme for mitigating smart grid vulnerability against faults and cyber-attacks. *Appl. Sci.* **2021**, *11*, 9972.
11. Rezaei, M.; Alharbi, S.A.; Razmjoo, A.; Mohamed, M.A. Accurate location planning for a wind-powered hydrogen refueling station: Fuzzy VIKOR method. *Int. J. Hydrogen Energy* **2021**, *46*, 33360–33374.
12. Slootweg, J.G.; Kling, W.L. The impact of large scale wind power generation on power system oscillations. *Electr. Power Syst. Res.* **2003**, *67*, 9–20.
13. Sanchez-Gasca, J.J.; Miller, N.W.; Price, W.W. A modal analysis of a two-area system with significant wind power penetration. In Proceedings of the IEEE PES Power Systems Conference and Exposition, New York, NY, USA, 10–13 October 2004; Volume 2, pp. 1148–1152.
14. Hagstrøm, E.; Norheim, I.; Uhlen, K. Large-scale wind power integration in Norway and impact on damping in the Nordic grid. *Wind Energy* **2005**, *8*, 375–384.
15. Wu, F.; Zhang, X.P.; Godfrey, K.; Ju, P. Small signal stability analysis and optimal control of a wind turbine with doubly fed induction generator. *IET Gener. Transm. Distrib.* **2007**, *1*, 751–760.
16. Tsourakis, G.; Nomikos, B.M.; Vournas, C.D. Effect of wind parks with doubly fed asynchronous generators on small-signal stability. *Electr. Power Syst. Res.* **2009**, *79*, 190–200.
17. Yang, L.; Xu, Z.; Østergaard, J.; Dong, Z.Y.; Wong, K.P.; Ma, X. Oscillatory Stability and Eigenvalue Sensitivity Analysis of A DFIG Wind Turbine System. *IEEE Trans. Energy Convers.* **2011**, *26*, 328–339.
18. Li, H.; Liu, S.; Ji, H.; Yang, D.; Yang, C.; Chen, H.; Zhao, B.; Hu, Y.; Chen, Z. Damping control strategies of inter-area low-frequency oscillation for DFIG-based wind farms integrated into a power system. *Int. J. Electr. Power Energy Syst.* **2014**, *61*, 279–287.
19. Singh, M.; Allen, A.J.; Muljadi, E.; Gevorgian, V.; Zhang, Y.; Santoso, S. Interarea Oscillation Damping Controls for Wind Power Plants. *IEEE Trans. Sustain. Energy* **2015**, *6*, 967–975.
20. Effatnejad, R.; Zare, A.; Choopani, K.; Effatnejad, M. DFIG-based damping controller design to damp low frequency oscillations in power plant industry. In Proceedings of the 2016 International Conference on Industrial Informatics and Computer Systems (CIICS), Sharjah, United Arab Emirates, 13–15 March 2016.
21. Du, W.; Bi, J.; Cao, J.; Wang, H.F. A Method to Examine the Impact of Grid Connection of the DFIGs on Power System Electromechanical Oscillation Modes. *IEEE Trans. Power Syst.* **2016**, *31*, 3775–3784.
22. Li, C.; Zhang, W.; Liu, R. Forced low frequency oscillation of wind-integrated power systems. In Proceedings of the 2016 IEEE Power & Energy Society Innovative Smart Grid Technologies Conference (ISGT), Minneapolis, MN, USA, 6–9 September 2016.
23. Zhang, X.; Fu, Y.; Wang, S.; Wang, Y. Effects of two-area variable inertia on transient stabilisation in interconnected power system with DFIG-based wind turbines. *IET Renew. Power Gener.* **2017**, *11*, 696–706.
24. Liu, C.; Cai, G.; Ge, W.; Yang, D.; Liu, C.; Sun, Z. Oscillation Analysis and Wide-Area Damping Control of DFIGs for Renewable Energy Power Systems Using Line Modal Potential Energy. *IEEE Trans. Power Syst.* **2018**, *33*, 3460–3471.
25. Simon, L.; Swarup, K.S.; Ravishankar, J. Wide area oscillation damping controller for DFIG using WAMS with delay compensation. *IET Renew. Power Gener.* **2019**, *13*, 128–137.
26. Eshkaftaki, A.A.; Rabiee, A.; Kargar, A.; Boroujeni, S.T. An Applicable Method to Improve Transient and Dynamic Performance of Power System Equipped With DFIG-Based Wind Turbines. *IEEE Trans. Power Syst.* **2020**, *35*, 2351–2361.
27. Kundur, P. *Power System Stability and Control*; McGraw-Hill: New York, NY, USA, 1994.
28. Zhao, S.; Chang, X.; He, R.; Ma, Y. Borrow damping phenomena and negative damping effect of PSS control. *Proc. CSEE* **2004**, *24*, 7–11.

29. Chen, J.; Tao, J.; Mohamed, M.A.; Wang, M. An adaptive TLS-ESPRIT algorithm based on an S-G filter for analysis of low frequency oscillation in wide area measurement systems. *IEEE Access* **2019**, *7*, 47644–47654.
30. Tan, H.; Ren, Z.; Yan, W.; Wang, Q.; Mohamed, M.A. A wind power accommodation capability assessment method for multi-energy microgrids. *IEEE Trans. Sustain. Energy* **2021**, *12*, 2482–2492.

sustainability

Article

Management and Policy Modeling of the Market Using Artificial Intelligence

Qunpeng Fan

School of Management, Jilin Normal University, Siping 136000, China; qunpengfan@sina.com

Abstract: This paper investigates the market management and modeling based on advanced artificial intelligence. The proposed model deploys the combination of the support vector machine and fuzzy set theory to provide a practical and powerful prediction model for the market price over the next day. A realistic and effective model is then introduced to model the market players, such as the renewable energy sources of solar and wind turbines, as well as the fossil-fueled sources of micro turbines and fuel cells. In order to provide an optimal management program, it introduces a stochastic framework based on the point estimate method and adaptive grey wolf optimization algorithm (GWO). The proposed optimization methods use an adaptive strategy to choose the most fitting modification for enhancing the GWO performance. A realistic scenario is simulated to demonstrate the model's effectiveness and impression on the real market management. The results clearly show the effectiveness of the prediction and management model. The prediction results show the superiority of the proposed model by RMSE of 2.9643 compared to the 3.217 for SVR, 3.2364 for ANN and 3.0621 for the grey model. Moreover, the optimal MAPE is 2.7453 by the proposed method, which is much better than the 3.052 by SVR, 3.1552 by ANN and 2.9285 by the grey model. From point of view of optimization, the most fitting power dispatch has been attained with the total cost of 300.8632 over 24 h.

Keywords: market price prediction; modeling and management; optimization; renewable energy sources

1. Introduction

The introduction of the electricity and energy, also called power, market is a system which makes power purchasing and sales possible through bids to offer [1,2]. In this way, the demand and supply regulations are deployed for clearing the market and setting the price. The market price idea would provide a competitive environment in which electricity consumers can play a more active role in deciding what to buy and how much to pay [3–5]. Therefore, it is clear that the price of energy would be a motivating term that would increase demand if preserved low enough [6,7]. It is proved that the power market can support the social welfare of the system by decreasing the power price and letting the cheap generation players produce more power than the expensive units [8–10]. Therefore, the analysis and modeling of the market is a very critical and significant task, which requires special work and focus to enhance it [11]. Figure 1 shows the way that the market is cleared and its point of common, called MCP, is determined as a tradeoff between demand and supply. One of the most significant issues for the consumer is the estimation or prediction of the next day's market price for making clear policies and strategies. In recent years, much research has been implemented, a list of which is described in the rest of the introduction.

In [12], the market price is forecasted using a hybrid generative adversarial network model and the results are compared with the some of the most well-known algorithms in the field. It is shown in [13] that the market price and market demand both have high nonlinearity and non-stationary characteristics which, if combined with seasonality, would create a big problem for accurate prediction. In [14], a heuristic flower pollination approach is introduced in conjunction with machine learning to boost the prediction of the market

Citation: Fan, Q. Management and Policy Modeling of the Market Using Artificial Intelligence. *Sustainability* **2022**, *14*, 8503. https://doi.org/10.3390/su14148503

Academic Editor: Mohamed A. Mohamed

Received: 7 May 2022
Accepted: 23 June 2022
Published: 11 July 2022

price accuracy. The proposed model considers the feature selection based on a hybrid regression model. In [15], authors assess the impact of renewable units on the market price and forecasting accuracy. A novel bootstrap gradient-based architecture is deployed for the short-term market price forecasting. In [16], a two-step market price forecasting tool is developed which uses deep learning with a loss function. The loss function would penalize any loss of spike in the price and thus some oversampling might be seen. Through the spike calibration approach, the prediction accuracy is improved in the market. The Brazil market price during the years 1998 to 2004 is assessed in [17]. It proposes a theoretical formulation based on Markov, switching to predict the market price and showing how a generator entry might be affected. In [18], the generator output power is forecasted to be an indirect way for estimating the market price. It uses machine learning models and probability sampling for analyzing the Italian electricity market. It is claimed that the proposed method does need to access all market transactions. In [19], the market price over the years January 2017–March 2020 are analyzed to forecast the price. It is shown in their research that the dominant retailers can play a critical role in controling the market price and its growth over the year. In [20], a hybrid Bayesian extreme learning and minimum redundancy are combined to forecast the market price based on the sequence segmentation. Two market price datasets from Ontario, Canada and New York are used to check the accuracy of the proposed model hourly. The authors in [21] try to address the weather effects on market behavior, including on demand and supply, using the fractional Brownian motion. A temperature based stochastic prediction model is developed which can help much to control the market fluctuations.

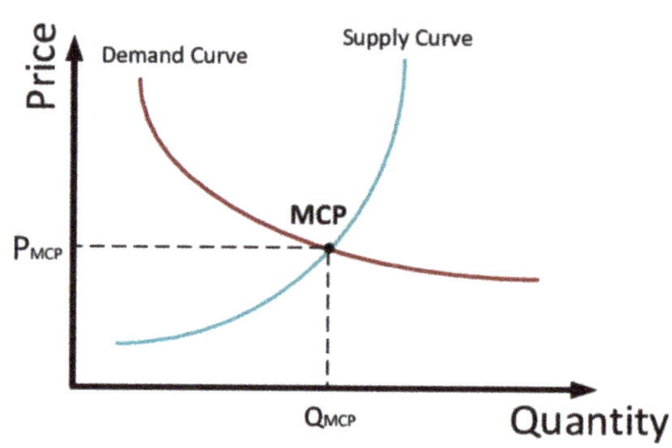

Figure 1. The market clearing price and the way it is determined.

According to the above explanations, it is well perceived that market modeling and analysis, especially prediction of the market price, are significant tasks which can benefit both the customers and the suppliers. This paper tries to provide market management and prediction based on artificial intelligence (AI) [22]. The proposed AI model uses support vector regression (SVR) [23] to avoid the possible overfitting concerns. It also suggests the fuzzy set theory to find a way for adjusting the kernel function parameters. In addition, the market price is then used as an input to a smart market, including several renewable or fossil fuel-based generators. An objective function is introduced to manage the power suppliers and minimize the cost of generation. Considering the very complex and nonlinear fluctuations in the market price, we need to make use of a novel smart optimization algorithm based on grey wolf optimization (GWO). It is a heuristic optimization method which has shown great abilities over the other algorithms in the literature [24,25]. Therefore, the main contributions of the paper can be shown as below:

- Proposing a fuzzy SVR-based prediction model for the market price
- Introducing a novel optimization algorithm called modified GWO for the optimal scheduling of units in an economic way

- Providing a comprehensive comparison with the most well-known algorithms in the area such ANN, Grey model, SVR, GA, PSO and original GWO

The analysis is implemented in a physical scenario to determine the effectiveness and influence of the proposed model. The simulation results obviously advocate the capabilities of the proposed prediction and management models.

The paper follows a structure organized as follows: Section 2 describes the fuzzy SVR for prediction of the market price. Section 3 describes the market model based on renewable sources and players. In Section 3, the GWO algorithm is explained in detail. Simulation results are then discussed in Section 4. At the end, the main conclusions are summarized.

2. Fuzzy Support Vector Regression

This section proposes a fuzzy support vector machine approach to forecast the electricity market price over the next day. The original idea of support vector machine is rooted in the nonlinear intrinsic structure of the dataset, which forces us to provide a tool for handling this nonlinearity. In this way, a support vector is able to assign a linear function on the nonlinear problem in a higher dimension. To describes this method, the sample points are shown by $\{(x_i, y_i)\}^M$, wherein M represents the size of the sample. In order to make a nonlinear mapping in the higher dimension, $\Lambda(.)$ is considered to form the input limited dimension to the output space with high dimension as $\Lambda(.) : \Re^n \rightarrow \Re^{n_h}$. This is the core idea of the SVR, which has made it a popular and successful method in machine learning. Figure 2 shows this idea from 2D to 3D conversion. As shown in Figure 2, a nonlinear relationship which is not detectable in 2D can be easily separated by a linear function in a 3D space. The linear function, also called the SVR function, may be formulated as below:

$$f(x) = W^T \Lambda(x) + b \tag{1}$$

In the above equation, the symbols W ($W \in \Re^{n_h}$) and b ($b \in \Re$) are the model coefficients, which, if adjusted well, make the final model trustable. The training process is through optimizing the error between the real data $W^T \Lambda(x) + b$ and the simulated data y as below:

$$R_{SVR} = \frac{1}{M} \sum_{i=1}^{M} \mathrm{E}_\varepsilon(y_i, W^T \Lambda(x_i) + b) \tag{2}$$

In this equation, $\mathrm{E}_\varepsilon(y_i, f(x_i))$ plays the role of the ε-insensitive loss function that is interpreted as below:

$$\mathrm{E}_\varepsilon(y, f(x)) = \begin{cases} |f(x) - y| - \varepsilon; & |f(x) - y| \geq \varepsilon \\ 0; & \text{Else} \end{cases} \tag{3}$$

Other than the training error, the SVR minimizes the structural complexity, which is shown by multiplication of weighting factors, as below:

$$Min_{W,b,\xi^*,\xi} R_\varepsilon(W, \xi^*, \xi) = \frac{1}{2} W^T W + C \sum_{i=1}^{M} (\xi_i^* + \xi_i) \tag{4}$$

The first item in this formulation represents the model structure's complexity, which, if it becomes too high, the overfitting problem appears. In addition, the parameter C is the constant value employed in the model structure for balancing between the error and complexity. The variables ξ_i^* / ξ_i are errors below or above the allowed values of $-\varepsilon$ and

+ ε, respectively. The dual format of the objective function in the new structure may be represented by (5), which is solvable linearly:

$$\begin{array}{ll} y_i - W^T\Lambda(x_i) - b \leq \varepsilon + \xi_i^*; & i = 1, \ldots, M \\ -y_i + W^T\Lambda(x_i) + b \leq \varepsilon + \xi_i; & i = 1, \ldots, M \\ \xi_i^* \geq 0; & i = 1, \ldots, M \\ \xi_i \geq 0; & i = 1, \ldots, M \end{array} \quad (5)$$

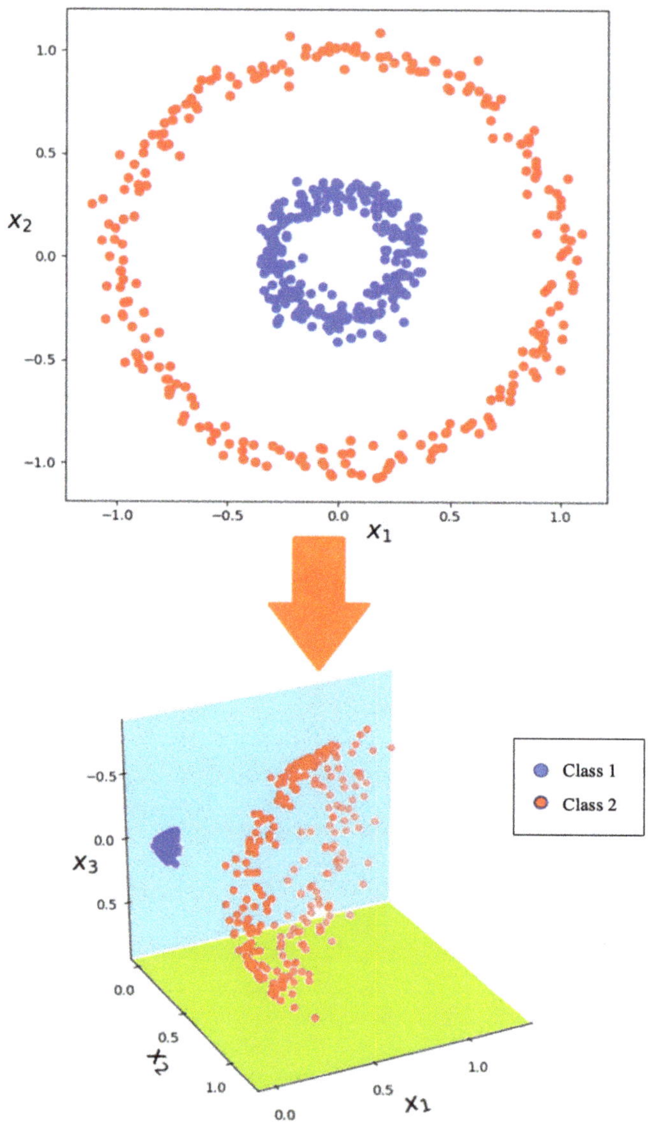

Figure 2. The higher dimension concept in the SVR.

A linear solution of the above equation would easily determine the optimal value of the weight factor, as below:

$$W = \sum_{i=1}^{M} (\beta_i^* - \beta_i)\Lambda(x_i) \tag{6}$$

In this model, the new parameters β_i^* and β_i are called the Lagrangian multipliers. Deploying β_i^* and β_i provides the dual format of the model, as below:

$$f(x) = \sum_{i=1}^{M} (\beta_i^* - \beta_i)K(x_i, x) + b \tag{7}$$

$$K(x_i, x) = \Lambda(x_i)^\circ\Lambda(x_j) \tag{8}$$

In (8), the term $K(x_i, x)$ is named a kernel function. Theoretically, the kernel function meets Mercer's condition [26].

A significant point is that the kernel function can much help the mapping process, due to its role in the connection of the input and output in the dataset. Fundamentally, the kernel function is computed by the inner multiplication of $\varphi(x_i)$ and $\varphi(x_j)$. There are different types of kernels, such as the Gaussian or RBF, which can be used. This article proposes a fuzzy model for the kernel function, which can enhance the model performance by suitably adjusting the fuzzy membership function. Figure 3 shows the trapezoidal fuzzy function considered as the kernel function in our model. The input feature of the function is the same as the kernel function in (7). The way that the output is determined instead can help us much in finding a better nonlinear mapping.

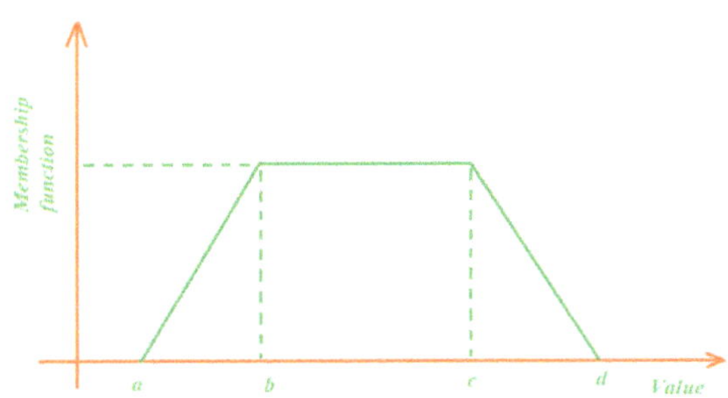

Figure 3. Fuzzy membership function used as the kernel function.

3. Market Economic Problem Formulation

In this section, an economic model is developed for the market players, including the consumers (loads) and generators (renewable and non-renewable sources), to make sure that the analysis can mitigate the total cost which is beneficial to both players. In order to model the market players, we first try to give a cost function which focuses on the cost of generation by the private generators, utility and battery storage. Therefore, we have considered the two renewable sources of wind turbines and solar units, and two other dispatchable units, i.e., micro turbines and fuel cells. The main goal is that the total power generation would cost the customers less based on the market hourly price. Therefore, the cost function is formulated as in (9), where the first term is the generators' power cost, the second term is the cost of On/OFF, the third term is the storage cost and the last term is the utility cost:

$$\begin{aligned} Min\ \ f(X) = \sum_{t=1}^{T} Cost^t = \sum_{t=1}^{T} \{ \sum_{i=1}^{N_g} [u_i(t)\, p_{Gi}(t) B_{Gi}(t)\ +\ \mathrm{D_{Gi}} \times \\ \max(0, u_i(t-1) - u_i(t)) + \sum_{j=1}^{N_s} [u_j(t) p_{sj}(t) B_{sj}(t)] \\ + p_{Grid}(t) B_{Grid}(t) \} \end{aligned} \tag{9}$$

where T shows the time interval length (here 24 h), N_g shows the number of generators, $u_i(t)$ is the status of a generator, $p_{Gi}(t)$ is the generator power output, $B_{Gi}(t)$ is the bidding of the generator, D_{Gi} is the On/Off switching cost, N_s is the number of storages, $u_j(t)$ is storage status, $p_{sj}(t)$ is the storage power value, $B_{sj}(t)$ is the storage power price, $p_{Grid}(t)$ is the main utility power value, and $B_{Grid}(t)$ is the grid price.

The control vector which is used for optimizing (9) contains some terms, as below:

$$\begin{array}{l} X = [P_g, U_g]_{1\times 2nT}; \\ P_g = [P_G, P_s]\ ; n = N_g + N_s + 1 \\ P_G = [P_{G,1}, P_{G,2}, \ldots, P_{G,N_g}]; \\ P_s = [P_{s,1}, P_{s,2}, \ldots, P_{s,N_s}] \\ P_{G,i} = [P_{G,i}(1), P_{G,i}(2), \ldots, P_{G,i}(T)];\ \ i = 1, 2, \ldots, N_g + 1 \\ P_{s,j} = [P_{s,j}(1), P_{s,j}(2), \ldots, P_{s,j}(T)];\ \ j = 1, 2, \ldots, N_s \\ U_g = [u_1, u_2, \ldots, u_n], \\ u_k \in \{0, 1\} \\ u_k = [u_k(1), u_k(2), \ldots, u_k(T)];\ \ k = 1, 2, \ldots, n \end{array} \tag{10}$$

where P_G is the generator output power and P_s is the storage output power.

The above cost function is optimized by meeting some operation limits which are explained in detail. Each of the limits is explained here:

(1) the total demand should be provided by the generators and storage, and the utility as active suppliers, to clear the market, as below:

$$\sum_{i=1}^{N_g} P_{G,i}(t) + \sum_{j=1}^{Ns} P_{s,j}(t) + P_{Grid}(t) = \sum_{l=1}^{N_L} P_{L,l}(t) \tag{11}$$

In the above, $P_{L,l}(t)$ is the load demand at time t and N_L is the number of load levels in the system.

(2) each dispatchable generator or unit is allowed to produce in its limited range, as below:

$$\begin{array}{l} P_{Gi,\min}(t) \le P_{Gi}(t) \le P_{Gi,\max}(t) \\ P_{grid,\min}(t) \le P_{Grid}(t) \le P_{grid,\max}(t) \\ P_{sj,\min}(t) \le P_{sj}(t) \le P_{sj,\max}(t) \end{array} \tag{12}$$

In (12), the min/max indices technically show the least and highest possible values.

(3) The battery storage can only charge/discharge according to its limited charging and discharging rate, as well as the energy storage limit:

$$\Delta_{ess}(t) = \Delta_{ess}(t-1) + \eta_{charge} P_{charge} - \frac{1}{\eta_{discharge}} P_{discharge} \tag{13}$$

$$\begin{cases} \Delta_{ess,\min} \le \Delta_{ess}(t) \le \Delta_{ess,\max} \\ P_{charge}(t) \le P_{charge,\max} \\ P_{discharge}(t) \le P_{discharge,\max} \end{cases} \tag{14}$$

where $\Delta_{ess}(t)$ is the energy stored at time t, η_{charge} is the charging efficiency, $\eta_{discharge}$ is the discharging efficiency, P_{charge} is the charging power and $P_{discharge}$ is the discharging value.

4. Adaptive Grey Wolf Optimization Method

In order to solve the problem formulation described in the last section, we need to make use of a suitable and reliable optimization algorithm. This paper proposes an adaptive grey wolf optimization (AGWO) to make sure that the optimization success is guaranteed. The original GWO is a heuristic method which uses the evolutionary habit of wolves in hunting prey. The GWO is a newly introduced algorithm which has some key features which make it a very powerful optimizer for nonlinear problems. Some of the main features of GWO include: being a simple concept, ease of implementation, powerful local search operators, providing an appropriate balance between local and global searches, and its sub-division ability. These animals live in a group and try to make use of the experience of leaders to make the hunt a successful process. Meanwhile, the leaders can consult with each other to obtain the most promising solution. Other than that, the leaders in some circumstances listen to the wolves to make sure that the maximum efficiency is attained. This shows a low level of democracy, which can boost the optimization process. The main leader of the group is called alpha α. This is the wolf with the smartest decisions in hunting. The next dominant wolf is called beta β, which can play the role of a leader for the group in the absence of α. Moreover, in the presence of α, beta can help him through consultation. In each group, the wolf with the least quality is called omega ω, which are assumed to be workers and need to obey the leader. The remaining wolf members in the group are called delta δ. Figure 4 provides a simple diagram showing the position of these members in the GWO.

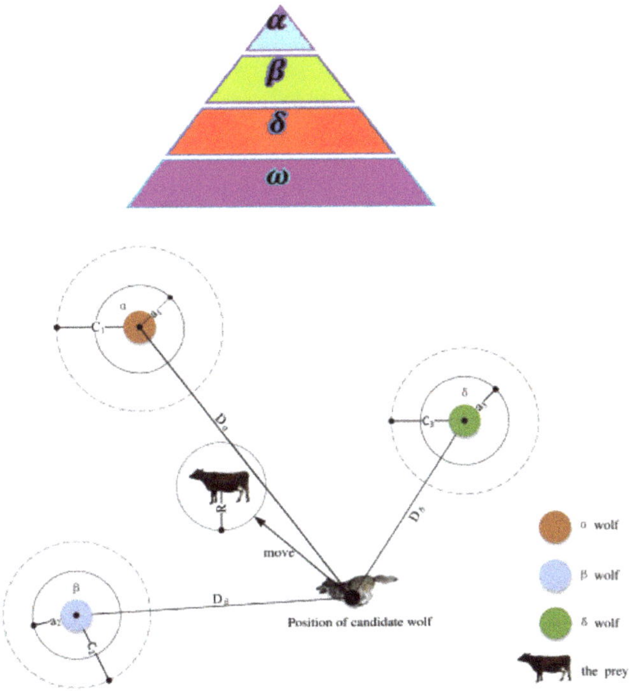

Figure 4. GWO algorithm structure and concept.

In order to mathematically formulate the idea of grey wolf hunting, we first produce the group of wolves, control vector, in a matrix. After calculating the cost objective for them, the population is sorted in a downward order and the alpha, beta, delta and omega members are determined. In the improvization stage, the wolves hunt a prey which is simulated in a circle with dimension:

$$D = \left| \vartheta_1 \times X_p(k) - X(k) \right| \quad (15)$$

where ϑ_1 is a random coefficient, k shows the epoch number and X_p is the prey position. The position of each wolf is updated according to (15) as below:

$$X(k+1) = \frac{1}{3} \sum_{p=\alpha,\beta,\delta} (X_p(k) - \vartheta_2 \times D_p) \tag{16}$$

where ϑ_1 is a random coefficient. The GWO suggests to update the coefficients ϑ_1 and ϑ_2 in a recursive way. Therefore, it suggests a recursive formulation to update it, as below:

$$\vartheta_1 = 2a\lambda_1 - a \tag{17}$$

$$\vartheta_2 = 2a\lambda_2 \tag{18}$$

$$a = 2 - 2(k/\text{kmax}) \tag{19}$$

$$\lambda_1 \& \lambda_2 = rand \tag{20}$$

In the above equations, the parameter a is a decreasing value in the domain (0,2). More explanations on the original GWO can be found in [19,20].

This paper proposes an adaptive formulation to enhance the search process in the GWO. The adaptive mechanism can choose the best modification in an iteration according to its success. First, the modifications are explained:

- Modification 1: This modification uses an exchange mechanism to combine two optimal solutions in the genes (like GA) in each vector randomly. Each grey wolf would attend this operator and exchange data with the leader alpha X_α:

$$\begin{array}{l} x_{i,j}^{new} = \begin{cases} x_{\alpha,j}; \ \vartheta_3 < \vartheta_4 \\ x_{i,j}^{rand}; \ \vartheta_3 > \vartheta_4 \end{cases} \\ X_\alpha = [x_{\alpha,1}, \ldots, x_{\alpha,d}] \\ X_i^{rand} = [x_{i,1}^{rand}, \ldots, x_{i,d}^{rand}] \end{array} \tag{21}$$

  where ϑ_3/ϑ_4 are random variables in the range (0,1).
- Modification 2: This modification would try to enhance the mean of the grey wolf population using a motivating operator as below:

$$X_i^{new} = X_i^{old} + L_F(X_\alpha - X_\beta) \tag{22}$$

  where L_F is an integer value of either 1 or 2.

Each grey wolf is allowed to pick one of these two modifications which best fits its situation. The probability of success for each modification is determined by an n index called Prb_θ = 0.5 & θ = 1,2. Moreover, a storage is considered whose value is zero at the beginning, but as time passes, it is updated as below:

$$Acum_\theta = Acum_\theta + \frac{\kappa_l}{n_{Mod_\theta}} l = 1, \ldots, n_{Mod_\theta} \tag{23}$$

$$\kappa_j = \frac{Log(N-j+1)}{\sum\limits_{i=1}^{n} Log(i)}; \ j = 1, \ldots, N \tag{24}$$

In the above, κ_j is the weight each solution X_j and N is the size of the population. The probability of any modification method is updated as below:

$$Prb_\theta = (1-\varepsilon) \times Prb_\theta + \varepsilon \times \frac{Acum_\theta}{k}, \ (\theta = 1,2) \tag{25}$$

where ε is a constant value for balancing the two modifications. In the last step, the Prb_θ is normalized as follows:

$$Prb_\theta = \frac{Prb_\theta}{(\sum_{\theta=1}^{2} Prb_\theta)} \tag{26}$$

5. Simulation Results

In this section, a test system is used for examining the proposed model. The test system structure is shown in Figure 5. As is seen here, there is a wind turbine, a fuel cell, a micro turbine and a solar panel installed in the system as the generators. In addition, battery is considered to play the role of an active player as ancillary services. The utility, also as a big supplier, can attend the market and sell its power. The total power generation of the wind and solar units are plotted in Figures 6 and 7, respectively. Furthermore, the market demand is considered as shown in Figure 8. It is worth noting that this demand can also be predicted, but since the paper's focus is on the market price, it is ignored in this research, though the solution procedure is the same.

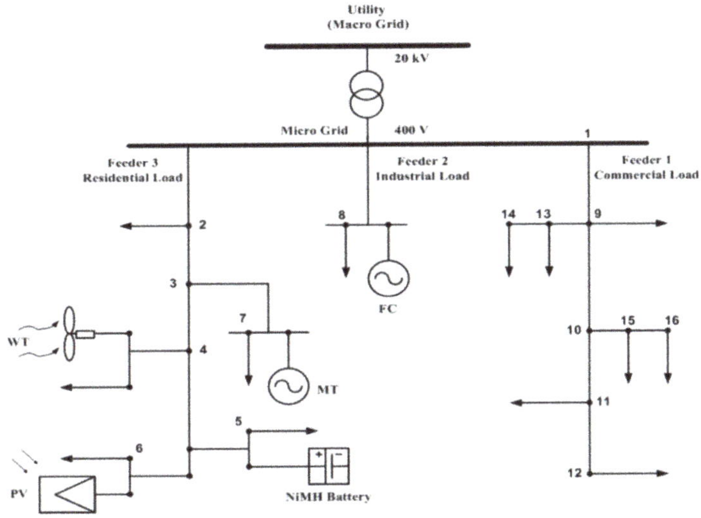

Figure 5. Single line diagram of the MG test system [27].

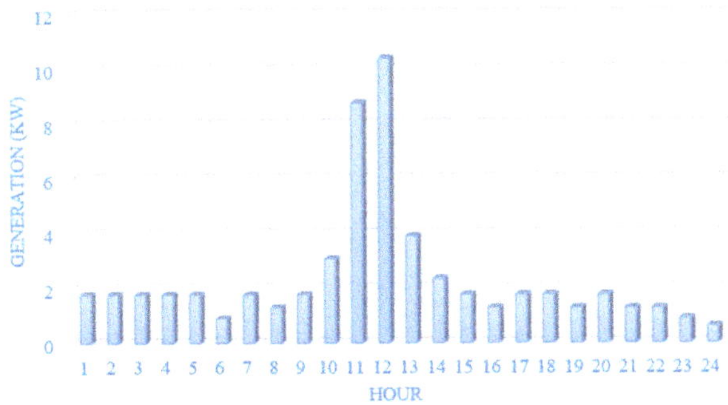

Figure 6. Wind turbine output power.

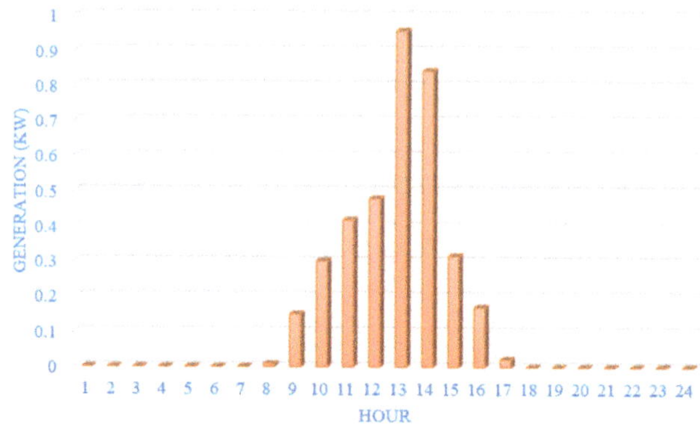

Figure 7. Solar panel output power.

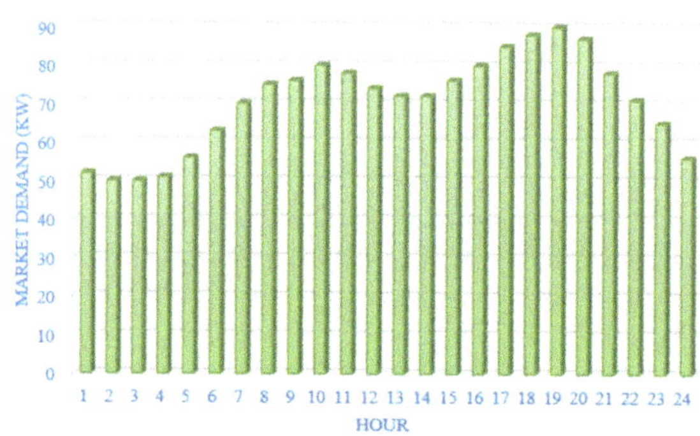

Figure 8. Market demand production.

Table 1 summarizes the test system data, including the capacity and prices of each generator. In order to show first the accuracy of the proposed prediction model for market price forecasting, the historical data over a year of Tehran Electricity market are considered as the benchmark [28]. The recorded data include the hourly market price over 480 h in 2018, from which 90% is used as the training set, 5% as the validation set and 5% as the test set. Therefore, we consider the next 24 h as the prediction time horizon. The comaprisons are made according to MAPE and RMSE, as described in [29]. For the sake of simplicity, the formualtions are not provided here but can be found in [29]. Table 2 shows the prediction results using the proposed fuzzy SVR (FSVR), the original support vector machine, artificial neural network (ANN) and the grey model. As can be seen in this table, the proposed FSVR shows a higher training result, as reflected in the MAPE and RMSE. The proposed model could enhance the RMSE rather than the SVR, ANN and grey model by 7.8%, 8.4% and 9.2%, respectively. A complete comaprison is provided in Figure 9. Almost similar iprovements can be seen for the MAPE. The same conclusion might be made for the RMSE, of which the low value shows the very high robustness. Figure 10 shows the forecast curve plotted in the same figure with the real data. The appropriate accuracy can be deduced from the following nature of the forecast signal that can be seen here.

Table 1. Data of the test system.

Type	Min Power (kW)	Max Power (kW)	Offer (EUR/kWh)	Start-Up/Shut-Down Cost (EURct)
Micro turbine	6	30	0.457	0.96
Fuel cell	3	30	0.294	1.65
Solar unit	0	25	2.584	0
Wind unit	0	15	1.073	0
Battery	−30	30	0.38	0
Utility	−30	30	-	-

Table 2. The results of prediction by the fuzzy SVR for the market price.

Type	Grey Model	ANN	SVR	FSVR
MAPE	2.9285	3.1552	3.052	2.7453
RMSE	3.0621	3.2364	3.217	2.9643

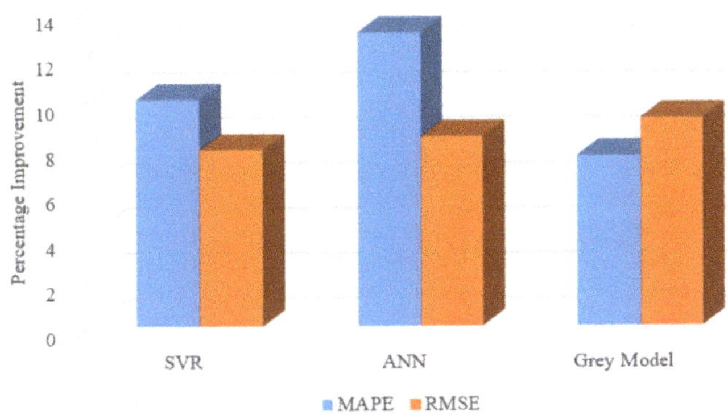

Figure 9. The percentage improvement made by the proposed prediction model compared to SVR, ANN and Grey model.

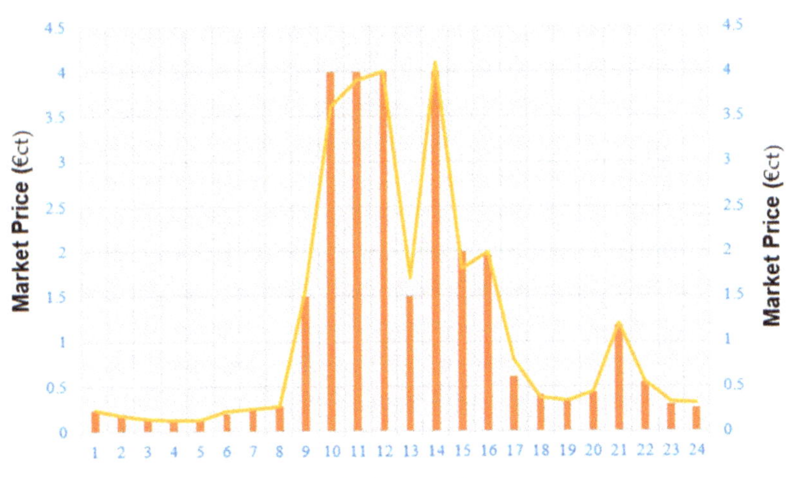

Figure 10. Forecast market price value versus real data.

Considering the market price, as shown in Figure 10, the cost minimization problem is solved using the AGWO algorithm. It is assumed that the battery has an initial charge of zero. The optimal scheduling is for 24 h (a full day). However, in order to see the search ability of the algorithm, the optimization process is repeated 25 times. Through this repetition, the best solution, the average of the optimal solutions and standard deviation values, would be achieved. This is a well-defined approach for assessing the stability of the algorithm. The simulation results are repeated 25 times and the final results are provided in Table 3. Accordingly, it is seen here that the proposed algorithm has discovered a more optimal solution than the other algorithms. The performance of the proposed algorithm is compared with the PSO, GA, FSAPSO and original GWO, and the results clearly show the superiority of the algorithm. Not only the best solution, but also the worst solution and standard deviation values, are still optimal compared with the other algorithms. These results show the robustness of the algorithm. The very time saving nature of the AGWO may be deduced from the last column, showing the CPU time. This is especially important if the operation time interval is reduced in the management (such as to 10 min or less). The results evidently reveal the high quality and search features of the algorithm for solving the problem. The optimal power generation and charging values of each unit is plotted as a bar diagram in Figure 11. According to this bar diagram, the battery is trying charge during the first hours and discharge during the mid-hour times. Moreover, the fuel cell, as a cheap unit, is attending with the maximum capacity. The utility is changing to a buyer of the power at the middle of the day to let the market increase its profit. The renewable wind and solar energy sources are also producing as forecasted in their initial power curves.

Table 3. Simulation results of optimizing cost function over 25 times.

Method	Best Solution (EURct)	Worst Solution (EURct)	Standard Deviation (EURct)	CPU Time (s)
GA [27]	335.0707	345.4849	20.9302	17.451
PSO [27]	329.4922	341.9583	15.2483	16.263
FSAPSO [27]	328.6594	336.5502	12.8409	15.685
GWO	320.3640	334.3312	8.84420	16.862
AGWO	300.8632	305.8774	0.75270	9.1673

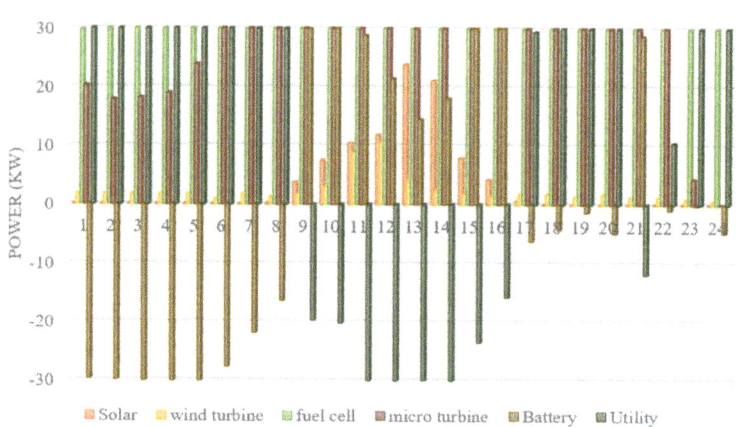

Figure 11. Output Power of RESs evaluated by the proposed method.

6. Conclusions

The focus of this work is to study market management and modeling based on AI. The proposed model uses the fuzzy-based SVR for the prediction of the market price with high accuracy. A management model based on cost minimization and unit dispatch is developed in the study, which can help scheduling the units. A novel adaptive-modified GWO is also devised which can help to minimize the operation cost over the next 24 h. The adaptive formulation is such that it allows the algorithm to chose the best modification according to its success rate and probability. The simulation results show that the proposed model has an appropriate performance for market price prediction with low MAPE and RMSE. Furthermore, it is seen that the test system model can suitably optimize the total cost of operations, including the solar and wind renewable energy sources, as well the fossil fuel-based models, such as fuel cell, micro turbine and battery. The AGWO's superiority over the set of algorithms is well proved.

Funding: This research received no external funding.

Institutional Review Board Statement: Not applicable.

Informed Consent Statement: Not applicable.

Data Availability Statement: This study did not report any data.

Conflicts of Interest: The authors declare no conflict of interest.

References

1. Mohamed, M.A.; Mirjalili, S.; Dampage, U.; Salmen, S.H.; Al Obaid, S.; Annuk, A. A Cost-Efficient-Based Cooperative Allocation of Mining Devices and Renewable Resources Enhancing Blockchain Architecture. *Sustainability* **2021**, *13*, 10382. [CrossRef]
2. Zeng, L.; Xia, T.; Elsayed, S.; Ahmed, M.; Rezaei, M.; Jermsittiparsert, K.; Dampage, U.; Mohamed, M. A Novel Machine Learning-Based Framework for Optimal and Secure Operation of Static VAR Compensators in EAFs. *Sustainability* **2021**, *13*, 5777. [CrossRef]
3. Abera, T.; Dereje, F.; Larago, B. Mobile Backhaul in Release 8 and Beyond: Benefits, Challenges, Market Status and Impact Analysis. *Int. J. Wirel. Microw. Technol.* **2020**, *10*, 9–15. [CrossRef]
4. Beni, S.A.; Sheikh-El-Eslami, M.K. Market power assessment in electricity markets based on social network analysis. *Comput. Electr. Eng.* **2021**, *94*, 107302. [CrossRef]
5. Yildirim, C.; Kasman, A.; Hamid, F.S. Impact of foreign ownership on market power: Do regional banks behave differently in ASEAN countries? *Econ. Model.* **2021**, *105*, 105654. [CrossRef]
6. Kamala, H. Development of an Effective Method of Data Collection for Advertising and Marketing on the Internet. *Int. J. Math. Sci. Comput.* **2021**, *7*, 1–11. [CrossRef]
7. Hossain, M.A.; Pota, H.R.; Squartini, S.; Abdou, A.F. Modified PSO algorithm for real-time energy management in grid-connected microgrids. *Renew. Energy* **2019**, *136*, 746–757. [CrossRef]
8. Rizwan, M.; Waseem, M.; Liaqat, R.; Sajjad, I.A.; Dampage, U.; Salmen, S.H.; Obaid, S.A.; Mohamed, M.A.; Annuk, A. SPSO Based Optimal Integration of DGs in Local Distribution Systems under Extreme Load Growth for Smart Cities. *Electronics* **2021**, *10*, 2542. [CrossRef]
9. Bui, V.-H.; Hussain, A.; Kim, H.-M. A Multiagent-Based Hierarchical Energy Management Strategy for Multi-Microgrids Considering Adjustable Power and Demand Response. *IEEE Trans. Smart Grid* **2018**, *9*, 1323–1333. [CrossRef]
10. Fan, W.; Huang, L.; Cong, B.; Tan, Z.; Xing, T. Research on an optimization model for wind power and thermal power participating in two-level power market transactions. *Int. J. Electr. Power Energy Syst.* **2021**, *134*, 107423. [CrossRef]
11. Liu, Y.-W.; Li, L.-L.; Tseng, M.-L.; Lim, M.K.; Helmi Ali, M. Optimal scheduling of combined cooling, heating, and power microgrid based on a hybrid gray wolf optimizer. *J. Ind. Prod. Eng.* **2022**, *39*, 277–292. [CrossRef]
12. Mahedy, A.; Abdelsalam, A.; Mohamed, R.; El-Nahry, I. Utilizing Neural Networks for Stocks Prices Prediction in Stocks Markets. *Int. J. Inf. Technol. Comput. Sci.* **2020**, *12*, 1–7. [CrossRef]
13. Amjady, N.; Daraeepour, A. Mixed price and load forecasting of electricity markets by a new iterative prediction method. *Electr. Power Syst. Res.* **2009**, *79*, 1329–1336. [CrossRef]
14. Sahoo, S.; Swain, S.; Subburaj, V. Novel Gaussian flower pollination algorithm with IoT for unit price prediction in peer-to-peer energy trading market. *Energy Rep.* **2021**, *7*, 8265–8276. [CrossRef]
15. Bhatia, K.; Mittal, R.; Tripathi, M.M. An ensemble approach for electricity price forecasting in markets with renewable energy resources. *Util. Policy* **2021**, *70*, 101185. [CrossRef]
16. Shi, W.; Wang, Y.; Ma, J. An effective Two-Stage Electricity Price forecasting scheme. *Electr. Power Syst. Res.* **2021**, *199*, 107416. [CrossRef]
17. Daglish, T.; de Bragança, G.G.F.; Romano, T. Pricing effects of the electricity market reform in Brazil. *Energy Econ.* **2021**, *97*, 105197. [CrossRef]
18. Flammini, M.G.; Prettico, G.; Chicco, G. Reducing fossil fuel-based generation: Impact on wholesale electricity market prices in the North-Italy bidding zone. *Electr. Power Syst. Res.* **2021**, *194*, 107095. [CrossRef]

19. Correa-Giraldo, M.; Garcia-Rendon, J.J.; Perez, A. Strategic behaviors and transfer of wholesale costs to retail prices in the electricity market: Evidence from Colombia. *Energy Econ.* **2021**, *99*, 105276. [CrossRef]
20. Shao, Z.; Zheng, Q.; Liu, C. Modeling and forecasting the electricity clearing price: A novel BELM based pattern classification framework and a comparative analytic study on multi-layer BELM and LSTM. *Energy Econ.* **2020**, *86*, 104648. [CrossRef]
21. Prabakaran, S.; Garcia, I.C.; Mora, J.U. A temperature stochastic model for option pricing and its impacts on the electricity market. *Econ. Anal. Policy* **2020**, *68*, 58–77. [CrossRef]
22. Prasenjit, C.; Banerjee, I.; Rahaman, H. Load management scheme for energy holes reduction in wireless sensor networks. *Comput. Electr. Eng.* **2015**, *48*, 343–357.
23. Xiang, Z.; Cao, J.; Guo, Q.; Wen, T. A novel network security algorithm based on improved support vector machine from smart city perspective. *Comput. Electr. Eng.* **2018**, *65*, 67–78.
24. Liu, M.; Luo, K.; Chen, S. A stock selection algorithm hybridizing grey wolf optimizer and support vector regression. *Expert Syst. Appl.* **2021**, *179*, 115078. [CrossRef]
25. Khalid, A.; Annuk, A.; Dampage, U.; Mohamed, M.A. Effective Energy Management via False Data Detection Scheme for the Interconnected Smart Energy Hub–Microgrid System under Stochastic Framework. *Sustainability* **2021**, *13*, 11836.
26. Figuera, C.; Barquero-Pérez, Ó.; Caamaño, A.J. Spectrally adapted Mercer kernels for support vector nonuniform interpolation. *Signal Process.* **2014**, *94*, 421–433. [CrossRef]
27. Baziar, A.; Kavousi-Fard, A. Considering uncertainty in the optimal energy management of renewable micro-grids including storage devices. *Renew. Energy* **2013**, *59*, 158–166. [CrossRef]
28. IREMA Website. Available online: https://www.irema.ir/market-data/statistical-report/daily-data/price/average (accessed on 30 March 2021).
29. Kavousifard, A.; Samet, H. A novel method based on modified shuffled frog leaping algorithm and artificial neural network for power system load prediction. *Emerg. Intell. Technol. Ind.* **2012**, *369*, 35–46.

sustainability MDPI

Article

Parameters Identification of Photovoltaic Cell and Module Models Using Modified Social Group Optimization Algorithm

Habib Kraiem [1,2,*], Ezzeddine Touti [1,3,*], Abdulaziz Alanazi [1], Ahmed M. Agwa [1,4], Tarek I. Alanazi [5], Mohamed Jamli [3] and Lassaad Sbita [2]

1 Department of Electrical Engineering, College of Engineering, Northern Border University, Arar 73222, Saudi Arabia
2 Processes, Energy, Environment and Electrical Systems, National Engineering School of Gabes, University of Gabes, Gabes 6029, Tunisia
3 Laboratory of Industrial Systems and Renewable Energies, National Higher Engineering School of Tunis, Tunis 1008, Tunisia
4 Department of Electrical Engineering, Faculty of Engineering, Al-Azhar University, Cairo 11651, Egypt
5 Department of Physics, College of Science, Northern Border University, Arar 73222, Saudi Arabia
* Correspondence: alhabeeb.kareem@nbu.edu.sa (H.K.); esseddine.touti@nbu.edu.sa (E.T.)

Abstract: Photovoltaic systems have become more attractive alternatives to be integrated into electrical power systems. Therefore, PV cells have gained immense interest in studies related to their operation. A photovoltaic module's performance can be optimized by identifying the parameters of a photovoltaic cell to understand its behavior and simulate its characteristics from a given mathematical model. This work aims to extract and identify the parameters of photovoltaic cells using a novel metaheuristic algorithm named Modified Social Group Optimization (MSGO). First, a comparative study was carried out based on various technologies and models of photovoltaic modules. Then, the proposed MSGO algorithm was tested on a monocrystalline type of panel with its single-diode and double-diode models. Then, it was tested on an amorphous type of photovoltaic cell (hydrogenated amorphous silicon (a-Si: H)). Finally, an experimental validation was carried out to test the proposed MSGO algorithm and identify the parameters of the polycrystalline type of panel. All obtained results were compared to previous research findings. The present study showed that the MSGO is highly competitive and demonstrates better efficiency in parameter identification compared to other optimization algorithms. The Individual Absolute Error (IAE) obtained by the MSGO is better than the other errors for most measurement values in the case of single- and double-diode models. Relatedly, the average fitness function obtained by the MSGO algorithm has the fastest convergence rate.

Keywords: photovoltaic cells; modeling; parameters estimation; MSGO algorithm; optimization

Citation: Kraiem, H.; Touti, E.; Alanazi, A.; Agwa, A.M.; Alanazi, T.I.; Jamli, M.; Sbita, L. Parameters Identification of Photovoltaic Cell and Module Models Using Modified Social Group Optimization Algorithm. *Sustainability* **2023**, *15*, 10510. https://doi.org/10.3390/su151310510

Academic Editor: Mohamed A. Mohamed

Received: 9 May 2023
Revised: 17 June 2023
Accepted: 21 June 2023
Published: 4 July 2023

1. Introduction

In recent decades, due to their inexhaustibility, non-polluting nature, and highly adaptable properties to decentralized generation, renewable energies have been the ecological alternative to fossil fuels and nuclear energy [1,2]. For these reasons, advanced technologies are currently being developed to benefit from these types of energy sources. Photovoltaic (PV) panels, which generate electricity using the sun's energy as a renewable energy source, are one of the most prevalent forms of renewable energy [3]. Solar energy is growing exponentially. Its main characteristic is to be a form of decentralized production, making it possible to meet strong demand from citizens and local authorities and to produce energy where it is consumed. Consequently, significant losses can be avoided during energy transportation. The PV industry has been overgrowing in recent years [4] because it is not only inexhaustible but also silent and non-disturbing for residents, unlike wind turbines which cause visual and acoustic disturbances. In addition, the market to produce electricity from solar energy is proliferating [5,6]. In this context, the importance of photovoltaic generators connected to the electricity distribution network is growing rapidly [7]. Hence, assessing and studying the performance of the photovoltaic module, which is the fundamental component of these generators, appears to be highly significant [8]. The manufacturers typically tend to provide only limited operational data for PV panels. These data are only available under standard conditions of 1000 W/m^2 irradiation, 25 °C cell temperature, and air mass of 1.5 [9]. Therefore, it is essential to understand each cell element's physical properties and electrical characteristics before developing an equivalent circuit for a photovoltaic cell. Performance evaluation of PV modules and the design of energy systems are derived from the electrical characteristic current–voltage (I–V) of the modules under different radiation levels and different temperatures of the PV cell [10,11]. There are three forms of solar cell technologies available on the market: amorphous, monocrystalline, and polycrystalline [12]. Monocrystalline and polycrystalline cells are found in rigid panels. The difference between the two types is mainly based on their efficiency. To achieve maximum performance, crystalline panels should be installed perpendicular to the sun's rays. Generally abbreviated a-Si, amorphous silicon is the non-crystallized allotropic variety of silicon; crystalline structures of the a-Si are formed from disordered atoms that are not arranged regularly. Thin layers of amorphous silicon can be deposited at low temperatures on a wide variety of substrates. Hence, a wide range of microelectronic applications can be envisaged. The advantage of amorphous silicon cells is that they are environmentally friendly because they do not use toxic heavy metals, such as cadmium or lead. Compared to amorphous cells, crystalline panels do not perform as well in partial shadowing, and they lose a tiny percentage of their output as the temperature rises over 25 °C. Various equations can be used to model PV cells and modules approximated to differing degrees of accuracy from the actual device. This modeling offers essential advantages, such as ease of use, thanks to the equivalent electrical circuit and the popularization of the system properties. Therefore, the understanding of complex phenomena will be facilitated. Therefore, solar cells are considered power generators and will be modeled by equivalent circuits and electric models. The most commonly used are the single-diode model [13], the model with two diodes [14], and the one with three diodes [15]. Each of these models has some unknown parameters that characterize and describe the behavior of a PV generator. In addition, the behavior of PV generators is influenced by various parameters related to electrical modeling [16]. The power output of a photovoltaic (PV) cell is influenced by several factors such as the orientation of the panels, quality factor, kind of material, absorbent layer, and optical window. The optimal orientation of panels should be perpendicular to the sun's direction to maximize the power output. The quality factor of the cell is a measure of its efficiency to convert sunlight into electricity, and it involves a trade-off between efficiency and cost. The choices of material and the thickness and composition of the absorbent layer also play a significant role in determining the power output. Additionally, optimizing the optical window requires a balance between light transmission and absorption by the window. The PV cell's performance is interdependent on various parameters, such as efficiency, open-circuit voltage, fill factor, short-circuit current, and maximum power point. These parameters are interdependent, and there are constraints between them that must be considered to optimize the cell's performance. Hence, understanding the constraints between the PV cell parameters is vital for designing efficient PV systems.

In order to optimize the various characteristics and simulate the behavior of a PV generator, it is crucial to identify the physical mechanisms at play within it. The complexity of the model is determined by the number of parameters that need to be identified. The ideal model includes a current source for solar power input and a diode for the PN junction, but additional components can be added to better represent the PV cell's behavior in specific operational situations. Various methods of parameter identification have been studied in the literature, including numerical, analytical, deterministic, and metaheuristic methods. Numerical methods utilize mathematical algorithms to iteratively optimize parameter values using measured or simulated data. These methods employ numerical optimization techniques, including iterative algorithms and metaheuristic approaches, to minimize the discrepancy between model predictions and observed data. Numerical methods offer flexibility in handling complex and nonlinear problems [17,18]. Analytical methods involve analyzing mathematical formulas to identify the parameters of PV models. These methods are characterized by their short execution time and simplicity. However, their solutions are not precise [19,20]. The deterministic methods have major drawbacks, such as the high sensitivity to the initial hypotheses and the tendency of these algorithms to converge to the local optimum [21]. Moreover, they depend on the convexity of the model [22]. However, the models of photovoltaic cells are multimodal and characterized by nonlinearities. Recently, metaheuristic methods seem to be good potential alternatives for extracting parameters from PV models [23]. Indeed, they overcome the shortcomings of the analytical and deterministic methods already cited. In the following, we mention some of the most popular metaheuristic methods: Genetic Algorithm (GA) [24], artificial bee colony algorithm (ABC) [25], differential evolution algorithm (DE) [26], bird mating optimization (BMO) [27], Ant Lion Optimizer (ALO) [28], bacterial foraging optimization (BFO) [29], gray wolf optimization (GWO) [30], whale optimization algorithm (WOA) [31], Slime Mould Algorithm (SMA) [32], Sal Swarm Algorithm (SSA) [33], and Coyote Optimization Algorithm (COA) [15].

The primary objective of this study is to investigate and analyze the efficiency of a novel algorithm called Modified Social Group Optimization (MSGO) [34] for the extraction and identification of the parameters of photovoltaic cells. To provide a comprehensive assessment, a comparative study was conducted, incorporating various technologies and models of photovoltaic panel cells.

In the initial phase of this investigation, the proposed algorithm was applied to the monocrystalline photovoltaic panel of RTC France Company, considering both single- and double-diode cell models. The outcomes obtained through the MSGO algorithm were compared with results from previous studies utilizing alternative metaheuristic algorithms, such as the Nelder–Mead method and modified particle swarm optimization (NM–MPSO) [35], Levenberg–Marquardt algorithm combined with Simulated Annealing (LMSA) [36], ABC [21], biogeography-based optimization algorithm with mutation strategies (BBO-M) [37], improved adaptive differential evolution (Rcr-IJADE) [38], artificial bee swarm optimization algorithm (ABSO) [39], and chaotic asexual reproduction optimization (CARO) [40]. All these algorithms were tested on the same photovoltaic panel, under identical lighting and temperature conditions (temperature of 33 °C and irradiation of 1000 W/m^2).

Subsequently, the proposed algorithm was also evaluated on a flexible photovoltaic panel composed of hydrogenated amorphous silicon (a-Si: H). The obtained results were compared with the findings presented by authors from [41], who based their research on the optimization algorithms Quasi-Newton Method (QNM) and the Self-Organizing Migrating Algorithm (SOMA).

To validate the results obtained by the proposed MSGO algorithm, an experimental study was performed on the TITAN-12-50 panel, utilizing polycrystalline cells [42]. Finally, the paper concludes with a comparative analysis between different optimization algorithms employed for photovoltaic parameter extraction. The results obtained through the proposed MSGO algorithm are compared with those derived from other algorithms such as the WOA, SSA, Sine Cosine Algorithm (SCA), Virus Colony Search Algorithm (VCS), Gravitational Search Algorithm (GSA), and Ant Lion Optimizer (ALO). Throughout the remainder of this paper, three sections are described: Section 2 introduces PV models and problem formulation. Section 3 details the proposed MSGO algorithm. Section 4 treats the study of the MSGO algorithm efficiency by testing various pieces of technology and PV cell models. In the last section, the obtained results are compared with those given in previous studies.

2. Mathematical PV Model Analysis

The evaluation of the PV module performance and the power system design is based on the current–voltage electrical characteristic of the modules under different radiation levels and various temperatures of the PV cells. It is possible to model PV cells and modules by means of equations that approximate the physical cell to varying degrees. Several electrical models are proposed in the literature for simulating PV cells under different conditions. The model's complexity varies depending on the number of parameters (R_s, R_{sh}, etc.) to be considered. Every model is basically refinements of the ideal model, which consists of a diode that represents the PN junction and a current source that represents incident solar power.

It is possible to add several additional elements to provide a better representation of the behavior of PV cells in some operating areas [43]. Single-diode models (SDMs), double-diode models (DDMs), and three-diode models (TDMs) are the most used models. Figure 1a represents the single-diode model, which is regarded as the most popular model. It is widely used because of its simplicity. It also provides high precision and simplicity in the power generation quadrant.

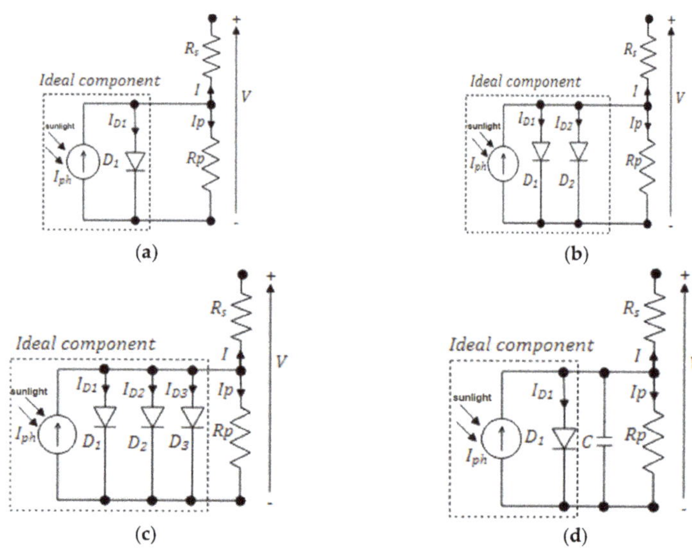

Figure 1. Solar cell models: (**a**) SDM, (**b**) DDM, (**c**) TDM, and (**d**) Dynamic Model.

The single-diode model has undergone various advancements that have led to the development of more accurate models, such as the Bishop model, which explains the behavior of the PV cell under reverse polarization. The double-diode model, shown in Figure 1b, considers losses due to various resistances and devices in the different electric components that constitute the circuit [44]. An enhanced model considers the effects of grain boundaries and leakage currents. This model involves three diodes as it is shown in Figure 1c. Although this model meets most of the physical requirements of solar cells, it involves computing nine parameters that require exceptionally high numerical execution. In addition, dynamic models are proposed by introducing the capacity to model the dynamic behavior of the PV cell. This model type is shown in Figure 1d. All these models differ in the number of parameters required for computing the I–V characteristic [45].

I_{ph} is the photo-generated current source; I_{D1}, I_{D2}, I_{D3} are the currents of diodes D_1, D_2, and D_3; R_p is the shunt resistance; R_s is the series resistance; I is the output current; and V is the output voltage.

2.1. Mathematical Development

2.1.1. Crystalline Cells

From the equivalent circuit (Figure 1a), it is evident that the current produced by the solar cell is equal to that produced by the current source (I_{ph}), minus that which flows through the diode (I_d), minus that which flows through the shunt resistor (I_p).

$$I = I_{ph} - K_i I_d - I_p \tag{1}$$

where $K_i = \begin{cases} K_1 = \begin{bmatrix} 1 & 0 & 0 \end{bmatrix}, \; i = 1 \text{ in case of SDM} \\ K_2 = \begin{bmatrix} 1 & 1 & 0 \end{bmatrix}, \; i = 2 \text{ in case of DDM} \\ K_3 = \begin{bmatrix} 1 & 1 & 1 \end{bmatrix}, \; i = 3 \text{ in case of TDM} \end{cases}$, where $I_d = \begin{bmatrix} I_{d1} \\ I_{d2} \\ I_{d3} \end{bmatrix}$, and $I_p = \frac{V+R_s I}{R_p}$.

The current in the jth diode is given by

$$I_{dj} = I_{sdj}\left(e^{\left(\frac{q(V+R_S I)}{n_j KT}\right)} - 1\right), \; (j = 1,2,3) \tag{2}$$

Then, the current given in Equation (1) is given by Equation (3), where I_{SDM}, I_{DDM}, and I_{TDM} are the total output current when considering the SDM, DDM, and TDM, respectively.

$$\begin{cases} I_{SDM} = I_{ph} - I_{sd}\left(e^{\left(\frac{q(V+R_S I)}{n_1 KT}\right)} - 1\right) - \frac{V+R_s I}{R_p} \\ I_{DDM} = I_{ph} - I_{sd1}\left[e^{\frac{q(V+R_S I)}{n_1 KT}} - 1\right] - I_{sd2}\left[e^{\frac{q(V+R_S I)}{n_2 KT}} - 1\right] - \frac{V+R_s I}{R_p} \\ I_{TDM} = I_{ph} - I_{sd1}\left[e^{\frac{q(V+R_S I)}{n_1 KT}} - 1\right] - I_{sd2}\left[e^{\frac{q(V+R_S I)}{n_2 KT}} - 1\right] - I_{sd3}\left[e^{\frac{q(V+R_S I)}{n_3 KT}} - 1\right] - \frac{V+R_s I}{R_p} \end{cases} \tag{3}$$

where n_1, n_2, and n_3 are the ideality factors of the diodes D_1, D_2, and D_3; K is the Boltzmann constant (1.380649×10^{-23} Joule/Kelvin); T is the temperature of the PV panel (Kelvin); and q is the charge of the electron ($1.602176634 \times 10^{-19}$ Coulomb).

The TDM does not seem suitable for fast computations and has complex nonlinear analytic expressions; therefore, this model will be excluded from the parametric identification tests.

2.1.2. Amorphous Silicon Cell

Equation (4) defines the current–voltage characteristic for an amorphous silicon cell:

$$I = I_{ph}\left(1 - \frac{d_i^2}{\mu_{eff}[V_b - (V + IR_s)]}\right) - I_s\left[\exp\left(\frac{V + IR_s}{aV_T} - 1\right)\right] - \frac{V + IR_s}{R_{sh}} \tag{4}$$

d_i denotes the width of the ith layer in the (a-Si) p_i_n diode, μ_{eff} represents the mobility-lifetime product of the electron and hole, and V_{bi} is the built-in field voltage.

The diode reverse saturation current and the photo-generated current of an (a-Si) cell under constant light and temperature are given, respectively, by

$$I_{ph} = qAg(x)(L_p + L_n + W) \tag{5}$$

$$I_0 = J_s \times A = \left(\frac{qD_p p_{n0}}{L_p} + \frac{qD_n n_{p0}}{L_n}\right) \times A \tag{6}$$

where A is the p_n junction area, L_p is the carrier diffusion length of the p-type area, L_n is the carrier diffusion length of the n-type area, W is the depletion layer, D_p and D_n are the holes and electrons diffusion coefficient, p_{n0} and n_{p0} are the minority carrier concentration in the P region and N region, and $g(x)$ is the electron hole formation ratio.

2.2. The Objective Functions

The term objective function is used in mathematical optimization and operations research to refer to a function that acts as a criterion for identifying the best solution to an optimization problem.

The objective function of the SDM may be written as

$$f_{SDM}(V,\ I,\ X) = I - X(1) + X(2)\left[e^{\left(\frac{q(V+X(3)I)}{X(4)KT}\right)} - 1\right] + \frac{V + X(3)I}{X(5)} \tag{7}$$

$$X = \begin{bmatrix} I_{ph} & I_{sd} & R_s & n_1 & R_p \end{bmatrix}$$

For the DDM, the error function is expressed by

$$f_{DDM}(V, I, X) = I - X(1) + X(2)\left[e^{\left(\frac{q(V+X(3)I)}{X(4)KT}\right)} - 1\right] + X(5)\left[e^{\left(\frac{q(V+X(3)I)}{X(6)KT}\right)} - 1\right] + \frac{V + X(3)I}{X(7)} \tag{8}$$

where $X = \begin{bmatrix} I_{ph} & I_{sd_1} & R_s & n_1 & I_{sd_2} & n_2 & R_p \end{bmatrix}$.

Whereas, the error function for the TDM is defined by

$$\begin{aligned} f_{TDM}(V, I, X) &= I - X(1) + X(2)\left[e^{\left(\frac{q(V+X(3)I)}{X(4)KT}\right)} - 1\right] + X(5)\left[e^{\left(\frac{q(V+X(3)I)}{X(6)KT}\right)} - 1\right] \\ &+ X(7)\left[e^{\left(\frac{q(V+X(3)I)}{X(8)KT}\right)} - 1\right] + \frac{V+X(3)I}{X(9)} \end{aligned} \tag{9}$$

where $X = \begin{bmatrix} I_{ph} & I_{sd_1} & R_s & n_1 & I_{sd_2} & n_2 & I_{sd_3} & n_3 & R_p \end{bmatrix}$.

It is necessary to use *Ne* samples (data points number) to widen the scope of the search and reach the global optimum. Equation (10) gives us a description of the cost function:

$$\text{RMSE}(x) = \sqrt{\frac{1}{Ne}\sum_{C=1}^{Ne}\left(f_M^C(V^C, I^C, x)\right)^2} \tag{10}$$

3. Procedure of Social Group Optimization for PV Parameters Estimation

The past twenty years have seen a remarkable rise in interest in metaheuristic optimization algorithms. The research work developed has enabled the appearance of new algorithms which are generally based on the following:

- A new idea inspired by a natural, physical, chemical phenomena;
- A modification of an existing algorithm to improve its performances;
- The hybridization of two methods allows the strengths to merge and the weaknesses to be eliminated of the two algorithms.

However, no algorithm can be adapted to all types of problems. In 2016, a new metaheuristic optimization algorithm appeared, known as Social Group Optimization (SGO) [46]. To solve complex problems, the new algorithm was inspired by the social behavior of individuals in groups. Each individual's knowledge is mapped by its fitness. The algorithm contains two phases. The first phase is called the improving phase in which each individual interacts with the best person (best solution) to improve his knowledge by interacting. The second phase is named the acquiring phase, during which the individuals acquire knowledge when they interact with the best person and randomly selected individuals simultaneously. A comparative study is carried out to show the performance of the new method. Detailed information on the SGO algorithm can be found in the following articles [47,48]. The SGO algorithm is described with the following:

P_i, (i = 1, 2, 3, ..., N): Pi is the social group persons, and N is the total number of people in the social group.

P_{ij}, (j = 1, 2, ..., D): D is the traits number related to a person which allows us to determine the dimensions of a person.
f_i, (i = 1, 2, ..., N) is their corresponding fitness value.

Improving phase

In each social group, the role of the best person (P_{best}) is to propagate knowledge between all persons. As a result, others in the group enhance their knowledge.

$$[minvalue, index] = \min\{f(Pi),\ i = 1,2,3\ldots\ldots\ldots N\}$$

$$\text{gbest} = \text{P}(\text{index}, :)$$

The following algorithm (Algorithm 1) can be used to calculate how often each person's knowledge is updated:

Algorithm 1 Improving phase

```
for i = 1: N
    for j = 1: D
        Pnew_ij = c * Pij + r * ( Pbest(j) − Pij)
    end for
end for
```

r: random number, and $r \in [0, 1]$. If *Pnew* provides higher fitness than P_{old}, it is accepted [34]. c is the parameter of self-introspection $c \in [0, 1]$.

Acquiring Phase

In the acquiring phase (Algorithm 2), a person acquires new knowledge by interacting with other persons of the group. The interaction can be with the best person (*Pbest*) or randomly with other persons who have more knowledge. To acquire knowledge, a person always interacts with the *Pbest* and with any other person of the group who has more knowledge than him. The ability to obtain a quantity of knowledge from another person is defined by the self-awareness probability (SAP). The modified acquiring phase is computed as

$$[value,\ index\_num] = \min\{f(Pi),\ i = 1,2,3,\ 4\ldots\ldots\ldots N\}$$

$$Pbest = P(index\_num, :)$$

where *Pi* is the updated value at the completion of the improving phase.

Algorithm 2 Acquiring phase

```
for i = 1: N
  Randomly select one person P_r where i ≠ r
      If f(P_i) < f(P_r)
        If rand > SAP
          for j = 1: D
          Pnew_i,j = P_i,j + rand_1 * (P_i,j − P_r,j) + rand_2 * (best_p (j) − P_i,j)
        end for
    else
          for j = 1: D
          Pnew_i,j = lb + rand_2 * (ub − lb)
          end for
        end if
    else
        for j = 1: D
        Pnew_i,j = P_i,j + rand_1 * (P_r,j − P_r,j) + rand_2 * (best_p (j) − P_i,j)
        end for
    end if
end for
```

Pnew is accepted if it provides a higher level of fitness than *P*.

The general steps to use the MSGO algorithm to extract parameters of a PV cell include:

Step 1: Define the objective function which describes the behavior of the PV cell under different conditions. This function takes input parameters, such as the cell's temperature, irradiance, and voltage, and outputs a value that represents the cell's performance. The goal is to find the values of these input parameters that maximize the output value of the objective function.

Step 2: Define the parameter space. The parameter space defines the objective function constraints for each parameter, such as the temperature may range from −10 °C to 100 °C, the irradiance may range from 0 W/m^2 to 1000 W/m^2, and the voltage may range from 0 V to 1 V. All other range parameters are declared in Equations (15)–(17).

Step 3: Initialize the population. The population (P_i) is a set of solutions that are randomly generated within the parameter space. Each solution corresponds to a set of input parameters that are used to evaluate the objective function.

Step 4: Evaluate the fitness. The fitness is a measure of how well each solution performs with respect to the objective function. The fitness function takes as input the output value of the objective function and returns a scalar value that represents the quality of the solution. The higher the fitness, the better the solution.

Step 5: Update the population. The MSGO algorithm updates the population in two phases: the improving phase and the acquiring phase. In the improving phase, each individual interacts with the best person in the social group to improve its knowledge. In the acquiring phase, each individual acquires knowledge by interacting with the best person and randomly selected individuals. The updating of each person's knowledge can be calculated using the formula described in the improving phase.

Step 6: Repeat steps 4 and 5 until convergence. The optimization process continues until the fitness values converge to a satisfactory level or the maximum number of iterations is reached. The best solution found during the optimization process corresponds to the set of input parameters that maximizes the output value of the objective function. These parameters can be used to characterize the behavior of the PV cell under the given conditions.

4. Results

The technical details of the software and hardware used for the extraction of the various simulation results are given in Table 1.

Table 1. Software and hardware details.

Hardware and Software	Setting
CPU	Intel (R) Core (TM) i7—7500U
Frequency	2.9 GHz
RAM	12 Gb
Simulation software	Matlab R2018b
Operating System	Windows 10

The adjustable parameters of the MSGO algorithm include: the population size is 40, and the maximum number of iterations is 3000. However, the parameters of other comparative algorithms are given in references cited in the first section.

All simulation work was conducted under the following solar irradiance and temperature conditions: 1000 W/m^2 and 33 °C. The obtained results of our parameter identification algorithm were compared to other optimization algorithms to determine the accuracy of the fitted curve between the MSGO algorithm values and experimental data. Table 2 summarizes the comparison work established in this paper.

Table 2. Summary of comparative study.

Types of Tested Panels	PV Cell Model	Examined Algorithms during the Comparison
RTC France Company monocrystalline	Single diode	NM–MPSO, LMSA, ABC, BBO-M, Rcr-IJADE
	Double diode	NM–MPSO, Rcr-IJADE, ABSO, CARO, ABC
Hydrogenated Amorphous Silicon a-Si: H	Single diode	SOMA, QNM
TITAN-12-50 Polycrystalline	Double diode	SCA, ALO, GSA, VCS, WOA, SSA

4.1. MSGO Implementation

To verify the accuracy of the fitted curve obtained by the MSGO algorithm using experimental data, a comparison is made against other algorithms. Tables 3 and 4 present a statistical analysis of the contrasted results for each model. The statistical errors used to demonstrate the performance of the proposed algorithm are presented below.

Table 3. Extracted parameters in case of an SDM.

Algorithm	I_{ph} (A)	I_0 (μA)	n	R_s (Ω)	R_p (Ω)
MSGO	0.7607877	0.31058918	1.47725615	0.0365470	52.88998
BBO-M	0.760781	0.318743	1.479842	0.036422	53.36226
Rcr-IJADE	0.760775	0.323022	1.481183	0.036376	53.718525
LMSA	0.760781	0.318492	1.479764	0.036433	53.326441
CARO	0.760792	0.317243	1.481681	0.036443	53.08930
ABC	0.76082	0.325155	1.481731	0.036443	53.64332
NM–MPSO	0.760781	0.323065	1.481202	0.036384	53.72221

Table 4. The estimated data and the resulted IAE obtained by the proposed algorithm **MSGO** compared with other algorithms in the case of an SDM.

	V_{exp} (V)	I_{exp} (A)	I_{est} (A)	MSGO	R_{cr}-IJADE	BBO-M	ABC	LMSA	NM–MPSO
1	−0.2057	0.764	0.764149248	0.00014925	0.00009559	0.000006	0.0001	0.000115762	0.000087
2	−0.1291	0.762	0.762293808	0.00029381	0.00066611	0.000604	0.0006	0.000680672	0.000662
3	−0.0588	0.7605	0.761373566	0.00087357	0.00085473	0.000817	0.0008	0.000863281	0.000854
4	0.0057	0.7605	0.7601543024	0.00034570	0.00035034	0.000364	0.0003	0.000346856	0.000346
5	0.0646	0.76	0.759038854	0.00096115	0.00094298	0.000946	0.0009	0.000953669	0.000945
6	0.1185	0.759	0.758010563	0.00098944	0.00095528	0.000943	0.0009	0.000973813	0.000957
7	0.1678	0.757	0.757045517	0.00004552	0.00009510	0.000120	0.0001	0.0000690271	0.000091
8	0.2132	0.757	0.756084674	0.00091533	0.00084950	0.000817	0.0008	0.000886778	0.000858
9	0.2545	0.7555	0.755022264	0.00047774	0.00041823	0.000361	0.0004	0.000445307	0.000413
10	0.2924	0.754	0.753597432	0.00040257	0.00032967	0.000276	0.0003	0.000370139	0.000336
11	0.3269	0.7505	0.751327686	0.00082769	0.00089542	0.000953	0.0008	0.000858429	0.000888
12	0.3585	0.7465	0.747306479	0.00080648	0.00085737	0.000914	0.0008	0.000827345	0.000848
13	0.3873	0.7385	0.740087107	0.00158711	0.00160420	0.001668	0.0016	0.00160213	0.001596
14	0.4137	0.728	0.727430948	0.00056905	0.00059912	0.000583	0.0006	0.000616337	0.000604
15	0.4373	0.7065	0.707034237	0.00053424	0.00044631	0.000485	0.0004	0.000492923	0.000452
16	0.459	0.6755	0.675413782	0.00008622	0.00019600	0.000230	0.0002	0.000182486	0.000206
17	0.4784	0.632	0.631018287	0.00098171	0.00110900	0.001271	0.0012	0.001194906	0.001117
18	0.496	0.573	0.572202755	0.00079724	0.00091027	0.001112	0.0011	0.001026552	0.00092
19	0.5119	0.499	0.499575662	0.00057566	0.00049902	0.000563	0.0005	0.000638902	0.00049
20	0.5265	0.413	0.413530488	0.00053049	0.00049030	0.000612	0.0006	0.00065758	0.000492
21	0.5398	0.3165	0.31721586	0.00071586	0.00071532	0.000985	0.001	0.000992379	0.000718
22	0.5521	0.212	0.212079153	0.00007915	0.00010468	0.000142	0.0001	0.000112783	0.000102
23	0.5633	0.1035	0.102706638	0.00079336	0.00078397	0.001254	0.0012	0.001305993	0.000779
24	0.5736	−0.01	−0.009221842	0.00077816	0.00075437	0.001268	0.0013	0.001228583	0.000751
25	0.5833	−0.123	−0.12427906	0.00127906	0.00137750	0.002537	0.0024	0.002545248	0.001381
26	0.59	−0.21	−0.209015291	0.00098471	0.00080320	0.001469	0.0015	0.001522512	0.000807

- The Individual Absolute Error (IAE) is defined by

$$IAE = |I_{measured} - I_{estimated}| \tag{11}$$

- The Median Absolute Error (MAE) is expressed as

$$MAE = \sum_{i=1}^{m} \frac{|I_{measured} - I_{estimated}|}{m} \tag{12}$$

- The Residual Sum of Squares (SSE) is defined by

$$SSE = \sum_{i=1}^{N_e} (I_{measured} - I_{estimated})^2 \tag{13}$$

- The Root Mean Square Error (RMSE) is given by

$$RMSE = \sqrt{\frac{1}{N_e} \sum_{i=1}^{N_e} (I_{measured} - I_{estimated})^2} \tag{14}$$

The equations below present the objective function constraints for each model, both for single and double diodes. The objective function constraints for the SDM is given by

$$\begin{cases} 0 \leq I_{ph} \leq 1\text{ A} \\ 0 \leq I_{01} \leq 1 \times 10^{-7}\text{ A} \\ 1 \leq n_1 \leq 2 \\ 0 \leq R_s \leq 0.8\ \Omega \\ 0 \leq R_p \leq 100\ \Omega \end{cases} \tag{15}$$

The objective function constraints for the DDM are

$$\begin{cases} 0 \leq I_{ph} \leq 1\text{ A} \\ 0 \leq I_{01} \leq 1 \times 10^{-7}\text{A} \\ 0 \leq I_{02} \leq 1 \times 10^{-7}\text{A} \\ 1 \leq n_1 \leq 2 \\ 1 \leq n_2 \leq 2 \\ 0 \leq R_s \leq 0.5\ \Omega \\ 0 \leq R_p \leq 100\ \Omega \end{cases} \tag{16}$$

The objective function constraints for (a-Si: H) are

$$\begin{cases} 0 \leq I_{ph} \leq 1\ \mu\text{A} \\ 0 \leq d \leq 10 \times 10^{-8}\text{ m} \\ 0 \leq \mu_{eff} \leq 10\ (\text{cm}^2/V) \\ 0 \leq V_{bi} \leq 1.5\text{ V} \\ 0 \leq R_s \leq 0.5\ \Omega \\ 0 \leq I_s \leq 5 \times 10^{-14} \\ 1 \leq a \leq 2.5 \\ 0 \leq R_{sh} \leq 50\ \Omega \end{cases} \tag{17}$$

4.1.1. A Comparative Study of Extraction Parameters for the SDM

The extracted parameters using the MSGO algorithm for the SDM are presented in Table 3. These values are compared to those obtained by other algorithms such as **BBO-M**, **R_{cr}-IJADE**, **LMSA**, **CARO**, **ABC**, and **NM–MPSO**.

The estimated current values obtained by the proposed algorithm and the resulted IAE are given in Table 4. These results are compared to those obtained by the **R_{cr}-IJADE,**

BBO-M, **ABC**, **LMSA**, and **NM–MPSO** algorithms. As it is shown, the IAE (IAT) obtained by the MSGO algorithm is better than the other errors for most measurement values.

The various statistical errors already defined by Equations (11)–(14) are presented in Table 5 and compared with those obtained by the other algorithms. One can remark that the MSGO error IAT (IAE) has the lowest value, which proves the robustness of the used algorithm.

Table 5. Statistical results for the SDM.

	MSGO	BBO-M	R_{cr}-IJADE	LMSA	CARO	ABC	NM–MPSO
IAT	0.01738	0.0213	0.017704	0.0215	0.0182	0.0205	0.0177
RMSE	7.21×10^{-4}	9.86×10^{-4}	7.75×10^{-4}	9.86×10^{-4}	9.87×10^{-4}	9.49×10^{-4}	7.75×10^{-4}
SSE	1.355×10^{-5}	2.529×10^{-5}	1.562×10^{-5}	2.529×10^{-5}	2.531×10^{-5}	2.343×10^{-5}	1.563×10^{-5}
MAE	6.68×10^{-4}	8.19×10^{-4}	6.81×10^{-4}	8.27×10^{-4}	6.98×10^{-4}	7.88×10^{-4}	6.81×10^{-4}

The different IAE results given in Table 4 are illustrated in Figure 2. One can notice that the IAE obtained by the MSGO algorithm (red color) is the lowest error for most values.

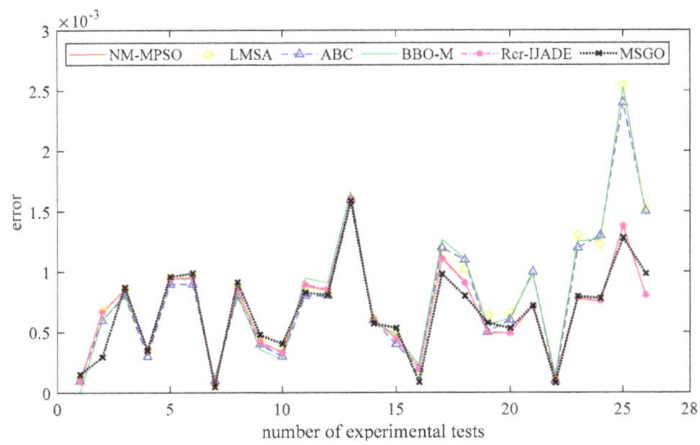

Figure 2. Calculated errors IAE obtained with the MSGO, R_{cr}-IJADE, BBO-M, ABC, LMSA, and NM–MPSO algorithms: case of SDM.

In order to assess the precision of the extracted parameters, one compares the I–V and P–V characteristics obtained from the estimated parameters using the MSGO method with the experimental data. Figure 3 illustrates this comparison, specifically for a single-diode case. These figures allow us to evaluate the quality of the parameter estimation process.

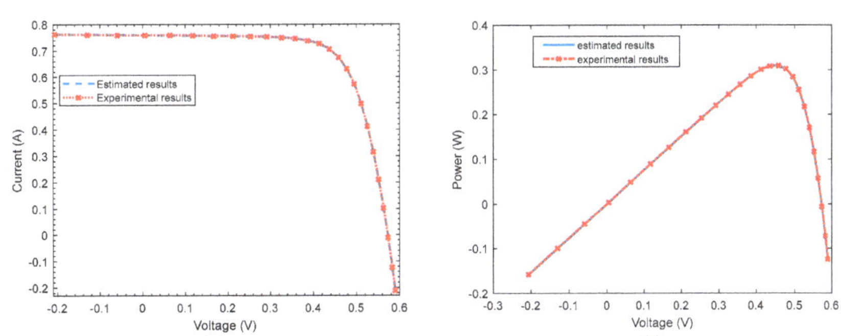

Figure 3. Experimental and estimated results obtained by the proposed **MSGO** algorithm in the case of an **SDM**: (**a**) Current_Voltage; (**b**) Power_Voltage.

The results depicted in Figure 3 indicate that the reconstructed SDM aligns well with the experimental data.

4.1.2. A Comparative Study of Extraction Parameters for the DDM

The extracted parameters using the MSGO algorithm for the DDM are presented in Table 6. These values are compared to those obtained by other algorithms, such as **BBO-M**, **R_{cr}-IJADE, LMSA**, **CARO**, **ABC**, and **NM–MPSO**.

Table 6. Extracted parameters in the case of double-diode model.

Algorithm	I_{ph} (A)	I_{01} (μA)	I_{02} (μA)	n_1	n_2	R_s (Ω)	R_p (Ω)
MSGO	0.7607	0.1465	0.6300	1.4190	1.8075	0.0371	54.7897
R_{cr}-IJADE	0.760821	0.225974	0.749347	1.451017	2.0000	0.036740	55.485443
CARO	0.760752	0.293151	0.090982	1.473383	1.77322	0.036414	54.39674
ABSO	0.760783	0.267135	0.381914	1.465125	1.98152	0.036572	54.62193
ABC	0.760825	0.040712	0.287433	1.449541	1.48852	0.036445	53.78046
NM–MPSO	0.760782	0.224761	0.755245	1.45054	1.99998	0.036752	55.52967

The estimated current values obtained by the proposed algorithm and the resulted IAE are given in Table 7. These results are compared to those obtained by the **R_{cr}-IJADE, BBO-M, ABC, LMSA**, and **NM–MPSO** algorithms. As it is shown, the IAE (IAT) obtained by the MSGO algorithm is better than the other errors for the majority of measurement values.

Table 7. Estimated data and the resulted IAE obtained by the proposed algorithm **MSGO** compared with other algorithms.

	V_{exp} (V)	I_{exp} (A)	I_{est} (A)	MSGO	ABC	CARO	ABSO	R_{cr}-IJADE	NM–MPSO
1	−0.2057	0.764	0.76402813	0.0000281297	0.000092908	0.00031	0.000031	0.00009268	0.000023
2	−0.1291	0.762	0.762630936	0.000630936	0.0006	0.000629	0.000629	0.00065394	0.000598
3	−0.0588	0.7605	0.761348408	0.000848408	0.0008	0.000843	0.000843	0.00085755	0.000832
4	0.0057	0.7605	0.760170729	0.000329271	0.0003	0.000338	0.000338	0.00033747	0.00033
5	0.0646	0.76	0.759091932	0.000908068	0.0009	0.00092	0.00092	0.00094	0.000895
6	0.1185	0.759	0.758093995	0.000906005	0.0009	0.000919	0.000919	0.00094935	0.00088
7	0.1678	0.757	0.757150045	0.000150045	0.0001	0.000139	0.000139	0.00009635	0.000187
8	0.2132	0.757	0.756196358	0.000803642	0.0008	0.000807	0.000807	0.00085535	0.000757
9	0.2545	0.7555	0.755121281	0.000378719	0.0004	0.000368	0.000368	0.00041885	0.000323
10	0.2924	0.754	0.753659367	0.000340633	0.0003	0.000306	0.000306	0.00033126	0.000277
11	0.3269	0.7505	0.751329108	0.000829108	0.0008	0.000892	0.000892	0.00089511	0.000896
12	0.3585	0.7465	0.74723368	0.00073368	0.0008	0.000822	0.000822	0.00084939	0.000798
13	0.3873	0.7385	0.739947501	0.001447501	0.0016	0.001544	0.001544	0.00160214	0.0001495
14	0.4137	0.728	0.727258182	0.000741818	0.0006	0.000669	0.000669	0.00061216	0.000729
15	0.4373	0.7065	0.706880574	0.000380574	0.0004	0.000396	0.000396	0.00045162	0.000344
16	0.459	0.6755	0.675327647	0.000172353	0.0002	0.000235	0.000235	0.00019888	0.000259
17	0.4784	0.632	0.631016281	0.000983719	0.0012	0.001111	0.001111	0.00111234	0.0001099
18	0.496	0.573	0.572261627	0.000738373	0.0011	0.000886	0.000886	0.00092523	0.000845
19	0.5119	0.499	0.499644454	0.000644454	0.0005	0.000533	0.000533	0.00049417	0.000586
20	0.5265	0.413	0.413556808	0.000556808	0.0006	0.000525	0.000525	0.00049125	0.000571
21	0.5398	0.3165	0.317169103	0.000669103	0.001	0.00073	0.00073	0.00071918	0.000753
22	0.5521	0.212	0.211959679	0.0000403211	0.0001	0.00009	0.00009	0.00010831	0.00088
23	0.5633	0.1035	0.102545511	0.000954489	0.0012	0.000806	0.000806	0.00077968	0.000827
24	0.5736	−0.01	−0.009374369	0.000625631	0.0012	0.00073	0.00073	0.00075539	0.000711
25	0.5833	−0.123	−0.124360342	0.001360342	0.0025	0.00139	0.00139	0.00137667	0.0001388
26	0.59	−0.21	−0.209001839	0.000998161	0.0014	0.00083	0.00083	0.00080501	0.000865

The various statistical errors already defined by Equations (11)–(14) are presented in Table 3 and compared with those obtained by the other algorithms. One can remark that the MSGO error IAT (IEA) has the lowest value, which proves the robustness of the used algorithm.

The various IAE results given in Table 7 are illustrated in Figure 4. One can notice that the IAE obtained by the MSGO algorithm (red color) is the lowest error for most values.

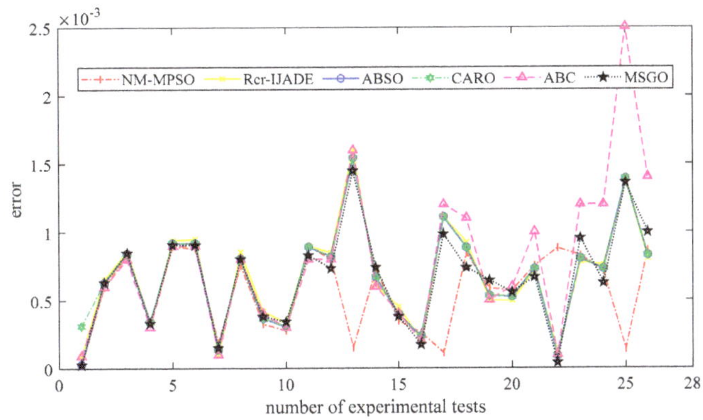

Figure 4. Calculated errors IAE obtained using the MSGO, ABC, CARO, ABSO, R_{cr}-IJADE, and NM–MPSO algorithms: case of DDM.

The I–V and P–V characteristics resulting from the parameters identified using the MSGO algorithm are compared to both experimental and estimated data to evaluate their quality. Figure 5 provides a comparison for a scenario involving two diodes, allowing us to determine the accuracy of the parameter estimation achieved through the MSGO algorithm. The results of the parameter identification using the MSGO algorithm and experimental data are compared with the estimated data to investigate the quality of the extracted parameters.

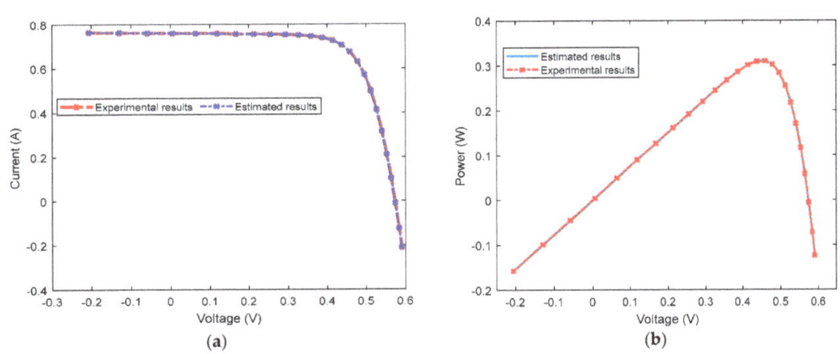

Figure 5. Experimental and estimated results obtained by the proposed MSGO algorithm in the case of a **DDM**: (**a**) Current_Voltage; (**b**) Power_Voltage.

Figure 6a,b illustrate the estimated I–V characteristic of the SDM and DDM compared with the experimental one. It is noted that there is a slight advantage of the DDM compared to the SDM, which is not clear enough in the figure. Upon closer inspection of the corresponding statistical results given in Tables 5 and 8, it becomes evident that the DDM outperforms the SDM by a small margin.

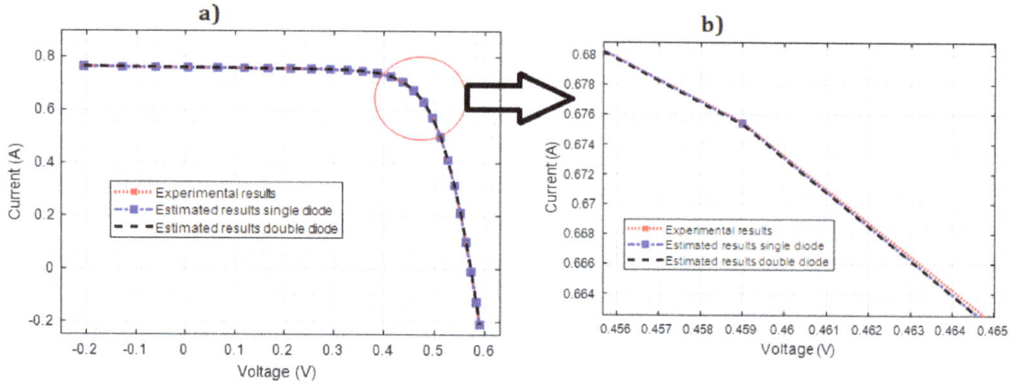

Figure 6. Experimental and estimated results: (**a**) I–V characteristic for SDM and DDM, and (**b**) Zoom of I–V characteristic in Figure 5a.

Table 8. Statistical results for the DDM (RTC France Company).

	MSGO	R_{cr}-IJADE	CARO	ABSO	ABC	NM–MPSO
Total IAE	0.0172	0.0177	0.0693	0.0178	0.0204	0.0174
RMSE	7.514×10^{-4}	7.754×10^{-4}	8.1×10^{-4}	7.682×10^{-4}	9.922×10^{-4}	7.581×10^{-4}
SSE	1.468×10^{-5}	1.56×10^{-5}	1.7×10^{-5}	1.53×10^{-5}	2.56×10^{-5}	1.49×10^{-5}
MAE	6.62×10^{-4}	6.81×10^{-4}	2.67×10^{-4}	6.83×10^{-4}	7.84×10^{-4}	6.68×10^{-4}

4.1.3. A Comparative Study of Extraction Parameters for the Amorphous PV Cell

The extracted parameters using the **MSGO** algorithm for the amorphous model are presented in Table 9. These values are compared to those obtained by other algorithms such as **QNM** and **SOMA**.

Table 9. Extracted parameters in case of amorphous cell.

Algorithm	I_{ph} (A)	d (10^{-8} m)	μ_{eff} (cm^2/V)	V_{bi} (V)	R_s (Ω)	I_s (A)	a	R_{sh} (Ω)
MSGO	0.3123	4.72	3.03	0.97	0.295	1.5	1.918	11.07
QNM	0.3043	5.8065	4.8812	0.9759	0.4242	3.0691	1.999	11.9138
SOMA	0.3181	4.9743	3.3277	0.9963	0.4706	3.0783	1.9931	13.9288

The various IAE results given by Table 10 are illustrated in Figure 7. One can notice that the IAE obtained by the MSGO algorithm (red color) is the lowest error for most values, as is confirmed in Table 11.

Table 10. Estimated data and resulted IAE obtained by the proposed algorithm **MSGO**.

	V_{exp} (V)	I_{exp} (A)	I_{est} (A)	IAE		
				MSGO	**QNM**	**SOMA**
1	1.525	0	−0.00328	0.00328	0.0041	0.00086
2	1.515	0.0158	0.0137831	0.002017	0.0058	0.0027
3	1.5	0.0302	0.0378181	0.007618	0.0003	0.0032
4	1.4775	0.0619	0.0701493	0.008249	0.0028	0.0004
5	1.47	0.0868	0.0798933	0.006907	0.0188	0.0153
6	1.445	0.1142	0.1085644	0.005636	0.0187	0.0138
7	1.37	0.1604	0.1623759	0.001976	0.0055	0.0075
8	0	0.3044	0.3042089	0.000191	0.0209	0.0026

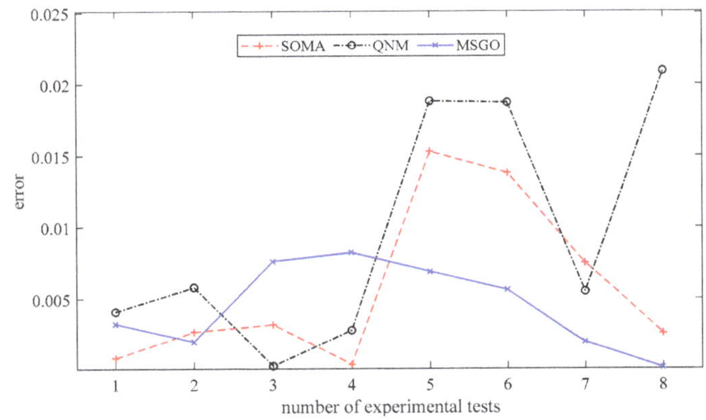

Figure 7. Estimated errors IAE.

Table 11. Statistical results for amorphous model.

	MSGO	SOMA	QNM
Total IAE	0.03587	0.04636	0.07690
RMSE	5.295×10^{-3}	7.952×10^{-3}	1.239×10^{-2}
SSE	2.243×10^{-4}	5.059×10^{-4}	1.228×10^{-3}
MAE	4.484×10^{-3}	5.796×10^{-3}	9.612×10^{-3}

4.2. Experimental Validation

The I–V and P–V characteristics of the TITAN-12-50 photovoltaic panel are implemented using the experimental test bench shown in Figure 8. The Parameter Specification of the TITAN-12-50 PV module is given in Table 12. To determine the various parameters of the photovoltaic generator, voltage and current measurements are required. These measurements are carried out using the LV25-P voltage sensor (Octapart, New York, NY, USA) and LA25-NP current sensor (Infineon, Munich, Germany). The solar sensor based on TLO82 is used to measure solar irradiation; meanwhile, the temperature is measured with an LM335 temperature sensor (ES Systems, Neo Psychico, Greece). The voltage and current are varied utilizing a variable resistor. The electronic oscilloscope scopiX, II (OX 7104) (TiePie, Sneek, The Netherlands) is used to display and record this variation.

Figure 8. Experimental test bench.

Table 12. Parameter Specification of the TITAN-12-50 PV module.

Characteristics	Value
I_{sc} (A)	3.2
V_{0c} (V)	21
P_{mpp} (W)	50
I_{mpp} (A)	2.9
V_{mpp} (V)	17.2
Cells number	32

Several experiments are conducted in this study to evaluate the I–V and P–V characteristics of the developed models under various lighting and temperature conditions. Various environmental factors affect the performance of a PV generator under real-life conditions. Consequently, four different tests are performed, and their data are recorded and presented in Figures 9 and 10.

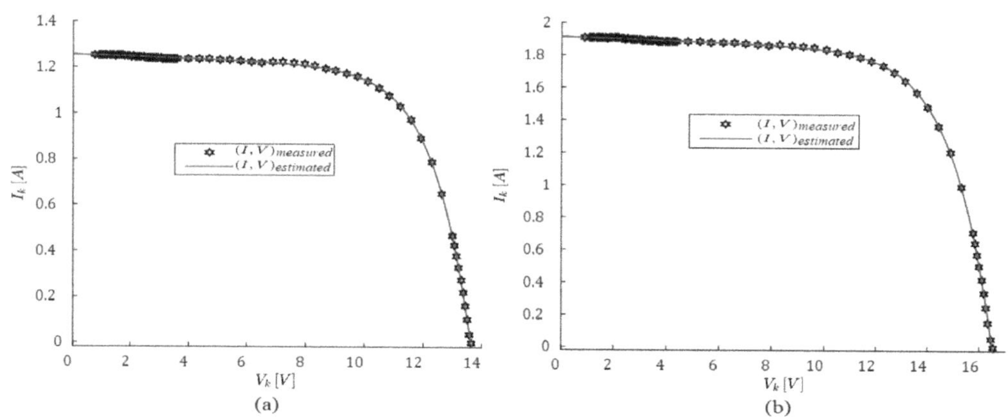

Figure 9. Comparison of I–V curves for TITAN-12-50 PV module using MSGO algorithm: (**a**) G = 360 W/m^2 and T = 18 °C; and (**b**) G = 556 W/m^2 and T = 20.5 °C.

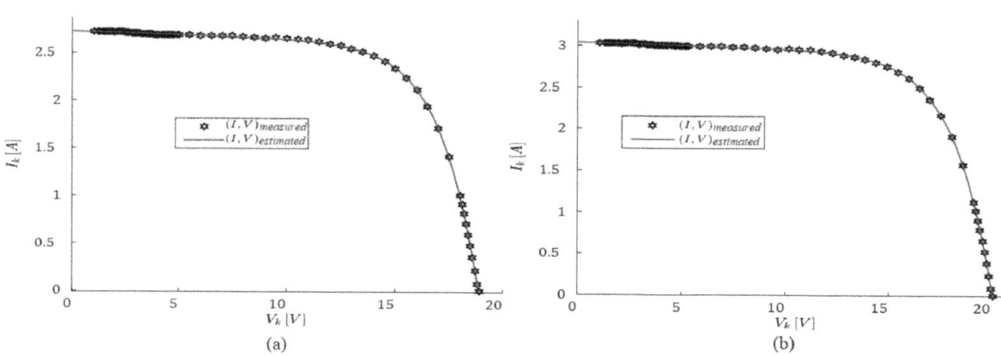

Figure 10. Comparison of I–V curves for TITAN-12-50 PV module using MSGO algorithm: (**a**) G = 810 W/m^2 and T = 22.60 °C; and (**b**) G = 900 W/m^2 and T = 23.80 °C.

According to the presented results in Figures 9 and 10, it can be noticed that the estimated current coincides with the measured current for the different cases of environmental factors.

This shows the effectiveness of the MSGO algorithm against changes in temperature and irradiation.

The convergence curves of various algorithms are illustrated in Figure 11. The average fitness functions ALO, WOA, VCS, GSA, SSA, and SCA are given in Ref. [40]. The MSGO results are presented in Figure 11 in black. In general, all algorithms exhibit an acceptable variation in the fitness function. The fastest convergence rate is seen in the MSGO results.

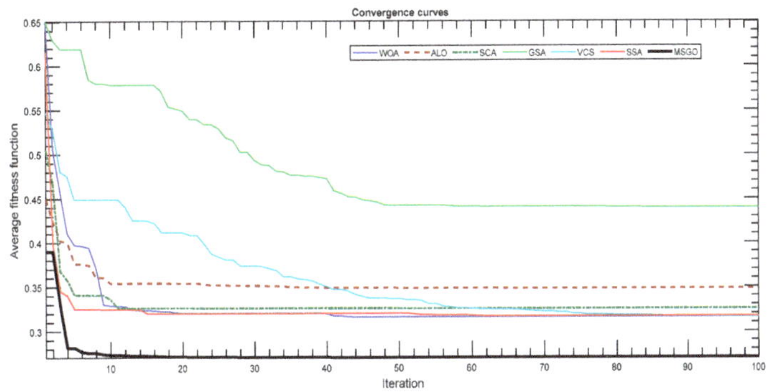

Figure 11. Average fitness functions.

All different parameters of the TITAN-12-50 solar panel extracted under G = 810 W/m^2 and T = 22.70 °C are illustrated in Table 13.

Table 13. Obtained RMSE values and estimated parameters of PV cells.

Parameters	I_{ph} (A)	R_s (Ω)	R_p (Ω)	α_1	α_2	I_{01} (μA)	I_{02} (μA)	RMS
MSGO	3.2810	0.3562	105.6204	1.4046	1.4046	4.48	4.48	1.3637×10^{-6}
SCA	2.74	0.169	72.000	1.456	1.200	9	9	1.3937×10^{-5}
ALO	2.733	0.489	50	1	1	0.6869	0.6786	1.5665×10^{-4}
GSA	2.716	0.818	140.659	1.013	1.058	0.6999	0.6405	4.8032×10^{-5}
VCS	2.734	0.333	70.189	1.003	1.002	0.6990	0.6993	1.6188×10^{-6}
WOA	2.75	0.351	90	1.60	1.48	7	7	3.6935×10^{-4}
SSA	2.722	0.174	98	1.2	1.3	7.8	7.8	1.5777×10^{-6}

Table 13 shows the extracted parameters of the TITAN-12-50 PV module at G = 810 W/m^2 and T = 22.70 °C. The MSGO parameter results are compared with previous results, given by Ref. [40], taking into account the same temperature and irradiation conditions. The proposed MSGO algorithm achieved the lowest RMSE value compared to all other algorithms by 1.3637×10^{-6}. After analyzing the data presented in Table 13, Figure 10, and Figure 11, it can be inferred that the MSGO algorithm exhibits several advantages, such as rapid convergence and minimal errors.

5. Discussion

The MSGO algorithm was developed to enhance the precision of solar cell parameter extraction. It was tested on monocrystalline, polycrystalline, and amorphous PV cells with the SDM and DDM to evaluate its performance. The MSGO algorithm was then compared to other techniques in the literature to determine its effectiveness. Results from statistical analysis and figures indicate that the MSGO algorithm is highly accurate and robust. Additionally, the results obtained from the MSGO algorithm are more promising than those of other previously proposed methods.

To further confirm the accuracy of the simulation model, experimental tests were conducted. The results of these tests indicate that the proposed MSGO algorithm is remarkably accurate, fast, and convergent, outperforming other similar algorithms proposed in the literature.

Author Contributions: All authors contributed to formulating the research idea, algorithm design, result analysis, writing, and reviewing the research. Conceptualization, H.K. and E.T.; data curation, E.T.; funding acquisition, E.T. and A.M.A.; investigation, M.J.; methodology, A.A. and A.M.A.; project administration, H.K.; resources, L.S.; software, T.I.A.; validation, A.A.; writing—original draft, H.K. and A.M.A.; writing—review and editing, T.I.A. All authors have read and agreed to the published version of the manuscript.

Funding: This work is funded by the Deanship of Scientific Research at Northern Border University under research group grant number (RG-NBU-2022-1215).

Institutional Review Board Statement: Not applicable.

Informed Consent Statement: Not applicable.

Data Availability Statement: Not applicable.

Acknowledgments: The authors extend their appreciation to the Deanship of Scientific Research at Northern Border University for funding this work through research group No. (RG-NBU-2022-1215). The authors gratefully thank the Prince Faisal bin Khalid bin Sultan Research Chair in Renewable Energy Studies and Applications (PFCRE) at Northern Border University for their support and assistance.

Conflicts of Interest: The authors declare no conflict of interest.

References

1. Ezzeddine, T. Reactive power analysis and frequency control of autonomous wind induction generator using particle swarm optimization and fuzzy logic. *Energy Explor. Exploit.* **2019**, *38*, 755–782. [CrossRef]
2. Touti, E.; Zayed, H.; Pusca, R.; Romary, R. Dynamic Stability Enhancement of a Hybrid Renewable Energy System in Stand-Alone Applications. *Computation* **2021**, *9*, 14. [CrossRef]
3. Kariem, H.; Touti, E.; Fetouh, T. The efficiency of PSO-based MPPT technique of an electric vehicle within the city. *Meas. Control.* **2020**, *53*, 461–473. [CrossRef]
4. Gul, M.; Kotak, Y.; Muneer, T. Review on recent trend of solar photovoltaic technology. *Energy Explor. Exploit.* **2016**, *34*, 485–526. [CrossRef]
5. Kihlström, V.; Elbe, J. Constructing Markets for Solar Energy—A Review of Literature about Market Barriers and Government Responses. *Sustainability* **2021**, *13*, 3273. [CrossRef]
6. Aminian, E.; Saffari, H. Experimental analysis of dropwise condensation heat transfer on a finned tube: Impact of pitch size. *Proc. Inst. Mech. Eng. Part A J. Power Energy* **2021**, *236*, 752–759. [CrossRef]
7. Belloni, M.; Conti, T.D.N. The Connection of Solar Generators to the Electricity Distribution Network as a Means of Mitigating Environmental Impacts by Dispensing the Banks of Accumulators. *Energy Power Eng.* **2019**, *11*, 392–397. [CrossRef]
8. Kumar, R.; Deshmukh, V.; Bharj, R.S. Performance enhancement of photovoltaic modules by nanofluid cooling: A comprehensive review. *Int. J. Energy Res.* **2020**, *44*, 6149–6169. [CrossRef]
9. Silva, P.R.; da Silva Jota, P.R.; Batista, A.P. PV Characterization System Outdoors—Case Study in Brazil. *J. Power Energy Eng.* **2017**, *05*, 119–132. [CrossRef]
10. Mokarram, M.; Mokarram, M.J.; Khosravi, M.R.; Saber, A.; Rahideh, A. Determination of the optimal location for constructing solar photovoltaic farms based on multi-criteria decision system and Dempster–Shafer theory. *Sci. Rep.* **2020**, *10*, 8200. [CrossRef]
11. Kraiem, H.; Aymen, F.; Yahya, L.; Triviño, A.; Alharthi, M.; Ghoneim, S.S.M. A Comparison between Particle Swarm and Grey Wolf Optimization Algorithms for Improving the Battery Autonomy in a Photovoltaic System. *Appl. Sci.* **2021**, *11*, 7732. [CrossRef]
12. Stutenbaeumer, U.; Mesfin, B. Equivalent model of monocrystalline, polycrystalline and amorphous silicon solar cells. *Renew. Energy* **1999**, *18*, 501–512. [CrossRef]
13. Anani, N.; Ibrahim, H.K. Adjusting the Single-Diode Model Parameters of a Photovoltaic Module with Irradiance and Temperature. *Energies* **2020**, *13*, 3226. [CrossRef]
14. Taherbaneh, M.; Rezaie, A.H.; Ghafoorifard, H.; Rahimi, K.; Menhaj, M.B.; Milimonfared, J.M. Evaluation of two-diode-model of a solar panel in a wide range of environmental conditions. *Int. J. Electron.* **2011**, *98*, 357–377. [CrossRef]
15. Qais, M.H.; Hasanien, H.M.; Alghuwainem, S.; Nouh, A.S. Coyote optimization algorithm for parameters extraction of three-diode photovoltaic models of photovoltaic modules. *Energy* **2019**, *187*, 116001. [CrossRef]
16. Premkumar, M.; Subramaniam, U.; Babu, T.S.; Elavarasan, R.M.; Mihet-Popa, L. Evaluation of Mathematical Model to Characterize the Performance of Conventional and Hybrid PV Array Topologies under Static and Dynamic Shading Patterns. *Energies* **2020**, *13*, 3216. [CrossRef]
17. Muhammadsharif, F.F.; Hashim, S.; Hameed, S.S.; Ghoshal, S.; Abdullah, I.K.; Macdonald, J.; Yahya, M.Y. Brent's algorithm based new computational approach for accurate determination of single-diode model parameters to simulate solar cells and modules. *Sol. Energy* **2019**, *193*, 782–798. [CrossRef]

18. Muhammadsharif, F.F.; Hashim, S. A Simple and Efficient Determination of the Ideality Factor of Solar Cells and Modules from the Knee Point of the Shunt Resistance Curve. *Arab. J. Sci. Eng.* **2023**, *48*, 8217–8225. [CrossRef]
19. DChan, D.S.H.; Phang, J.C.H. Analytical methods for the extraction of solar-cell single- and double-diode model parameters from I-V characteristics. *IEEE Trans. Electron Devices* **1987**, *34*, 286–293. [CrossRef]
20. Ibrahim, H.; Anani, N. Evaluation of Analytical Methods for Parameter Extraction of PV modules. *Energy Procedia* **2017**, *134*, 69–78. [CrossRef]
21. Oliva, D.; Cuevas, E.; Pajares, G.J.E. Parameter identification of solar cells using artificial bee colony optimization. *Energy* **2014**, *72*, 93–102. [CrossRef]
22. Narasimman, K.; Iniyan, S.; Goic, R. Experimental investigation of ridge concentrator PV-based solar water pumping system for small-scale applications. *Energy Sources Part A Recover. Util. Environ. Eff.* **2019**, *42*, 1844–1860. [CrossRef]
23. Fadlallah, S.O.; Serradj, D.E.B. Determination of the optimal solar photovoltaic (PV) system for Sudan. *Sol. Energy* **2020**, *208*, 800–813. [CrossRef]
24. Bastidas-Rodriguez, J.; Petrone, G.; Ramos-Paja, C.; Spagnuolo, G. A genetic algorithm for identifying the single diode model parameters of a photovoltaic panel. *Math. Comput. Simul.* **2017**, *131*, 38–54. [CrossRef]
25. Chen, X.; Xu, B.; Mei, C.; Ding, Y.; Li, K. Teaching–learning–based artificial bee colony for solar photovoltaic parameter estimation. *Appl. Energy* **2018**, *212*, 1578–1588. [CrossRef]
26. Yang, X.; Gong, W.; Wang, L. Comparative study on parameter extraction of photovoltaic models via differential evolution. *Energy Convers. Manag.* **2019**, *201*, 112113. [CrossRef]
27. Askarzadeh, A.; dos Santos Coelho, L. Determination of photovoltaic modules parameters at different operating conditions using a novel bird mating optimizer approach. *Energy Convers. Manag.* **2015**, *89*, 608–614. [CrossRef]
28. Kanimozhi, G.; Kumar, H. Modeling of solar cell under different conditions by Ant Lion Optimizer with LambertW function. *Appl. Soft Comput.* **2018**, *71*, 141–151. [CrossRef]
29. Subudhi, B.; Pradhan, R. Bacterial Foraging Optimization Approach to Parameter Extraction of a Photovoltaic Module. *IEEE Trans. Sustain. Energy* **2017**, *9*, 381–389. [CrossRef]
30. Saxena, A.; Sharma, A.; Shekhawat, S. Parameter extraction of solar cell using intelligent grey wolf optimizer. *Evol. Intell.* **2020**, *15*, 167–183. [CrossRef]
31. Elazab, O.S.; Hasanien, H.M.; Elgendy, M.A.; Abdeen, A.M. Parameters estimation of single- and multiple-diode photovoltaic model using whale optimisation algorithm. *IET Renew. Power Gener.* **2018**, *12*, 1755–1761. [CrossRef]
32. Kumar, C.; Raj, T.D.; Premkumar, M. A new stochastic slime mould optimization algorithm for the estimation of solar photovoltaic cell parameters. *Optik* **2020**, *223*, 165277. [CrossRef]
33. Ben Messaoud, R. Extraction of uncertain parameters of single and double diode model of a photovoltaic panel using Salp Swarm algorithm. *Measurement* **2019**, *154*, 107446. [CrossRef]
34. Naik, A.; Satapathy, S.C.; Abraham, A. Modified Social Group Optimization—A meta-heuristic algorithm to solve short-term hydrothermal scheduling. *Appl. Soft Comput.* **2020**, *95*, 106524. [CrossRef]
35. Hamid, N.F.A.; Rahim, N.A.; Selvaraj, J. Solar cell parameters identification using hybrid Nelder-Mead and modified particle swarm optimization. *J. Renew. Sustain. Energy* **2016**, *8*, 015502. [CrossRef]
36. Dkhichi, F.; Oukarfi, B.; Fakkar, A.; Belbounaguia, N. Parameter identification of solar cell model using Levenberg–Marquardt algorithm combined with simulated annealing. *Sol. Energy* **2014**, *110*, 781–788. [CrossRef]
37. Niu, Q.; Zhang, L.; Li, K. A biogeography-based optimization algorithm with mutation strategies for model parameter estimation of solar and fuel cells. *Energy Convers. Manag.* **2014**, *86*, 1173–1185. [CrossRef]
38. Gong, W.; Cai, Z. Parameter extraction of solar cell models using repaired adaptive differential evolution. *Sol. Energy* **2013**, *94*, 209–220. [CrossRef]
39. Askarzadeh, A.; Rezazadeh, A. Artificial bee swarm optimization algorithm for parameters identification of solar cell models. *Appl. Energy* **2013**, *102*, 943–949. [CrossRef]
40. Yuan, X.; He, Y.; Liu, L. Parameter extraction of solar cell models using chaotic asexual reproduction optimization. *Neural Comput. Appl.* **2014**, *26*, 1227–1239. [CrossRef]
41. Wang, Y.; Xi, J.; Han, N.; Xie, J. Modeling method research of flexible amorphous silicon solar cell. *Appl. Sol. Energy* **2015**, *51*, 41–46. [CrossRef]
42. Abbassi, R.; Abbassi, A.; Heidari, A.A.; Mirjalili, S. An efficient salp swarm-inspired algorithm for parameters identification of photovoltaic cell models. *Energy Convers. Manag.* **2018**, *179*, 362–372. [CrossRef]
43. Rodrigo, P.; Fernández, E.; Almonacid, F.; Pérez-Higueras, P. Models for the electrical characterization of high concentration photovoltaic cells and modules: A review. *Renew. Sustain. Energy Rev.* **2013**, *26*, 752–760. [CrossRef]
44. Bishop, J. Computer simulation of the effects of electrical mismatches in photovoltaic cell interconnection circuits. *Sol. Cells* **1988**, *25*, 73–89. [CrossRef]
45. Sulyok, G.; Summhammer, J. Extraction of a photovoltaic cell's double-diode model parameters from data sheet values. *Energy Sci. Eng.* **2018**, *6*, 424–436. [CrossRef]
46. Elazab, O.S.; Hasanien, H.M.; Alsaidan, I.; Abdelaziz, A.Y.; Muyeen, S.M. Parameter Estimation of Three Diode Photovoltaic Model Using Grasshopper Optimization Algorithm. *Energies* **2020**, *13*, 497. [CrossRef]

47. Satapathy, S.; Naik, A. Social group optimization (SGO): A new population evolutionary optimization technique. *Complex Intell. Syst.* **2016**, *2*, 173–203. [CrossRef]
48. Naik, A.; Satapathy, S.C.; Ashour, A.S.; Dey, N. Social group optimization for global optimization of multimodal functions and data clustering problems. *Neural Comput. Appl.* **2016**, *30*, 271–287. [CrossRef]

sustainability MDPI

Article

A New Five-Port Energy Router Structure and Common Bus Voltage Stabilization Control Strategy

Xianyang Cui [1,2], Yulong Liu [1,2], Ding Yuan [1,2], Tao Jin [1,2,*] and Mohamed A. Mohamed [3,*]

1 College of Electrical Engineering and Automation, Fuzhou University, Fuzhou 350108, China
2 Fujian Province University Engineering Research Center of Smart Distribution Grid Equipment, Fuzhou 350108, China
3 Electrical Engineering Department, Faculty of Engineering, Minia University, Minia 61519, Egypt
* Correspondence: jintly@fzu.edu.cn (T.J.); dr.mohamed.abdelaziz@mu.edu.eg (M.A.M.)

Abstract: Multi-port energy routers are a core device that integrates distributed energy sources and enables energy-to-energy interconnections. For the energy routing system, the construction of its topology, the establishment of internal model switching and the control of common bus voltage stability are the key elements of the research. In this paper, a five-port energy router structure is proposed, including a PV port, an energy storage port, a grid-connected port, a DC load port, and an AC load port. Among them, the energy storage port and the grid-connected port involve bidirectional energy flow, which are the core ports of control. For the system state, a model switching strategy is proposed based on the topology and the port energy flow direction. When the external conditions change, the system can be stabilized by means of a quick response from the energy storage port. When the energy storage is saturated, the state is switched, and the grid-connected port works to achieve system stability. The rapid stabilization of the bus voltage and the free flow of energy are achieved by combining the fast response of the model predictive control with the properties of multiple model switching. Finally, the feasibility of this energy router topology and control strategy is verified by building simulations in MATLAB.

Keywords: energy internet; energy router; common bus voltage; cooperative control; model transformation

Citation: Cui, X.; Liu, Y.; Yuan, D.; Jin, T.; Mohamed, M.A. A New Five-Port Energy Router Structure and Common Bus Voltage Stabilization Control Strategy. *Sustainability* **2023**, *15*, 2958. https://doi.org/10.3390/su15042958

Academic Editor: Pablo García Triviño

Received: 21 December 2022
Revised: 1 February 2023
Accepted: 2 February 2023
Published: 6 February 2023

1. Introduction

With the shortage of energy and the increasing problem of environmental pollution, distributed energy has gradually come into people's view, and the energy internet (EI), with electricity at its core, has come to be paid more attention [1,2]. The energy router (ER) is the core equipment for energy internet architecture, effectively integrating distributed energy sources, realizing AC–DC free conversion and multi-directional flow of energy, and effectively reducing the impact brought about by distributed energy sources and AC–DC load access to the power system [3–5], and the popular use of energy routers is an essential trend for the future development of the energy internet. For interconnected systems of distributed energy sources, a reasonable energy router architecture, suitable energy scheduling and coordination control strategy are the key factors for determining the stable operation of the system [6,7]. ER can not only solve the problem of distributed energy grid connection, but also provide a plug-and-play interface for different devices. ER can achieve the goal of power system control integration. Therefore, research on ER is very urgent.

Depending on the demand and the location, energy routers with different classifications are also used, and can mainly be divided into terminal energy routers, regional energy routers, and backbone energy routers [8–10]. Normally, energy routers applied to the backbone and area are relatively mature, while those applied to the terminal have been less thoroughly researched. Energy routers for terminal applications are widely used, and can effectively utilize distributed energy and meet the development goal of "double carbon", having high research significance.

Therefore, energy routers have accordingly been studied in terms of overall architecture [11], control strategy [12,13], internal communication design [14], and key parameters [15], but so far, no expert scholars have proposed a general energy router architecture and a control strategy that can satisfy various operating conditions. Ref. [16] proposed a coordinated control strategy for AC–DC hybrid ER based on energy storage and voltage stabilization that used multiple sets of converters to provide a wealth of AC–DC ports, and also classified the external ports for consideration, but its module control strategy was relatively simple, and the dynamic performance required improvement. Ref. [17] proposed a multi-LAN port energy router that could easily connect distributed power sources and AC–DC loads to the backbone grid, but did not consider the access of energy storage systems, and the energy router architecture was not complete. Ref. [18] provided the design idea of a community energy router, giving different operating states of the energy router under distribution network failure. However, no specific feasible topology was given. In [19], a power allocation strategy for seamless switching conditions was proposed, but it lacked precise consideration of the battery. In [20], a two-level modular ER structure that could be applied to DC microgrid clusters was proposed to realize the function of flexible interconnection of microgrids. In [21], the impact on the energy router under grid fault conditions and the balancing strategy of the capacitor voltage under various conditions were analyzed, but only situations under single-phase grid faults were considered, without considering two-phase or multiple three-phase faults.

The study of ER is at this stage no longer limited to the ER itself, as the grid-connected port is connected to the distribution grid. Therefore, research on the power quality of the distribution grid port and the energy allocation and optimization of the distribution grid side is also meaningful. In ref. [22], the authors proposed a hierarchical grid model to enhance the resilience of DC-MGs. In refs. [23,24], the researchers studied energy hubs. These include wind power, solar power, and energy storage. The management of distributed energy sources and the improvement of distribution network reliability, respectively, were studied. However, these papers are all macroscopic studies, but do not explain the details of ER.

Due to the nature of the energy router, the presence of a grid-connected port is necessary. When the energy router is connected to the grid, if it is not established with a battery or the battery is not properly controlled and directly connected to the grid, it will result in high network loss. The battery is able to play the role of a bridge for energy circulation, carry out energy buffering, and can also help the energy router to stabilize the bus voltage. For the regulation of the energy router, the use of the battery, the state of charge of the battery, and the comprehensive deployment of the battery discharge power also need to be considered [25–28]. Refs. [29,30] mention a variety of optimization algorithms that serve as a theoretical basis for the future expansion of ER and its integration with power systems, and the future expansion of power systems will be studied in depth on the basis of the algorithms. Refs. [31,32] propose optimization algorithms in combination with biology, and these algorithms are better able to optimize the energy allocation strategy of energy routers and improve their efficiency.

On the basis of the existing research, the study of ER has primarily focused on structure as well as on control. Therefore, in this paper, we design a five-port energy router topology in consideration of these two points and propose a corresponding control strategy for this structure.

The first point is the design of the topology. In accordance with the requirements, an ER including PV, energy storage, grid-connected, AC, and DC ports was proposed. All ports are connected by a DC bus, and are unified and controlled by a central processor. The topologies of different ports are designed to realize the energy flow.

The second point is the implementation of the control strategy. The control strategy is divided into two layers, which are the independent control of the ports and the overall coordinated control. For the port functions, each port designs an independent control strategy to maintain stability. At the same time, the top-level control is designed to divide the working modes, and the different ports operate in a coordinated fashion under different working modes to finally work stably.

2. Energy Router Topology Creation and Analysis

In the original model, the ER system integrates the PV, the AC load, and the DC load. In response to the need to provide various ports involved in energy routers, this paper proposes a five-port energy router for a wide range of applications. The router architecture is shown in Figure 1, and the five groups of ports are the PV port, the energy storage port, the grid-connected port, the DC load port, and the AC load port. All ports of the energy router converge into a DC common bus (CB), and U_{CB} is the common bus voltage. i_1~i_5 are the output current values of the five ports, which can represent the interaction with the common bus energy handover.

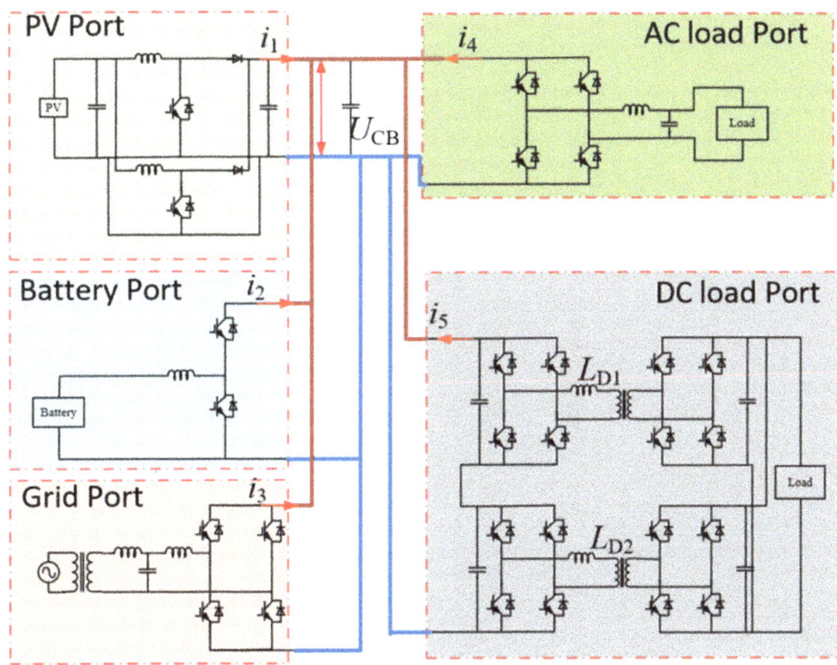

Figure 1. The architecture of five-port ER.

Port 1 is the PV port structure, considering the low voltage level of solar power (PV) and the high voltage level of the common bus, the boost topology is used, and considering the switching tube current stress and ripple, this port adopts the interleaved parallel boost structure. i1 is the PV port and common bus interaction current, the PV port and common bus interaction power $P_1 = i_1 U_{CB}$, for the energy storage port, the adopted structure is a two-level converter; i_2 is the storage port, and the common bus current storage port interaction power $P_2 = i_2 U_{CB}$, where the port plays the role of stabilizing the DC bus voltage. For the grid-connected port, the structure of a two-level converter combined with an LCL filter is used, which is connected to the grid to realize the interaction of energy with the grid, and the interaction power is $P_3 = i_3 U_{CB}$, for the AC load port, the structure of a two-level converter and an LC filter is used, and the interaction power with the common bus is $P_4 = i_4 U_{CB}$; for the DC load port, a parallel input structure is used. For the DC load port, the structure used is an input parallel-type double active bridge converter (ISOP DAB), and the interaction power of the port with the common bus is $P_5 = i_5 U_{CB}$.

The system has a total of five groups of ports, ignoring port losses, and the real-time power balance of the system using the common bus is shown below.

$$P_1\Delta t + P_2\Delta t + P_3\Delta t + P_4\Delta t + P_5\Delta t + P_{non}\Delta t = 0 \quad (1)$$

$$P_{non}\Delta t = \frac{1}{2}C_{CB}U_{CB}^2 \quad (2)$$

$$\sum_{k=1}^{5} i_k U_{CB} + \frac{1}{2}C_{CB}U_{CB}^2 = 0 \quad (3)$$

where U_{CB} is the current common bus voltage value, C_{CB} is the DC bus capacitance value, i_k is the interaction power generated by each port with the DC bus, and P_{non} is the unbalanced power of the system at the current moment. Due to the changes in PV, load, etc., fluctuations in P_{non} will be caused, and will be accompanied by fluctuations in U_{CB}. Therefore, the overall coordination control starts from the stable control of U_{CB}, and stable operation of the system is finally realized by controlling the energy interaction status of the energy storage port and the grid-connected port with the main line.

For the energy router, the five groups of ports involved are divided into two parts, consisting of the functional port, for the unidirectional flow of energy, and the voltage stabilization port, for the bidirectional flow of energy. Among them, the PV, AC and DC load ports are the functional ports, and the grid-connected energy storage ports are the voltage stabilization ports.

The structural design of the PV port adopts the structure presented in Figure 2.

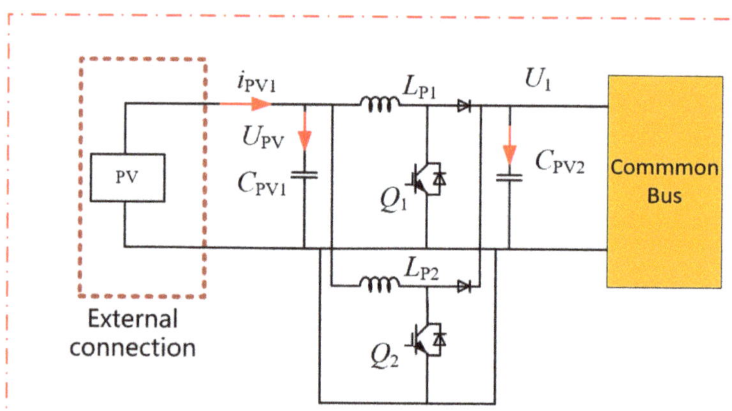

Figure 2. The architecture of the PV port.

The function of the PV port is to incorporate solar energy into the energy router system, and the criterion of its work is the full utilization of light energy without considering factors such as energy flow system stability, so the design and control of the PV port start from the efficient acquisition of energy without involving other factors. Considering the characteristics of low PV voltage level, multiple parallel PV cells, and high DC bus voltage level, a boost converter is used as the port, and considering the expandability and the current stress of the switching devices, the boost converter structure adopts an interleaved parallel boost circuit. When the power is the same, the current stress of the staggered parallel booster structure switch tube is lower than that of the ordinary booster circuit; the current ripple is also smaller, which is advantageous for the stability of the common bus voltage and current control.

For the design of the AC load port, considering the controllability of the system and the overall control difficulty, the AC port is not involved in internal voltage stabilization, and is only used as a functional port to supply the load, so the control method and basic structure are the same as those of the conventional inverter. It is sufficient to ensure the output stability and realize the basic external functions. Figure 3 shows the topology of the AC load port, with the output connected to the outside through an LC filter.

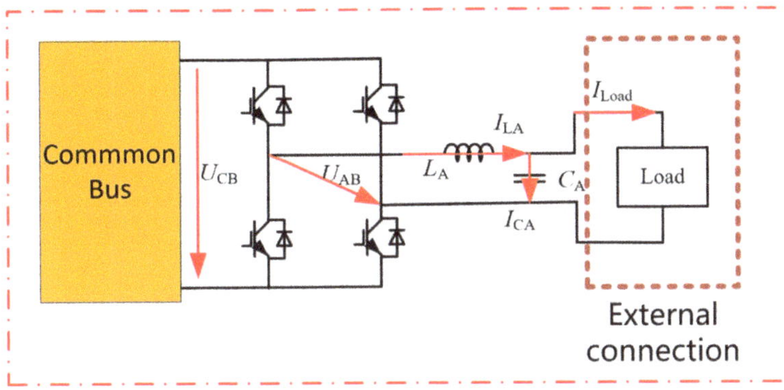

Figure 3. Topology of AC load port.

For the DC port shown in Figure 4, the overall situation is similar to that of the AC port, which is also only used as a functional port to provide DC output without participating in system regulation, and only needs to ensure output stability and achieve basic functions. Considering the widespread use of electric vehicles, the DC port usually needs to provide a greater amount of power, and in terms of electrical isolation, the output of the two modules is connected in series with the output of the parallel-type double active bridge converter (ISOP DAB) to realize the external connection. This structure can effectively increase the output power and reduce the device's current stress.

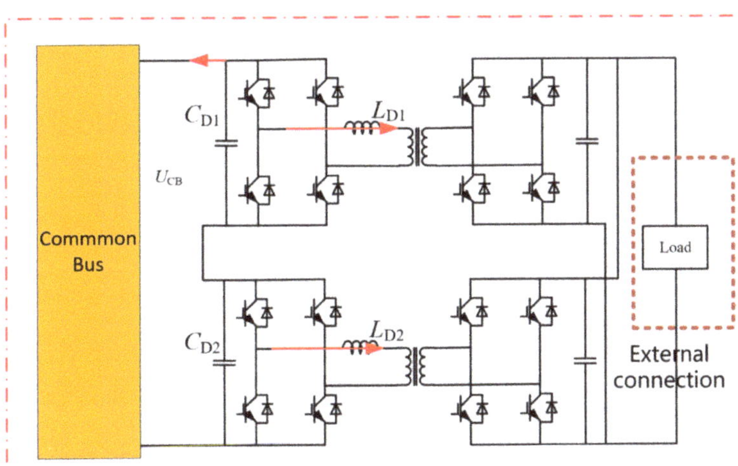

Figure 4. Topology of the DC load port.

For a two-module ISOP-type DAB structure, as described in Figure 5, the modules at the input are voltage-divided, and the voltage at the output is automatically equalized. Due to the manufacturing process, the parameters between the modules are not exactly the same, and the power balance between the modules needs to be considered. When the input voltage equalization loop is in stable operation, the system achieves power equalization among modules and current equalization at the output with $I_{21} = I_{22}$. When the input voltage of the system is disturbed, the input voltage equalization loop can be controlled so that the input voltage of each module receives equalized control. When the output current of the module is disturbed and increases, the output voltage of the module decreases, because the input voltage equalization loop has already achieved system power balance.

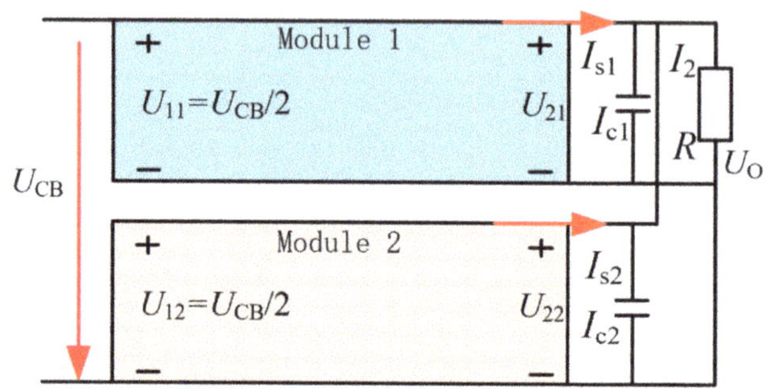

Figure 5. Equivalent diagram of ISOP DAB converter.

The output voltage drop leads to the current flowing to the load becoming smaller and the current flowing to the inductor becoming larger, at which time the inductor is charged and the voltage increases, forming negative feedback regulation, and the system output voltage restores a stable value again. As can be observed, the input voltage equalization loop can simultaneously control the system input voltage waveform and the system output current fluctuation, and ensure the power balance between the system modules from the input side and the output side at the same time.

The grid-connected port is different from the two load ports and the PV port in that it assumes the function of bidirectional energy flow, and is thus an important structure for realizing the energy routing function by obtaining energy from the grid to provide loads when the system energy is insufficient and feeding the excess energy back to the grid when the system energy is excessive. For the grid-connected ports, the structure used is the single-phase rectifier structure shown in Figure 6.

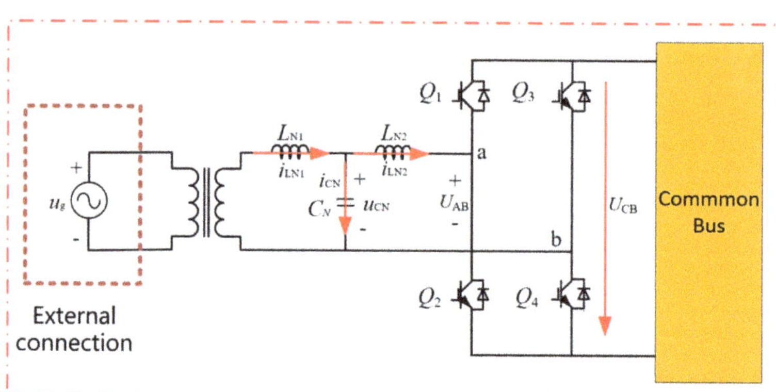

Figure 6. Topology of the grid-connected port.

The grid-connected port is connected to the grid through a set of LCL filters. As a third-order system, LCL filters are small in size and have good high-frequency filtering performance, but the structure has resonant peaks, which can reduce the stability of the system. In the control, the resonant peaks are suppressed by adding virtual resistors.

The energy storage port plays the role of stabilizing the DC bus voltage, connecting the battery, and buffering the energy, and is the most critical port in terms of the control of the energy router. This port needs to realize the bidirectional flow of energy, and also requires rapid control that is able to quickly sense the fluctuation of the common bus voltage and make adjustments in the control. Therefore, a bidirectional buck–boost structure, as shown in Figure 7, was chosen, and its control adopts the faster model predictive control.

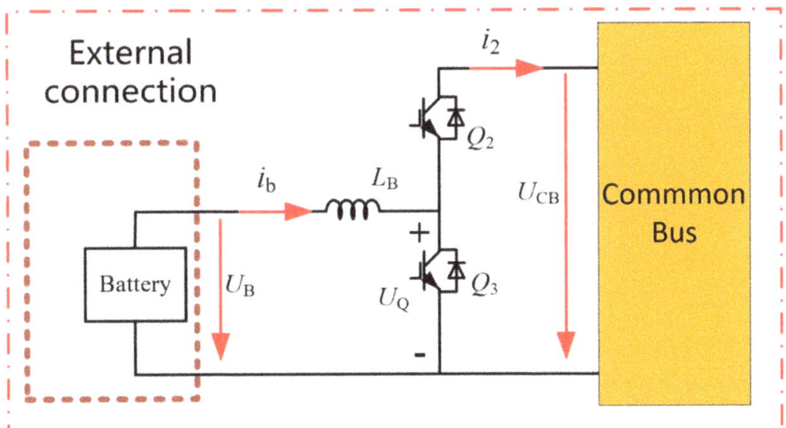

Figure 7. Structure of the energy storage part.

The port realizes the bidirectional flow of energy by controlling the opening and closing of switch tubes Q_2 and Q_3. When Q_3 is cut off, the opening and closing of Q_2 are controlled, meaning that the circuit works in the boost mode, whereby energy is transferred from the battery to the DC bus. When Q_2 is cut off, the opening and closing of Q_3 are controlled, and the circuit works in the buck mode, whereby energy is transferred from the DC bus to the battery, thus realizing the bidirectional flow of energy.

An overview of the energy flow directions and functions of the five ports is provided in Table 1. Obviously, the energy of the first two ports flows in both directions, and the functions are more complex and have more important positions. The last three ports only function to perform energy transfer, and do not play a dominant role in the control.

Table 1. ER ports and functional description.

Port	Direction of Energy Flow	Function
Battery Port	Bidirectional	1. Coordinated energy flow 2. Stabilize UCB 3. Coordinated energy storage batteries
Grid port	Bidirectional	1. Coordinated energy flow 2. Stabilize UCB 3. Integrating ER with the grid
PV Port	Unidirectional external to ER	Transferring solar energy to the ER.
AC load port	Unidirectional ER to external	Provide power for AC load
DC load port	Unidirectional ER to external	Provide power for the DC load

3. Energy Router Control Strategy Research

3.1. Overall Control Architecture and Control Strategy

For the control of the energy router, the most effective current is achieved using the hierarchical control strategy; in this control strategy, the port controller performs control using local information, and on this basis maintains two-way contact with the central controller. The control architecture is shown in Figure 8, where the overall control architecture is divided into four layers, comprising (from top to bottom) the top-level control platform, the information collection platform, decision control and the bottom-level control platform. Information enters the controller at the bottom and flows upwards, and control signals enter the port at the top and flow downwards.

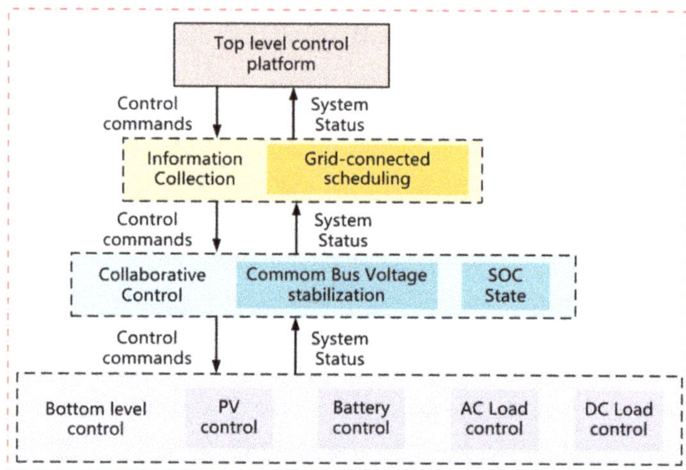

Figure 8. Hierarchical control architecture of five-port ER.

For the proposed five-port energy router topology and the hierarchical control architecture used in this paper, the bottom-level control layer and the cooperative control layer are the core of the study. The bottom-level control layer corresponds to the independent control of the five ports, and the cooperative control performs overall control of the ports' operating states by means of bus voltage and battery charge state. The cooperative control layer includes two points, bus voltage stabilization and battery charge state adjustment, both of which are closely related to the energy storage ports. Therefore, this study focuses on the energy storage port in the design and control of the bottom-level port, and establishes an overall control strategy for the port that is able to realize reasonable control of the bus voltage and stability, as well as the battery charge state, so as to finally realize the free flow among multiple energy sources and the stable operation of the energy router. The energy router works stably.

3.2. Five-Port Control Strategy Design

The first is the control of the PV port, which is a unidirectional functional port, and its control goal is to inject as much light energy as possible into the common bus. For this system, in order to ensure that the load attains maximum power, proper load matching is required. When the load resistance is equal to the internal resistance of the power supply system, the load receives the maximum power, with such a load matching process being referred to as maximum power point tracking (MPPT) [33,34]. Figure 9 shows the process of MPPT for the PV port. By collecting the power at the current moment and comparing it with the power at the previous moment, the switching tubes are controlled by judging the change in power, while using a staggered parallel boost structure, where the control signals of the two groups of switching tubes are the same, but the delta carrier signals are staggered by half a cycle to achieve harmonic offset.

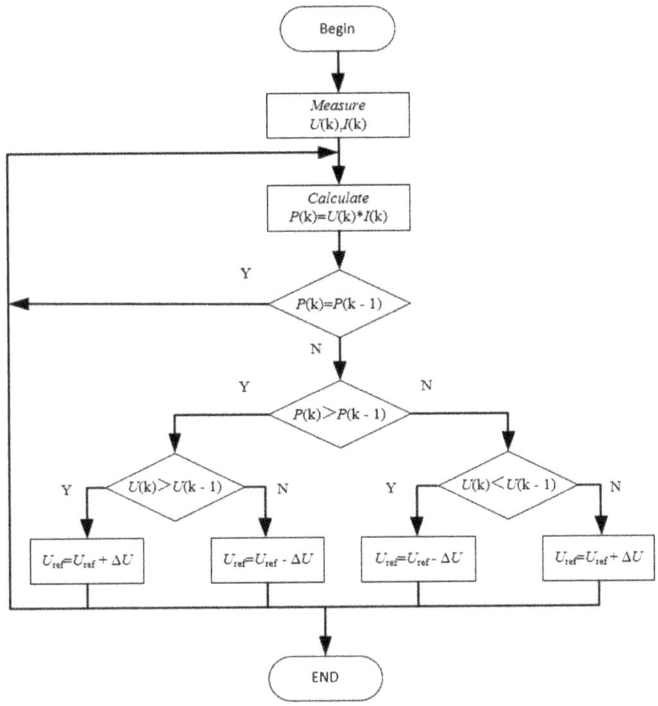

Figure 9. PV module structure and control block diagram.

For the AC load port, as can be seen from the structure topology diagram in Figure 3, the output is connected to the load using an LC filter. It is not a bad idea to set the load as a resistive load with resistance R. Then, in the complex frequency domain, we have:

$$\begin{cases} I_{LA}(s) = \frac{U_{AB}(s)-U_c(s)}{L_A s} \\ (I_{L1}(s) - I_{Load}(s))\frac{1}{Cs} = Uc(s) \\ I_{Load}(s) = \frac{U_c(s)}{R} \end{cases} \tag{4}$$

Using bipolar SPWM modulation, in one switching cycle:

$$\overline{U_{AB}} = (2D-1)U_{CB} \tag{5}$$

$$D = \frac{1}{2}(1 + \frac{V_m}{V_{tri}}) \tag{6}$$

where D is the duty cycle, V_m is the reference sine wave, which is the input to the modulator, and V_{tri} is the peak delta carrier wave.

$$\overline{U_{ab}} = (2D-1)U_{CB} \tag{7}$$

$$\overline{U_{ab}} = \left[2 \times \frac{1}{2}\left(1 + \frac{V_m}{V_{tri}}\right) - 1\right]U_{CB} = \frac{V_m}{V_{tri}}U_{CB} \tag{8}$$

Then, the transfer function of the AC port can be obtained as follows:

$$G_0(s) = \frac{U_0(s)}{V_m(s)} = \frac{1}{L_A C s^2 + \frac{L_A}{R} s + 1} \frac{U_{CB}}{V_{tri}} \tag{9}$$

For the AC load port, a dual closed-loop voltage and current control are used, and the control strategy is shown in Figure 10.

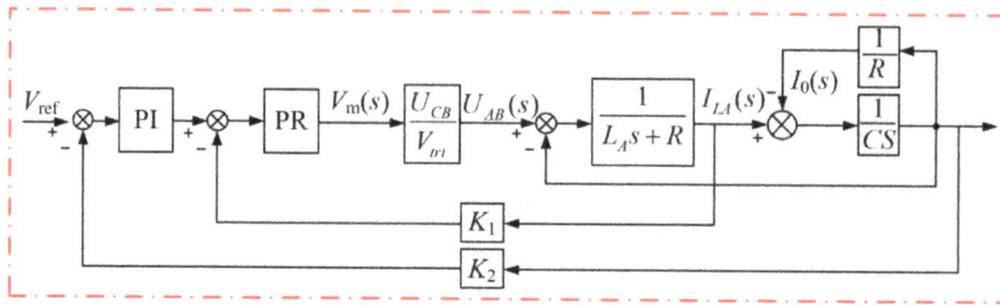

Figure 10. AC load port control structure diagram.

The voltage outer loop is the error signal obtained by comparing the reference value with the sampled value of the output voltage, and then the current reference value of the inner loop is generated by the PI controller; the current reference value of the inner loop and the inductor current obtained from the actual sampling are used to generate the current error, and the modulated wave is obtained from the proportional resonance of the inner loop. The feedforward of the reference value is also added to improve the time response characteristics.

For the DC port, an ISOP DAB topology is used, as shown in Figure 4. For this structure, a centralized dual-loop voltage equalization control strategy is used, which is based on the principle whereby a voltage loop is used to control the output voltage, and an independent input voltage loop is used to equalize the input voltage. When the input voltage of a module in the system is disturbed and increases, it can be seen from Figure 11 that the offset control amount of the module also increases, so the output phase shift duty cycle increases, the output power increases, the system input side capacitor is in a discharged state, and the voltage drops, thus stabilizing the output voltage [35].

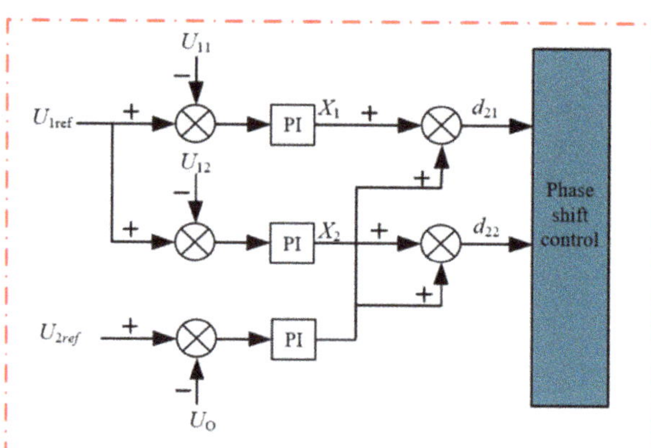

Figure 11. Block diagram of the double-loop voltage equalization control system for DAB.

For the grid-connected port, a single-phase rectifier structure is used, as shown in Figure 6: the grid side and the LCL filter and rectifier are isolated using a transformer, and the LCL filter circuit is connected to the DC bus side using the single-phase rectifier.

For the single-phase rectifier, the mathematical model in the complex frequency domain is as follows:

$$\begin{cases} L_{N1}s \cdot i_{LN1}(s) = -u_{CN}(s) + u_g(s) \\ C_{N}s \cdot u_{CN}(s) = u_g(s) - i_{LN2}(s) \\ L_{N2}s \cdot i_{LN2}(s) = u_{CN}(s) - U_{AB}(s) \end{cases} \tag{10}$$

For SVPWM, the following can be obtained:

$$\frac{U_{AB}(s)}{U_m(s)} = \frac{U_{CB}}{U_{tri}} \tag{11}$$

where $U_{AB}(s)$ is the actual output voltage, $U_m(s)$ is the modulating waveform, U_{CB} is the DC bus voltage, and U_{tri} is the carrier peak. The actual grid current is obtained, as well as the reference value for the purpose of comparison, and the modulating signal is obtained using a proportional resonant controller, which obtains the capacitor current reference value i_{CN_ref} and then compares it with the actual capacitor current i_{CN}.

With respect to the resonance of the LCL filter, the passive damping method is used to suppress the system resonance, and the control strategy with full feedback of the grid voltage is adopted in consideration of the influence that the voltage fluctuation of the grid has on the system. Figure 12 shows the grid voltage full-feedback control block diagram of the grid-connected port.

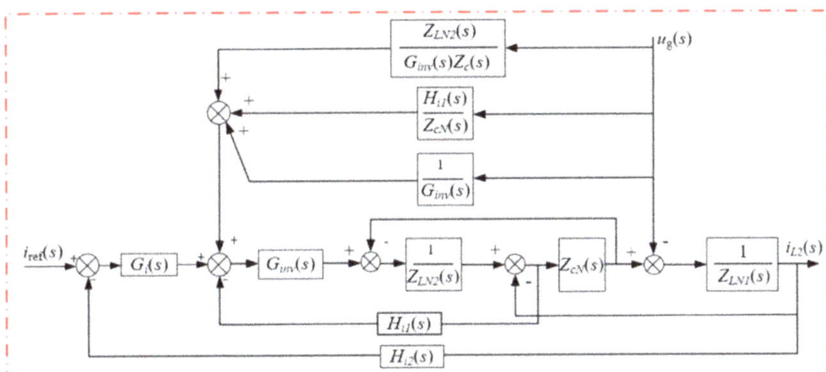

Figure 12. Grid voltage full-feedback control block diagram of the grid-connected port.

$H_{i1}(s)$ and $H_{i2}(s)$ in Figure 12 are the current sensor gain and $G_i(s)$ is the transfer function of the outer loop, respectively. Figure 13 shows the block diagram of the grid-connected port control, which is used to build the grid-connected port and realize grid-connected port control of the ER.

The energy storage port of the ER can play the role of energy buffer and dissipate the unbalanced energy in a timely fashion, thus playing a role in stabilizing the voltage of the common bus. As a core port, the sensitive control of the energy storage port is of great significance. Model predictive control, as a mature control strategy, has excellent dynamic performance, while the control structure is simple and can be combined with different control strategies such as PI to achieve even better control performance by combining their advantages.

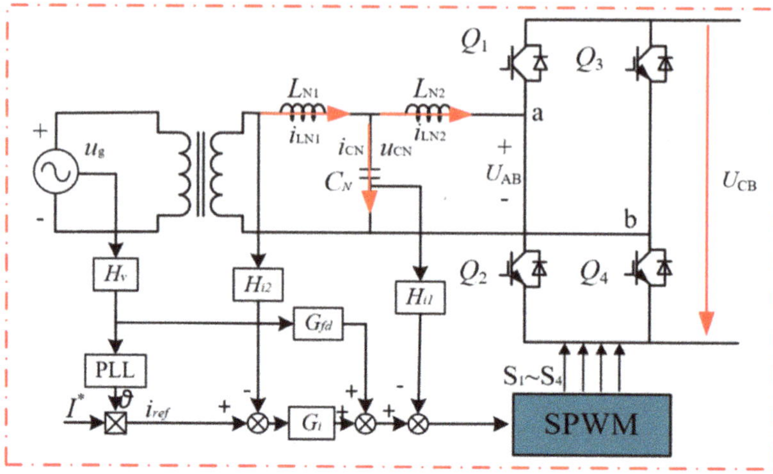

Figure 13. Control block diagram of the structure of the grid-connected port.

As can be observed from the energy storage port structure diagram depicted in Figure 7,

$$U_Q = L_B \frac{di_b}{dt} + U_b \tag{12}$$

Using a discretized form, the transformation of Equation (7) yields

$$i_L(k+1) = (U_Q - U_b)\frac{T_s}{L_B} + i_L(k) \tag{13}$$

Using the sensor, we can get i_l at moment K. On the basis of prediction, we can get i_{lK+1}. Let's assume that the converter is working in the buck mode at this time; then, we have

$$i_L(k+1) = (DU_{CB} - U_B)\frac{T_s}{L_B} + i_L(k) \tag{14}$$

Usually, for digital control systems, there is a certain delay time, which is expressed as follows:

$$i_L(k+1) = (D_{old}U_{CB} - U_B)\frac{T_s}{L} + i_L(k) \tag{15}$$

where D_{old} is the duty cycle of the previous moment. Generally, a constant voltage is considered for a given sampling period; therefore, we have

$$D = \frac{(i_{L\_ref} - i_L(k+1))L}{T_s U_{CB}} + \frac{U_B}{U_{CB}} \tag{16}$$

The block diagram of the energy storage converter control is shown in Figure 14. First, the U_{CB} and the given reference voltage U_{CB_ref} are sent to the proportional-integral PI for comparison, at which time a reference current is output for comparison with the actual inductor current and sent to the model predictive controller in order to generate the duty cycle, and the output is compared with the carrier waveform to generate the control signal. When the bus voltage fluctuates, it can be adjusted by this model predictive control strategy.

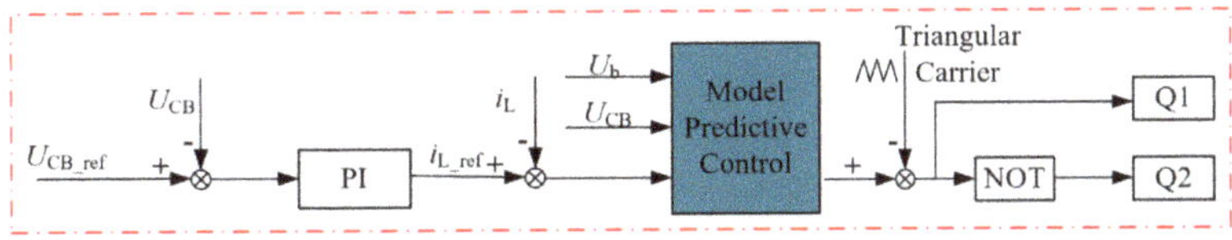

Figure 14. Control block diagram of energy storage converter.

3.3. Five-Port Energy Router Cooperative Control Strategy

In the energy router, the grid-connected port and the energy storage port are bidirectional ports, which, in addition to their basic functions, also assume the role of stabilizing the common bus voltage, U_{CB}, which are now defined as voltage stabilization ports. The PV port and the DC and AC load ports are unidirectional ports, which realize the basic function of the energy router is to transmit energy in one direction, which will be defined as the functional port.

The cooperative control of the five ports of the energy router starts with the current common bus voltage U_{CB}, compares it with the theoretical U_{CB_ref}, and also analyzes the state of charge (SOC) of the battery, combining the current PV output with the actual consumption, and divides the work of the energy router into 12 working models, which can be classified into five groups of working models, and the cooperative control of the functional ports and the voltage stabilization ports. The synergistic control of the functional ports and voltage stabilization ports makes it possible to realize the stable operation of the energy router. Figure 15 presents a flowchart of the model discrimination used in the volume router, and Table 2 presents the detailed model classification criteria.

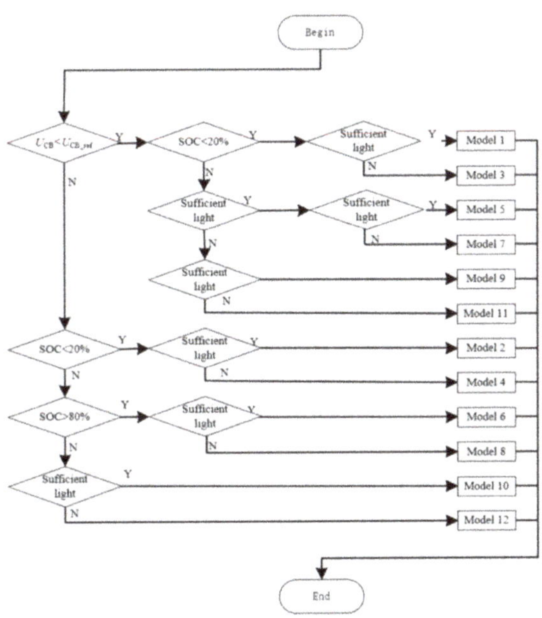

Figure 15. Flowchart of model switching of energy router.

According to Table 2, the common bus, energy storage and PV states are delineated as models A, B, C, D and Z.

Model A is an "off-grid charging model"; in this state, the energy supply of storage is insufficient and PV is sufficient. While the energy supply comes from the PV port, the storage port receives energy from the bus, and the grid-connected port does not participate in the work.

Table 2. The working model of ER.

Model	U_{CB}	SOC	Illumination	Model
Model A	$U_{CB} < U_{CB\_ref}$	SOC < 20%	Sufficient	Model 1
Model A	$U_{CB} > U_{CB\_ref}$	SOC < 20%	Sufficient	Model 2
Model B	$U_{CB} < U_{CB\_ref}$	SOC < 20%	Insufficient	Model 3
Model B	$U_{CB} > U_{CB\_ref}$	SOC < 20%	Insufficient	Model 4
Model C	$U_{CB} < U_{CB\_ref}$	SOC > 80%	Sufficient	Model 5
Model C	$U_{CB} > U_{CB\_ref}$	SOC > 80%	Sufficient	Model 6
Model D	$U_{CB} < U_{CB\_ref}$	SOC > 80%	Insufficient	Model 7
Model D	$U_{CB} > U_{CB\_ref}$	SOC > 80%	Insufficient	Model 8
Model Z	$U_{CB} < U_{CB\_ref}$	20% < SOC < 80%	Sufficient	Model 9
Model Z	$U_{CB} > U_{CB\_ref}$	20% < SOC < 80%	Sufficient	Model 10
Model Z	$U_{CB} < U_{CB\_ref}$	20% < SOC < 80%	Insufficient	Model 11
Model Z	$U_{CB} > U_{CB\_ref}$	20% < SOC < 80%	Insufficient	Model 12

Model B is the "grid-connected charging model", where the energy router storage port is "non-stocked" and the PV is insufficient, and it is necessary to get energy from the grid to meet the requirements of load consumption.

Model C is a "grid-connected power generation model", where the energy router is full of energy storage and the PV is sufficient to feed the acquired energy back to the grid.

Model D is an "off-grid discharge model", where the energy router has sufficient energy storage and the PV is insufficient, and the energy router supplies the AC/DC port consumption from its own energy storage.

Model Z is the "intermediate model", which is the middle model among Models A–D, indicating that the energy router is self-sufficient and works normally, and this intermediate state is off grid, which is the ideal working state, and the energy router is regarded as a "micro grid" without an external connection. The energy is generated and consumed by the system itself, without any interaction with the outside world.

The five-port energy router transitions between the above five working models, and the different working models represent the different working states of the energy storage and grid-connected ports, thus realizing the stable overall control.

4. Simulation Verification and Analysis

In order to verify that the energy router topology proposed in this paper and the corresponding control strategy are feasible, an energy router model based on model predictive control was built using the MATLAB/Simulink platform, and its main parameter settings are shown in Table 3.

Considering the battery storage state and the limitation of simulation time, the initial storage value of the battery, SOC, in the energy router "Model Z" was set to 60%, and its range was set to 59.993 < SOC < 60.009. The initial light intensity of the simulation was set to 1000 w/m^2, and the output power of the PV port was about 10 kW at this time. The initial light intensity of the simulation was set to 1000 w/m^2, and the output power of the PV port was about 10 kW at this time. The light dropped at 1.1 s and then dreopped again to 0 at 1.3 s, after which the PV port stops working.

The system functional port states are as follows: the initial state is set to have only the AC port working, the load power is about 1 kW, and the initial working model of the system is the Z-intermediate model. The system puts the DC port into operation at 0.6 s, the light starts to drop at 1.1 s, and decreases to 0 at 1.3 s, the DC load port < at 1.6 s, the AC port increases the load at 1.9 s, and the load further increases at the AC port at 2.2 s.

Table 3. Parameters of ER.

Parameters	Value
Common Bus voltage U_{CB_ref}	400 V
PV port inductors L_{P1}, L_{P2}	5 mH
PV module open circuit voltage	309 V
PV port capacitor C_{P1}	3300 μF
Energy storage inductor L_B	5 mH
Grid port inductor L_{N1}	0.6 mH
Grid port inductor L_{N2}	0.2 mH
Grid port capacitor C_N	10 μF
AC load port inductor L_A	5 mH
AC load port capacitor C_A	6 μF
DC load inductor L_{D1}	50 mH
DC load inductor L_{D2}	47 mH
DC load capacitor C_{D1}	470 μF
DC load capacitor C_{D2}	500 μF
Initial light intensity I_0	1 kW/m^2
Initial SOC	60%

The system voltage regulator ports work as follows: 0–0.8 s, only the energy storage port works, and the battery is charged; 0.8–1.2 s, the grid-connected port works and outputs energy to the grid; 1.2–2.9 s, the energy storage port works, and the battery is discharged. Then, 2.9 s later, the grid-connected port works, and obtaining energy from the grid.

The light intensity, battery SOC, and common bus voltage U_{CB} waveforms are plotted in Figure 16. The plot characterizes the expected battery storage state with the change in external model change and the throwing of AC and DC loads, with the common bus changing with the load.

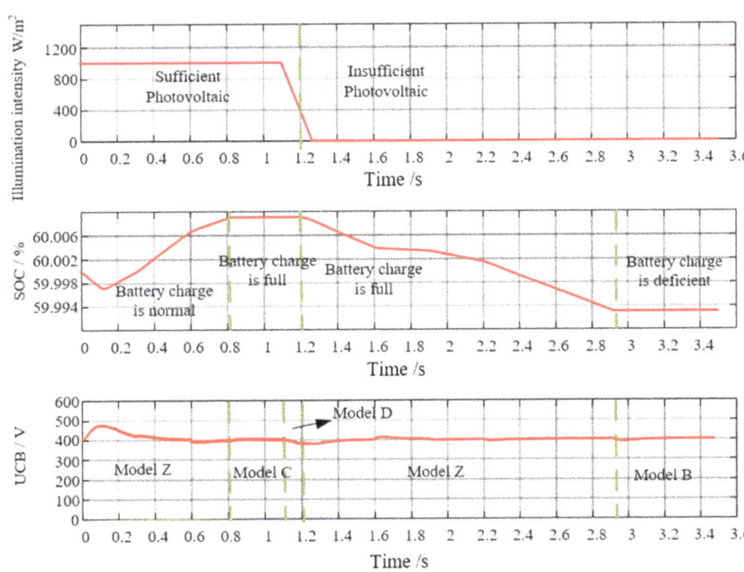

Figure 16. Lighting intensity, energy storage, common bus voltage.

As can be seen in Figure 16, the simulation is started, and the system is in model Z at 0–0.8 s. At 0.8 s, the animal battery is fully charged, and the system enters model C. At 1.1 s, the PV drops to 0, and the system enters model D. The system stabilizes and re-enters model Z. The battery SOC drops further until the system enters model C at 2.9 s.

Different model switching methods correspond to different converter operating states, and the steady-state operation of the energy router can be observed on the basis of the DC bus voltage, which fluctuates with load switching and model change, but can be stabilized quickly.

Figure 17 shows the PV port power curve, which reaches the light inflection point at 1.1 s, after which the PV power gradually decreases to 0, which is consistent with the light situation in Figure 16, and the PV port works normally.

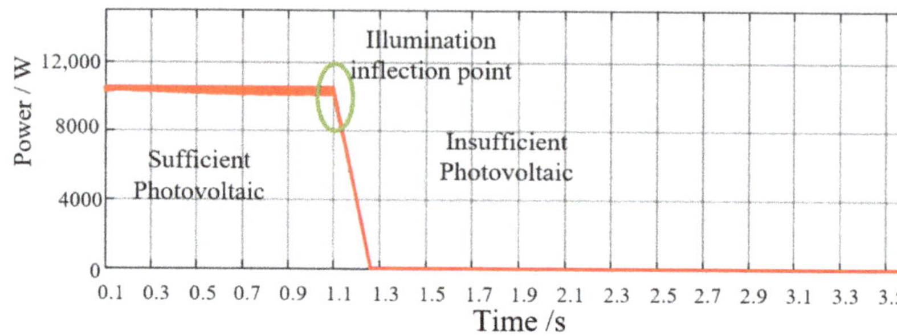

Figure 17. PV port power curve.

Figure 18 shows the power fluctuation of the AC and DC load ports and the bus voltage fluctuation. The four graphs represent the AC port load power, DC port load power, total load power, and the fluctuation of bus voltage when the load is switched. The AC load increases from 1 kW to 3 kW to 5 kW, and its output power is rapidly stabilized, while the DC load increases from 0 to 5 kW and then decreases. From the power fluctuation graph and the DC bus voltage fluctuation graph, it can be seen that the AC and DC load ports are able to realize basic functions, and the bus voltage can quickly return to stability when the AC and DC loads are added and removed.

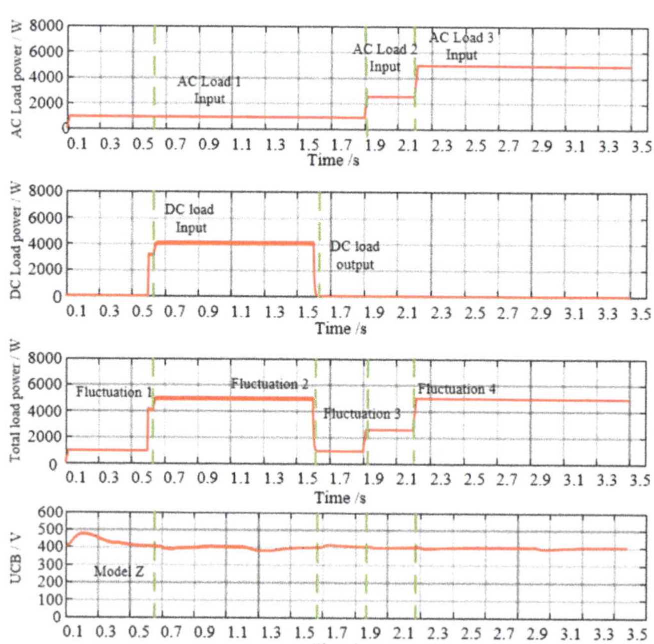

Figure 18. Power fluctuation and common bus voltage diagram.

Figure 19 shows the voltage and current power of the grid-connected port of the energy router; there is no energy exchange at the power port before 0.8 s, and at 0.8 s, the power port delivers energy to the grid, corresponding to model C in Figure 14; the corresponding battery is full of electricity, and the PV port continues to generate energy, at which time the excess energy is fed back to the grid. At 1.1 s, the PV exits, and the AC/DC port is supplied with energy from the battery, at which point the grid-connected port stops working. At 2.9 s, the storage port stops working, and the load continues to consume energy, at which point it enters model B. The grid-connected port receives energy from the grid in order to provide the required load for consumption.

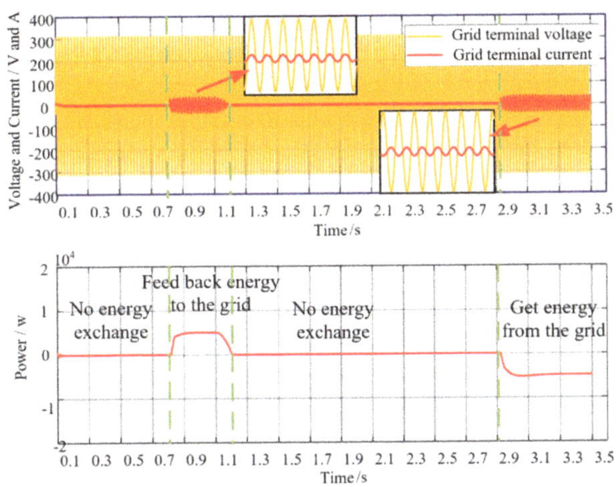

Figure 19. Voltage, current and power diagram of grid-connected port.

Figure 20 shows the voltage at the input of the ISOP DAB and the division of the CD1, CD2, and DC load voltage is implemented at 0.6 s. Due to the different input parameters of the ISOP DAB, the voltage division at the input is different. Under the double closed-loop voltage division control designed as shown in Figure 11, the two sets of input voltages are rapidly equalized, while at the same time, due to the change in the model, it is possible to satisfactorily follow the voltage at the input.

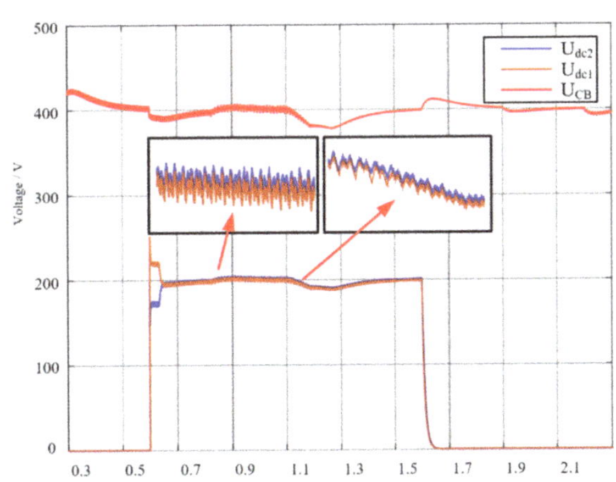

Figure 20. Capacitive voltage at the DAB input.

Figure 21 presents a spectral analysis of the grid-connected port, where the harmonic distortion rate is only 3.02%. Using the energy router topology and the corresponding control method proposed in this paper, the output current waveform is good, and the impact on the grid is small.

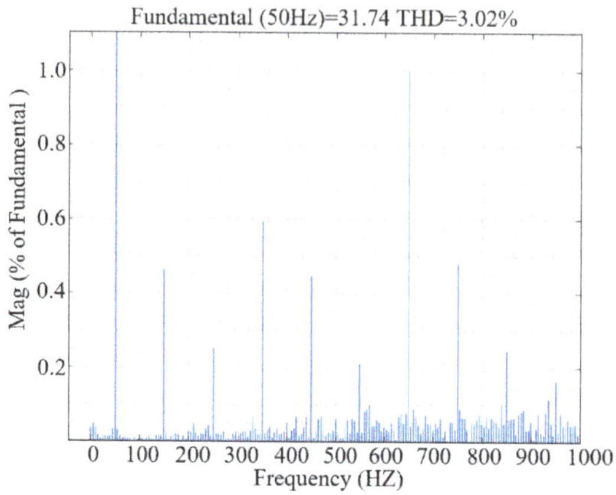

Figure 21. Spectral analysis of the grid-connected port.

The study of ERs is divided into analysis at the module level and at the system level. The study in this article also starts from these two levels.

As shown in Table 1, at the module level, the system is divided into five ports. The first port is the energy storage port. As shown in Table 1, the functions of the energy storage port are divided into three parts. These three functions are verified in Figure 16. Figure 16 shows the change in the SOC of the energy storage port and the corresponding change in the ER operating mode. The second is the grid-connected port, as shown in Table 1, the functions of which can be divided into three parts. These functions are demonstrated in Figure 19, as well as in Figure 21. Figure 19 shows the direction of energy flow in the grid-connected port and the role played by the energy storage port in stabilizing the bus voltage. From Figure 21, it can be observed that the grid-connected port not only fulfills the function of energy coordination, but also has less impact on the distribution network. The third is the photovoltaic port; Figures 16 and 17 show the functionality of the photovoltaic port with different current light intensities and different levels of output power. Finally, there are two load ports, and Figure 18 shows the stability of the system when the two load ports are working in alternation.

System-level studies are effectively equivalent to all ports working in coordination, with energy flowing freely and on demand. All ports are able to work properly throughout this process, thus reflecting the stable operation of the system.

Figure 22 shows the working status of different ports at different points in time, as well as the coordinated work of all ports, which makes it possible to finally realize the free flow of energy.

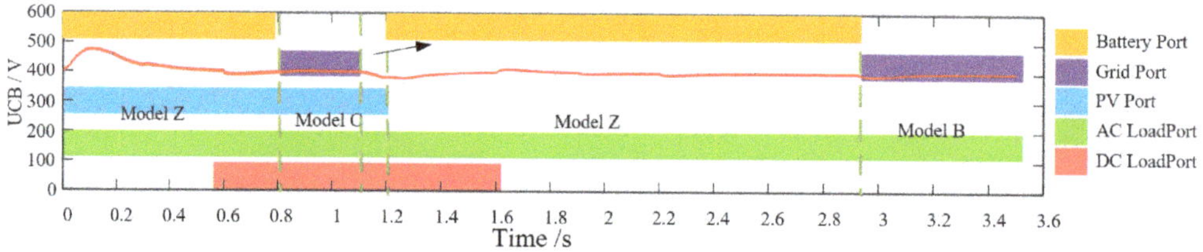

Figure 22. Model switching and port coordination diagram.

5. Conclusions

In this paper, a five-port energy router topology was designed on the basis of the basic function of the energy router, where the five external groups of ports correspond to the five groups of topologies, and a corresponding control strategy is designed for each port. At the same time, an energy router model switching strategy was designed according to the flow of energy and the storage of that state. Through the switching of different models, the external state changes can be satisfied, while stability is achieved for the voltage of the bottom bus. The feasibility of this energy router was subsequently verified. Compared with previous studies, the innovations presented in this study are as follows:

(1) At the module control level, model predictive control was applied to the module control of ER, resulting in the common bus voltage having a faster stabilization speed. At the module topology level, for high-power modules, such as photovoltaic modules, a staggered parallel structure was used in order to reduce the current stress and improve feasibility. Finally, a reasonable and feasible five-port topology and control structure was designed.

(2) A corresponding modal division strategy was proposed for the internal energy storage state and energy flow direction of the energy router, and through the division of this strategy, reasonable control of the energy storage port and the grid-connected port is realized, and the effective switching of the working modalities reflects the characteristics of the interactive flow of energy, which has a certain practical value in engineering applications.

Author Contributions: Conceptualization, X.C., Y.L., D.Y., T.J. and M.A.M.; Methodology, X.C., Y.L., D.Y., T.J. and M.A.M.; Software, X.C. and Y.L.; Validation, X.C., D.Y., T.J. and M.A.M.; Formal analysis, M.A.M.; Investigation, D.Y. and T.J.; Writing—original draft, X.C., Y.L., T.J. and M.A.M.; Writing—review and editing, X.C., Y.L., D.Y., T.J. and M.A.M.; Supervision, T.J. and M.A.M. All authors have read and agreed to the published version of the manuscript.

Funding: This work was supported by the Chinese National Natural Science Foundation (grant number 51977039) and the central government guiding local science and technology development project under grant number 2021L3005.

Data Availability Statement: Not applicable.

Conflicts of Interest: The authors declare no conflict of interest.

References

1. Huang, A.Q.; Crow, M.L.; Heydt, G.T.; Zheng, J.P.; Dale, S.J. The Future Renewable Electric Energy Delivery and Management (FREEDM) System: The Energy Internet. *Proc. IEEE* **2010**, *99*, 133–148. [CrossRef]
2. Wen, J.; Zhou, B.; Wei, L. Preliminary study on an energy storage grid for future power system in China. *Power Syst. Prot. Control* **2022**, *50*, 1–10.
3. Guo, H.; Wang, F.; Zhang, L.; Luo, J. Technologies of energy router-based smart distributed energy network. *Proc. CSEE* **2016**, *36*, 3314–3324.
4. Liu, Y.; Yuan, D.; Gong, Z.; Jin, T.; Mohamed, M.A. Adaptive spectral trend based optimized EWT for monitoring the parameters of multiple power quality disturbances. *Int. J. Electr. Power Energy Syst.* **2023**, *146*, 108797. [CrossRef]
5. Zhao, D.; Hu, N.; Fu, J.; Ben, J.S.; Zi, N.Z. Research on the practice and road map of enhancing the flexibility of a new generation power system in China. *Power Syst. Prot. Control* **2020**, *53*, 1–9.
6. Loh, P.C.; Li, D.; Chai, Y.K.; Blaabjerg, F. Autonomous operation of hybrid microgrid with AC and DC subgrids. *IEEE Trans. Power Electron.* **2012**, *28*, 2214–2223. [CrossRef]
7. Zheng, L.; Chunming, T.; Fan, X.; Jun, G.; Songhui, X. The power control of power electronic transformer in hybrid AC-DC microgrid. *Trans. China Electrotech. Soc.* **2015**, *30*, 50–57.
8. Wang, S.; Su, S.; Wang, H.; Ouyang, Z.; Wang, H.; Zhou, W. An energy routing control method for a terminal energy router based on switching system theory. *Power Syst. Prot. Control* **2022**, *50*, 81–90.
9. Chen, R.; Yang, Y.; Jin, T. A hierarchical coordinated control strategy based on multi-port energy router of urban rail transit. *Prot. Control Mod. Power Syst.* **2022**, *7*, 15. [CrossRef]
10. Cui, X.; Jin, T.; Dai, Q. Research on Key Technologies of Energy Router for Urban Rail Transit. In Proceedings of the 2022 IEEE 5th International Conference on Electronics Technology (ICET), Chengdu, China, 16 May 2022; pp. 1250–1254.

11. Bu, Z.; Sun, X.; Teng, J.; Zhao, W.; Li, X.; Wang, B. A Compact Energy Router Scheme Based on MMC. *Proc. CSEE* **2022**, *42*, 4536–4546.
12. Li, P.; Sheng, W.; Duan, Q.; Li, Z.; Zhu, C.; Zhang, X. A Lyapunov Optimization-Based Energy Management Strategy for Energy Hub with Energy Router. *IEEE Trans. Smart Grid* **2020**, *11*, 4860–4870. [CrossRef]
13. Tu, C.; Lan, Z.; Luan, S.; Xiao, F. A Coordinated Power Control Strategy of Three-port DC Energy Router Based on Droop Phase-shift. *Power Syst. Technol.* **2019**, *43*, 4105–4114.
14. Hussain, S.M.S.; Aftab, M.A.; Nadeem, F.; Ali, I.; Ustun, T.S. Optimal Energy Routing in Microgrids with IEC 61850 Based Energy Routers. *IEEE Trans. Ind. Electron.* **2019**, *67*, 5161–5169. [CrossRef]
15. Cao, Y.; Yuan, L.; Zhu, S.; Huang, R.; Feng, G.; Zhao, Z. Parameter Design of Energy Router Orienting Energy Internet. *Power Syst. Technol.* **2015**, 3094–3101. [CrossRef]
16. Li, Z.; Sheng, W.; Duan, Q. Coordinated control strategy of AC/DC hybrid power router based on energy storage and voltage stabilization. *Autom. Electr. Power Syst.* **2019**, 121–129. [CrossRef]
17. Shi, L.; Liu, G. Research on switching control strategy for the multi-LAN-port energy router. *J. Electr. Power Sci. Technol.* **2015**, 3094–3101. [CrossRef]
18. Xu, M.; Tai, N.; Huang, W. Energy router design based on community energy network. *Power Syst. Prot. Control* **2016**, *44*, 177–183.
19. Li, S.; Wu, C.; Jiang, X.; Pan, P.; Ke, Z. Coordinated Control of Multiple Operation Conditions for Multi-port Energy Router in Energy Internet Framework. *Autom. Electr. Power Syst.* **2020**, *44*, 32–45.
20. Tu, C.; Xiao, F.; Lan, Z.; Guo, Q.; Shuai, Z. Analysis and Control of a Novel Modular-Based Energy Router for DC Microgrid Cluster. *IEEE J. Emerg. Sel. Top. Power Electron.* **2018**, *7*, 331–342. [CrossRef]
21. Xiao, Q.; He, J.; Wang, H.; Jia, H. Voltage balance control strategy for DC capacitor of electrical energy rounter under power grid failure. *Autom. Electr. Power Syst.* **2018**, *42*, 20–25.
22. Mansouri, S.A.; Nematbakhsh, E.; Ahmarinejad, A.; Jordehi, A.R.; Javadi, M.S.; Marzband, M. A hierarchical scheduling framework for resilience enhancement of decentralized renewable-based microgrids considering proactive actions and mobile units. *Renew. Sustain. Energy Rev.* **2022**, *168*, 112854. [CrossRef]
23. Nasir, M.; Jordehi, A.R.; Tostado-Véliz, M.; Tabar, V.S.; Mansouri, S.A.; Jurado, F. Operation of energy hubs with storage systems, solar, wind and biomass units connected to demand response aggregators. *Sustain. Cities Soc.* **2022**, *83*, 103974. [CrossRef]
24. Matin, S.A.A.; Mansouri, S.A.; Bayat, M.; Jordehi, A.R.; Radmehr, P. A multi-objective bi-level optimization framework for dynamic maintenance planning of active distribution networks in the presence of energy storage systems. *J. Energy Storage* **2022**, *52*, 104762. [CrossRef]
25. Zhang, Y.; Du, G.; Lei, Y.; Zheng, H. Current status and prospects of control strategy for a DC micro grid hybrid energy storage system. *Power Syst. Prot. Control.* **2021**, *49*, 177–187.
26. Ramos, F.; Pinheiro, A.; Nascimento, R.; de Araujo Silva, W., Jr.; Mohamed, M.A.; Annuk, A.; Marinho, M.H.N. Development of Operation Strategy for Battery Energy Storage System into Hybrid AC Microgrids. *Sustainability* **2022**, *14*, 13765. [CrossRef]
27. Nascimento, R.; Ramos, F.; Pinheiro, A.; Junior, W.d.A.S.; Arcanjo, A.M.C.; Filho, R.F.D.; Mohamed, M.A.; Marinho, M.H.N. Case Study of Backup Application with Energy Storage in Microgrids. *Energies* **2022**, *15*, 9514. [CrossRef]
28. Chen, W.; Liu, B.; Nazir, M.S.; Abdalla, A.N.; Mohamed, M.A.; Ding, Z.; Bhutta, M.S.; Gul, M. An Energy Storage Assessment: Using Frequency Modulation Approach to Capture Optimal Coordination. *Sustainability* **2022**, *14*, 8510. [CrossRef]
29. Dehghani, M.; Shiraz University of Technology; Montazeri, Z.; Givi, H.; Guerrero, J.; Dhiman, G.; University of Shahreza; Aalborg University; Government Bikram College of Commerce. Darts Game Optimizer: A New Optimization Technique Based on Darts Game. *Int. J. Intell. Eng. Syst.* **2020**, *13*, 286–294. [CrossRef]
30. Dehghani, M.; Montazeri, Z.; Dhiman, G.; Malik, O.; Morales-Menendez, R.; Ramirez-Mendoza, R.; Dehghani, A.; Guerrero, J.; Parra-Arroyo, L. A Spring Search Algorithm Applied to Engineering Optimization Problems. *Appl. Sci.* **2020**, *10*, 6173. [CrossRef]
31. Dhiman, G.; Oliva, D.; Kaur, A.; Singh, K.K.; Vimal, S.; Sharma, A.; Cengiz, K. BEPO: A novel binary emperor penguin optimizer for automatic feature selection. *Knowl. Based Syst.* **2020**, *211*, 106560. [CrossRef]
32. Dhiman, G. ESA: A hybrid bio-inspired metaheuristic optimization approach for engineering problems. *Eng. Comput.* **2019**, *37*, 323–353. [CrossRef]
33. Ali, A.I.M.; Alaas, Z.M.; Sayed, M.A.; Almalaq, A.; Farah, A.; Mohamed, M.A. An Efficient MPPT Technique-Based Single-Stage Incremental Conductance for Integrated PV Systems Considering Flyback Central-Type PV Inverter. *Sustainability* **2022**, *14*, 12105. [CrossRef]
34. Rao, C.; Hajjiah, A.; El-Meligy, M.A.; Sharaf, M.; Soliman, A.T.; Mohamed, M.A. A Novel High-Gain Soft-Switching DC-DC Converter with Improved P&O MPPT for Photovoltaic Applications. *IEEE Access* **2021**, *9*, 58790–58806. [CrossRef]
35. Mohamed, M.A.; Awwad, E.M.; El-Sherbeeny, A.M.; Nasr, E.A.; Ali, Z.M. Optimal scheduling of reconfigurable grids considering dynamic line rating constraint. *IET Gener. Transm. Distrib.* **2020**, *14*, 1862–1871. [CrossRef]

sustainability

MDPI

Article

Cost Estimation Process of Green Energy Production and Consumption Using Probability Learning Approach

Jian Xiao [1,*] and Wei Hou [2,3]

[1] Dong Fureng Economic & Social Development School, Wuhan University, Wuhan 430072, China
[2] Chakrabongse Bhuvanarth International Institute for Interdisciplinary Studies, Rajamangala University of Technology Tawan-Ok, Bangkok 10400, Thailand; houwei9999@gmail.com
[3] State Grid Beijing Electric Power Corporation, Beijing 100031, China
* Correspondence: hhbbj1108@163.com

Abstract: With electric vehicle (EV) charging, green energy production costs could be reduced, and smart grid (SG) reliability improved. Nevertheless, the vast number of EVs could adversely affect the stability of the voltage and cost of operation. The present study designs a new security-based system based on a new EV participation charging method for a decentralized blockchain-enabled SG system. It is aimed at minimizing the level of power alternation in the electrical network and the total charging costs of EVs as mobile systems. In the first step, the power alternation level issue of the SG is formulated based on the capacity of EV batteries, the rate of charging, and EV users' charging behavior. Next, a new adaptive blockchain-based EV participation (AdBEV) method is proposed, using the Iceberg order execution algorithm for improving EV discharging and charging schedules. Simulated outcomes demonstrate that the suggested method is superior to the genetic algorithm method when it comes to reducing power fluctuation levels and total charging cost.

Keywords: smart grid; modern transportation system; hybrid blockchain technology; EV mobility; machine learning

Citation: Xiao, J.; Hou, W. Cost Estimation Process of Green Energy Production and Consumption Using Probability Learning Approach. *Sustainability* **2022**, *14*, 7091. https://doi.org/10.3390/su14127091

Academic Editor: Mohamed A. Mohamed

Received: 23 April 2022
Accepted: 27 May 2022
Published: 9 June 2022

1. Introduction

The introduction of electric vehicles (EVs) brings the idea of vehicle-to-grid (V2G) and grid-to-vehicle (G2V) into the energy market, which will convert the overloaded network into a valuable source. EVs are supplied with proper control methods to fatten the load, shave the peak, and regulate the frequency. EV charging and discharging via a bidirectional power flow in a smart grid (SG) is expected to decrease the cost of subsidiary (reserved) energy production and improve the system's robustness [1]. In more detail, as an example, when many EVs begin charging simultaneously throughout peak energy consumption periods, the large power generators must begin the subsidiary generators (with a shorter response time) to supplement the energy demand, and this results in a delay for the subsidiary generators to begin producing energy. Ten to fifteen min is needed for energy to be provided. Ref [2] proposes game theory structures such as the energy grid, EVs, and smart communities, providing insight into possible optimization techniques for achieving efficient and intelligent energy management. In spite of this, large penetrations of EVs into the electrical system raise issues regarding the impact on electrical system stability and operational costs [3].

Refs [4,5] examined a centralized information center operator as the aggregator for gathering energy consumption demand and further commanding the transfer of energy in the traditional EV discharging/charging method. Utilizing the aggregator, control methods are used for controlling power flow during peak and off-peak hours, respectively. Since the behavior of EV users is uncontrollable, EVs are considered a widely distributed power load on the power grids. This makes it possible to schedule the exchange of energy, even if a predetermined scheduling process does not consider every factor affecting the electrical grid [6]. Although SG aggregators are designed for decentralizing traditional energy grids and supporting micro-distributed renewable generators, they compromise the purpose of the SGs [7]. As a result of the centralized-based system, decision-making is delayed and grid participants' autonomy is undermined, in that users cannot control the discharging or charging of their processes. The current average cost of electricity that users are paying at the time of consumption is generally not indicative of the actual wholesale costs. When operating costs vary, it is difficult to adjust to fluctuating power demand [8,9].

Several control methods for EV discharging and charging techniques have been presented before to control the duration and amount of energy transferred in order to enhance grid performance. Smart charging planning for low-voltage residential grids based on the state-of-charge (SOC) amounts like battery residual and battery capacity, thus impacting the entire grid offloading, is presented in ref [5]. In ref [4], automatic generation control signals are applied for regulating EV discharging and charging schedules for enhancing the frequency regulation service's efficiency is examined by considering the dynamic departure time and arrival time of EVs. Ref [10] examined an aggregated based on the optimization model of the EV charging process by considering the stochastic characteristics of the charging process and applied the Genetic Algorithm (GA) for determining the parameters of the system model. The proposed EV discharging/charging schedule algorithms in [5] assumed static parameters for the EV's accessible charging time. It is important to note that in [4,10] the suggested charging layout was on the basis of parameter estimations and theoretical calculations which did not completely account for the flexibility of EVs. Thus, it would be difficult for the blockchain (BC) platform to be adapted to the kinds of battery and consumer behavior associated with EVs.

Due to a large number of EV charging and dispatching demands, a BC idea has been developed for providing peer-to-peer transferred structures, which utilize decentralized storage for whole transferred information data. Therefore, BC technology will enable a non-trusting network to eliminate the operational costs of the intermediaries, which would result in a cheaper, faster, and more reliable way of reflecting fluctuating wholesale costs to consumers. BC technology can also shift high-load equipment to off-peak hours for reducing energy prices and assist in reducing the overloaded peaks [11,12]. Therefore, ref [13] examined a new scheme to trade energy on the basis of BC technology for adopting the decentralized and competitive environment of local energy production, however, the BC was only utilized as a data storage warehouse for transaction recording. A detailed analysis of the economic assessment of the market mechanism for local energy trading can be found in ref [14]. According to ref [15], BCs and decentralized consensus methods were applied for coordinating the planning of distributed power sources in a microgrid and guaranteeing fair payments with no need for a centralized aggregator.

In order to charge EVs on a large scale, simple control methods are required for reducing operating delay and investment costs. As a result, an adaptive EV discharging/charging planning algorithm is presented in this study according to the BC platform, called Adaptive BC-based EV Participation (AdBEV) for executing the data sharing and decision-making method.

As a method of solving the problem, the Iceberg order management algorithm [16] is widely applied in the digital financial trading market and is assumed a powerful machine learning approach that can be used for managing EV discharging and charging demands. In summary, the study makes the following contributions:

(1) A new EV participation method, which can introduce the idea of decentralized EV discharging and charging on a BC-enabled SG system is presented in this study.
(2) A discharging and charging plan issue related to EVs on the BC-enabled SG network is formulated for the EV charging schedule.
(3) The AdBEV layout on the basis of the iceberg order algorithm, which executes the optimal order process for matching the SG electricity discharging and charging demand is proposed in order to reduce the power fluctuation level.

(4) AdBEV has been shown to have a better power fluctuation level in comparison with the benchmark method that utilizes GA. Besides, it has been shown that the suggested algorithm achieves low computation prices in comparison with the benchmark layout in the Ethereum platform.

Following are the remaining parts of this study. The system model according to the matching imbalances of the EV consumer profile related to discharging/charging is built in Section 2. The problem formulation and the algorithm definition are presented in Section 3. The simulation outcomes and the AdBEV vs. GA comparison are demonstrated in Section 4. Section 5 presents the conclusions.

2. System Model

As an extension to the system model developed in [17], a residential district that has a substation transformer whose maximum power capacity is P_{max}, is considered here. On the supplier side, there are traditional extend power plants, distributed micro renewable generators, and storage. In addition, the user power loads, such as residential areas and hospitals, are linked to the public BC energy exchange platform that allows for the transmission, encryption, and storage of information about electricity supply and demand. Figure 1 illustrates the architecture of the power network including the public BC platform to trade power.

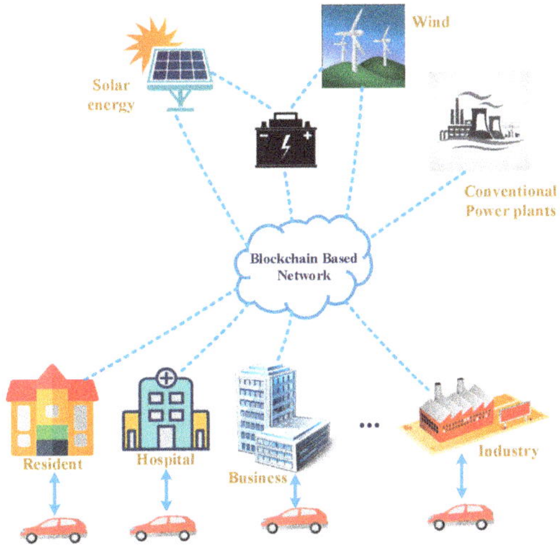

Figure 1. The SG diagram contains the public BC trading platform.

This model assumes that EVs can publish and transmit the discharging or charging orders to the SG public BC trading platform. EVs are able to be charged and discharged using a programmable charging installation. Using this design, the grid operator will be able to switch the power transmission for the EV instantly on/off (presuming the switches are complicated). Figure 2 illustrates the process for processing the transaction in the BC platform. Driver owners initiate the power orders including buying and selling, and once the identity of the initiator has been verified, orders can be entered on the BC-based trading platform. After that, the AdBEV layout is used to process the orders, which are then published in the open order book. Peers in the grid transact and verify the matched orders. Both parties must confirm orders before they can be stored in a distributed manner.

$$X_{i,t} = \begin{cases} 1, & if\ EV_i\ is\ connescted\ at\ t \\ 0, & otherwise \end{cases} \tag{1}$$

where X represents the EV status matrix as Equation (2). It should be noted that in our research, we considered 100 households, where every household is assumed to own an EV. Hence, the diminution of this matrix of our research is 100×1.

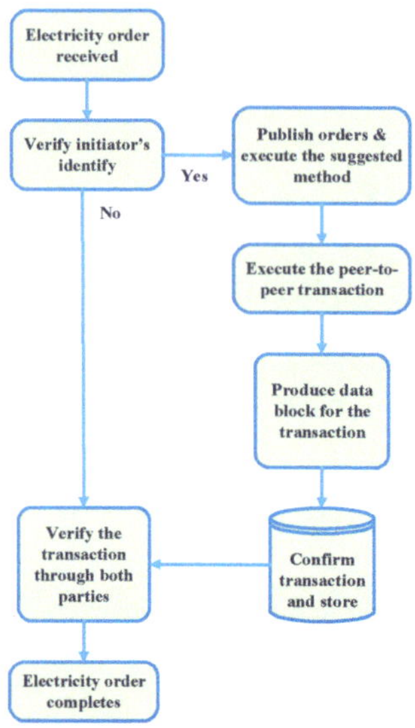

Figure 2. The process for transferred execution in the BC-enabled SG network.

Battery residual for every EV is (SOC_{ini}) and the predicted SOC when charging determines the power demand for EV is (SOC_{exp}). The formula looks like this:

$$P_{EV}(t) = \sum_{i=1}^{I} \left(X_{i,t} \left(SOC_{exp}(i) \pm SOC_{ini}(i) \right) \right) \quad (2)$$

When scheduling EV discharging/charging, it is essential to take into consideration how many EVs remain linked to the system in order to determine the maximum waiting time for the order. According to this study, the duration of the charging, and the amount of energy transferred, are combined to predict the hourly charging demand pattern throughout the day for achieving minimal energy fluctuations and consumer satisfaction.

Profiles of vehicle activity related to the normal residential EV charging demands are modeled using [18]. The pattern of the charging demand on a daily basis can be inferred by combining the duration of the charging and the amount of energy transferred. Equation (3) can be used to model the amount of energy transferred during EV charging based on the balance between complexity and precision, and this is expressed using a time segment vector $\eta(t)$ comprised of the percentage of stay-on-line EV in Equation (4).

$$k_j(t) = \sum_{j=1}^{J} \sum_{t=T_f}^{T_s} \left(a_j \sin\left(b_j t + c_j \right) \right) + \varepsilon \quad (3)$$

$$\overrightarrow{\eta(t)} = \left[k_1(t), k_2(t), \cdots, k_j(t) \right]\% \quad (4)$$

The T_f refers to the first time step and while T_s refers to the second time step elapsed that can enable the formation of a certain fitting in j^{th} order of sine series. The j shows the order for the total of sine series derived from the Matlab curve fitting toolbox, in which J indicates the model complexity of the sum of the sines. The parameters in the sum of the sine series are shown by the a_j, b_j, and c_j. In view of the charging load profile (LP) indication that a specific number of EVs should remain online during the procedure, Equation (4) is rewritten accordingly:

$$N_{car}(t) = \Omega \cdot \left[k_1(t), k_2(t), \cdots, k_j(t)\right] \tag{5}$$

Here, the whole number of EVs in an area is shown by Ω. The power demand for one-time segment changes during every iteration of order execution. Accordingly, Equation (2) is reformulated with the amount of power exchange Q_i according to the order category:

$$P_{EV}(f) = \sum_{i=1}^{I} \left(X_{i,f} Q_i\right) \tag{6}$$

Therefore, the whole residential load is described as the total demand for EV discharging/charging and LP excluding EV for formulating EV charging problems.

$$P_{total}(t) = P_{home}(t) + P_{EV}(t), \qquad t \in T \tag{7}$$

$$P_{total}(t) + \varsigma(t) \leq P_{sub}, \quad \forall t, \tag{8}$$

$$V_{min} \leq V(t) \leq V_{max}, \quad \forall t, \tag{9}$$

Here, the power load excluding EV is shown by P_{home}. Aggregating the above random process (7) with the suggested algorithm can solve the EV discharging/charging scheduling problem. In order to enhance the operation of the electric grid, it is advisable not to exceed the peak transformer substation load following the implementation of EV discharging and charging to the residential electricity demand. In limitation (8), the error item $\varsigma(t)$ shows power losses or branch overloaded plus the total power load should not exceed the substation power capacity P_{sub}. Hence, there is a maximum number of EV $\max(N_{car}(t))$, which is employed in the system to prevent exceeding the substation capacity. The levels of the voltage in buses have been limited, and cannot exceed the minimum and maximum limits in limitation (9).

3. Formulation of the Problem and Explanation of the Algorithm

EV discharging and charging demand is scheduled to minimize the effect of consuming or injecting excessive amounts of energy on the network. The paper examines an adaptive discharging/charging method for different kinds of EVs for flattening the LP of transformer substations in distribution networks.

A. Formulation of the problem

As input data, the daily 30 min power interchange order book profile of a residential grid has been used, with EV charging/discharging demands on a daily basis as the decision parameter. A schedule that is adaptive for filling in the gaps in the residential LP is essential for equalizing the level of alternation of the entire system. Accordingly, the total function of utility for measuring the level of alternations in 2-time parts of a network is:

$$P_{PFL} = \sum_{t=1}^{T} \|P_{total}(t) - P_{total}(t-1)\| \tag{10}$$

Here, the total power alternation level during 24 h with 30 min temporal resolution is shown by P_{PFL}, $P_{total}(t)$, and the overall power in the transformer at hour t and $t-1$ is represented by $P_{total}(t-1)$. This system is aimed at minimizing the power alternation level criteria P_{PFL} of the total energy network using a set of variables to be optimized for related i and its related SOC_{exp}, as shown below:

$$\min_{\forall SOC_{\exp}(i) \in I} P_{PFL} \tag{11}$$

$$S{\cdot}T{\cdot}\sum_{t=1}^{T}\sum_{i=1}^{I}(SOC_{exp}(i) \pm (SOC_{ini}(i))) = \sum_{t=1}^{T}P_{EV}(t) \tag{12}$$

$$SOC_{ini}(i) \in (0,1), \forall i, \tag{13}$$

$$SOC_{exp}(i) \leq P_{max}, \forall i, \tag{14}$$

$$X_{i,t} = \{0,1\}, \quad \forall i, \quad \forall t, \tag{15}$$

Using Equation (12), the charging and discharging power from EVs is limited to the order demand based on the existing EV number i and the obtained $SOC_{exp}(i)$. In Equation (13), the initial SOC is set between (0,1). Despite the fact that this limitation might lead to a less flattened power LP, users' requirements are met as a consequence of this limitation. The limitation (14) ensures that the maximum SOC following charging is not greater than the EV battery capacity P_{max} for every EV i. As explained in (13), an EV could have just 2 states, one being connected to the network and one being disconnected.

As Equation (15) has a binary constraint for EV connection state $X_{i,t}$, the formulation of the issue is a mixed combinatorial nonconvex issue. The optimal solution to this problem is not a systematic method with an effective computational method. The optimization problem can be to determine how many EVs should be used for transferring energy (discharging/charging) to reduce the total power fluctuations. It should be noted that the required infrastructures to deploy the proposed technique are including, but are not limited to, open cloud, reliable and scalable hardware, updated software, a stable internet connection, databases, battery, several energy resources, different levels of load demand, EV, point of sales terminals, crypto automated teller machines (ATMs), etc.

B. ADBEV layout

This part proposes an AdBEV layout for solving this issue by utilizing the power exchange book as a trading system. A scheduling result in the previous time slot can affect the power demand in the next one as well. The grid system needs to be kept at a minimum power fluctuation so as to provide the total power charging demand of EVs.

a. Normal electricity exchange order

In the case of a small quantity power exchange order, the demand has been formatted as an input that can be sent to the power exchange stand book Std_{in} via a vector as shown below:

$$\vec{Q_i} = (\gamma, Id_i, \sigma_i, Q_i) \tag{16}$$

Here, the specific identifier for the discharging/charging initiators is shown by the Id_i in which they are EVs or other equipment, the confirmed agent price for the power order is represented by the σ_i, the power demand amount of the order is shown by the Q_i, and γ presents a matrix showing if it is a power discharging or charging order:

$$\gamma = \begin{cases} 1, & charging\ order \\ 0, & discharging\ order \end{cases} \tag{17}$$

To produce every matched trade, the solution must be used in the existing book Std_{in} for each inserted order message, and whole non-error output must be directed to the Std_{out}. The trade information format can be described in the following way:

$$\vec{T_i} = \left(Id_{sell}, Id_{buy}, \sigma_m, Q_m\right) \tag{18}$$

Here, the Id_{buy} and Id_{sell} indicate the matched power buy and sell order identifier respectively, this matched price in pence is shown by the σ_m and the matched amount for the order is represented by the Q_m. The solutions must display the existing full order book in the previous format upon receiving an order message and upon receiving every matching entry in the book.

b. ICEBERG power interchange structure

An extend user can hold an overall exchange demand (ϕ_0) and intend to liquidate it before time T_{max}, in which case the peak size ϕ_P and a limit $\overline{S}$ to the demand of the iceberg are assigned. The latter exceeds the initial optimal bid price S_0 on the charging side, as shown below:

$$S_0 < \overline{S} \tag{19}$$

The first portion of the order cannot be quickly executed, and the reverse is also true.

For ensuring the smooth running of the iceberg power exchange and the advantage achieved by participants, choosing the right price is critical. Based on ref [16], a jump-diffusion process is used to model the optimal charging price S_t. This paper aimed at building an electricity exchange market for EV consumers, to find a guide price related to every period of time, when $S_t < \overline{S}$, the extensively applied geometric Brownian motion for the stock price is adopted for modeling the real-time power price on a daily basis:

$$dS_t = \mu S_t dt + \sigma S_t dW_t\,, \quad with\ S_t < \overline{S} dS_t = \mu S_t dt + \sigma S_t dW_t\,, \quad with\ S_t < \overline{S}, \tag{20}$$

Here, it is possible to set the percentage drift μ and percentage volatility σ as constants, and a Wiener process is shown by W_t. So, the next equations allow obtaining the optimal iceberg price S_t for a certain top price value S_0:

$$S_t = S_0 \exp\left(\left(\mu - \frac{\sigma^2}{2}\right)t + \sigma W_t\right) \tag{21}$$

$$E(S_t) = S_0 e^{\mu t} \tag{22}$$

In Equation (22), the iceberg format is given by the vector integrating the order optimal price σ_{S_t} and the overall demand ϕ_i. In theory, Q_{pi} shows the peak for a single trading period that will never exceed ϕ_i:

$$\vec{Q_i} = (\gamma, Id_i, \sigma_{S_t}, \phi_i, Q_{pi}) \tag{23}$$

Based on the priority function, both normal and iceberg power exchanges must be depicted in the order book:

$$f(P_{1p}(n);\ P_{2t}\ (n)) = \alpha Rank(Pr) + \beta Rank(T) \tag{24}$$

Here, the ranking for the price is shown by $Rank(Pr)$, and the ranking for generation time is indicated by $Rank(T)$, and $\alpha = 10\beta$ for building a price-competitive market. In Figure 3, the algorithm first determines the normal and iceberg orders after initializing the demand for power discharging and charging. Following that, entire orders are sorted based on the ranking function $f(P_{1p}(n);\ P_{2t}\ (n))$ for matching them for exchange. Following the execution of entire orders in every trade frame period, provides a respective response for the orders.

Figure 4 illustrates the optimal order method for matching the power discharging and charging demand for three different scenarios. The algorithm can execute whole matching orders in the order of price ranking provided that the power demand is satisfied in a manner that minimizes power fluctuations. Those orders whose priority amounts are the highest are executed initially, even when the overall amount of power sell orders is lower compared to the overall amount of power buy orders. Afterward, the iceberg execution strategy has been applied to the unmatched electricity demand orders for matching the orders in the one-time frame $(t \in T)$. Iceberg orders that equal 1 $\left(N_{Q_i} = 1\right)$ are held until the next cycle with a similar priority value, or, they are allocated new priorities. When there are a larger number of sell orders, the iceberg execution is applied to passively wait for the next cycle execution. Order books are built and updated for every time frame $(t \in T)$.

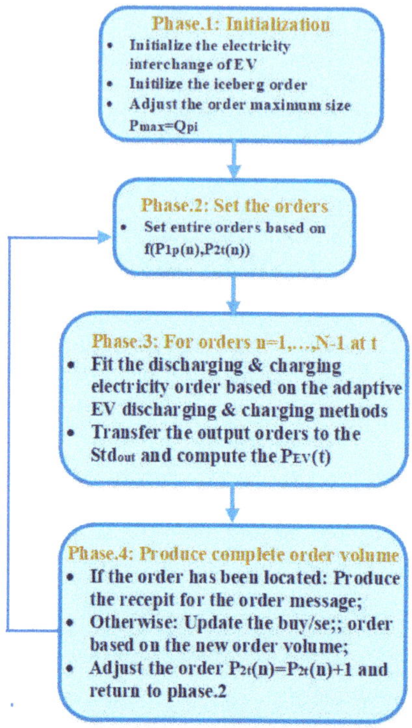

Figure 3. Electricity trading network with order book initialization.

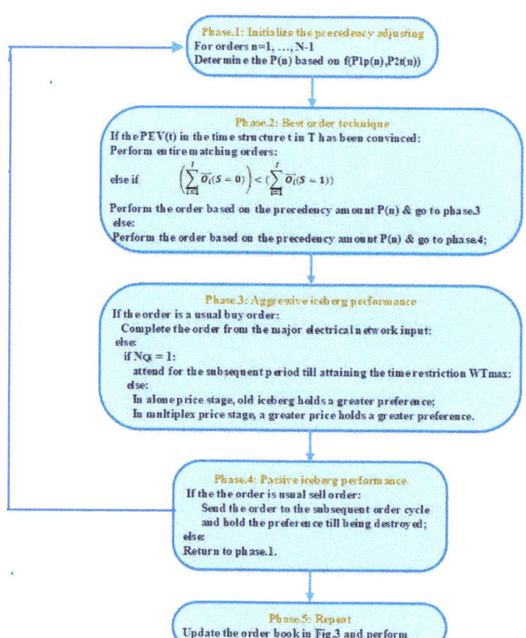

Figure 4. Adaptive BC-based EV participation plan.

Because AdBEV can optimize charging patterns in day-ahead markets, when one day all the time slots have been allocated, the suggested scheduling layout can execute just once a day according to the prior EV arrival pattern and residential LPs.

4. Simulation Outcomes

A. Experiment setup

AdBEV is evaluated using the residential region substation transformer with a $P_{max} = 250$ kVA serving 100 households. Every household is assumed to own an EV. Taking into account the different kinds of EVs in the market, 2 kinds of the most common battery sizes of 8.8 kWh (Toyota Prius), and 60 kWh (Tesla model S) are selected. In terms of charging rates, various manufacturers have various connector kinds, with slow charging and fast charging compatible with 3 kW (16 A), and 43 kW (63 A) chargers, respectively. Based on the charging port availability, the number of EVs for fast and slow charging has been assigned at a ratio of 1:4 in the simulation procedure. There are 2 parts to the EV charge connection state, the first one is between 06:00 and 18:00, and the other one is between 18:30 and 05:30 (+1). EV batteries have an initial residual voltage (SOC_{ini}) created at random, while the battery level (SOC_{exp}) following charging has been assigned to 80% for battery protection, whereas the SOC following discharging has been assigned to 50% to make it easier to use.

The Ethereum platform is used for implementing the developed algorithm by taking into account the distributed trading platform for the power interchange market. Platforms such as Ethereum allow consumers running distributed applications on public BCs [19]. The Solidity language with version 0.4.0 has been applied for deploying the suggested smart contract for executing the AdBEV layout. As a result, gas consumption from the Ethereum platform can provide a direct indication of the algorithm's operational difficulty. Consumers of the public BC platform must pay *gas* costs for executing smart contract commands in the Ethereum platform [20]. The cost of executing a complicated algorithm vastly increases the trade cost for the power, as the number of peers in the grid increases.

The simulation consists of 2 kinds of orders: iceberg and normal orders based on Equations (5) and (16). Applying iceberg orders can decrease the power load fluctuations caused by large trading volume orders. For the simulation, the iceberg and normal orders have a ratio of about 1:1. Furthermore, the peak size of the trading quantity (Q) for every order has been set at 4 kW. Table 1 shows the simulation of a partial order book. The system model determines data architecture (described in Section 2). Orders on the buy-side have been ranked in ascending order based on the cost, while they are ranked in descending order on the sell-side that is comparable to a stock exchange market with price competition. Those iceberg orders are highlighted with bold figures in the quantity (Q) column. This exchange process with the order input is simulated for calculating the total cost fluctuation based on the real-time price.

Table 1. Electricity interchange order book.

S(Buy/Sell)	Buy	Buy	Buy	...	Sell	Sell	Sell
Q (kW)	2.14	4.0	1.07	...	4.0	0.44	1.80
Price (pence)	13	12	11		17	15	14
Id	0	70	12		0	18	48

Prices of power exchange markets vary based on the iceberg order execution algorithm, in which the drift of the optimal bid cost is taken as constant. It is assumed that a simple scenario is used for keeping the setup tractable for exposition: the optimal bid cost shows a zero-drift $\mu = 0$ before iceberg orders are submitted. As a result, the original cost fluctuation interval has been assigned at $\sigma_i(10;\ 30)$ based on the local area and as a consequence, the order price σ_t has been adjusted throughout particular times of the day for simulating the retail power costs as shown in Figure 5. Every hour's price has been determined from the average price of whole deal orders for every time interval from the order book.

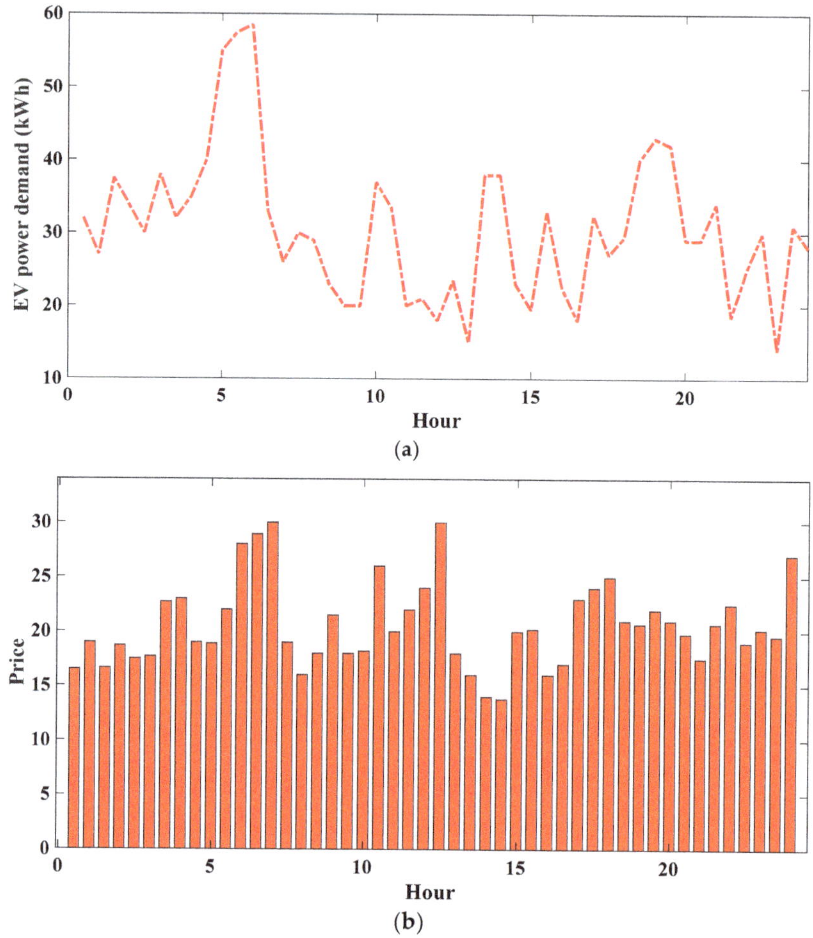

Figure 5. Mean generated trading cost within a day: (**a**) EV power demand; (**b**) price.

B. Power alteration amount minimization

Figure 6 illustrates the impact of the algorithm on the power load alternation. Here is a comparison of the daily half-hourly resolution of LP in a residential area excluding each EV charging scheduling optimization (top panel) and with scheduling method utilizing GA (bottom panel). The daily profile of the domestic residential from the Elexon report is used in the simulation [21]. Charging demand of EV and residential load is represented by the red dashed line v in every timeframe. Because of the high number of charging requirements during this period, the overall electricity consumption at 20:30 exceeds the power capacity, with $P_{PFL} = 1.15$ by Equations (7)–(9). As described in [17], following the application of a planning algorithm, the load of charging has been moved to the off-peak time and the discharging characteristics have been taken into account.

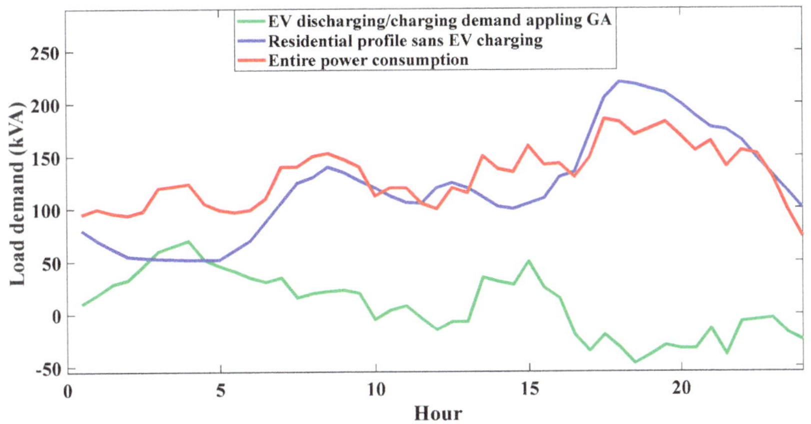

(**a**)

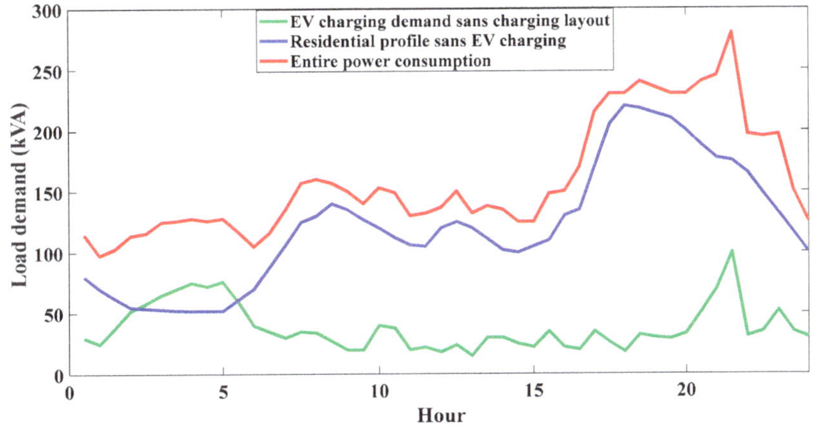

(**b**)

Figure 6. A comparison of the domestic 30 min LP. (**a**) LP applying charging/discharging scheme; (**b**) LP sans charging method.

The suggested AdBEV scheduling layout allows EVs for generating discharging and charging orders for the market based on their states of connection, battery capacities, and discharging/charging limitations. By using the proposed algorithm, the total power demand fluctuation level can be reduced, with the optimized outcome P^*_{PFL} with the algorithm being 0.63, determined through Equations (7) to (9), which results in a reduction of 25.9 by comparison with the P_{PFL} applying the GA scheduling algorithm. Figure 7 illustrates how the index P is better suited to flatten consumption loads by applying the optimized EV charging layout in ref [17]. In addition, by increasing the number of EVs in the system, the P^*_{PFL} would rise linearly since the SOC is assumed to be normally distributed [10]. The total utility function Equation (8) shows that the power fluctuation level P_{PFL} has been aggregated with the absolute difference of power consumption over two consecutive hours, in which the ability to minimize power fluctuation increases linearly as $P_{EV}(t)$ increases.

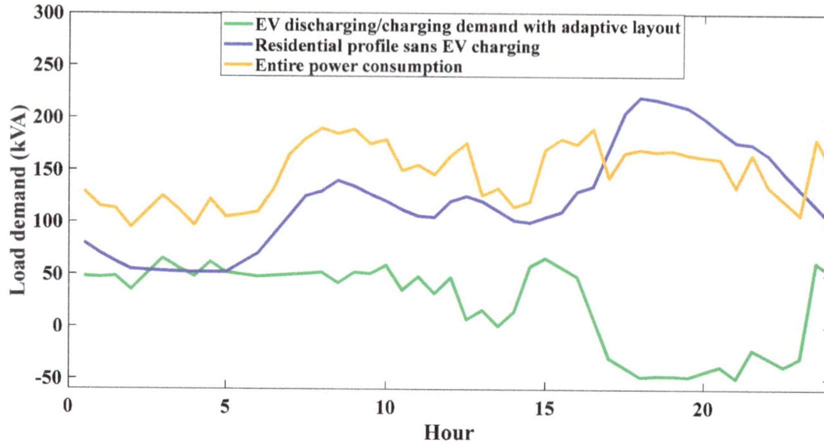

Figure 7. Domestic half-hourly LP applying the offered adaptive EV discharging/charging demand matching scheme.

C. Discussion and analyzing computation cost

The computational complexity should be noted because calculation complexity directly affects exchange performance and cost. A *gas* computation can correspond to a low-level operation in the Ethereum Virtual Machine, in which every opcode can have its own *gas*. The operators *add* applies 3 *gas*, whereas multiplication 2 integers apply 5 *gas*. The total gas cost is 21,000 *gas* as a base despite not having interacted with the contract, and every *gas* resulting from running a contract plus 21,000 *gas* is included in the overall *gas*. As shown in Figure 8, the theoretical computation cost for Ethereum based on the number of peers for every system is modeled [22–24]. As it is inferred from Figure 5, the power cost goes up from 6:00 to 8:00, 11:00 to 13:00, and 17:00 to 19:00 corresponding to the higher energy demand P_{EV} for EV charging periods. In Figure 6, it is seen that when the GA scheduling algorithm is used, the peak time power consumption is reduced with less energy fluctuation index at $P_{PFL} = 0.85$. In comparison to the *PPFL* index with no scheduling algorithm, there is a 26.1% reduction in the index, and the transformer load has been reduced. It is deduced from the results that as the number of EVs in the system increases, the P^*_{PFL} will linearly since the SOC is assumed to be normally distributed. Moreover, it is observed that the total utility function, as calculated and represented in Equation (8), shows that the power fluctuation level P_{PFL} has been aggregated with the absolute difference in power consumption over two consecutive hours. The simulation results show that the proposed technique not only helps the policymakers to reduce the price but also provides a dynamic and transparent environment that can mitigate possible fraud within the system. The appropriate performance and high quality of the proposed model are well proved in the results.

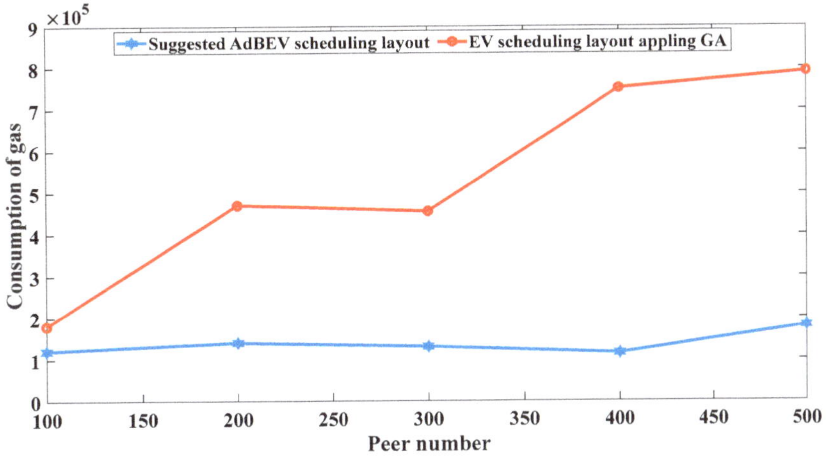

Figure 8. Comparison of calculation cost among the EV scheduling layout applying GA and the AdBEV.

5. Conclusions

An AdBEV scheduling layout for minimizing power fluctuations in order to provide a reliable and safe trading platform is examined in the present study. The EV stay-online model has been modeled for controlling the accessibility of discharging/charging amounts in the market. Most scheduling demands are processed using the iceberg order execution algorithm. The outcomes of the simulation are satisfactory and suggest that the proposed hybrid blockchain and machine learning algorithm can significantly lower power fluctuation levels, while also maximizing EV driver usefulness. The discharging and charging procedure can be controlled dynamically by adapting the very popular EV battery kinds and charging rates. AdBEV can provide insights into the construction of buildings in the transactional power market in the context of BC technology for improving decentralizing of the SGs. By reducing gas consumption, the suggested algorithm increases the performance of order trading. The future requires finding a balance between the on- and off-chain complexity, whilst taking advantage of BC's decentralized nature. It is worth noting that the main shortcoming of the proposed technique is the required space for the BL technique. Indeed, after a long time, storing all these transactions on a database requires a huge amount of space. Moreover, in the case of an attack, the retrieval of data would be challenging.

Author Contributions: Data curation, W.H.; Investigation, J.X. and W.H.; Project administration, W.H.; Resources, J.X.; Supervision, J.X.; Visualization, J.X. All authors have read and agreed to the published version of the manuscript.

Funding: This research received no external funding.

Institutional Review Board Statement: Not applicable.

Informed Consent Statement: Not applicable.

Data Availability Statement: This study did not report any data.

Conflicts of Interest: The authors have no conflict of interest to declare.

References

1. Alnowibet, K.; Annuk, A.; Dampage, U.; Mohamed, M.A. Effective Energy Management via False Data Detection Scheme for the Interconnected Smart Energy Hub–Microgrid System under Stochastic Framework. *Sustainability* **2021**, *13*, 11836. [CrossRef]
2. Mohamed, M.A.; Mirjalili, S.; Dampage, U.; Salmen, S.H.; Al Obaid, S.; Annuk, A. A cost-efficient-based cooperative allocation of mining devices and renewable resources enhancing blockchain architecture. *Sustainability* **2021**, *13*, 10382. [CrossRef]

3. Wang, B.; Dabbaghjamanesh, M.; Kavousi-Fard, A.; Mehraeen, S. Cybersecurity enhancement of power trading within the networked microgrids based on blockchain and directed acyclic graph approach. *IEEE Trans. Ind. Appl.* **2019**, *55*, 7300–7309. [CrossRef]
4. Taherzadeh, E.; Javadi, S.; Dabbaghjamanesh, M. New optimal power management strategy for series plug-in hybrid electric vehicles. *Int. J. Automot. Technol.* **2018**, *19*, 1061–1069. [CrossRef]
5. Dabbaghjamanesh, M.; Kavousi-Fard, A.; Dong, Z.Y. A novel distributed cloud-fog based framework for energy management of networked microgrids. *IEEE Trans. Power Syst.* **2020**, *35*, 2847–2862. [CrossRef]
6. Angelim, J.H.; Affonso, C.D.M. Probabilistic assessment of voltage quality on solar-powered electric vehicle charging station. *Electr. Power Syst. Res.* **2020**, *189*, 106655. [CrossRef]
7. Dehghani, M.; Kavousi-Fard, A.; Dabbaghjamanesh, M.; Avatefipour, O. Deep learning based method for false data injection attack detection in AC smart islands. *IET Gener. Transm. Distrib.* **2020**, *14*, 5756–5765. [CrossRef]
8. Rajić, M.N.; Maksimović, R.M.; Milosavljević, P.; Pavlović, D. Energy management system application for sustainable development in wood industry enterprises. *Sustainability* **2019**, *12*, 76. [CrossRef]
9. Shojaei, F.; Rastegar, M.; Dabbaghjamanesh, M. Simultaneous placement of tie-lines and distributed generations to optimize distribution system post-outage operations and minimize energy losses. *CSEE J. Power Energy Syst.* **2020**, *7*, 318–328.
10. Aghajan-Eshkevari, S.; Azad, S.; Nazari-Heris, M.; Ameli, M.T.; Asadi, S. Charging and Discharging of Electric Vehicles in Power Systems: An Updated and Detailed Review of Methods, Control Structures, Objectives, and Optimization Methodologies. *Sustainability* **2022**, *14*, 2137. [CrossRef]
11. Dabbaghjamanesh, M.; Wang, B.; Kavousi-Fard, A.; Hatziargyriou, N.D.; Zhang, J. Blockchain-based stochastic energy management of interconnected microgrids considering incentive price. *IEEE Trans. Control. Netw. Syst.* **2021**, *8*, 1201–1211. [CrossRef]
12. Dabbaghjamanesh, M.; Wang, B.; Kavousi-Fard, A.; Mehraeen, S.; Hatziargyriou, N.D.; Trakas, D.N.; Ferdowsi, F. A novel two-stage multi-layer constrained spectral clustering strategy for intentional islanding of power grids. *IEEE Trans. Power Deliv.* **2019**, *35*, 560–570. [CrossRef]
13. Aslam, S.; Javaid, N.; Khan, F.A.; Alamri, A.; Almogren, A.; Abdul, W. Towards efficient energy management and power trading in a residential area via integrating a grid-connected microgrid. *Sustainability* **2018**, *10*, 1245. [CrossRef]
14. Antal, C.; Cioara, T.; Antal, M.; Mihailescu, V.; Mitrea, D.; Anghel, I.; Salomie, I.; Raveduto, G.; Bertoncini, M.; Croce, V.; et al. Blockchain based decentralized local energy flexibility market. *Energy Rep.* **2021**, *7*, 5269–5288. [CrossRef]
15. Vilathgamuwa, M.; Mishra, Y.; Yigitcanlar, T.; Bhaskar, A.; Wilson, C. Mobile-Energy-as-a-Service (MEaaS): Sustainable Electromobility via Integrated Energy–Transport–Urban Infrastructure. *Sustainability* **2022**, *14*, 2796. [CrossRef]
16. Zeng, L.; Xia, T.; Elsayed, S.; Ahmed, M.; Rezaei, M.; Jermsittiparsert, K.; Dampage, U.; Mohamed, M. A novel machine learning-based framework for optimal and secure operation of static VAR compensators in EAFs. *Sustainability* **2021**, *13*, 5777. [CrossRef]
17. Ghaffari, S.; Ashkaboosi, M. Applying Hidden Markov Model Baby Cry Signal Recognition Based on Cybernetic Theory. *IJEIR* **2016**, *5*, 243–247.
18. Lehtinen, O.; Pitkäniemi, S.; Weckman, A.; Aikio, M.; Mabano, M.; Lehtonen, M. Electric Vehicle Charging Loads in Residential Areas of Apartment Houses. In Proceedings of the 2020 21st International Scientific Conference on Electric Power Engineering (EPE), Prague, Czech Republic, 19–21 October 2020; IEEE: Piscataway, NJ, USA, 2020; pp. 1–6.
19. Ashkaboosi, M.; Ashkaboosi, F.; Nourani, S.M. *The Interaction of Cybernetics and Contemporary Economic Graphic Art as "Interactive Graphics"*; University Library of Munich: Munich, Germany, 2016.
20. Badruddoja, S.; Dantu, R.; He, Y.; Upadhayay, K.; Thompson, M. Making smart contracts smarter. In Proceedings of the 2021 IEEE International Conference on Blockchain and Cryptocurrency (ICBC), Sydney, Australia, 3–6 May 2021; IEEE: Piscataway, NJ, USA; pp. 1–3.
21. Rizwan, M.; Waseem, M.; Liaqat, R.; Sajjad, I.A.; Dampage, U.; Salmen, S.H.; Al Obaid, S.; Mohamed, M.A.; Annuk, A. SPSO Based Optimal Integration of DGs in Local Distribution Systems under Extreme Load Growth for Smart Cities. *Electronics* **2021**, *10*, 2542. [CrossRef]
22. Shahab, S.; Lades, L.K. Sludge and transaction costs. *Behav. Public Policy* **2021**, 1–22. [CrossRef]
23. Cavusoglu, S.S.; Macário, R. Minimum delay or maximum efficiency? Rising productivity of available capacity at airports: Review of current practice and future needs. *J. Air Transp. Manag.* **2021**, *90*, 101947.
24. Foryś, I.; Głuszak, M.; Konowalczuk, J. Compensation due to land use restrictions: The case of limited use area in the vicinity of Polish airports. *Oeconomia Copernic.* **2019**, *10*, 649–667. [CrossRef]

sustainability

MDPI

Article

Coordinated Design of Type-2 Fuzzy Lead–Lag-Structured SSSCs and PSSs for Power System Stability Improvement

Prabodh Khampariya [1], Sidhartha Panda [2], Hisham Alharbi [3], Almoataz Y. Abdelaziz [4,*] and Sherif S. M. Ghoneim [3,*]

1 Department of EEE, Sri Satya Sai University of Technology & Medical Sciences, Sehore 466001, India; khampariya5@gmail.com
2 Department of Electrical Engineering, Veer Surendra Sai University of Technology, Odisha 768018, India; prof.dr.spanda@gmail.com
3 Electrical Engineering Department, College of Engineering, Taif University, P.O. Box 11099, Taif 21944, Saudi Arabia; h.alharbi@tu.edu.sa
4 Faculty of Engineering and Technology, Future University in Egypt, Cairo 11835, Egypt
* Correspondence: almoatazabdelaziz@hotmail.com (A.Y.A.); s.ghoneim@tu.edu.sa (S.S.M.G.)

Abstract: This work suggests a type-2 fuzzy lead–lag (T2FLL) controller structure for flexible AC transmission system (FACTS)-based damping controllers and power system stabilizers (PSSs) for power system stability improvement. The values of the suggested controller are optimized by a hybrid adaptive differential evolution and pattern search algorithm (hADE-PS) method. Initially, a single-machine infinite-bus (SMIB) system with lead–lag (LL)-structured FACTS and PSS controllers is considered, and the dominance of the hADE-PS method is established over the original differential evolution (DE), genetic algorithm (GA), and particle swarm optimization (PSO). The supremacy of T2FLL over the lead–lag (LL) controller is established under different large and small disturbance conditions, as well as varied loading conditions and fault positions. Lastly, the effectiveness of T2FLL is evaluated in a multimachine power system (MMPS). It is demonstrated that the suggested T2FLL offers better performance than the LL controller under various large and small disturbance conditions by providing significantly more damping to all modes of oscillations.

Keywords: power system stability; power system stabilizer; static synchronous series compensator; type-2 fuzzy lead–lag controller; differential evolution; pattern search

Citation: Khampariya, P.; Panda, S.; Alharbi, H.; Abdelaziz, A.Y.; Ghoneim, S.S.M. Coordinated Design of Type-2 Fuzzy Lead-Lag-Structured SSSCs and PSSs for Power System Stability Improvement. *Sustainability* **2022**, *14*, 6656. https://doi.org/10.3390/su14116656

Academic Editors: J. C. Hernandez and Mohamed A. Mohamed

Received: 18 April 2022
Accepted: 26 May 2022
Published: 29 May 2022

1. Introduction

Low-frequency oscillations are detected after disturbances in power systems that are joined by weak transmission lines. Under such circumstances, a lack of adequate system damping may result in sustained oscillations and, in turn, lead to the separation of the system [1,2]. To handle this issue, power system stabilizers (PSSs) are extensively favored by utilities. However, PSSs may not provide sufficient damping, and additional controllers are required. In this regard, recently developed flexible AC transmission system (FACTS) devices can be employed to provide the required damping. The static synchronous series compensator (SSSC) is an effective series FACTS device for power flow control and, hence, can be designed to provide additional damping [3].

Various soft-computing-based schemes have been suggested for damping controller design. Methodologies such as particle swarm optimization (PSO) [4,5], genetic algorithm (GA) [6,7], oppositional cuckoo algorithm [8], simulated annealing (SA) [9], tabu search (TS) [10], and bacteria foraging algorithm (BFA) [11] have been proposed in the literature. When SSSCs and PSSs are present in a power system, both should be coordinately designed. In the literature, PSSs are coordinately designed with other controllers such as PSS and SSSC by GA [12], PSS and SVC by GA [13], PSS and SSSC by hybrid PSO and gravitational search algorithm (GSA) [14], and PSS and thyristor-controlled series compensator (TCSC) [15–17].

It is observed in the literature that lead–lag (LL) controllers are commonly employed by researchers as structures of damping controllers. Recently, fuzzy logic controllers (FLCs) have been proposed by researchers, as FLCs can operate with imprecise inputs and handle nonlinearity. In [18], an improved control strategy has been proposed using a fuzzy PID regulator for a hybrid system. An FLC-based hybrid scheme is projected in [19] for power oscillation damping with FACTS and PSS controllers. In [20], a mixture of fuzzy-neural schemes is employed to capture maximum power. A hybrid firefly swarm-based type-2-based fractional-order (FO) fuzzy PID-structured PSS has been suggested in [21] for stability improvement of power systems.

Different optimization techniques such as the cuckoo search (CS) algorithm [22] and BAT search algorithm [23] have been employed in the literature for the optimal design of PSS parameters, and the superiority of the approach for system damping was demonstrated for various loading conditions. Numerous optimization-technique-based FACTS devices such as CS-based static synchronous compensator (STATCOM) [24], flower-pollination-based FACTS damping controller for a multimachine power system (MMPS) [25], and gravitational search algorithm (GSA)-based SSSCs [26] have been proposed for power system stability improvement. It is also demonstrated in the literature that when PSS and FACTS devices are present, they should be designed in a coordinated manner for achieving improved system performance [27].

From the above survey, it is evident that many optimization methods have been used for various power system problems, but no technique is exactly suited for all types of problems, and there are opportunities for performance enhancement by proposing new and modified techniques. Differential evolution (DE) is an easy but capable optimization method [28]. DE effectiveness relies on the choice of DE parameters such as crossover constant (*Cr*) and scaling factor (*F*) [28]. Throughout the search procedure, appropriate F and Cr parameters must be engaged in place of employing constant parameters during the evolution. An adaptive method is suggested here, where an adaptive mechanism is used for the selection of suitable *F* and *Cr* parameters during the optimization procedure. As DE is a global search method, it is intended for exploring the search area. To improve search capabilities, hybrid schemes have been recently projected for engineering design problems. A hybrid DE–PS procedure was suggested in [29] to tune the modified integral-derivative (MID) structure for frequency regulation. Fuzzy PID values are tuned by many hybrid optimization liaisons and GSAs for different systems in [30]. A hybrid-shuffled frog-leaping and PS-tuned PID has been proposed for frequency control in [31]. Recently, a hybrid-modified DE and PS (hMDE-PS)-based SSSC controller was proposed under communication constraints for stability improvement [32]. The superiority of hMDE-PS over other variants of DE such as JADE [33], CoDE [34], and SHADE [35] has been demonstrated for benchmark test functions. A hybrid adaptive DE and PS (hADE-PS)-based fractional-order fuzzy PID (FOFPID) structure for frequency regulation of power systems was reported in [36], where the performance of the FOFPID controller was equated with a double-derivative PI and PID controller for assessing frequency regulation performance using transfer function models of power systems.

Based on the above facts, this work proposes a novel scheme by employing a hybrid ADE and PS (hADE-PS) method for the coordinated design of a type-2 fuzzy lead–lag (T2FLL)-structured FACTS controller and PSS for power oscillation damping. The effectiveness of hADE-PS is compared with other popular algorithms such as GA, PSO, and the original DE. The novelties of the current work are:

i. A new controller structure, known as type-2 fuzzy lead–lag (T2FLL), is proposed in this paper for PSS- and SSSC-based controllers for power system stability improvement.
ii. The design task is taken as an optimization problem, and controller parameters are optimized by a recently proposed hADE-PS method.

iii. Technique-wise, the hDE-PS method is compared with GA, PSO, and DE methods, and controller-wise, T2FLL is compared with type-1 fuzzy lead–lag (T1FLL) and widely used lead–lag (LL) controllers.
iv. Various disturbance scenarios and changed loading/fault locations are simulated for both single-machine infinite-bus (SMIB) and MMPS, and it is seen that improved damping is attained with T2FLL related to T1FLL and LL controllers for all scenarios.

This paper is structured as follows: Section 2 introduces the systems under investigation. The projected control approach is described in Section 3. An overview of the hybrid adaptive DE and PS technique is provided in Section 4. Results are discussed in Section 5. Conclusions and future work are provided in Section 6.

2. Systems under Investigation

2.1. The SMIB System

Initially, SMIB system exposed in Figure 1 is taken. In Figure 1, the transformer is represented by *T*; the infinite bus and generator terminal voltages are denoted by V_S and V_T, respectively; while V_1 and V_2 are the bus voltages, the SSSC converter output voltage and the DC voltage source are represented by V_{cnv} and V_{DC}, respectively; the line current is *I*; the real powers in the transmission lines and one line are each represented by P_L and P_{L1}, respectively. The generator is provided with a turbine and governor, an excitation system, and a PSS. The excitation system contains a voltage regulator and exciter [37]. The system parameters are specified in Appendix A.

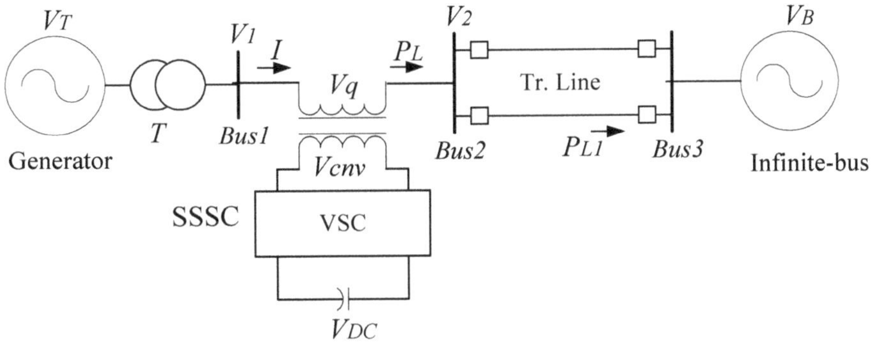

Figure 1. SMIB system.

2.2. Kundur's Test System (Four-Machine Two-Area)

The multi-machine power system (MMPS) contains 11 buses and 2 areas, joined by a weak transmission line as revealed in Figure 2. Loads and shunt capacitors are assumed at buses 7 and 9. The 2 areas are joined by 2 lines of 220 km. It was precisely planned to investigate low-frequency oscillations [38]. The test system follows the characteristics of usual systems in real operation. The load is characterized as constant impedances. The power transfer from area 1 to area 2 is 413 MW. The system is stressed and oscillates in the occurrence of any disturbance. The system details are taken from [38] and given in Appendix A.

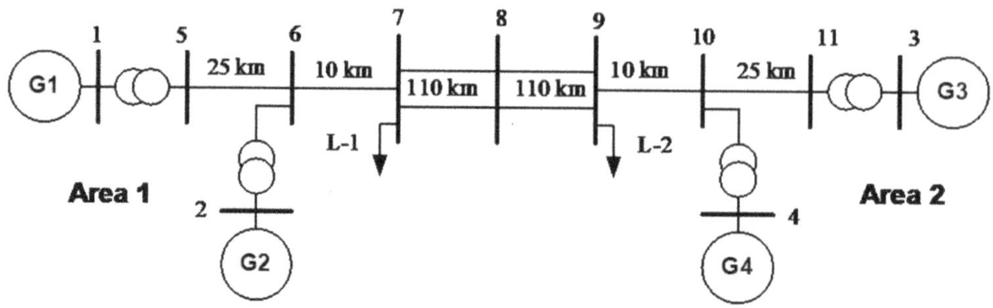

Figure 2. Kundur's 4-machine 2-area system.

3. The Proposed Approach

3.1. Type-2 Fuzzy Logic Overview

When uncertainties are present, the conventional fuzzy logic control (FLC) may not be effective to improve the performance of the system. However, the twin membership function (MF) grounded type-2 FLC (T2F) provides improved system performance. The present study proposes a type-2 fuzzy lead–lag (T2FLL) controller as a damping controller. The MFs of the T2FLL engage upper and lower MFs (UMF and LMF). The type-2 fuzzy act includes fuzzification, knowledge base, type reducer (TR), and defuzzification.

Fuzzification is the initial phase of FLC. It processes inputs and creates the necessary prearranged fuzzy sets by the MFs. The linguistic parameters engaged for MFs are Extreme Positive (EXP), Least Positive (LP), Zero (ZER), Least Negative (LN), and Extreme Negative (EXN).

The T2F fuzzy set (*FS*) is formulated as:

$$FS = ((Var, a), \mu_U(Var, a)), vVar \in P, va \in J_{Var}[0,1] \tag{1}$$

where $\mu_U(Var, a)$ is the UMF, *Var* is the main variable, *a* is the extra parameter in area J_{Var}

The universe of discourse is formulated as:

$$FS = \int_{Var \in P} \int_{a \in J_{Var[0,1]}} \frac{\mu_E(Var, a)}{(Var, a)} \tag{2}$$

where, $\int \int$ = Union on *ACE* and *a*

Now the membership expression is:

$$\mu_{\bar{U}}(Var, a) = \overline{FOU(U)} vVar \in P, va \in J_{Var}[0,1] \tag{3}$$

where J_{Var} is expressed as:

$$JVar = [\mu_{\overline{U}}(Var, a), \mu_{\underline{U}}(Var, a)] vVar \in P, va \in J_{Var}[0,1] \tag{4}$$

The MF associated with type-I FLC inspires to develop MFs of T2F.

The rule base and interface engine form the knowledge base. The rule base is demonstrated in Table 1. Individually, *Var* and *dAVar* are the inputs to T2F and y is the output.

Table 1. Rule base of fuzzy controller.

$\dot{e}$ \ e	**EXN**	**LN**	**ZER**	**LP**	**EXP**
EXN	EXN	EXN	LN	LN	ZER
LN	EXN	LN	EXN	ZER	LP
ZER	LN	LN	ZER	LP	LP
LP	LN	ZER	LP	LP	EXP
EXP	ZER	LP	LP	EXP	EXP

The characteristic of the type-2 FLC is

$$LMF : forVar = \underline{LN}; dVar = \underline{Z}; Y = \underline{LN} \tag{5}$$

$$UMF : forVar = \overline{LN}; dVar = \overline{Z}; Y = \overline{LN} \tag{6}$$

where $\underline{LN}$, $\underline{Z}$ and $\overline{LN}$, $\overline{Z}$ are related to LMF and UMF, respectively.

The related FS firing forte is

$$\underline{f^s} = \min(\mu_{\overline{US}}(Var, a), \mu_{\overline{US}}(dVar, a)) \tag{7}$$

$$\overline{f^s} = \max(\mu_{\overline{US}}(Var, a), \mu_{\overline{US}}(dVar, a)) \tag{8}$$

$$F^S = [\underline{f^s}, \overline{f^s}] \tag{9}$$

where $\underline{f^s}$ and $\overline{f^s}$ are related to LMF and UMF, respectively.

Type reduction (TR) is used to modify type-2 to type-1 FS for defuzzification. The schemes for defuzzification are centroid, the center of sets (SOC), and the center of sums. Here, SOC is used as it is reported as the best scheme. The outputs are:

$$Y_{\cos} = \sum_{s=1}^{25} \frac{F^s Y^s}{F^s} = [Y_{m1}, Y_{m2}] \tag{10}$$

$$Y_{m1} = \frac{\sum\limits_{s=1}^{25} \underline{f^s y^s}}{\sum\limits_{s=1}^{25} \underline{fy}} \tag{11}$$

$$Y_{m2} = \frac{\sum\limits_{s=1}^{25} \overline{f^s y^s}}{\sum\limits_{s=1}^{25} \overline{fs}} \tag{12}$$

where Y_{m1} and Y_{m2} are associated with MFs of type-1 FLC. The output of type-2 FLC $Y_{\cos}$ is found by averaging. The recommended T2FLL is set bearing in mind all characteristics of lead–lag controller and T2F.

3.2. Structure of T2FLL Controller

The configuration of the proposed type-2 fuzzy lead–lag (T2FLL) SSSC-based controller is revealed in Figure 3. It contains a sensor block, transport delay block, scaling factor blocks for fuzzy input, gain, a washout, and lead–lag (LL) blocks. The washout tie constant is generally taken as 10 s [38]. The LL blocks deliver the suitable phase lead to balance for the lag among output and input. The output of the controller is the V_q.

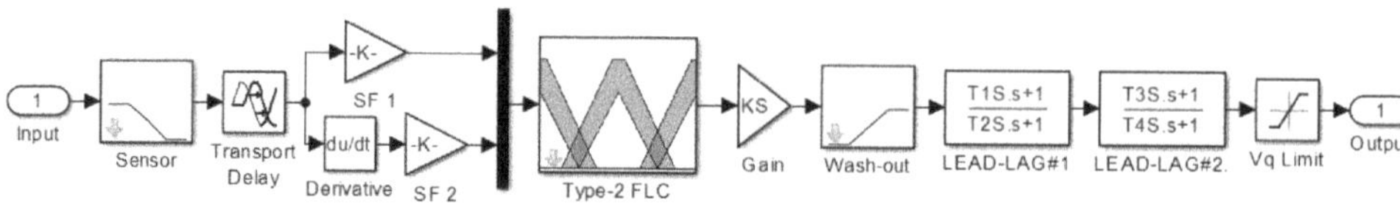

Figure 3. SSSC-based T2FLL controller.

Figure 4 shows the structure of the T2FLL-based PSS. It consists of a gain, washout, and LL block. The speed variation is fed as input to the PSS. The PSS output (*Vs*) is used to change the excitation voltage.

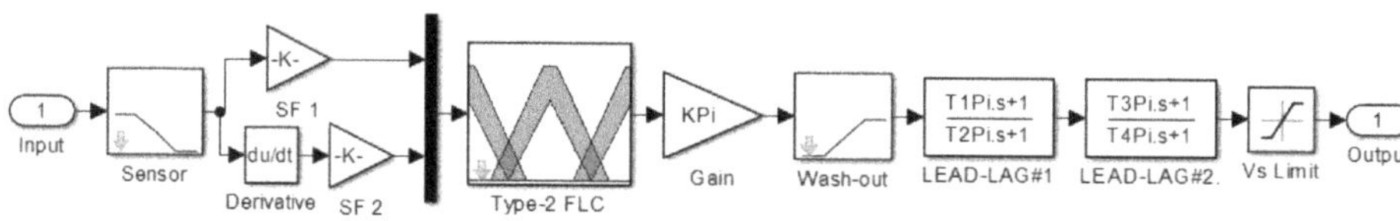

Figure 4. PSS-based T2FLL controller.

Speed deviations are selected as T2FLL controller input signals. A sensor time constant of 15 ms is assumed. A transport delay of 50 ms is assumed for the SSSC controller.

3.3. Optimization Problem

An integral time multiplied absolute error (ITAE) criterion is employed for the SMIB system. For MMPS, an ITAE criterion using local and inter-area modes of oscillations is selected. They are represented by:

For SMIB:

$$J = \int_{t=0}^{t=t_{sim}} |\Delta\omega| \cdot t \cdot dt \tag{13}$$

For MMPS:

$$J = \int_{t=0}^{t=t_{sim}} \left(\sum|\Delta\omega_L| + \sum|\Delta\omega_I|\right) \cdot t \cdot dt \tag{14}$$

where $\Delta\omega$ is the speed deviance in the SMIB system, $\Delta\omega_L$ and $\Delta\omega_I$ are the speed deviations of MMPS related to local and inter-area modes, respectively, and t_{sim} is the simulation time.

Therefore, the design task is formulated as an optimization task as:

$$\text{Minimize } J \tag{15}$$

Subject to

$$\begin{array}{l} KSFi_{\min} \leq KSFi \leq KSFi_{max}(KSFi = \text{Scaling factors}) \\ Ki_{\min} \leq Ki \leq Ki_{max}(Ki = \text{Gains}) \\ Ti_{\min} \leq Ti \leq Ti_{max}(Ti = \text{Time constants}) \end{array} . \tag{16}$$

The limits of the parameters are [0.01–100] for *Ki*, [0.01–10] for *Ti*, and [0.01–100] for *KFSi*. In the present work, the hADE-PS algorithm determines T2FLLvalues.

4. Overview of Hybrid Adaptive DE and PS Technique

The differential evolution (DE) method is a simple, competent but efficient method [28]. Encouraged by the evolution approach, the DE method is found to be a capable method for complex engineering problems. The stages of DE are:

4.1. Mutation

The mutant vector $w^j\big|_i$ is to be performed for each parent vector $y^j\big|_i$ on each generation '*i*' is:

$$w^j\Big|_i = y^{q_1}\Big|_i + F(y^{q_2}\Big|_i - y^{q_3}\Big|_i) \tag{17}$$

where $q_1 \neq q_2 \neq q_3 \neq j$. F indicates the scaling factor.

4.2. Crossover

It is a vital method for the expansion of population diversity. In Binary crossover, the trial vector $v^j|_i = [v_1^j\big|_i, v_2^j\big|_i, \cdots, v_n^j\big|_i]$ can be generated through the target vector $y^j\big|_i$ and its mutant vector $w^j\big|_i$. As:

$$v_l^j|_i = \begin{cases} w_l^j\Big|_i \; if \; rand\ (0,1) \leq Cr \\ y_l^j\Big|_i \; otherwise \end{cases} \tag{18}$$

where $l \in 1, 2, \cdots, n$. For every vector $y^j\big|_i$ and $w^j\big|_i$, the trial vector $v^j\big|_i$ can be generated as:

$$v^j\Big|_i = y^j\Big|_i + F_j(w^j\Big|_i - y^j\Big|_i) \tag{19}$$

4.3. Selection

The selection operation is engaged to maintain a fixed population size in all generations. The vector having the best fitness value replaces the fewer fit vectors.

The equation representing this process is:

$$y^j|_{i+1} = \begin{cases} v^j|_i \; if \; f(v^j|_i) \leq f(y^j|_i) \\ y^j|_i \; otherwise \end{cases} \tag{20}$$

DE is a competent method but relies on its algorithm parameters such as F and Cr. Appropriate choice of F and Cr is vital for the performance of DE. Numerous researchers have recommended different schemes to choose F and Cr values [39].

Liu and Lampinen [40] indicated the control parameters value to evaluate the robustness as well as the effectiveness of the DE algorithm. Gamperle et al. [41] have suggested the values such as 0.6 and [0.3, 0.9] for F and Cr to control the parameters. On the other hand, Ali and Torn [42] have tested their results on the values such as 0.5 and [0.4, 1] for F and Cr. Moreover, different self-adaptive approaches are proposed to select 'Cr &F' during the experimentation for effective results [43].

Selecting correct F and Cr parameters is a problem-dependent task. The projected scheme uses an adaptive approach to select F and Cr values during the optimization process. For generation i, the triangular factors such as $\wedge F$, and $\wedge Cr$ have been used for evaluating $f_F(y)\big|_i$ and $f_{Cr}(y)\big|_i$ for each child in the population. For Triangular Distribution factor of F ($\wedge F$) and Cr ($\wedge Cr$), where $\wedge F$ = [0.1, 1, 2] and $\wedge Cr$ = [0.1, 0.4, 1] are engaged in the present study.

To select F^i and Cr^i, mx (maximum), mn (minimum), and md (median) are engaged in a generation as indicated in Equations. (21) and (22).

$$F^i = \begin{cases} F_{mn} + \sqrt{rand\,V * (F_{mx} - F_{mn})/(F_{md} - F_{mn})} \\ if\ rand\,V < (F_{mx} - F_{mn})/(F_{md} - F_{mn}) \\ F_{mx} - \sqrt{(1 - rand\,V) * (F_{mx} - F_{mn})/(F_{mx} - F_{md})} \\ otherwise \end{cases} \tag{21}$$

$$Cr^i = \begin{cases} Cr_{mn} + \sqrt{rand\,V * (Cr_{mx} - Cr_{mn})/(Cr_{md} - Cr_{mn})} \\ if\ rand\,V < (Cr_{mx} - Cr_{mn})/(Cr_{md} - Cr_{mn}) \\ Cr_{mx} - \sqrt{(1 - rand\,V) * (Cr_{mx} - Cr_{mn})/(Cr_{mx} - Cr_{md})} \\ otherwise \end{cases} \tag{22}$$

Here, *rand V* has been selected in an arbitrary distribution with a limit [0, 1].

Before performing the mutation scheme, the F^i and Cr^i values are set which affects the mutation, cross-over, and selection process for the newly generator vector $y^j|_{i+1}$. Due to the easy generation of adaptive values for *F* and *Cr*, it can maintain the time complexity as compared to conventional DE.

The pattern search (PS) method is shown to be useful in numerous hybrid approaches [23,25]. The information regarding the PS algorithm is available in the literature [23,25]. At first, the ADE technique is executed and PS is then executed considering the final results of ADE as the starting points. So, the benefit of hybridizing the ADE and PS algorithms is that the hybrid algorithm will have enhanced exploring and exploiting capability.

5. Outcomes

5.1. SMIB System

The SMIB system is developed in MATLAB / SIMULINK as displayed in Figure 5. In the present study, the ode23tb (stiff/TR-BDF2) solver which is a variable-step with 1 cycle of the fundamental frequency as the maximum time step is used as in the literature [4,5]. The system is developed using the Sim Power Systems (SPS) library of SIMULINK. The dynamics of all windings are specified in the rotor reference frame (q-d frame). The Hydraulic Turbine and Governor (HTG) and Excitation system are present in the subsystem represented by the "Generator Control System" (Reg_M).

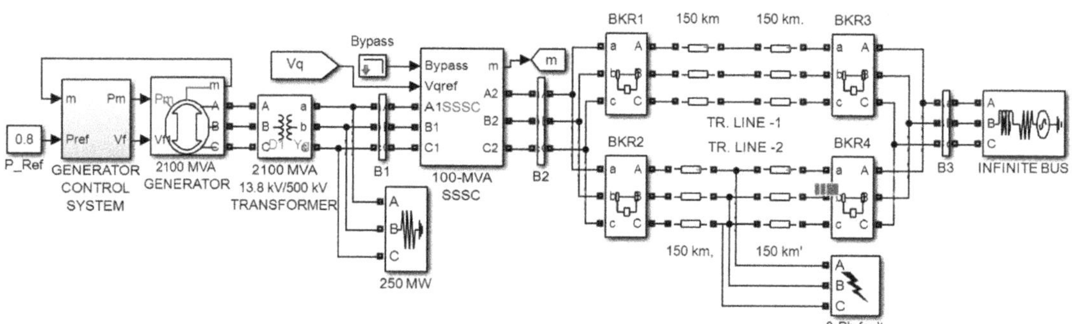

Figure 5. MATLAB/SIMULINK model of SMIB system.

The system model is constructed in the SIMULINK environment and optimization routines are specified in a separate m file. For ITAE evaluation, the model is simulated considering a severe disturbance. The *J* value is transferred to the optimization routine by transferring it to the workspace and used by the optimization program to minimize *J* value.

To validate the advantage of the hADE-PS approach, the LL controller is initially considered. For ITAE calculation, a 3-phase fault of 5 cycles is assumed at the midpoint of one transmission line. The lines are opened for 5 cycles to clear the fault. To authenticate the better performance of the hADE-PS method, the LL values are optimized by the hADE-PS, DE, PSO, and GA methods. It is worth mentioning here that, the parameters of the compared algorithm are taken from the related reference papers [44,45].

One important challenge of a hybrid global search (GS) and local search (LS) is the tradeoff between GS and LS because the cost of LS can be rather high. If LS is applied at the first stage, several solutions in the final generation may lie very close to each other at the end of LS, which makes it harder for the second stage GS method to maintain diversity in its population. This may lead the hybrid algorithm to converge to a local optimum. Therefore, the LS is performed at the second stage using the best solution provided by the GS method. This action tends to make the hybrid algorithm more computationally efficient and helps to prevent converging to a local optimum [29,31,32,36]. As the LS method is used to improve the solution quality (fine-tuning) but not to explore the search space to find independently the global optimum, more iterations (about 90%) are assigned to the GS method and fewer iterations (about 10%) are assigned to the LS method [29–32].

For the application of DE, PSO, and GA, 50 populations and 100 generations are employed and for the hADE-PS technique, 90 generations are assigned to the ADE technique and 10 generations are assigned to the PS technique. The parameters employed for each technique are as follows: GA: Tournament selection, Crossover and mutation rates 0.9 and 0.1; PSO: Inertia weight reduces from 0.9 to 0.2, social and cognitive components 2.0; DE: DE/best/2/bin strategy, step size and crossover probability 0.2 and 0.6. The PS is executed with a mesh size of 1, mesh expansion factor of 2 and mesh contraction factor of 0.5. The maximum number of objective function evaluations is set to 10. All the techniques are executed 10 times, and optimal values (as per the least ITAE value) obtained are taken as controller parameters. The outcomes are delivered in Table 2. It is obvious from Table 2 that, with lead–lag structure, the least ITAE values (33.7926×10^{-3}) are obtained with the hADE-PS technique compared to GA (44.3216×10^{-3}), PSO (40.9781×10^{-3}) and DE (37.7718×10^{-3}) techniques. So with the identical system and controller structure, the percentage reduction in ITAE value with hADE-PS compared to GA, PSO, and DE techniques are 23.56%, 17.53%, and 10.53%, respectively. This demonstrates the dominance of hADE-PS over GA, PSO, and DE techniques. The speed deviation response is revealed in Figure 6, from which it is evident that the best system response is attained with hADE-PS related to GA, PSO, and DE.

Table 2. Optimized lead–lag-structured PSS and SSSC parameters for SMIB.

Technique	Controller	K_i	T_{1i}	T_{2i}	T_{3i}	T_{4i}	J Value $\times 10^{-3}$
GA	SSSC	74.0740	0.4286	0.0511	0.4523	1.1098	44.3216
	PSS	19.8294	0.1415	0.0011	0.0630	1.2736	
PSO	SSSC	38.5588	1.0653	0.3218	0.6751	0.7496	40.9781
	PSS	24.2563	1.5546	1.4223	0.0134	0.4930	
DE	SSSC	50.1218	0.8175	1.1379	1.4961	0.4326	37.7718
	PSS	16.0860	0.1553	0.7317	1.7974	1.7974	
hADE-PS	SSSC	71.5265	1.2006	1.3051	0.0011	0.0012	33.7926
	PSS	18.2475	0.0011	0.5204	0.0011	0.4981	

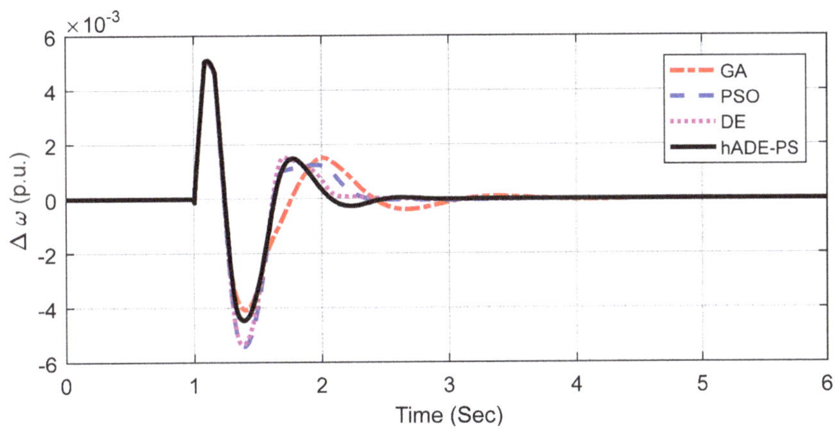

Figure 6. $\Delta\omega$ response with GA-, PSO-, DE-, and hADE-PS-tuned lead–lag controller.

In the next step, the proposed Type-2 fuzzy lead–lag controller is implemented and hADE-PS-tuned parameters are specified in Table 3. For comparison, the results of the type-1 fuzzy lead–lag (T1FLL) controller are also provided in Table 3. The type-1 fuzzy part, i.e., membership functions, range, etc. is adopted from reference [36]. It can be noticed that with the same hADE-PS technique, the ITAE value is further reduced to 31.9342 $\times 10^{-3}$ with T1FLL, and the least ITAE of 28.6961 $\times 10^{-3}$ is attained with the T2FLL controller. So there is a reduction of 10.13% and 15.08% in J value with T2FLL related to the T1FLL and LL controller, respectively.

Table 3. hADE-PS optimized T2FLL- and T1FLL-structured PSS and SSSC parameters for SMIB.

Controller	KSF_1	KSF_2	K_i	T_{1i}	T_{2i}	T_{3i}	T_{4i}	J Value $\times 10^{-3}$
				T2FLL controller				
SSSC	0.0105	0.1824	79.5279	1.9974	1.5774	1.0292	0.4604	28.6961
PSS	0.0104	0.7241	49.9350	0.0622	1.6995	0.0012	1.9974	
				T1FLL controller				
SSSC	1.9093	1.4684	14.2391	0.0154	0.4538	1.9974	0.9935	31.9342
PSS	0.3209	1.2285	30.5106	0.2421	1.7827	0.4172	1.9974	

To explore the viability of T2FLL, several contingencies are taken. These scenarios are widely used in literature to evaluate the performance of and are given as follows [46–48].

5.1.1. Scenario 1: Large Disturbance Condition

Initially, the performance is investigated at nominal loading (Pe = 0.8 p.u. $\delta_0 = 48.5°$). A 3-phase fault of 5 cycles is applied at the midpoint of one transmission line. The fault is removed by opening the line for 5 cycles. The responses with the hADE-PS-tuned LL, T1FLL, and T2FLL controllers are revealed in Figure 7a–e. For comparison, the responses without any control are also shown in Figure 7a–e. Figure 7a–e displays the speed deviation response ($\Delta\omega$) in p.u., power angle response (δ) in degrees, tie-line power in the line (P_L) in MW SSSC output voltage (V_q) in p.u., and PSS output (V_s) in p.u. under above severe disturbance.

Figure 7a–e demonstrate that the proposed controller tuned by hADE-PS has a superior damping performance compared to both the T1FLL and LL controller. Figure 7a–c also indicate that T2FLL offers improved response with less overshoot/undershoot compared to both the T1FLL and LL controller.

5.1.2. Scenario 2: Small Disturbance Condition

The efficiency of the T2FLL is also studied at small disturbance conditions by removing load at bus 1 for 100 ms at t = 1 s. Figure 8 displays the $\Delta\omega$ response, from which it is seen that T2FLL designed at large disturbance conditions, suppresses power system oscillations efficiently at the small disturbance, and T2FLL also delivers a better transient response compared to the T1FLL and LL in this case.

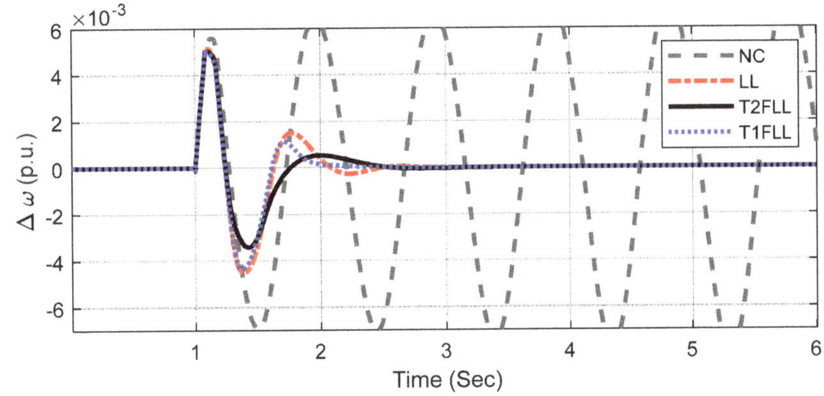

(**a**)

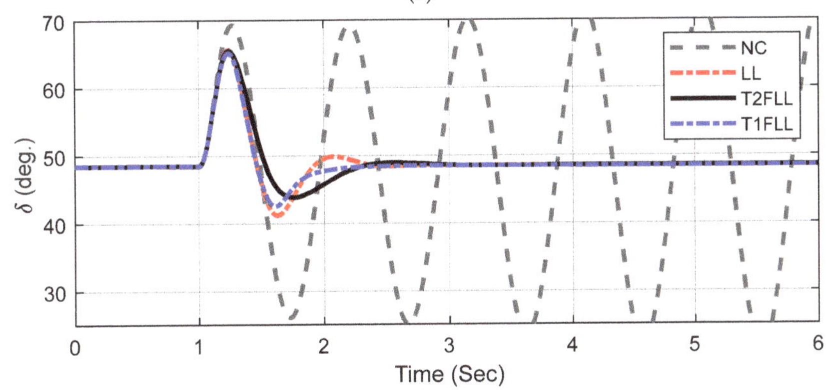

(**b**)

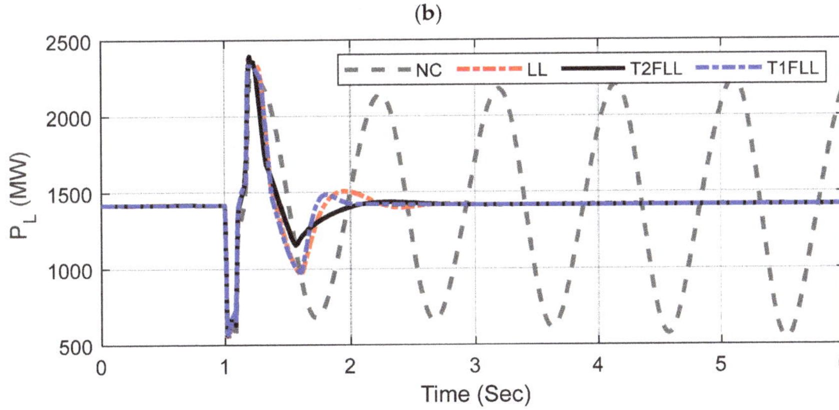

(**c**)

Figure 7. *Cont.*

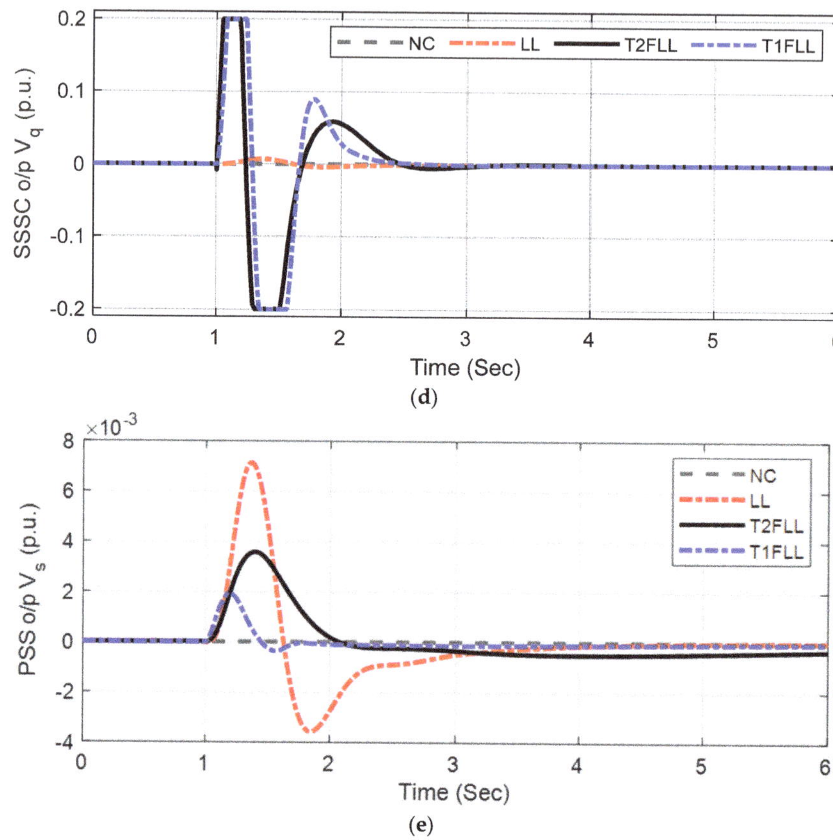

Figure 7. System response of SMIB for Scenario 1: (**a**) $\Delta\omega$, (**b**) $\Delta\delta$, (**c**) P_L, (**d**) SSSC output, (**e**) PSS output.

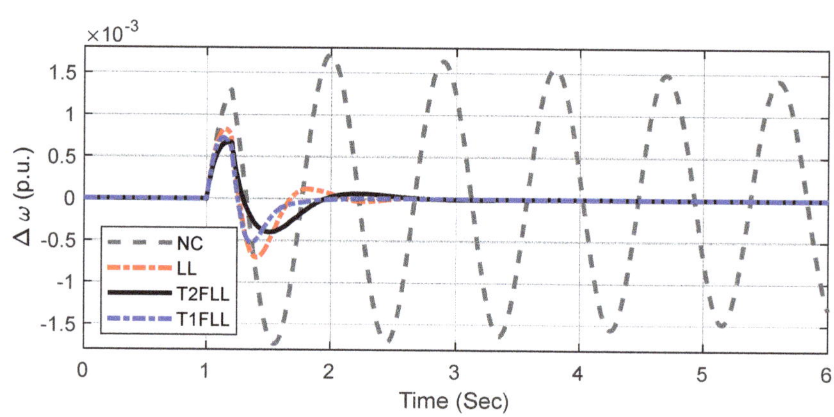

Figure 8. $\Delta\omega$ response of SMIB for Scenario 2.

5.1.3. Scenario 3: Changed Loading Condition and Fault Location

Controller efficiency is also verified in changed fault location and loading conditions. A self-clearing, 5-cycle, 3-phase fault is assumed at bus 1 in light loading (Pe = 0.5 p.u., δ_0 = 38.2°) condition. Under this contingency, $\Delta\omega$ variation is revealed in Figure 9. It is noticed that oscillations are damped quickly with T2FLL in comparison with the T1FLL and LL controllers and without the controllers.

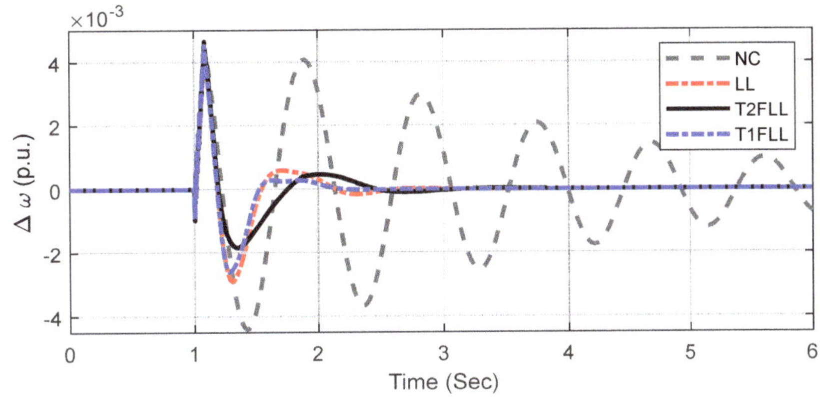

Figure 9. $\Delta\omega$ response of SMIB for Scenario 3.

The performance for the above three scenarios is gathered in Table 4 for improved clarification of enhancement by the proposed method. It can be observed that for different scenarios, the least performance indexes are found with T2FLL related to the T1FLL and LL controller.

Table 4. Performance comparison under various scenarios for SMIB.

Scenario/Controller		ISE ($\times 10^{-6}$)	ITAE ($\times 10^{-2}$)	ITSE ($\times 10^{-3}$)	IAE ($\times 10^{-6}$)	Overshoots in $\Delta\omega$ ($\times 10^{-3}$)	Undershoots in $\Delta\omega$ ($\times 10^{-3}$)
Scenario 1	NC	256.6942	244.0563	43.0972	1493.7607	7.786	−8.6124
	LL	7.7379	3.3792	2.3878	10.1409	5.0881	−4.4641
	T1FLL	6.8863	3.1934	2.0871	8.8434	4.998	−4.3516
	T2FLL	5.8701	2.86961	2.0413	7.4721	5.0363	−3.3831
Scenario 2	NC	9.3642	42.5216	8.1817	45.9841	1.7148	1.7312
	LL	0.1951	0.4901	0.3586	0.2514	0.8321	−0.7049
	T1FLL	0.1668	0.4862	0.3398	0.2163	0.7148	−0.5189
	T2FLL	0.1162	0.4618	0.3142	0.1528	0.6732	−0.3972
Scenario 3	NC	19.6749	30.5829	9.0338	48.6916	4.6224	−4.3954
	LL	2.6978	1.825	1.3121	3.3519	4.5324	−2.8929
	T1FLL	2.4873	1.8011	1.2967	3.0572	4.5324	−2.2581
	T2FLL	2.2042	1.8169	1.2816	2.73604	4.5324	−1.8272

5.2. Extension to Multi-Machine System (MMPS)

The MMPS presented in Figure 2 is taken in the next stage and constructed in the MATLAB/SIMULINK setting as revealed in Figure 10. The MATLAB/SIMULINK model of Area 1 is revealed in Figure 11. In the same way, Area 2 is also modeled. The location of series FACTS devices constitutes a major step in the employment of FACTS devices in power systems. Depending on the required objective, sensitivity analysis can be performed to decide the best location of SSSC in a large power system. Keeping in view of our objective of power system stability enhancement, the SSSC is incorporated between bus 1 and bus 2, as presented in Figure 10. To design the proposed T2FLL controller, a similar approach as used in the SMIB case is followed. The objective function is defined in Equation (14) for a multi-machine system. The tuned parameters for LL, T1FLL, and T2FLL are provided in Tables 5 and 6, respectively.

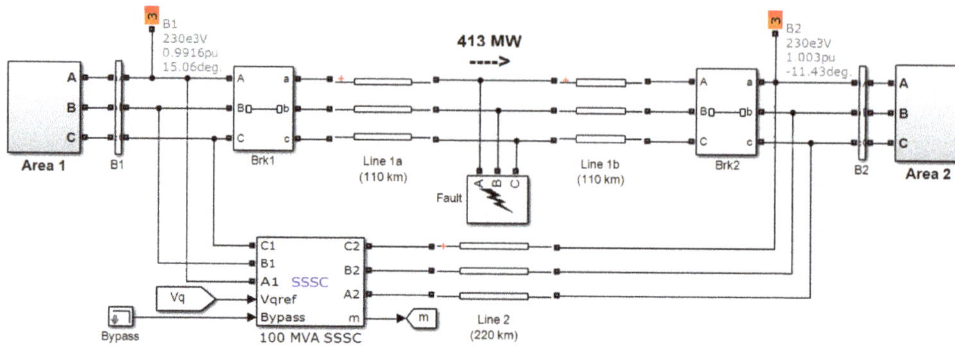

Figure 10. MATLAB/SIMULINK model of Kundur's 4-machine 2-area system.

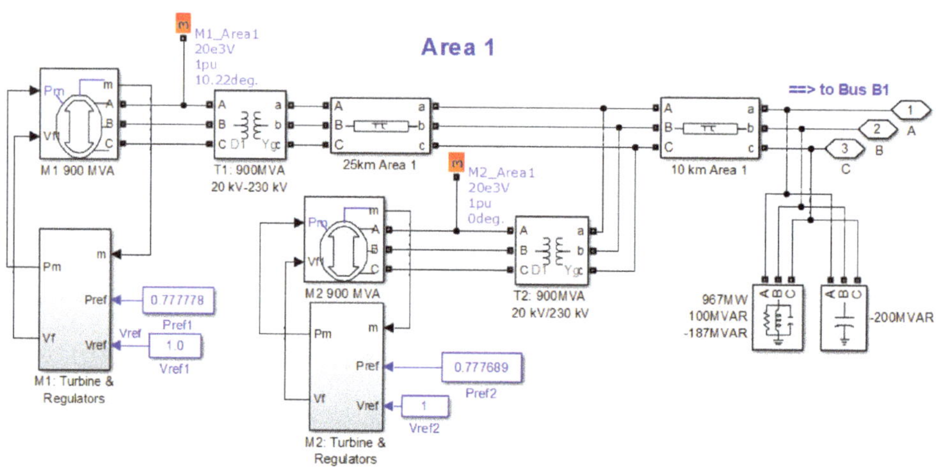

Figure 11. MATLAB/SIMULINK model of Area 1 of Kundur's 4-machine 4-area system.

Table 5. Optimized lead–lag-structured PSS and SSSC parameters for multi-machine system.

Controller	K_i	T_{1i}	T_{2i}	T_{3i}	T_{4i}
SSSC	92.8460	0.0885	0.0308	4.5494	7.4982
PSS1	46.8908	0.0430	0.0182	2.7794	8.8044
PSS2	48.0950	0.0114	0.0155	3.9735	3.5212
PSS3	45.8967	0.0712	0.0128	3.7749	3.2167
PSS4	27.6544	0.0974	0.0319	3.8884	3.2087

Table 6. Optimized T2FLL-structured PSS and SSSC parameters for multi-machine system.

Controller	KSF_1	KSF_2	K_i	T_{1i}	T_{2i}	T_{3i}	T_{4i}
			T2FLL controller				
SSSC	0.0102	0.0148	41.0552	0.0990	0.0064	1.5070	2.9408
PSS1	0.7591	0.0126	24.6432	0.0193	0.0066	2.4321	2.3374
PSS2	1.9797	0.0468	16.5071	0.0394	0.0104	1.9728	1.5423
PSS3	0.0765	0.0284	37.0997	0.0992	0.0059	2.4474	3.0812
PSS4	1.2245	0.2568	11.0542	0.0155	0.0051	1.7857	0.9991
			T1FLL controller				
SSSC	0.2323	0.5164	2.7141	0.0103	0.0101	4.9493	2.0462
PSS1	1.0206	0.0871	11.7451	0.0194	0.0066	3.2127	3.8817
PSS2	1.5428	0.4353	40.6019	0.0415	0.0383	2.3425	3.8154
PSS3	0.9495	0.0100	49.4935	0.0121	0.0060	4.2398	4.4935
PSS4	1.4140	0.0492	35.8082	0.0115	0.0239	2.4098	3.6268

To investigate the viability of T2FLL, several contingencies are taken. These scenarios are widely used in the literature to evaluate the performance and are given as follows [46–48].

5.2.1. Scenario 1: Three Phase Fault Disturbance Condition

Initially, the performance of the controller is investigated under a large disturbance scenario. A 3-phase fault (self-clearing) of 5 cycles is assumed at the middle of one transmission line at t = 0 s. The local and inter-area modes of oscillations with the hADE-PS-tuned LL, T1FLL, and T2FLL controllers are shown in Figure 12a,b. For comparison, the responses without any control are also presented in Figure 12a,b.

It is obvious from Figure 12a,b that the system loses stability without control, and stability is maintained with all the controllers. It is also evident from Figure 12a,b that improved system damping performance with fewer overshoots/undershoots is attained with the proposed T2FLL controller compared to the T1FLL and LL controllers.

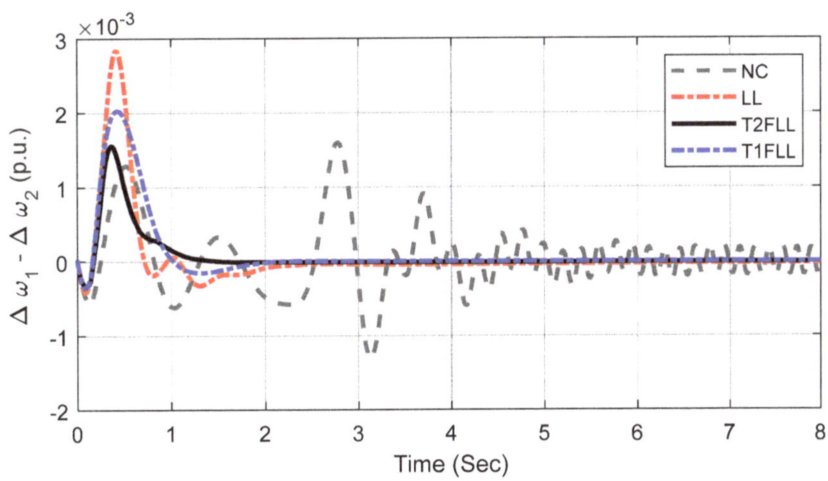

(a)

Figure 12. *Cont.*

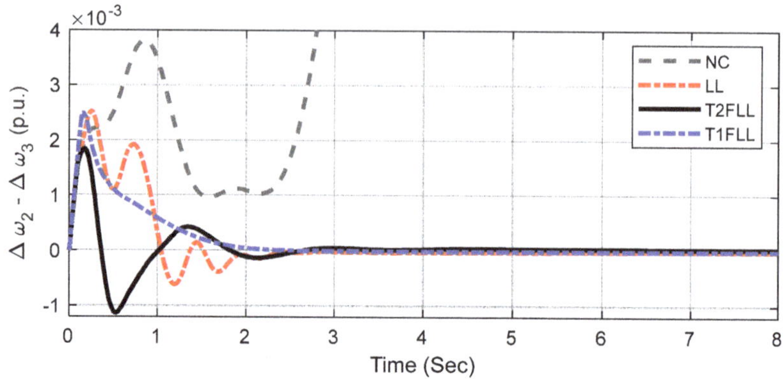

(**b**)

Figure 12. System response of MMPS system for Scenario 1: (**a**) local mode; (**b**) inter-area mode. (**a**) local mode of oscillation of MMPS for Scenario 1. (**b**) Inter-area mode of oscillation of MMPS for Scenario 1.

5.2.2. Scenario 2: Line Outage Disturbance Condition

In this scenario, one of the two tie-lines is removed at t = 0 sec and reclosed after 5 cycles. The responses are exposed in Figure 15a,b. It is evident from Figure 15a,b that in this scenario also, the system loses stability without control, and improved responses are attained with the T2FLL controller related to the T1FLL and LL controllers.

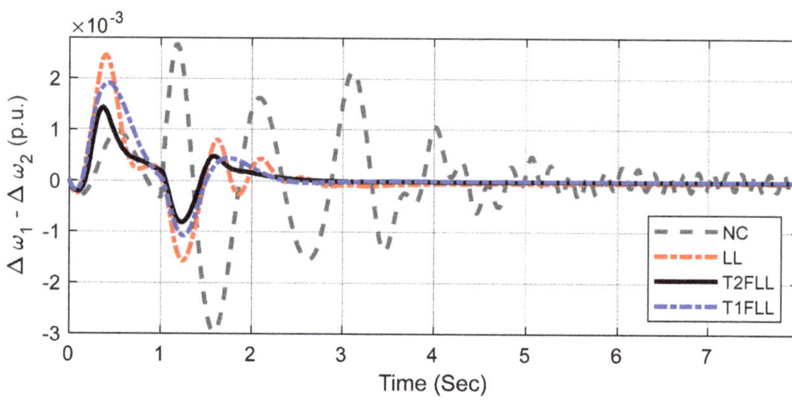

(**a**)

Figure 13. *Cont.*

Figure 14. *Cont.*

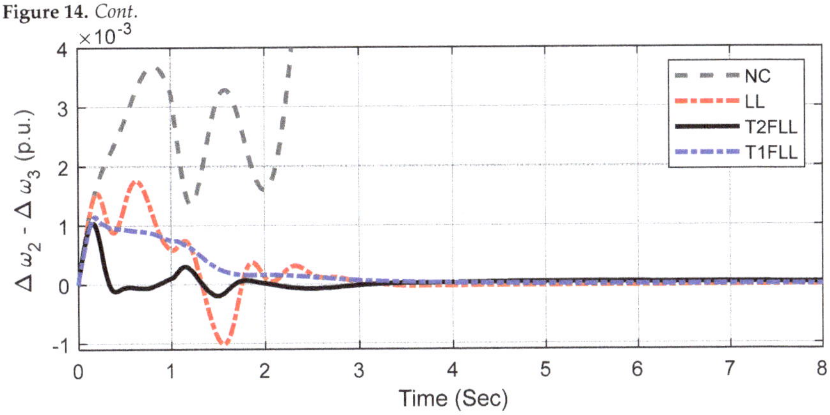

(**b**)

Figure 15. System response of MMPS system for Scenario 2: (**a**) local mode; (**b**) inter-area mode. (**a**) local mode of oscillation of MMPS system for Scenario 2. (**b**) Inter-area mode of oscillation of MMPS system for Scenario 2.

5.2.3. Scenario 3: Small Disturbance Condition

To simulate the small disturbance condition, the voltage reference of M1 is increased by 5% for 12 cycles, and the responses are revealed in Figure 16a,b. It is obvious from Figure 16 that both the LL and T2FLL controllers maintain stability and T2FLL is superior to the T1FLL and LL controller, providing significantly more damping to all modes. For better illustration, objective function values for MMPS for the above scenarios are presented in Table 7. It can be noticed from Table 7 that fewer J values are attained with T2FLL for all the scenarios compared to T1FLL and LL controllers.

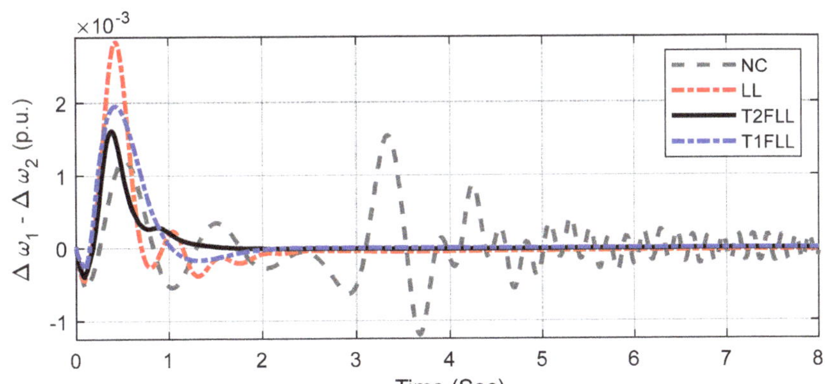

(**a**)

Figure 16. *Cont.*

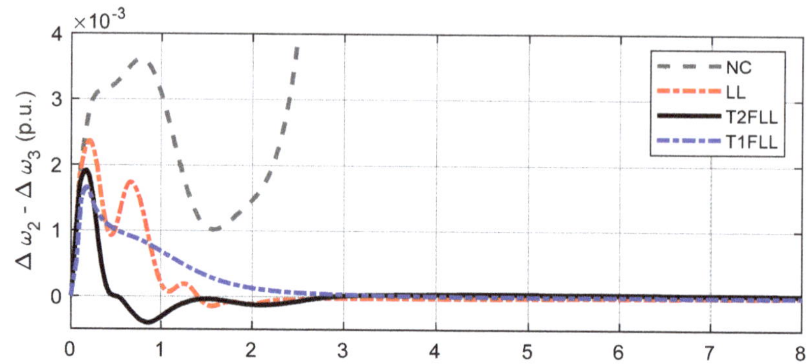

(**b**)

Figure 16. System response of MMPS system for Scenario 3: (**a**) local mode; (**b**) inter-area mode. (**a**) Local mode of oscillation of MMPS system for Scenario 3. (**b**) Inter-area mode of oscillation of MMPS system for Scenario 3.

Table 7. Objective function value comparison under various scenario for MMPS.

Scenario/Controller	Scenario 1 ($\times 10^{-3}$)	Scenario 2 ($\times 10^{-3}$)	Scenario 3 ($\times 10^{-3}$)
NC	20,806.0798	21,201.1951	21,040.6094
LL	14.5224	12.8586	17.4648
T1FLL	12.6361	11.8245	13.8912
T2FLL	9.9952	10.6961	10.5976

6. Conclusions

In this present work, type-2 fuzzy lead–lag (T2FLL)-structured SSSC and PSS damping controller are suggested for stability improvement of power systems. The controllers are coordinated and designed using a hybrid adaptive DE-PS (hADE-PS) technique.

Initially, a SMIB system with a lead–lag controller is considered, and the dominance of hADE-PS as related to DE, PSO, and GA is demonstrated. It is noticed that for the SMIB system with LL controller, the percentage reduction in ITAE value with hADE-PS related to GA, PSO, and DE methods are 23.56%, 17.53%, and 10.53%, respectively.

Next, the T2FLL controller is designed and it is found that there is a reduction of 10.13% and 15.08% in J value with T2FLL related to the T1FLL and LL controller, respectively. Various disturbance scenarios such as large disturbance, small disturbance, and changed loading/fault location are simulated and it is seen that improved damping is attained with T2FLL related to the T1FLL and LL controllers for all the scenarios.

The study is also applied to an MMPS and it is noticed that though all controllers maintain synchronism under various scenarios, an improved system response is attained with T2FLL compared to others. For all the scenarios, the proposed T2FLL is superior to the T1FLL and LL controller, providing significantly more damping to all modes.

Extension of the presented control scheme to a large scale power system is the future scope of this work. Additionally, new hybrid optimization techniques could be implemented for improving the system performance further.

Author Contributions: Conceptualization, P.K. and S.P.; data curation, P.K.; formal analysis, P.K., S.P. and A.Y.A.; funding acquisition, H.A., A.Y.A. and S.S.M.G.; investigation, P.K., S.P., A.Y.A. and S.S.M.G.; methodology, P.K.; project administration, H.A., A.Y.A. and S.S.M.G.; resources, S.P. and S.S.M.G.; software, S.P.; supervision, S.P. and A.Y.A.; validation, P.K., S.P., H.A. and A.Y.A.; visualization, P.K., S.P. and S.S.M.G.; writing—original draft, P.K.; writing—review and editing,

S.P., H.A., A.Y.A. and S.S.M.G. All authors have read and agreed to the published version of the manuscript.

Funding: This work was supported by Taif University Researchers Supporting Project number (TURSP-2020/34), Taif University, Taif, Saudi Arabia.

Institutional Review Board Statement: Not applicable.

Informed Consent Statement: Not applicable.

Data Availability Statement: Not applicable.

Acknowledgments: The authors appreciate Taif University Researchers Supporting Project number (TURSP-2020/34), Taif University, Taif, Saudi Arabia for supporting this work.

Conflicts of Interest: The authors declare no conflict of interest.

Appendix A

SMIB Parameters

Generator: Nominal power (S_B),voltage (V_B) and frequency (f): 2100 MVA, 13.8 kV, 60 Hz
Parameters: Stator resistance (R_S): 2.8544 × 10^{-3}, Reactances: X_d = 1.305, X_q = 0.474, X'_d = 0.296, X'_q = 0.243, X''_d = 0.252, X''_q = 0.18, T_d = 1.01s, T'_d = 0.053s, T''_{qo} = 0.1s.
Inertia constant (H) and pole pairs (p): 3.7 s and 32
Excitation System:
Gains (K_A) and time constant (T_A) of regulator: 200 and 0.001 s
Gains (K_e) and time constant (T_e) of exciter: 1 and 0 s
Gains (K_f) and time constant (T_f) of damping filter: 0.001 and 0.1 s
Low-pass filter time constant (T_{LP}): = 0.02 s,
Transient gain reduction (T_b, T_c): 0
Regulator output limits (E_{fmax}/E_{fmin}) and gain (K_p): 7/0 and 0
Hydraulic Turbine and Governor:
Gains (K_a) and time constant (T_a) of Servo-motor: 3.33 and 0.07 s
Limits of Gate opening (G_{max}/G_{min} and V_{gmax}/V_{gmin}): 0.97518/0.01 and 0.01/−0.1 pu/s
Permanent droop (R_p): = 0.05, Hydraulic turbine: β = 0, T_w = 2.67 sPID regulator (K_p, Ki, Kd, T_d): 1.163, 0.105, 0, 0.01 s
Transformer:
Nominal power (S_B) = 2100 MVA, Winding connection: D_1/Y_g, Primary and secondary voltage (V_1/V_2): 13.8/500 kV, Resistance (R): 0.002 p.u., Inductances (L_1/L_2): 0/0.12, Magnetization resistance (R_m) and reactance (L_m): 500 Ω
Transmission line:
Line length and no. of phases: 300 km, 3-Ph, Resistance per unit length (R_1/R_0): 0.02546/0.3864 Ω/ km, Inductance per unit length (L_1/L_0): 0.9337 × 10^{-3} /4.1264 × 10^{-3} H/km, Capacitance per unit length (C_1/C_0): 12.74 × 10^{-9}/7.751 × 10^{-9} F/ km
Load at Bus2:
250 MW (500 kV, 60 Hz, Y-grounded)

Kundur's 4-machine 2-area system

textbfGenerator: Nominal powers: 900 MVA each, Nominal voltages: 20 kV each, frequency (f): 60 Hz
Parameters: Stator resistance (R_S): 2.8544 × 10^{-3}, Reactances: X_d = 1.8, X_q = 1.7, X'_d = 0.3, X'_q = 0.55, X''_d = 0.25, X''_q = 0.25, T'_{do} = 8 s, T''_{do} = 0.03 s, T'_{qo} = 0.4 s, T''_{qo} = 0.05 s, Stator resistance (R_S): 0.0025 Ω
Excitation Systems: Each same as SMIB system
Steam Turbine and Governor:
Regulator Gain (K_p) = 1, Permanent drooop (R_p): 0.05, Dead zone (D_z): 0, Speed relay and servo-motor time constants (T_{sr}/T_{sm}): 0.001/0.15 s, Limits of gate opening (G_{max}/G_{min} and V_{gmax}/V_{gmin}): 4.496/0 and 0.01/−0.1 pu/s, time constants of steam turbine (T_1, T_2, T_3, T_4): 0, 10, 3.3, 0.5 s, Turbine torque fractions (F_1, F_2, F_3, F_4): 0, 0.36, 0.36, 0.28
Transformers:

Nominal powers = 900 MVA each, Winding connection: D_1/Y_g, Primary and secondary voltage (V_1/V_2): 20/230 kV, Resistance (R): 1×10^{-6}, Inductances (L_1/L_2): 0/0.15, Magnetization resistance (R_m) and reactance (L_m): 500 Ω
Transmission lines:
Distributed parameter line (110 km line sections) and *PI section line* (10 km and 25 km line sections) Line length and n0. of phases: 300 km, 3-Ph, Resistance per unit length (R_1/R_0): 0.0529/1.61 Ω/km, Inductance per unit length (L_1/L_0): 0.0014/0.0061 H/km, Capacitance per unit length (C_1/C_0): $8.7749 \times 10^{-9}/5.2489 \times 10^{-9}$ F/km
Loads:
Area-1: Active power: 967 MW, Inductive reactive power: 100 MVAR, Capacitive reactive power: 387 MVAR
Area-2: Active power: 1767 MW, Inductive reactive power: 100 MVAR, Capacitive reactive power: 537 MVAR
SSSC Data
Converter rating: = 100 MVA, Nominal voltage: 500 kV, Frequency (*f*): 60 Hz, Maximum rate of change in reference voltage = 3 pu/s, Converter impedances (R/L): 0.00533/0.16, DC link voltage: 40 kV, DC link equivalent capacitance: 375×10^{-6} F, Injected Voltage regulator gains (K_P/K_I): 0.00375/0.1875, DC Voltage regulator gains (K_P/K_I): $0.1 \times 10^{-3}/20 \times 10^{-3}$, Limits of injected voltage: ± 0.2

References

1. Yu, Y.N. *Electric Power System Dynamics*; Academic Press: New York, NY, USA, 1983.
2. Sauer, P.W.; Pai, M.A. *Power System Dynamics and Stability*; Prentice-Hall: Englewood Cliffs, NJ, USA, 1998.
3. Gyugyi, L.; Schauder, C.D.; Sen, K.K. Static synchronous series compensator: A solid-state approach to the series compensation of transmission lines. *IEEE Trans. Power Deliv.* **1997**, *12*, 406–417. [CrossRef]
4. Panda, S.; Ardil, C. Robust coordinated design of multiple power system stabilizers using particle swarm optimization technique. *Int. J. Electr. Syst. Sci. Eng.* **2008**, *1*, 41–48.
5. Panda, S.; Padhy, N.P. Robust power system stabilizer design using particle swarm optimization technique. *Int. J. Electr. Syst. Sci. Eng.* **2008**, *1*, 2260–2267.
6. Abdel-Magid, Y.L.; Abido, M.A.; Al-Baiyat, S.; Mantawy, A.H. Simultaneous stabilization of multi-machine stabilizers via genetic algorithm. *IEEE Trans. Power Syst.* **1999**, *14*, 1428–1439. [CrossRef]
7. Sebaa, K.; Boudour, M. Optimal locations and tuning of robust power system stabilizer using genetic algorithms. *Int. J. Electr. Power Syst. Res.* **2009**, *79*, 406–416. [CrossRef]
8. Paul, S.; Roy, P.K. Oppositional cuckoo optimization algorithm for optimal tuning of power system stabilizers. In Proceedings of the Michael Faraday IET International Summit, Kolkata, India, 12–13 September 2015; pp. 176–181.
9. Abido, M.A. Robust design of multi-machine power system stabilizers using simulated annealing. *IEEE Trans Energy Convers.* **2000**, *15*, 297–304. [CrossRef]
10. Abdel-Magid, Y.L.; Abido, M.A.; Mantawy, A.H. Robust tuning of power system stabilizers in multi-machine power systems. *IEEE Trans Power Syst.* **2000**, *15*, 735–740. [CrossRef]
11. Mishra, S.; Tripathy, M.; Nanda, J. Multi-machine power system stabilizer design by rule based bacteria foraging. *Int. J. Electr. Power Syst. Res.* **2007**, *77*, 1595–1607. [CrossRef]
12. Jolfaei, M.G.; Sharaf, A.M.; Shariatmadar, S.M.; Poudeh, M.B. A hybrid PSS–SSSC GA-stabilization scheme for damping power system small signal oscillations. *Electr. Power Energy Syst.* **2016**, *75*, 337–344. [CrossRef]
13. Falehi, A.D.; Rostami, M.; Doroudi, A.; Ashrafian, A. Optimization and coordination of SVC-based supplementary controllers and PSSs to improve power system stability using a genetic algorithm. *Turk. J. Elec. Eng. Comp. Sci.* **2012**, *20*, 639–654.
14. Khadanga, R.K.; Satapathy, J.K. Time delay approach for PSS and SSSC based coordinated controller design using hybrid PSO–GSA algorithm. *Int. J. Electr. Power Energy Syst.* **2015**, *71*, 262–273. [CrossRef]
15. Tripathy, M.; Mishra, S. Coordinated tuning of PSS and TCSC to improve hopf bifurcation margin in multi-machine power system by a modified bacteria foraging algorithm. *Int. J. Electr. Power Energy Syst.* **2015**, *66*, 97–109. [CrossRef]
16. Ali, E.S.; Abd-Elazim, S.M. Coordinated design of PSSs and TCSC via bacterial swarm optimization algorithm in a multi-machine power system. *Int. J. Electr. Power Energy Syst.* **2012**, *36*, 84–92. [CrossRef]
17. Shayeghi, H.; Safari, A.; Shayanfar, H.A. PSS and TCSC damping controller coordinated design using PSO in multi-machine power system. *Energy Convers Manag.* **2010**, *51*, 2930–2937. [CrossRef]
18. Dadfar, S.; Wakil, K.; Khaksar, M.; Rezvani, A.; Miveh, M.R.; Gandomkar, M. Enhanced control strategies for a hybrid battery/photovoltaic system using FGS-PID in grid-connected mode. *Int. J. Hydrog. Energy* **2019**, *44*, 14642–14660. [CrossRef]
19. Khaksar, M.; Rezvani, A.; Moradi, M.H. Simulation of novel hybrid method to improve dynamic responses with PSS and UPFC by fuzzy logic controller. *Neural Comput. Appl.* **2018**, *29*, 837–853. [CrossRef]

20. Rezvani, A.; Esmaeily, A.; Etaati, H.; Mohammadinodoushan, M. Intelligent hybrid power generation system using new hybrid fuzzy-neural for photovoltaic system and RBFNSM for wind turbine in the grid connected mode. *Front. Energy* **2019**, *13*, 131–148. [CrossRef]
21. Ray, P.K.; Paital, S.R.; Mohanty, A.; Eddy, F.Y.S.; Krishnan, A.; Gooi, H.B.; Amaratunga, G.A.J. A hybrid firefly-swarm optimized fractional order interval type-2 fuzzy PID-PSS for transient stability improvement. *IEEE Trans. Ind. Appl.* **2019**, *55*, 6486–6498. [CrossRef]
22. Elazim, S.M.A.; Ali, E.S. Optimal power system stabilizers design via cuckoo search algorithm. *Int. J. Electr. Power Energy Syst.* **2016**, *75*, 99–107. [CrossRef]
23. Ali, E.S. Optimization of power system stabilizers using BAT search algorithm. *Int. J. Electr. Power Energy Syst.* **2014**, *61*, 683–690. [CrossRef]
24. Elazim, S.M.A.; Ali, E.S. Optimal location of STATCOM in multimachine power system for increasing loadability by cuckoo search algorithm. *Int. J. Electr. Power Energy Syst.* **2016**, *80*, 240–251. [CrossRef]
25. Abdelaziz, A.Y.; Ali, E.S. Static VAR compensator damping controller design based on flower pollination algorithm for a multi-machine power system. *Electr. Power Compon. Syst.* **2015**, *43*, 1268–1277. [CrossRef]
26. Abd Elazim, S.M.; Ali, E.S. Optimal SSSC design for damping power systems oscillations via gravitational search algorithm. *Int. J. Electr. Power Energy Syst.* **2016**, *82*, 161–168. [CrossRef]
27. Elazim, S.M.A.; Ali, E.S. Coordinated design of PSSs and SVC via bacteria foraging optimization algorithm in a multi-machine power system. *Int. J. Electr. Power Energy Syst.* **2012**, *41*, 44–53. [CrossRef]
28. Stron, R.; Price, K. Differential evolution—A simple and efficient adaptive scheme for global optimization over continuous spaces. *J. Glob. Optim.* **1997**, *11*, 341–359. [CrossRef]
29. Sahu, R.K.; Gorripotu, T.S.; Panda, S. A hybrid DE–PS algorithm for load frequency control under deregulated power system with UPFC and RFB. *Ain Shams Eng. J.* **2015**, *6*, 893–911. [CrossRef]
30. Mohanty, P.; Sahu, R.K.; Panda, S. A novel hybrid many optimizing liaisons gravitational search algorithm approach for AGC of power systems. *Autom. J. Control. Meas. Electron. Comput. Commun.* **2020**, *61*, 158–178. [CrossRef]
31. Khadanga, R.K.; Kumar, A.; Panda, S. A hybrid shuffled frog-leaping and pattern search algorithm for load frequency controller design of a two-area system composing of PV grid and thermal generator. *Int. J. Numer. Modeling* **2020**, *33*, e2694. [CrossRef]
32. Rout, B.D.; Pati, B.B.; Panda, S. A hybrid modified differential evolution-pattern search approach for SSSC based damping controller design under communication constraints. *Int. J. Syst. Assur. Eng. Manag.* **2018**, *9*, 962–971. [CrossRef]
33. Zhang, J.; Sanderson, A.C. JADE: Adaptive differential evolution with optional external archive. *IEEE Trans Evol. Comput.* **2009**, *13*, 945–958. [CrossRef]
34. Wang, Y.; Cai, Z.; Zhang, Q. Differential evolution with composite trial vector generation strategies and control parameters. *IEEE Trans Evol. Comput.* **2011**, *15*, 55–66. [CrossRef]
35. Tanabe, R.; Fukunaga, A. Success-history based parameter adaptation for differential evolution. In Proceedings of the IEEE CEC, Cancun, Mexico, 20–23 June 2013; pp. 71–78.
36. Mohanty, P.; Sahu, R.K.; Sahoo, D.K.; Panda, S. Adaptive differential evolution and pattern search tuned fractional order fuzzy PID for frequency control of power systems. *Int. J. Model. Simul.* **2022**, *42*, 240–254. [CrossRef]
37. SimPowerSystems 4.3 User's Guide. Available online: http://www.mathworks.com/products/simpower/ (accessed on 3 March 2022).
38. Kundur, P. *Power System Stability and Control*; McGraw-Hill: Irvine, CA, USA, 1994; p. 813, Example 12.6.
39. Chintu, J.M.; Sahu, R.K.; Panda, S. Adaptive differential evolution tuned hybrid fuzzy PD-PI controller for automatic generation control of power systems. *Int. J. Ambient. Energy* **2019**, *17*, 1–6. [CrossRef]
40. Liu, J.; Lampinen, J. A fuzzy adaptive differential evolution algorithm. *Soft Comput.* **2005**, *9*, 448–462. [CrossRef]
41. Gamperle, R.; Muller, S.D.; Koumoutsakos, P. A parameter study for differential evolution. *Adv. Intell. Syst. Fuzzy Syst. Evol. Comput.* **2002**, *10*, 293–298.
42. Ali, M.M.; Torn, A. Population set-based global optimization algorithms: Some modifications and numerical studies. *Comput. Oper. Res.* **2004**, *31*, 1703–1725. [CrossRef]
43. Biswas, P.P.; Suganthan, P.N.; Wu, G.; Amaratunga, G.A. Parameter estimation of solar cells using datasheet information with the application of an adaptive differential evolution algorithm. *Renew. Energy* **2019**, *132*, 425–438. [CrossRef]
44. Panda, S.; Padhy, N.P. Comparison of Particle Swarm Optimization and Genetic Algorithm for FACTS-based Controller Design. *Appl. Soft Comput.* **2008**, *8*, 1418–1427. [CrossRef]
45. Mohanty, B.; Panda, S.; Hota, P.K. Controller parameters tuning of differential evolution algorithm and its application to load frequency control of multi-source power system. *Int. J. Electr. Power Energy Syst.* **2014**, *54*, 77–85. [CrossRef]
46. Swain, S.C.; Panda, S.; Mahapatra, S. A multi-criteria optimization technique for SSSC based power oscillation damping controller design. *Ain Shams Eng. J.* **2016**, *7*, 553–565. [CrossRef]
47. Panda, S.; Yegireddy, N. Multi-input single output SSSC based damping controller design by a hybrid improved differential evolution-pattern search approach. *ISA Trans.* **2015**, *58*, 173–185. [CrossRef] [PubMed]
48. Panda, S.; Kiran, S.H.; Dash, S.S.; Subramani, C. A PD-type multi input single output SSSC damping controller design employing hybrid improved differential evolution-pattern search approach. *Appl. Soft Comput.* **2015**, *32*, 532–543. [CrossRef]

sustainability

MDPI

Article

Development of Operation Strategy for Battery Energy Storage System into Hybrid AC Microgrids

Felipe Ramos [1], Aline Pinheiro [2], Rafaela Nascimento [1], Washington de Araujo Silva Junior [1], Mohamed A. Mohamed [3,*], Andres Annuk [4] and Manoel H. N. Marinho [1,*]

[1] Polytechnic School of Pernambuco, University of Pernambuco (UPE), Recife 50720-001, PE, Brazil
[2] Energy Storage Division, Edson Mororó Moura Institute of Technology (ITEMM), Recife 51020-280, PE, Brazil
[3] Department of Electrical Engineering, Faculty of Engineering, Minia University, Minia 61519, Egypt
[4] Institute of Forestry and Engineering, Estonian University of Life Sciences, 51006 Tartu, Estonia
* Correspondence: dr.mohamed.abdelaziz@mu.edu.eg (M.A.M.); marinho75@poli.br (M.H.N.M.)

Abstract: With continuous technological advances, increasing competitiveness of renewable sources, and concerns about the environmental impacts of the energy matrix, the use of hybrid microgrids has been promoted. These generation microsystems, historically composed basically of fossil fuels as the main source, have experienced an energy revolution with the introduction of renewable and intermittent sources. However, with the introduction of these uncontrollable sources, the technical challenges to system stability, low diesel consumption, and security of supply increase. The main objective of this work is to develop an operation and control strategy for energy storage systems intended for application in hybrid microgrids with AC coupling. Throughout the work, a bibliographic review of the existing applications is carried out, as well as a proposal for modification and combination to create a new control strategy. This strategy, based on optimized indirect control of diesel generators, seeks to increase generation efficiency, reduce working time, and increase the introduction of renewable sources in the system. As a result, there is a significant reduction in diesel consumption, a decrease in the power output variance of the diesel generation system, and an increase in the average operating power, which ensures effective control of hybrid plants.

Keywords: renewable Resources; microgrids; generation efficiency; energy management; battery energy storage system

Citation: Ramos, F.; Pinheiro, A.; Nascimento, R.; de Araujo Silva Junior, W.; Mohamed, M.A.; Annuk, A.; Marinho, M.H.N. Development of Operation Strategy for Battery Energy Storage System into Hybrid AC Microgrids. *Sustainability* **2022**, *14*, 13765. https://doi.org/10.3390/su142113765

Academic Editor: Alberto-Jesus Perea-Moreno

Received: 11 September 2022
Accepted: 11 October 2022
Published: 24 October 2022

1. Introduction

The hybridization of microgrids integrated with photovoltaic generation systems plays an important role as an alternative to reduce diesel consumption and consequent reduction in operating costs [1]. In this context, energy storage systems act as enablers and enhancers of this integration, since several instability factors can be found when only the photovoltaic system is introduced, mainly due to the lack of control of the solar source [2].

This work has as its main proposal the development of a parameterizable operating strategy for energy storage systems applied to hybrid microgrids.

The textual body of the work is organized into five sections, and in Section 2—Theoretical Reference, the definition of microgrids, their main components, and classifications are presented. Furthermore, a detailed description of the Battery Energy Storage System (BESS) applications associated with the scope of this work or the use in isolated systems is provided. In Section 3—Methodology, the applications that make up the proposed control strategy, the changes made to the traditional methodology, and the way in which they are combined are described. At the end, the proposed system for validation is detailed, as well as the analysis methods. In Section 4—Results, the effectiveness of the operation is demonstrated through the analysis of the main energy parameters of the microgrid. In addition, a sensitivity analysis regarding the integration of the PV array (Photovoltaic) and

BESS is performed. Finally, in Section 5—Conclusions, the results are consolidated and future work is proposed.

2. Theoretical Reference

2.1. Literature Review

In this section, a bibliographic review is carried out covering topics related to the operation of microgrids in their various configurations.

In [3], an operation strategy proposal was made that uses a day-ahead forecast model to optimize the response to demand, seeking the lowest overall cost of the operation. The proposal, which relies on the connection of the electricity grid, resulted in a sensitivity analysis indicating a microgrid composed of BESS, PV generation, and wind power, reaching lower cost levels when compared to the traditional model of single supply through the distributor.

In [4], a study was carried out based on the application of BESS and renewable energy sources for the short-term energy market, basing the operation on generation and cost forecast models. As a result, there is a cost reduction in the global operation of the system, proving the effectiveness of the proposed algorithm.

A review of the applications of hybrid energy storage systems, based on the combination of batteries and supercapacitors, was presented in [5], focusing on renewable power smoothing strategies, voltage and frequency control, lifetime and optimization, among others. Furthermore, a comparison between storage technologies based on batteries and capacitors is developed.

A comparison of operating strategies for microgrids was presented in [6]. As a base case, the traditional model of load following the electrical network is used, where load fluctuations are absorbed by the main source. Two additional strategies for controlling the supply of the network are proposed: the first is the stabilization of a constant daily supply power and the second is the stabilization of a constant annual supply power. The proposed cases presented a lower total cost of supply than the base case, although they also presented a greater waste of energy.

A comparative study of three operating strategies for microgrids was carried out in [7]: i. Load following, ii. Cycle charging, and iii. Combined dispatch. The strategy iii. operates as a combination of the two previous strategies, aiming at the best option depending on the conditions. The evaluated microgrid is composed of PV generation, Diesel generation, and BESS, based on the Homer (Hybrid Optimization of Multiple Energy Resources) software. In response, for the proposed scenario, the combined dispatch strategy is proven to be more effective, with lower CO_2 emissions and lower global cost of supply.

In [8], a study similar to the one discussed in [7] is presented, evaluating the operating strategies available in the Homer software, these being i. Load following, ii. Cycle charging, and iii. Combined dispatch. Additionally, studies are carried out regarding the storage technology used, with analyses for Lead-Acid, Lithium-Ion, Vanadium Flux, Supercapacitors, and Reversible Hydroelectric Batteries.

In [9], a control algorithm for supercapacitors in dynamic frequency regulation applications was developed. Based on this use, an option is presented to reduce the traditional load-sharing strategies, with sudden variations being absorbed by the storage system. The article also shows that although the supercapacitor was used as a base, other technologies can be used to perform the function.

In [10], a study of PV insertion in electrical networks was developed, applying two strategies that make it possible to increase the share of PV generation. The first strategy is based on changing the consumption profile, using variable pricing for energy depending on resource availability. The second strategy uses storage systems as sources of generation stabilization and peak absorption, allowing insertions above 50% of the total energy balance.

A case study of a microgrid composed of diesel, PV, wind, and BESS generation was presented by [11]. The strategy, based on BESS's SoC (State-of-Charge) steps, ensures that the operation of the diesel generator is minimized and at rated power, being used to recharge the BESS at times of low renewable generation. Optimization and dimensioning studies based on the global cost of the operation are also presented.

More recently, [12] carried out a case study to complement a ship's energy generation system with fuel cells and BESS. In the proposed strategy, which uses Homer as a simulation platform, the scenario with the lowest possible CO_2 emission is sought, so that the ship's original energy generator is maintained. The study also shows that the result of the reduction and optimal scenario for the insertion of additional elements varies according to the load profile considered.

The literature review contributed to the consolidation of the knowledge base and an investigation of the current technological stage of the topic. Of great importance for the construction of the work methodology, this review also made it possible to identify the gaps in the current proposals for microgrid control. As described in this section, a proposal was not identified in the evaluated works that combined the applications of power smoothing applied to Load and PV, optimal dispatch of BESS to increase the efficiency of diesel generation, sharing of the rotating reserve, and use of the surplus energy generated by PV, which is the focus of this work.

2.2. Microgrids

Microgrids can be defined as independent generation structures composed minimally of one or more generation sources and interconnected loads. These microgrids can have in their generation matrix elements such as renewable energy sources, mainly solar and wind, storage systems, and generation based on fossil fuels. In addition to the generation and consumption components, the importance of the control and protection elements of the system is also highlighted, which guarantees the integrity and correct functioning of the integrated systems [13].

According to [13], microgrids can be classified in different ways, and the main ones can be cited: i. Generation capacity, ii. Type of installation and load, iii. Network connection and supply nature. Other classifications may consider the number of phases, voltage level, and other technical aspects.

Microgrids that work in AC (Alternate Current), DC (Direct Current), and hybrid models of work, both continuous and alternating, can also be classified according to the type of load to be supplied.

For direct current microgrids, the main highlight is the greater efficiency due to reduced energy conversion processes, in addition to facilitating the process of integrating renewable sources. On the other hand, it is important to consider that most loads used today were designed and operated only with AC supply, bringing the need for at least one DC to AC conversion system.

For DC microgrids, the main advantage is the high level of maturity of the global electricity sector on the subject since traditional electrical networks are developed in AC and there are no major complications for connection to the grid or grid conventional electrical, if necessary. However, for the system to function as desired, a robust control and operation structure is required, combined with a good management system.

In addition to the aforementioned options, there is also the possibility of integrating hybrid systems from the point of view of load supply and connection, using part of the system in direct current and part of the system in AC.

Another relevant point in the classification and design of microgrids refers directly to the control modes applied to this structure. For master-slave controls, a structure is used where only one of the generation sources represents the voltage and frequency parameters, whereas the others act as current sources, following the standards established by the main source.

With the use of Peer-to-Peer topology, the figure of the main source is not necessary, since all installed sources work in automatic load sharing mode. In this scenario, all sources are configured with an operating droop curve, where active and reactive power is automatically generated through voltage and frequency parameters at the connection point.

For this classification scenario there are also hybrid models, where part of the sources work in Slave mode and part of the sources in Peer-to-Peer mode. For the system analyzed in this work, the term hybrid refers to the fact that different sources of generation and control are considered, unlike traditional microgrids that are based on only one generation source.

2.3. BESS Applications

According to [14], it is possible to differentiate applications of energy storage systems between five main categories as shown in Figure 1, so that applications related to use in microgrids are directly or indirectly related to one of these proposed items.

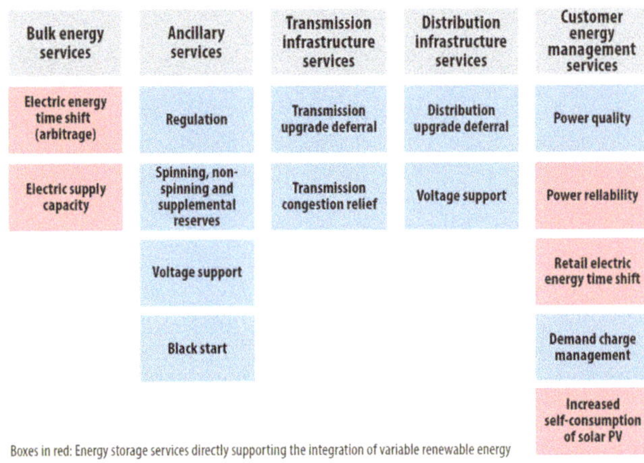

Figure 1. BESS Applications.

2.3.1. Energy Arbitrage

This application, used both in isolated and grid-connected systems, consists of storing energy in periods of high generation and use in periods of low resource availability and/or high consumption. With this strategy it is possible to store energy in periods when energy is cheaper and use it in periods when production is more expensive [15]. In addition, especially in microgrids with high insertion of renewable sources, energy waste is common due to the imbalance between generation and consumption, thus, storage plays a fundamental role in the technical and economic feasibility of these systems [16].

Illustrating this behavior, Figure 2 shows a typical operation of a storage system connected to the electricity grid, charging between 01:00 and 10:00 and discharging between 17:00 and 20:00.

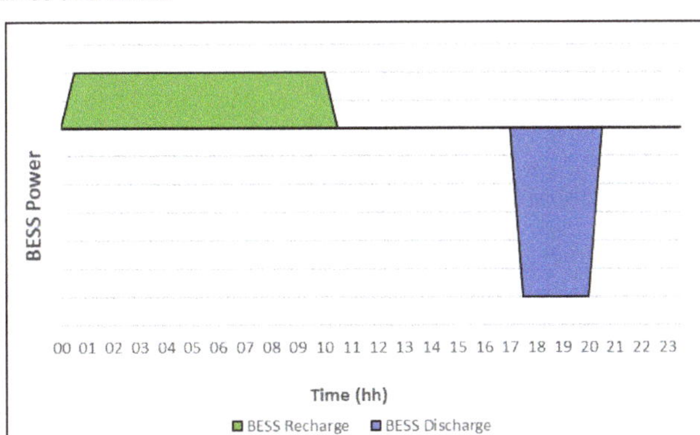

Figure 2. A typical operation of a storage system connected to the electricity grid.

In this work, a method of performing the energy displacement function is proposed, based on intelligent discharge control. This method is detailed in Section 4.

2.3.2. Increase Electric Supply Capacity

In order to meet consumption peaks, this functionality is characterized by the complementation of the storage system to power not available in the generation system, using the stored energy to supply the excess part of consumption [17]. This application makes it possible to install generation systems smaller than the peak consumption required for service, especially in cases where these do not occur continuously and frequently [18].

2.3.3. Upgrade Deferral

This application is used in both transmission and distribution, and characterized by the inhibition or postponement of the need for investments in electrical infrastructure reinforcements.

Usually, the repowering of electrical systems occurs due to sporadic peaks in demand, lasting for a few hours. In this scenario, the investment in storage systems can be reduced when compared to the traditional method of meeting these peaks [19].

Figure 3 shows the typical operation of a storage system performing the investment postponement function.

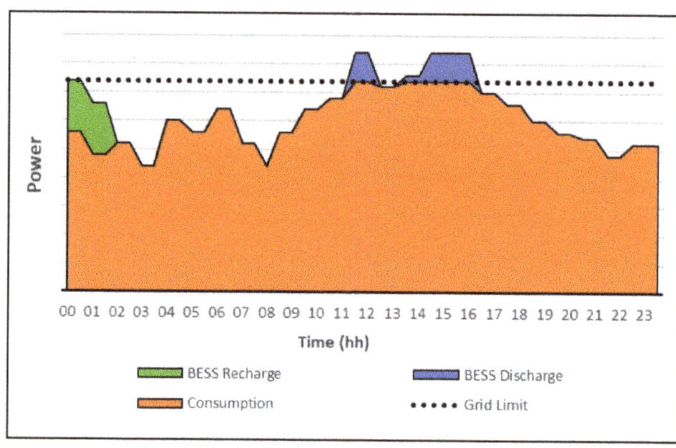

Figure 3. Upgrade Deferral Application.

2.3.4. Black Start

This function is characterized by the capacity of the storage system to restore the electrical grid after a complete system collapse, serving as a voltage and frequency source until synchronization and connection of other generation sources [20].

2.3.5. Load Leveling

This application, typically used in transmission and distribution systems, aims to ensure that energy consumption at a given point is always within pre-established maximum and minimum limits, increasing the predictability of energy supply [21]. If consumption is lower than the established limit, the storage system recharges; if consumption is greater than the established limit, discharge is carried out [22].

Figure 4 shows the typical operation of a storage system performing the leveling function of a hypothetical load.

In this work, a method of performing the load-leveling function is proposed, based on the combination of power smoothing algorithms. This method will be detailed in Section 4.

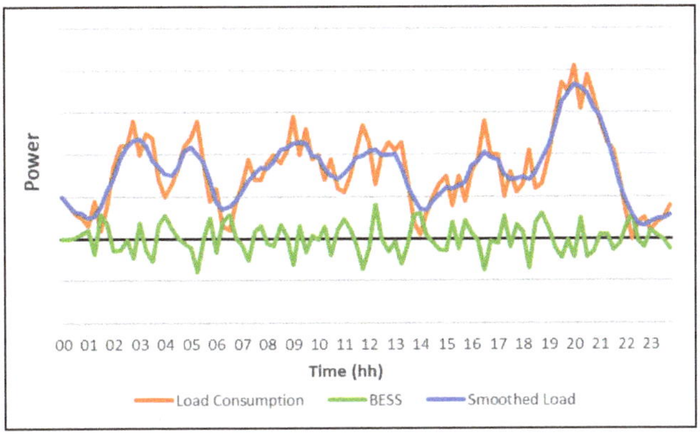

Figure 4. Load Leveling.

2.3.6. Power Smoothing

Renewable energy sources are becoming increasingly present in the world's electricity generation. For example, according to [23], photovoltaic solar energy has increased its capacity by almost nine times in 9 years, from 381.6 GW in 2010 to 3.4 TW in 2019. On the other hand, energy fluctuations caused by weather conditions bring several technical challenges related to power quality, protection, dispatch control, and reliability, reducing the possibility of massive insertion of this source [24].

An alternative to maintain the stability of the network is to smooth the output of the intermittent source with the use of energy storage systems, so that, providing or absorbing power, it stabilizes the variation in the connection point to an acceptable range [25].

Figure 5 shows the typical operation of a storage system connected to a photovoltaic system, performing power smoothing.

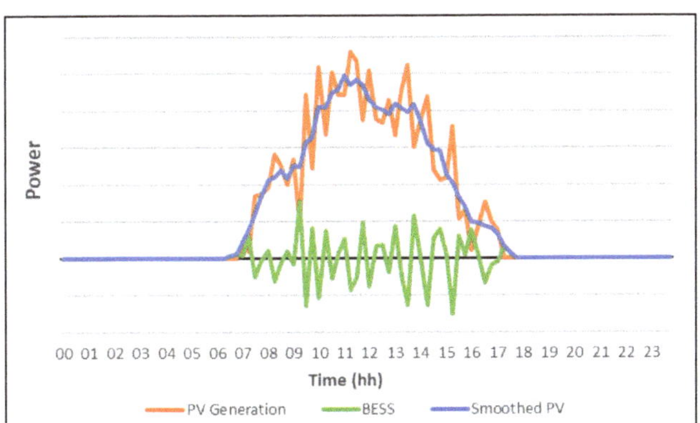

Figure 5. Typical operation of a storage system connected to a photovoltaic system.

The main ways of performing this control are described below:

Moving Average: This technique is widely used in stock analysis, smoothing out volatility and making it easier to see the price trend of a security. It can also be applied to smooth power outputs, being known for its computational simplicity. The basic operation of this algorithm is to sum the recent output power values and then divide by the number of periods in the calculation average.

The only input parameter used for this technique is the analyzed time interval, so that, regardless of the weather conditions, longer intervals will generate smoother curves. However, the greater the gap and smoothing, the greater the energy storage system and its utilization [26].

Double Moving Average: In order to enhance the power smoothing of the single moving average method, the double moving average applies the above technique twice, the first being on the original data and the second on the response data. Similar to the previous scenario, although this strategy produces greater smoothing, the energy and utilization of the storage system is also intensified [27].

Exponential Moving Average: Similar to the simple moving average algorithm, this strategy uses historical data to generate its smoothed output. However, a multiplication constant is also used, which has greater weight for more recent data.

Moving Median: This smoothing algorithm applies a similar concept to the simple moving average, however, using the median instead of the arithmetic average. From a statistical point of view, the moving median can reach the power trend more quickly; however, in certain scenarios, this behavior can be negative, since the algorithm can present sudden transitions.

Ramp Rate: Unlike previous algorithms, the ramp rate method calculates the power variation and uses the storage system only if the variation exceeds the established threshold, for example, 10% per minute. This proves attractive for most scenarios as it minimizes battery usage and sizing. The percentage change is calculated by the difference between the current output and the previous output, divided by the rated power of the plant. This variation can remain within the established limits or exceed positively or negatively [28].

In this work, a method of performing the power smoothing function based on the modification of the ramp rate control algorithm is proposed. This method will be detailed in Section 4.

2.3.7. Clipping Recapture

With the traditional oversizing of DC power against the AC power of photovoltaic systems, it is common for an effect known as "clipping" to occur, where the power available on the DC bus exceeds the rated power of the inverters to which the bus is connected. In this way, the output power of the system is limited, even if the photovoltaic modules have more power available for generation.

This oversizing is generally used for several technical and economic reasons, which are beyond the scope of this work; however, it is possible to direct the additional energy to the storage system and later use, aiming to maximize the gains of the generation system as a whole.

This application has a relatively simple operating logic, since whenever the DC power exceeds the PV inverter's transformation capacity, the excess power is directed to the batteries, which are connected to the photovoltaic module's busbars through a DC/ AC [29].

Figure 6 shows the typical operation of a storage system performing the clipping utilization function.

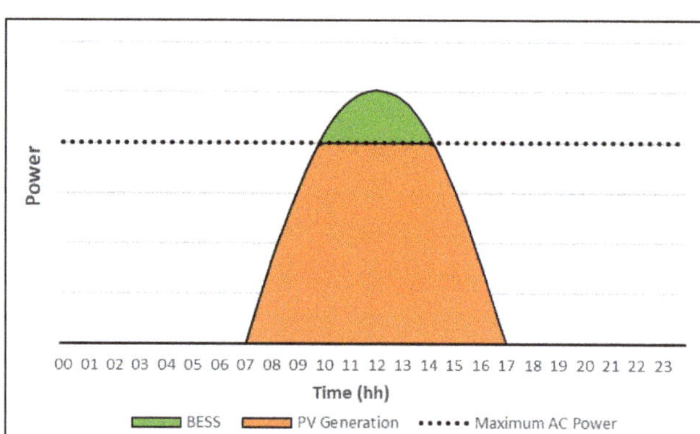

Figure 6. Typical operation of a storage system performing the clipping utilization function.

2.3.8. Capacity Reserve

The capacity reserve can be defined as energy available for use in sudden load variations or partial failures in the generation system. These reserves can be divided into a few categories, which are mainly related to the time of operation [30].

Spinning Reserve: Generation capacity connected to the grid with a response between seconds to a few minutes, sufficient to compensate for load increases or partial drops of generation systems. This service can be provided in a simple way by storage systems, since the response to commands sent takes place on a millisecond scale, depending on the priority of the application [31].

Supplementary Reserve: Generation capacity used after the entry of the rotating reserve, which may be disconnected from the network, with a response in tens of minutes [32].

In this work, a method of performing the capacity reservation function based on the load state of the storage system is proposed. This method will be detailed in Section 4.

2.3.9. Frequency and Voltage Regulation

Aimed at guaranteeing the stability of the network, these functions have as their main objective the guarantee of the dynamic balance of power between generation and consumption. To carry out this application, the storage system performs fast and high-power operations, discharging in situations where the frequency has values lower than those established and recharging at times when the frequency has high values. In a similar way to frequency regulation, with in voltage regulation it is possible to control the bus where the system is connected, but with the use of reactive power [33].

Figure 7 shows the typical operation of a storage system performing the frequency control function.

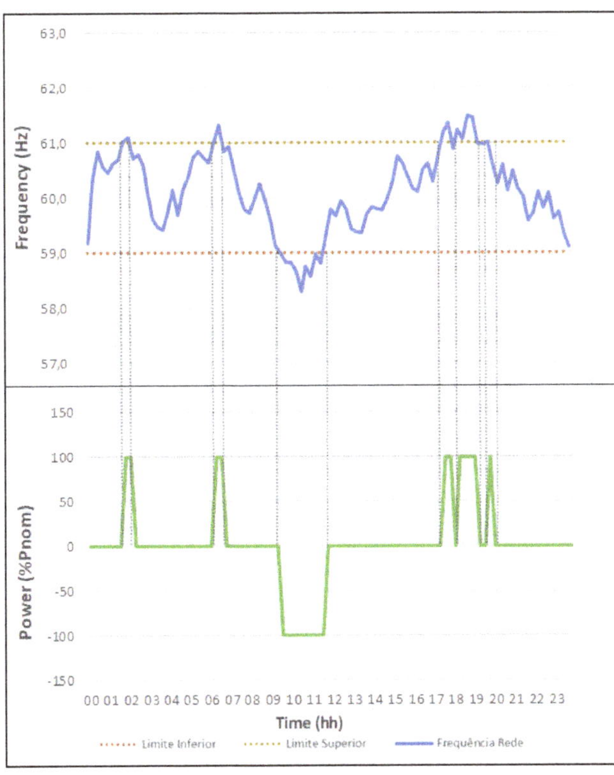

Figure 7. Typical operation of a storage system performing the frequency control function.

2.3.10. Virtual Inertia

With the massive penetration of non-rotating energy sources in electrical power systems, it is expected that part of the traditional generation systems will lose space, reducing their participation. However, traditional generation systems operate predominantly with synchronous machines, which in their working principle present mechanical inertia, hence being able to respond to frequency variations instantaneously.

This effect of substitution of generation sources causes the weakening of electrical networks, since most power generation systems based on power electronics cannot perform these functions.

An alternative to solve this problem is the use of energy storage systems operating as a voltage source, performing the virtual inertia function in parallel to the electrical grid, associated with control techniques for "damping" the system response.

The main difference between this functionality and traditional frequency regulation is in the connection/operation mode of the storage system, which directly affects the response time. For frequency regulation, the system is connected as a current source, in this way, in an event of frequency variation, it is necessary for the system to perform the measurement, process the information, and then act to correct it. For the virtual inertia function, as the system remains in voltage source mode, the response is instantaneous to frequency variations, whether positive or negative [34].

Figure 8 shows the frequency control levels for traditional electrical systems, as follows:
Mech Inertia: Mechanical inertia, provided by synchronous generators;
Inertia Emu.: Virtual inertia, provided by BESS, topic 2.3.10;
Primary Control: Frequency Regulation, topic 2.3.9;
Secondary Control: Rotating Reserve, topic 2.3.8;
Tertiary Control: Supplementary Reserve, topic 2.3.8;

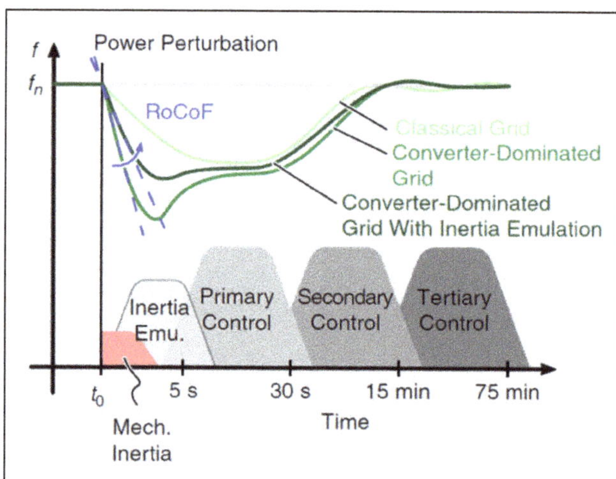

Figure 8. Energy Reserve.

3. Methodology

The methodology used in this work consists of the combination of several applications of storage systems aimed at microgrids, with the main objectives of enhancing the diesel economy generated by the photovoltaic system and minimizing the power variations imposed on the Diesel Genset. For the proposed study, it was considered that the Gensets always operate in load sharing mode, where all generated power is divided equally between the connected machines, respecting the minimum operating limit of the machines. The following topics describe the operating modes of each of the storage system applications.

3.1. Applications

3.1.1. Excess Energy Recapture

In traditional isolated systems, the balance between generation and consumption is carried out through the control of the Genset's, so that load variations are compensated by an increase or decrease in the generation. However, with the application of photovoltaic systems, due to the lack of control of the main source, it is possible the occurrence of scenarios where the generation exceeds the load consumption, causing a waste of energy. This scenario intensifies as the power of the PV system increases.

Figure 9 shows the PV generation and consumption curves for a hypothetical day, highlighting the effect of surplus energy, which can be used to recharge the storage system and later use.

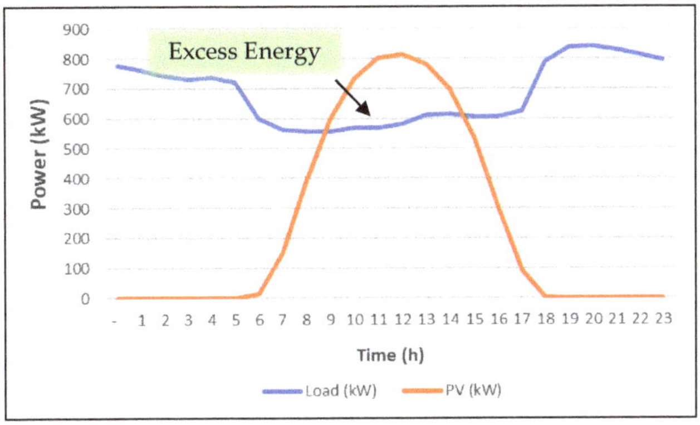

Figure 9. Microgrid with PV Generation.

Storage systems are an alternative for the use of surplus energy and subsequent use, enhancing the economy of diesel since more energy comes from a renewable source. As the PV system grows, more surplus energy is generated, bringing greater viability to the application of storage systems. Since this is the application with the highest economic return for BESS in the microgrid, priority is given to its execution over other functionalities.

In line with the strategy of prioritizing the recharge of the storage system with surplus energy, a controlled discharge is proposed in two scenarios. The first scenario occurs when more than one Genset is in operation; in this way, the calculation of the required discharge power of the storage system for the sequential shutdown of the machines is performed. In this way, based on the power and energy available in BESS, as many machines as possible are turned off, ensuring that the machines that remain on continue their operation at the optimum point of efficiency.

This strategy aligns the decrease in diesel consumption by reducing the energy demanded by the Genset's with the increase in the generation efficiency of the remaining Genset's, seeking the optimal point of operation.

For example, if three Gensets are operating at 80% of their power (3 × 80% = 240%), the power required to turn off a machine will be equal to 40% of its rated power. If the storage system has enough power and energy to shut down a machine, it will discharge the equivalent of 40% of the Genset power, causing its shutdown, keeping the two remaining machines at their optimal operating point. If the storage system has available power and energy equivalent to 140% of the Genset power, it will shut down two machines, keeping the remaining machine at its optimal operating point.

The second algorithm control scenario occurs when only one Genset is in operation. In the proposed configuration, at least one Genset must remain connected and generating, in order to guarantee the voltage and frequency of the network. Based on this, it is not possible to discharge the storage system to the point of causing the total shutdown of the Diesel generation system.

However, it is proposed to configure a target power of operation of the Genset until the storage system can perform the discharge. Ideally, this target power should not present low Genset efficiency values.

For example, if the configured target power is 75% of the rated power and the generator is operating at 100% of the rated power, the storage will discharge with a power equivalent to 25%. In another hypothetical scenario, using the same target power setting of 75%, but operating the Genset at 50% of rated power, the storage system does not perform unloading or recharging operations.

It is worth mentioning that this configured parameter is different from the minimum operating power of the Genset. The minimum operating power is an immutable value, so continuous operation below the minimum power can cause irreversible damage to the machine.

3.1.2. Power Smoothing of Photovoltaic Generation

As pointed out in topic 2.3.6, the power smoothing of photovoltaic generation is an important service provided by storage systems. In this work, two modifications are proposed to the commonly used methods.

In traditional microgrids, power generation is based on Gensets; in this way, since this is the main source, all load variations are absorbed by this set. Based on this, it is proposed to apply power smoothing in the sum of the load consumption data with the PV generation, creating a real smoothing scenario for the Genset's, unifying the power smoothing and load leveling applications in a single algorithm.

This modification aims to focus the application of the storage system on reducing the variations imposed on the Genset and not on the output of the PV system. As an example, if the PV system has a positive variation in generation and the load also presents an increase in consumption, this variation against the Genset can be cancelled without interference from storage, depending on the amplitude of the events. The opposite is also valid, since a drop in generation can be potentiated with an increase in consumption, generating a high-power ramp in the Genset's.

Figures 10–16 are possible to observe, for a hypothetical day: i. PV generation; ii. load consumption; iii. load consumption + PV generation; iv. smoothed load consumption + PV generation; v. load consumption + smoothed PV generation; saw. smoothed load consumption and PV generation; vii. Comparison of smoothing.

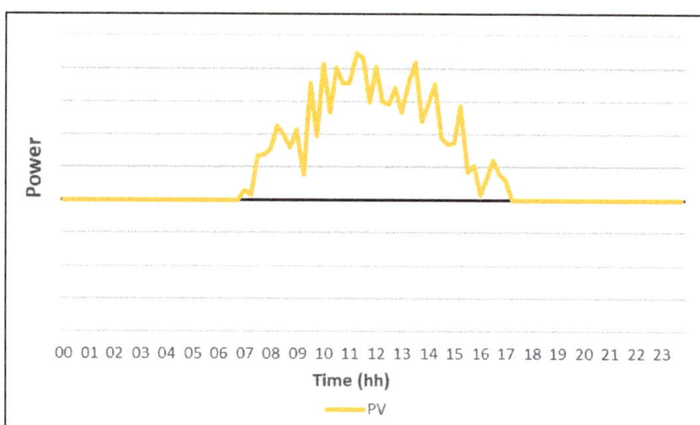

Figure 10. PV Generation.

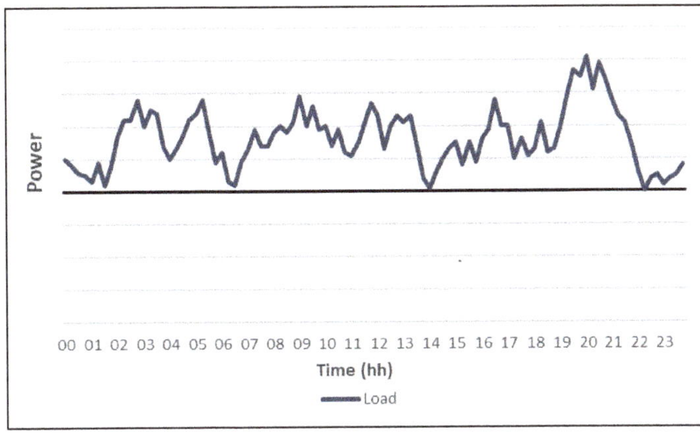

Figure 11. Load Power.

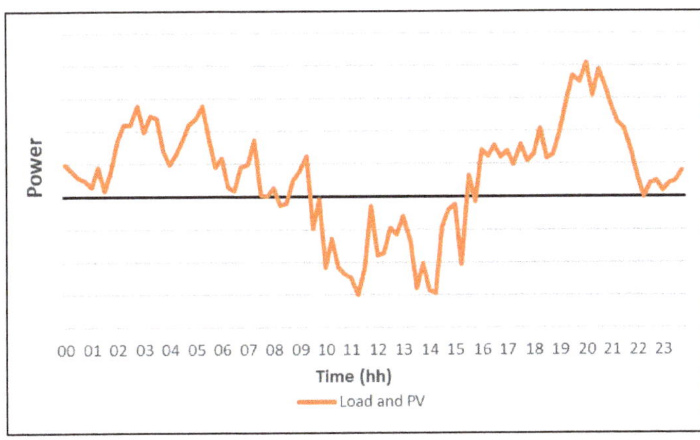

Figure 12. Load and PV Power.

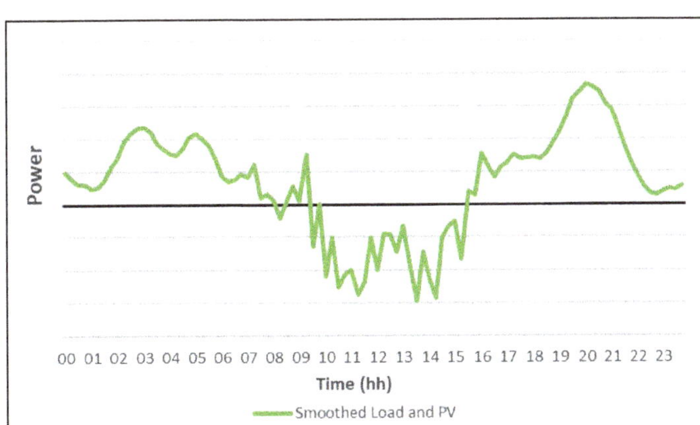

Figure 13. Smoothed Load and Natural PV.

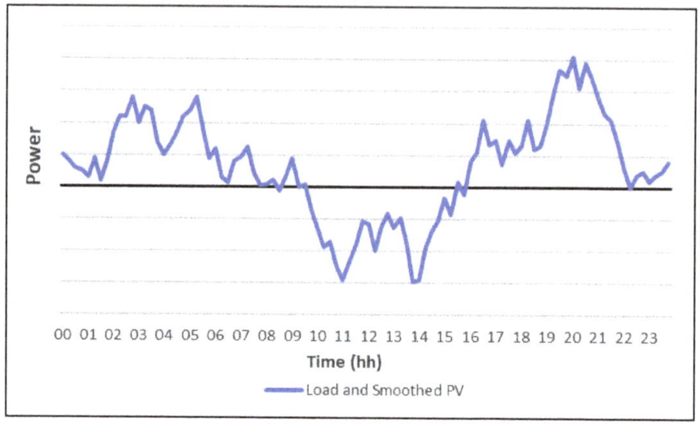

Figure 14. Natural Load and Smoothed PV.

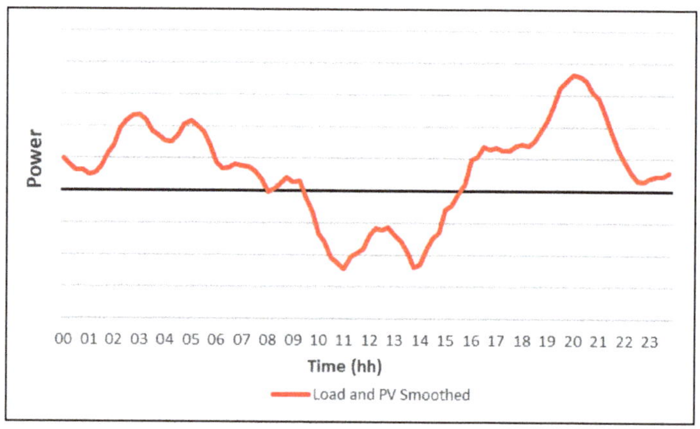

Figure 15. Load and Smoothed PV.

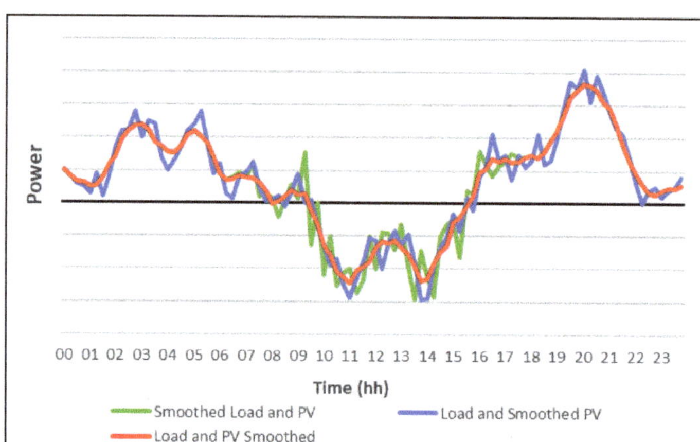

Figure 16. Smoothing Comparison.

Another point of great importance for the correct application of smoothing methods, as studied by Pinheiro, A. in [35], refers to the SoC level of the storage system. Depending on load and PV generation variations, scenarios may occur where the consumption increase always happens abruptly and the drops in soft variations, causing an imbalance in the SoC at the beginning and end of the day, since BESS would only carry out operations to smooth out the abrupt rise in consumption, without carrying out recharging operations, as the drop in consumption did not exceed the threshold rate.

Based on this, the proposed smoothing method is a modification of the ramp rate strategy, recharging or discharging the system to reach the desired SoC with energy from the Gensets, without extrapolating the designated threshold rate.

In other words, if the Load + PV set decreases consumption in relation to the previous smoothed power, enough to exceed the threshold rate, the battery bank must consume part of this power, recharging to keep the variation seen by the Genset within the admissible limit, and vice versa. However, if the variation is within the limits (that is, no limit exceeded), the BESS can perform the function of returning to the ideal SoC, recharging or discharging, without extrapolating the limit rate of change.

3.1.3. Capacity Reserve

As presented in 2.3.8, isolated generation systems must work with an available reserve power so that any sudden load increases or partial failures in the generation system can be met without violating the established voltage and frequency values. However, due to this reserve percentage, it is likely that the generators will start to operate outside the optimal consumption ranges, since this value is found close to the rated power of the machine.

As storage systems are capable of providing power and energy quickly, this power reserve can be shifted from the Genset's to the storage system, so that the optimal range of consumption of the Genset can be reached more easily, since there is greater use of available power.

In this work, it is proposed to perform the capacity reserve function of the generation system in the storage system. For the execution of this functionality, the power available by the storage system is verified, conflicting with the state of load and power limits of the equipment. If full power is not available, the diesel generation system returns to the execution of the functionality, which can also be shared, with part of the reserve coming from storage and part from the Genset's.

3.2. *Proposed System*

In the proposed microgrid, the PV system, five Gensets of equal power, and the storage and load system are connected to the AC bus, as illustrated in Figure 17.

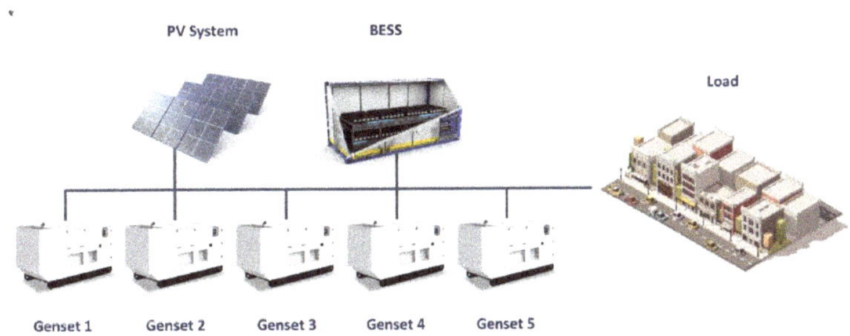

Figure 17. AC Microgrid Topology.

PV system: With peak power being an input variable, the simulated PV system has an average monthly generation of 120 kWh/kWp/month and a monthly generation profile as shown in Figure 18. To perform the simulations, the generation profile was determined as equivalent to 1 kWp, so multipliers on top of the base profile were applied to reach the peak power target. Figure 18 presents the monthly generation for a simulated 100 kWp system.

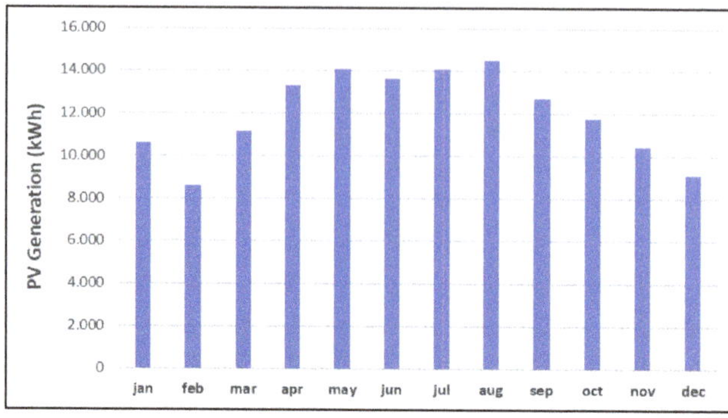

Figure 18. PV Monthly Generation Profile.

Genset: The simulated Gensets have a nominal power of 500 kW and a minimum operating power of 125 kW, with the five machines having the same configuration. The consumption curve, shown in Figure 19, is based on the data presented in [36].

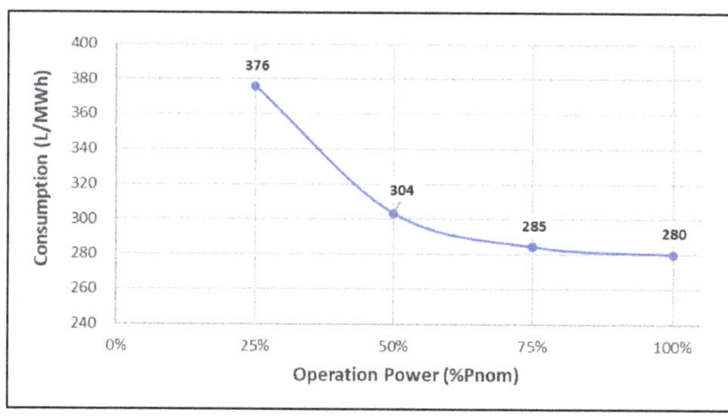

Figure 19. Consumption Curve of Diesel Genset.

Storage System: With useful energy being an input variable, the simulated storage system has a round-trip efficiency of 85% and a power/useful energy ratio of 1.

Load: The base load for the simulations has a peak consumption of 1.25 MW and an average consumption of 500 MWh/month. The typical consumption curve is shown in Figure 20 and the monthly consumption is shown in Figure 21.

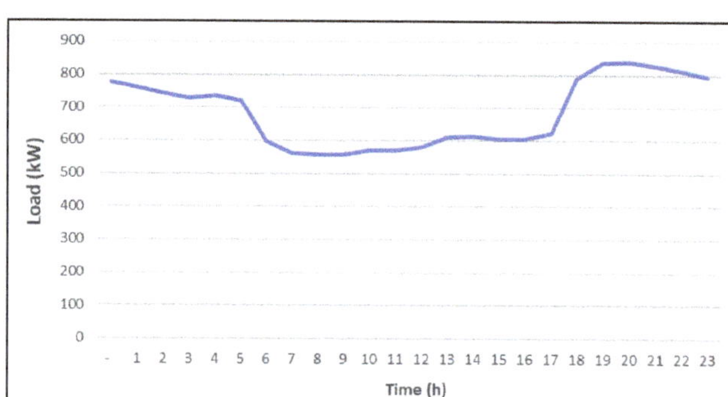

Figure 20. Daily Consumption Profile.

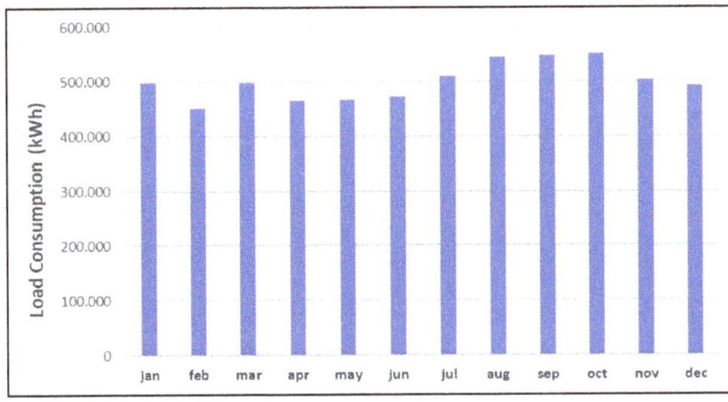

Figure 21. Monthly Consumption Profile.

3.3. *Simulations and Results Analysis*

The simulations were carried out for a period of one year of consumption and generation, with a time interval of 5 min between samples, thus, dynamic stability studies are not contemplated in this work.

For the amplitude of the subsystems, analyses were carried out with photovoltaic systems from 0 to 4 MWp, in steps of 500 kWp, and storage system from 0 to 6 MWh, in steps of 500 kWh.

For the analysis of results, the microgrids operation indicators were standardized, as follows:

1. Diesel consumption (L);
2. Surplus Energy (kWh);
3. Energy Recharged by BESS (kWh);
4. PV generation (kWh);
5. PV share in the energy matrix (%)
6. Hours of Operation of Gensets (h);
7. Average operating power of Gensets (kW);
8. Percentage of Gensets power variation within the established limits (%);
9. Variation in the average power of the Gensets;
10. Distribution of power variation bands for Gensets.

4. Results

4.1. *Energy Balance*

In this section, the energy balance of the microgrid will be analyzed, presenting the effectiveness of the implementation of the hybrid system to reduce diesel consumption, increase the PV share, energy recharged by BESS, and the sensitivity analysis with the variation of these components.

Seeking to simplify the understanding of the behavior of the microgrid and the results achieved by the operation algorithm, Figures 22–24 show the standard operation data for scenarios of low, medium, and high penetration of renewables.

In the scenario of low penetration of renewables (Figure 22), it is possible to observe a decrease in the operating power of the Genset in relation to the consumption of the load during the period of higher PV generation, followed also by a slight reduction, where the discharge of the BESS is found. This reduction in operating power is due to PV participation in the generation matrix and storage as an element for absorbing excess energy.

In this scenario, it is possible to point out that even if the PV generation does not exceed the consumption of the load, there is surplus energy in the microgrid, since the minimum generation to be exported by the Gensets is 125 kW.

In the scenario of medium penetration of renewables (Figure 23), it is possible to observe that the PV system has greater participation in the microgrid, with more surplus energy being generated. In the same way, BESS recharges and discharges more expressively, extending the time in which the generator operates with less power than the load consumption.

For this result, it is possible to observe that close to 11:00 am, BESS decreases the recharge power, even if there is excess energy in the microgrid. This happens because the state of charge has reached the maximum value and there is no more storage capacity, causing this excess energy to be wasted.

In the scenario of high penetration of renewables (Figure 24), the surplus energy due to high generation during the day is even more evident. The PV overtakes the load during almost the entire generation period and is followed by a controlled discharge of the storage system, causing the generator to operate at close to its minimum power throughout the entire period.

Similar to the previous analysis, around 11:00 a.m. the charging power of BESS decreases, since charging is complete and there is no space for storage.

As the main parameter to be evaluated in economic analyses, since it corresponds to the component with the highest operating cost in a microgrid, diesel consumption is presented in Figure 25. The operation analysis is performed for systems PV from 0 to 4 MWp, in steps of 500 kWp and BESS from 0 to 6 MWh, in steps of 500 kWh.

At points of lower PV insertion (0 and 500 kWp), it is noted that the change in the size of the storage system does not present significant differences in diesel consumption. This is mainly due to the low surplus energy generated in these scenarios, the main source of BESS recharge. As the PV system grows, the more effective the storage system is, as there is more surplus energy available for recharging.

As an analysis of the operating trends, it is highlighted that the increase in the PV system presents greater effectiveness for the relative reduction in diesel consumption in the first steps (0 to 1 MWp). This is because most of the energy generated in these scenarios is used by the microgrid. With the increase in the PV system, it is possible to observe a stabilization trend in the reduction in diesel consumption, so that this stabilization behavior is observed earlier in scenarios with lower insertion of BESS.

As a complement to the PV trend analysis, it is noteworthy that the first step of inserting storage (0 to 500 kWh) presents a greater percentage reduction in diesel consumption, since the power smoothing and capacity reservation services pass to be performed.

In addition, with the increase in BESS, it is possible to notice that in the final stages (5.5 MWh to 6 MWh) the reduction in diesel loses percentage effectiveness, also reaching a stabilization trend. This behavior is mainly due to the high energy stored and the need to keep at least one generator in operation to guarantee the voltage and frequency reference, even at minimum power.

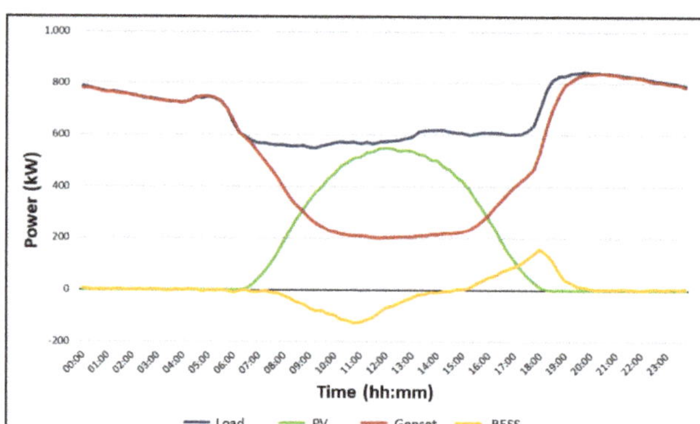

Figure 22. Average Operation–Low Renewables Penetration Rate (1 MWp–0.5 MWh).

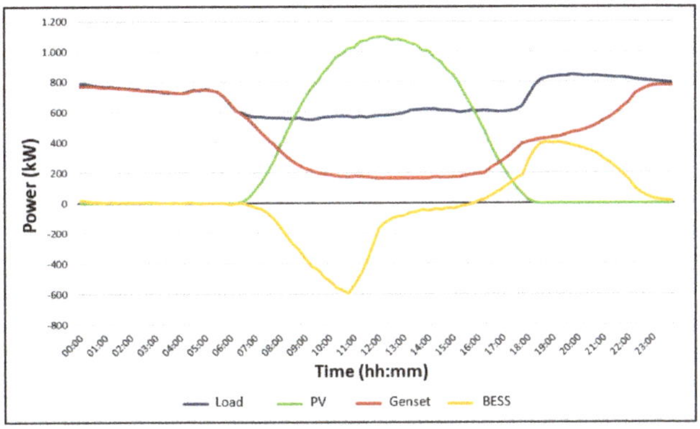

Figure 23. Average Operation–Medium Renewables Penetration Rate (2 MWp–2 MWh).

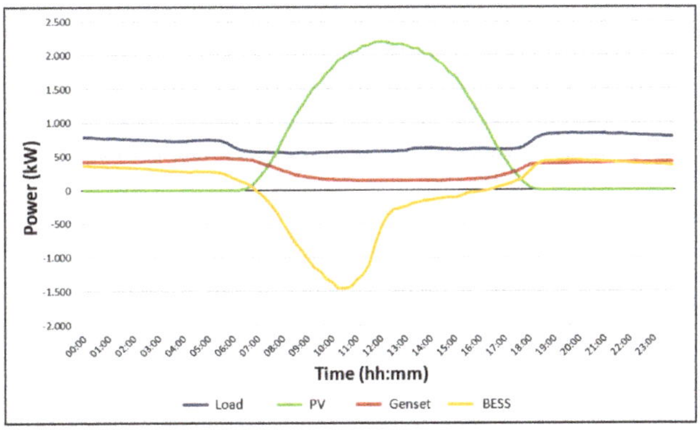

Figure 24. Average Operation–High Renewables Penetration Rate (4 MWp–6 MWh).

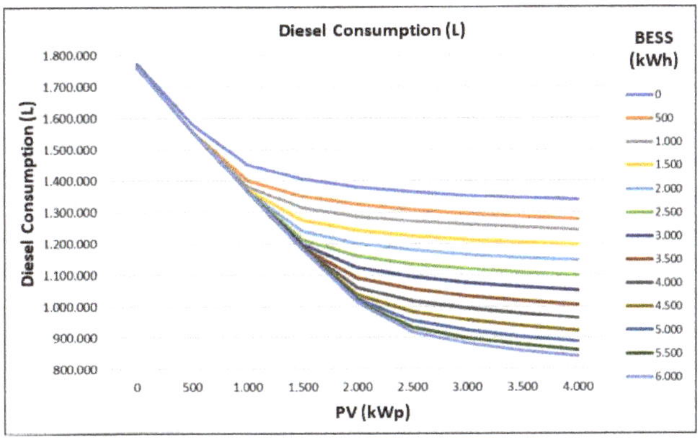

Figure 25. Diesel Consumption vs. PV and BESS insertion.

Complementary to the analyses carried out for diesel consumption, Figure 26 shows the surplus energy of the microgrid with the increase in PVS (Photovoltaic System) and BESS. For the lower PV insertion scenarios (0 to 500 kWp) there is no surplus energy, since the load consumption remains above the PV generation + Genset minimum generation. However, it is noted that the increase in PVS is accompanied by an increase in excess energy, even if minimized with the insertion of BESS.

The trend of constant increase in excess energy, even with the insertion of larger BESS is noticeable, explained by the need to keep at least one Genset in operation, even at minimum power.

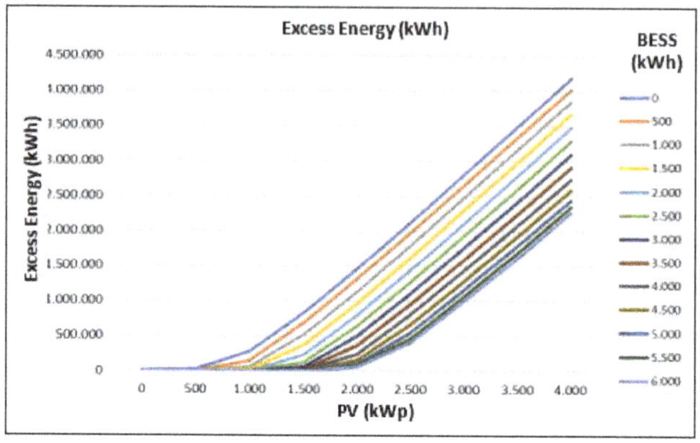

Figure 26. Excess Energy vs. PV and BESS insertion.

For the recharged energy of the BESS (Figure 27), it is noted that the stabilization of the parameter is reached earlier for smaller storage capacities. Similar to previous analyses, this occurs due to the limitation of BESS recharge to absorb excess energy from the plant. This analysis becomes important, mainly for economic aspects, as it details the real need for storage capacity through BESS recharge.

Taking the 1 MWp PVS as an example, it is possible to infer that from the point of view of energy use, there is no considerable difference between using a 2 MWh or 6 MWh BESS, since the energy recharged by both systems is close. In this scenario, using a higher capacity system would bring an extra cost to the system, which could lead to the economic unfeasibility of the investment.

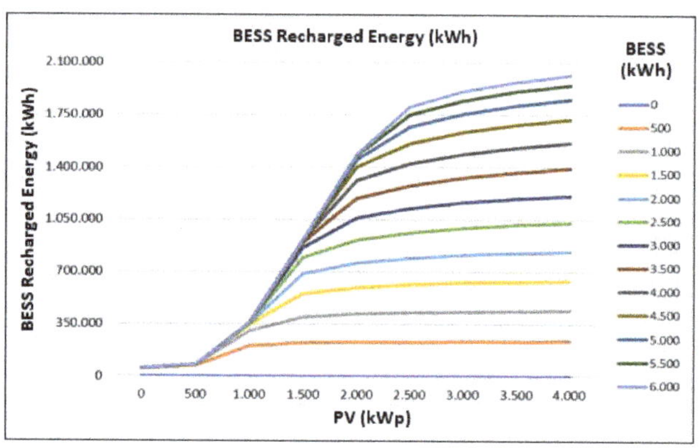

Figure 27. BESS Recharged Energy vs. PV and BESS Integration.

Adding to the analysis of the energy balance of the microgrid, Figure 28 shows the PV participation in meeting the load in view of the increase in the generation and storage system. Additionally, the percentage of PV generation was inserted in relation to the total consumption of the load, in order to illustrate the low use of generated energy for larger systems.

With the low penetration of PVS, it is clear that the PV share follows the percentage generation compared to consumption, as there is low surplus energy in these scenarios. As the PV insertion increases, the greater the dispersion between generation and use, even with the increase in BESS, and a trend towards stabilization of participation is noted.

As already pointed out in previous analyses, this stabilization takes place far from the point of 100% renewable penetration due to the need to guarantee the operation of at least one Genset. It is important to emphasize that different load profiles will bring different results regarding the penetration of renewables in the system, being the analysis of varied loads outside the scope of this work.

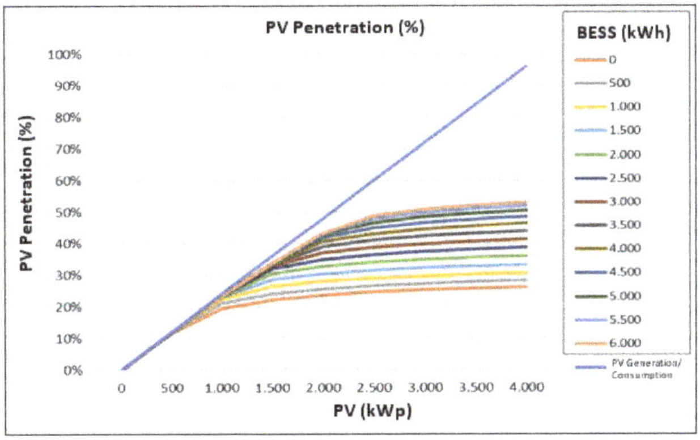

Figure 28. Renewable Penetration vs. PV and BESS Integration.

4.2. Capacity Reserve

One of the functions developed for the operation algorithm refers to the capacity reservation of the microgrid. In this proposal, the reserve power is transferred to BESS, if available. This feature aims to reduce operating time and increase the average working power of the Genset's.

To exemplify the proposed strategy, considering a scenario with exclusive generation by Genset's and 30% of the capacity reserve, for a load of 500 kW, a minimum available power of 650 kW will be required. As the generator simulated for this microgrid has a maximum power of 500 kW, two generators would be needed working in parallel. If this functionality is performed by another system, this load could be serviced with only one generator connected.

In the scenario where the reserve function is not performed by the Genset, the operation is based on one generator, with a working power of 500 kW, compared to two generators with a working power of 250 kW. Note, for this example, that performing the capacity reserve function reduces the operating time by half and the operating power remains at the optimal point, bringing better efficiency in the generated L/kWh ratio and lower maintenance costs, since the machine spends less time in operation.

Figure 29 shows the graph with the two extremes of the BESS analyzes (0.5 MWh and 6 MWh). For the system with a capacity of 0.5 MWh, it is possible to achieve a reduction of approximately 900 h of operation for smaller PVSs (0.5 to 1 MWp) and 1500 h for the PVS of 4 MWp.

For the system with a capacity of 6 MWh, the same reduction is achieved for scenarios with lower penetration of renewables and a reduction of approximately 1000 h considering PVS of 6 MWp.

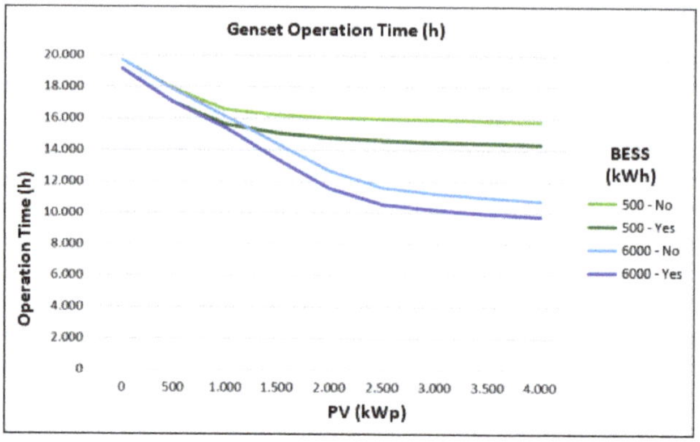

Figure 29. Diesel Genset Operating Hours vs. Capacity Reserve (500 kWh e 6000 kWh).

As a consequence of the reduction in the reserve percentage of the Genset's, an increase in the average operating power of the machines is observed. Figure 30 shows the average power data for the 500 kWh BESS. Of note is an average power increase of 15 kW for scenarios with lower PV penetration and 28 kW for the scenario with higher penetration.

Figure 31 shows the average power data for the 6 MWh BESS. The same reduction of 15 kW for smaller PVSs and 41 kW for the scenario of higher penetration stands out.

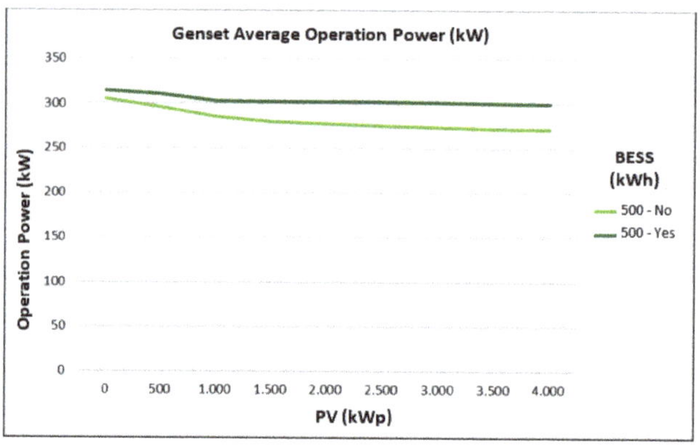

Figure 30. Average Diesel Genset Operating Power vs. Capacity Reserve (500 kWh).

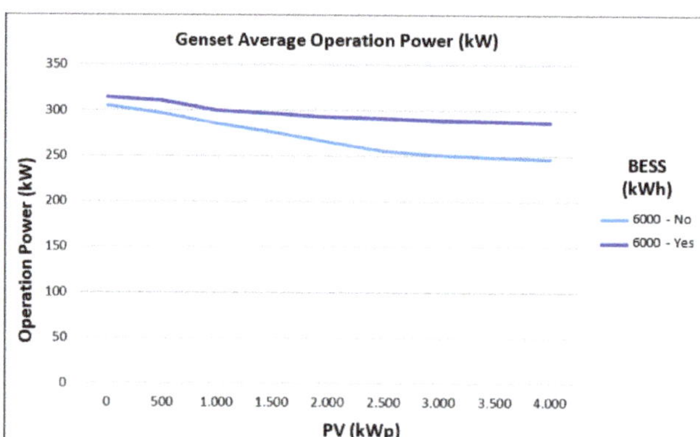

Figure 31. Average Diesel Genset Operating Power vs. Capacity Reserve (6000 kWh).

4.3. Power Smoothing

The power smoothing functionality has a main objective to limit the variation imposed on the Genset, which is caused by the load or PVS. This variation minimization seeks to increase the voltage and frequency stability of the network (parameters not analyzed in this work) and minimize the chance of damage to the Genset's due to the inertia found in the motors, which contrast with abrupt load variations.

In short, when executing the functionality, BESS analyzes the working power of the Genset in the previous instant and controls the maximum variation in ±1%/minute. It should be noted that this parameter can be configured according to the needs of the Gensets used.

Figure 32 shows the power variation data within the limits of ±1%/min for a BESS of 500 kWh. It is possible to notice that, without the execution of the functionality, the limits present a high rate of violation, remaining below 80% for low penetration of renewables (0.5 to 1.5 MWp) and close to 83% for higher penetrations.

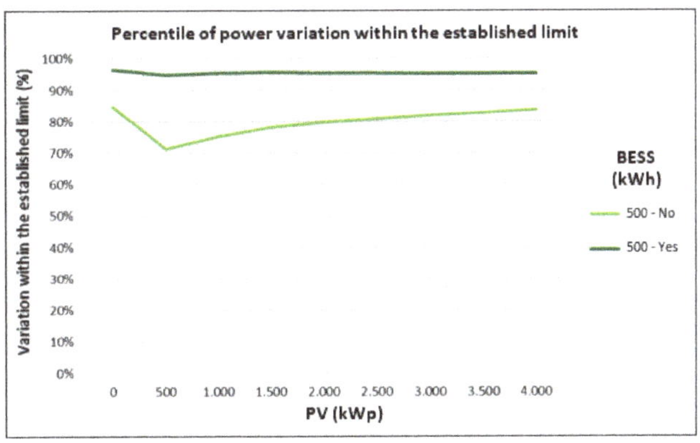

Figure 32. Power Variation of the Diesel Genset within the Established Limit (500 kWh).

The increase in data with a variation rate within the limits for larger PVSs is basically due to the generation limitation at times when the generation exceeds the load consumption. As illustrated in Figures 23 and 24, in high-generation scenarios, the Genset output is kept fixed at the minimum power, making the observed variation null. With the functionality running, all scenarios are kept above 95%, proving the effectiveness of the proposed solution even for low storage capacity.

Figure 33 shows the power variation data within the limits of ±1%/min for a BESS of 6 MWh. In a similar way to the behavior of the lower capacity system, the effectiveness of the functionality stands out, guaranteeing 95% of the variation within the limit established for low PV penetration (0.5 MWp) and above 98% for medium and large systems (2 to 4 MWp).

Without functionality running, data within limits remains below 80% for low penetration of renewables (0.5 to 1.0 MWp) and increases up to 94% as PVS increases. As explained in the previous analysis, the increase in the percentage of data within the established variation rate is due to the limitation of PV generation to guarantee the minimum operating power of the Genset.

As a consequence of BESS operation, keeping the power variation within the established limit of ±1%/min, it is also possible to observe a considerable reduction in the average power variation, in kW/min. Figure 34 shows the power variation data, in kW/min, for the 500 kWh BESS, with and without the power smoothing function. For systems with low penetration of renewables it is possible to observe a reduction of approximately 20% of the average variation (10.4 kW to 8.4 kW). For the highest PV insertion point (4 MWp) a reduction of approximately 29% of the average variation is achieved (10.1 kW to 7.2 kW).

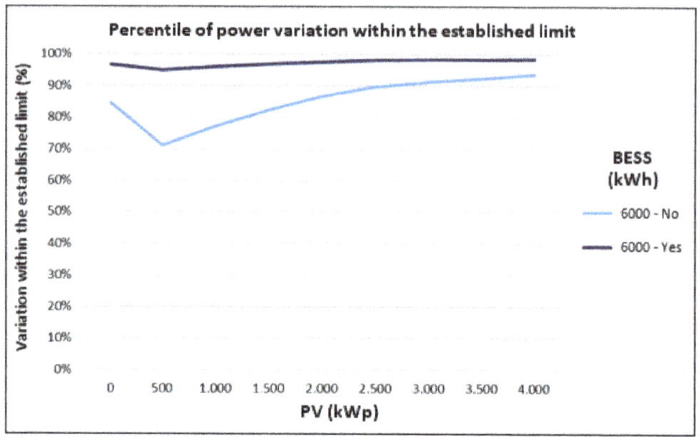

Figure 33. Power Variation of the Diesel Genset within the Established Limit (6000 kWh).

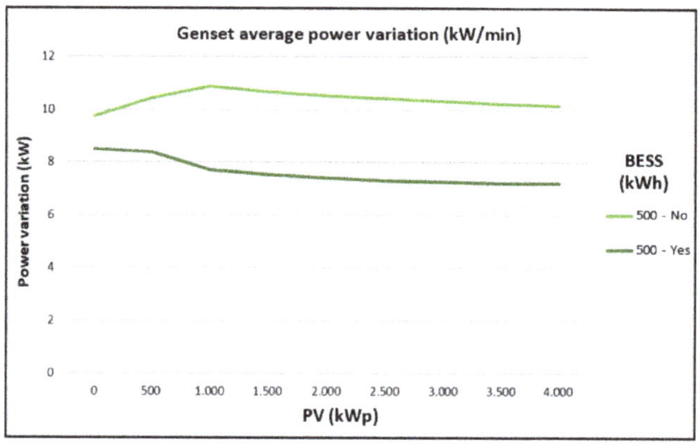

Figure 34. Average Power Variation of Diesel Genset vs. PV Power Smoothing (500 kWh).

For the analysis of the system with a BESS of 6 MWh, shown in Figure 35, it is possible to observe a reduction of the same 20% in low PV insertion (10.4 kW to 8.4 kW) and of 22% for higher insertions (4.3 kW to 3.3 kW). Additionally, noteworthy is the natural reduction in the average power variation for scenarios with greater PV insertion, since, due to the phenomenon presented in the previous analyses, the larger the PV + BESS set, the longer the generator remains at minimum operating power, or with the output power controlled by the action of BESS, according to the strategy of using surplus energy and optimized discharge, presented in Figure 34.

Figure 36 shows the power variation distribution for the simulation performed, 2 MWp–500 kWh. This representation illustrates the scenario with and without the application of the power smoothing strategy, where BESS performs the variation limitation if it has a modulus greater than 1%. As seen from this scenario, it is possible to detect that without the application of the smoothing strategy, there are variations that violate these limits.

It is also noted from Figure 36 that all data with a population outside the allowed variation zone of −1% to 1% had a reduction in their occurrence, due to the application of the smoothing strategy.

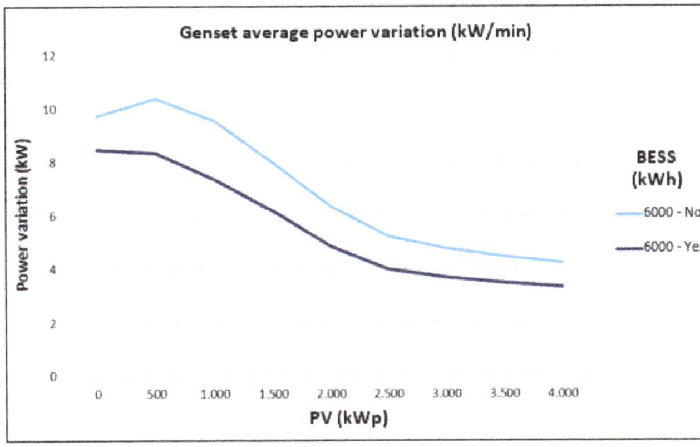

Figure 35. Average Power Variation of Diesel Genset vs. PV Power Smoothing (6000 kWh).

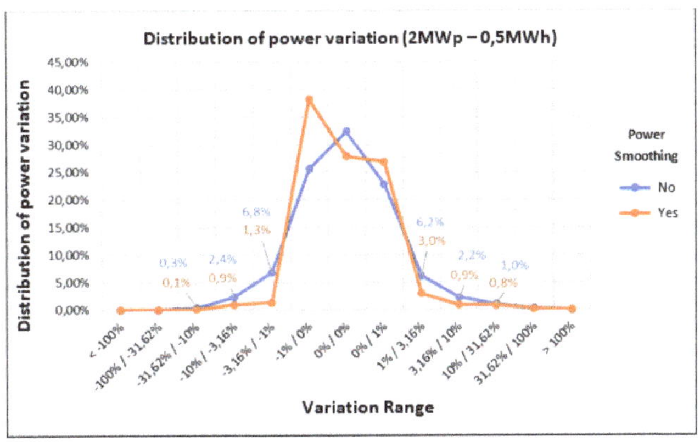

Figure 36. Distribution of Power Variation of Diesel Genset (2 MWp–500 kWh).

5. Conclusions

Throughout this work, in which the control strategy for microgrids was developed and applied, it was possible to verify the effectiveness of the proposal with reductions in diesel consumption with values greater than 50%, reduction in Genset operating time by up to 50%, increase in the average operating power, and guarantee of power variation within the limits with the confidence of up to 99%, all results being variable according to the level of PV + BESS penetration. The proposed strategy contributes with a new perspective of BESS applications in microgrids, enabling the realization of simultaneous and combined functionalities, such as intelligent discharge control, Genset power smoothing, and sharing of the rotating reserve. The proposed study was limited to the construction and demonstration of the effectiveness of the strategy for applications in microgrids with AC coupling, with the guarantee of at least one generator working as a voltage and frequency reference source. For future work, a detailed economic analysis of each of the components and of the complete microgrid is suggested, in order to evaluate the combination of PVS, BESS, and Genset that best relates to the technical and economic benefits, being able to use sensitivity matrices for different estimates of the cost of components, diesel, and maintenance.

Author Contributions: Conceptualization, F.R., A.P., R.N., W.d.A.S.J., M.A.M. and M.H.N.M.; data curation, F.R., W.d.A.S.J., M.A.M. and M.H.N.M.; formal analysis, R.N. and M.H.N.M.; investigation, F.R., A.P., W.d.A.S.J., M.A.M., A.A. and M.H.N.M.; methodology, R.N., M.A.M. and M.H.N.M.; resources, A.P., R.N. and W.d.A.S.J.; software, F.R. and M.H.N.M.; supervision, M.A.M. and M.H.N.M.; validation, A.A.; visualization, A.P. and A.A.; writing—original draft, F.R., A.P., R.N., W.d.A.S.J., M.A.M. and M.H.N.M.; writing—review and editing, F.R., A.P., R.N., W.d.A.S.J., M.A.M., A.A. and M.H.N.M. All authors have read and agreed to the published version of the manuscript.

Funding: This research received no external funding.

Institutional Review Board Statement: Not applicable.

Informed Consent Statement: Not applicable.

Data Availability Statement: Not applicable.

Conflicts of Interest: The authors declare no conflict of interest.

References

1. Dayalan, S.; Gul, S.S.; Rathinam, R.; Fernandez Savari, G.; Aleem, S.H.A.; Mohamed, M.A.; Ali, Z.M. Multi-Stage Incentive-Based Demand Response Using a Novel Stackelberg–Particle Swarm Optimization. *Sustainability* **2022**, *14*, 10985. [CrossRef]
2. Chen, W.; Liu, B.; Nazir, M.S.; Abdalla, A.N.; Mohamed, M.A.; Ding, Z.; Bhutta, M.S.; Gul, M. An Energy Storage Assessment: Using Frequency Modulation Approach to Capture Optimal Coordination. *Sustainability* **2022**, *14*, 8510. [CrossRef]
3. Peng, F.; Su, H.; Li, P.; Song, G.; Zhao, J. *An Interactive Operation Strategy for Microgrid Cooperated with Distribution System Based on Demand Response*; IEEE: New York, NY, USA, 2015; ISBN 9781467371063.
4. Zhang, Y.; Liu, B.; Zhang, T.; Guo, B. An Intelligent Control Strategy of Battery Energy Storage System for Microgrid Energy Management under Forecast Uncertainties. *Int. J. Electrochem. Sci.* **2014**, *9*, 4190–4204.
5. Khalid, M. A Review on the Selected Applications of Battery-Supercapacitor Hybrid Energy Storage Systems for Microgrids. *Energies* **2019**, *12*, 4559. [CrossRef]
6. Huang, Y.; Masrur, H.; Shigenobu, R.; Hemeida, A.M.; Mikhaylov, A.; Senjyu, T. A Comparative Design of a Campus Microgrid Considering a Multi-Scenario and Multi-Objective Approach. *Energies* **2021**, *14*, 2853. [CrossRef]
7. Aziz, A.S.; Tajuddin, M.F.N.; Adzman, M.R.; Ramli, M.A.M.; Mekhilef, S. Energy Management and Optimization of a PV/Diesel/Battery Hybrid Energy System Using a Combined Dispatch Strategy. *Sustainability* **2019**, *11*, 683. [CrossRef]
8. Arévalo-Cordero, P.; Benavides, D.J.; Espinoza, J.L.; Hernández-Callejo, L.; Jurado, F. Optimal Energy Management Strategies to Reduce Diesel Consumption for a Hybrid Off-Grid System. *Rev. Fac. Ing.* **2020**, *98*, 47–58. [CrossRef]
9. Delille, G.; François, B.; Malarange, G. Dynamic Frequency Control Support: A Virtual Inertia Provided by Distributed Energy Storage to Isolated Power Systems. In Proceedings of the 2010 IEEE PES Innovative Smart Grid Technologies Conference Europe (ISGT Europe), Gothenburg, Sweden, 11–13 October 2010; pp. 1–8.
10. Denholm, P.; Margolis, R.M. Evaluating the Limits of Solar Photovoltaics (PV) in Electric Power Systems Utilizing Energy Storage and Other Enabling Technologies. *Energy Policy* **2007**, *35*, 4424–4433. [CrossRef]
11. Zhao, B.; Zhang, X.; Li, P.; Wang, K.; Xue, M.; Wang, C. Optimal Sizing, Operating Strategy and Operational Experience of a Stand-Alone Microgrid on Dongfushan Island. *Appl. Energy* **2014**, *113*, 1656–1666. [CrossRef]
12. Vieira, G.T.T.; Pereira, D.F.; Taheri, S.I.; Khan, K.S.; Salles, M.B.C.; Guerrero, J.M.; Carmo, B.S. Optimized Configuration of Diesel Engine-Fuel Cell-Battery Hybrid Power Systems in a Platform Supply Vessel to Reduce CO_2 Emissions. *Energies* **2022**, *15*, 2184. [CrossRef]
13. Mohamed, M.A. A relaxed consensus plus innovation based effective negotiation approach for energy cooperation between smart grid and microgrid. *Energy* **2022**, *252*, 123996. [CrossRef]
14. Ralon, P.; Taylor, M.; Ilas, A.; Diaz-Bone, H.; Kairies, K. *Electricity Storage and Renewables: Costs and Markets to 2030*; International Renewable Energy Agency: Abu Dhabi, United Arab Emirates, 2017; ISBN 9789292600389.
15. Mohamad, F.; Teh, J.; Lai, C.M. Optimum allocation of battery energy storage systems for power grid enhanced with solar energy. *Energy* **2021**, *223*, 120105. [CrossRef]
16. Ramos, F.O.; Neto, M.M.B.; Pinheiro, A.L.; Junior, W.A.S.; Lima, R.N. A Real Case Analysis of a Battery Energy Storage System for Energy Time Shift, Demand Management, and Reactive Control. In Proceedings of the 2021 IEEE PES Innovative Smart Grid Technologies Conference-Latin America (ISGT Latin America), Quito, Ecuador, 15 September 2021.
17. Metwaly, M.K.; Teh, J. Optimum network ageing and battery sizing for improved wind penetration and reliability. *IEEE Access* **2020**, *8*, 118603–118611. [CrossRef]
18. Ribeiro, P.F.; Johnson, B.K.; Crow, M.L.; Arsoy, A.; Liu, Y. Energy Storage Systems for Advanced Power Applications. *Proc. IEEE* **2001**, *89*, 1744–1756. [CrossRef]
19. Fidalgo, J.N.; Couto, M.; Fournié, L. The worth of network upgrade deferral in distribution systems—Truism or myth? *Electr. Power Syst. Res.* **2016**, *137*, 96–103. [CrossRef]

20. Liu, L.; Wang, Y.; Zhao, Y. The Energy Storage System Control Research Based on Black-Start. In Proceedings of the 2014 China International Conference on Electricity Distribution (CICED), Shenzhen, China, 23 September 2014; pp. 1472–1476.
21. Mielczarski, W. Impact of Energy Storage on Load Balancing. In Proceedings of the 2018 15th International Conference on the European Energy Market (EEM), Lodz, Poland, 27–29 June 2018; Institute of Electrical and Electronics Engineers (IEEE): New York, NY, USA, 2018; pp. 1–5.
22. Izidio, D.M.F.; de Mattos Neto, P.S.G.; Barbosa, L.; de Oliveira, J.F.L.; da Marinho, M.H.N.; Rissi, G.F. Evolutionary Hybrid System for Energy Consumption Forecasting for Smart Meters. *Energies* **2021**, *14*, 1794. [CrossRef]
23. IRENA. *Renewable Capacity Statistics 2020*; International Renewable Energy Agency: Abu Dhabi, United Arab Emirates, 2020; ISBN 9789292602390.
24. de Mattos Neto, P.S.G.; de Oliveira, J.F.L.; Domingos, D.S.; Siqueira, H.V.; Marinho, M.H.N.; Madeiro, F. An Adaptive Hybrid System Using Deep Learning for Wind Speed Forecasting. *Inf. Sci.* **2021**, *581*, 495–514. [CrossRef]
25. Mohamed, M.A.; Almalaq, A.; Abdullah, H.M.; Alnowibet, K.A.; Alrasheedi, A.F.; Zaindin, M.S.A. A distributed stochastic energy management framework based-fuzzy-PDMM for smart grids considering wind park and energy storage systems. *IEEE Access* **2021**, *9*, 46674–46685. [CrossRef]
26. Haque, M.E.; Sakib Khan, M.N.; Islam Sheikh, M.R. *Smoothing Control of Wind Farm Output Fluctuations by Proposed Low Pass Filter, and Moving Averages*; International Conference on Electrical & Electronics Engineering: New York, NY, USA, 2015; ISBN 9781467378192.
27. Alevizakos, V.; Chatterjee, K.; Koukouvinos, C.; Lappa, A. A Double Moving Average Control Chart. *Commun. Stat. Simul. Comput.* **2020**, *37*, 1696–1708. [CrossRef]
28. Martins, J.; Spataru, S.; Sera, D.; Stroe, D.I.; Lashab, A. Comparative Study of Ramp-Rate Control Algorithms for PV with Energy Storage Systems. *Energies* **2019**, *12*, 1342. [CrossRef]
29. Mohamed, M.A.; Abdullah, H.M.; Al-Sumaiti, A.S.; El-Meligy, M.A.; Sharaf, M.; Soliman, A.T. Towards energy management negotiation between distributed AC/DC networks. *IEEE Access* **2020**, *8*, 215438–215456. [CrossRef]
30. Al-Shamma'a, A.A.; Hussein Farh, H.M.; Noman, A.M.; Al-Shaalan, A.M.; Alkuhayli, A. Optimal Sizing of a Hybrid Renewable Photovoltaic-Wind System-Based Microgrid Using Harris Hawk Optimizer. *Int. J. Photoenergy* **2022**, *2022*, 4825411. [CrossRef]
31. Teh, J.; Lai, C.M. Reliability impacts of the dynamic thermal rating and battery energy storage systems on wind-integrated power networks. *Sustain. Energy Grids Netw.* **2019**, *20*, 100268. [CrossRef]
32. Al-Shamma'a, A.A.; Alturki, F.A.; Farh, H.M. Techno-economic assessment for energy transition from diesel-based to hybrid energy system-based off-grids in Saudi Arabia. *Energy Transit.* **2020**, *4*, 31–43. [CrossRef]
33. Zeng, L.; Xia, T.; Elsayed, S.K.; Ahmed, M.; Rezaei, M.; Jermsittiparsert, K.; Dampage, U.; Mohamed, M.A. A novel machine learning-based framework for optimal and secure operation of static VAR compensators in EAFs. *Sustainability* **2021**, *13*, 5777. [CrossRef]
34. Unamuno, E.; Paniagua, J.; Barrena, J.A. Unified Virtual Inertia for Ac and Dc Microgrids. *IEEE Electrif. Mag.* **2019**, *7*, 56–68. [CrossRef]
35. Pinheiro, A.L.; Ramos, F.O.; Neto, M.M.B.; Lima, R.N.; Bezerra, L.G.S.; Junior, W.A.S. A Review and Comparison of Smoothing Methods for Solar Photovoltaic Power Fluctuation Using Battery Energy Storage Systems. In Proceedings of the 2021 IEEE PES Innovative Smart Grid Technologies Conference-Latin America (ISGT Latin America), Quito, Ecuador, 15 September 2021.
36. Caterpillar Inc. *C18 500kW Diesel Generator Sets*; Caterpillar Inc.: Peoria, IL, USA, 2019.

sustainability

MDPI

Article

A Vector Inspection Technique for Active Distribution Networks Based on Improved Back-to-Back Converters

Weiming Zhang [1], Hui Fan [1], Jiangbo Ren [2], Xianzhi Wang [1], Tiecheng Li [1] and Yibo Wang [3,*]

[1] State Grid Hebei Electric Power Research Institute, Shijiazhuang 050021, China
[2] State Grid Hebei Electric Power Co., Ltd., Shijiazhuang 050021, China
[3] Wuhan Kemov Electric Co., Ltd., Wuhan 430223, China
* Correspondence: yibowangwuh@gmail.com

Abstract: In this paper, an improved back-to-back converter is proposed, and the converter is used as a test power source for vector inspection of relay protection in an active distribution network, which effectively solves the problem that the output voltage and current of the test power source cannot be continuously and stably adjusted. Firstly, a three-phase back-to-back cascade converter is established to analyze the impedance characteristics of its DC terminal. Then a feedforward voltage is added to the inverter to improve the input impedance characteristics of the inverter. Secondly, the system stability and parameter stability of the improved back-to-back converter are analyzed. Finally, the improved converter is used as the test power source for vector inspection of relay protection in the active distribution network. The simulation results show that the stability of the improved back-to-back converter system is greatly improved. The experiment shows that the vector check technology based on an improved back-to-back converter can effectively check the vector of relay protection in an active distribution network and find various installation problems.

Keywords: active distribution network; relay protection; vector inspection; back-to-back converter; stability

Citation: Zhang, W.; Fan, H.; Ren, J.; Wang, X.; Li, T.; Wang, Y. A Vector Inspection Technique for Active Distribution Networks Based on Improved Back-to-Back Converters. *Sustainability* **2023**, *15*, 750. https://doi.org/10.3390/su15010750

Academic Editor: Mohamed A. Mohamed

Received: 25 November 2022
Revised: 23 December 2022
Accepted: 27 December 2022
Published: 31 December 2022

1. Introduction

Vector inspection of relay protection is important to enact before the power grid is put into operation. Vector error will lead to false-action or non-action of relay protection in a normal operation state and fault state. Therefore, relevant power grid regulations stipulate that the correctness of the relay protection vector must be strictly checked comprehensively before newly installed equipment or devices with great changes in the circuit are put into operation [1]. At the same time, the relay protection vector inspection technology, applied before commissioning, reduces the safety risk caused by a system switching operation during the commissioning process and reduces the operation workload [2].

In recent years, vector inspection technology that has first been applied in newly built high-voltage substations and transmission networks is put into operation before relay protection [3,4], but there are some difficulties in its application to distribution networks. Compared with the high-voltage substation and transmission network, the line distance of the distribution network is shorter and the transformer capacity is smaller. As a result, the output voltage and output current of the test power source cannot be adjusted continuously and stably, which cannot satisfy the requirements of relay protection vector inspection of the distribution network [5,6].

Back-to-back converters have the advantages of four-quadrant operation, low current harmonic content, and controllable DC voltage, which attracts more and more scholars' attention and research. In this paper, the control strategy and stability of back-to-back converters are analyzed and studied. Ref. [7] adopts a no-difference beat current control and active power feedback compensation mechanism to improve the harmonic current of the back-to-back converter system and grid-connection reliability. Ref. [8] proposed a coordinated control strategy of two-end converters, which can enhance the stable and reliable operation of the back-to-back MMC-HVDC transmission system. Ref. [9] uses the harmonic state space method to study the stability of back-to-back converters in wind turbines. Impedance matching theory points out that the stability of a cascade system not only depends on the stability of a single converter itself, but also the impedance matching between converters will affect the stability of the system to a large extent [10]. The impedance matching analysis method can be applied to improve and evaluate the stability of AC grid-connected inverters [11–13]. In refs. [14,15], the output impedance of the grid-connected inverter is constructed by virtual impedance correction to improve grid-connected stability. Refs. [16,17] proposed an active full bridge converter (DAB)—inverter cascade topology. The control link is designed according to the DC impedance of the cascade system to improve the impedance characteristics of the system and enhance its stability of the system. Ref. [18] studies the dynamic interaction between photovoltaic grid-connected inverters and power grids based on impedance analysis. None of the above references have been able to analyze the stability of the back-to-back converter from an impedance perspective and propose an improvement plan.

To solve these problems, an improved AC/DC-DC/AC cascade converter is proposed, and the converter is used as the test power source for relay protection vector inspection of the active distribution network in this paper. The main innovations of this paper are as follows:

(1) Adding a feedforward voltage to the inverter end of the converter can effectively improve the input impedance characteristics of the inverter, and thus the system stability of the converter;
(2) The vector inspection technology based on the improved back-to-back converter is applied to the vector inspection of the active distribution network for the first time, which can accurately find various installation problems of relay protection.

This paper is organized as follows: in Section 2, the three-phase back-to-back cascaded converter is established to analyze the impedance characteristics of its DC terminal; in Section 3, the system stability and parameter stability of the improved back-to-back converter are analyzed; in Section 4, the principle and flow of distribution network vector inspection are introduced; in Section 5, the improved converter is used as the test power source for vector inspection of relay protection, and vector inspection of relay protection of the active distribution network is carried out; in Section 6, conclusions are drawn.

2. Impedance Characteristics of the Converter

This section analyzes the system impedance of the three-phase back-to-back converter, the output impedance of the rectifier at the DC terminal, and the input impedance of the inverter. When analyzing the input impedance of the inverter, feedforward voltage is added to the inverter end, which can effectively improve the input impedance characteristics of the inverter.

2.1. Converter System Impedance

Firstly, we set up the back-to-back cascading converter topology as shown in Figure 1. The converter on both terminals is composed of three pairs of switch tubes in a three-phase bridge circuit. e_1 is the grid voltage at the rectifier end, L_1 and r_1 are the filter inductance and parasitic resistance of the rectifier converter, respectively, C is DC filter capacitor, u_{dc} and i_{dc} are DC voltage and current, respectively, L_2 and r_2 are the filter inductance and parasitic resistance of the inverter converter, respectively, and e_2 is the grid-connected voltage of the inverter.

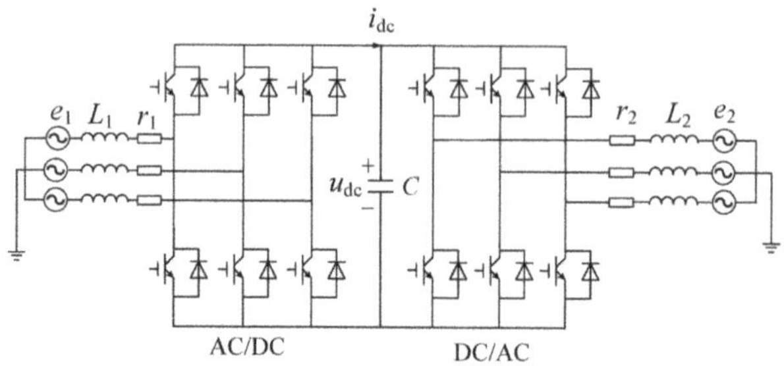

Figure 1. Topology of back-to-back converters.

The block diagram of the impedance principle of a back-to-back converter cascade system is shown in Figure 2. G_A, G_B are transfer functions of rectifier A and inverter B, respectively [19].

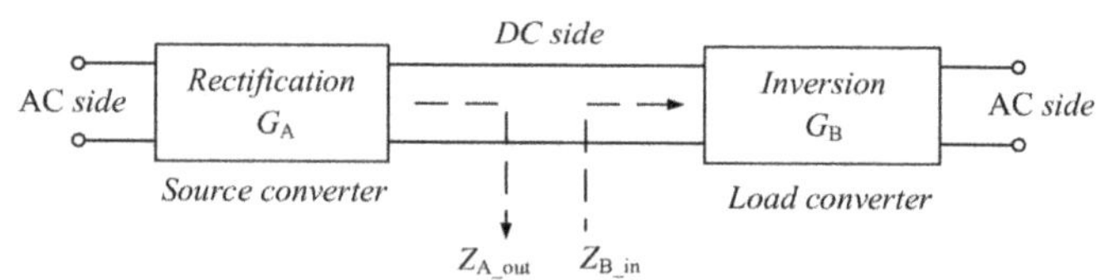

Figure 2. Schematic block diagram of cascade system impedance.

The transfer function G of the converter cascade system is:

$$G = G_A G_B \frac{1}{1+T_m} \quad (1)$$

As can be seen from the transfer function of the cascade system, the system stability is not only dependent on the stability of a single system but also related to the impedance matching between subsystems. According to the principle of automatic control, if the open loop transfer function of a system is stable, the system is stable. Therefore, T_m can be regarded as the open loop transfer function of the cascade system, which is called the minimum loop ratio. Thus, on the premise that all subsystems are stable, the stability of the whole cascade system can be judged by the stability of the T_m. T_m is related to the control mode of the subsystem. If A is the voltage source converter and B is the current source converter, then T_m is the output impedance of rectifier A compared with the input impedance of inverter B [20], as shown in Equation (2):

$$T_m = \frac{Z_{A\_out}}{Z_{B\_in}} \quad (2)$$

where Z_{Aout} represents the output impedance of source converter A, and Z_{Ain} represents the input impedance of load converter B.

If A and B have opposite characteristics—that is, A is the power converter and B is the voltage source converter—then T_m is the input impedance ratio of inverter B to the output impedance of rectifier A [21], as shown in Equation (3):

$$T_m = \frac{Z_{B\_in}}{Z_{A\_out}} \quad (3)$$

Input impedance Z_{in} and output impedance Z_{out} can be obtained by constructing a small signal model of the system at the static operating point, and the calculation method is shown in Equations (4) and (5):

$$Z_{in} = \frac{\hat{u}_{in}}{\hat{i}_{in}} \tag{4}$$

$$Z_{out} = -\frac{\hat{u}_{out}}{\hat{i}_{out}} \tag{5}$$

where $\hat{u}_{in}$ and $\hat{i}_{in}$ are the input voltage and current small signal disturbance, respectively. Similarly, $\hat{u}_{out}$ and $\hat{i}_{out}$ are the input voltage and current small signal disturbance, respectively.

2.2. Rectifier Impedance

The control block diagram of the voltage and current double closed-loop control rectifier is established in the *dq* coordinate system, as shown in Figure 3. In the figure, i_{id} and i_{iq}, respectively, represent axis *d*- and *q*-axis components of grid-connected current at the rectifier terminal. d_{id} and d_{iq}, respectively, represent the *d* and *q* axis components of the rectifier switch duty cycle. e_{id} and e_{iq}, respectively, represent the components of axis *d* and axis *q* of rectifier grid-connected voltage. The symbol "*" represents the given value of the corresponding variable, and the symbol "^" below represents the small signal disturbance of the corresponding variable [22].

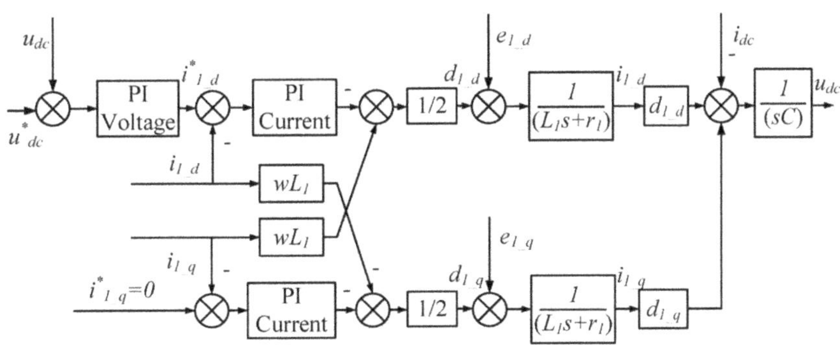

Figure 3. Control block diagram of the rectifier in *dq* coordinate system.

The small signal model of the rectifier is established by the small signal analysis method, as shown in Equation (6):

$$L_1 \frac{d}{dx}\begin{bmatrix} \hat{i}_{1\_d} \\ \hat{i}_{1\_q} \end{bmatrix} = \begin{bmatrix} -r_1 & 0 \\ 0 & -r_1 \end{bmatrix}\begin{bmatrix} \hat{i}_{1\_d} \\ \hat{i}_{1\_q} \end{bmatrix} + \begin{bmatrix} \hat{e}_{1\_d} \\ \hat{e}_{1\_q} \end{bmatrix} - \left(\begin{bmatrix} \hat{d}_{1\_d} \\ \hat{d}_{1\_q} \end{bmatrix} U_{dc} + \begin{bmatrix} D_{1\_d} \\ D_{1\_q} \end{bmatrix} \hat{u}_{dc} \right) \tag{6}$$

where D_{1d} and D_{1q} are, respectively, the *d*- and *q*-axis components of the DC duty ratio of the rectifier system, and U is the steady-state DC voltage.

According to the power balance principle, when the loss in the power transmission process is ignored, the power at the input and output sides of the rectifier should be equal.

$$P = 1.5(e_{1\_d} i_{1\_d} + e_{1\_q} i_{1\_q}) = u_{dc} i_{dc} \tag{7}$$

Assuming that the three-phase input voltage is symmetrical and balanced without disturbance, while $\hat{e}_{1d} = 0$, $\hat{e}_{1q} = 0$, small signal analysis is performed on Equation (7), and the results are shown in Equation (8):

$$1.5E_{1\_d}\hat{i}_{1\_d} = \hat{u}_{dc} I_{dc} + U_{dc}\hat{i}_{dc} \tag{8}$$

where E_{1d} is the voltage of the d-axis of the rectifier system grid in steady state and I_{dc} is the steady-state current.

Let the current transfer function from voltage to axis D be $G_{u_{dc}i_{1\_d}} = \hat{i}_{1\_d}/\hat{u}_{dc}$, then the small signal equation of power balance is shown in Equation (9), and Equation (10) is the output impedance of the DC terminal of the rectifier:

$$1.5E_{1\_d}G_{u_{dc}i_{1\_d}}\hat{u}_{dc} = \hat{u}_{dc}I_{dc} + U_{dc}\hat{i}_{dc} \tag{9}$$

$$Z_{rec\_out} = -\frac{\hat{u}_{dc}}{\hat{i}_{dc}} = \frac{-U_{dc}}{1.5E_{1\_d}G_{u_{dc}i_{1\_d}} - I_{dc}} \tag{10}$$

Since the rectifier in this paper adopts voltage and current double closed-loop control, the axis components of duty cycle disturbance d and q are, respectively:

$$\begin{bmatrix} \hat{i}_{1\_d} \\ \hat{i}_{1\_q} \end{bmatrix} = \frac{1}{2}\begin{bmatrix} H_{1\_i}(-H_u\hat{u}_{dc} - \hat{i}_{1\_d}) \\ -H_{1\_i}\hat{i}_{1\_q} \end{bmatrix} = \begin{bmatrix} -\frac{H_{1\_i}}{2} & 0 \\ 0 & -\frac{H_{1\_i}}{2} \end{bmatrix}\begin{bmatrix} \hat{i}_{1\_d} \\ \hat{i}_{1\_q} \end{bmatrix} + \begin{bmatrix} -\frac{H_{1\_i}H_u}{2} \\ 0 \end{bmatrix}\hat{u}_{dc} \tag{11}$$

where $H_{1\_i}(s)$ is the transfer function of the current loop PI regulator in rectifier control and $H_u(s)$ is the transfer function of the voltage loop PI regulator.

Substitute Equation (11) into Equation (10), then the rectifier model is as follows.

$$M_1\begin{bmatrix} \hat{i}_{1\_d} \\ \hat{i}_{1\_q} \end{bmatrix} = \begin{bmatrix} \hat{e}_{1\_d} \\ \hat{e}_{1\_q} \end{bmatrix} + \begin{bmatrix} -D_{1\_d} + H_{1\_i}H_u/2 \\ -D_{1\_q} \end{bmatrix}\hat{u}_{dc} \tag{12}$$

where $M_1 = \begin{bmatrix} L_1s + r_1 - H_{1\_i}/2 & 0 \\ 0 & L_1s + r_1 - H_{1\_i}/2 \end{bmatrix}$. Then, the transfer function of DC voltage to d-axis current is shown in Equation (13):

$$G_{u_{dc}i_{1\_d}} = \frac{\hat{i}_{1\_d}}{\hat{u}_{dc}} = \frac{-D_{1\_d} + H_{1\_i}H_u/2}{L_1s + r_1 - H_{1\_i}/2} \tag{13}$$

By substituting Equation (13) into Equation (10), the expression of output impedance Z_{rec_out} at the DC terminal of the rectifier can be obtained.

2.3. Inverter Impedance

We establish the control block diagram of the inverter in the converter in the dq coordinate system, as shown in Figure 4.

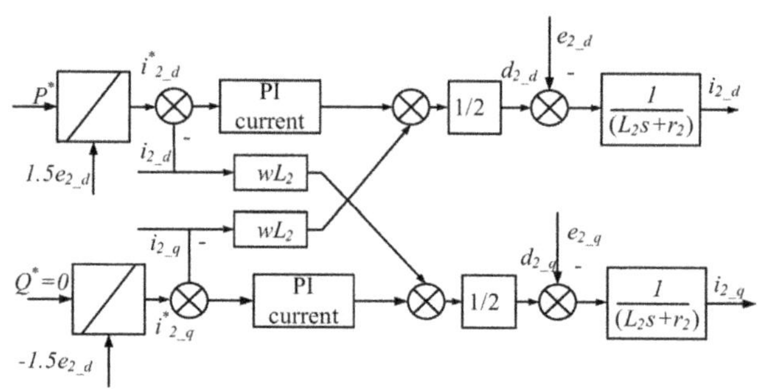

Figure 4. Control block diagram of an inverter in dq coordinate system.

In this figure, P and Q represent active and reactive power respectively, e_{2_d} and e_{2_q}, respectively, represent axis d and axis q components of the grid-connected voltage of the inverter system, i_{2_d} and i_{2_q}, respectively, represent d- and q-axis components of inductance current, and d_{2_d} and d_{2_q}, respectively, represent the *d*- and *q*-axis components of the inverter switch duty cycle [23].

Similar to the derivation of the rectifier model, the small signal model of the inverter is shown in Equation (14):

$$L_2\frac{d}{dx}\begin{bmatrix}\hat{i}_{2\_d}\\ \hat{i}_{2\_q}\end{bmatrix}=\begin{bmatrix}-r_2 & 0\\ 0 & -r_2\end{bmatrix}\begin{bmatrix}\hat{i}_{2\_d}\\ \hat{i}_{2\_q}\end{bmatrix}+\begin{bmatrix}\hat{e}_{2\_d}\\ \hat{e}_{2\_q}\end{bmatrix}-\left(\begin{bmatrix}\hat{d}_{2\_d}\\ \hat{d}_{2\_q}\end{bmatrix}U_{dc}+\begin{bmatrix}D_{2\_d}\\ D_{2\_q}\end{bmatrix}\hat{u}_{dc}\right) \tag{14}$$

where D_{2_d}, D_{2_q} are the DC duty cycle *d*- and *q*-axis components of the inverter system. Similar to rectifier analysis, the loss in the process of power transmission is ignored, and the impedance of the DC terminal is obtained according to the power balance equation, as shown in Equation (15):

$$Z_{inv\_in}=\frac{\hat{u}_{dc}}{\hat{i}_{dc}}=\frac{U_{dc}}{1.5E_{2\_d}G_{u_{dc}i_{1\_d}}-I_{dc}} \tag{15}$$

where E_{2_d} is the grid *d*-axis voltage of the inverter system at a steady state.

The inverter adopts constant power control, so the duty cycle disturbance is shown in Equation (16):

$$\begin{bmatrix}\hat{d}_{2\_d}\\ \hat{d}_{2\_q}\end{bmatrix}=\begin{bmatrix}-H_{2\_i}\hat{i}_{2\_d}\\ -H_{2\_i}\hat{i}_{2\_q}\end{bmatrix}=-\begin{bmatrix}-H_{2\_i}/2 & 0\\ 0 & -H_{2\_i}/2\end{bmatrix}\begin{bmatrix}\hat{i}_{2\_d}\\ \hat{i}_{2\_q}\end{bmatrix} \tag{16}$$

where $H_{2\_i}(s)$ is the transfer function of the *PI* regulator of the current loop in the control loop of the inverter system. Substitute Equation (16) into Equation (14), then the small signal model of the inverter is shown in Equation (17):

$$M_2\begin{bmatrix}\hat{i}_{2\_d}\\ \hat{i}_{2\_q}\end{bmatrix}=-\begin{bmatrix}\hat{e}_{2\_d}\\ \hat{e}_{2\_q}\end{bmatrix}+\begin{bmatrix}D_{2\_d}\\ D_{2\_q}\end{bmatrix}\hat{u}_{dc} \tag{17}$$

where $M_2=\begin{bmatrix}L_2s+r_2-H_{2\_i}/2 & 0\\ 0 & L_2s+r_2-H_{2\_i}/2\end{bmatrix}$. Then, the current transfer function $G_{u_{dc}i_{2\_d}}$ of the inverter DC terminal voltage to axis D is shown in Equation (18):

$$G_{u_{dc}i_{2\_d}}=\frac{\hat{i}_{2\_d}}{\hat{u}_{dc}}=\frac{D_{2\_d}}{L_2s+r_2+H_{2\_i}/2} \tag{18}$$

After substituting Equation (18) into Equation (15), the input impedance Z_{inv_in} expression of the DC inverter can be obtained.

2.4. Improved Inverter Impedance

The DC terminal voltage error is used as a given compensation for the inverter *d*-axis power loop, and the improved inverter transfer function $G_{u_{dc}i_d}$ block diagram is shown in Figure 5. The transfer function of inverter DC terminal voltage to *d*-axis current can be directly derived from Figure 5, as shown in Equation (19):

$$G'_{u_{dc}i_d}=\frac{\hat{u}_{dc}}{\hat{i}_{dc}}=\frac{D_d+kH_i/2}{L_2s+r_2+H_{2\_i}/2} \tag{19}$$

where k is the voltage feedforward coefficient and $H_i(s)$ is the transfer function of the current loop *PI* regulator. Substitute Equation (19) into Equation (18) to obtain the improved input impedance of the DC inverter Z'_{inv_in}:

$$Z'_{inv\_in}=\frac{\hat{u}_{dc}}{\hat{i}_{dc}}=\frac{U_{dc}}{1.5E_dG'_{u_{dc}i_d}-I_{dc}} \tag{20}$$

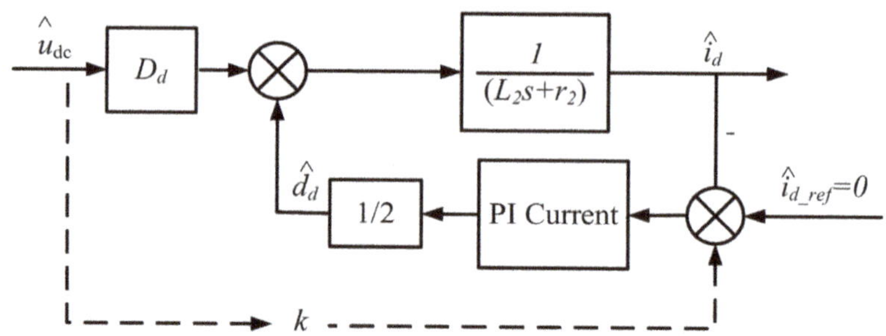

Figure 5. Improved inverter transfer function block diagram.

3. Stability Analysis

The stability of the improved back-to-back converter directly determines the feasibility of its application in relay protection vector inspection of the distribution network. Therefore, this section analyzes the stability of the improved back-to-back converter from two aspects of subsystem and parameter influence.

3.1. Subsystem Stability Analysis

From the above analysis, the mathematical models of the output impedance Z_{rec_out} of the rectifier and the input impedance Z_{inv_in} of the inverter before and after optimization have been obtained. Because the rectifier acts as a voltage source converter, the minimum loop ratio of the cascade system is the ratio of the rectifier's output impedance to the inverter's input impedance. Under the condition of a given power of 5 kW, the output impedance of the rectifier at the DC terminal, the input impedance of the inverter, and the porter diagram of the optimized input impedance of the inverter are shown in Figure 6.

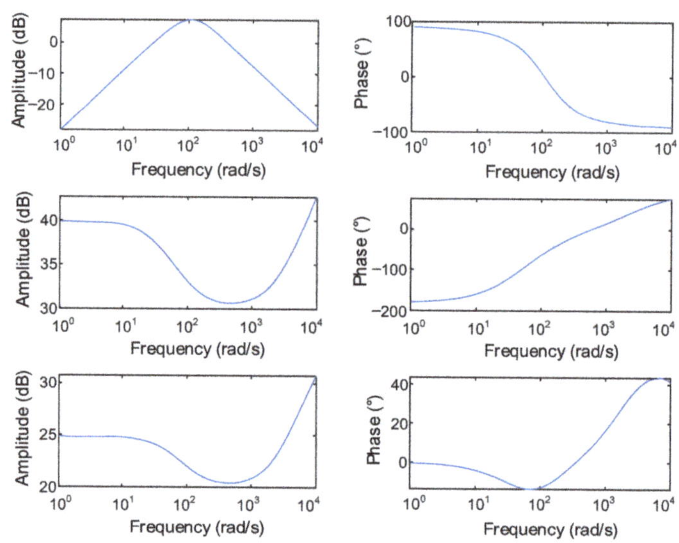

Figure 6. Bode diagram of input/output impedance of converter.

Figure 6 shows that the output impedance amplitude of the DC rectifier is smaller, and the phase ranges from 90° in the low-frequency band to −90° in the high-frequency band. Compared with the output impedance, the input impedance amplitude of the DC inverter is larger. Its low-frequency phase is −180° showing negative impedance, which is also an important reason for the instability of the cascade system. After the voltage feedforward is introduced into the inverter, the impedance amplitude is reduced and the low-frequency phase is 0°, which shows resistance. Theoretically, it is beneficial to system stability.

3.2. Influence of Parameters on System Stability

First, three parameters K_1, K_2, and K_3 are introduced, which represent the common gain coefficients of the rectifier current loop PI controller, the voltage loop PI controller, and the inverter power loop PI controller, respectively. Equation (21) is the transfer function of the *PI* controller after introducing the common gain coefficient of the PI parameter:

$$H(s) = K_i(k_p + \frac{k_i}{s}) \tag{21}$$

where k_p, k_i are the ratio and integral coefficient.

When the common gain coefficient K_1 of the rectifier current loop parameter changes, the bode diagram of the output impedance of the DC terminal is shown in Figure 7a. It can be seen from the figure that when the coefficient K_1 changes from 0.1 to 1.5, the amplitude and phase of the output impedance do not change. That is to say, the current loop parameters of the rectifier under the voltage and current double closed-loop control have little effect on the output impedance characteristics, and they are not the dominant factor affecting the instability of the system.

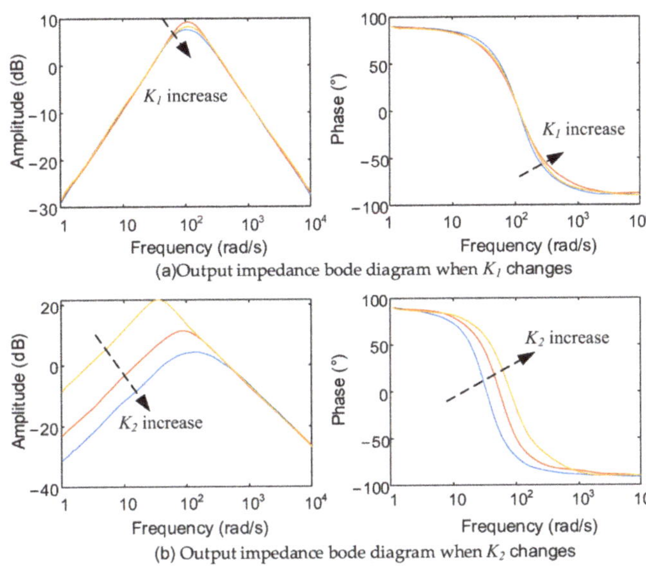

Figure 7. Output impedance bode diagram when coefficients K_1 and K_2 change. (**a**) Output impedance bode with K_1 changes. (**b**) Output impedance bode with K_2 changes.

When the common gain coefficient K_2 of the voltage loop *PI* parameter of the rectifier changes, the bode diagram of the output impedance of the DC terminal is shown in Figure 7b. In general, the increase in the voltage loop PI parameter will lead to the increase in high-frequency resonant amplitude in the system, which is not conducive to system stability. In this paper, the influence of impedance on system stability is studied under the premise of system stability. It can be seen from Figure 7 that when the coefficient K_2 increases from 0.1 to 1.5, the impedance amplitude of low-frequency rectifier decreases, and the capacitive impedance characteristic of high-frequency rectifier weakens. The increase in coefficient leads to the reduction of impedance amplitude, which is beneficial to the cascade system impedance matching. Therefore, the stability of back-to-back converters will be positively affected.

When the feedforward voltage is introduced into the power loop of the inverter, only the input impedance of the inverter is affected, and it has nothing to do with the rectifier side. Therefore, the change in the input impedance of the optimized inverter is analyzed. Figure 8 shows the input impedance bode diagram before and after optimization when the coefficient K_3 changes.

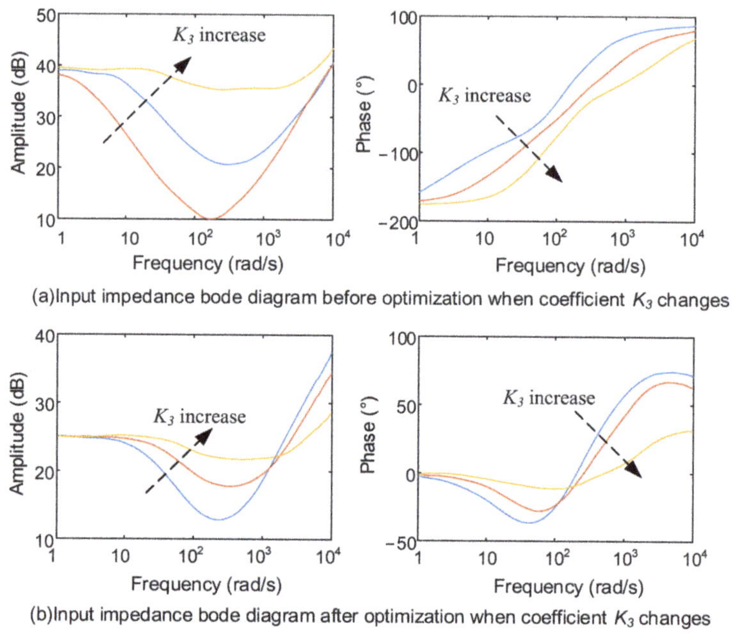

Figure 8. Input impedance bode diagram before and after optimization when coefficient K_3 changes. (**a**) Input impedance bode diagram before optimization with K_3 changes. (**b**) Input impedance bode diagram after optimization with K_3 changes.

When the common gain coefficient K_3 of the inverter power loop *PI* parameter changes, the bode diagram of the input impedance of the DC terminal is shown in Figure 8a. When the coefficient K_3 changes from 0.1 to 1.5, the high-frequency input impedance amplitude increases, and the negative impedance characteristic of low-frequency inverter increases. Figure 8b is the bode diagram of the optimized input impedance at the DC terminal when the coefficient K_3 changes. By comparing Figure 8a,b, it is obvious that the amplitude of the input impedance after optimization decreases. In addition, the low frequency band changes from the negative impedance characteristic which endangers the stability of the system to the pure resistive characteristic which is beneficial to the stability of the system. It can be seen that the input impedance of inverter with feedforward voltage is beneficial to the stability of cascade system.

Figure 9 shows the bode diagram of the input impedance of the DC inverter before optimization when the power changes. When the power changes from 1 kW to 6 kW, the amplitude of the input impedance of the low-frequency inverter decreases, and the negative impedance characteristic increases. The decrease in the input impedance amplitude can easily cause instability in the cascade system, so the stability of the system can be improved by increasing the amplitude of the input impedance.

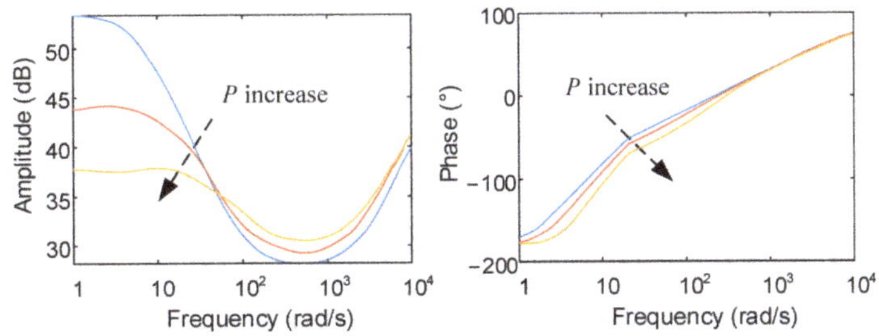

Figure 9. Input impedance bode diagram before optimization when power changes.

4. Principle of Vector Inspection

Vector inspection of relay protection includes secondary voltage value, secondary current value, the angle between secondary voltage, the angle between secondary current, and the correspondence between angle value and amplitude, etc. Vector inspection mainly checks the current transformer ratio range, polarity selection, voltage, and current secondary circuit wiring correctness. The test device for relay protection vector inspection of the distribution network is shown in Figure 10, which includes the current converter, test device, simulated load, and protection device.

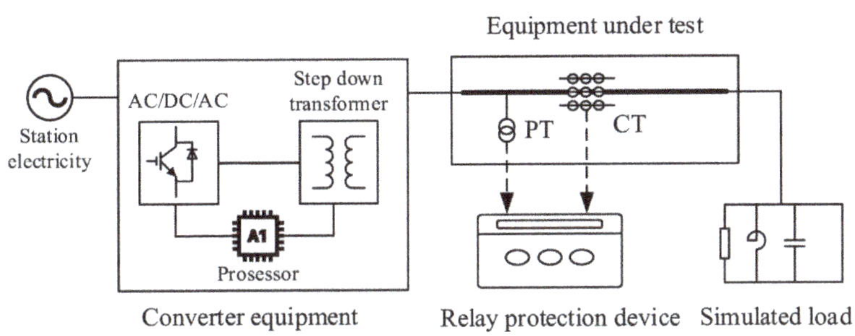

Figure 10. Vector inspection test device structure.

The input end of the current transformer is connected to 380 V mains power, and the output end is connected to the reception test equipment. The output end of the equipment to be tested is connected to the analog load, and the protection device is connected to the equipment to be tested through the voltage transformer and current transformer.

The current converter includes *AC/DC-DC/AC* cascade converter module, step-down transformer module, and processor module. The converter module can output continuous, stable, and adjustable current and voltage. Step-down transformers are used for voltage regulation. The processor controls and adjusts the converter module according to the output voltage and current of the converter. The whole converter equipment is required to be able to output the required load current stably and quickly respond to the required pulse current. Its nominal design is 50 kVA, 100 A/150 V phase voltage.

The analog load includes parallel test inductance, inductive analog load, capacitive analog load, and analog grounding line, which can form a variety of loops with the equipment to be tested. The test inductance adopts a fixed value of 0.5 Ω. The inductive analog load is for star connection access, which has an adjustable range of 0–1.0 Ω. The capacitive analog load is for angular connection access, which has an adjustable range of 50 μF 200 μF. The dynamic stability of the analog ground cable must be at least 150 A/min.

The impedance of the test system load is the sum of the impedance of the device under test and the impedance of the simulated load. During the test, we adjust the simulated load to compensate for the test equipment and adjust the system load to the inductive or capacitive load of 1 Ω.

The relay protection vector inspection process of the whole distribution network is as follows:

- According to the type of relay protection, select different locations to connect to the converter equipment and simulated loads;
- Adjust the output voltage and current of the converter equipment and calculate the impedance of the test system load;
- Subtract the simulated load impedance from the system load impedance to obtain an estimated impedance value of the device under test;
- According to the estimated impedance value of the device under test and the test type, adjust the converter device and the simulated load to compensate the system load impedance to the preset value;
- Connect the protection device to the test system and start the test. Check the correctness of the vector according to the secondary voltage and secondary current collected by the protection device.

5. Simulation and Experimental Verification

In this section, the stability of the improved converter is verified by simulation. Then, the improved converter is used as the test power source to verify the relay protection vector inspection technology in the distribution network.

5.1. Converter Stability Verification

For the above theoretical analysis, this paper built a three-phase back-to-back converter cascade system on the MATLAB/SIMULINK simulation platform. Table 1 shows the main circuit parameters in the simulation, and Table 2 shows the *PI* parameters of each control link. In accordance with the theoretical analysis, the rectifier adopts the DC voltage and filter inductor current double closed-loop control strategy, while the inverter adopts the constant power control mode. Figures 11–13, respectively, show the changes in DC terminal voltage, rectifier, and inverter power of the cascade system before and after optimization when the coefficients K_1-K_3 change. Figure 14 shows the changes in DC terminal voltage, rectifier, and inverter power when output power changes.

Table 1. Main circuit parameters of the back-to-back converter cascade system.

Parameter	Value
The effective value of grid voltage(e)/V	220
Rectifier side inductance (L1)/mH	2
Rectifier side inductance parasitic resistance (r1)/Ω	0.16
Capacitance (C)/mF	2.35
Switching frequency (fs)/k Hz	10
Inverter inductance (L2)/mH	3
Inverter inductance parasitic resistance (r2)/Ω	0.16
Specifies the DC terminal voltage (UDC)/V	700

Table 2. Controller parameters.

PI Control Loop	Scale Parameter	Integral Parameter
Rectifier current loop	50.87	103.49
Rectifier voltage loop	0.58	42.2
Inverter current loop	15	1520

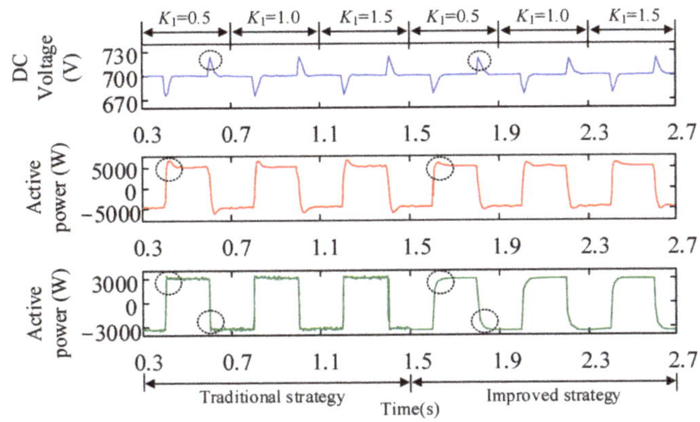

Figure 11. Waveform of DC terminal voltage, rectifier and inverter power output when coefficient K_1 changes.

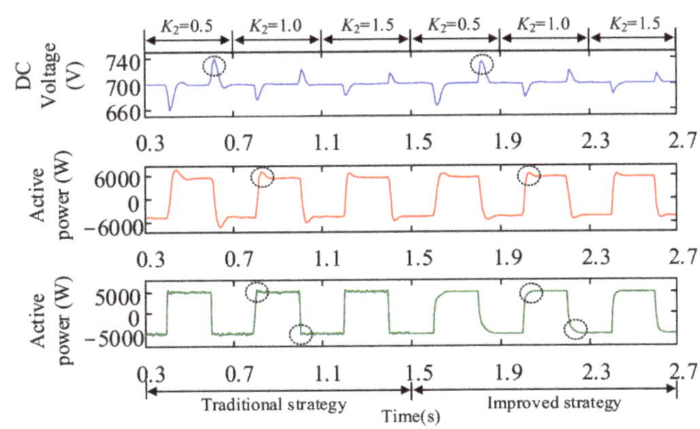

Figure 12. Waveform of DC terminal voltage, rectifier, and inverter power output when coefficient K_2 changes.

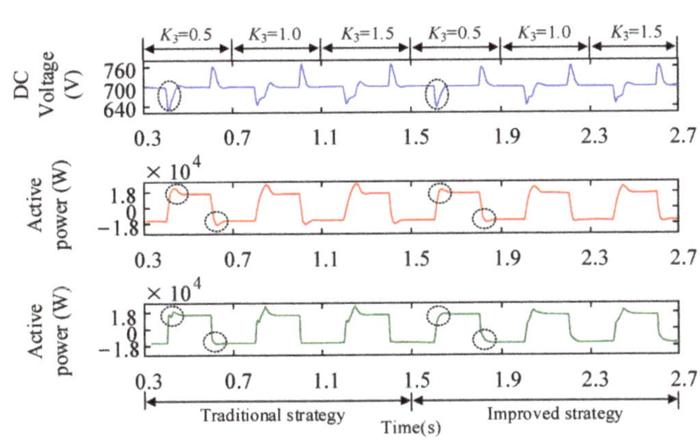

Figure 13. Waveform of DC terminal voltage, rectifier and inverter power output when coefficient K_3 changes.

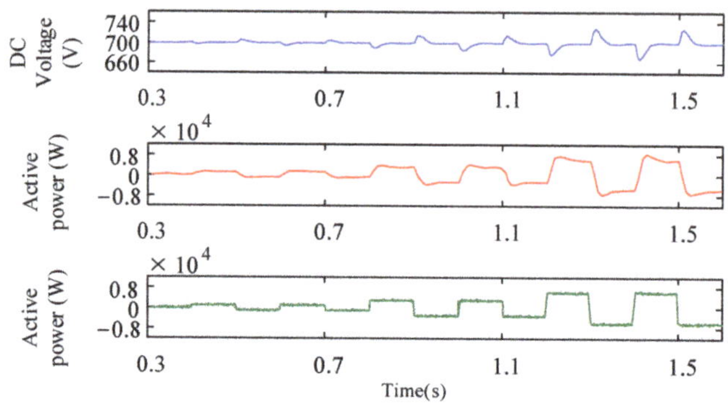

Figure 14. Waveform of DC terminal voltage, rectifier, and inverter power output when power changes.

Figure 11 shows the waveforms of the DC terminal voltage, rectifier, and inverter power output optimized when coefficient K_1 is 0.5, 1.0, and 1.5, respectively. In each case, when the coefficient K_1 changes, the system voltage and power are not greatly disturbed. Comparing the two cases, it is obvious that the peak value of the output voltage of the optimized system is smaller. The power transition process is relatively gentle, which has a favorable effect on the system.

Figure 12 shows the waveforms of DC terminal voltage, rectifier, and inverter power output when coefficient K_2 is 0.5, 1.0, and 1.5, respectively. As K_2 increases, the peak value of DC terminal voltage gradually weakens, and the short-term impact on the system gradually decreases. At the same time, the inverter power output waveform gradually flattens with the increase in the coefficient. According to the above analysis, the increase in the K_2 coefficient reduces the output impedance of the rectifier, which is conducive to system stability. The input impedance amplitude of the DC inverter decreases after the feedforward voltage is introduced. Regulating voltage is no longer only the responsibility of the rectifier. At this time, the inverter is also sharing part of the work of regulating voltage, and the system's adjustability has also been improved.

Figure 13 shows the waveforms of DC terminal voltage, rectifier, and inverter power output when coefficient K_3 is 0.1, 0.5, and 0.9, respectively. With the increase in the coefficient, DC terminal voltage distortion and voltage peak value gradually increase. At the same time, the harmonic wave of the rectifier and inverter output power becomes larger under forward power. The increase in inverter power loop parameters enhances the negative impedance characteristics of the inverter, which is not conducive to the stability of the cascade system. Compared with the waveform before and after optimization, the introduction of feedforward voltage effectively reduces the voltage harmonics of the DC terminal, and the output power distortion rate of the rectifier and inverter decreases.

Figure 14 shows the DC terminal voltage, rectifier, and inverter power output of the improved converter when output power P is 1 kW, 3 kW, and 7 kW, respectively. As the power increases, the voltage spike generated by the DC terminal voltage at the power mutation increases. The power variation of the rectifier and inverter gradually becomes uneven, and the stability of the converter system becomes worse.

5.2. Distribution Network Vector Inspection Application

First, the analog load is adjusted so that the end of the device to be tested is connected to the three-phase test inductance, the impedance of which is X_S. Adjust the output voltage and current of the converter. The test device automatically reads current, voltage, and active power values, then the impedance of the test system load is calculated as X_L. The system load impedance minus the analog load impedance, the estimated value of the impedance to be tested equipment $X_d = X_L - X_s$. Under different test devices, the estimated impedance values of the devices to be tested are shown in Table 3.

Table 3. Main circuit parameters of the back-to-back converter cascade system.

Type	System Load Impedance X_L	Analog Load Impedance X_s	Estimated Impedance X_d
Bus	0.5037 Ω	0.5 Ω	0.0037 Ω
Line	0.7524 Ω	0.5 Ω	0.2524 Ω
Distribution converter	18.5495 Ω	0.5 Ω	18.0495 Ω

Secondly, according to the estimated impedance value of the equipment to be tested, the simulation load is adjusted to compensate for the system load impedance to 1 Ω. When adjusting the current converter device for vector inspection, the current output by the converter device should meet certain requirements. Its size is related to the primary current of the device to be tested, as shown in Table 4.

Table 4. Output current requirements of converter equipment.

Primary Current of the Device	Current Converter Output Current
Rated current ≥ 500 A	≥10 A
Rated current ≥ 1000 A	≥20 A
Rated current ≥ 2000 A	≥40 A
Fault current	≤100 A

5.2.1. Bus and Feeder Vector Inspection

The wiring diagram of the vector inspection of the 10 kV bus and feeder is shown in Figure 15. The power source of the test system comes from the station power, and the currency converter is connected to the equipment under test from the incoming line of the bus I. The simulated load is connected to the device under test from feeder F12. The protection device is connected to the device to be tested through the voltage transformer and current transformer. We adjust the system load to inductive 1 Ω, and then adjust the output current and output voltage of the converter equipment. The relay protections that can be tested are bus low voltage overcurrent protection, feeder F12 overcurrent protection, and quick break protection. The inspection results are shown in Table 5.

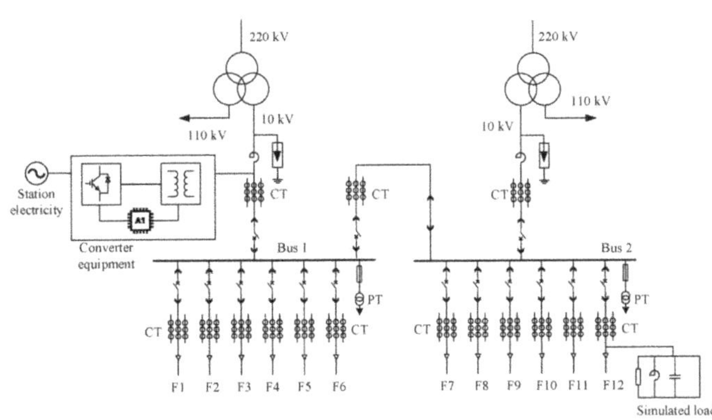

Figure 15. Relay protection vector inspection wiring diagram of feeder and bus.

Table 5. Busbar protection test data.

Transformer	U Phase	V Phase	W Phase
PT of bus I	1.274V∠0°	1.271V∠239°	1.281V∠118°
Incoming CT of bus I	0.082A∠270°	0.082A∠150°	0.082A∠31°
Bus coupler CT	0.082A∠270°	0.082A∠150°	0.082A∠31°

It can be seen from Tables 5 and 6 that the secondary voltage and secondary current of bus protection and feeder F12 protection are both greater than 0.02 A and 0.2 V, and the secondary current lags almost 90° behind the secondary voltage. Therefore, it can be judged that the vectors of bus protection and feeder F12 protection are correct. Similarly, by changing the position of the converter device and the simulated load, vector inspections can be carried out on other protections of the bus and feeder.

Table 6. Feeder protection test data.

Transformer	U Phase	V Phase	W Phase
PT for bus II	0.358A∠5°	0.354A∠242°	0.359A∠124°
CT of feeder 12	0.082A∠270°	0.082A∠150°	0.082A∠31°

5.2.2. Distribution Vector Inspection

The wiring diagram for vector inspection of the distribution transformer is shown in Figure 16. The power source of the test system is from the station electricity, and the currency converter is connected to the equipment to be tested from the 1# incoming line of the distribution transformer box, and the analog load is connected to the equipment to be tested from the 1# outgoing line of the distribution transformer. Vector inspection can be carried out for overcurrent protection and quick break protection of the 1# distribution transformer. The current converter is connected to the equipment to be tested from the 2# incoming line of the distribution transformer box, and the analog load is connected to the equipment to be tested from the 2# outgoing line of the distribution transformer. Vector inspection can be carried out for overcurrent protection and quick break protection of the 2# distribution transformer. The inspection results are shown in Table 7.

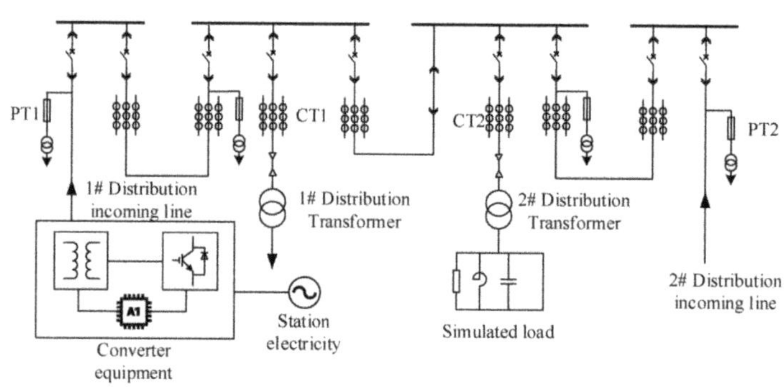

Figure 16. Vector inspection wiring diagram of relay protection of distribution transformer.

Table 7. Test data of transformer protection.

Transformer	U Phase	V Phase	W Phase
1# Incoming cabinet PT1	1.238V∠2°	1.236V∠237°	1.242V∠121°
1# Distribution converter CT1	0.075A∠272°	0.071A∠153°	0.078A∠29°
2# Incoming cabinet PT2	1.242V∠3°	1.248V∠247°	1.261V∠119°
2# Distribution converter CT2	0.062A∠264°	0.064A∠159°	0.068A∠27°

As can be seen from Table 7, the secondary voltage and secondary current of distribution transformer overcurrent protection and quick break protection are both greater than 0.02 and 0.2 V. Moreover, the secondary current lags behind the secondary voltage by almost 90°, so the vector of distribution protection is judged to be correct. Similarly, vector inspection can be carried out on other measurement and control devices of the distribution transformer box by changing the position of the current converter and the simulated load. It is not discussed here.

5.2.3. Installation Problems Were Found in Application

The test device constructed in this paper was used to carry out vector inspection on relay protection of a 10 kV distribution network in a certain area, and several installation problems were found, as shown in Table 8. This shows that the vector inspection technology based on improved back-to-back converters can accurately carry out vector inspection of relay protection in the active distribution network and find various installation problems.

Table 8. Installation problems were found in the application.

Installation Problem	Annotation
The cable connection of the digital input optical fiber of the voltage combination unit is incorrect	In the process of a line protection vector inspection test, an error in voltage phase sequence was found, which was considered a phase sequence error of line interval combination unit voltage input optical fiber. It belonged to misconnection.
The remote module of the voltage transformer is incorrectly connected	During the protection vector inspection test of a certain line, the protection device shows no voltage of a certain phase, and the quality problem of the remote module is determined by the inspection
The polarity of the busbar current is incorrectly configured	In the process of vector inspection of bus protection, the polarity of the current is wrong. After checking, the polarity of the bus current is wrong, which belongs to the fixed value configuration error

6. Conclusions

This paper presents an improved back-to-back converter which is used as a test power source for relay protection vector inspection. The simulation results show that adding feed-forward voltage to the inverter can effectively improve the input impedance characteristics of the inverter, thus improving the system stability of the converter. At the same time, the vector inspection technology based on the improved back-to-back converter can meet the requirements of the distribution network and find various installation problems of the relay protection. Due to the limited laboratory conditions, this paper only uses simulation to verify the stability of the improved converter. In future work, experiments will be used to further verify the stability of the improved converter.

Author Contributions: Conceptualization, W.Z., J.R., X.W., T.L., Y.W. and H.F.; software, H.F., T.L.; validation, W.Z. and Y.W.; formal analysis, W.Z., J.R., X.W. and Y.W.; investigation, H.F.; data curation, X.W., T.L. and Y.W.; writing—original draft preparation, H.F., X.W., T.L. and Y.W.; writing—review and editing, H.F., J.R., T.L. and Y.W.; visualization, J.R.; supervision, Y.W.; All authors have read and agreed to the published version of the manuscript.

Funding: This research was funded by Science and Technology Project of State Grid Hebei Electric Power Co., Ltd. (kj2022-001).

Institutional Review Board Statement: Not applicable.

Informed Consent Statement: Not applicable.

Data Availability Statement: Not applicable.

Conflicts of Interest: The authors declare no conflict of interest.

References

1. Li, T. Research on vector inspection technology of relay protection before operation. *Hebei Electr. Power Technol.* **2013**, *32*, 2.
2. Zhang, B.; Hao, X.; Cao, S.; Li, T.; Zhao, C. Research on vector inspection technology of relay protection before operation using capacitor compensation. *Hebei Electr. Power Technol.* **2013**, *32*, 1–3.
3. Wang, Q.; Jin, T.; Mohamed, M.A. A Fast and Robust Fault Section Location Method for Power Distribution Systems Considering Multisource Information. *IEEE Syst. J.* **2021**, *16*, 1954–1964. [CrossRef]
4. Rao, C.; Hajjiah, A.; El-Meligy, M.A.; Sharaf, M.; Soliman, A.T.; Mohamed, M.A. A novel high-gain soft-switching DC-DC converter with improved P&O MPPT for photovoltaic applications. *IEEE Access* **2021**, *9*, 58790–58806.
5. Ma, H.; Liu, Z.; Li, M.; Wang, B.; Si, Y.; Yang, Y.; Mohamed, M.A. A two-stage optimal scheduling method for active distribution networks considering uncertainty risk. *Energy Rep.* **2021**, *7*, 4633–4641. [CrossRef]
6. Wang, Q.; Jin, T.; Mohamed, M.A. An innovative minimum hitting set algorithm for model-based fault diagnosis in power distribution network. *IEEE Access* **2019**, *7*, 30683–30692. [CrossRef]
7. Zhou, S. *Research on Harmonic Elimination and Reliability of Three-Level back to back Converter*; Anhui university: Hefei, China, 2017.
8. Yang, Y.; Yang, J.; He, Z.; Li, Q.; Xu, W. Control strategy of MMC based back-to-back HVDC transmission system. *Autom. Electr. Power Syst.* **2017**, *41*, 120–124.
9. Kwon, J.; Wang, X.; Bak, C.L.; Blaabjerg, F. Analysis of harmonic coupling and stability in back-to-back converter systems for wind turbines using Harmonic State Space (HSS). In Proceedings of the Energy Conversion Congress and Exposition, Montreal, QC, Canada, 20–24 September 2015; IEEE: Manhattan, NY, USA, 2015; pp. 730–737.
10. Gao, J.; Zhao, J.; Qu, K.; Li, F. Reconstruction of impedance-based stability criteria in weak grid. *Power Syst. Technol.* **2017**, *41*, 2762–2768.
11. Cespedes, M.; Sun, J. Impedance modeling and analysis of grid-connected voltage-source converters. *IEEE Trans. Power Electron.* **2013**, *29*, 1254–1261. [CrossRef]
12. Xu, J.; Zhang, B.; Qian, Q.; Meng, X.; Xie, S. Robust control and design based on impedance-based stability criterion for improving stability and harmonics rejection of inverters in weak grid. In Proceedings of the Applied Power Electronics Conference and Exposition, Tampa, FL, USA, 26–30 March 2017; IEEE: Manhattan, NY, USA, 2017; pp. 3619–3624.
13. Lissandron, S.; Dalla Santa, L.; Mattavelli, P.; Wen, B. Experimental validation for impedance-based small-signal stability analysis of single-phase interconnected power systems with grid-feeding inverters. *IEEE J. Emerg. Sel. Top. Power Electron.* **2016**, *4*, 103–115. [CrossRef]
14. Yang, D.; Ruan, X.; Wu, H. Virtual impedance method to improve LCL grid-inverter's adaptability in weak power grid. *Proc. CSEE* **2014**, *34*, 2327–2335.
15. Wei, S.; Tian, Y.; Wang, Y. Virtual Inductance Control for Enhancing Stability of LCL Inverter in Weak Grid. *High Volt. Eng.* **2019**, *45*, 1827–1834.
16. Tian, Y.; Loh, P.C.; Deng, F.; Chen, Z.; Sun, X.; Hu, Y. Impedance coordinative control for cascaded converter in bidirectional application. *IEEE Trans. Ind. Appl.* **2015**, *52*, 4084–4095. [CrossRef]
17. Tian, Y.; Loh, P.C.; Deng, F.; Chen, Z.; Hu, Y. DC-Link voltage coordinated proportional control for cascaded converter with zero steady-state error and reduced system type. *IEEE Trans. Power Electron.* **2015**, *31*, 3177–3188. [CrossRef]
18. Chen, X.; Zhang, Y.; Wang, Y. Research on the dynamic interaction of photovoltaic grid-connected inverter and power grid based on impedance analysis. *Proc. CSEE* **2014**, *34*, 459–4567.
19. Raza, M.; Prieto-araujo, E.; Gomis-bellmunt, O. Small signal stability analysis of offshore AC network having multiple VSC-HVDC system. *IEEE Trans. Power Deliv.* **2018**, *33*, 830–839. [CrossRef]
20. Zhang, X.; Ruan, X.; Zhong, Q.C. Improving the stability of cascaded DC/DC converter systems via shaping the input impedance of the load converter with a parallel or series virtual impedance. *IEEE Trans. Ind. Electron.* **2015**, *62*, 7499–7512. [CrossRef]
21. Aapro, A.; Messo, T.; Roinila, T.; Suntio, T. Effect of active damping on output impedance of three-phase grid-connected converter. *IEEE Trans. Ind. Electron.* **2017**, *64*, 7532–7541. [CrossRef]
22. Alenius, H.; Roinila, T. Impedance-based stability analysis of paralleled grid-connected rectifiers: Experimental case study in a data center. *Energies* **2020**, *13*, 2109. [CrossRef]
23. Yadav, A.; Chandra, S.; Bajaj, M.; Sharma, N.K.; Ahmed, E.M.; Kamel, S. A Topological Advancement Review of Magnetically Coupled Impedance Source Network Configurations. *Sustainability* **2022**, *14*, 3123. [CrossRef]

sustainability

MDPI

Article

Experimental Investigation of an Adaptive Fuzzy-Neural Fast Terminal Synergetic Controller for Buck DC/DC Converters

Badreddine Babes [1], Noureddine Hamouda [1], Fahad Albalawi [2], Oualid Aissa [3], Sherif S. M. Ghoneim [2,*] and Saad A. Mohamed Abdelwahab [4,5]

1 Research Center in Industrial Technologies (CRTI), P.O. Box 64, Cheraga 16014, Algeria; elect_babes@yahoo.fr (B.B.); n.hamouda@crti.dz (N.H.)
2 Department of Electrical Engineering, College of Engineering, Taif University, P.O. Box 11099, Taif 21944, Saudi Arabia; f.albiloi@tu.edu.sa
3 LPMRN Laboratory, Faculty of Sciences and Technology, University of Bordj Bou Arreridj, El Anseur 34000, Algeria; oualid.aissa@univ-bba.dz
4 Electrical Department, Faculty of Technology and Education, Suez University, Suez 43533, Egypt; saad.abdelwahab@suezuniv.edu.eg
5 Department of Computers & Systems Engineering, High Institute of Electronic Engineering, Ministry of Higher Education, Bilbis-Sharqiya 44621, Egypt
* Correspondence: s.ghoneim@tu.edu.sa

Abstract: This study proposes a way of designing a reliable voltage controller for buck DC/DC converter in which the terminal attractor approach is combined with an enhanced reaching law-based Fast Terminal Synergetic Controller (FTSC). The proposed scheme will overcome the chattering phenomena constraint of existing Sliding Mode Controllers (SMCs) and the issue related to the indefinite time convergence of traditional Synergetic Controllers (SCs). In this approach, the FTSC algorithm will ensure the proper tracking of the voltage while the enhanced reaching law will guarantee finite-time convergence. A Fuzzy Neural Network (FNN) structure is exploited here to approximate the unknown converter nonlinear dynamics due to changes in the input voltage and loads. The Fuzzy Neural Network (FNN) weights are adjusted according to the adaptive law in real-time to respond to changes in system uncertainties, enhancing the increasing the system's robustness. The applicability of the proposed controller, i.e., the Adaptive Fuzzy-Neural Fast Terminal Synergetic Controller (AFN-FTSC), is evaluated through comprehensive analyses in real-time platforms, along with rigorous comparative studies with an existing FTSC. A dSPACE ds1103 platform is used for the implementation of the proposed scheme. All results confirm fast reference tracking capability with low overshoots and robustness against disturbances while comparing with the FTSC.

Keywords: synergetic control (SC) law; fuzzy neural network (FNN) approximator; fast terminal synergetic controller (FTSC); finite-time convergence; DC/DC buck converter

Citation: Babes, B.; Hamouda, N.; Albalawi, F.; Aissa, O.; Ghoneim, S.S.M.; Abdelwahab, S.A.M. Experimental Investigation of an Adaptive Fuzzy-Neural Fast Terminal Synergetic Controller for Buck DC/DC Converters. *Sustainability* **2022**, *14*, 7967. https://doi.org/10.3390/su14137967

Academic Editors: Luigi Aldieri and Mohamed A. Mohamed

Received: 2 June 2022
Accepted: 26 June 2022
Published: 29 June 2022

1. Introduction

1.1. Motivation

The buck DC/DC converter is a common choice in many applications in electric power supply systems, including DC motor drives, DC microgrids, automobiles, sailboats, airplanes, solar systems, and many more. Its mathematical model is a nonlinear and time-varying system based on state-space equations. These equations must be treated around the steady-operating point with a small-signal linearization approach before being studied with PID controllers. When a severe disturbance causes a substantial divergence from the typical operating point, it is difficult for a standard buck DC/DC converter to achieve perfect stability [1]. In light of this, adjusting the output voltage of the buck DC/DC converter is a challenging task. Realizing a robust control strategy for buck DC/DC converter is a critical challenge for better understanding the design aspect of the regulator and stability problem [2]. Furthermore, it can aid to enhance transient and steady-state conditions under a variety of disturbances. To overcome some of the previously mentioned challenges, this work suggests an AFN-FTSC algorithm for the output voltage control without complete converter model knowledge. FNN model, and universal approximator, are utilized in an adaptive scheme to approximate the converter nonlinear dynamics, while the FTSC algorithm is utilized to guarantee robustness through an easy-to-implement chatter-free continuous control law [1].

1.2. Related Work

To provide better performance for buck DC/DC converters, numerous intelligent and nonlinear control techniques have been introduced over the last decade. In paper [3], the performance of the buck DC/DC converter is evaluated by variable structure controllers in which the output voltage is tracked and regulated in a good way; however, the robustness of these controlling techniques is not acceptable against noise, chattering, parametric variations, and disturbances, which makes them practically unsuitable [4]. Otherwise, because of the switching properties of buck DC/DC converter, digital discrete control methods like repetitive [5], dead beat [6], and internal model [7] controllers have been recommended for buck DC/DC converter. The main advantage of a repetitive controller is low output distortion; however, it has some limitations such as poor tracking, large memory capacity, poor performance, and slow dynamics. Nevertheless, the internal model and dead beat controller can provide output voltage tracking without ringing and overshoot; however, the complex controlling structure of the internal model and dead beat controllers have influenced their efficiency.

In paper [8], a feedback linearization scheme is utilized for the buck DC/DC converter to control the output voltage, and model predictive controllers are presented in [9], and [10] to achieve the same goal. Both feedback linearizing and model predictive controllers provide similar advantages such as reliable reference tracking, constant switching frequency, and fast dynamic response; these controllers do not perform well under the variation in parameters due to their requirements of using precise parametric information during the implementation, and it is quite impossible to precisely know the parameters of the system as many of these vary with operating scenarios. An adaptive model predictive controller is introduced in [11] by capturing linear parameter variations so that parameter sensitivity can be alleviated. However, the inclusion of such parameter variations complicates the process of solving the optimization problem for the model predictive scheme. The Synergetic Controller (SC) is another useful alternative way of controlling the buck DC/DC converters, which can also ensure the robustness against parameter variations and disturbances by using a generalized state-space averaged representation of the system. The SC inherently alleviates the effects of external disturbances, nonlinearities, and uncertainties [12]. For all these features, SCs are extensively used for power electronic applications, including the buck DC/DC converter [13]. The standard SC scheme uses the linear manifold as a macro-variable as presented in [14] to guarantee the asymptotic convergence of the error. However, such a macro-variable makes the system static during the fluctuations in the load voltage, and therefore, it results in poor voltage tracking during transients in the system [14]. Although the gain parameters for such traditional SCs can be adjusted to ensure better transients, there are no guarantees that the system variable reaches their equilibriums in a finite time. These problems can be overcome by using SC schemes such as the Non-Singular Terminal Synergetic Control and FTSC that ensure finite-time convergence, as presented in [15,16]. However, these methods slow down the dynamic response by reducing the likelihood of overshoots during load changes and other transients during the starting. The dynamic response for regulating the output voltage during transients is accelerated in [13] using an adaptive FTSC based on dual Radial Basis Function Neural Networks (RBFNN-FTSC) control. Moreover, all adaptive SC approaches in the existing literature on the buck DC/DC converters use classical adaptive law, for which there is a possibility to induce the chattering causing ripples at the output voltage during their practical or real-time implementations. In recent years, FNN models are a hybrid approach that combines the learning ability of a neural network with the noise-handling capabilities of fuzzy logic (FL). They are commonly used in adaptive control to cope with uncertainties and attain improved performance standards. [17,18]. Although the FNNs have extensively been utilized in different adaptive controls, the applications of FNNs for uncertain nonlinear systems are relatively new and only a few applications can be seen throughout the literature. Based on the author's knowledge, the applications of AFN-FTSC for DC/DC buck converters have not been investigated until now. Motivated by the existing literature, the considered controller design process in the current research investigates a new digital controller based on AFN-FTSC for buck DC/DC converters without the previously mentioned problems.

1.3. Novelty and Principal Contributions

The main innovative features of the proposed scheme are:

(i) It is the first time a FNN model has been developed based on FTSC for DC/DC buck converters;
(ii) Unlike closely related work, the proposed AFN-FTSC is created by incorporating the macro-variable to tackle the chattering effects, decrease the time of convergence, simplify the expression of the controller, and ensure a fast transient reaction, low steady-state error, and high output voltage tracking accuracy;
(iii) A Lyapunov stability theorem is rigorously used to demonstrate the overall stability of the system and to obtain the updated rules for the FNN weights.
(iv) The removal of demand for an accurate model by using a FNN approximator to estimate an unknown buck DC/DC converter functions;
(v) Experiments are carried out to demonstrate efficiency while assisting to achieve desired goals under a variety of operating situations.
(vi) It simultaneously ensures the higher output voltage tracking accuracy, swift transient responses, and less impact by disturbances and uncertainties due to the use of adaptive time-varying reaching law.

1.4. Organization of the Paper

The remaining part of the study is structured as follows. The basic requirements for SC and FNNs are briefly discussed in Section 2. The representation of the DC/DC buck converter is introduced in Section 3. The steps for designing the AFN-FTSC algorithm are presented in Section 4, including the stability and reachability analysis. Afterward, results from experimental tests are presented in Section 5 to illustrate the efficacy and practicality of the designed AFN-FTSC algorithm. Finally, the conclusions and some recommendations for further investigation are given in Section 6.

2. Preliminaries

2.1. Principle of Synergetic Control (sc)

SC algorithm is effectively a nonlinear control strategy based on the principle of directional self-organization theory and the usage of the standard nonlinear characters of dynamic systems. The fundamental idea of the SC algorithm is as follows:

1. Invariant manifold is created in the state-space of a controllable system. On this attractor, we guarantee the organization of the preferred static and dynamic behavior of the controllable system. The design of the attractor is the indication of a directed self-organization principle.
2. The most important premise in the theory of SC is the principle of compression–decompression of the phase flow of the controllable objects.
3. The designer's necessities are given in the form of an affine system that describes the preferred operating modes of the controlled systems.

Consider the nonlinear affine dynamical system of degree n given by:

$$\dot{x} = \frac{dx}{dt} = f(x, u, t) \quad (1)$$

where $x \in R^n$, $u \in R^m$, indicating the state-variable and control-variables, respectively. The $f(.)$ indicates a continuous nonlinear function. The SC design structure starts with defining a macro-variable φ as a function of the state variables [19]:

$$\varphi = \varphi(x, t) \quad (2)$$

The primary goal of the SC is to force the system to function on a pre-determined manifold $\varphi = 0$. Develop a SC that forces system variables to approach the desired manifold exponentially, with an evolution constraint that may be expressed as:

$$\kappa\dot{\varphi} + \varphi = 0, \kappa > 0 \quad (3)$$

where φ is the derivative of the aggregated macro-variables, and κ is a positive integer that enables the designer to choose the rate of convergence to the selected attractor. The solution of (3) gives the following function for $\varphi(t)$:

$$\varphi(t) = \varphi_0 e^{-t/\kappa} \quad (4)$$

It affirms that $\varphi(t) \rightarrow 0$ at $t \rightarrow \infty$, which implies that $\varphi(t)$ is attracted to $\varphi = 0$ from any starting location φ_0 (Figure 1). Since the time constant $\kappa > 0$, the macro-variable φ will decay exponentially with a speed determined by κ. As long as the system is stable, the smaller value of κ is faster than the macro-variable decays. When φ approaches zero, the system converges to the manifold and then functions on the manifold without interruption.

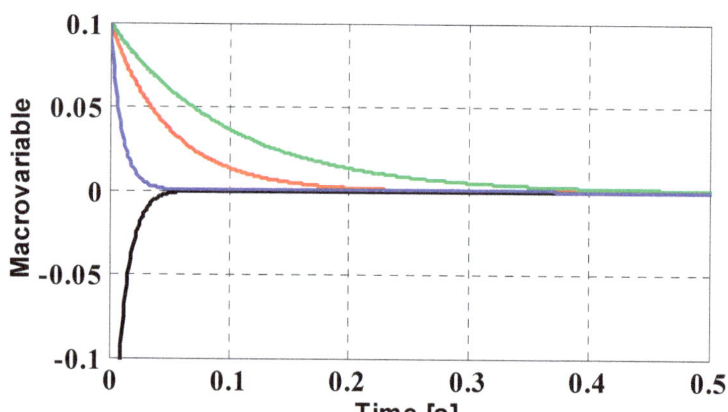

Figure 1. Convergence of macro-variable for various starting locations to the attractor.

Considering the differentiation chain provided by:

$$\dot{\varphi} = \frac{d\varphi(x,t)}{dt} = \frac{\partial\varphi(x,t)}{\partial x} \cdot \frac{dx(t)}{dt} \quad (5)$$

When (1) and (3) are substituted into (5), we get (6):

$$\kappa\frac{d\varphi(x,t)}{\partial x} f(x, u, t) + \varphi(x, t) = 0 \quad (6)$$

Resolving (6) for the control vector u gives the SC law as:

$$u = g(x, \varphi(x,t), \kappa, t) \tag{7}$$

The SC law (7) permits state trajectories to follow certain paths (3). Proper selection of the macro-variable (2) and judicious manifolds ensure the necessary performance and stability [20]. It is worth noting that SC law (7) can be rewritten as a solution to a Kolesnikov's problem (8). Choose the performance index as follows [21]:

$$J_{\Sigma} = \int_0^{\infty} \Gamma(\dot{\varphi}, \varphi)dt = \int_0^{\infty} \left(\sum_{k=1}^{m} \kappa_k^2 \dot{\varphi}_k{}^2 + \varphi_k{}^2 \right) dt \tag{8}$$

A Lyapunov function candidate as in (9) can be used to determine the system's asymptotic stability:

$$V(t) = 0.5 * \varphi(x,t)^2 \tag{9}$$

The following inequality is fulfilled when the condition of $\kappa > 0$ is met:

$$\dot{V}(t) = \varphi(x,t)\dot{\varphi}(x,t) = -\frac{1}{\kappa}\varphi(x,t)^2 < 0 \tag{10}$$

Figure 2 illustrates the system's phase profile as well as the SC law's stability features, and also shows the convergence to the manifold. The steady-state operating point is the origin if the error is equal to 0. A straight line through the origin with a slope $(-1/\kappa)$ is represented by Equation (3). The operational point of the system converges to the straight line (the control manifold) and then moves along it to the origin.

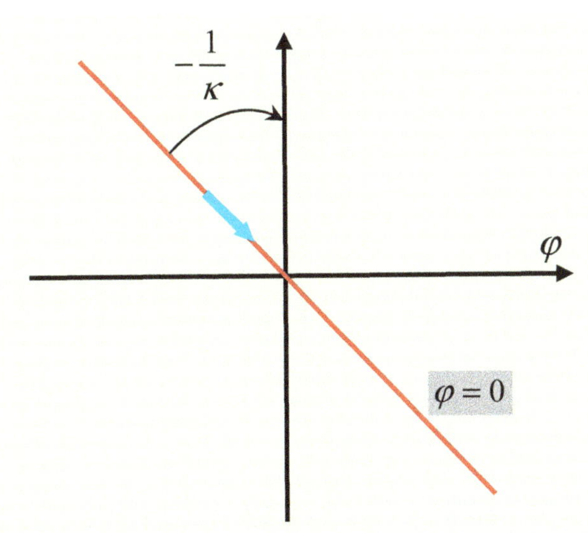

Figure 2. Geometric description of SC in the phase plane.

The schematic representation, illustrating the suggested controller's computations, is shown in Figure 3.

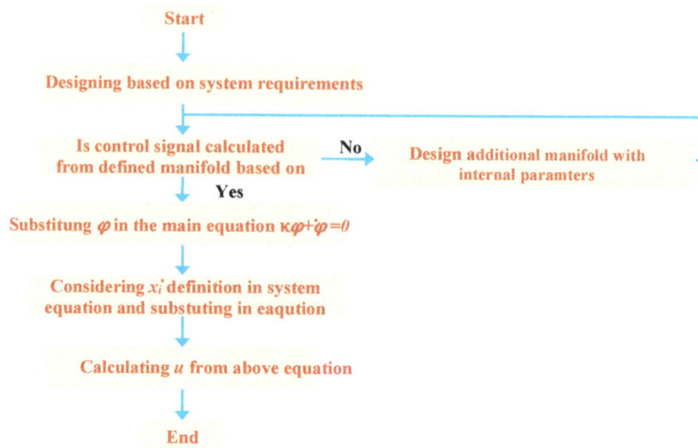

Figure 3. Schematic diagram of the proposed SC algorithm.

2.2. Principle of Fnn Approximator

It is known that FNNs are universal estimators and have exceptional functions in controller design and identification. The properties of the FNN, such as accelerating the learning process and finer granularity to make decisions [22], make it appropriate for real-time control systems and are capable of enhancing the control precision, stability, and flexibility of the system. The FNN is an implementation of fuzzy logic (FL) with trainable parameters similar to an ANN, which comprises four layers of neurons. They are the input layer-1; the membership layer-2; the rule layer-3; and the output layer-4, respectively, all of which are depicted in Figure 4. The key characteristic of the FNN is its ability to describe the human-like reasoning and successfully cope with experience by using conditional fuzzy IF-THEN rules to build a mapping from an input vector x to an output vector y, which may be expressed as:

$$R^l : Ifx_1isA_1^l(x_1), x_2isA_2^l(x_2), \ldots, x_nisA_n^l(x_n)Theny_1isB_1^l, y_2isB_2^l, \ldots, y_misB_m^l \quad (11)$$

where $L = 1, 2, \ldots$, with L representing the overall number of IF-THEN rules, $x = [x_1, x_2, \ldots, x_n]^T$ and $y = [y_1, y_2, \ldots, y_m]^T$ characterize the input and output parameter vectors of FNN. A_1^l *and* B_1^l are the linguistic variables of the fuzzy sets, expressed by their membership function vectors $\mu A_i{}^l(x_i)$ and $\mu B_i{}^l(y)$, respectively.

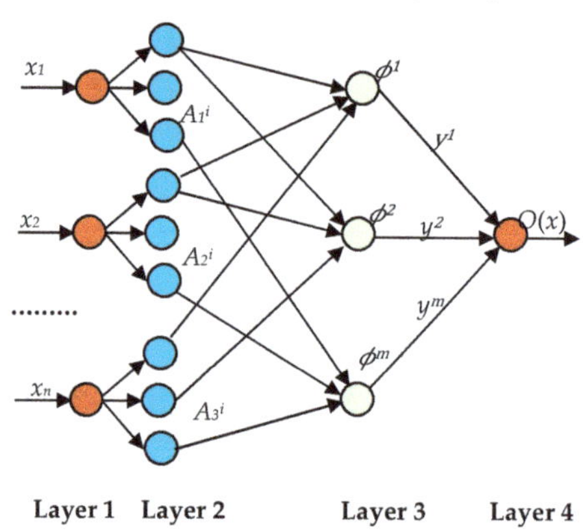

Figure 4. Structure of a FNN approximator with four layers.

Based on the singleton fuzzifier, product inference model, and center-average defuzzifier, the final output $O(x)$ can be obtained by the FNN estimator as in [22]:

$$o(x) = \frac{\sum_{l=1}^{L}\left(\prod_{i=1}^{n}\mu_{A_i^l}(x_i)\right)\overline{y}_j^l}{\sum_{l=1}^{L}\left(\prod_{i=1}^{n}\mu_{A_i^l}(x_i)\right)} \tag{12}$$

where $\overline{y}_j^l$ is the point in at which $\mu_{B_j^l}(y_j)$ attains its optimal value $\left(\mu_{B_j^l}(y_j) = 1 \text{ considering}\right)$.

If $\overline{y}_j^l$ are selected as the free parameter vector, the FNN approximator (12) becomes an adaptive FNN. Then, it can be expressed in compact form as:

$$y_j(x) = \sum_{l=1}^{L}\overline{y}_j^l = w^T\phi(x) \tag{13}$$

where $w = \left[\overline{y}_j^1, \overline{y}_j^2, \ldots, \overline{y}_j^L\right]^T \in R^L$ is called the regulating parameter vector, and $\phi(x) = [\phi_1(x), \phi_2(x), \ldots, \phi_L(x)]^T \in R^L$ is the vector of fuzzy primary function, or the fuzzy basis function (FBFs) is obtained by (14):

$$\phi_j(x) = \frac{\prod_{i=1}^{n}\mu_{A_i^l}(x_i)}{\sum_{j=1}^{L}\left(\prod_{i=1}^{n}\mu_{A_i^l}(x_i)\right)} \tag{14}$$

Remark 1. *It ought to be indicated that the complexity of the designed AFN-FTSC algorithm is dependent on the complexity of FNN approximator. Suppose that the number of FNN inputs is n, and the number of fuzzy variables is m.*

The complexity of the AFN-FTSC is composed of the fuzzy layer (layer 2) and fuzzy inference layer (layer 3). The complexity of the fuzzy layer is $O(n \times m)$, and the complexity for the fuzzy inference production layer is $O(m^2)$. The total complexity of the AFN-FTSC algorithm is $O(n \times m) + O(m^2)$.

3. Mathematical Model of Buck DC/DC Converter

Figure 5 shows a buck DC/DC converter that connects an input DC voltage V_{in} with a DC load R_o having its output voltage as V_o. The buck DC/DC converter encompasses an adjustable switch Q, diode D, an inductor L with i_L as the current flowing through it, a capacitor C with i_C as the current flowing through it, and a load resistor R_O where i_R is the load current.

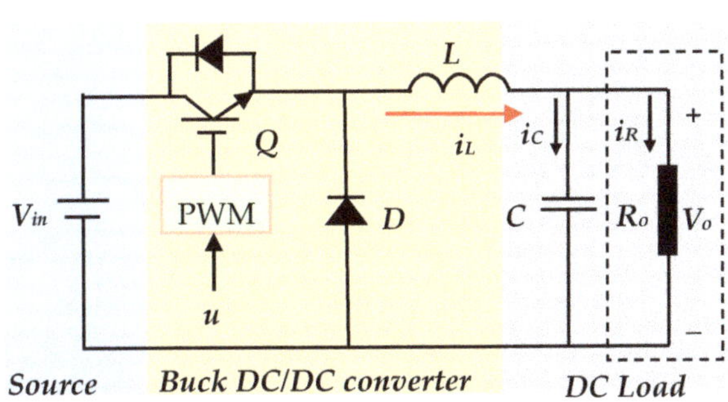

Figure 5. The circuit model of buck DC/DC converter.

Due to the nonlinear behavior of the diode and MOSFET/IGBT switch, the buck DC/DC converter has a nonlinear structure. Moreover, there are two charging elements an inductor and capacitor for which the mathematical model for the system will be a second-order one.

The dynamic of the system in Figure 5 will be represented by the change in the voltage V_o across R_o, which is also the voltage across C. Similarly, i_L will change as it flows through L. Since the switch will be in two modes, ON and OFF, there will be two stages.

When the switch is turned ON in the first stage, equations representing the state-space model can be stated in the following way [23]:

$$\begin{cases} \frac{di_L}{dt} = -\frac{1}{L}V_o + \frac{V_{in}}{L}u \\ \frac{dV_o}{dt} = \frac{1}{C}i_L - \frac{1}{RC}V_o \end{cases} \tag{15}$$

In the second stage, the switch is turned OFF. In this case, the diode D is in forwarding bias and the circuit is then equivalent to a parallel RLC circuit.

For this circuit, the equation representing the state-space model can be written as follows:

$$\begin{cases} \frac{di_L}{dt} = -\frac{1}{L}V_o + \frac{V_{in}}{L}u \\ \frac{dV_o}{dt} = \frac{1}{C}i_L - \frac{1}{RC}V_o \end{cases} \tag{16}$$

At this stage, the complete dynamical average model can be obtained by combining (15) and (16) as:

$$\begin{cases} \frac{di_L}{dt} = -\frac{1}{L}V_o + \frac{V_{in}}{L}u \\ \frac{dV_o}{dt} = \frac{1}{C}i_L - \frac{1}{RC}V_o \end{cases} \tag{17}$$

Where u is the control input, which is recognized as 0 when the switch Q is turned OFF and 1 when it is turned ON, considering the capacitor voltage and its derivative as system state variables x_1 and x_2, respectively. Equation (18) represents the final model of the buck DC/DC converter, which is utilized to design the proposed composite controller as discussed in the section below:

$$\begin{cases} \dot{x}_1 = x_2 \\ \dot{x}_2 = -\frac{x_1}{LC} - \frac{x_2}{RC} + \frac{V_{in}}{LC}u \end{cases} \tag{18}$$

4. Proposed Robust AFN-FTSC Design

This section focuses on determining the switching control input u using the proposed AFN-FTSC law for system (18) as a function of state coordinates (e_1, e_2), which gives the desired value of converter output voltage $V_{o\text{-}ref}$ without the prior modeling information and is subject to the requirement that the voltage error converges to origin asymptotically in finite-time t_r, i.e., Further analyses are presented to comprehensively analyze the stability and reachability of the proposed method. The determination of the switching control law is presented below.

4.1. Determination of the Control Law

Since the regulation of V_0 is the main task, the design process needs to start by defining the error for this output voltage.

(1) Step 1:

Based on the design criterion, the tracking error (e_1) for V_0 can be expressed as:

$$e_1 = x_1 - V_{o-ref} \tag{19}$$

$$\dot{e}_1 = e_2 = x_2 - \dot{V}_{o-ref} \tag{20}$$

where V_{0-ref} is the reference output voltage and its dynamic will be as:

$$\dot{e}_1 = e_2 = \frac{i_L}{C} - \frac{V_o}{RC} - \dot{V}_{o-ref} \tag{21}$$

(2) Step 2:

As seen in [24], a special variable referred to as the nonlinear macro-variable function must be added as follows:

$$\varphi = \dot{e}_1 + ae_1 + be_1{}^{p/q} \quad (22)$$

where a and b are the macro-variable parameters, p and q are positive odd constants, which satisfy the following criterion: $1 < p/q < 2$.

(3) Step 3:

A constraint imposing preferred dynamics to the macro-variable is chosen as:

$$\kappa\dot{\varphi} + \varphi = 0 \quad (23)$$

where κ is a positive parameter that obliges the designer select speed convergence to the desired attractor.

(4) Step 4:

To determine the control input while satisfying the synergetic mode reaching condition, $\varphi = 0$, the dynamics of the system (22) may be described as:

$$\dot{e}_1 + ae_1 + be_1{}^{p/q} = 0 \quad (24)$$

At this point, the reaching time t_r for e_1 can be obtained by integrating Equation (20), which may be calculated as:

$$\dot{\varphi} = \dot{e}_2 + a\dot{e}_1 + b\frac{p}{q}e_1{}^{(p/q)-1}\dot{e}_1 \quad (25)$$

where $e_1(0)$ is the initial value of $e_1(t)$. The time derivative of φ can be determined as:

$$\dot{\varphi} = \dot{e}_2 + a\dot{e}_1 + b\frac{p}{q}e_1{}^{(p/q)-1}\dot{e}_1 \quad (26)$$

By combining (23) and (26), we get (27):

$$\dot{e}_2 + a\dot{e}_1 + b\frac{p}{q}e_1{}^{(p/q)-1}\dot{e}_1 = -\frac{1}{\kappa}\varphi \quad (27)$$

Rewriting Equation (27), can be expressed as:

$$\dot{e}_2 = -\frac{x_1}{LC} - \frac{x_2}{RC} + \frac{V_{in}}{LC}u - \ddot{V}_{o-ref} \quad (28)$$

We specified the following abstractions for the controller design:

$$g(x) = \frac{V_{in}}{LC} \text{ and } f(x) = -\frac{x_1}{LC} - \frac{x_2}{RC}$$

Solving for the FTSC law, U_{FTSC}, guides to an enhanced reaching law (29):

$$u_{FTSC} = \frac{1}{g(x)}\left[\ddot{V}_{o-ref} - f\left(x\right) + \frac{1}{\kappa}\varphi - a\,\dot{e}_1 - b\frac{p}{q}e_1{}^{(p/q)-1}\dot{e}_1\right] \quad (29)$$

The next subsection follows the overall stability analysis with this SC law.

4.2. Stability Analysis

The Lyapounov theorem can be used to assess the stability of the system as in (30):

$$V_1 = 0.5 * \varphi^2 \quad (30)$$

The time derivative of Equation (30), using Equations (23) and (27), can be written as:

$$\dot{V}_1 = \varphi\dot{\varphi} = \varphi\left(\ddot{e}_2 + a\dot{e}_1 + b\frac{p}{q}e_1{}^{(p/q)-1}\dot{e}_1\right) = \varphi\left(-\frac{1}{\kappa}\varphi\right) = -\frac{1}{\kappa}\varphi^2 \leq 0 \tag{31}$$

As a result, the regulator (29) may satisfy the system's overall stability criteria.

4.3. Approximation of f(x) and g(x) Based on FNN

Sadly, in the practical application, the nonlinear system terms $f(x)$ and $g(x)$ are unknown and difficult to determine precisely. As a consequence, there is no way to apply the conventional FTSC law (29), of which the terms $f(x)$ and $g(x)$ are known to be scalar. To tackle this difficulty and guarantee global stability, a FNN model using the universal approximation approach is developed in the following theorem to approximate the nonlinear and unknown terms $f(x)$ and $g(x)$ in (29) through adaptive laws.

Theorem 1. *Considering system (18) with unknown terms f(x) and g(x), and design the macro-variable in the form of (22). If the updated AFN-FTSC controller $u_{AFN\text{-}FTSC}$ is constructed as (32):*

$$u_{AFN-FTSC} = \frac{1}{\hat{g}(x|w_g)}\left[\ddot{V}_{o-ref} - \hat{f}\left(x|w_f\right) + \frac{1}{\kappa}\varphi - a\dot{e}_1 b\frac{p}{q}e_1{}^{(p/q)-1}\dot{e}_1\right] \tag{32}$$

Next, the tracking error and its first derivative converge to zero in finite-time, where $\hat{f}\left(x|w_f\right)$ and $\hat{g}(x|w_g)$ represent the FNN estimates of the terms $f(x)$ and $g(x)$. These functions can be estimated through Equations (33) and (34):

$$\hat{f}\left(x|w_f\right) = w_f^T\phi_f(x) \tag{33}$$

$$\hat{g}(x|w_g) = w_g^T\phi_g(x) \tag{34}$$

where $\phi_f(x)$ and $\phi_g(x)$ are the transfer functions from the input layer to the rule layer, w_f and w_g are the connection weights of FNN. By updating the network weights, the system uncertainties can be estimated adaptively.

Theorem 2. *Considering the nonlinear model (18) with the input signal (32), and if the FNN-based adaptive laws are designed as:*

$$\dot{\hat{w}}_f = -\mu_1\varphi\phi_f(x) \tag{35}$$

$$\dot{\hat{w}}_g = -\mu_2\varphi\phi_g(x) \tag{36}$$

where η_1 and η_2 are arbitrary positive parameters.

Assumption 1. *Suppose x be a member of a compact set $R^n = \{x \in R^n\text{: } |x| \leq K_x < +\infty\}$, with K_x is a constant. The ideal FNN weights $w_f{}^*$ and $w_g{}^*$ are located in the convex area shown below:*

$$\Delta_f = \left\{w_f \in R^n : \left|w_f\right| \leq H_f\right\} \tag{37}$$

$$\Delta_g = \left\{w_g \in R^n : \left|w_g\right| \leq H_g\right\} \tag{38}$$

where H_f and H_g are designed parameters, and the radius Δ_f and Δ_g are limitations for w_f and w_g. The universal approximation theory states that there is an ideal FNN weights $w_f{}^*$ and $w_g{}^*$ satisfies:

$$w_f^* = \underset{w_f \in \Delta_f}{argmin}\left\{\sup_{x \in R^n}\left|\hat{f}\left(x|w_f\right) - f(x)\right|\right\} \tag{39}$$

$$w_g^* = \underset{w_g \in \Delta_g}{argmin}\left\{ \sup_{x \in R^n} |\hat{g}(x|w_g) - g(x)| \right\} \tag{40}$$

Consequently, $f(x)$ and $g(x)$ are approximated to arbitrary accuracy by the FNN approximators (35) and (36) as the below assumption.

Assumption 2 (See [25]). *For each given real smooth variable f(x) and g(x) defined on a compact set $x \in R^n$ and for any arbitrary $\varepsilon_f > 0$ and $\varepsilon_g > 0$, there exists a FNN approximator and in the formula of (41) and (42) so that:*

$$\sup_{x \in R^n} \left|\hat{f}\left(x|w_f\right) - f(x)\right| < \varepsilon_f \tag{41}$$

$$\sup_{x \in R^n} |\hat{g}(x|w_g) - g(x)| < \varepsilon_g \tag{42}$$

Remark 2. *By using the FNN approximator designed above, the FNN gains $\phi_f(x)$ and $\phi_g(x)$ can be scheduled adaptively according to the variety of x_1 and x_2. Then, the minimum approximation error can be obtained as:*

$$\varepsilon = \left[\hat{f}\left(x|w_f^*\right) - f(x)\right] + \left[\hat{g}\left(x|w_g^*\right) - g(x)\right]u \tag{43}$$

where ε is bounded by a positive constant $\varepsilon \leq \varepsilon_{max}$. The dynamic of the macro-variable is calculated by replacing (43) into (23):

$$\dot{\varphi} = \left[\hat{f}\left(x|w_f^*\right) - f(x)\right] + \left[\hat{g}\left(x|w_g^*\right) - g(x)\right]u - \frac{1}{\kappa}\varphi + \varepsilon \tag{44}$$

Leading, after some straightforward manipulations, to:

$$\dot{\varphi} = \left(w_f^{*T} - w_f^T\right)\phi(x) + \left(w_g^{*T} - w_g^T\right)\phi(x)u - \frac{1}{\kappa}\varphi + \varepsilon \tag{45}$$

From the above expression, one may write:

$$\dot{\varphi} = \hat{w}_f^T\phi(x) + \hat{w}_g^T\phi(x)u - \frac{1}{\kappa}\varphi + \varepsilon \tag{46}$$

where $\hat{w}_f = w_f^* - w_f$ and $\hat{w}_g = w_g^* - w_g$ are the error between w and the ideal weight w^*.

Stability demonstration: Let us choose the Lyapunov function candidate as:

$$V = 0.5 * \varphi\left(\varphi^2 + \frac{1}{\mu_1}\hat{w}_f^T\hat{w}_f + \frac{1}{\mu_2}\hat{w}_g^T\hat{w}_g\right) \tag{47}$$

The time derivative of Equation (47) gives:

$$\dot{V} = \varphi\dot{\varphi} + \frac{1}{\mu_1}\hat{w}_f^T\dot{\hat{w}}_f + \frac{1}{\mu_2}\hat{w}_g^T\dot{\hat{w}}_g \tag{48}$$

Substituting (46) into (48), we can get (49):

$$\dot{V} = \varphi\left(\hat{w}_f^T f(x) + \hat{w}_g^T f(x)u - \frac{1}{\kappa}\varphi + \varepsilon\right) + \frac{1}{\mu_1}\hat{w}_f^T\dot{\hat{w}}_f + \frac{1}{\mu_2}\hat{w}_g^T\dot{\hat{w}}_g \tag{49}$$

We utilize the following Equations (50) and (51):

$$\dot{\hat{w}}_f = \dot{w}_f \tag{50}$$

$$\dot{\tilde{w}}_g = \dot{w}_g \tag{51}$$

Substituting (50) and (51) into (49) leads to:

$$\dot{V} = -\frac{1}{\kappa}\varphi^2 + \varphi\varepsilon + \frac{1}{\mu_1}\tilde{w}_f^T\left(\mu_1\varphi f(x) + \dot{w}_f\right) + \ldots + \frac{1}{\mu_2}\tilde{w}_g^T\left(\mu_2\varphi f(x)u + \dot{w}_g\right) \tag{52}$$

The parameter vector $\dot{w}_f$ and $\dot{w}_g$ is adapted according to the (35) and (36) laws. Then, Equations (23) and (52) bring about the following result:

$$\dot{V} \leq -\frac{1}{\kappa}\varphi^2 + \varphi\varepsilon \tag{53}$$

Remark 3. *The FNN approximation error* ε *should preferably be equal to zero, in this situation* $\dot{V} \leq -\frac{1}{\kappa}\varphi^2 \leq 0$ *emphasizes the definiteness of the negative, thus the regulated system is stable. In reality, the estimation error is always present, hence the stability of the system cannot be assured. To finish the analysis and check the asymptotic convergence of the tracking error, we must demonstrate that* $\varphi \to 0$ *as* $t \to \infty$*. Besides, suppose that* $\|\varphi\| \leq \tau$*, then Equations (53) can be reformulated as:*

$$\dot{V} \leq -\frac{1}{\kappa}\tau|\varphi| + \tau|\varepsilon| \tag{54}$$

Integration on both sides of Equation (54) provides:

$$\int_0^t |\varphi| dt \leq \frac{\kappa}{\tau}(|V(0)| - |V(t)|) + \kappa \int_0^t |\varepsilon| dt \tag{55}$$

Equation (42) means that all the signals are homogeneously bounded in the closed-loop system. Then, we do have $\varphi \in L_1$, from (41), we recognize that the macro-variable is limited and every term in (42) is limited, hence $(\varphi, \dot{\varphi}) \in L_\infty$, making use of the Barbalat lemma [26]. We can conclude that the tracking error converges to zero asymptotically, which confirms the stability condition of a closed-loop system. Consequently, the stable control performance of the buck DC/DC converter can be ensured without the necessity of system information. Figure 6 shows the overall schematic diagram of the developed AFN-FTSC algorithm.

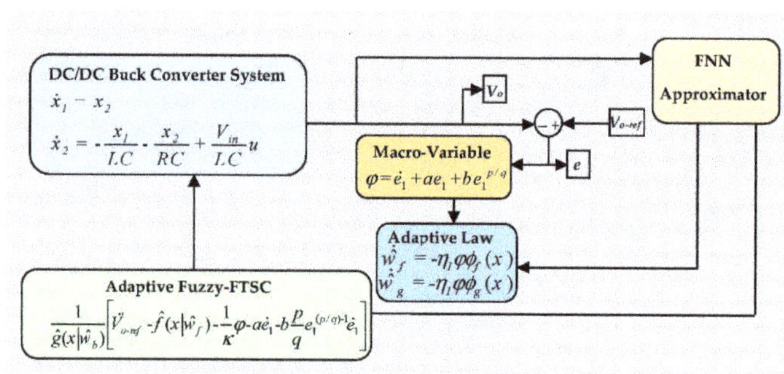

Figure 6. Block diagram for proposed AFN-FTSC controller.

5. Controller Performance Evaluation

Since the satisfactory control performance will require properly selecting user-defined parameters (please note that the proposed method is less sensitive than existing control methods), some discussions have been included in the following subsection.

5.1. Selection of the User-Defined Parameters

One of the key goals to design and implement new controllers is to achieve a faster response during the transient [27]. For the proposed AFN-FTSC scheme, the larger values of these user-defined parameters will ensure fast-transient performance. However, such large user-defined parameters or gains are not always used, as these degrade the performance of the controller by amplifying the overshoot and providing the undesirable disturbance rejection. Therefore, it is essential to make a trade-off among various performance factors while tuning these gains. These performance factors must include the desired criteria for tracking (e.g., fewer or no offsets, small overshoots, fast settling time, etc.), the appropriate level of the disturbance rejection, and robustness against parametric constraints. Since the major focus of this work is on the disturbance rejection and resilience against uncertainties, tall user-defined parameters are carefully chosen to ensure an appropriate tracking performance. For the proposed AFN-FTSC algorithm, it is found that the most sensitive parameter is (p/q), which is basically associated with the voltage tracking error. The dynamic performance depicting the output voltage with the proposed AFN-FTSC is depicted in Figure 7 for various values of (p/q). This figure shows the minimum value of the voltage overshoot when the value of (p/q) is equal to 2; however, the settling time is much higher than other values of (p/q). At the same time, the voltage overshoot and the settling time are 0.15 V and 9.44 ms, respectively, when the value of (p/q) is 1.5, and this is more acceptable as compared to other values in terms of both voltage overshoot and settling time.

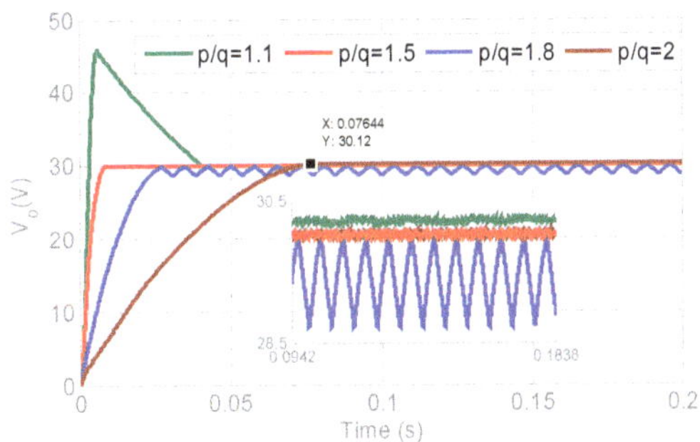

Figure 7. The output voltage of the converter for different values of (p/q) with the designed controller.

The voltage overshoot and settling time for different values of (p/q) with the designed AFN-FTSC are detailed in Table 1.

Table 1. Parameter sensitivity analysis.

p/q	1.1	1.5	1.8	2.0
Voltage overshoot (V)	15.85	0.13	0.11	0.07
Settling-time (ms)	42.96	9.44	27.60	75.44

To ensure a fair comparison, the acquired results are compared to that of the existing FTSC while maintaining the same value of the most sensitive gain parameter, i.e., (p/q = 1.5) for both controllers. Similarly, the values of other user-defined parameter values can be determined, and these are presented in Table 2.

Table 2. Regulator Specifications.

Regulator	Parameters	Gain Value
AFN-FTSC	κ	0.005
	a	200
	b	300
	p	3
	q	2
	μ_1	500
	μ_2	800
FTSC	κ	0.005
	a	200
	b	300
	p	3
	q	2

In the FNN estimator, we can specify a set of five decision variables regularly distributed on a universe of discourse $[-1,1]$, as follows:

$$\mu_f^k = \mu_g^k = exp[-(x + 4 - 1.6(k-1)^2], k = 1, \ldots, 5 \tag{56}$$

i = 5, which means that there are 25 rules to estimate the unknown terms. The selected 25 rules can cover the entire space and approximate any non-linear function. The member function degree is illustrated in Figure 8.

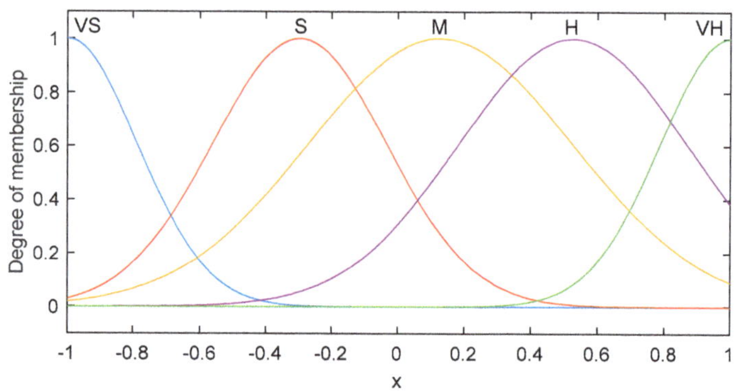

Figure 8. Membership function of x.

The FNN weights are selected as follows:

$$\begin{bmatrix} w_f \\ w_g \end{bmatrix} = \begin{bmatrix} w_f^1 & w_f^2 & w_f^3 & w_f^4 & w_f^5 \\ w_g^1 & w_g^2 & w_g^3 & w_g^4 & w_g^5 \end{bmatrix} \tag{57}$$

The initial values for the FNN weights are selected in a random way, and the vector of fuzzy basis functions was constructed by (12). Using these gain parameters, the designed AFN-FTSC and existing FTSC are implemented in the following under different operating points of the buck DC/DC converter.

5.2. Experimental Validation

In this part, the efficacy of the suggested AFN-FTSC technique for regulating the output voltage of a buck DC/DC converter is demonstrated through a real-time implementation on dSPACE 1103 platforms (Figure 9). All system parameters are carefully chosen by keeping the practical conditions in mind, which are given in Table 3.

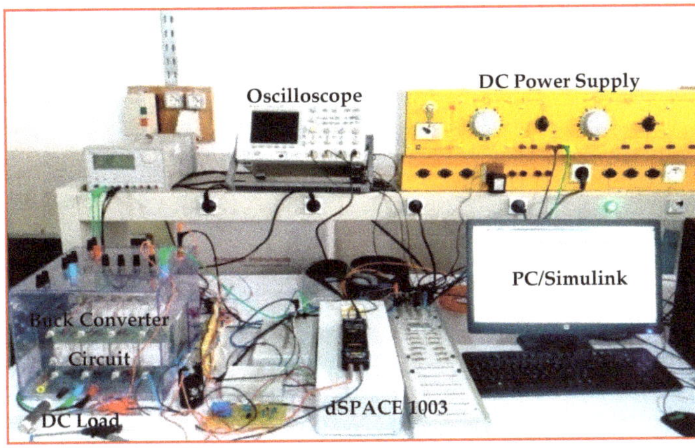

Figure 9. The experimental prototype of the buck DC/DC converter.

Table 3. System Parameters.

Parameter	Value
V_{in}	100 V
R_o	40/80/40 Ω
V_{o_red}	20/30/50 V
L	7 mH
C	800 μF
f_s	20 kHz

To show the efficacy of the designed AFN-FTSC algorithm, variations in the output reference voltage, input supply voltage, and load resistance are done based on the values in Table 3. However, the nominal output voltage of the converter is set at 50 V with an input voltage of 100 V during the standard conditions, i.e., while considering no disturbances. To deploy the AFN-FTSC algorithm for the buck DC/DC converter, the sampling frequency, and the switching frequency are considered as 200 kHz and 20 kHz, respectively. The superiority of the AFN-FTSC is analyzed against an existing FTSC as proposed in [14] in terms of tracking error, overshoot, and settling time. The performance of the designed AFN-FTSC algorithm is validated under a variety of operating scenarios: variations in input supply voltage, load resistance, and output reference voltage as demonstrated through the following four cases:

- Controller performance under variations in the load resistance,
- Controller performance under variations in the input supply voltage,
- Controller performance under start-up transient,
- Controller performance under variations in the output reference voltage,

Case I: Output voltage regulation with variations in the load resistance.

In this scenario, the test is conducted to justify the load disturbance rejection capacity of the proposed AFN-FTSC algorithm while making a comparison with the FTSC algorithm. For this purpose, the effects of only variations in the load resistance are considered, while the output reference voltage remains constant at 50 V. In this test, the load resistance of the buck DC/DC converter is decreased from 80 Ω to 40 Ω, conversely. With such changes in the load resistance, the dynamic responses of the Buck DC/DC converter are represented by two states: output voltage (V_o) and inductor current (i_L), as illustrated in Figure 10. The output voltage quickly settles down to its reference value without affecting the steady-state behavior while the designed one is used. However, the overshoots are a bit higher with higher settling times, especially at the instant of changes, i.e., at the decrease in load resistance when the existing FTSC is used. Hence, Figure 10a provides an observation that the designed AFN-FTSC controller can stabilize the output voltage very quickly to its desired value without having any significant impact even with a large variation in the load resistance. Figure 10a also confirms the faster settling time with less transient with the AFN-FTSC controller during the post-disturbance operation while making its comparison with the FTSC controller.

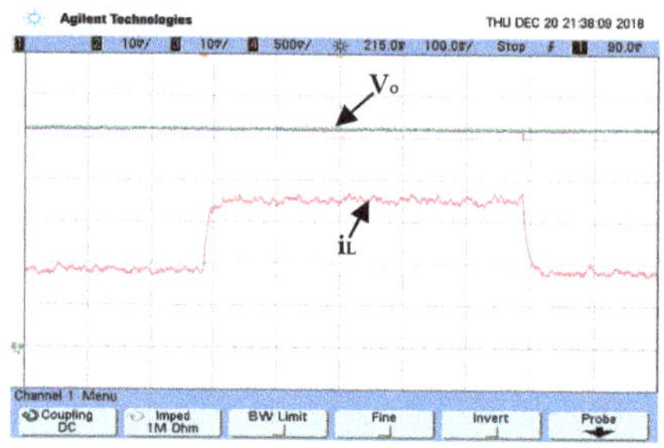

(a)

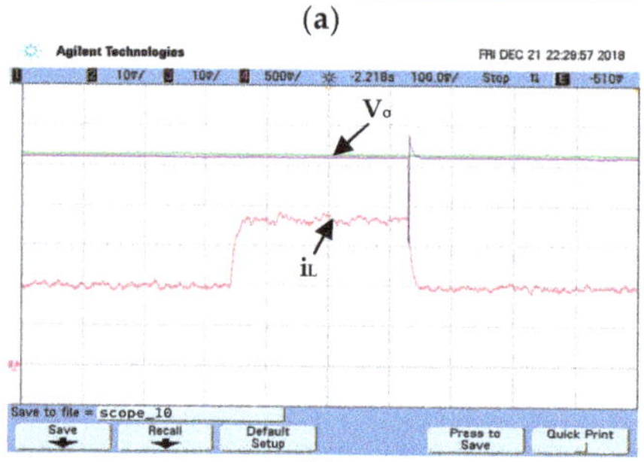

(b)

Figure 10. Dynamic performance of the converter under load-resistance variations (V_o: 10 V/div, i_L: 0.25 A/div and *time*: 100 ms/div): (**a**) AFN-FTSC controller, (**b**) FTSC controller.

Several quantitative factors such as the peak value, overshoot, undershoot, and settling time are calculated to validate the performance of the designed AFN-FTSC against the FTSC, where these values for both controllers are captured in Table 4. The designed AFN-FTSC outperforms the existing FTSC in all aspects that can also be found in Table 4. The next case study analyzes and compares the performance of the designed and existing controllers against variations in the input supply voltage.

Table 4. Peak value, overshoot, undershoot, and settling time with variations in the load.

Controller	Peak Value (V)	Overshoot (%)	Undershoot (%)	Settling Time (ms)
AFN-FTSC	0	0	6.66	2
FTSC	8	16	40	20

Case II: Output voltage regulation with variations in the input supply voltage

This test considers the variation in the input supply voltage to show the superiority of the designed AFN-FTSC over the existing FTSC algorithm. The input voltage is rated at 0 V at the start of the test, and it is suddenly raised to 100 V at t = 0.25 s. However, the output reference voltage is kept constant at 30 V. The corresponding dynamic responses of different states for the converter are shown in Figure 11. As illustrated in Figure 11a, the designed AFN-FTSC can efficiently eliminate the effects of fluctuations in the input as evidenced by the proper tracking of the output reference voltage. On the other hand, the existing FTSC fails to realize the desired voltage tracking performance when responding to changes in input supply voltage. The result in Figure 11b shows that the FTSC cannot cope with changes in the input supply voltage, however, the designed AFN-FTSC does not experience any problem to maintain the desired performance.

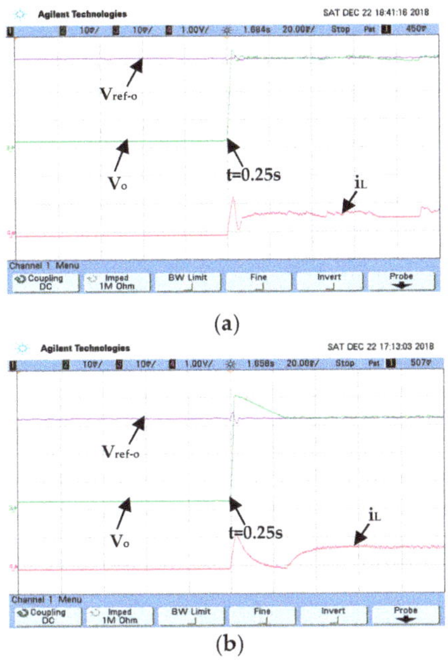

Figure 11. Dynamic performance of the converter under input supply voltage variations (V_o: 10 V/div, i_L: 0.5 A/div and *time*: 20 ms/div): (**a**) AFN-FTSC controller, (**b**) FTSC controller.

The peak value, overshoot, and settling time are calculated for this scenario and presented in Table 5, which clearly demonstrates the effectiveness of the AFN-FTSC over the FTSC. From Table 5, it can be seen that the output voltage recovery time of the AFN-FTSC is 5 ms, whereas it is 28 ms for the existing FTSC. Furthermore, from Table 5, it can be seen that the voltage fluctuation with the designed AFN-FTSC is 3 V, while it is 11 V with the existing FTSC. In comparison to the existing FTSC controller, the designed AFN-FTSC controller has a shorter settling time and lower voltage changes, as shown by the data analysis.

Table 5. Peak value, overshoot, and settling time with variations in the input voltage.

Control Strategy	Peak Value (V)	Overshoot (%)	Undershoot (%)	Settling Time (ms)
AFN-FTSC	3	10	0	5
FTSC	11	36.66	0	28

In the next case, the converter's reference output voltage variation is regarded as an external perturbation to the converter, which is discussed in the following subsection.

Case III: Output voltage regulation with variations in the output reference voltage.

This test counts variations in the output reference voltage to demonstrate the robustness of both controllers. At the start of the test, the reference voltage is rated at 20 V and then suddenly changed to 30 V and vice-versa. Under this disturbance, the dynamic responses of both states (V_o and i_L) are depicted in Figure 12a,b. These responses clearly demonstrate the superior capability of the AFN-FTSC controller (Figure 12a) for tracking output reference voltage, as the existing FTSC controller fails to do so (Figure 12b).

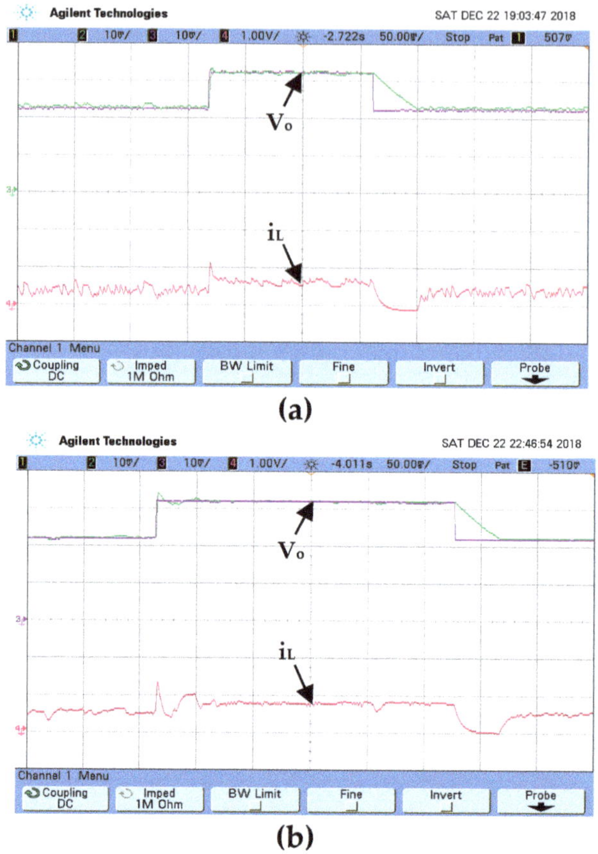

Figure 12. Dynamic performance of the converter under output reference voltage variations (V_o: 10 V/div, i_L: 0.5 A/div and *time*: 50 ms/div): (**a**) AFN-FTSC controller, (**b**) FTSC controller.

The peak value, overshoot, undershoot, and settling time are calculated for this scenario and presented in Table 6. This table clearly shows very satisfactory performance in terms of both settling time and overshoot with the designed AFN-FTSC over the FTSC.

Table 6. Peak value, overshoot, undershoot, and settling time with variations in the reference voltage.

Control Strategy	Peak Value (V)	Overshoot (%)	Undershoot (%)	Settling Time (ms)
AFN-FTSC	0	0	0	2
FTSC	2	6.66	0	40

The voltage tracking capability of each controller is evaluated using a sin-wave reference output voltage. As illustrated in Figure 13, the AFN-FTSC not only offers better precise results, but it also eliminates chattering.

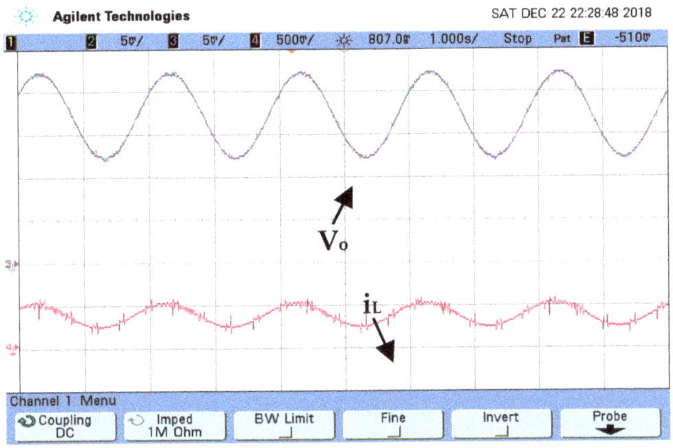

(a)

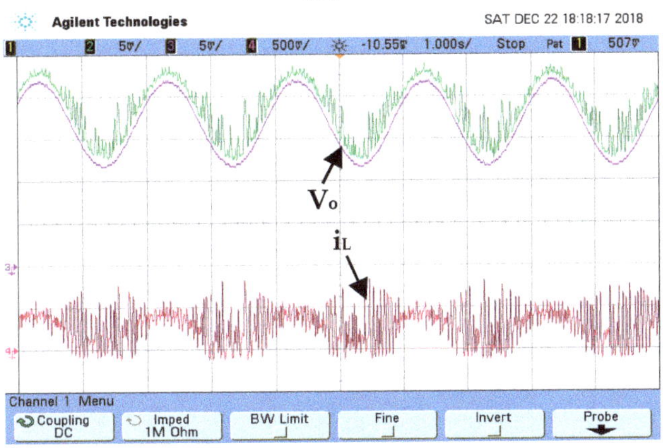

(b)

Figure 13. Dynamic performance of the converter under sinusoidal wave in the output voltage reference (V_o: 10 V/div, i_L: 0.5 A/div, and *time*: 1 s/div): (**a**) AFN-FTSC controller, (**b**) FTSC controller.

The voltage tracking capability of each controller is evaluated using a triangular-wave reference output voltage (Figure 14). The results indicate the advantage of the considered AFN-FTSC over the FTSC algorithm in terms of output voltage tracking. The studied AFN-FTSC displays a small tracking error (Figure 14a).

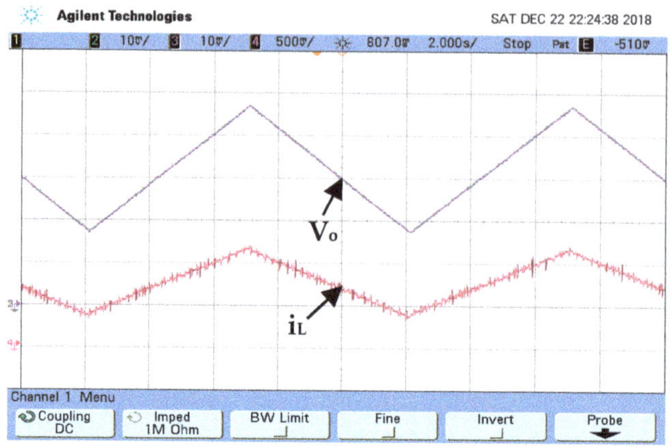

(a)

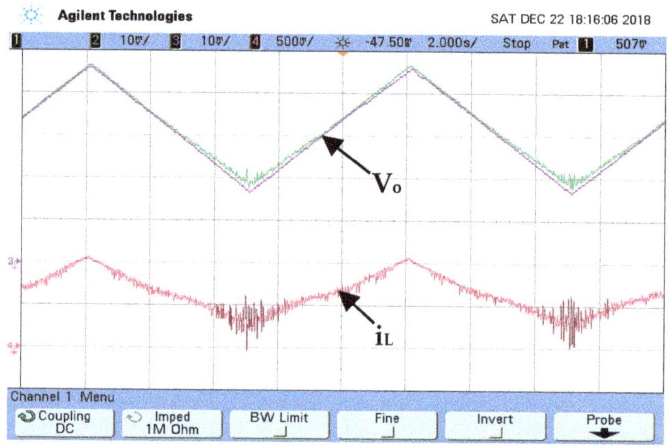

(b)

Figure 14. Dynamic responses of the converter with triangular-wave variations in the output voltage reference (V_o: 10 V/div, i_L: 0.25 A/div and *time*: 2 s/div): (**a**) AFN-FTSC controller, (**b**) FTSC controller.

Case IV: Output voltage regulation under nominal start-up in the reference voltage.

The purpose of this test is to evaluate the robustness of both controllers under nominal start-up in the output voltage reference. The output voltage of the converter, as presented in Figure 15, is observed for this case, which clearly shows the settling time as almost 5 ms with the AFN-FTSC. However, this settling-time is a bit larger, i.e., around 20 ms if the FTSC is used. Similarly, the FTSC exhibits an overshoot of 20%, while there are no overshoots when the AFN-FTSC is used. Therefore, it is also possible to deduce that the AFN-FTSC method provides a superior dynamic response to the existing FTSC method.

The peak value, overshoot, undershoot, and settling time are calculated for this case study and presented in Table 7. This table clearly shows very satisfactory performance in terms of both maximum overshoot and settling time with the suggested AFN-FTSC over the FTSC. Hence, the overall performance of the AFN-FTSC significantly dominates that of the FTSC for all cases.

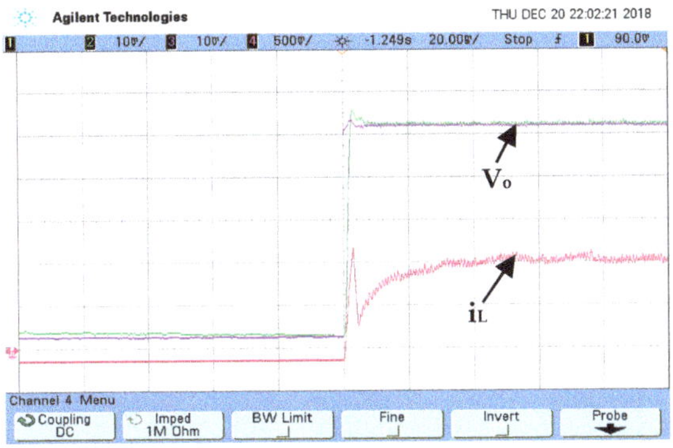

(a)

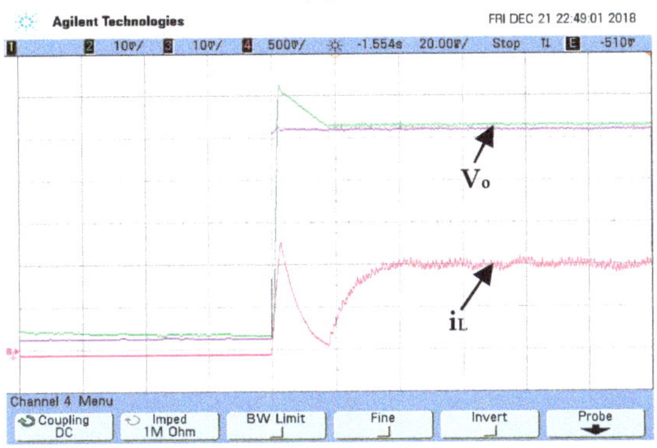

(b)

Figure 15. Dynamic responses of the converter under nominal start-up in the output voltage reference (V_o: 10 V/div, i_L: 0.25 A/div and *time*: 20 ms/div): (**a**) AFN-FTSC controller, (**b**) FTSC controller.

Table 7. Peak value, overshoot, undershoot, and settling time with variations in reference voltage.

Controller	Peak Value (V)	Overshoot (%)	Undershoot (%)	Settling Time (ms)
AFN-FTSC	3	2	0	5
FTSC	10	20	0	20

5.3. Experimental Validation

A comparative study between the proposed and existing controllers was performed, using integrals of error-based performance indices that are defined as the integral of absolute error (IAE), integral of squared error (ISE), and an integral of time-weighted squared error (*ITSE*). The mathematical expressions for these indices can be expressed as in [26]:

$$\mathrm{IAE} = \int_0^T |e(t)|dt, \mathrm{ISE} = \int_0^T e(t)^2 dt, \& ITSE = \int_0^T te(t)^2 dt \quad (58)$$

where T denotes simulation time with *e(t)* as the error signal. The smaller value of these indices indicates the better performance, as these indices basically represent the deviation from the ideal condition. All these three indices are calculated for both controllers and presented in Figure 16, which clearly demonstrates significantly smaller values for the AFN-FTSC algorithm when compared with the FTSC algorithm.

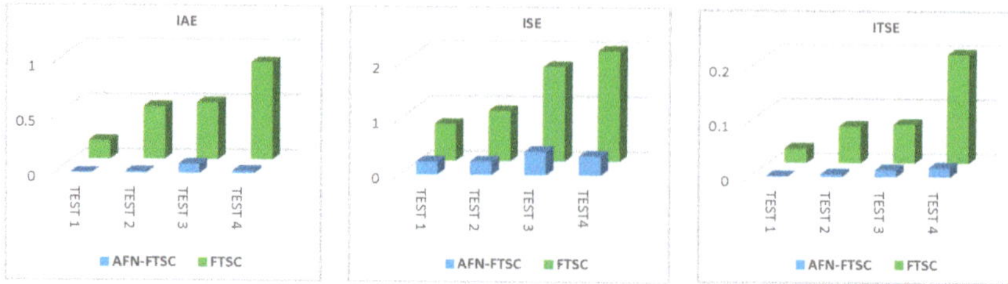

(**a**) (**b**) (**c**)

Figure 16. Integral-error-performance indices: (**a**) IAE, (**b**) ISE, and (**c**) *ITSE*.

6. Conclusions

In this study, a robust regulator is developed by combining the synergetic theory of control (STC) with a terminal attractor method to track the desired output voltage across the load connected to a buck DC/DC converter. The time-varying dynamic reaching law is employed to ensure the finite-time convergence of the tracking voltage errors and the FNN model guarantees the estimation of the unknown converter nonlinear dynamics. Experimental results are analyzed under various conditions including qualitative and quantitative ways. Comparisons were made with an existing FTSC, and the newly designed AFN-FTSC performs better in all conditions. Based on the analyses presented in this study, the key findings can be summarized as:

- The designed AFN-FTSC properly tracks the reference value of the output voltage with the minimum overshoot and faster settling time—even variations in the input voltage, load resistance, and reference voltage.
- The steady-state is significantly low under any operating scenario.

The impacts of model perturbations and external disturbances are not incorporated into the mathematical model, although these have been included during the analysis in different platforms. Future work will further improve the AFN-FTSC design process by capturing these impacts within the model.

Author Contributions: Conceptualization, B.B. and F.A.; Data curation, B.B., N.H. and O.A.; Formal analysis, N.H., O.A. and S.A.M.A.; Funding acquisition, F.A. and S.S.M.G.; Investigation, S.S.M.G.; Methodology, B.B., N.H., O.A. and S.A.M.A.; Project administration, B.B., F.A. and S.S.M.G.; Resources, B.B., F.A. and S.S.M.G.; Software, B.B., N.H. and O.A.; Supervision, B.B. and S.S.M.G.; Validation, N.H., O.A. and S.A.M.A.; Visualization, F.A. and S.A.M.A.; Writing—original draft, B.B., N.H. and O.A.; Writing—review & editing, F.A., S.S.M.G. and S.A.M.A. All authors have read and agreed to the published version of the manuscript.

Funding: This work funded by Taif University Researchers Supporting Project TURSP-2020/97, Taif University, Taif, Saudi Arabia.

Acknowledgments: The authors thank Taif University Researchers Supporting Project TURSP-2020/97, through Taif University, Taif, Saudi Arabia for supporting this work.

Conflicts of Interest: The authors declare no conflict of interest.

References

1. Babes, B.; Boutaghane, A.; Hamouda, N.; Mezaache, M. Design of a robust voltage controller for a dc-dc buck converter using fractional-order terminal sliding mode control strategy. In Proceedings of the International Conference on Advanced Electrical Engineering (ICAEE), Algiers, Algeria, 19–21 November 2019.
2. Amir, M.; Prajapati, A.K.; Refaat, S.S. Dynamic Performance Evaluation of Grid-Connected Hybrid Renewable Energy-Based Power Generation for Stability and Power Quality Enhancement in Smart Grid. *Front. Energy Res.* **2022**, *10*, 861282. [CrossRef]
3. Yan, Y.; Liu, J. Analysis of passivity-based sliding-mode control strategy in DC/DC converter. In Proceedings of the Chinese Control Conference, Harbin, China, 7–11 August 2006; pp. 171–174.
4. Young, D.S.; Hen, T.-W.; Santi, E.; Monti, A. Synergetic control approach for induction motor speed control. In Proceedings of the Annual Conference of IEEE Industrial Electronics Society, Busan, Korea, 2–6 November 2004. IECON 2004.

5. Dehri, K.; Nouri, A.S. A discrete repetitive adaptive sliding mode control for DC-DC buck converter. *Inst. Mech. Eng. Part I J. Syst. Control. Eng.* **2021**, *235*, 1698–1708. [CrossRef]
6. Chen, J.J.; Hwang, Y.S.; Lin, J.Y.; Ku, Y. A dead-beat-controlled fast-transient-response buck converter with active pseudo-current-sensing techniques. *IEEE Trans. Very Large Scale Integr. (VLSI) Syst.* **2019**, *27*, 1751–1759. [CrossRef]
7. Kumar, V.I.; Kapat, S. Mixed-signal hysteretic internal model control of buck converters for ultra-fast envelope tracking. In Proceedings of the IEEE Applied Power Electronics Conference and Exposition (APEC), Long Beach, CA, USA, 20–24 March 2016; pp. 3224–3230.
8. Linares-Flores, J.; Hernandez Mendez, A.; Garcia-Rodriguez, C.; Sira-Ramirez, H. Robust nonlinear adaptive control of a boost converter via algebraic parameter identification. *IEEE Trans. Ind. Electron.* **2014**, *61*, 4105–4114. [CrossRef]
9. Xu, Q.; Yan, Y.; Zhang, C.; Dragicevic, T.; Blaabjerg, F. An offset-free composite model predictive control strategy for DC/DC buck converter feeding constant power loads. *IEEE Trans. Power Electron.* **2020**, *35*, 5331–5342. [CrossRef]
10. Hausberger, T.; Kugi, A.; Eder, A.; Kemmetmüller, W. High-speed nonlinear model predictive control of an interleaved switching DC/DC-converter. *Control. Eng. Pract.* **2020**, *103*, 104576. [CrossRef]
11. Albira, M.E.; Zohdy, M.A. Adaptive model predictive control for DC-DC power converters with parameters uncertainties. *IEEE Access* **2021**, *9*, 135121–135131. [CrossRef]
12. Hamouda, N.; Babes, B.; Boutaghane, A. Design and analysis of robust nonlinear synergetic controller for a PMDC motor driven wire-feeder system (WFS). In *Lecture Notes in Electrical Engineering*; Springer: Singapore, 2020; pp. 373–387.
13. Babes, B.; Boutaghane, A.; Hamouda, N. Design and real-time implementation of an adaptive fast terminal synergetic controller based on dual RBF neural networks for voltage control of DC–DC step-down converter. *Electr. Eng.* **2021**, *104*, 945–957. [CrossRef]
14. Hamouda, N.; Babes, B. A DC/DC Buck converter voltage regulation using an adaptive fuzzy fast terminal synergetic control. In *Lecture Notes in Electrical Engineering*; Springer: Singapore, 2020; pp. 711–721.
15. Hadjer, A.; Ameur, A.; Harmas, N.M. Adaptive non-singular terminal synergetic power system control using PSO. In Proceedings of the 8th International Conference on Modellin, Identification and Control (ICMIC-2016), Algiers, Algeria, 15–17 November 2016.
16. Babes, B.; Boutaghane, A.; Hamouda, N.; Mezaache, M.; Kahla, S. A robust adaptive fuzzy fast terminal synergetic voltage control scheme for DC/DC buck converter. In Proceedings of the International Conference on Advanced Electrical Engineering (ICAEE), Algiers, Algeria, 19–21 November 2019.
17. Wen, S.; Chen, M.Z.Q.; Zeng, Z.; Huang, T.; Li, C. Adaptive neural-fuzzy sliding-mode fault-tolerant control for uncertain nonlinear systems. *IEEE Trans. Syst. Man Cybern. Syst.* **2017**, *47*, 2268–2278. [CrossRef]
18. Chen, Z.; Li, Z.; Chen, C.L.P. Adaptive neural control of uncertain MIMO nonlinear systems with state and input constraints. *IEEE Trans. Neural Netw. Learn. Syst.* **2017**, *28*, 1318–1330. [CrossRef]
19. Santi, E.; Monti, A.; Proddutur, D.; Li, K.; Dougal, R.A. Synergetic control for power electronics applications: A comparison with the sliding mode approach. *J. Circuits Syst. Comput.* **2004**, *13*, 737–760. [CrossRef]
20. Shahgholian, G. Power system stabilizer application for load frequency control in hydro-electric power plant. *Int. J. Theor. Appl. Math.* **2017**, *3*, 148. [CrossRef]
21. Rubaai, A.; Young, P. Hardware/software implementation of fuzzy-neural-network self-learning control methods for brushless DC motor drives. *IEEE Trans. Ind. Appl.* **2016**, *52*, 414–424. [CrossRef]
22. Nettari, Y.; Kurt, S. Design of a new non-singular robust control using synergetic theory for DC-DC buck converter. *Electrica* **2018**, *18*, 292–329. [CrossRef]
23. Zerroug, N.; Harmas, M.N.; Benaggoune, S.; Bouchama, Z.; Zehar, K. DSP-based implementation of fast terminal synergetic control for a DC–DC Buck converter. *J. Frankl. Inst.* **2018**, *355*, 2329–2343. [CrossRef]
24. Ullah, N.; Shaoping, W. High performance direct torque control of electrical aerodynamics load simulator using adaptive fuzzy backstepping control. *Proc. Inst. Mech. Eng. Part G J. Aerosp. Eng.* **2015**, *229*, 369–383. [CrossRef]
25. Sastry, S.; Bodson, M. *Adaptive Control: Stability, Convergence, and Robustness*; Englewood Cliffs, N.J., Ed.; Prentice-Hall: Hoboken, NJ, USA, 1989.
26. Babes, B.; Mekhilef, S.; Boutaghane, A.; Rahmani, L. Fuzzy Approximation-Based Fractional-Order Nonsingular Terminal Sliding Mode Controller for DC–DC Buck Converters. *IEEE Trans. Power Electron.* **2022**, *37*, 2749–2760. [CrossRef]
27. Alanqar, A.; Durand, H.; Albalawi, F.; Cristofides, P.D. An economic model predictive control approach to integrated production management and process operation. *AIChE J.* **2017**, *63*, 1892–1906. [CrossRef]

MDPI
St. Alban-Anlage 66
4052 Basel
Switzerland
www.mdpi.com

Sustainability Editorial Office
E-mail: sustainability@mdpi.com
www.mdpi.com/journal/sustainability

www.ingramcontent.com/pod-product-compliance
Lightning Source LLC
Chambersburg PA
CBHW061452220926
51850CB00004B/67
9783725800872